Introductory Algebra

SECOND EDITION

Julie Miller
Daytona State College

Molly O'Neill
Daytona State College

Nancy Hyde
Formerly of Broward College

Higher Education

Boston   Burr Ridge, IL   Dubuque, IA   New York   San Francisco   St. Louis
Bangkok   Bogotá   Caracas   Kuala Lumpur   Lisbon   London   Madrid   Mexico City
Milan   Montreal   New Delhi   Santiago   Seoul   Singapore   Sydney   Taipei   Toronto

Higher Education

INTRODUCTORY ALGEBRA, SECOND EDITION

Published by McGraw-Hill, a business unit of The McGraw-Hill Companies, Inc., 1221 Avenue of the Americas, New York, NY 10020. Copyright © 2009 by The McGraw-Hill Companies, Inc. All rights reserved. Previous edition © 2007. No part of this publication may be reproduced or distributed in any form or by any means, or stored in a database or retrieval system, without the prior written consent of The McGraw-Hill Companies, Inc., including, but not limited to, in any network or other electronic storage or transmission, or broadcast for distance learning.

Some ancillaries, including electronic and print components, may not be available to customers outside the United States.

This book is printed on acid-free paper.

1 2 3 4 5 6 7 8 9 0 DOW/DOW 0 9 8

ISBN 978–0–07–340609–1
MHID 0–07–340609–0

ISBN 978–0–07–335701–0 (Annotated Instructor's Edition)
MHID 0–07–335701–4

ISBN 978–0–07–730341–9 (Hard Cover)
MHID 0–07–730341–5

ISBN 978–0–07–730340–2 (Hard Cover Annotated Instructor's Edition)
MHID 0–07–730340–7

Editorial Director: *Stewart K. Mattson*
Senior Sponsoring Editor: *David Millage*
Director of Development: *Kristine Tibbetts*
Senior Developmental Editor: *Emilie J. Berglund*
Marketing Manager: *Victoria Anderson*
Lead Project Manager: *Peggy J. Selle*
Senior Production Supervisor: *Sherry L. Kane*
Lead Media Project Manager: *Stacy A. Patch*
Designer: *Laurie B. Janssen*
Cover Illustration: *Imagineering Media Services, Inc.*
Lead Photo Research Coordinator: *Carrie K. Burger*
Supplement Producer: *Mary Jane Lampe*
Compositor: *Aptara®, Inc.*
Typeface: *10/12 Times Ten Roman*
Printer: *R. R. Donnelley Willard, OH*

The credits section for this book begins on page C-1 and is considered an extension of the copyright page.

Library of Congress Cataloging-in-Publication Data

Miller, Julie, 1962–
    Introductory algebra / Julie Miller, Molly O'Neill, Nancy Hyde. — 2nd ed.
        p. cm.
    Includes index.
    ISBN 978–0–07–340609–1—ISBN   0–07–340609–0 (hard copy : alk. paper)
    ISBN 978–0–07–730340–2 —ISBN   0–07–730340–7 (inst. ed. : hard copy : alk. paper)
    [etc.]
    1. Algebra—Textbooks.   I. O'Neill, Molly, 1953–   II. Hyde, Nancy.   III. Title.
QA152.3.M576 2009
512—dc22
                            2008026083

Letter from the Authors

Julie, Molly, and Nancy: "Dedicated to Getting Better Results"

Dear Colleagues,

We originally embarked on this textbook project because we were seeing a lack of student success in our developmental math sequence. In short, we were not getting the results we wanted from our students with the materials and textbooks that we were using at the time. The primary goal of our project was to create teaching and learning materials that would get better results.

At Daytona State College, our students were instrumental in helping us develop the clarity of writing, the step-by-step examples, and the pedagogical elements, such as Avoiding Mistakes, Concept Connections, and Problem Recognition Exercises, found in our textbooks. When our text and course redesign were implemented at Daytona State College in 2006, our student success rates in developmental courses improved by 20%. We think you will agree that these are the kinds of results we are all striving for in developmental mathematics courses.

This project has been a true collaboration with our Board of Advisors and colleagues in developmental mathematics around the country. We have been truly humbled by those of you who adopted the first edition and the over 400 colleagues around the country who partnered with us providing valuable feedback and suggestions through reviews, symposia, focus groups, and being on our Board of Advisors. You partnered with us to create materials that will help students get better results. For that we are immeasurably grateful.

As an author team, we have an ongoing commitment to provide the best possible materials for instructors and students. With your continued help and suggestions we will continue the quest to help all of our students get better results.

Sincerely,

Julie Miller
millerj@DaytonaState.edu

Molly O'Neill
oneillm@DaytonaState.edu

Nancy Hyde
nhyde@montanasky.com

Julie Miller

Julie Miller has been on the faculty of the Mathematics Department at Daytona State College for 19 years, where she has taught developmental and upper level courses. Prior to her work at Daytona State College, she worked as a software engineer for General Electric in the area of flight and radar simulation. Julie earned a bachelor of science in applied mathematics from Union College in Schenectady, New York, and a master of science in mathematics from the University of Florida. In addition to this textbook, she has authored several course supplements for college algebra, trigonometry, and precalculus, as well as several short works of fiction and nonfiction for young readers.

"My father is a medical researcher, and I got hooked on math and science when I was young and would visit his laboratory. I can remember using graph paper to plot data points for his experiments and doing simple calculations. He would then tell me what the peaks and features in the graph meant in the context of his experiment. I think that applications and hands-on experience made math come alive for me and I'd like to see math come alive for my students."

—Julie Miller

Molly O'Neill

Molly O'Neill is also from Daytona State College, where she has taught for 21 years in the Mathematics Department. She has taught a variety of courses from developmental mathematics to calculus. Before she came to Florida, Molly taught as an adjunct instructor at the University of Michigan–Dearborn, Eastern Michigan University, Wayne State University, and Oakland Community College. Molly earned a bachelor of science in mathematics and a master of arts and teaching from Western Michigan University in Kalamazoo, Michigan. Besides this textbook, she has authored several course supplements for college algebra, trigonometry, and precalculus and has reviewed texts for developmental mathematics.

"I differ from many of my colleagues in that math was not always easy for me. But in seventh grade I had a teacher who taught me that if I follow the rules of mathematics, even I could solve math problems. Once I understood this, I enjoyed math to the point of choosing it for my career. I now have the greatest job because I get to do math everyday and I have the opportunity to influence my students just as I was influenced. Authoring these texts has given me another avenue to reach even more students."

—Molly O'Neill

Nancy Hyde served as a full-time faculty member of the Mathematics Department at Broward College for 24 years. During this time she taught the full spectrum of courses from developmental math through differential equations. She received a bachelor of science degree in math education from Florida State University and master's degree in math education from Florida Atlantic University. She has conducted workshops and seminars for both students and teachers on the use of technology in the classroom. In addition to this textbook, she has authored a graphing calculator supplement for College Algebra.

Nancy Hyde

"I grew up in Brevard County, Florida, where my father worked at Cape Canaveral. I was always excited by mathematics and physics in relation to the space program. As I studied higher levels of mathematics I became more intrigued by its abstract nature and infinite possibilities. It is enjoyable and rewarding to convey this perspective to students while helping them to understand mathematics."

—Nancy Hyde

*D*edication

To Jonas, Erica, Nicolas, and Max
    —*Julie Miller*

In memory of my mother, Doris
    —*Molly O'Neill*

To my friends and colleagues at
Broward College
    —*Nancy Hyde*

Get Better Results with Miller/O'Neill/Hyde

About the Cover

A mosaic is made up of pieces placed together to create a unified whole. Similarly, an introductory algebra course provides an array of materials that together create a solid mathematical foundation for the developmental mathematics student.

The Miller/O'Neill/Hyde developmental mathematics series helps students to see the whole picture through better pedagogy and supplemental materials. In the second edition of their developmental mathematics series, Julie Miller, Molly O'Neill, and Nancy Hyde focused their efforts on guiding students successfully through core topics to build students' mathematical proficiency to get better results.

"We originally embarked on this textbook project because we were seeing a lack of student success in courses beyond our developmental sequence. We wanted to build a better bridge between developmental algebra and higher level math courses. Our goal has been to develop pedagogical features to help students achieve better results in mathematics."

—Julie Miller, Molly O'Neill, Nancy Hyde

How Will Miller/O'Neill/Hyde Help Your Students *Get Better Results?*

Better Clarity, Quality, and Accuracy

Julie Miller, Molly O'Neill, and Nancy Hyde know what students need to be successful in mathematics. Better results come from clarity in their exposition, quality of step-by-step worked examples, and accuracy of exercise sets, but it takes more than just great authors to build a textbook series to help students achieve success in mathematics. Our authors worked with a strong mathematical team of instructors from around the country to ensure clarity, quality, and accuracy.

> "The authors' writing style is clear and concise. The tone is appropriate for our course. Students will find it easy to read and follow."
>
> —Carla Kulinsky, *Salt Lake Community College*

Better Exercise Sets!

A comprehensive set of exercises are available for every student level. Julie Miller, Molly O'Neill, and Nancy Hyde worked with a national board of advisors from across the country to ensure the series will offer the appropriate depth and breadth of exercises for your students. New to this edition, **Problem Recognition Exercises** were created in direct response to student need and resulted in improved student performance on tests.

Our exercise sets help students progress from skill development to conceptual understanding. Student tested and instructor approved, the Miller/O'Neill/Hyde exercise sets will help your students get better results.

- ▶ **Problem Recognition Exercises**
- ▶ **Skill Practice Exercises**
- ▶ **Study Skills Exercises**
- ▶ **Mixed Exercises**
- ▶ **Expanding Your Skills Exercises**

> "The number, type, various levels of difficulty and quality of practice exercises are excellent in this book. If students go over most of the available resources in this book, they can excel at the next math level. No doubt."
>
> —Pauline Chow, *Harrisburg Area Community College*

Better Step-By-Step Pedagogy!

The second edition provides enhanced step-by-step learning tools available to help students *get better results*.

- ▶ **Worked Examples** provide an "easy-to-understand" approach, clearly guiding each student through a step-by-step approach to master each practice exercise for better comprehension.
- ▶ **TIPS** offer students extra cautious direction to help improve understanding through hints and further insight.
- ▶ **Avoiding Mistakes** boxes alert students to common errors and provide practical ways to avoid them.

  These learning aids will help students get better results by learning how to work through a problem using a clearly defined step-by-step methodology that has been class-tested and student approved.

> "These are useful reminders to the instructor as well as a resource for students."
>
> —Patricia Roux, *Delgado Community College*

Formula for Student Sucess

Step-by-Step Worked Examples

▶ Do you get the feeling that there is a disconnect between your students classwork and homework?

▶ Do your students have trouble finding worked examples that match the practice exercises?

▶ Would you like your students to see examples in the textbook that match the ones you use in class?

Miller/O'Neill/Hyde's worked examples offer a clear, concise methodology that replicates the mathematical processes used in the authors' classroom lectures!

"The wording is clear, consistent and concise in the worked out examples. Having seen the text itself, I know the use of pedagogy here gets my top grade!"
—Linda Murphy, *Northern Essex Community College*

PROCEDURE Solving a Proportion
Step 1 Set the cross products equal to each other.
Step 2 Divide both sides of the equation by the number being multiplied by the variable.
Step 3 Check the solution in the original proportion.

Example 4 Solving a Linear Equation

Solve the equation. $7 + 3 = 2(p - 3)$

— Skill Practice —
Solve the equation.
4. $12 + 2 = 7(3 - y)$

Solution:

$7 + 3 = 2(p - 3)$

$10 = 2p - 6$ **Step 1:** Simplify both sides of the equation by clearing parentheses and combining *like* terms.

Step 2: The variable terms are already on one side.

$10 + 6 = 2p - 6 + 6$ **Step 3:** Add 6 to both sides to collect the constant terms on the other side.

$16 = 2p$

$\dfrac{16}{2} = \dfrac{2p}{2}$ **Step 4:** Divide both sides by 2 to obtain a coefficient of 1 for p.

$8 = p$ **Step 5:** Check:

$7 + 3 = 2(p - 3)$

$10 \overset{?}{=} 2(8 - 3)$

$10 \overset{?}{=} 2(5)$

The solution is 8. $10 \overset{?}{=} 10$ ✔ True

Good step-by-step examples with easy to follow explanations."
—Lynn Beckett-Lemus, *El Camino College*

"All of the worked examples are good and easy to understand. There are plenty of examples given. Also, it appears that there is at least one example for each different particular type of exercise in each section – very good ☺"
—Mike Kirby, *Tidewater Community College*

Better Learning Tools

Avoiding Mistakes Boxes

Avoiding Mistakes boxes are integrated throughout the textbook to alert students to common errors and how to avoid them.

> "The Avoiding Mistakes is probably one of my favorite "supplements" to the text. This idea is an EXCELLENT tool. I wouldn't change anything with the exception of adding more when appropriate."
>
> —Mark Billiris, *St. Petersburg Community College*

If the slopes of two lines are known, then we can compare the slopes to determine if the lines are parallel, perpendicular, or neither.

Skill Practice

Determine if lines l_1 and l_2 are parallel, perpendicular, or neither.

9. l_1: $(-2, -3)$ and $(4, -1)$
   l_2: $(0, 2)$ and $(-3, 1)$

Avoiding Mistakes

You can check that two lines are perpendicular by checking that the product of their slopes is -1.

$$4\left(-\frac{1}{4}\right) = -1$$

Example 7 Determining If Lines Are Parallel, Perpendicular, or Neither

Lines l_1 and l_2 pass through the given points. Determine if l_1 and l_2 are parallel, perpendicular, or neither.

$$l_1: \ (2, -7) \text{ and } (4, 1) \qquad l_2: \ (-3, 1) \text{ and } (1, 0)$$

Solution:

Find the slope of each line.

$$l_1: \ \underset{(x_1, y_1)}{(2, -7)} \text{ and } \underset{(x_2, y_2)}{(4, 1)} \qquad l_2: \ \underset{(x_1, y_1)}{(-3, 1)} \text{ and } \underset{(x_2, y_2)}{(1, 0)}$$

$$m_1 = \frac{1 - (-7)}{4 - 2} \qquad\qquad m_2 = \frac{0 - 1}{1 - (-3)}$$

$$m_1 = \frac{8}{2} \qquad\qquad m_2 = \frac{-1}{4}$$

$$m_1 = 4 \qquad\qquad m_2 = -\frac{1}{4}$$

One slope is the opposite of the reciprocal of the other slope. Therefore, the lines are perpendicular.

TIP: The solution to a system of linear equations can be confirmed by graphing. The system from Example 2 is graphed here.

$x + y = 4$

$(3, 1)$

$-5x + 3y = -12$

Tips Boxes

Teaching Tips are usually only revealed in the classroom. Not anymore. Tip boxes offer students helpful hints and extra direction to help improve understanding and further insight.

> "The nice thing is that as I am reading through the text, I think to myself, "I would say this," and then the authors will say that very thing in a tip off to the side."
>
> —Barb Elzey, *Bluegrass Community & Technical College*

Concept Connection Boxes

Concept Connections help students understand the conceptual meaning of the problems they are solving—a vital skill in mathematics.

> "The concept connections seem to ask almost the same questions I ask my students when we are in class. I think they are helpful."
>
> —Lori Grady, *University of Wisconsin-Whitewater*

---

2. Dependent and Inconsistent Systems of Linear Equations

When two lines are drawn in a rectangular coordinate system, three geometric relationships are possible:

1. Two lines may intersect at *exactly one point*.

2. Two lines may intersect at *no point*. This occurs if the lines are parallel.

Concept Connections

3. If the linear equations in a system represent a pair of parallel lines, how many solutions does the system have?

4. If the linear equations in a system represent a pair of perpendicular lines, how many solutions does the system have?

New to this Edition

▶ Do your students have trouble with problem solving?

▶ Do you want to help students overcome math anxiety?

▶ Do you want to help your students improve performance on math assessments?

Problem Recognition Exercises!

Problem Recognition Exercises present a collection of problems that look similar to a student upon first glance, but are actually quite different in the manner of their individual solutions. Students sharpen critical thinking skills and better develop their "solution recall" to help them distinguish the method needed to solve an exercise—an essential skill in mathematics. Problem Recognition Exercises, tested in a developmental mathematics classroom, were created in direct response to student need to improve performance in testing where different problem types are mixed.

"Excellent problems. The PRE's review all different exercise possibilities and require students to identify what they see."

—Patricia Roux, *Delgado Community College*

"I think this is a great way for students to have to think through the different topics. These problems are significantly different than they have been presented previously, so the student cannot use the visual clues they have used previously."

—Ron Koehn, *Southwestern Oklahoma State University*

Problem Recognition Exercises

Polynomial Expressions and Polynomial Equations

For Exercises 1–36, factor each expression and solve each equation.

1. **a.** $x^2 + 6x - 7$
   b. $x^2 + 6x - 7 = 0$

2. **a.** $c^2 + 8c + 12$
   b. $c^2 + 8c + 12 = 0$

3. **a.** $2y^2 + 7y + 3$
   b. $2y^2 + 7y + 3 = 0$

4. **a.** $3x^2 - 8x + 5$
   b. $3x^2 - 8x + 5 = 0$

5. **a.** $5q^2 + q - 4 = 0$
   b. $5q^2 + q - 4$

6. **a.** $6a^2 - 7a - 3 = 0$
   b. $6a^2 - 7a - 3$

7. **a.** $a^2 - 64 = 0$

8. **a.** $v^2 - 100 = 0$
   b. $v^2 - 100$

9. **a.** $4b^2 - 81$
   b. $4b^2 - 81 = 0$

11. **a.** $8x^2 + 16x + 6 = 0$
    b. $8x^2 + 16x + 6$

12. **a.** $12y^2 + 40y + 32 = 0$
    b. $12y^2 + 40y + 32$

14. **a.** $k^3 + 5k^2 - 14k$
    b. $k^3 + 5k^2 - 14k = 0$

15. **a.** $b^3 + b^2 - 9b - 9 = 0$
    b. $b^3 + b^2 - 9b - 9$

$= 0$

17. $2s^2 - 6s + rs - 3r$

18. $6t^2 + 3t + 10tu + 5u$

20. $2b^3 - 50b = 0$

21. $2x^3 - 4x^2 + 2x = 0$

23. $7c^2 - 2c + 3 = 7(c^2 + c)$

24. $3z(2z + 4) = -7 + 6z^2$

25. $8w^3 + 27$

26. $1000q^3 - 1$

27. $5z^2 + 2z = 7$

28. $4h^2 + 25h = -6$

29. $3b(b + 6) = b - 10$

30. $3y^2 + 1 = y(y - 3)$

31. $5(2x - 3) - 2(3x + 1) = 4 - 3x$

32. $11 - 6a = -4(2a - 3) - 1$

33. $4s^2 = 64$

34. $81v^2 = 36$

35. $(x - 3)(x - 4) = 6$

36. $(x + 5)(x + 9) = 21$

New and Improved Applications!

Class-Tested and Student Approved!

New and improved applications have been developed by an advisory team. The Miller/O'Neill/Hyde Board of Advisor Team partnered with our authors to bring you the *best applications* from every region of the country! These applications include real data and topics which are more relevant and interesting to today's student.

> "The application problems and examples are MUCH better than most texts. The real-life examples seem to drive much of the development of the text."
>
> —John Close, *Salt Lake Community College*

Objective 2: Applications Involving Principal and Interest

15. Shanelle invested $10,000, and at the end of 1 yr, she received $805 in interest. She invested part of the money in an account earning 10% simple interest and the remaining money in an account earning 7% simple interest. How much did she invest in each account? **(See Example 2.)**

	10% Account	7% Account	Total
Principal invested			
Interest earned			

16. $12,000 was invested in two accounts, one earning 12% simple interest and the other earning 8% simple interest. If the total interest at the end of 1 yr was $1240, how much was invested in each account?

	12% Account	8% Account	Total
Principal invested			
Interest earned			

17. Troy borrowed a total of $12,000 in two different loans to help pay for his new Chevy Silverado. One loan charges 9% simple interest, and the other charges 6% simple interest. If he is charged $810 in interest after 1 yr, find the amount borrowed at each rate.

18. Blake has a total of $4000 to invest in two accounts. One account earns 2% simple interest, and the other earns 5% simple interest. How much should be invested in both accounts to earn exactly $155 at the end of 1 yr?

NEW Group Activities!

Each chapter concludes with a Group Activity selected by objective to promote classroom discussion and collaboration—helping students not only to solve problems but to explain their solutions for better mathematical mastery. Group Activities are great for instructors and adjuncts—bringing a more interactive approach to teaching mathematics! All required materials, activity time, and suggested group sizes are provided in the directions of the activity. Activities include: Investigating Probability, Tracking Stocks, Using Card Games with Fractions, and more!

> "What I liked the most was how the applications required students to figure out the appropriate operation to use—I also appreciate the variety of applications from those dealing with simplifying fractions to the area of rectangles and composite figures."
>
> —Vernon Bridges, *Durham Technical Community College*

Group Activity

Modeling a Linear Equation

Materials: Yardstick or other device for making linear measurements

Estimated Time: 15–20 minutes

Group Size: 3

1. The members of each group should measure the length of their arms (in inches) from elbow to wrist. Record this measurement as x and the person's height (in inches) as y. Write these values as ordered pairs for each member of the group. Then write the ordered pairs on the board.

2. Next, copy the ordered pairs collected from all groups in the class and plot the ordered pairs. (This is called a "scatter diagram.")

Height Versus Arm Length

What makes this new edition special?

Better Development!

Question: How do you build a better developmental mathematics textbook series?

Answer: Employ a developmental mathematics instructor from the classroom to become a McGraw-Hill editor!

Emilie Berglund joined the developmental mathematics team at McGraw-Hill bringing her extensive classroom experience to the Miller/O'Neill/Hyde textbook series. A former developmental mathematics instructor at Utah Valley State College, Ms. Berglund has won numerous teaching awards and has served as the beginning algebra course coordinator for the department. Ms. Berglund's experience teaching developmental mathematics students from the Miller/O'Neill/Hyde textbook translates into more well-developed pedagogy throughout the textbook series and can be seen in everything from the updated Worked Examples to the Exercise Sets.

Listening to You . . .

This textbook has been reviewed by over 300 teachers across the country. Our textbook is a commitment to your students, providing a clear explanation, concise writing style, step-by-step learning tools, and the best exercises and applications in developmental mathematics. **How do we know? You told us so!**

Teachers *Just Like You* are saying great things about the Miller/O'Neill/Hyde developmental mathematics series:

"The authors have an excellent style that is very conducive to this level of learning."
—Barb Elzey, *Bluegrass Community & Technical College*

"It reads as a teacher who tries explaining using everyday language and everyday examples."
—Pat Rome, *Delgado Community College*

"The exercise sets for each section have many and varied problems. The examples are well-constructed and students find them useful. Then—in addition to the needed topics, numerous practice exercises, and helpful examples—the writing style is good with explanations that are accurate, clear, and sufficient!"
—Charlotte Newsom, *Tidewater Community College*

"I feel like they wrote a book of my notes. The topics are presented as I have been doing in my class for ten years."
—Brenda *Norman-Tidewater Community College*

"These authors present the material in a writing style that is geared to developmental students. The material is well explained and the paragraphs are in small, manageable blocks of information. The authors' use of bullets and other techniques in the explanation style is helpful for students and does not "overwhelm" them."
—Susan D. Caire, *Delgado Community College*

"The topics are clear and understandable. It is probably the most complete textbook I have ever reviewed in terms of clarity and understandability. Nothing needs to change."
—Sonny Kirby, *Gadsden State College*

360° Development Process

McGraw-Hill's 360° Development Process is an ongoing, never-ending, market-oriented approach to building accurate and innovative print and digital products. It is dedicated to continual large-scale and incremental improvement driven by multiple customer feedback loops and checkpoints. The process is initiated during the early planning stages of our new products, and is intensified during development and production. Then the process begins again upon publication in anticipation of the next edition.

A key principle in the development of any mathematics text is its ability to adapt to teaching specifications in a universal way. The only way to do so is by contacting those universal voices—and learning from their suggestions. We are confident that our book has the most current content the industry has to offer, thus pushing our desire for accuracy to the highest standard possible. In order to accomplish this, we have moved through an arduous road to production. Extensive and open-minded advice is critical in the production of a superior text.

Here is a brief overview of the initiatives included in the *Introductory Algebra*, Second Edition, 360° Development Process:

Board of Advisors

A hand-picked group of trusted teachers active in the Introductory Algebra course served as chief advisors and consultants to the authors and editorial team with regards to manuscript development. The Board of Advisors reviewed parts of the manuscript; served as a sounding board for pedagogical, media, and design concerns; consulted on organizational changes; and attended a focus group to confirm the manuscript's readiness for publication.

Basic College Mathematics

Vernon Bridges, *Durham Technical Community College*

Lynette King, *Gadsden State Community College*

Sharon Morrison, *St. Petersburg College*

Deanna Murphy, *Lane County Community College*

Rod Oberdick, *Delaware Technical and Community College*

Matthew Robinson, *Tallahassee Community College*

Pat Rome, *Delgado Community College–City Park*

Introductory Algebra

Mark Billiris, *St. Petersburg Community College*

Pauline Chow, *Harrisburg Community College*

John Close, *Salt Lake Community College*

Barbara Elzy, *Bluegrass Community College*

Lori Grady, *University of Wisconsin-Whitewater*

Mike Kirby, *Tidewater Community College*

Patricia Roux, *Delgado Community College*

Intermediate Algebra

Susan Dimick, *Spokane Community College*

Sue Duff, *Guilford Technical Community College*

Alicia Giovinazzo, *Miami-Dade Community College*

Charlotte Newsome, *Tidewater Community College*

Ena Salter, *Manatee Community College*

Would you like to inquire about becoming a BOA member?
If so, email the editor, David Millage at david_millage@mcgraw-hill.com.

Acknowledgments and Reviewers

The development of this textbook series would never have been possible without the creative ideas and feedback offered by many reviewers. We are especially thankful to the following instructors for their careful review of the manuscript.

Symposia

Every year McGraw-Hill conducts a general mathematics symposium which is attended by instructors from across the country. These events are an opportunity for editors from McGraw-Hill to gather information about the needs and challenges of instructors teaching these courses. This information helped to create the book plan for *Introductory Algebra.* They also offer a forum for the attendees to exchange ideas and experiences with colleagues they might have not otherwise met.

Advisors Symposium

Pauline Chow, *Harrisburg Area Community College*
Alina Coronel, *Miami-Dade College*
Anabel Darini, *Suffolk County Community College*
Susan Dimick, *Spokane Community College*
Barbara Elzey-Miller, *Bluegrass Community & Technical College*
Lori Grady, *University of Wisconsin–Whitewater*

Lynette King, *Gadsden State Community College*
Mike Kirby, *Tidewater Community College*
Charlotte Newsom, *Tidewater Community College*
Matthew Robinson, *Tallahassee Community College*
Patricia Rome, *Delgado Community College*
Suzanne Rosenberger, *Harrisburg Area College*
Patricia Roux, *Delgado Community College*

Focus Groups

In addition to the symposia, we held two specific focus groups for this book—on the overall project and on the art. These selected mathematics professors provided ideas on improvements and suggestions for fine tuning the content, pedagogy, and problems.

Santa Fe Focus Group Attendees

Rajeed Carriman, *Miami-Dade College*
Nancy Chell, *Anne Arundel Community College*
Alina Coronel, *Miami-Dade College–Kendall*
Sarah Fallis, *Tarrant County College*
Nancy Graham, *Rose State College*
Jane Hammontree, *Tulsa Community College*
Greta Harris-Hardland, *Tarrant County College*
Kristie Johnson, *Tarrant County College*

Tina Levy, *Diablo Valley College*
Barbara Lott, *Seminole Community College*
Diane McHugh, *Longview Community College*
Valarie Melvin, *Cape Fear Community College*
Janis Orinson, *Central Piedmont Community College*
Mari Peddycoart, *Lone Star College–Kingwood*
Trudy Streiliein, *Northern Virginia Community College*
Jane Wyatt, *Longview Community College*

Class Tests

Six student class tests provided the editorial team with an understanding of how content and the design of a textbook impacts a student's homework and study habits in the general mathematics course area.

Chicago Digital Focus Group Attendees

Antonio Alfonso, *Miami-Dade College*
Eric Bennett, *Lansing Community College*
David DelRossi, *Tallahassee Community College*
Maria DeLucia, *Middlesex County College*
Patricia D'Emidio, *Montclair State University*
Brandie Faulkner, *Tallahassee Community College*
Mary Lou Hammond, *Spokane Community College*
Nicole Lloyd, *Lansing Community College*

Bill Morrow, *Delaware Technical College*
Mari Peddycoart, *Lone Star College–Kingwood*
Adelaida Quesada, *Miami-Dade College–Kendall*
Patricia Roux, *Delgado Community College*
Sharon Sledge, *San Jacinto College*
Beverly Vredevelty, *Spokane Falls Community College*
Kathryn Wetzel, *Amarillo College*
Bridget Young, *Suffolk County Community College*

Get Better Results

Manuscript Review Panels

Over 400 teachers and academics from across the country reviewed the various drafts of the manuscript to give feedback on content, design, pedagogy, and organization. This feedback was summarized by the book team and used to guide the direction of the text.

Special "*thank you*" to our Manuscript Class-Testers

Vernon Bridges, *Durham Technical Community College*
Susan Dimick, *Spokane Community College*
Lori Grady, *University of Wisconsin–Whitewater*

Rod Oberdick, *Delaware Technical and Community College*
Matthew Robinson, *Tallahassee Community College*
Pat Rome, *Delgado Community College–City Park*

Reviewers of the Miller/O'Neill/Hyde Developmental Mathematics Series

Darla Aguilar, *Pima Community College–Desert Vista*
Ebrahim Ahmadizadeh, *Northampton Community College*
Sara Alford, *North Central Texas College*
Theresa Allen, *University of Idaho*
Sheila Anderson, *Housatonic Community College*
Lane Andrew, *Arapahoe Community College*
Jan Archibald, *Ventura College*
Yvonne Aucoin, *Tidewater Community College–Norfolk*
Eric Aurand, *Mohave Community College*
Sohrab Bakhtyari, *St. Petersburg College*
Anna Bakman, *Los Angeles Trade Technical*
Andrew Ball, *Durham Technical Community College*
Russell Banks, *Guilford Technical Community College*
Suzanne Battista, *St. Petersburg College*
Kevin Baughn, *Kirtland Community College*
Sarah Baxter, *Gloucester County College*
Lynn Beckett-Lemus, *El Camino College*
Edward Bender, *Century College*
Emilie Berglund, *Utah Valley State College*
Rebecca Berthiaume, *Edison College–Fort Myers*
John Beyers, *Miami Dade College–Hialeah*
Leila Bicksler, *Delgado Community College–West Bank*
Norma Bisulca, *University of Maine–Augusta*
Kaye Black, *Bluegrass Community and Technical College*
Deronn Bowen, *Broward College–Central*
Timmy Bremer, *Broome Community College*
Donald Bridgewater, *Broward College*
Peggy Brock, *TVI Community College*
Kelly Brooks, *Pierce College*
Susan D. Caire, *Delgado Community College–West Bank*
Peter Carlson, *Delta College*
Judy Carter, *North Shore Community College*
Veena Chadha, *University of Wisconsin–Eau Claire*
Zhixiong Chen, *New Jersey City University*
Tyrone Clinton, *Saint Petersburg College–Gibbs*
Eugenia Cox, *Palm Beach Community College*
Julane Crabtree, *Johnson Community College*
Mark Crawford, *Waubonsee Community College*

Natalie Creed, *Gaston College*
Anabel Darini, *Suffolk County Community College–Brentwood*
Antonio David, *Del Mar College*
Ron Davis, *Kennedy-King College–Chicago*
Laurie Delitsky, *Nassau Community College*
Patti D'Emidio, *Montclair State University*
Bob Denton, *Orange Coast College*
Robert Diaz, *Fullerton College*
Robert Doran, Eileen, *Palm Beach Community College*
Deborah Doucette, *Erie Community College–North Campus—Williamsville*
Thomas Drucker, *University of Wisconsin–Whitewater*
Michael Dubrowsky, *Wayne Community College*
Barbara Duncan, *Hillsborough Community College–Dale Mabry*
Jeffrey Dyess, *Bishop State Community College*
Elizabeth Eagle, *University of North Carolina–Charlotte*
Sabine Eggleston, *Edison College–Fort Myers*
Lynn Eisenberg, *Rowan-Cabarrus Community College*
Barb Elzey, *Bluegrass Community and Technical College*
Nerissa Felder, *Polk Community College*
Mark Ferguson, *Chemeketa Community College*
Diane Fisher, *Lousiana State University–Eunice*
David French, *Tidewater Community College–Chesapeake*
Dot French, *Community College of Philadelphia*
Deborah Fries, *Wor-Wic Community College*
Robert Frye, *Polk Community College*
Jesse M. Fuson, *Mountain State University*
Patricia Gary, *North Virginia Community College–Manassas*
Calvin Gatson, *Alabama State University*
Donna Gerken, *Miami Dade College–Kendall*
Mehrnaz Ghaffarian, *Tarrant County College South*
Mark Glucksman, *El Camino College*
Judy Godwin, *Collin County Community College*
William Graesser, *Ivy Tech Community College*
Victoria Gray, *Scott Community College*

Reviewers of the Miller/O'Neill/Hyde Developmental Mathematics Series *(continued)*

Edna Greenwood, *Tarrant County College–Northwest*

Kimberly Gregor, *Delaware Technical Community College–Wilmington*

Vanetta Grier-Felix, *Seminole Community College*

Kathy Grigsby, *Moraine Valley Community College*

Joseph Guiciardi, *Community College of Allegheny County–Monroeville*

Susan Haley, *Florence-Darlington Technical College*

Mary Lou Hammond, *Spokane Community College*

Joseph Harris, *Gulf Coast Community College*

Lloyd Harris, *Gulf Coast Community College*

Mary Harris, *Harrisburg Area Community College–Lancaster*

Susan Harrison, *University of Wisconsin–Eau Claire*

Kristen Hathcock, *Barton County Community College*

Marie Hoover, *University of Toledo*

Linda Hoppe, *Jefferson College*

Joe Howe, *St. Charles County Community College*

Juan Jimenez, *Springfield Technical Community College*

Jennifer Johnson, *Delgado Community College*

Yolanda Johnson, *Tarrant County College South*

Shelbra Jones, *Wake Technical Community College*

Joe Jordan, *John Tyler Community College*

Cheryl Kane, *University of Nebraska–Lincoln*

Ismail Karahouni, *Lamar University*

Mike Karahouni, *Lamar University–Beaumont*

Joanne Kawczenski, *Luzerne County Community College*

Elaine Keane, *Miami Dade College–North*

Miriam Keesey, *San Diego State University*

Joe Kemble, *Lamar University–Beaumont*

Patrick Kimani, *Morrisville State College*

Sonny Kirby, *Gadsden State Community College*

Vicky Kirkpatrick, *Lane Community College*

Marcia Kleinz, *Atlantic Cape Community College*

Ron Koehn, *Southwestern Oklahoma State University*

Jeff Koleno, *Lorain County Community College*

Rosa Kontos, *Bergen Community College*

Randa Kress, *Idaho State University*

Gayle Krzemie, *Pikes Peak Community College*

Gayle Kulinsky, Carla, *Salt Lake Community College*

Linda Kuroski, *Erie Community College*

Catherine Laberta, *Erie Community College–North Campus—Williamsville*

Joyce Langguth, *University of Missouri–St. Louis*

Betty Larson, *South Dakota State University*

Katie Lathan, *Tri-County Technical College*

Kathryn Lavelle, *Westchester Community College*

Patricia Lazzarino, *North Virginia Community College–Manassas*

Julie Letellier, *University of Wisconsin–Whitewater*

Mickey Levendusky, *Pima Community College*

Barbara Little, *Central Texas College*

David Liu, *Central Oregon Community College*

Maureen Loiacano, *Montgomery College*

Wanda Long, *St. Charles County Community College*

Kerri Lookabill, *Mountain State University*

Jessica Lowenfield, *Nassau Community College*

Diane Lussier, *Pima Community College*

Mark Marino, *Erie Community College–North Campus—Williamsville*

Dorothy Marshall, *Edison College–Fort Myers*

Diane Masarik, *University of Wisconsin–Whitewater*

Louise Mataox, *Miami Dade College*

Cindy McCallum, *Tarrant County College South*

Joyce McCleod, *Florida Community College–South Campus*

Roger McCoach, *County College of Morris*

Stephen F. McCune, *Austin State University*

Ennis McKenna, Hazel, *Utah Valley State College*

Harry McLaughlin, *Montclair State University*

Valerie Melvin, *Cape Fear Community College*

Richard Moore, *St. Petersburg College–Seminole*

Elizabeth Morrison, *Valencia Community College*

Sharon Morrison, *St. Petersburg College*

Shauna Mullins, *Murray State University*

Linda Murphy, *Northern Essex Community College*

Michael Murphy, *Guilford Technical Community College*

Kathy Nabours, *Riverside Community College*

Roya Namavar, *Rogers State University*

Tony Nelson, *Tulsa Community College*

Melinda Nevels, *Utah Valley State College*

Charlotte Newsom, *Tidewater Community College–Virginia Beach*

Brenda Norman, *Tidewater Community College*

David Norwood, *Alabama State University*

Rhoda Oden, *Gadsden State Community College*

Tammy Payton, *North Idaho College*

Melissa Pedone, *Valencia Community College–Osceola*

Shirley Pereira, *Grossmont College*

Pete Peterson, *John Tyler Community College*

Suzie Pickle, *St. Petersburg College*

Sheila Pisa, *Riverside Community College–Moreno Valley*

Marilyn Platt, *Gaston College*

Richard Ponticelli, *North Shore Community College*

Tammy Potter, *Gadsden State Community College*

Joel Rappaport, *Florida Community College*

Sherry Ray, *Oklahoma City Community College*

Angelia Reynolds, *Gulf Coast Community College*

Suellen Robinson, *North Shore Community College*

Jeri Rogers, *Seminole Community College–Oviedo*

Trisha Roth, *Gloucester County College*

Richard Rupp, *Del Mar College*

Dave Ruszkiewicz, *Milwaukee Area Technical College*

Nancy Sattler, *Terra Community College*

Vicki Schell, *Pensacola Junior College*

Nyeita Schult, *St. Petersburg College*

Wendiann Sethi, *Seton Hall University*

Dustin Sharp, *Pittsburg Community College*

Marvin Shubert, *Hagerstown Community College*

Plamen Simeonov, *University of Houston–Downtown*

Carolyn Smith, *Armstrong Atlantic State University*

Melanie Smith, *Bishop State Community College*

John Squires, *Cleveland State Community College*

Sharon Staver, Judith, *Florida Community College–South Campus*

Sharon Steuer, *Nassau Community College*

Trudy Streilein, *North Virginia Community College–Annandale*

Gretchen Syhre, *Hawkeye Community College*

Katalin Szucs, *Pittsburg Community College*

Mike Tiano, *Suffolk County Community College*

Stephen Toner, *Victor Valley College*

Mary Lou Townsend, *Wor-Wic Community College*

Susan Twigg, *Wor-Wic Community College*

Matthew Utz, *University of Arkansas–Fort Smith*

Joan Van Glabek, *Edison College–Fort Myers*

John Van Kleef, *Guilford Technical Community College*

Diane Veneziale, *Burlington County College–Pemberton*

Andrea Vorwark, *Metropolitan Community College–Maple Woods*

Edward Wagner, *Central Texas College*

David Wainaina, *Coastal Carolina Community College*

James Wang, *University of Alabama*

Richard Watkins, *Tidewater Community College–Virginia Beach*

Sharon Wayne, *Patrick Henry Community College*

Leben Wee, *Montgomery College*

Betty Vix Weinberger, *Delgado Community College–West Bank*

Christine Wetzel-Ulrich, *Northampton Community College*

Jackie Wing, *Angelina College*

Michelle Wolcott, *Pierce College*

Deborah Wolfson, *Suffolk County Community College–Brentwood*

Mary Wolyniak, *Broome Community College*

Rick Woodmansee, *Cosumnes River College*

Susan Working, *Grossmont College*

Karen Wyrick, *Cleveland State Community College*

Alan Yang, *Columbus State Community College*

William Young, Jr, *Century College*

Vasilis Zafiris, *University of Houston*

Vivian Zimmerman, *Prairie State College*

Special thanks go to Jon Weerts for preparing the *Instructor's Solutions Manual* and the *Student's Solution Manual* and to Rebecca Hubiak for her work ensuring accuracy. Many thanks to Cindy Reed for her work in the video series, and to Ethel Wheland for advising us on the Instructor Notes.

Finally, we are forever grateful to the many people behind the scenes at McGraw-Hill without whom we would still be on page 1. To our developmental editor (and math instructor extraordinaire), Emilie Berglund, thanks for your day-to-day support and understanding of the world of developmental mathematics. To David Millage, our sponsoring editor and overall team captain, thanks for keeping the train on the track. Where did you find enough hours in the day? To Torie Anderson and Sabina Navsariwala, we greatly appreciate your countless hours of support and creative ideas promoting all of our efforts. To our director of development and champion,

Kris Tibbetts, thanks for being there in our time of need. To Pat Steele, where would we be without your watchful eye over our manuscript? To our editorial director, Stewart Mattson, we're grateful for your experience and energizing new ideas. Thanks for believing in us. To Jeff Huettman and Amber Bettcher, we give our greatest appreciation for the exciting technology so critical to student success. To Peggy Selle thanks for keeping watch over the whole team as the project came together. Thank you to our wonderful designer Laurie Janssen, not only did Laurie help develop a better textbook series by delivering a clean, clear design framework for the mathematics content, Laurie also designed the best covers of the Miller/O'Neill/Hyde series to date.

Most importantly, we give special thanks to all the students and instructors who use *Introductory Algebra* in their classes.

Supplements For the Instructor

Instructor's Resource Manual

The *Instructor's Resource Manual* (*IRM*), written by the authors, is a printable electronic supplement available through Mathzone. The *IRM* includes discovery-based classroom activities, worksheets for drill and practice, materials for a student portfolio, and tips for implementing successful cooperative learning. Numerous classroom activities are available for each section of text and can be used as a complement to the lectures or can be assigned for work outside of class. The activities are designed for group or individual work and take about 5–10 minutes each. With increasing demands on faculty schedules, these ready-made lessons offer a convenient means for both full-time and adjunct faculty to promote active learning in the classroom.

MathZone www.mathzone.com

McGraw-Hill's **MathZone** is a complete online tutorial and course management system for mathematics and statistics, designed for greater ease of use than any other system available. Available with selected McGraw-Hill textbooks, the system enables instructors to **create and share courses and assignments** with colleagues and adjuncts with only a few clicks of the mouse. All assignments, questions, e-professors, online tutoring, and video lectures are directly tied to **text-specific** materials.

MathZone courses are customized to your textbook, but you can edit questions and algorithms, import your own content, and **create** announcements and due dates for assignments.

MathZone has **automatic grading** and reporting of easy-to-assign, algorithmically generated homework, quizzing, and testing. All student activity within **MathZone** is automatically recorded and available to you through a **fully integrated grade book** that can be downloaded to Excel.

MathZone offers:

- **Practice exercises** based on the textbook and generated in an unlimited number for as much practice as needed to master any topic you study.
- **Videos** of classroom instructors giving lectures and showing you how to solve exercises from the textbook.
- **e-Professors** to take you through animated, step-by-step instructions (delivered via on-screen text and synchronized audio) for solving problems in the book, allowing you to digest each step at your own pace.
- **NetTutor,** which offers live, personalized tutoring via the Internet.

Instructor's Testing and Resource Online

Provides a wealth of resources for the instructor. Among the supplements is a **computerized test bank** utilizing Brownstone Diploma® algorithm-based testing software to create customized exams quickly. This user-friendly program enables instructors to search for questions by topic, format, or difficulty level; to edit existing questions or to add new ones; and to scramble questions and answer keys for multiple versions of a single test. Hundreds of text-specific, open-ended, and multiple-choice questions are included in the question bank. Sample chapter tests are also provided. CD available upon request.

New ALEKS Instructor Module

The new ALEKS Instructor Module features enhanced functionality and a stream-lined interface based on research with ALEKS instructors and homework management instructors. Paired with powerful assignment driven features, textbook integration, and extensive content flexibility, the new ALEKS Instructor Module simplifies administrative tasks and makes ALEKS more powerful than ever. Features include:

Gradebook Instructors can seamlessly track student scores on automatically graded assignments. They can also easily adjust the weighting and grading scale of each assignment.

Course Calendar Instructors can schedule assignments and reminders for students.

Automatically Graded Assignments Instructors can easily assign homework, quizzes, tests, and assessments to all or select students. Deadline extensions can also be created for select students.

Set-Up Wizards Instructors can use wizards to easily set up assignments, course content, textbook integration, etc.

Message Center Instructors can use the redesigned Message Center to send, receive, and archive messages; input tools are available to convey mathematical expressions via email.

Miller/O'Neill/Hyde Video Lectures on Digital Video Disk (DVD)

In the videos, qualified instructors work through selected problems from the text-book, following the solution methodology employed in the text. The video series is available on DVD or online as an assignable element of MathZone. The DVDs are closed-captioned for the hearing-impaired, are subtitled in Spanish, and meet the Americans with Disabilities Act Standards for Accessible Design. Instructors may use them as resources in a learning center, for online courses, and to provide extra help for students who require extra practice.

Annotated Instructor's Edition

In the *Annotated Instructor's Edition* (*AIE*), **answers to all exercises and tests appear adjacent to each exercise,** in a color used *only* for annotations. The *AIE* also contains **Instructor Notes** that appear in the margin. The notes may assist with lecture preparation. Also found in the *AIE* are icons within the Practice Exercises that serve to guide instructors in their preparation of homework assignments and lessons.

Another significant feature new to this edition is the inclusion of ***Classroom Examples*** for the instructor. In the Annotated Instructor's Edition of the text, we include references to even-numbered exercises at the end of the section for instructors to use as *Classroom Examples*. These exercises mirror the examples in the text. Therefore, if an instructor covers these exercises as classroom examples, then all the major objectives in that section will have been covered. This feature was added because we recognize the growing demands on faculty time, and to assist new faculty, adjunct faculty, and graduate assistants. Furthermore, because these exercises appear in the student edition of the text, students will not waste valuable class time copying down complicated examples from the board.

Instructor's Solutions Manual

The *Instructor's Solutions Manual* provides comprehensive, worked-out solutions to all exercises in the Chapter Openers; the Practice Exercises; the Problem Recognition Exercises; the end-of-chapter Review Exercises; the Chapter Tests; and the Cumulative Review Exercises.

Supplements For the Student

 www.mathzone.com

McGraw-Hill's MathZone is a powerful Web-based tutorial for homework, quizzing, testing, and multimedia instruction. Also available in CD-ROM format, Math-Zone offers:

Practice exercises based on the text and generated in an unlimited quantity for as much practice as needed to master any objective

Video clips of classroom instructors showing how to solve exercises from the text, step by step

e-Professor animations that take the student through step-by-step instructions, delivered on-screen and narrated by a teacher on audio, for solving exercises from the textbook; the user controls the pace of the explanations and can review as needed

NetTutor, which offers personalized instruction by live tutors familiar with the textbook's objectives and problem-solving methods

Every assignment, exercise, video lecture, and e-Professor is derived from the textbook.

ALEKS Prep for Developmental Mathematics

ALEKS Prep for Beginning Algebra and Prep for Intermediate Algebra focus on prerequisite and introductory material for Beginning Algebra and Intermediate Algebra. These prep products can be used during the first 3 weeks of a course to prepare students for future success in the course and to increase retention and pass rates. Backed by two decades of National Science Foundation funded research, ALEKS interacts with students much like a human tutor, with the ability to precisely assess a student's preparedness and provide instruction on the topics the student is most likely to learn.

ALEKS Prep Course Products Feature:

- Artificial Intelligence Targets Gaps in Individual Students Knowledge
- Assessment and Learning Directed Toward Individual Students Needs
- Open Response Environment with Realistic Input Tools
- Unlimited Online Access-PC & Mac Compatible

Free trial at www.aleks.com/free_trial/instructor

Student's Solutions Manual

The *Student's Solutions Manual* provides comprehensive, worked-out solutions to the odd-numbered exercises in the Practice Exercise sets; the Problem Recognition Exercises, the end-of-chapter Review Exercises, the Chapter Tests, and the Cumulative Review Exercises. Answers to the odd- and even-numbered entries to the Chapter Opener Puzzles are also provided.

New connect2Developmental Mathematics video Series!

Available on DVD and the MathZone website, these innovative videos bring essential Developmental Mathematics concepts to life! The videos take the concepts and place them in a real world setting so that students make the connection from what they learn in the classroom to real world experiences outside the classroom. Making use of 3D animations and lectures, Connect2Developmental Mathematics video series answers the age-old questions "Why is this important?" and "When will I ever use it?" The videos cover topics from Arithmetic and Basic Mathematics through the Algebra sequence, mixing student-oriented themes and settings with basic theory.

Video Lectures on Digital Video Disk

The video series is based on exercises from the textbook. Each presenter works through selected problems, following the solution methodology employed in the text. The video series is available on DVD or online as part of MathZone. The DVDs are closed-captioned for the hearing impaired, are subtitled in Spanish, and meet the Americans with Disabilities Act Standards for Accessible Design.

NetTutor

Available through MathZone, NetTutor is a revolutionary system that enables students to interact with a live tutor over the web. NetTutor's Web-based, graphical chat capabilities enable students and tutors to use mathematical notation and even to draw graphs as they work through a problem together. Students can also submit questions and receive answers, browse previously answered questions, and view previous sessions. Tutors are familiar with the textbook's objectives and problem-solving styles.

A COMMITMENT TO ACCURACY

You have a right to expect an accurate textbook, and McGraw-Hill invests considerable time and effort to make sure that we deliver one. Listed below are the many steps we take to make sure this happens.

Our Accuracy Verification Process

First Round

Step 1: Numerous **college math instructors** review the manuscript and report on any errors that they may find. Then the authors make these corrections in their final manuscript.

Second Round

Step 2: Once the manuscript has been typeset, the **authors** check their manuscript against the first page proofs to ensure that all illustrations, graphs, examples, exercises, solutions, and answers have been correctly laid out on the pages, and that all notation is correctly used.

Step 3: An outside, **professional mathematician** works through every example and exercise in the page proofs to verify the accuracy of the answers.

Step 4: A **proofreader** adds a triple layer of accuracy assurance in the first pages by hunting for errors, then a second, corrected round of page proofs is produced.

Third Round

Step 5: The **author team** reviews the second round of page proofs for two reasons: (1) to make certain that any previous corrections were properly made, and (2) to look for any errors they might have missed on the first round.

Step 6: A **second proofreader** is added to the project to examine the new round of page proofs to double check the author team's work and to lend a fresh, critical eye to the book before the third round of paging.

Fourth Round

Step 7: A **third proofreader** inspects the third round of page proofs to verify that all previous corrections have been properly made and that there are no new or remaining errors.

Step 8: Meanwhile, in partnership with **independent mathematicians,** the text accuracy is verified from a variety of fresh perspectives:

- The **test bank author** checks for consistency and accuracy as he/she prepares the computerized test item file.
- The **solutions manual author** works every exercise and verifies his/her answers, reporting any errors to the publisher.
- A **consulting group of mathematicians,** who write material for the text's MathZone site, notifies the publisher of any errors they encounter in the page proofs.
- A video production company employing **expert math instructors** for the text's videos will alert the publisher of any errors they might find in the page proofs.

Final Round

Step 9: The **project manager,** who has overseen the book from the beginning, performs a **fourth proofread** of the textbook during the printing process, providing a final accuracy review.

➡ What results is a mathematics textbook that is as accurate and error-free as is humanly possible, and our authors and publishing staff are confident that our many layers of quality assurance have produced textbooks that are the leaders in the industry for their integrity and correctness.

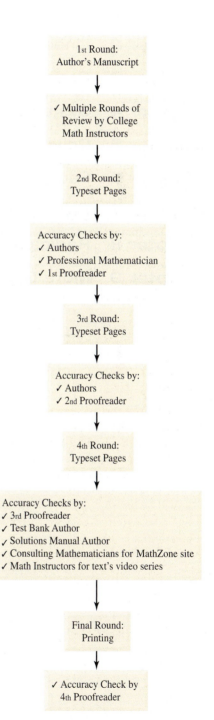

1st Round:
Author's Manuscript

✓ Multiple Rounds of
Review by College
Math Instructors

2nd Round:
Typeset Pages

Accuracy Checks by:
✓ Authors
✓ Professional Mathematician
✓ 1st Proofreader

3rd Round:
Typeset Pages

Accuracy Checks by:
✓ Authors
✓ 2nd Proofreader

4th Round:
Typeset Pages

Accuracy Checks by:
✓ 3rd Proofreader
✓ Test Bank Author
✓ Solutions Manual Author
✓ Consulting Mathematicians for MathZone site
✓ Math Instructors for text's video series

Final Round:
Printing

✓ Accuracy Check by
4th Proofreader

Brief Contents

Contents

Reference

Chapter R

This chapter is a reference chapter that provides a review of the basic operations on fractions, decimals, percents, and geometry. This chapter also provides a section of valuable study tips.

As you read through Chapter R, try to become familiar with the features of this textbook. Then match the feature in column B with its description in column A. Then fill in the numbered blanks with the matching letters to complete the riddle.

Column A

1. Allows you to check your work as you do your homework
2. Shows you how to avoid common errors
3. Provides an online tutorial and exercise supplement
4. Outlines key concepts for each section in the chapter
5. Provides exercises that allow you to distinguish between different types of problems
6. Offers helpful hints and insight
7. Offers practice exercises that go along with each example

Column B

a. Tips
c. ALEKS
l. Skill Practice Exercises
m. Problem Recognition Exercises
d. Answers to odd exercises
i. Chapter Summary
e. Avoiding Mistakes

Where do mathematicians shop? At the $\underline{\ \ }\ \underline{\ \ }\ \underline{\ \ }\ \underline{\ \ }\ \underline{\ \ }\ \underline{\ \ }\ \underline{\ \ }$.
12345677

| Section R.1 | Study Tips |

Objectives

1. Before the Course
2. During the Course
3. Preparation for Exams
4. Where to Go for Help

In taking a course in algebra, you are making a commitment to yourself, your instructor, and your classmates. Following some or all of the study tips below can help you be successful in this endeavor. The features of this text that will assist you are printed in blue.

1. Before the Course

- Purchase the necessary materials for the course before the course begins or on the first day.
- Obtain a three-ring binder to keep and organize your notes, homework, tests, and any other materials acquired in the class. We call this type of notebook a portfolio.
- Arrange your schedule so that you have enough time to attend class and to do homework. A common rule is to set aside at least 2 hours for homework for every hour spent in class. That is, if you are taking a 4-credit-hour course, plan on at least 8 hours a week for homework. A 4-credit-hour course will then take *at least* 12 hours each week—about the same as a part-time job. If you experience difficulty in mathematics, plan for more time.
- Communicate with your employer and family members the importance of your success in this course so that they can support you.
- Be sure to find out the type of calculator (if any) that your instructor requires.

2. During the Course

- Read the section in the text *before* the lecture to familiarize yourself with the material and terminology.
- Attend every class, and be on time.
- Take notes in class. Write down all of the examples that the instructor presents. Read the notes after class, and add any comments to make your notes clearer to you. Use a tape recorder to record the lecture if the instructor permits the recording of lectures.
- Ask questions in class.
- Read the section in the text *after* the lecture, and pay special attention to the Tip boxes and Avoiding Mistakes boxes.
- After you read an example, try the accompanying Skill Practice problem in the margin. The skill practice problem mirrors the example and tests your understanding of what you have read.
- Do homework every night. Even if your class does not meet every day, you should still do some work every night to keep the material fresh in your mind.
- Check your homework with the answers that are supplied in the back of this text. Correct the exercises that do not match, and circle or star those that you cannot correct yourself. This way you can easily find them and ask your instructor the next day.
- Write the definition and give an example of each Key Term found at the beginning of the Practice Exercises.
- The Problem Recognition Exercises are located in most chapters. These provide additional practice distinguishing among a variety of problem types. Sometimes the most difficult part of learning mathematics is retaining all that you learn. These exercises are excellent tools for retention of material.

- Form a study group with fellow students in your class, and exchange phone numbers. You will be surprised by how much you can learn by talking about mathematics with other students.
- If you use a calculator in your class, read the Calculator Connections boxes to learn how and when to use your calculator.
- Ask your instructor where you might obtain extra help if necessary.

3. Preparation for Exams

- Look over your homework. Pay special attention to the exercises you have circled or starred to be sure that you have learned that concept.
- Read through the Summary at the end of the chapter. Be sure that you understand each concept and example. If not, go to the section in the text and reread that section.
- Give yourself enough time to take the Chapter Test uninterrupted. Then check the answers. For each problem you answered incorrectly, go to the Review Exercises and do all of the problems that are similar.
- To prepare for the final exam, complete the Cumulative Review Exercises at the end of each chapter, starting with Chapter 2. If you complete the cumulative reviews after finishing each chapter, then you will be preparing for the final exam throughout the course. The Cumulative Review Exercises are another excellent tool for helping you retain material.

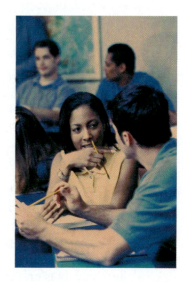

4. Where to Go for Help

- At the first sign of trouble, see your instructor. Most instructors have specific office hours set aside to help students. Don't wait until after you have failed an exam to seek assistance.
- Get a tutor. Most colleges and universities have free tutoring available.
- When your instructor and tutor are unavailable, use the Student Solutions Manual for step-by-step solutions to the odd-numbered problems in the exercise sets.
- Work with another student from your class.
- Work on the computer. Many mathematics tutorial programs and websites are available on the Internet, including the website that accompanies this text.

Section R.1   Practice Exercises

Boost your GRADE at ALEKS.com!

- Practice Problems
- Self-Tests
- NetTutor
- e-Professors
- Videos

Objective 1: Before the Course

1. To motivate yourself to complete a course, it is helpful to have a very clear reason for taking the course. List your goals for taking this course.

2. Budgeting enough time to do homework and to study for a class is one of the most important steps to success in a class. Use a weekly calendar to help you plan your time for your studies this week. Also write other obligations such as the time required for your job, for your family, for sleeping, and for eating. Be realistic when estimating the time for each activity.

3. Taking 12 credit-hours is the equivalent of a full-time job. Often students try to work too many hours while taking classes at school.

    a. Write down how many hours you work per week and the number of credit hours you are taking this term.

    number of hours worked per week _____

    number of credit-hours this term _____

    b. The table gives a recommended limit to the number of hours you should work for the number of credit hours you are taking at school. (Keep in mind that other responsibilities in your life such as your family might also make it necessary to limit your hours at work even more.) How do your numbers from part (a) compare to those in the table? Are you working too many hours?

Number of Credit-Hours	Maximum Number of Hours of Work per Week
3	40
6	30
9	20
12	10
15	0

4. It is important to establish a place where you can study—someplace that has few distractions and is readily available. Answer the questions about the space that you have chosen.

    a. Is there enough room to spread out books and paper to do homework?

    b. Is this space available anytime?

    c. Are there any distractions? Can you be interrupted?

    d. Is the furniture appropriate for studying? That is, is there a comfortable chair and good lighting?

5. Organization is an important ingredient to success. A calendar or pocket planner is a valuable resource for keeping track of assignments and test dates. Write the date of the first test in this class.

Objective 2: During the Course

6. Taking notes can help in many ways. Good notes provide examples for reference as you do your homework. Also, taking notes keeps your mind on track during the lecture and helps make you an active listener. Here are some tips to help you take better notes in class.

    a. In your next math class, take notes by drawing a vertical line about $\frac{3}{4}$ of the way across the paper as shown. On the left side, write down what your instructor puts on the board or overhead. On the right side, make your own comments about important words, procedures, or questions that you have.

    b. Revisit your notes as soon as possible after class to fill in the missing parts that you recall from lecture but did not have time to write down.

    c. Be sure that you label each page with the date, chapter, section, and topic. This will make your notes easier to study from when you study for a test.

7. Many instructors use a variety of styles to accommodate all types of learners. From the following list, try to identify the type of learner that best describes you.

Auditory Learner   Do you learn best by listening to your instructor's lectures? Do you tape the lecture so that you can listen to it as many times as you need? Do you talk aloud when doing homework or studying for a test?

Visual Learner   Do you learn best by seeing problems worked out on the board? Do you understand better if there is a picture or illustration accompanying the problem? Do you take notes in class?

Tactile Learner   Do you learn best with hands-on projects? Do you prefer having some sort of physical objects to manipulate? Do you prefer to move around the classroom in a lab situation?

Objective 3: Preparation for Exams

8. When taking a test, go through the test and work all the problems that you know first. Then go back and work on the problems that were more difficult. Give yourself a time limit for each problem (maybe 3 to 5 minutes the first time through the test). Circle the importance of each statement.

	not important	somewhat important	very important
a. Read through the entire test first.	1	2	3
b. If time allows, go back and check each problem.	1	2	3
c. Write out all steps instead of doing the work in your head.	1	2	3

9. One way to lessen test anxiety is to feel prepared for the exam. Check the ways that you think might be helpful in preparing for a test.

_____ Do the chapter test in the text.

_____ Get in a study group with your peers and go over the major topics.

_____ Write your own pretest.

_____ Use the online component that goes with your textbook.

_____ Write an outline of the major topics. Then find an example for each topic.

_____ Make flash cards with a definition, property, or rule on one side and an example on the other.

_____ Re-read your notes that you took in class.

_____ Write down the definitions of all key terms introduced in the chapter.

10. The following list gives symptoms of math anxiety.

• Experiencing loss of sleep and worrying about an upcoming exam.

• Mind becoming blank when answering a question or taking a test.

• Becoming nervous about asking questions in class.

• Experiencing anxiety that interferes with studying.

• Being afraid or embarrassed to let the instructor see your work.

• Becoming physically ill during a test.

• Having sweaty palms or shaking hands when asked a math question.

Have you experienced any of these symptoms? If so, how many and how often?

11. If you think that you have math anxiety, read the following list for some possible solutions. Check the activities that you can realistically try to help you overcome this problem.

_____ Read a book on math anxiety.

_____ Search the Web for helpful tips on handling math anxiety.

_____ See a counselor to discuss your anxiety.

_____ See your instructor to inform him or her about your situation.

_____ Evaluate your time management to see if you are trying to do too much. Then adjust your schedule accordingly.

Objective 4: Where to Go for Help

12. Does your college offer free tutoring? If so, write down the room number and the hours of the tutoring center.

13. Does your instructor have office hours? If so, write down your professor's office number and office hours.

14. Is there a supplement to your text? If so, find out its price and where you can get it.

15. Find out how to access the online tutoring available with this text.

Section R.2	Fractions

Objectives

1. Basic Definitions
2. Prime Factorization
3. Simplifying Fractions to Lowest Terms
4. Multiplying Fractions
5. Dividing Fractions
6. Adding and Subtracting Fractions
7. Operations on Mixed Numbers

1. Basic Definitions

The study of algebra involves many of the operations and procedures used in arithmetic. Therefore, we begin this text by reviewing the basic operations of addition, subtraction, multiplication, and division on fractions and mixed numbers.

In day-to-day life, the numbers we use for counting are

the **natural numbers**: $1, 2, 3, 4, \ldots$

and

the **whole numbers**: $0, 1, 2, 3, \ldots$

Whole numbers are used to count the number of whole units in a quantity. A fraction is used to express part of a whole unit. If a child gains $2\frac{1}{2}$ lb, the child has gained two whole pounds plus a portion of a pound. To express the additional half pound mathematically, we may use the fraction, $\frac{1}{2}$.

> **DEFINITION**   **A Fraction and Its Parts**
>
> **Fractions** are numbers of the form $\frac{a}{b}$, where $\frac{a}{b} = a \div b$ and b does not equal zero.
>
> In the fraction $\frac{a}{b}$, the **numerator** is a, and the **denominator** is b.

The denominator of a fraction indicates how many equal parts divide the whole. The numerator indicates how many parts are being represented. For instance, suppose Jack wants to plant carrots in $\frac{2}{5}$ of a rectangular garden. He can divide the garden into five equal parts and use two of the parts for carrots (Figure R-1).

5 equal parts

2 parts   The shaded region represents $\dfrac{2}{5}$ of the garden.

Figure R-1

DEFINITION **Proper Fractions, Improper Fractions, and Mixed Numbers**

1. If the numerator of a fraction is less than the denominator, the fraction is a **proper fraction**. A proper fraction represents a quantity that is less than a whole unit.
2. If the numerator of a fraction is greater than or equal to the denominator, then the fraction is an **improper fraction**. An improper fraction represents a quantity greater than or equal to a whole unit.
3. A **mixed number** is a whole number added to a proper fraction.

Proper Fractions: $\dfrac{3}{5}$ $\dfrac{1}{8}$

Improper Fractions: $\dfrac{7}{5}$ $\dfrac{8}{8}$

Mixed Numbers: $1\frac{1}{5}$ $2\frac{3}{8}$

Concept Connections

1. Write a proper or improper fraction associated with the shaded region.

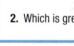

2. Which is greater, $\dfrac{3}{2}$ or $3\dfrac{1}{2}$?

2. Prime Factorization

To perform operations on fractions it is important to understand the concept of a factor. For example, when the numbers 2 and 6 are multiplied, the result (called the **product**) is 12.

$$2 \times 6 = 12$$

factors   product

The numbers 2 and 6 are said to be **factors** of 12. (In this context, we refer only to natural number factors.) The number 12 is said to be factored when it is written as the product of two or more natural numbers. For example, 12 can be factored in several ways:

$$12 = 1 \times 12 \qquad 12 = 2 \times 6 \qquad 12 = 3 \times 4 \qquad 12 = 2 \times 2 \times 3$$

A natural number greater than 1 that has only two factors, 1 and itself, is called a **prime number**. The first several prime numbers are 2, 3, 5, 7, 11, and 13. A natural number greater than 1 that is not prime is called a **composite number**. That is, a composite number has factors other than itself and 1. The first several composite numbers are 4, 6, 8, 9, 10, 12, 14, 15, and 16.

Avoiding Mistakes

The number 1 is neither prime nor composite.

Answers

1. $\dfrac{7}{4}$ 2. $3\dfrac{1}{2}$

Example 1 **Writing a Natural Number as a Product of Prime Factors**

Write each number as a product of prime factors.

a. 12   **b.** 30

Solution:

a. $12 = 2 \times 2 \times 3$ Divide 12 by prime numbers until only prime numbers are obtained.   Or use a factor tree

$$\begin{array}{r} 2\overline{)12} \\ 2\overline{)6} \\ 3 \end{array}$$

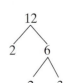

b. $30 = 2 \times 3 \times 5$

$$\begin{array}{r} 2\overline{)30} \\ 3\overline{)15} \\ 5 \end{array}$$

3. Simplifying Fractions to Lowest Terms

The process of factoring numbers can be used to reduce or simplify fractions to lowest terms. A fractional portion of a whole can be represented by infinitely many fractions. For example, Figure R-2 shows that $\frac{1}{2}$ is equivalent to $\frac{2}{4}, \frac{3}{6}, \frac{4}{8}$, and so on.

   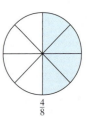

$$\frac{1}{2} \qquad \frac{2}{4} \qquad \frac{3}{6} \qquad \frac{4}{8}$$

Figure R-2

The fraction $\frac{1}{2}$ is said to be in **lowest terms** because the numerator and denominator share no common factor other than 1.

To simplify a fraction to lowest terms, we use the following important principle.

PROPERTY **Fundamental Principle of Fractions**

Suppose that a number, c, is a common factor in the numerator and denominator of a fraction. Then

$$\frac{a \times c}{b \times c} = \frac{a}{b} \times \frac{c}{c} = \frac{a}{b} \times 1 = \frac{a}{b}$$

To simplify a fraction, we begin by factoring the numerator and denominator into prime factors. This will help identify the common factors.

Example 2 Simplifying a Fraction to Lowest Terms

Simplify $\dfrac{45}{30}$ to lowest terms.

Solution:

$\dfrac{45}{30} = \dfrac{3 \times 3 \times 5}{2 \times 3 \times 5}$ Factor the numerator and denominator.

$= \dfrac{3}{2} \times \dfrac{3}{3} \times \dfrac{5}{5}$ Apply the fundamental principle of fractions.

$= \dfrac{3}{2} \times 1 \times 1$ Any nonzero number divided by itself is 1.

$= \dfrac{3}{2}$ Any number multiplied by 1 is itself.

Skill Practice

5. Simplify to lowest terms.
$\dfrac{20}{50}$

In Example 2, we showed numerous steps to reduce fractions to lowest terms. However, the process is often simplified. Notice that the same result can be obtained by dividing out the greatest common factor from the numerator and denominator. (The **greatest common factor** is the largest factor that is common to both numerator and denominator.)

$\dfrac{45}{30} = \dfrac{3 \times 15}{2 \times 15}$ The greatest common factor of 45 and 30 is 15.

$= \dfrac{3 \times \overset{1}{\cancel{15}}}{2 \times \underset{1}{\cancel{15}}}$ The symbol ╱ is often used to show that a common factor has been divided out.

$= \dfrac{3}{2}$ Notice that "dividing out" the common factor of 15 has the same effect as dividing the numerator and denominator by 15. This is often done mentally.

$\dfrac{\overset{3}{\cancel{45}}}{\underset{2}{\cancel{30}}} = \dfrac{3}{2}$ ⟵ 45 divided by 15 equals 3.
            ⟵ 30 divided by 15 equals 2.

Avoiding Mistakes

In Example 3, the common factor 14 in the numerator and denominator simplifies to 1. It is important to remember to write the factor of 1 in the numerator. The simplified form of the fraction is $\frac{1}{3}$.

Example 3 Simplifying a Fraction to Lowest Terms

Simplify $\dfrac{14}{42}$ to lowest terms.

Solution:

$\dfrac{14}{42} = \dfrac{1 \times 14}{3 \times 14}$ The greatest common factor of 14 and 42 is 14.

$= \dfrac{1 \times \overset{1}{\cancel{14}}}{3 \times \underset{1}{\cancel{14}}}$

$= \dfrac{1}{3}$ $\dfrac{\overset{1}{\cancel{14}}}{\underset{3}{\cancel{42}}} = \dfrac{1}{3}$ ⟵ 14 divided by 14 equals 1.
                        ⟵ 42 divided by 14 equals 3.

Skill Practice

6. Simplify to lowest terms.
$\dfrac{32}{12}$

Answers

5. $\dfrac{2}{5}$ 6. $\dfrac{8}{3}$

4. Multiplying Fractions

> **PROCEDURE   Multiplying Fractions**
>
> If b is not zero and d is not zero, then
>
> $$\frac{a}{b} \times \frac{c}{d} = \frac{a \times c}{b \times d}$$
>
> To multiply fractions, multiply the numerators and multiply the denominators.

Example 4   Multiplying Fractions

Multiply the fractions: $\dfrac{1}{4} \times \dfrac{1}{2}$

Solution:

$$\frac{1}{4} \times \frac{1}{2} = \frac{1 \times 1}{4 \times 2} = \frac{1}{8}$$

Multiply the numerators. Multiply the denominators.

Notice that the product $\frac{1}{4} \times \frac{1}{2}$ represents a quantity that is $\frac{1}{4}$ of $\frac{1}{2}$. Taking $\frac{1}{4}$ of a quantity is equivalent to dividing the quantity by 4. One-half of a pie divided into four pieces leaves pieces that each represent $\frac{1}{8}$ of the pie (Figure R-3).

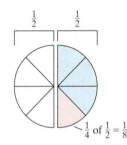

$\frac{1}{4}$ of $\frac{1}{2} = \frac{1}{8}$

Figure R-3

Example 5   Multiplying Fractions

Multiply the fractions.

a. $\dfrac{7}{10} \times \dfrac{15}{14}$ **b.** $\dfrac{2}{13} \times \dfrac{13}{2}$ **c.** $5 \times \dfrac{1}{5}$

Solution:

a. $\dfrac{7}{10} \times \dfrac{15}{14} = \dfrac{7 \times 15}{10 \times 14}$

Multiply the numerators. Multiply the denominators.

$$= \frac{\overset{3}{\cancel{105}}}{\underset{4}{\cancel{140}}}$$

Divide out the common factor, 35.

$$= \frac{3}{4}$$

> **TIP:** The same result can be obtained by dividing out common factors *before* multiplying.
>
> $$\frac{\overset{1}{\cancel{7}}}{\underset{2}{\cancel{10}}} \times \frac{\overset{3}{\cancel{15}}}{\underset{2}{\cancel{14}}} = \frac{3}{4}$$

b. $\dfrac{2}{13} \times \dfrac{13}{2} = \dfrac{2 \times 13}{13 \times 2} = \dfrac{\overset{1}{2} \times \overset{1}{13}}{\underset{1}{13} \times \underset{1}{2}} = \dfrac{1}{1} = 1$

Multiply $1 \times 1 = 1$.

Multiply $1 \times 1 = 1$.

c. $5 \times \dfrac{1}{5} = \dfrac{5}{1} \times \dfrac{1}{5}$

The whole number 5 can be written as $\frac{5}{1}$.

$= \dfrac{\overset{1}{5} \times 1}{1 \times \underset{1}{5}} = \dfrac{1}{1} = 1$

Multiply and simplify to lowest terms.

5. Dividing Fractions

Before we divide fractions, we need to know how to find the reciprocal of a fraction. Notice from Example 5 that $\frac{2}{13} \times \frac{13}{2} = 1$ and $5 \times \frac{1}{5} = 1$. The numbers $\frac{2}{13}$ and $\frac{13}{2}$ are said to be reciprocals because their product is 1. Likewise the numbers 5 and $\frac{1}{5}$ are reciprocals.

DEFINITION   The Reciprocal of a Number

Two nonzero numbers are **reciprocals** of each other if their product is 1. Therefore, the reciprocal of the fraction

$$\dfrac{a}{b} \text{ is } \dfrac{b}{a} \qquad \text{because} \qquad \dfrac{a}{b} \times \dfrac{b}{a} = 1$$

Number	Reciprocal	Product
$\dfrac{2}{15}$	$\dfrac{15}{2}$	$\dfrac{2}{15} \times \dfrac{15}{2} = 1$
$\dfrac{1}{8}$	$\dfrac{8}{1}$ (or equivalently 8)	$\dfrac{1}{8} \times 8 = 1$
$6\left(\text{or equivalently } \dfrac{6}{1}\right)$	$\dfrac{1}{6}$	$6 \times \dfrac{1}{6} = 1$

Concept Connections

Give the reciprocal.

11. $\dfrac{9}{4}$ **12.** 7   **13.** $\dfrac{1}{15}$

To understand the concept of dividing fractions, consider a pie that is half-eaten. Suppose the remaining half must be divided among three people, that is, $\frac{1}{2} \div 3$. However, dividing by 3 is equivalent to taking $\frac{1}{3}$ of the remaining $\frac{1}{2}$ of the pie (Figure R-4).

$$\dfrac{1}{2} \div 3 = \dfrac{1}{2} \cdot \dfrac{1}{3} = \dfrac{1}{6}$$

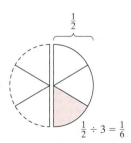

$\frac{1}{2} \div 3 = \frac{1}{6}$

Figure R-4

Answers

11. $\dfrac{4}{9}$ **12.** $\dfrac{1}{7}$ **13.** 15

This example illustrates that dividing two numbers is equivalent to multiplying the first number by the reciprocal of the second number.

PROCEDURE **Dividing Fractions**

Let a, b, c, and d be numbers such that b, c, and d are not zero. Then,

$$\frac{a}{b} \div \frac{c}{d} = \frac{a}{b} \times \frac{d}{c}$$

multiply

reciprocal

To divide fractions, multiply the first fraction by the reciprocal of the second fraction.

Skill Practice

Divide.

14. $\dfrac{12}{25} \div \dfrac{8}{15}$ **15.** $\dfrac{1}{4} \div 2$

Example 6 Dividing Fractions

Divide the fractions.

a. $\dfrac{8}{5} \div \dfrac{3}{10}$ **b.** $\dfrac{12}{13} \div 6$

Solution:

a. $\dfrac{8}{5} \div \dfrac{3}{10} = \dfrac{8}{5} \times \dfrac{10}{3}$ Multiply by the reciprocal of $\frac{3}{10}$, which is $\frac{10}{3}$.

$= \dfrac{8 \times \overset{2}{\cancel{10}}}{\underset{1}{\cancel{5}} \times 3} = \dfrac{16}{3}$ Multiply and simplify to lowest terms.

b. $\dfrac{12}{13} \div 6 = \dfrac{12}{13} \div \dfrac{6}{1}$ Write the whole number 6 as $\frac{6}{1}$.

$= \dfrac{12}{13} \times \dfrac{1}{6}$ Multiply by the reciprocal of $\frac{6}{1}$, which is $\frac{1}{6}$.

$= \dfrac{\overset{2}{\cancel{12}} \times 1}{13 \times \underset{1}{\cancel{6}}} = \dfrac{2}{13}$ Multiply and simplify to lowest terms.

---

6. Adding and Subtracting Fractions

PROCEDURE **Adding and Subtracting Fractions**

Two fractions can be added or subtracted if they have a common denominator. Let a, b, and c be numbers such that b does not equal zero. Then,

$$\frac{a}{b} + \frac{c}{b} = \frac{a+c}{b} \qquad \text{and} \qquad \frac{a}{b} - \frac{c}{b} = \frac{a-c}{b}$$

To add or subtract fractions with the same denominator, add or subtract the numerators and write the result over the common denominator.

Answers

14. $\dfrac{9}{10}$ **15.** $\dfrac{1}{8}$

Example 7   Adding and Subtracting Fractions with the Same Denominator

Add or subtract as indicated.

a. $\dfrac{1}{12} + \dfrac{7}{12}$ **b.** $\dfrac{13}{5} - \dfrac{3}{5}$

Solution:

a. $\dfrac{1}{12} + \dfrac{7}{12} = \dfrac{1+7}{12}$ Add the numerators.

$\qquad = \dfrac{8}{12}$

$\qquad = \dfrac{2}{3}$ Simplify to lowest terms.

> **TIP:** The sum $\frac{1}{12} + \frac{7}{12}$ can be visualized as the sum of the pink and blue sections of the figure.

b. $\dfrac{13}{5} - \dfrac{3}{5} = \dfrac{13-3}{5}$ Subtract the numerators.

$\qquad = \dfrac{10}{5}$ Simplify.

$\qquad = 2$ Simplify to lowest terms.

In Example 7, we added and subtracted fractions with the same denominators. To add or subtract fractions with different denominators, we must first become familiar with the idea of a least common multiple between two or more numbers. The **least common multiple (LCM)** of two numbers is the smallest whole number that is a multiple of each number. For example, the LCM of 6 and 9 is 18.

multiples of 6:   6, 12, 18, 24, 30, 36, …

multiples of 9:   9, 18, 27, 36, 45, 54, …

Listing the multiples of two or more given numbers can be a cumbersome way to find the LCM. Therefore, we offer the following method to find the LCM of two numbers.

PROCEDURE   Finding the LCM of Two Numbers

Step 1   Write each number as a product of prime factors.
Step 2   The LCM is the product of unique prime factors from *both* numbers. Use repeated factors the maximum number of times they appear in *either* factorization.

Skill Practice

23. Subtract. $2\frac{3}{4} - 1\frac{1}{3}$

Example 11 Operations on Mixed Numbers

Subtract. $5\frac{1}{3} - 2\frac{1}{4}$

Solution:

$5\frac{1}{3} - 2\frac{1}{4}$

$= \dfrac{16}{3} - \dfrac{9}{4}$ Write the mixed numbers as improper fractions.

$= \dfrac{16 \times 4}{3 \times 4} - \dfrac{9 \times 3}{4 \times 3}$ The LCD is 12. Multiply numerators and denominators by the missing factors from the denominators.

$= \dfrac{64}{12} - \dfrac{27}{12}$

$= \dfrac{37}{12} \text{ or } 3\frac{1}{12}$ Subtract the fractions.

TIP: An improper fraction can also be written as a mixed number. Both answers are acceptable. Note that

$$\frac{37}{12} = \frac{36}{12} + \frac{1}{12} = 3 + \frac{1}{12}, \text{ or } 3\frac{1}{12}$$

This can easily be found by dividing.

Skill Practice

24. Divide. $5\frac{5}{6} \div 3\frac{2}{3}$

Example 12 Operations on Mixed Numbers

Divide. $7\frac{1}{2} \div 3$

Solution:

$7\frac{1}{2} \div 3$

$= \dfrac{15}{2} \div \dfrac{3}{1}$ Write the mixed number and whole number as fractions.

$= \dfrac{\overset{5}{\cancel{15}}}{2} \times \dfrac{1}{\underset{1}{\cancel{3}}}$ Multiply by the reciprocal of $\frac{3}{1}$, which is $\frac{1}{3}$.

$= \dfrac{5}{2} \text{ or } 2\frac{1}{2}$ The answer may be written as an improper fraction or as a mixed number.

Avoiding Mistakes

Remember that when dividing (or multiplying) fractions, a common denominator is not necessary.

Answers

23. $\dfrac{17}{12}$ or $1\dfrac{5}{12}$ **24.** $\dfrac{35}{22}$ or $1\dfrac{13}{22}$

Section R.2 Practice Exercises

Objective 1: Basic Definitions

For Exercises 1–8, identify the numerator and denominator of the fraction. Then determine if the fraction is a proper fraction or an improper fraction.

1. $\dfrac{7}{8}$ 2. $\dfrac{2}{3}$ 3. $\dfrac{9}{5}$ 4. $\dfrac{5}{2}$

5. $\dfrac{6}{6}$ 6. $\dfrac{4}{4}$ 7. $\dfrac{12}{1}$ 8. $\dfrac{5}{1}$

For Exercises 9–16, write a proper or improper fraction associated with the shaded region of each figure.

9.

10.

11.

12.

13.   14.   15.   16.

For Exercises 17–20, write both an improper fraction and a mixed number associated with the shaded region of each figure.

17.   18.   19.   20.

21. Explain the difference between the set of whole numbers and the set of natural numbers.

22. Explain the difference between a proper fraction and an improper fraction.

23. Write a fraction that simplifies to $\frac{1}{2}$. (Answers may vary.)

24. Write a fraction that simplifies to $\frac{1}{3}$. (Answers may vary.)

111. $3\dfrac{1}{2} - 1\dfrac{7}{8}$

112. $5\dfrac{1}{3} - 2\dfrac{3}{4}$

113. $1\dfrac{1}{6} + 3\dfrac{3}{4}$

114. $4\dfrac{1}{2} + 2\dfrac{2}{3}$

115. $1 - \dfrac{7}{8}$

116. $2 - \dfrac{3}{7}$

117. A board $26\dfrac{3}{8}$ in. long must be cut into three pieces of equal length. Find the length of each piece.

$26\dfrac{3}{8}$ in.

118. A futon, when set up as a sofa, measures $3\dfrac{5}{6}$ ft wide. When it is opened to be used as a bed, the width is increased by $1\dfrac{3}{4}$ ft. What is the total width of this bed?

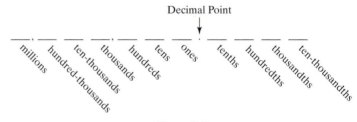

$3\dfrac{5}{6}$ ft

119. A plane trip from Orlando to Detroit takes $2\dfrac{3}{4}$ hr. If the plane traveled for $1\dfrac{1}{6}$ hr, how much time remains for the flight?

120. Antonio bought $3\dfrac{3}{4}$ lb of smoked turkey for sandwiches. If he made 10 sandwiches, how much turkey did he put in each sandwich?

121. José ordered two seafood platters for a party. One platter has $1\dfrac{1}{2}$ lb of shrimp, and the other has $\dfrac{3}{4}$ lb of shrimp. How many pounds of shrimp does he have altogether?

122. Ayako took a trip to the store $5\dfrac{1}{2}$ miles away. If she rode the bus for $4\dfrac{5}{6}$ miles and walked the rest of the way, how far did she have to walk?

123. If Tampa, Florida, averages $6\dfrac{1}{4}$ in. of rain during each summer month, how much total rain would be expected in June, July, August, and September?

124. Pete started working out and found that he lost approximately $\dfrac{3}{4}$ in. off his waistline every month. How much would he lose around his waist in 6 months?

Section R.3   **Decimals and Percents**

Objectives

1. Introduction to a Place Value System
2. Converting Fractions to Decimals
3. Converting Decimals to Fractions
4. Converting Percents to Decimals and Fractions
5. Converting Decimals and Fractions to Percents
6. Applications of Percents

1. Introduction to a Place Value System

In a **place value** number system, each digit in a numeral has a particular value determined by its location in the numeral (Figure R-5).

Figure R-5

For example, the number 197.215 represents

$$(1 \times 100) + (9 \times 10) + (7 \times 1) + \left(2 \times \dfrac{1}{10}\right) + \left(1 \times \dfrac{1}{100}\right) + \left(5 \times \dfrac{1}{1000}\right)$$

Each of the digits 1, 9, 7, 2, 1, and 5 is multiplied by 100, 10, 1, $\frac{1}{10}$, $\frac{1}{100}$, and $\frac{1}{1000}$, respectively, depending on its location in the numeral 197.215.

By obtaining a common denominator and adding fractions, we have

$$197.215 = 100 + 90 + 7 + \frac{200}{1000} + \frac{10}{1000} + \frac{5}{1000}$$

$$= 197 + \frac{215}{1000} \quad \text{or} \quad 197\frac{215}{1000}$$

Because 197.215 is equal to the mixed number $197\frac{215}{1000}$, we read 197.215 as one hundred ninety-seven *and* two hundred fifteen thousandths. The decimal point is read as the word *and*.

If there are no digits to the right of the decimal point, we usually omit the decimal point. For example, the number 7125. is written simply as 7125 without a decimal point.

2. Converting Fractions to Decimals

In Section R.2, we learned that a fraction represents part of a whole unit. Likewise, the digits to the right of the decimal point represent a fraction of a whole unit. In this section, we will learn how to convert a fraction to a decimal number and vice versa.

> **PROCEDURE** **Converting a Fraction to a Decimal**
>
> To convert a fraction to a decimal, divide the numerator of the fraction by the denominator of the fraction.

Example 1 **Converting Fractions to Decimals**

Convert each fraction to a decimal. **a.** $\frac{7}{40}$ **b.** $\frac{2}{3}$

Solution:

a. $\frac{7}{40} = 0.175$

$$\begin{array}{r} 0.175 \\ 40\overline{)7.000} \\ \underline{40} \\ 300 \\ \underline{280} \\ 200 \\ \underline{200} \\ 0 \end{array}$$

The number 0.175 is said to be a **terminating decimal** because there are no nonzero digits to the right of the last digit, 5.

b. $\frac{2}{3} = 0.666\ldots$

$$\begin{array}{r} 0.666\ldots \\ 3\overline{)2.00000} \\ \underline{18} \\ 20 \\ \underline{18} \\ 20 \\ \underline{18} \\ 2\ldots \end{array}$$

The pattern 0.666... continues indefinitely. Therefore, we say that this is a **repeating decimal**.

For a repeating decimal, a horizontal bar is often used to denote the repeating pattern after the decimal point. Hence, $\frac{2}{3} = 0.\overline{6}$.

Sometimes it is useful to round a decimal number to a desired place value. This is demonstrated in Example 2.

Skill Practice

Round to the indicated place value.
5. 0.624 hundredths place
6. 1.$\overline{62}$ ten-thousandths place

Example 2 Rounding Decimal Numbers

Round the numbers to the indicated place value.

a. 0.175 to the hundredths place   b. 0.$\overline{54}$ to the thousandths place

Solution:

To round decimal numbers to a given place value, we actually look at the digit to the right of that position. If the digit to the right is 5 or greater, we round up. If the digit to the right is less than 5, we truncate the number beyond the given place value.

hundredths place

a. 0.17**5**   The digit to the right of the hundredths place is 5 or greater.

≈ 0.18   Round up.

b. 0.$\overline{54}$ = 0.545454...   This is a repeating decimal. We write out several digits.

thousandths place

0.545**4**54...   The digit to the right of the thousandths place is less than 5.

≈ 0.545

3. Converting Decimals to Fractions

For a terminating decimal, use the word name to write the number as a fraction or mixed number.

Skill Practice

Convert to a fraction.
7. 0.107
8. 11.25

Example 3 Converting Terminating Decimals to Fractions

Convert each decimal to a fraction.

a. 0.0023   b. 50.06

Solution:

a. 0.0023 is read as twenty-three ten-thousandths. Thus,

$$0.0023 = \frac{23}{10,000}$$

b. 50.06 is read as fifty and six hundredths. Thus,

$$50.06 = 50\frac{6}{100}$$

$$= 50\frac{3}{50}$$ Simplify the fraction to lowest terms.

$$= \frac{2503}{50}$$ Write the mixed number as a fraction.

Answers

5. 0.62   6. 1.6263
7. $\frac{107}{1000}$ 8. $\frac{45}{4}$

Repeating decimals also can be written as fractions. However, the procedure to convert a repeating decimal to a fraction requires some knowledge of algebra. Table R-1 shows some common repeating decimals and an equivalent fraction for each.

Table R-1

$0.\overline{1} = \dfrac{1}{9}$	$0.\overline{4} = \dfrac{4}{9}$	$0.\overline{7} = \dfrac{7}{9}$
$0.\overline{2} = \dfrac{2}{9}$	$0.\overline{5} = \dfrac{5}{9}$	$0.\overline{8} = \dfrac{8}{9}$
$0.\overline{3} = \dfrac{3}{9} = \dfrac{1}{3}$	$0.\overline{6} = \dfrac{6}{9} = \dfrac{2}{3}$	$0.\overline{9} = \dfrac{9}{9} = 1$

4. Converting Percents to Decimals and Fractions

The concept of percent (%) is widely used in a variety of mathematical applications. The word *percent* means "per 100." Therefore, we can write percents as fractions.

$6\% = \dfrac{6}{100}$ A sales tax of 6% means that 6 cents in tax is charged for every 100 cents spent.

$91\% = \dfrac{91}{100}$ The fact that 91% of the population is right-handed means that 91 people out of 100 are right-handed.

The quantity $91\% = \dfrac{91}{100}$ can be written as $91 \times \dfrac{1}{100}$ or as 91×0.01.

Notice that the % symbol implies "division by 100" or, equivalently, "multiplication by $\frac{1}{100}$." Thus, we have the following rule to convert a percent to a fraction (or to a decimal).

PROCEDURE Converting a Percent to a Decimal or Fraction

Replace the % symbol by $\div\ 100$ (or equivalently $\times \dfrac{1}{100}$ or $\times\ 0.01$).

Example 4 Converting Percents to Decimals

Convert the percents to decimals.

a. 78%   **b.** 412%   **c.** 0.045%

Solution:

a. $78\% = 78 \times 0.01 = 0.78$

b. $412\% = 412 \times 0.01 = 4.12$

c. $0.045\% = 0.045 \times 0.01 = 0.00045$

TIP: Multiplying by 0.01 is equivalent to dividing by 100. This has the effect of moving the decimal point two places to the left.

Skill Practice

Convert the percent to a decimal.
9. 29%
10. 3.5%
11. 100%

Answers
9. 0.29   **10.** 0.035   **11.** 1.00

Example 5 **Converting Percents to Fractions**

Convert the percents to fractions.

   a. 52%   **b.** $33\frac{1}{3}$%   **c.** 6.5%

Solution:

a. $52\% = 52 \times \dfrac{1}{100}$ Replace the % symbol by $\frac{1}{100}$.

$\qquad\quad = \dfrac{52}{100}$ Multiply.

$\qquad\quad = \dfrac{13}{25}$ Simplify to lowest terms.

b. $33\dfrac{1}{3}\% = 33\dfrac{1}{3} \times \dfrac{1}{100}$ Replace the % symbol by $\frac{1}{100}$.

$\qquad\quad = \dfrac{100}{3} \times \dfrac{1}{100}$ Write the mixed number as a fraction $33\frac{1}{3} = \frac{100}{3}$.

$\qquad\quad = \dfrac{100}{300}$ Multiply the fractions.

$\qquad\quad = \dfrac{1}{3}$ Simplify to lowest terms.

c. $6.5\% = 6.5 \times \dfrac{1}{100}$ Replace the % symbol by $\frac{1}{100}$.

$\qquad\quad = \dfrac{65}{10} \times \dfrac{1}{100}$ Write 6.5 as an improper fraction.

$\qquad\quad = \dfrac{65}{1000}$ Multiply the fractions.

$\qquad\quad = \dfrac{13}{200}$ Simplify to lowest terms.

5. Converting Decimals and Fractions to Percents

To convert a percent to a decimal or fraction, we replace the % symbol by ÷ 100. To convert a decimal or fraction to a percent, we reverse this process.

PROCEDURE **Converting Decimals and Fractions to Percents**

Multiply the fraction or decimal by 100%.

Example 6 **Converting Decimals to Percents**

Convert the decimals to percents.

a. 0.92 **b.** 10.80 **c.** 0.005

Solution:

a. $0.92 = 0.92 \times 100\% = 92\%$ Multiply by 100%.

b. $10.80 = 10.80 \times 100\% = 1080\%$ Multiply by 100%.

c. $0.005 = 0.005 \times 100\% = 0.5\%$ Multiply by 100%.

Skill Practice

Convert the decimal to a percent.

15. 0.56

16. 4.36

17. 0.002

Example 7 **Converting Fractions to Percents**

Convert the fractions to percents.

a. $\dfrac{2}{5}$ **b.** $\dfrac{5}{3}$

Solution:

a. $\dfrac{2}{5} = \dfrac{2}{5} \times 100\%$ Multiply by 100%.

$= \dfrac{2}{5} \times \dfrac{100}{1}\%$ Write the whole number as a fraction.

$= \dfrac{2}{\underset{1}{5}} \times \dfrac{\overset{20}{100}}{1}\%$ Multiply fractions.

$= \dfrac{40}{1}\%$ or 40% Simplify.

b. $\dfrac{5}{3} = \dfrac{5}{3} \times 100\%$ Multiply by 100%.

$= \dfrac{5}{3} \times \dfrac{100}{1}\%$ Write the whole number as a fraction.

$= \dfrac{500}{3}\%$ Multiply fractions.

$= \dfrac{500}{3}\%$ or $166.\overline{6}\%$ The value $\frac{500}{3}$ can be written in decimal form by dividing 500 by 3.

Skill Practice

Convert the fraction to a percent.

18. $\dfrac{7}{8}$ **19.** $\dfrac{5}{6}$

6. Applications of Percents

Many applications involving percents involve finding a percent of some base number. For example, suppose a textbook is discounted 25%. If the book originally cost $60, find the amount of the discount.

In this example, we must find 25% of $60. In this context, the word *of* means multiply.

$$25\% \text{ of } \$60$$
$$\downarrow \quad \downarrow \quad \downarrow$$
$$0.25 \times 60 = 15 \qquad \text{The amount of the discount is } \$15.$$

Answers

15. 56% **16.** 436% **17.** 0.2%

18. 87.5% **19.** $83.\overline{3}\%$ or $83\frac{1}{3}\%$

Note that the *decimal form* of a percent is always used in calculations. Therefore, 25% was converted to 0.25 *before* multiplying by $60.

Example 8 **Applying Percentages**

Shauna received a raise, so now her new salary is 105% of her old salary. Find Shauna's new salary if her old salary was $36,000 per year.

Solution:

The new salary is 105% of $36,000.

$$1.05 \times 36,000 = 37,800 \qquad \text{The new salary is \$37,800 per year.}$$

In some applications, it is necessary to convert a fractional part of a whole to a percent of the whole.

Example 9 **Finding a Percentage**

Union College in Schenectady, New York, accepts approximately 520 students each year from 3500 applicants. What percent does 520 represent? Round to the nearest tenth of a percent.

Solution:

$$\frac{520}{3500} \approx 0.149 \qquad \text{Convert the fractional part of the total number of applicants to decimal form. (\textit{Note:} Rounding the decimal form of the quotient to the thousandths place gives us the nearest tenth of a percent.)}$$

$$= 0.149 \times 100\% \qquad \text{Convert the decimal to a percent.}$$

$$= 14.9\% \qquad \text{Simplify.}$$

Approximately 14.9% of the applicants to Union College are accepted.

Answers

20. $3.12   **21.** 88%

Calculator Connections

Topic: Approximating Repeating Decimals on a Calculator

Calculators can display only a limited number of digits on the calculator screen. Therefore, repeating decimals and terminating decimals with a large number of digits will be truncated or rounded to fit the calculator display. For example, the fraction $\frac{2}{3} = 0.\overline{6}$ may be entered into the calculator as [**2**] [**÷**] [**3**]. The result may appear as 0.6666666667 or as 0.6666666666. The fraction $\frac{2}{11}$ equals the repeating decimal $0.\overline{18}$. However, the calculator converts $\frac{2}{11}$ to the terminating decimal 0.1818181818.

```
2/3
        .6666666667
2/11
        .1818181818
```

Calculator Exercises

Without using a calculator, find a repeating decimal to represent each of the following fractions. Then use a calculator to confirm your answer.

1. $\dfrac{4}{9}$ **2.** $\dfrac{7}{11}$ **3.** $\dfrac{3}{22}$ **4.** $\dfrac{5}{13}$

Section R.3   Practice Exercises

**Boost *your* GRADE at
ALEKS.com!**   **ALEKS**   • Practice Problems   • e-Professors
version 3.0   • Self-Tests   • Videos
• NetTutor

Objective 1: Introduction to a Place Value System

For Exercises 1–8, write the name of the place value for the underlined digit.

1. 481.24

2. 1345.42

3. 2912.032

4. 4208.03

5. 2.381

6. 8.249

7. 21.413

8. 82.794

9. The first 10 Roman numerals are: I, II, III, IV, V, VI, VII, VIII, IX, X. Is this numbering system a place value system? Explain your answer.

10. Write the number in decimal form. $3(100) + 7(10) + 6 + \dfrac{1}{100} + \dfrac{5}{1000}$

Objective 2: Converting Fractions to Decimals

For Exercises 11–18, convert each fraction to a terminating decimal or a repeating decimal. **(See Example 1.)**

11. $\dfrac{7}{10}$

12. $\dfrac{9}{10}$

13. $\dfrac{9}{25}$

14. $\dfrac{3}{25}$

15. $\dfrac{11}{9}$

16. $\dfrac{16}{9}$

17. $\dfrac{7}{33}$

18. $\dfrac{2}{11}$

For Exercises 19–26, round the decimal to the given place value. **(See Example 2.)**

19. 214.059; tenths

20. 1004.165; hundredths

21. 39.26849; thousandths

22. 0.059499; thousandths

23. 39,918.2; thousands

24. 599,621.5; thousands

25. $0.\overline{72}$; hundredths

26. $0.3\overline{4}$; thousandths

Objective 3: Converting Decimals to Fractions

For Exercises 27–38, convert each decimal to a fraction or a mixed number. **(See Example 3.)**

27. 0.45

28. 0.65

29. 0.181

30. 0.273

31. 2.04

32. 6.02

33. 13.007

34. 12.003

35. $0.\overline{5}$ (*Hint*: Refer to Table R-1)

36. $0.\overline{8}$

37. $1.\overline{1}$

38. $2.\overline{3}$

Objective 4: Converting Percents to Decimals and Fractions

For Exercises 39–48, convert each percent to a decimal and to a fraction. **(See Examples 4–5.)**

39. The sale price is 30% off of the original price.

40. An HMO (health maintenance organization) pays 80% of all doctors' bills.

41. The building will be 75% complete by spring.

42. Chan plants roses in 25% of his garden.

43. The bank pays $3\frac{3}{4}\%$ interest on a checking account.

44. A credit union pays $4\frac{1}{2}\%$ interest on a savings account.

45. Kansas received 15.7% of its annual rainfall in 1 week.

46. Social Security withholds 6.2% of an employee's gross pay.

47. The world population in 2008 was 270% of the world population in 1950.

48. The cost of a home is 140% of its cost 10 years ago.

Objective 5: Converting Decimals and Fractions to Percents

49. Explain how to convert a decimal to a percent.   50. Explain how to convert a percent to a decimal.

For Exercises 51–62, convert the decimal to a percent. **(See Example 6.)**

51. 0.05	52. 0.06	53. 0.90	54. 0.70
55. 1.2	56. 4.8	57. 7.5	58. 9.3
59. 0.135	60. 0.536	61. 0.003	62. 0.002

For Exercises 63–74, convert the fraction to a percent. **(See Example 7.)**

63. $\dfrac{3}{50}$	64. $\dfrac{23}{50}$	65. $\dfrac{9}{2}$	66. $\dfrac{7}{4}$
67. $\dfrac{5}{8}$	68. $\dfrac{1}{8}$	69. $\dfrac{5}{16}$	70. $\dfrac{7}{16}$
71. $\dfrac{5}{6}$	72. $\dfrac{4}{15}$	73. $\dfrac{14}{15}$	74. $\dfrac{5}{18}$

Objective 6: Applications of Percents

75. A suit that costs $140 is discounted by 30%. How much is the discount? **(See Example 8.)**

76. Louise completed 40% of her task that takes a total of 60 hr to finish. How many hours did she complete?

77. Tom's federal taxes amount to 27% of his income. If Tom earns $12,500 per quarter, how much will he pay in taxes for that quarter?

78. A tip of $7 is left for a meal that costs $56. What percent of the cost does the tip represent?

79. Jamie paid $5.95 in sales tax on a textbook that costs $85. Find the percent of the sales tax. **(See Example 9.)**

80. Sue saves $37.50 each week out of her paycheck of $625. What percent of her paycheck does her savings represent?

For Exercises 81–84, refer to the graph. The pie graph shows a family budget based on a net income of $2400 per month.

81. Determine the amount spent on rent.

82. Determine the amount spent on car payments.

83. Determine the amount spent on utilities.

84. How much more money is spent than saved?

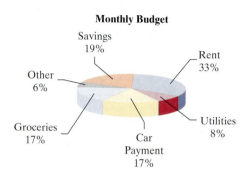

Monthly Budget

Savings 19%
Rent 33%
Other 6%
Utilities 8%
Groceries 17%
Car Payment 17%

 85. By the end of the year, Felipe will have 75% of his mortgage paid. If the mortgage was originally for $90,000, how much will have been paid at the end of the year?

 86. A certificate of deposit (CD) earns 4% interest in 1 year. If Mr. Patel has $12,000 invested in the CD, how much interest will he receive at the end of the year?

Introduction to Geometry   Section R.4

1. Perimeter

Objectives

1. Perimeter
2. Area
3. Volume
4. Angles
5. Triangles

In this section, we present several facts and formulas that may be used throughout the text in applications of geometry. One of the most important uses of geometry involves the measurement of objects of various shapes. We begin with an introduction to perimeter, area, and volume for several common shapes and objects.

Perimeter is defined as the distance around a figure. If we were to put up a fence around a field, the perimeter would determine the amount of fencing. For example, in Figure R-6 the distance around the field is 300 ft. For a polygon (a closed figure constructed from line segments), the perimeter is the sum of the lengths of the sides. For a circle, the distance around the outside is called the **circumference**.

100 ft
50 ft
50 ft
Perimeter = 300 ft
100 ft

Figure R-6

Rectangle

w w
ℓ
ℓ

$P = 2\ell + 2w$

Square

s s
s
s

$P = 4s$

Triangle

a b
c

$P = a + b + c$

Circle

r

Circumference: $C = 2\pi r$

$d = 2r$

For a circle, r represents the length of a radius—the distance from the center to any point on the circle. The length of a diameter, d, of a circle is twice that of a radius. Thus, $d = 2r$. The number π is a constant equal to the circumference of a circle divided by the length of a diameter. That is, $\pi = \frac{C}{d}$. The value of π is often approximated by 3.14 or $\frac{22}{7}$.

Example 1 **Finding Perimeter and Circumference**

Find the perimeter or circumference as indicated. Use 3.14 for π.

a. Perimeter of the rectangle

3.1 ft

5.5 ft

b. Circumference of the circle

6 cm

Solution:

a. $P = 2\ell + 2w$

$\quad = 2(5.5\text{ ft}) + 2(3.1\text{ ft})$ Substitute $\ell = 5.5$ ft and $w = 3.1$ ft.

$\quad = 11\text{ ft} + 6.2\text{ ft}$

$\quad = 17.2\text{ ft}$ The perimeter is 17.2 ft.

b. $C = 2\pi r$

$\quad = 2(3.14)(6\text{ cm})$ Substitute 3.14 for π and $r = 6$ cm.

$\quad = 6.28(6\text{ cm})$

$\quad = 37.68\text{ cm}$ The circumference is 37.68 cm.

TIP: If a calculator is used to find the circumference of a circle, use the π key to get a more accurate answer.

2. Area

The **area** of a geometric figure is the number of square units that can be enclosed within the figure. In applications, we would find the area of a region if we were laying carpet or putting down sod for a lawn. For example, the rectangle shown in Figure R-7 encloses 6 square inches (6 in.^{2}).

The formulas used to compute the area for several common geometric shapes are given here:

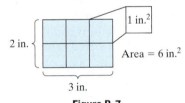

2 in.

3 in.

1 in.^{2}

Area = 6 in.^{2}

Figure R-7

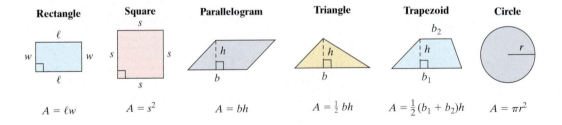

Rectangle	Square	Parallelogram	Triangle	Trapezoid	Circle
$A = \ell w$	$A = s^2$	$A = bh$	$A = \frac{1}{2}bh$	$A = \frac{1}{2}(b_1 + b_2)h$	$A = \pi r^2$

Example 2   Finding Area

Find the area.

Solution:

$A = bh$ The figure is a parallelogram.

$= (4\frac{1}{4} \text{ in.})(2\frac{1}{2} \text{ in.})$ Substitute $b = 4\frac{1}{4}$ in. and $h = 2\frac{1}{2}$ in.

$= \left(\frac{17}{4} \text{ in.}\right)\left(\frac{5}{2} \text{ in.}\right)$

$= \frac{85}{8} \text{ in.}^2$ or $10\frac{5}{8} \text{ in.}^2$

> **TIP:** Notice that the units of area are given in square units such as square inches (in.²), square feet (ft²), square yards (yd²), square centimeters (cm²), and so on.

Skill Practice

Find the area.

3.

Example 3   Finding Area

Find the area.

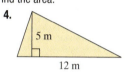

Solution:

$A = \frac{1}{2}(b_1 + b_2)h$ The figure is a trapezoid.

$= \frac{1}{2}(16 \text{ yd} + 10 \text{ yd})(3 \text{ yd})$ Substitute $b_1 = 16$ yd, $b_2 = 10$ yd, and $h = 3$ yd.

$= \frac{1}{2}(26 \text{ yd})(3 \text{ yd})$

$= (13 \text{ yd})(3 \text{ yd})$

$= 39 \text{ yd}^2$ The area is 39 yd².

Skill Practice

Find the area.

4.

> **TIP:** Notice that several of the formulas presented thus far involve multiple operations. The order in which we perform the arithmetic is called the **order of operations** and is covered in detail in Section 1.2. We will follow these guidelines in the order given below:
>
> **1.** Perform operations within parentheses first.
>
> **2.** Evaluate expressions with exponents.
>
> **3.** Perform multiplication or division in order from left to right.
>
> **4.** Perform addition or subtraction in order from left to right.

Answers

3. $\frac{9}{16}$ cm²   **4.** 30 m²

Example 4   **Finding Area of a Circle**

Find the area of a circular fountain if the radius is 25 ft. Use 3.14 for π.

25 ft

Solution:

$A = \pi r^2$

$\approx (3.14)(25 \text{ ft})^2$

$= (3.14)(625 \text{ ft}^2)$ Substitute 3.14 for π and $r = 25$ ft.

$= 1962.5 \text{ ft}^2$ The area of the fountain is 1962.5 ft^2.

3. Volume

The **volume** of a solid is the number of cubic units that can be enclosed within a solid. The solid shown in Figure R-8 contains 18 cubic inches (18 in.³). In applications, volume might refer to the amount of water in a swimming pool.

1 in.³

Volume = 18 in.³

2 in.

3 in.

3 in.

Figure R-8

The formulas used to compute the volume of several common solids are given here:

Rectangular Solid   **Cube**   **Right Circular Cylinder**

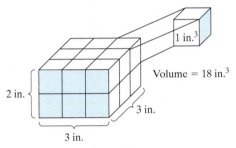

h

w

ℓ

s

s

s

h

r

$V = \ell w h$ $V = s^3$ $V = \pi r^2 h$

TIP: Notice that the volume formulas for the three figures just shown are given by the product of the area of the base and the height of the figure:

$V = \ell w h$ $V = s \cdot s \cdot s$ $V = \pi r^2 h$

Area of Rectangular Base   Area of Square Base   Area of Circular Base

---

Right Circular Cone   **Sphere**

h

r

r

$V = \frac{1}{3}\pi r^2 h$ $V = \frac{4}{3}\pi r^3$

Example 5	**Finding Volume**

Find the volume.

$1\frac{1}{2}$ ft

$1\frac{1}{2}$ ft

$1\frac{1}{2}$ ft

Solution:

$V = s^3$ The object is a cube.

$= (1\frac{1}{2} \text{ ft})^3$ Substitute $s = 1\frac{1}{2}$ ft.

$= \left(\frac{3}{2} \text{ ft}\right)^3$

$= \left(\frac{3}{2} \text{ ft}\right)\left(\frac{3}{2} \text{ ft}\right)\left(\frac{3}{2} \text{ ft}\right)$

$= \frac{27}{8} \text{ ft}^3$, or $3\frac{3}{8} \text{ ft}^3$

TIP: Notice that the units of volume are cubic units such as cubic inches (in.^{3}), cubic feet (ft^3), cubic yards (yd^3), cubic centimeters (cm^3), and so on.

Skill Practice

6. Find the volume.

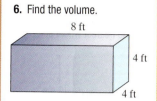

8 ft

4 ft

4 ft

Example 6	**Finding Volume**

Find the volume. Round to the nearest whole unit.

$h = 12$ cm

$r = 4$ cm

Solution:

$V = \frac{1}{3}\pi r^2 h$ The object is a right circular cone.

$\approx \frac{1}{3}(3.14)(4 \text{ cm})^2(12 \text{ cm})$ Substitute 3.14 for π, $r = 4$ cm, and $h = 12$ cm.

$= \frac{1}{3}(3.14)(16 \text{ cm}^2)(12 \text{ cm})$

$= 200.96 \text{ cm}^3$

$\approx 201 \text{ cm}^3$ Round to the nearest whole unit.

Skill Practice

7. Find the volume. Use 3.14 for π. Round to the nearest whole unit.

$r = 2$ in.

Example 7	**Finding Volume in an Application**

An underground gas tank is in the shape of a right circular cylinder.

a. Find the volume of the tank. Use 3.14 for π.

b. Find the cost to fill the tank with gasoline if gasoline costs \$35/ft^3.

1 ft

10 ft

Answers

6. 128 ft^3 **7.** 33 in.^{3}

Skill Practice

Refer to the figure. Assume that lines L_1 and L_2 are parallel.

12. Given that $m(\angle 3) = 23°$, find $m(\angle 2)$, $m(\angle 4)$, $m(\angle 7)$, and $m(\angle 8)$.

Example 8 Finding Unknown Angles in a Diagram

Find the measure of each angle and explain how the angle is related to the given angle of 70°.

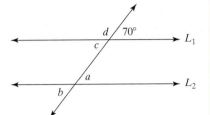

a. $\angle a$ **b.** $\angle b$

c. $\angle c$ **d.** $\angle d$

Solution:

a. $m(\angle a) = 70°$ $\angle a$ is a corresponding angle to the given angle of 70°.

b. $m(\angle b) = 70°$ $\angle b$ and the given angle of 70° are alternate exterior angles.

c. $m(\angle c) = 70°$ $\angle c$ and the given angle of 70° are vertical angles.

d. $m(\angle d) = 110°$ $\angle d$ is the supplement of the given angle of 70°.

5. Triangles

Triangles are categorized by the measures of the angles (Figure R-12) and by the number of equal sides or angles (Figure R-13).

- An **acute triangle** is a triangle in which all three angles are acute.
- A **right triangle** is a triangle in which one angle is a right angle.
- An **obtuse triangle** is a triangle in which one angle is obtuse.

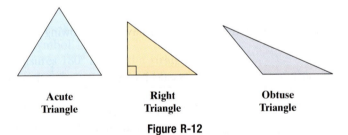

Figure R-12

- An **equilateral triangle** is a triangle in which all three sides (and all three angles) are equal in measure.
- An **isosceles triangle** is a triangle in which two sides are equal in measure (the angles opposite the equal sides are also equal in measure).
- A **scalene triangle** is a triangle in which no sides (or angles) are equal in measure.

Figure R-13

Answer

12. $m(\angle 2) = 23°$; $m(\angle 4) = 157°$; $m(\angle 7) = 23°$; $m(\angle 8) = 157°$

The following important property is true for all triangles.

> **PROPERTY Sum of the Angles in a Triangle**
> The sum of the measures of the angles of a triangle is 180°.

Example 9 Finding Unknown Angles in a Diagram

Find the measure of each angle in the figure.

a. ∠a

b. ∠b

c. ∠c

d. ∠d

e. ∠e

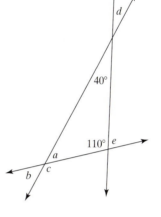

For Exercises 13–17, refer to the figure. Find the measure of the indicated angle.

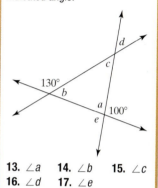

13. ∠a   **14.** ∠b   **15.** ∠c
16. ∠d   **17.** ∠e

Solution:

a. $m(\angle a) = 30°$ The sum of the angles in a triangle is 180°.

b. $m(\angle b) = 30°$ ∠a and ∠b are vertical angles and are equal.

c. $m(\angle c) = 150°$ ∠c and ∠a are supplementary angles (∠c and ∠b are also supplementary).

d. $m(\angle d) = 40°$ ∠d and the given angle of 40° are vertical angles.

e. $m(\angle e) = 70°$ ∠e and the given angle of 110° are supplementary angles.

Answers

13. 80°   **14.** 50°   **15.** 50°
16. 50°   **17.** 100°

Section R.4 Practice Exercises

Objective 1: Perimeter

1. Identify which of the following units could be measures of perimeter.

   a. Square inches (in.^{2})   **b.** Meters (m)   **c.** Cubic feet (ft^3)

   d. Cubic meters (m^3)   **e.** Miles (mi)   **f.** Square centimeters (cm^2)

   g. Square yards (yd^2)   **h.** Cubic inches (in.^{3})   **i.** Kilometers (km)

2. Identify which of the following units could be measures of circumference.

   a. Square inches (in.^{2})   **b.** Meters (m)   **c.** Cubic feet (ft^3)

   d. Cubic meters (m^3)   **e.** Miles (mi)   **f.** Square centimeters (cm^2)

   g. Square yards (yd^2)   **h.** Cubic inches (in.^{3})   **i.** Kilometers (km)

The Set of Real Numbers

1

Chapter 1

In Chapter 1, we present operations on real numbers. The skills that you will learn in this chapter are particularly important as you continue in algebra. Pay careful attention to the order of operations.

Use what you learn about operations on real numbers and the order of operations to complete the puzzle. Fill in each blank box with one of the four basic operations, +, −, •, or ÷ so that the statement is true both going across and going down. Pay careful attention to the order of operations.

−10		2		−3	=	−11
12		−2		3	=	−18
6		4		10	=	−8
=		=		=		=
8		0		1	=	−1

Section 1.1 Sets of Numbers and the Real Number Line

Objectives

1. **The Set of Real Numbers**
2. **Inequalities**
3. **Opposite of a Real Number**
4. **Absolute Value of a Real Number**

1. The Set of Real Numbers

The numbers we work with on a day-to-day basis are all part of the set of **real numbers**. The real numbers encompass zero, all positive, and all negative numbers, including those represented by fractions and decimal numbers. The set of real numbers can be represented graphically on a horizontal number line with a point labeled as 0. Positive real numbers are graphed to the right of 0, and negative real numbers are graphed to the left of 0. Zero is neither positive nor negative. Each point on the number line corresponds to exactly one real number. For this reason, this number line is called the *real number line* (Figure 1-1).

Figure 1-1

Skill Practice

1. Plot the numbers on a real number line.
$\{-1, \frac{3}{4}, -2.5, \frac{10}{3}\}$

Example 1 Plotting Points on the Real Number Line

Plot the points on the real number line that represent the following real numbers.

a. -3 **b.** $\frac{3}{2}$ **c.** -4.7 **d.** $\frac{16}{5}$

Solution:

a. Because -3 is negative, it lies three units to the left of 0.

b. The fraction $\frac{3}{2}$ can be expressed as the mixed number $1\frac{1}{2}$, which lies half-way between 1 and 2 on the number line.

c. The negative number -4.7 lies $\frac{7}{10}$ units to the left of -4 on the number line.

d. The fraction $\frac{16}{5}$ can be expressed as the mixed number $3\frac{1}{5}$, which lies $\frac{1}{5}$ unit to the right of 3 on the number line.

TIP: The natural numbers are used for counting. For this reason, they are sometimes called the "counting numbers."

In mathematics, a well-defined collection of elements is called a **set**. "Well-defined" means the set is described in such a way that it is clear whether an element is in the set. The symbols { } are used to enclose the elements of the set. For example, the set {A, B, C, D, E} represents the set of the first five letters of the alphabet.

Several sets of numbers are used extensively in algebra and are *subsets* (or part) of the set of real numbers.

DEFINITION Natural Numbers, Whole Numbers, and Integers

The set of **natural numbers** is $\{1, 2, 3, \ldots\}$
The set of **whole numbers** is $\{0, 1, 2, 3, \ldots\}$
The set of **integers** is $\{\ldots -3, -2, -1, 0, 1, 2, 3, \ldots\}$

Answer

1.

Notice that the set of whole numbers includes the natural numbers. Therefore, every natural number is also a whole number. The set of integers includes the set of whole numbers. Therefore, every whole number is also an integer.

Fractions are also among the numbers we use frequently. A number that can be written as a fraction whose numerator is an integer and whose denominator is a nonzero integer is called a *rational number*.

DEFINITION Rational Numbers

The set of **rational numbers** is the set of numbers that can be expressed in the form $\frac{p}{q}$, where both p and q are integers and q does not equal 0.

We also say that a rational number $\frac{p}{q}$ is a *ratio* of two integers, p and q, where q is not equal to zero.

Example 2   Identifying Rational Numbers

Show that the following numbers are rational numbers by finding an equivalent ratio of two integers.

a. $\dfrac{-2}{3}$ **b.** -12 **c.** 0.5 **d.** $0.\overline{6}$

Solution:

a. The fraction $\frac{-2}{3}$ is a rational number because it can be expressed as the ratio of -2 and 3.

b. The number -12 is a rational number because it can be expressed as the ratio of -12 and 1, that is, $-12 = \frac{-12}{1}$. In this example, we see that an integer is also a rational number.

c. The terminating decimal 0.5 is a rational number because it can be expressed as the ratio of 5 and 10. That is, $0.5 = \frac{5}{10}$. In this example, we see that a terminating decimal is also a rational number.

d. The repeating decimal $0.\overline{6}$ is a rational number because it can be expressed as the ratio of 2 and 3. That is, $0.\overline{6} = \frac{2}{3}$. In this example, we see that a repeating decimal is also a rational number.

Skill Practice

Show that each number is rational by finding an equivalent ratio of two integers.

2. $\dfrac{3}{7}$ **3.** -5

4. 0.3 **5.** $0.\overline{3}$

TIP: Any rational number can be represented by a terminating decimal or by a repeating decimal.

Some real numbers, such as the number π, cannot be represented by the ratio of two integers. These numbers are called irrational numbers and in decimal form are nonterminating, nonrepeating decimals. The value of π, for example, can be approximated as $\pi \approx 3.1415926535897932$. However, the decimal digits continue forever with no repeated pattern. Another example of an irrational number is $\sqrt{3}$ (read as "the positive square root of 3"). The expression $\sqrt{3}$ is a number that when multiplied by itself is 3. There is no rational number that satisfies this condition. Thus, $\sqrt{3}$ is an irrational number.

DEFINITION Irrational Numbers

The set of **irrational numbers** is a subset of the real numbers whose elements cannot be written as a ratio of two integers.

Note: An irrational number cannot be written as a terminating decimal or as a repeating decimal.

Answers

2. ratio of 3 and 7
3. ratio of -5 and 1
4. ratio of 3 and 10
5. ratio of 1 and 3

The set of real numbers consists of both the rational and the irrational numbers. The relationship among these important sets of numbers is illustrated in Figure 1-2:

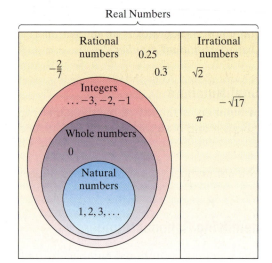

Figure 1-2

Example 3   **Classifying Numbers by Set**

Check the set(s) to which each number belongs. The numbers may belong to more than one set.

	Natural Numbers	Whole Numbers	Integers	Rational Numbers	Irrational Numbers	Real Numbers
5						
$\dfrac{-47}{3}$						
1.48						
$\sqrt{7}$						
0						

Solution:

	Natural Numbers	Whole Numbers	Integers	Rational Numbers	Irrational Numbers	Real Numbers
5	✔	✔	✔	✔ (ratio of 5 and 1)		✔
$\dfrac{-47}{3}$				✔ (ratio of −47 and 3)		✔
1.48				✔ (ratio of 148 and 100)		✔
$\sqrt{7}$					✔	✔
0		✔	✔	✔ (ratio of 0 and 1)		✔

Answers

6. Integers, rational numbers, real numbers
7. Rational numbers, real numbers
8. Irrational numbers, real numbers
9. Natural numbers, whole numbers, integers, rational numbers, real numbers
10. Whole numbers, integers, rational numbers, real numbers

2. Inequalities

The relative size of two real numbers can be compared using the real number line. Suppose a and b represent two real numbers. We say that a is less than b, denoted $a < b$, if a lies to the left of b on the number line.

$a < b$

We say that a is greater than b, denoted $a > b$, if a lies to the right of b on the number line.

$a > b$

Table 1-1 summarizes the relational operators that compare two real numbers a and b.

Table 1-1

Mathematical Expression	Translation	Example
$a < b$	a is less than b.	$2 < 3$
$a > b$	a is greater than b.	$5 > 1$
$a \leq b$	a is less than or equal to b.	$4 \leq 4$
$a \geq b$	a is greater than or equal to b.	$10 \geq 9$
$a = b$	a is equal to b.	$6 = 6$
$a \neq b$	a is not equal to b.	$7 \neq 0$
$a \approx b$	a is approximately equal to b.	$2.3 \approx 2$

The symbols $<$, $>$, $\leq$, $\geq$, and $\neq$ are called *inequality signs*, and the expressions $a < b$, $a > b$, $a \leq b$, $a \geq b$, and $a \neq b$ are called **inequalities**.

Example 4   Ordering Real Numbers

The average temperatures (in degrees Celsius) for selected cities in the United States and Canada in January are shown in Table 1-2.

Table 1-2

City	Temp (°C)
Prince George, British Columbia	-12.5
Corpus Christi, Texas	13.4
Parkersburg, West Virginia	-0.9
San Jose, California	9.7
Juneau, Alaska	-5.7
New Bedford, Massachusetts	-0.2
Durham, North Carolina	4.2

Plot a point on the real number line representing the temperature of each city. Compare the temperatures between the following cities, and fill in the blank with the appropriate inequality sign: $<$ or $>$.

Skill Practice

Fill in the blanks with the appropriate inequality sign: $<$ or $>$.

11. -11 _____ 20

12. -3 _____ -6

13. 0 _____ -9

14. -6.2 _____ -1.8

Answers

11. $<$ **12.** $>$ **13.** $>$ **14.** $<$

Solution:

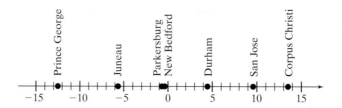

a. Temperature of San Jose $\boxed{<}$ temperature of Corpus Christi

b. Temperature of Juneau $\boxed{>}$ temperature of Prince George

c. Temperature of Parkersburg $\boxed{<}$ temperature of New Bedford

d. Temperature of Parkersburg $\boxed{>}$ temperature of Prince George

3. Opposite of a Real Number

To gain mastery of any algebraic skill, it is necessary to know the meaning of key definitions and key symbols. Two important definitions are the *opposite* of a real number and the *absolute value* of a real number.

> **DEFINITION**   The Opposite of a Real Number
>
> Two numbers that are the same distance from 0 but on opposite sides of 0 on the number line are called **opposites** of each other. Symbolically, we denote the opposite of a real number a as $-a$.

Skill Practice

18. Find the opposite of 224.

19. Find the opposite of -3.4.

Example 5   Finding the Opposite of a Real Number

a. Find the opposite of 5.   **b.** Find the opposite of $-\frac{4}{7}$.

Solution:

a. The opposite of 5 is -5.   **b.** The opposite of $-\frac{4}{7}$ is $\frac{4}{7}$.

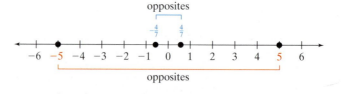

Skill Practice

20. Evaluate $-(2.8)$.

21. Evaluate $-\left(-\dfrac{1}{5}\right)$.

Example 6   Finding the Opposite of a Real Number

a. Evaluate $-(0.46)$.   **b.** Evaluate $-\left(-\dfrac{11}{3}\right)$.

Solution:

a. $-(0.46) = -0.46$ The expression $-(0.46)$ represents the opposite of 0.46.

b. $-\left(-\dfrac{11}{3}\right) = \dfrac{11}{3}$ The expression $-(-\frac{11}{3})$ represents the opposite of $-\frac{11}{3}$.

Answers

15. $<$ **16.** $>$ **17.** $>$

18. -224 **19.** 3.4   **20.** -2.8

21. $\dfrac{1}{5}$

4. Absolute Value of a Real Number

The concept of absolute value will be used to define the addition of real numbers in Section 1.3.

DEFINITION Informal Definition of the Absolute Value
of a Real Number

The **absolute value** of a real number a, denoted $|a|$, is the distance between a and 0 on the number line.

Note: The absolute value of any real number is positive or zero.

For example, $|3| = 3$ and $|-3| = 3$.

Example 7   Finding the Absolute Value of a Real Number

Evaluate the absolute value expressions.

a. $|-4|$ **b.** $\left|\frac{1}{2}\right|$ **c.** $|-6.2|$ **d.** $|0|$

Solution:

a. $|-4| = 4$ −4 is 4 units from 0 on the
number line.

b. $\left|\frac{1}{2}\right| = \frac{1}{2}$ $\frac{1}{2}$ is $\frac{1}{2}$ unit from 0 on the
number line.

c. $|-6.2| = 6.2$ −6.2 is 6.2 units from 0 on the number line.

d. $|0| = 0$ 0 is 0 units from 0 on the
number line.

The absolute value of a number a is its distance from 0 on the number line. The definition of $|a|$ may also be given symbolically depending on whether a is negative or nonnegative.

DEFINITION Absolute Value of a Real Number

Let a be a real number. Then

 1. If a is nonnegative (that is, $a \geq 0$), then $|a| = a$.

 2. If a is negative (that is, $a < 0$), then $|a| = -a$.

Answers

22. 99 **23.** $\dfrac{7}{8}$

24. 1.4 **25.** 1

This definition states that if *a* is a nonnegative number, then |*a*| equals *a* itself. If *a* is a negative number, then |*a*| equals the opposite of *a*. For example:

|9| = 9   Because 9 is positive, then |9| equals the number 9 itself.

|−7| = 7   Because −7 is negative, then |−7| equals the opposite of −7, which is 7.

Example 8   Comparing Absolute Value Expressions

Determine if the statements are true or false.

a. |3| ≤ 3   **b.** −|5| = |−5|

Solution:

a. |3| ≤ 3   True. The symbol ≤ means "less than *or* equal to."
Since |3| is equal to 3, then |3| ≤ 3 is a true statement.

b. −|5| = |−5| False. On the left-hand side, −|5| is the opposite of |5|. Hence
−|5| = −5. On the right-hand side, |−5| = 5. Therefore, the
original statement simplifies to −5 = 5, which is false.

Calculator Connections

Topic: Approximating Irrational Numbers on a Calculator

Scientific and graphing calculators approximate irrational numbers by using rational numbers in the form of terminating decimals. For example, consider approximating π and $\sqrt{3}$.

Scientific Calculator:

Enter: or **2nd** π **Result:** | 3.141592654 |

Enter: 3 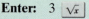   **Result:** | 1.732050808 |

Graphing Calculator:

Enter: **2nd** π **ENTER**

Enter: **2nd** $\sqrt{}$ 3 **ENTER**

```
π
              3.141592654
√(3)
              1.732050808
```

Note that when writing approximations, we use the symbol, ≈.

$$\pi \approx 3.141592654 \quad \text{and} \quad \sqrt{3} \approx 1.732050808$$

Calculator Exercises

Use a calculator to approximate the irrational numbers. Remember to use the appropriate symbol, ≈, when expressing answers.

1. $\sqrt{12}$ **2.** $\sqrt{99}$ **3.** $4 \cdot \pi$ **4.** $\sqrt{\pi}$

Section 1.1 Practice Exercises

Boost your GRADE at ALEKS.com!

ALEKS version 3.0

- Practice Problems
- Self-Tests
- NetTutor
- e-Professors
- Videos

Study Skills Exercises

1. In this text, we will provide skills for you to enhance your learning experience. In the first four chapters, each set of Practice Exercises begins with an activity that focuses on one of eight areas: learning about your course, using your text, taking notes, doing homework, taking an exam (test and math anxiety), managing your time, recognizing your learning style, and studying for the final exam.

   Each activity requires only a few minutes and will help you to pass this course and become a better math student. Many of these skills can be carried over to other disciplines and help you to become a model college student. To begin, write down the following information:

   a. Instructor's name
   b. Instructor's office number
   c. Instructor's telephone number
   d. Instructor's e-mail address
   e. Instructor's office hours
   f. Days of the week that the class meets
   g. The room number in which the class meets
   h. Is there a lab requirement for this course? How often must you attend lab and where is it located?

2. Define the key terms:

   a. real numbers
   b. set
   c. natural numbers
   d. whole numbers
   e. integers
   f. rational numbers
   g. irrational numbers
   h. inequality
   i. opposite
   j. absolute value

Objective 1: The Set of Real Numbers

3. Plot the numbers on a real number line: $\{1, -2, -\pi, 0, -\frac{5}{2}, 5.1\}$ **(See Example 1.)**

4. Plot the numbers on a real number line: $\{3, -4, \frac{1}{8}, -1.7, -\frac{4}{3}, 1.75\}$

For Exercises 5–20, describe each number as (a) a terminating decimal, (b) a repeating decimal, or (c) a nonterminating, nonrepeating decimal. Then classify the number as a rational number or as an irrational number. **(See Example 2.)**

5. 0.29

6. 3.8

7. $\dfrac{1}{9}$

8. $\dfrac{1}{3}$

9. $\dfrac{1}{8}$

10. $\dfrac{1}{5}$

11. 2π

12. 3π

13. -0.125

14. -3.24

15. -3

16. -6

17. $0.\overline{2}$

18. $0.\overline{6}$

19. $\sqrt{6}$

20. $\sqrt{10}$

21. List three numbers that are real numbers but not rational numbers.

22. List three numbers that are real numbers but not irrational numbers.

 23. List three numbers that are integers but not natural numbers.

24. List three numbers that are integers but not whole numbers.

25. List three numbers that are rational numbers but not integers.

For Exercises 26–32, let $A = \{-\frac{3}{2}, \sqrt{11}, -4, 0.\overline{6}, 0, \sqrt{7}, 1\}$ **(See Example 3.)**

26. Are all of the numbers in set A real numbers? **27.** List all of the rational numbers in set A.

28. List all of the whole numbers in set A.

29. List all of the natural numbers in set A.

30. List all of the irrational numbers in set A.

31. List all of the integers in set A.

 32. Plot the real numbers from set A on a number line. (*Hint:* $\sqrt{11} \approx 3.3$ and $\sqrt{7} \approx 2.6$)

Objective 2: Inequalities

33. The LPGA Samsung World Championship of women's golf scores for selected players are given in the table. Compare the scores and fill in the blank with the appropriate inequality sign: $<$ or $>$. **(See Example 4.)**

a. Kane's score _____ Pak's score.

b. Sorenstam's score _____ Davies' score.

c. Pak's score _____ McCurdy's score.

d. Kane's score _____ Davies' score.

LPGA Golfers	Final Score with Respect to Par
Annika Sorenstam	7
Laura Davies	−4
Lorie Kane	0
Cindy McCurdy	3
Se Ri Pak	−8

34. The elevations of selected cities in the United States are shown in the figure. Compare the elevations and fill in the blank with the appropriate inequality sign: $<$ or $>$. (A negative number indicates that the city is below sea level.)

a. Elevation of Tucson _____ elevation of Cincinnati.

b. Elevation of New Orleans _____ elevation of Chicago.

c. Elevation of New Orleans _____ elevation of Houston.

d. Elevation of Chicago _____ elevation of Cincinnati.

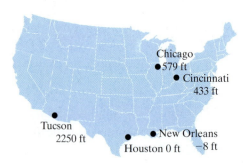

Objective 3: Opposite of a Real Number

For Exercises 35–42, find the opposite of each number. **(See Example 5.)**

35. 18

36. 2

37. −6.1

38. −2.5

39. $-\dfrac{5}{8}$

40. $-\dfrac{1}{3}$

41. $\dfrac{7}{3}$

42. $\dfrac{1}{9}$

The opposite of a is denoted as $-a$. For Exercises 43–50, simplify. **(See Example 6.)**

43. $-(-3)$ **44.** $-(-5.1)$ **45.** $-\left(\dfrac{7}{3}\right)$ **46.** $-(-7)$

47. $-(-8)$ **48.** $-(36)$ **49.** $-(72.1)$ **50.** $-\left(\dfrac{9}{10}\right)$

Objective 4: Absolute Value of a Real Number

For Exercises 51–62, simplify. **(See Example 7.)**

51. $|-2|$ **52.** $|-7|$ **53.** $|-1.5|$ **54.** $|-3.7|$

55. $-|-1.5|$ **56.** $-|-3.7|$ **57.** $\left|\dfrac{3}{2}\right|$ **58.** $\left|\dfrac{7}{4}\right|$

59. $-|10|$ **60.** $-|20|$ **61.** $-\left|-\dfrac{1}{2}\right|$ **62.** $-\left|-\dfrac{11}{3}\right|$

For Exercises 63–64, answer true or false. If a statement is false, explain why.

63. If n is positive, then $|n|$ is negative.   **64.** If m is negative, then $|m|$ is negative.

For Exercises 65–88, determine if the statements are true or false. Use the real number line to justify the answer. **(See Example 8.)**

65. $5 > 2$ **66.** $8 < 10$ **67.** $6 < 6$ **68.** $19 > 19$

69. $-7 \geq -7$ **70.** $-1 \leq -1$ **71.** $\dfrac{3}{2} \leq \dfrac{1}{6}$ **72.** $-\dfrac{1}{4} \geq -\dfrac{7}{8}$

73. $-5 > -2$ **74.** $6 < -10$ **75.** $8 \neq 8$ **76.** $10 \neq 10$

77. $|-2| \geq |-1|$ **78.** $|3| \leq |-1|$ **79.** $\left|-\dfrac{1}{9}\right| = \left|\dfrac{1}{9}\right|$ **80.** $\left|-\dfrac{1}{3}\right| = \left|\dfrac{1}{3}\right|$

81. $|7| \neq |-7|$ **82.** $|-13| \neq |13|$ **83.** $-1 < |-1|$ **84.** $-6 < |-6|$

85. $|-8| \geq |8|$ **86.** $|-11| \geq |11|$ **87.** $|-2| \leq |2|$ **88.** $|-21| \leq |21|$

Expanding Your Skills

89. For what numbers, a, is $-a$ positive?   **90.** For what numbers, a, is $|a| = a$?

Exponents, Square Roots, and the Order of Operations

Objectives

1. **Variables and Expressions**
2. **Evaluating Algebraic Expressions**
3. **Exponential Expressions**
4. **Square Roots**
5. **Order of Operations**
6. **Translations**

Concept Connections

1. Why is π not a variable?

---

1. Variables and Expressions

A **variable** is a symbol or letter such as x, y, and z, used to represent an unknown number. **Constants** are values that do not vary such as the numbers 3, -1.5, $\frac{2}{7}$, and π. An algebraic **expression** is a collection of variables and constants under algebraic operations. For example, $\frac{3}{x}$, $y + 7$, and $t - 1.4$ are algebraic expressions.

The symbols used to show the four basic operations of addition, subtraction, multiplication, and division are summarized in Table 1-3.

Table 1-3

Operation	Symbols	Translation
Addition	$a + b$	**sum** of a and b a plus b b added to a b more than a a increased by b the total of a and b
Subtraction	$a - b$	**difference** of a and b a minus b b subtracted from a a decreased by b b less than a a less b
Multiplication	$a \times b, a \cdot b, a(b), (a)b, (a)(b), ab$ (*Note:* We rarely use the notation $a \times b$ because the symbol, $\times$, might be confused with the variable, x.)	**product** of a and b a times b a multiplied by b
Division	$a \div b, \dfrac{a}{b}, a/b, b\overline{)a}$	**quotient** of a and b a divided by b b divided into a ratio of a and b a over b a per b

2. Evaluating Algebraic Expressions

The value of an algebraic expression depends on the values of the variables within the expression.

Skill Practice

Evaluate the algebraic expressions when $x = 5$ and $y = 2$.

2. $20 - y$ **3.** xy

Example 1 **Evaluating an Algebraic Expression**

Evaluate the algebraic expression when $p = 4$ and $q = \frac{3}{4}$.

a. $100 - p$ **b.** pq

Solution:

a. $100 - p$

$100 - (\ \ )$ When substituting a number for a variable, use parentheses.

$= 100 - (4)$ Substitute $p = 4$ in the parentheses.

$= 96$ Subtract.

Answers

1. The symbol π is not a variable because its value does not vary. It is always equal to the irrational number 3.14159....

2. 18 **3.** 10

b. pq

$= (\ )(\ )$	When substituting a number for a variable, use parentheses.
$= (4)\left(\dfrac{3}{4}\right)$	Substitute $p = 4$ and $q = \frac{3}{4}$.
$= \dfrac{\overset{1}{4}}{1} \cdot \dfrac{3}{\underset{1}{4}}$	Write the whole number as a fraction.
$= \dfrac{3}{1}$	Multiply fractions.
$= 3$	Simplify.

3. Exponential Expressions

In algebra, repeated multiplication can be expressed using exponents. The expression, $4 \cdot 4 \cdot 4$ can be written as

In the expression 4^3, 4 is the base, and 3 is the exponent, or power. The exponent indicates how many factors of the base to multiply.

DEFINITION   Definition of b^n

Let b represent any real number and n represent a positive integer. Then,

$$b^n = \underbrace{b \cdot b \cdot b \cdot b \ldots \cdot b}_{n \text{ factors of } b}$$

b^n is read as "b to the nth power."

b is called the **base**, and n is called the **exponent**, or **power**.

b^2 is read as "b squared," and b^3 is read as "b cubed."

The exponent, n, is the number of times the base, b, is used as a factor.

TIP: A number or variable with no exponent shown implies that there is an exponent of 1. That is, $b = b^1$.

Example 2   **Evaluating Exponential Expressions**

Translate the expression into words and then evaluate the expression.

a. 2^5 **b.** 5^2 **c.** $\left(\dfrac{3}{4}\right)^3$ **d.** 1^6

Solution:

a. The expression 2^5 is read as "two to the fifth power."
$2^5 = (2)(2)(2)(2)(2) = 32$

b. The expression 5^2 is read as "five to the second power" or "five, squared."
$5^2 = (5)(5) = 25$

Skill Practice

Evaluate.

4. 4^3 **5.** 2^4

6. $\left(\dfrac{2}{3}\right)^2$ **7.** $(1)^7$

Answers

4. 64   **5.** 16   **6.** $\dfrac{4}{9}$ **7.** 1

c. The expression $\left(\frac{3}{4}\right)^3$ is read as "three-fourths to the third power" or "three-fourths, cubed."

$$\left(\frac{3}{4}\right)^3 = \left(\frac{3}{4}\right)\left(\frac{3}{4}\right)\left(\frac{3}{4}\right) = \frac{27}{64}$$

d. The expression 1^6 is read as "one to the sixth power."
$$1^6 = (1)(1)(1)(1)(1)(1) = 1$$

4. Square Roots

The inverse operation to squaring a number is to find its **square roots**. For example, finding a square root of 9 is equivalent to asking "what number(s) when squared equals 9?" The symbol, $\sqrt{}$, (called a *radical sign*) is used to find the *principal* square root of a number. By definition, the principal square root of a number is nonnegative. Therefore, $\sqrt{9}$ is the nonnegative number that when squared equals 9. Hence $\sqrt{9} = 3$ because 3 is nonnegative and $(3)^2 = 9$.

Skill Practice

Evaluate.
8. $\sqrt{81}$ **9.** $\sqrt{100}$
10. $\sqrt{1}$ **11.** $\sqrt{\dfrac{9}{25}}$

Example 3 **Evaluating Square Roots**

Evaluate the square roots.

a. $\sqrt{64}$ **b.** $\sqrt{121}$ **c.** $\sqrt{0}$ **d.** $\sqrt{\dfrac{4}{9}}$

Solution:

a. $\sqrt{64} = 8$ Because $(8)^2 = 64$

b. $\sqrt{121} = 11$ Because $(11)^2 = 121$

c. $\sqrt{0} = 0$ Because $(0)^2 = 0$

d. $\sqrt{\dfrac{4}{9}} = \dfrac{2}{3}$ Because $\dfrac{2}{3} \cdot \dfrac{2}{3} = \dfrac{4}{9}$

A perfect square is a number whose square root is a rational number. If a number is not a perfect square, its square root is an irrational number that can be approximated on a calculator.

TIP: To simplify square roots, it is advisable to become familiar with the following perfect squares and square roots.

$0^2 = 0 \rightarrow \sqrt{0} = 0$	$7^2 = 49 \rightarrow \sqrt{49} = 7$
$1^2 = 1 \rightarrow \sqrt{1} = 1$	$8^2 = 64 \rightarrow \sqrt{64} = 8$
$2^2 = 4 \rightarrow \sqrt{4} = 2$	$9^2 = 81 \rightarrow \sqrt{81} = 9$
$3^2 = 9 \rightarrow \sqrt{9} = 3$	$10^2 = 100 \rightarrow \sqrt{100} = 10$
$4^2 = 16 \rightarrow \sqrt{16} = 4$	$11^2 = 121 \rightarrow \sqrt{121} = 11$
$5^2 = 25 \rightarrow \sqrt{25} = 5$	$12^2 = 144 \rightarrow \sqrt{144} = 12$
$6^2 = 36 \rightarrow \sqrt{36} = 6$	$13^2 = 169 \rightarrow \sqrt{169} = 13$

Answers
8. 9   **9.** 10
10. 1   **11.** $\dfrac{3}{5}$

5. Order of Operations

When algebraic expressions contain numerous operations, it is important to evaluate the operations in the proper order. Parentheses ( ), brackets [ ], and braces { } are used for grouping numbers and algebraic expressions. It is important to recognize that operations must be done within parentheses and other grouping symbols first. Other grouping symbols include absolute value bars, radical signs, and fraction bars.

PROCEDURE Order of Operations

Step 1 Simplify expressions within parentheses and other grouping symbols first. These include absolute value bars, fraction bars, and radicals. If imbedded parentheses are present, start with the innermost parentheses.

Step 2 Evaluate expressions involving exponents, radicals, and absolute values.

Step 3 Perform multiplication or division in the order that they occur from left to right.

Step 4 Perform addition or subtraction in the order that they occur from left to right.

Example 4 Applying the Order of Operations

Simplify the expressions.

a. $17 - 3 \cdot 2 + 2^2$ **b.** $\dfrac{1}{2}\left(\dfrac{5}{6} - \dfrac{3}{4}\right)$

Skill Practice

Simplify the expressions.

12. $14 - 3 \cdot 2 + 3^2$

13. $\dfrac{13}{4} - \dfrac{1}{4}(10 - 2)$

Solution:

a. $17 - 3 \cdot 2 + 2^2$

$= 17 - 3 \cdot 2 + 4$ Simplify exponents.

$= 17 - 6 + 4$ Multiply before adding or subtracting.

$= 11 + 4$ Add or subtract from left to right.

$= 15$

b. $\dfrac{1}{2}\left(\dfrac{5}{6} - \dfrac{3}{4}\right)$ Subtract fractions within the parentheses.

$= \dfrac{1}{2}\left(\dfrac{10}{12} - \dfrac{9}{12}\right)$ The least common denominator is 12.

$= \dfrac{1}{2}\left(\dfrac{1}{12}\right)$

$= \dfrac{1}{24}$ Multiply fractions.

Answers

12. 17   **13.** $\dfrac{5}{4}$

Skill Practice

Simplify the expressions.
14. $1 + 2 \cdot 3^2 \div 6$
15. $|-20| - \sqrt{20 - 4}$
16. $60 - 5[(7 - 4) + 2^2]$

Example 5 **Applying the Order of Operations**

Simplify the expressions.

a. $25 - 12 \div 3 \cdot 4$

b. $6.2 - |-2.1| + \sqrt{15 - 6}$

c. $28 - 2[(6 - 3)^2 + 4]$

Solution:

a. $25 - 12 \div 3 \cdot 4$ Multiply or divide in order from left to right.

$= 25 - 4 \cdot 4$ Notice that the operation $12 \div 3$ is performed first (not $3 \cdot 4$).

$= 25 - 16$ Multiply $4 \cdot 4$ before subtracting.

$= 9$ Subtract.

Avoiding Mistakes

In Example 5(a), division is performed before multiplication because it occurs first as we read from left to right.

b. $6.2 - |-2.1| + \sqrt{15 - 6}$

$= 6.2 - |-2.1| + \sqrt{9}$ Simplify within the square root.

$= 6.2 - (2.1) + 3$ Simplify the absolute value and square root.

$= 4.1 + 3$ Add or subtract from left to right.

$= 7.1$ Add.

c. $28 - 2[(6 - 3)^2 + 4]$

$= 28 - 2[(3)^2 + 4]$ Simplify within the inner parentheses first.

$= 28 - 2[(9) + 4]$ Simplify exponents.

$= 28 - 2[13]$ Add within the square brackets.

$= 28 - 26$ Multiply before subtracting.

$= 2$ Subtract.

Skill Practice

Translate each English phrase to an algebraic expression.

17. The product of 6 and y
18. The difference of the square root of t and 7
19. Twelve less than x
20. Twelve less x
21. One more than two times x
22. Five subtracted from the absolute value of w.

6. Translations

Example 6 **Translating from English Form to Algebraic Form**

Translate each English phrase to an algebraic expression.

a. The quotient of x and 5

b. The difference of p and the square root of q

c. Seven less than n

d. Seven less n

e. Eight more than the absolute value of w

f. x subtracted from 18

Answers

14. 4   **15.** 16   **16.** 25   **17.** $6y$
18. $\sqrt{t} - 7$ **19.** $x - 12$
20. $12 - x$ **21.** $2x + 1$
22. $|w| - 5$

Solution:

a. $\dfrac{x}{5}$ or $x \div 5$ The quotient of x and 5

b. $p - \sqrt{q}$ The difference of p and the square root of q

c. $n - 7$ Seven less than n

d. $7 - n$ Seven less n

e. $|w| + 8$ Eight more than the absolute value of w

f. $18 - x$ x subtracted from 18

> **Avoiding Mistakes**
>
> Recall that "a less than b" is translated as $b - a$. Therefore, the statement "seven less than n" must be translated as $n - 7$, not $7 - n$.

Example 7   **Translating from English Form to Algebraic Form**

Translate each English phrase into an algebraic expression. Then evaluate the expression for $a = 6$, $b = 4$, and $c = 20$.

a. The product of a and the square root of b

b. Twice the sum of b and c

c. The difference of twice a and b

Solution:

a. The product of a and the square root of b

     $a\sqrt{b}$

     $= (\ \ )\sqrt{(\ \ )}$ Use parentheses to substitute a number for a variable.

     $= (6)\sqrt{(4)}$ Substitute $a = 6$ and $b = 4$.

     $= 6 \cdot 2$ Simplify the radical first.

     $= 12$ Multiply.

b. Twice the sum of b and c

     $2(b + c)$ To compute "twice the sum of b and c," it is necessary to take the sum first and then multiply by 2. To ensure the proper order, the sum of b and c must be enclosed in parentheses. The proper translation is $2(b + c)$.

     $= 2((\ \ ) + (\ \ ))$ Use parentheses to substitute a number for a variable.

     $= 2((4) + (20))$ Substitute $b = 4$ and $c = 20$.

     $= 2(24)$ Simplify within the parentheses first.

     $= 48$ Multiply.

c. The difference of twice a and b

     $2a - b$

     $= 2(\ \ ) - (\ \ )$ Use parentheses to substitute a number for a variable.

     $= 2(6) - (4)$ Substitute $a = 6$ and $b = 4$.

     $= 12 - 4$ Multiply first.

     $= 8$ Subtract.

Skill Practice

Translate each English phrase to an algebraic expression. Then evaluate the expression for $x = 3$, $y = 9$, $z = 10$.

23. The quotient of the square root of y and x.

24. One-half the sum of x and y.

25. The difference of z and twice x.

Answers

23. $\dfrac{\sqrt{y}}{x}$; 1   **24.** $\dfrac{1}{2}(x + y)$; 6

25. $z - 2x$; 4

Calculator Connections

Topic: Evaluating Exponential Expressions on a Calculator

On a calculator, we enter exponents greater than the second power by using the key labeled y^x or $\wedge$. For example, evaluate 2^4 and 10^6:

Scientific Calculator:

Enter: 2 y^x 4 =   Result: 16

Enter: 10 y^x 6 =   Result: 1000000

Graphing Calculator:

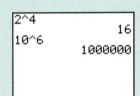

```
2^4
             16
10^6
       1000000
```

Topic: Applying the Order of Operations on a Calculator

Most calculators also have the capability to enter several operations at once. However, it is important to note that fraction bars and radicals require user-defined parentheses to ensure that the proper order of operations is followed. For example, evaluate the following expressions on a calculator:

a. $130 - 2(5 - 1)^3$ **b.** $\dfrac{18 - 2}{11 - 9}$ **c.** $\sqrt{25 - 9}$

Scientific Calculator:

Enter: 130 − 2 × (5 − 1) y^x 3 =   Result: 2

Enter: (18 − 2) ÷ (11 − 9) =   Result: 8

Enter: (25 − 9) √   Result: 4

Graphing Calculator:

```
130-2*(5-1)^3
              2
(18-2)/(11-9)
              8
√(25-9)
              4
```

Calculator Exercises

Simplify the expression without the use of a calculator. Then enter the expression into the calculator to verify your answer.

1. $\dfrac{4 + 6}{8 - 3}$ **2.** $110 - 5(2 + 1) - 4$ **3.** $100 - 2(5 - 3)^3$

4. $3 + (4 - 1)^2$ **5.** $(12 - 6 + 1)^2$ **6.** $3 \cdot 8 - \sqrt{32 + 2^2}$

7. $\sqrt{18 - 2}$ **8.** $(4 \cdot 3 - 3 \cdot 3)^3$ **9.** $\dfrac{20 - 3^2}{26 - 2^2}$

Section 1.2   Practice Exercises

Study Skills Exercises

1. Sometimes you may run into a problem with homework or you find that you are having trouble keeping up with the pace of the class. A tutor can be a good resource.

   a. Does your college offer tutoring?

   b. Is it free?

   c. Where would you go to sign up for a tutor?

2. Define the key terms:

   a. variable   **b.** constant   **c.** expression   **d.** base

   e. exponent   **f.** square root   **g.** order of operations

Review Exercises

3. Which of the following are rational numbers. $\left\{-4, 5.\overline{6}, \sqrt{29}, 0, \pi, 4.02, \frac{7}{9}\right\}$

4. Evaluate. $|-56|$

5. Evaluate. $|9.2|$

6. Evaluate. $-|-14|$

7. Find the opposite of 19.

8. Find the opposite of -34.2.

Objective 2: Evaluating Algebraic Expressions

For Exercises 9–14, evaluate the expressions for the given substitutions. **(See Example 1.)**

9. $y - 3$ when $y = 18$

10. $3q$ when $q = 5$

11. $\dfrac{15}{t}$ when $t = 5$

12. $8 + w$ when $w = 12$

13. $5 + 6d$ when $d = \dfrac{2}{3}$

14. $\dfrac{6}{5}h - 1$ when $h = 10$

Objective 3: Exponential Expressions

For Exercises 15–20, write each of the products using exponents.

15. $\dfrac{1}{6} \cdot \dfrac{1}{6} \cdot \dfrac{1}{6} \cdot \dfrac{1}{6}$

16. $10 \cdot 10 \cdot 10 \cdot 10 \cdot 10 \cdot 10$

17. $a \cdot a \cdot a \cdot b \cdot b$

18. $7 \cdot x \cdot x \cdot y \cdot y$

19. $5c \cdot 5c \cdot 5c \cdot 5c \cdot 5c$

20. $3 \cdot w \cdot z \cdot z \cdot z \cdot z$

21. **a.** For the expression $5x^3$, what is the base for the exponent 3?

    b. Does 5 have an exponent? If so, what is it?

22. **a.** For the expression $2y^4$, what is the base for the exponent 4?

    b. Does 2 have an exponent? If so, what is it?

For Exercises 23–30, write each expression in expanded form using the definition of an exponent.

23. x^3 **24.** y^4 **25.** $(2b)^3$ **26.** $(8c)^2$

27. $10y^5$ **28.** x^2y^3 **29.** $2wz^2$ **30.** $3a^3b$

For Exercises 31–38, simplify the expressions. **(See Example 2.)**

31. 6^2 **32.** 5^3 **33.** $\left(\dfrac{1}{7}\right)^2$ **34.** $\left(\dfrac{1}{2}\right)^5$

35. $(0.2)^3$ **36.** $(0.8)^2$ **37.** 2^6 **38.** 13^2

Objective 4: Square Roots

For Exercises 39–50, simplify the square roots. **(See Example 3.)**

39. $\sqrt{81}$ **40.** $\sqrt{64}$ **41.** $\sqrt{4}$ **42.** $\sqrt{9}$

43. $\sqrt{144}$ **44.** $\sqrt{49}$ **45.** $\sqrt{16}$ **46.** $\sqrt{36}$

47. $\sqrt{\dfrac{1}{9}}$ **48.** $\sqrt{\dfrac{1}{64}}$ **49.** $\sqrt{\dfrac{25}{81}}$ **50.** $\sqrt{\dfrac{49}{100}}$

Objective 5: Order of Operations

For Exercises 51–80, use the order of operations to simplify the expressions. **(See Examples 4–5.)**

51. $8 + 2 \cdot 6$ **52.** $7 + 3 \cdot 4$ **53.** $(8 + 2) \cdot 6$ **54.** $(7 + 3) \cdot 4$

55. $4 + 2 \div 2 \cdot 3 + 1$ **56.** $5 + 12 \div 2 \cdot 6 - 1$ **57.** $81 - 4 \cdot 3 + 3^2$ **58.** $100 - 25 \cdot 2 - 5^2$

59. $\dfrac{1}{4} \cdot \dfrac{2}{3} - \dfrac{1}{6}$ **60.** $\dfrac{3}{4} \cdot \dfrac{2}{3} + \dfrac{2}{3}$ **61.** $\left(\dfrac{11}{6} - \dfrac{3}{8}\right) \cdot \dfrac{4}{5}$ **62.** $\left(\dfrac{9}{8} - \dfrac{1}{3}\right) \cdot \dfrac{3}{4}$

63. $3[5 + 2(8 - 3)]$ **64.** $2[4 + 3(6 - 4)]$ **65.** $10 + |-6|$ **66.** $18 + |-3|$

67. $21 - |8 - 2|$ **68.** $12 - |6 - 1|$ **69.** $2^2 + \sqrt{9} \cdot 5$ **70.** $3^2 + \sqrt{16} \cdot 2$

71. $\sqrt{9 + 16} - 2$ **72.** $\sqrt{36 + 13} - 5$ **73.** $[4^2 \cdot (6 - 4) \div 8] + [7 \cdot (8 - 3)]$

74. $(18 \div \sqrt{4}) \cdot \{[(9^2 - 1) \div 2] - 15\}$ **75.** $48 - 13 \cdot 3 + [(50 - 7 \cdot 5) + 2]$

76. $80 \div 16 \cdot 2 + (6^2 - |-2|)$ **77.** $\dfrac{7 + 3(8 - 2)}{(7 + 3)(8 - 2)}$ **78.** $\dfrac{16 - 8 \div 4}{4 + 8 \div 4 - 2}$

79. $\dfrac{15 - 5(3 \cdot 2 - 4)}{10 - 2(4 \cdot 5 - 16)}$ **80.** $\dfrac{5(7 - 3) + 8(6 - 4)}{4[7 + 3(2 \cdot 9 - 8)]}$

81. The area of a rectangle is given by $A = \ell w$, where ℓ is the length of the rectangle and w is the width. Find the area for the rectangle shown.

160 ft

360 ft

82. The perimeter of a rectangle is given by $P = 2\ell + 2w$. Find the perimeter for the rectangle shown.

83. The area of a trapezoid is given by $A = \frac{1}{2}(b_1 + b_2)h$, where b_1 and b_2 are the lengths of the two parallel sides and h is the height. A window is in the shape of a trapezoid. Find the area of the trapezoid with dimensions shown in the figure.

84. The volume of a rectangular solid is given by $V = \ell wh$, where ℓ is the length of the box, w is the width, and h is the height. Find the volume of the box shown in the figure.

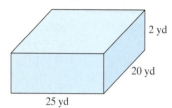

Objective 6: Translations

For Exercises 85–96, translate each English phrase into an algebraic expression. **(See Example 6.)**

85. The product of 3 and x

86. The sum of b and 6

87. The quotient of x and 7

88. Four divided by k

89. The difference of 2 and a

90. Three subtracted from t

91. x more than twice y

92. Nine decreased by the product of 3 and p

93. Four times the sum of x and 12

94. Twice the difference of x and 3

95. Q less than 3

96. Fourteen less than t

For Exercises 97–104, translate the English phrase into an algebraic expression. Then evaluate the expression for $x = 4$, $y = 2$, and $z = 10$. **(See Example 7.)**

97. Two times y cubed

98. Three times z squared

99. The absolute value of the difference of z and 8

100. The absolute value of the difference of x and 3

101. The product of 5 and the square root of x

102. The square root of the difference of z and 1

103. The value x subtracted from the product of y and z

104. The difference of z and the product of x and y

Expanding Your Skills

For Exercises 105–108, use the order of operations to simplify the expressions.

105. $\dfrac{\sqrt{\frac{1}{9}} + \frac{2}{3}}{\sqrt{\frac{4}{25}} + \frac{3}{5}}$

106. $\dfrac{5 - \sqrt{9}}{\sqrt{\frac{4}{9}} + \frac{1}{3}}$

107. $\dfrac{|-2|}{|-10| - |2|}$

108. $\dfrac{|-4|^2}{2^2 + \sqrt{144}}$

109. Some students use the following common memorization device (mnemonic) to help them remember the order of operations: the acronym PEMDAS or **P**lease **E**xcuse **M**y **D**ear **A**unt **S**ally to remember **P**arentheses, **E**xponents, **M**ultiplication, **D**ivision, **A**ddition, and **S**ubtraction. The problem with this mnemonic is that it suggests that multiplication is done before division and similarly, it suggests that addition is performed before subtraction. Explain why following this acronym may give incorrect answers for the expressions:

   a. $36 \div 4 \cdot 3$

   b. $36 - 4 + 3$

110. If you use the acronym **P**lease **E**xcuse **M**y **D**ear **A**unt **S**ally to remember the order of operations, what must you keep in mind about the last four operations?

111. Explain why the acronym **P**lease **E**xcuse **D**r. **M**ichael **S**mith's **A**unt could also be used as a memory device for the order of operations.

Section 1.3 Addition of Real Numbers

1. Addition of Real Numbers and the Number Line

Adding real numbers can be visualized on the number line. To add a positive number, move to the right on the number line. To add a negative number, move to the left on the number line. The following example may help to illustrate the process.

On a winter day in Detroit, suppose the temperature starts out at 5 degrees Fahrenheit (5°F) at noon, and then drops 12° two hours later when a cold front passes through. The resulting temperature can be represented by the expression $5° + (−12°)$. On the number line, start at 5 and count 12 units to the left (Figure 1-3). The resulting temperature at 2:00 P.M. is −7°F.

Figure 1-3

Skill Practice

Use the number line to add the numbers.

1. $-2 + 4$
2. $-2 + (-3)$
3. $5 + (-6)$

TIP: Note that we move to the left when we add a negative number, and we move to the right when we add a positive number.

| Example 1 | Using the Number Line to Add Real Numbers |

Use the number line to add the numbers.

a. $-5 + 2$ **b.** $-1 + (-4)$ **c.** $4 + (-7)$

Solution:

a. $-5 + 2 = -3$

Start at −5, and count 2 units to the right.

b. $-1 + (-4) = -5$

Start at −1, and count 4 units to the left.

c. $4 + (-7) = -3$

Start at 4, and count 7 units to the left.

2. Addition of Real Numbers

When adding large numbers or numbers that involve fractions or decimals, counting units on the number line can be cumbersome. Study the following example to determine a pattern for adding two numbers with the *same* sign.

$1 + 4 = 5$

$-1 + (-4) = -5$

PROCEDURE Adding Numbers with the *Same* Sign

To add two numbers with the *same* sign, add their absolute values and apply the common sign.

Study the following example to determine a pattern for adding two numbers with *different* signs.

$1 + (-4) = -3$

$-1 + 4 = 3$

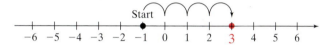

PROCEDURE Adding Numbers with *Different* Signs

To add two numbers with *different* signs, subtract the smaller absolute value from the larger absolute value. Then apply the sign of the number having the larger absolute value.

Example 2 **Adding Real Numbers with the Same Sign**

Add.

a. $-12 + (-14)$ **b.** $-8.8 + (-3.7)$ **c.** $-\dfrac{4}{3} + \left(-\dfrac{6}{7}\right)$

Solution:

a. $-12 + (-14)$ 　First find the absolute value of the addends.
　　　　↓　　↓　　　$|-12| = 12$ and $|-14| = 14$.

$= -(12 + 14)$ 　Add their absolute values and apply the
　　　　　　　　common sign (in this case, the common sign is
common sign is negative 　　negative).

$= -26$ 　　　The sum is -26.

b. $-8.8 + (-3.7)$ 　First find the absolute value of the addends.
　　　　↓　　↓　　　$|-8.8| = 8.8$ and $|-3.7| = 3.7$.

$= -(8.8 + 3.7)$ 　Add their absolute values and apply the common
　　　　　　　　sign (in this case, the common sign is negative).
common sign is negative

$= -12.5$ 　　　The sum is -12.5.

Skill Practice

Add the numbers.

4. $-5 + (-25)$

5. $-14.8 + (-9.7)$

6. $-\dfrac{1}{2} + \left(-\dfrac{5}{8}\right)$

Answers

4. -30 　**5.** -24.5 　**6.** $-\dfrac{9}{8}$

c. $-\dfrac{4}{3} + \left(-\dfrac{6}{7}\right)$ The least common denominator (LCD) is 21.

$= -\dfrac{4 \cdot 7}{3 \cdot 7} + \left(-\dfrac{6 \cdot 3}{7 \cdot 3}\right)$ Write each fraction with the LCD.

$= -\dfrac{28}{21} + \left(-\dfrac{18}{21}\right)$ Find the absolute value of the addends.

$$\left|-\dfrac{28}{21}\right| = \dfrac{28}{21} \text{ and } \left|-\dfrac{18}{21}\right| = \dfrac{18}{21}.$$

$= -\left(\dfrac{28}{21} + \dfrac{18}{21}\right)$ Add their absolute values and apply the common sign (in this case, the common sign is negative).

↑ common sign is negative

$= -\dfrac{46}{21}$ The sum is $-\dfrac{46}{21}$.

Skill Practice

Add.

7. $-15 + 16$

8. $6 + (-6)$

Example 3 **Adding Real Numbers with Different Signs**

Add.   **a.** $12 + (-17)$ **b.** $-8 + 8$

Solution:

a. $12 + (-17)$ First find the absolute value of the addends.
$|12| = 12$ and $|-17| = 17$.

The absolute value of -17 is greater than the absolute value of 12. Therefore, the sum is negative.

$= -(17 - 12)$ Next, subtract the smaller absolute value from the larger absolute value.

↑ Apply the sign of the number with the larger absolute value.

$= -5$

b. $-8 + 8$ First find the absolute value of the addends.
$|-8| = 8$ and $|8| = 8$.

$= (8 - 8)$ The absolute values are equal. Therefore, their difference is 0. The number zero is neither positive nor negative.

$= 0$

Skill Practice

Add.

9. $27.3 + (-18.1)$

10. $-\dfrac{9}{10} + \dfrac{2}{5}$

Example 4 **Adding Real Numbers with Different Signs**

Add.   **a.** $-10.6 + 20.4$ **b.** $\dfrac{2}{15} + \left(-\dfrac{4}{5}\right)$

Answers

7. 1   **8.** 0   **9.** 9.2   **10.** $-\dfrac{1}{2}$

Solution:

a. $-10.6 + 20.4$ First find the absolute value of the addends. $|-10.6| = 10.6$ and $|20.4| = 20.4$.

The absolute value of 20.4 is greater than the absolute value of -10.6. Therefore, the sum is positive.

$= +(20.4 - 10.6)$ Next, subtract the smaller absolute value from the larger absolute value.

Apply the sign of the number with the larger absolute value.

$= 9.8$

b. $\dfrac{2}{15} + \left(-\dfrac{4}{5}\right)$ The least common denominator is 15.

$= \dfrac{2}{15} + \left(-\dfrac{4 \cdot 3}{5 \cdot 3}\right)$ Write each fraction with the LCD.

$= \dfrac{2}{15} + \left(-\dfrac{12}{15}\right)$ Find the absolute value of the addends.

$\left|\dfrac{2}{15}\right| = \dfrac{2}{15}$ and $\left|-\dfrac{12}{15}\right| = \dfrac{12}{15}$.

The absolute value of $-\dfrac{12}{15}$ is greater than the absolute value of $\dfrac{2}{15}$. Therefore, the sum is negative.

$= -\left(\dfrac{12}{15} - \dfrac{2}{15}\right)$ Next, subtract the smaller absolute value from the larger absolute value.

Apply the sign of the number with the larger absolute value.

$= -\dfrac{10}{15}$ Subtract.

$= -\dfrac{2}{3}$ Simplify by reducing to lowest terms. $-\dfrac{\overset{2}{\cancel{10}}}{\underset{3}{\cancel{15}}} = -\dfrac{2}{3}$

3. Translations

Example 5   **Translating Expressions Involving the Addition of Real Numbers**

Translate each English phrase into an algebraic expression. Then simplify the result.

a. The sum of $-12, -8, 9$, and -1

b. Negative three-tenths added to $-\dfrac{7}{8}$

c. The sum of -12 and its opposite

Solution:

a. The sum of $-12, -8, 9$, and -1

$-12 + (-8) + 9 + (-1)$

$= -20 + 9 + (-1)$ Add from left to right.

$= -11 + (-1)$

$= -12$

b. Negative three-tenths added to $-\frac{7}{8}$

$$-\frac{7}{8} + \left(-\frac{3}{10}\right)$$

$$= -\frac{35}{40} + \left(-\frac{12}{40}\right) \qquad \text{The common denominator is 40.}$$

$$= -\frac{47}{40} \qquad \text{The numbers have the same signs. Add their absolute values and keep the common sign. } -\left(\frac{35}{40} + \frac{12}{40}\right).$$

c. The sum of -12 and its opposite

$$-12 + (12)$$

$$= 0 \qquad \text{Add.}$$

TIP: The sum of any number and its opposite is 0.

4. Applications Involving Addition of Real Numbers

Example 6 **Adding Real Numbers in Applications**

a. A running back on a football team gains 4 yd. On the next play, the quarterback is sacked and loses 13 yd. Write a mathematical expression to describe this situation and then simplify the result.

b. A student has $120 in her checking account. After depositing her paycheck of $215, she writes a check for $255 to cover her portion of the rent and another check for $294 to cover her car payment. Write a mathematical expression to describe this situation and then simplify the result.

Solution:

a. $4 + (-13)$ The loss of 13 yd can be interpreted as adding -13 yd.

$\quad = -9$ The football team has a net loss of 9 yd.

b. $120 + 215 + (-255) + (-294)$ Writing a check is equivalent to adding a negative amount to the bank account.

$= 335 + (-255) + (-294)$ Use the order of operations. Add from left to right.

$= \quad 80 + (-294)$

$= \quad\quad -214$ The student has overdrawn her account by $214.

Answer
14. $32.00 + 2.15 + (-3.28)$; $30.87 per share

Study Skills Exercise

1. It is very important to attend class every day. Math is cumulative in nature, and you must master the material learned in the previous class to understand today's lesson. Because this is so important, many instructors have an attendance policy that may affect your final grade. Write down the attendance policy for your class.

Review Exercises

Plot the points in set A on a number line. Then for Exercises 2–7 place the appropriate inequality (<, >) between the numbers.

$$A = \left\{ -2, \frac{3}{4}, -\frac{5}{2}, 3, \frac{9}{2}, 1.6, 0 \right\}$$

2. $-2 \ \square \ 0$

3. $\dfrac{9}{2} \ \square \ \dfrac{3}{4}$

4. $-2 \ \square \ -\dfrac{5}{2}$

5. $0 \ \square \ -\dfrac{5}{2}$

6. $\dfrac{3}{4} \ \square \ 1.6$

7. $\dfrac{3}{4} \ \square \ -\dfrac{5}{2}$

8. Evaluate the expressions.

   a. $-(-8)$ **b.** $-|-8|$

Objective 1: Addition of Real Numbers and the Number Line

For Exercises 9–16, add the numbers using the number line. **(See Example 1.)**

9. $-2 + (-4)$

10. $-3 + (-5)$

11. $-7 + 10$

12. $-2 + 9$

13. $6 + (-3)$

14. $8 + (-2)$

15. $2 + (-5)$

16. $7 + (-3)$

Objective 2: Addition of Real Numbers

For Exercises 17–70, add. **(See Examples 2–4.)**

17. $-19 + 2$

18. $-25 + 18$

19. $-4 + 11$

20. $-3 + 9$

21. $-16 + (-3)$

22. $-12 + (-23)$

23. $-2 + (-21)$

24. $-13 + (-1)$

25. $0 + (-5)$

26. $0 + (-4)$

27. $-3 + 0$

28. $-8 + 0$

29. $-16 + 16$

30. $11 + (-11)$

31. $41 + (-41)$

32. $-15 + 15$

33. $4 + (-9)$

34. $6 + (-9)$

35. $7 + (-2) + (-8)$

36. $2 + (-3) + (-6)$

37. $-17 + (-3) + 20$

38. $-9 + (-6) + 15$

39. $-3 + (-8) + (-12)$

40. $-8 + (-2) + (-13)$

41. $-42 + (-3) + 45 + (-6)$

42. $36 + (-3) + (-8) + (-25)$

43. $-5 + (-3) + (-7) + 4 + 8$

44. $-13 + (-1) + 5 + 2 + (-20)$

45. $23.81 + (-2.51)$

46. $-9.23 + 10.53$

47. $-\frac{2}{7} + \frac{1}{14}$

48. $-\frac{1}{8} + \frac{5}{16}$

49. $\frac{2}{3} + \left(-\frac{5}{6}\right)$

50. $\frac{1}{2} + \left(-\frac{3}{4}\right)$

51. $-\frac{7}{8} + \left(-\frac{1}{16}\right)$

52. $-\frac{1}{9} + \left(-\frac{4}{3}\right)$

53. $-\frac{1}{4} + \frac{3}{10}$

54. $-\frac{7}{6} + \frac{7}{8}$

55. $-2.1 + \left(-\frac{3}{10}\right)$

56. $-8.3 + \left(-\frac{9}{10}\right)$

57. $\frac{3}{4} + (-0.5)$

58. $-\frac{3}{2} + 0.45$

59. $8.23 + (-8.23)$

60. $-7.5 + 7.5$

61. $-\frac{7}{8} + 0$

62. $0 + \left(-\frac{21}{22}\right)$

63. $-\frac{3}{2} + \left(-\frac{1}{3}\right) + \frac{5}{6}$

64. $-\frac{7}{8} + \frac{7}{6} + \frac{7}{12}$

65. $-\frac{2}{3} + \left(-\frac{1}{9}\right) + 2$

66. $-\frac{1}{4} + \left(-\frac{3}{2}\right) + 2$

67. $-47.36 + 24.28$

68. $-0.015 + (0.0026)$

69. $-0.000617 + (-0.0015)$

70. $-5315.26 + (-314.89)$

71. State the rule for adding two numbers with different signs.

72. State the rule for adding two numbers with the same signs.

For Exercises 73–80, evaluate the expression for $x = -3$, $y = -2$, and $z = 16$.

73. $x + y + \sqrt{z}$

74. $2z + x + y$

75. $y + 3\sqrt{z}$

76. $-\sqrt{z} + y$

77. $|x| + |y|$

78. $z + x + |y|$

79. $-x + y$

80. $x + (-y) + z$

Objective 3: Translations

For Exercises 81–90, translate the English phrase into an algebraic expression. Then simplify the result.
(See Example 5.)

81. The sum of -6 and -10

82. The sum of -3 and 5

83. Negative three increased by 8

84. Twenty-one increased by 4

85. Seventeen more than -21

86. Twenty-four more than -7

87. Three times the sum of -14 and 20

88. Two times the sum of 6 and -10

89. Five more than the sum of -7 and -2

90. Negative six more than the sum of 4 and -1

Objective 4: Applications Involving Addition of Real Numbers

91. The temperature in Minneapolis, Minnesota, began at −5°F (5° below zero) at 6:00 A.M. By noon the temperature had risen 13°, and by the end of the day, the temperature had dropped 11° from its noontime high. Write an expression using addition that describes the change in temperatures during the day. Then evaluate the expression to give the temperature at the end of the day.

92. The temperature in Toronto, Ontario, Canada, began at 4°F. A cold front went through at noon, and the temperature dropped 9°. By 4:00 P.M. the temperature had risen 2° from its noontime low. Write an expression using addition that describes the changes in temperature during the day. Then evaluate the expression to give the temperature at the end of the day.

93. During a football game, the Nebraska Cornhuskers lost 2 yd, gained 6 yd, and then lost 5 yd. Write an expression using addition that describes the team's total loss or gain and evaluate the expression. **(See Example 6.)**

94. During a football game, the University of Oklahoma's team gained 3 yd, lost 5 yd, and then gained 14 yd. Write an expression using addition that describes the team's total loss or gain and evaluate the expression.

95. Yoshima has $52.23 in her checking account. She writes a check for groceries for $52.95. **(See Example 6.)**

  a. Write an addition problem that expresses Yoshima's transaction.

  b. Is Yoshima's account overdrawn?

96. Mohammad has $40.02 in his checking account. He writes a check for a pair of shoes for $40.96.

  a. Write an addition problem that expresses Mohammad's transaction.

  b. Is Mohammad's account overdrawn?

97. In the game show *Jeopardy*, a contestant responds to six questions with the following outcomes:

$$\$100, \$200, -\$500, \$300, \$100, -\$200$$

  a. Write an expression using addition to describe the contestant's scoring activity.

  b. Evaluate the expression from part (a) to determine the contestant's final outcome.

98. A company that has been in business for 5 yr has the following profit and loss record.

  a. Write an expression using addition to describe the company's profit/loss activity.

  b. Evaluate the expression from part (a) to determine the company's net profit or loss.

Year	Profit/Loss ($)
1	−50,000
2	−32,000
3	−5000
4	13,000
5	26,000

Section 1.4 Subtraction of Real Numbers

Objectives

1. Subtraction of Real Numbers
2. Translations
3. Applications Involving Subtraction
4. Applying the Order of Operations

1. Subtraction of Real Numbers

In Section 1.3, we learned the rules for adding real numbers. Subtraction of real numbers is defined in terms of the addition process. For example, consider the following subtraction problem and the corresponding addition problem:

$$6 - 4 = 2 \quad \Leftrightarrow \quad 6 + (-4) = 2$$

In each case, we start at 6 on the number line and move to the left 4 units. That is, adding the opposite of 4 produces the same result as subtracting 4. This is true in general. To subtract two real numbers, add the opposite of the second number to the first number.

> **PROCEDURE** Subtracting Real Numbers
>
> If a and b are real numbers, then $a - b = a + (-b)$.

$$\left. \begin{array}{l} 10 - 4 = 10 + (-4) = 6 \\ -10 - 4 = -10 + (-4) = -14 \end{array} \right\} \quad \text{Subtracting 4 is the same as adding } -4.$$

$$\left. \begin{array}{l} 10 - (-4) = 10 + (4) = 14 \\ -10 - (-4) = -10 + (4) = -6 \end{array} \right\} \quad \text{Subtracting } -4 \text{ is the same as adding 4.}$$

Skill Practice

Subtract.

1. $1 - (-3)$
2. $-2 - 2$
3. $-6 - (-11)$
4. $8 - 15$

Example 1 Subtracting Integers

Subtract the numbers.

a. $4 - (-9)$ **b.** $-6 - 9$ **c.** $-11 - (-5)$ **d.** $7 - 10$

Solution:

a. $4 - (-9)$

$\qquad = 4 + (9) = 13$

Change subtraction to addition. Take the opposite of -9.

b. $-6 - 9$

$\qquad = -6 + (-9) = -15$

Change subtraction to addition. Take the opposite of 9.

c. $-11 - (-5)$

$\qquad = -11 + (5) = -6$

Change subtraction to addition. Take the opposite of -5.

d. $7 - 10$

$\qquad = 7 + (-10) = -3$

Change subtraction to addition. Take the opposite of 10.

Answers

1. 4   **2.** -4 **3.** 5   **4.** -7

Example 2 Subtracting Real Numbers

a. $\dfrac{3}{20} - \left(-\dfrac{4}{15}\right)$ **b.** $-2.3 - 6.04$

Solution:

a. $\dfrac{3}{20} - \left(-\dfrac{4}{15}\right)$ The least common denominator is 60.

$\dfrac{9}{60} - \left(-\dfrac{16}{60}\right)$ Write equivalent fractions with the LCD.

$\dfrac{9}{60} + \left(\dfrac{16}{60}\right)$ Rewrite subtraction in terms of addition.

$\dfrac{25}{60}$ Add.

$\dfrac{5}{12}$ Reduce to lowest terms.

b. $-2.3 - 6.04$

$-2.3 + (-6.04)$ Rewrite subtraction in terms of addition.

-8.34 Add.

Skill Practice

Subtract.

5. $\dfrac{1}{6} - \left(-\dfrac{7}{12}\right)$

6. $-7.5 - 1.5$

---

2. Translations

Example 3 Translating Expressions Involving Subtraction

Write an algebraic expression for each English phrase and then simplify the result.

a. The difference of -7 and -5

b. 12.4 subtracted from -4.7

c. -24 decreased by the sum of -10 and 13

d. Seven-fourths less than one-third

Solution:

a. The difference of -7 and -5

$-7 - (-5)$

$= -7 + (5)$ Rewrite subtraction in terms of addition.

$= -2$ Simplify.

b. 12.4 subtracted from -4.7

$-4.7 - 12.4$

$= -4.7 + (-12.4)$ Rewrite subtraction in terms of addition.

$= -17.1$ Simplify.

Skill Practice

Write an algebraic expression for each phrase and then simplify.

7. 8 less than -10

8. -7.2 subtracted from -8.2

9. 10 more than the difference of -2 and 3

10. Two-fifths decreased by four-thirds

TIP: Recall that "b subtracted from a" is translated as $a - b$. In Example 3(b), -4.7 is written first and then 12.4.

Answers

5. $\dfrac{3}{4}$ **6.** -9

7. $-10 - 8;\ -18$

8. $-8.2 - (-7.2);\ -1$

9. $(-2 - 3) + 10;\ 5$

10. $\dfrac{2}{5} - \dfrac{4}{3};\ -\dfrac{14}{15}$

c. -24 decreased by the sum of -10 and 13

$$-24 - (-10 + 13)$$

$$= -24 - (3) \qquad \text{Simplify inside parentheses.}$$

$$= -24 + (-3) \qquad \text{Rewrite subtraction in terms of addition.}$$

$$= -27 \qquad \text{Simplify.}$$

d. Seven-fourths less than one-third

$$\frac{1}{3} - \frac{7}{4}$$

$$= \frac{1}{3} + \left(-\frac{7}{4}\right) \qquad \text{Rewrite subtraction in terms of addition.}$$

$$= \frac{4}{12} + \left(-\frac{21}{12}\right) \qquad \text{The common denominator is 12.}$$

$$= -\frac{17}{12} \quad \text{or} \quad -1\frac{5}{12}$$

3. Applications Involving Subtraction

Skill Practice

11. During Harold's first round on *Jeopardy*, he got the $100, $200, and $400 questions correct but he got the $300 and $500 questions incorrect. Determine Harold's score for this round.

Example 4 Using Subtraction of Real Numbers in an Application

During one of his turns on *Jeopardy*, Harold selected the category "Show Tunes." He got the $200, $600, and $1000 questions correct, but he got the $400 and $800 questions incorrect. Write an expression that determines Harold's score. Then simplify the expression to find his total winnings for that category.

Solution:

$$200 + 600 + 1000 - 400 - 800$$

$$= 200 + 600 + 1000 + (-400) + (-800) \qquad \begin{array}{l}\text{Add the positive numbers.}\\\text{Add the negative numbers.}\end{array}$$

$$= 1800 + (-1200)$$

$$= 600 \qquad \text{Harold won \$600.}$$

Answer

11. -100, Harold lost $100.

Example 5	**Using Subtraction of Real Numbers in an Application**

The highest recorded temperature in North America was 134°F, recorded on July 10, 1913, in Death Valley, California. The lowest temperature of −81°F was recorded on February 3, 1947, in Snag, Yukon, Canada.

Find the difference between the highest and lowest recorded temperatures in North America.

Solution:

$134 - (-81)$

$= 134 + (81)$ Rewrite subtraction in terms of addition.

$= 215$ Add.

The difference between the highest and lowest temperatures is 215°F.

Skill Practice

12. The record high temperature for the state of Montana occurred in 1937 and was 117°F. The record low occurred in 1954 and was −70°F. Find the difference between the highest and lowest temperatures.

4. Applying the Order of Operations

Example 6	**Applying the Order of Operations**

Simplify the expressions.

a. $-6 + \{10 - [7 - (-4)]\}$ **b.** $5 - \sqrt{35 - (-14)} - 2$

Skill Practice

Simplify the expressions.
13. $-11 - \{8 - [2 - (-3)]\}$
14. $(12 - 5)^2 + \sqrt{4 - (-21)}$

Solution:

a. $-6 + \{10 - [7 - (-4)]\}$ Work inside the inner brackets first.

$= -6 + \{10 - [7 + (4)]\}$ Rewrite subtraction in terms of addition.

$= -6 + \{10 - (11)\}$ Simplify the expression inside braces.

$= -6 + \{10 + (-11)\}$ Rewrite subtraction in terms of addition.

$= -6 + (-1)$

$= -7$ Add.

b. $5 - \sqrt{35 - (-14)} - 2$ Work inside the radical first.

$= 5 - \sqrt{35 + (14)} - 2$ Rewrite subtraction in terms of addition.

$= 5 - \sqrt{49} - 2$

$= 5 - 7 - 2$ Simplify the radical.

$= 5 + (-7) + (-2)$ Rewrite subtraction in terms of addition.

$= -2 + (-2)$ Add from left to right.

$= -4$

Answers

12. 187°F   **13.** −14   **14.** 54

Example 7 Applying the Order of Operations

Simplify the expressions.

a. $\left(-\dfrac{5}{8} - \dfrac{2}{3}\right) - \left(\dfrac{1}{8} + 2\right)$

b. $-6 - |7 - 11| + (-3 + 7)^2$

Solution:

a. $\left(-\dfrac{5}{8} - \dfrac{2}{3}\right) - \left(\dfrac{1}{8} + 2\right)$ Work inside the parentheses first.

$= \left[-\dfrac{5}{8} + \left(-\dfrac{2}{3}\right)\right] - \left(\dfrac{1}{8} + 2\right)$ Rewrite subtraction in terms of addition.

$= \left[-\dfrac{15}{24} + \left(-\dfrac{16}{24}\right)\right] - \left(\dfrac{1}{8} + \dfrac{16}{8}\right)$ Get a common denominator in each parentheses.

$= \left(-\dfrac{31}{24}\right) - \left(\dfrac{17}{8}\right)$ Add fractions in each parentheses.

$= \left(-\dfrac{31}{24}\right) + \left(-\dfrac{17}{8}\right)$ Rewrite subtraction in terms of addition.

$= -\dfrac{31}{24} + \left(-\dfrac{51}{24}\right)$ Get a common denominator.

$= -\dfrac{82}{24}$ Add.

$= -\dfrac{41}{12}$ Reduce to lowest terms.

b. $-6 - |7 - 11| + (-3 + 7)^2$ Simplify within absolute value bars and parentheses first.

$= -6 - |7 + (-11)| + (-3 + 7)^2$ Rewrite subtraction in terms of addition.

$= -6 - |-4| + (4)^2$

$= -6 - (4) + 16$ Simplify absolute value and exponent.

$= -6 + (-4) + 16$ Rewrite subtraction in terms of addition.

$= -10 + 16$ Add from left to right.

$= 6$

Calculator Connections

Topic: Operations with Signed Numbers on a Calculator

Most calculators can add, subtract, multiply, and divide signed numbers. It is important to note, however, that the key used for the negative sign is different from the key used for subtraction. On a scientific calculator, the $\boxed{+/-}$ key or $\boxed{+\circ-}$ key is used to enter a negative number or to change the sign of an existing number. On a graphing calculator, the $\boxed{(-)}$ key is used. These keys should not be confused with the $\boxed{-}$ key which is used for subtraction. For example, try simplifying the following expressions.

a. $-7 + (-4) - 6$ **b.** $-3.1 - (-0.5) + 1.1$

Scientific Calculator:

Enter: 7 +o- + (4 +o-) − 6 =   **Result:**   −17

Enter: 3.1 +o- − (0.5 +o-) + 1.1 =   **Result:**   −1.5

Graphing Calculator:

```
-7+(-4)-6
              -17
-3.1-(-0.5)+1.1
              -1.5
```

Calculator Exercises

Simplify the expression without the use of a calculator. Then use the calculator to verify your answer.

1. $-8 + (-5)$ **2.** $4 + (-5) + (-1)$ **3.** $627 - (-84)$ **4.** $-0.06 - 0.12$

5. $-3.2 - (-14.5)$ **6.** $-472 + (-518)$ **7.** $-12 - 9 + 4$ **8.** $209 - 108 + (-63)$

Section 1.4   Practice Exercises

Study Skills Exercise

1. Some instructors allow the use of calculators. What is your instructor's policy regarding calculators in class, on the homework, and on tests?

Helpful Hint: If you are not permitted to use a calculator on tests, it is a good idea to do your homework in the same way, without a calculator.

Review Exercises

For Exercises 2–5, translate each English phrase into an algebraic expression.

2. The square root of 6   **3.** The square of x

4. Negative seven increased by 10   **5.** Two more than $-b$

For Exercises 6–8, simplify the expression.

6. $4^2 - 6 \div 2$ **7.** $1 + 36 \div 9 \cdot 2$ **8.** $14 - |10 - 6|$

Objective 1: Subtraction of Real Numbers

For Exercises 9–14, fill in the blank to make each statement correct.

9. $5 - 3 = 5 + $ _____   **10.** $8 - 7 = 8 + $ _____   **11.** $-2 - 12 = -2 + $ _____

12. $-4 - 9 = -4 + $ _____   **13.** $7 - (-4) = 7 + $ _____   **14.** $13 - (-4) = 13 + $ _____

For Exercises 15–60, subtract. **(See Examples 1–2.)**

15. $3 - 5$

16. $9 - 12$

17. $3 - (-5)$

18. $9 - (-12)$

19. $-3 - 5$

20. $-9 - 12$

21. $-3 - (-5)$

22. $-9 - (-5)$

23. $23 - 17$

24. $14 - 2$

25. $23 - (-17)$

26. $14 - (-2)$

27. $-23 - 17$

28. $-14 - 2$

29. $-23 - (-23)$

30. $-14 - (-14)$

31. $-6 - 14$

32. $-9 - 12$

33. $-7 - 17$

34. $-8 - 21$

35. $13 - (-12)$

36. $20 - (-5)$

37. $-14 - (-9)$

38. $-21 - (-17)$

39. $-\dfrac{6}{5} - \dfrac{3}{10}$

40. $-\dfrac{2}{9} - \dfrac{5}{3}$

41. $\dfrac{3}{8} - \left(-\dfrac{4}{3}\right)$

42. $\dfrac{7}{10} - \left(-\dfrac{5}{6}\right)$

43. $\dfrac{1}{2} - \dfrac{1}{10}$

44. $\dfrac{2}{7} - \dfrac{3}{14}$

45. $-\dfrac{11}{12} - \left(-\dfrac{1}{4}\right)$

46. $-\dfrac{7}{8} - \left(-\dfrac{1}{6}\right)$

47. $6.8 - (-2.4)$

48. $7.2 - (-1.9)$

49. $3.1 - 8.82$

50. $1.8 - 9.59$

51. $-4 - 3 - 2 - 1$

52. $-10 - 9 - 8 - 7$

53. $6 - 8 - 2 - 10$

54. $20 - 50 - 10 - 5$

55. $-36.75 - 14.25$

56. $-84.21 - 112.16$

57. $-112.846 + (-13.03) - 47.312$

58. $-96.473 + (-36.02) - 16.617$

59. $0.085 - (-3.14) + 0.018$

60. $0.00061 - (-0.00057) + 0.0014$

Objective 2: Translations

For Exercises 61–70, translate each English phrase into an algebraic expression. Then evaluate the expression. **(See Example 3.)**

61. Six minus -7

62. Eighteen minus -1

63. Eighteen subtracted from 3

64. Twenty-one subtracted from 8

65. The difference of -5 and -11

66. The difference of -2 and -18

67. Negative thirteen subtracted from -1

68. Negative thirty-one subtracted from -19

69. Twenty less than -32

70. Seven less than -3

Objective 3: Applications Involving Subtraction

71. On the game, *Jeopardy*, Jasper selected the category "The Last." He got the first four questions correct (worth $200, $400, $600, and $800) but then missed the last question (worth $1000). Write an expression that determines Jasper's score. Then simplify the expression to find his total winnings for that category. **(See Example 4.)**

72. On Courtney's turn in *Jeopardy*, she chose the category "Birds of a Feather." She already had $1200 when she selected a Double Jeopardy question. She wagered $500 but guessed incorrectly (therefore she lost $500). On her next turn, she got the $800 question correct. Write an expression that determines Courtney's score. Then simplify the expression to find her total winnings for that category.

73. In Ohio, the highest temperature ever recorded was 113°F and the lowest was −39°F. Find the difference between the highest and lowest temperatures. (*Source: Information Please Almanac*) **(See Example 5.)**

74. In Mississippi, the highest temperature ever recorded was 115°F and the lowest was −19°F. Find the difference between the highest and lowest temperatures. (*Source: Information Please Almanac*)

75. The highest mountain in the world is Mt. Everest, located in the Himalayas. Its height is 8848 meters (m) (29,028 ft). The lowest recorded depth in the ocean is located in the Marianas Trench in the Pacific Ocean. Its "height" relative to sea level is −11,033 m (−36,198 ft). Determine the difference in elevation, in meters, between the highest mountain in the world and the deepest ocean trench. (*Source: Information Please Almanac*)

76. The lowest point in North America is located in Death Valley, California, at an elevation of −282 ft (−86 m). The highest point in North America is Mt. McKinley, Alaska, at an elevation of 20,320 ft (6194 m). Find the difference in elevation, in feet, between the highest and lowest points in North America. (*Source: Information Please Almanac*)

Objective 4: Applying the Order of Operations

For Exercises 77–96, perform the indicated operations. **(See Examples 6–7.)**

77. $6 + 8 - (-2) - 4 + 1$

78. $-3 - (-4) + 1 - 2 - 5$

79. $-1 - 7 + (-3) - 8 + 10$

80. $13 - 7 + 4 - 3 - (-1)$

81. $2 - (-8) + 7 + 3 - 15$

82. $8 - (-13) + 1 - 9$

83. $-6 + (-1) + (-8) + (-10)$

84. $-8 + (-3) + (-5) + (-2)$

85. $-4 - \{11 - [4 - (-9)]\}$

86. $15 - \{25 + 2[3 - (-1)]\}$

87. $-\dfrac{13}{10} + \dfrac{8}{15} - \left(-\dfrac{2}{5}\right)$

88. $\dfrac{11}{14} - \left(-\dfrac{9}{7}\right) - \dfrac{3}{2}$

89. $\left(\dfrac{2}{3} - \dfrac{5}{9}\right) - \left(\dfrac{4}{3} - (-2)\right)$

90. $\left(-\dfrac{9}{8} - \dfrac{1}{4}\right) - \left(-\dfrac{5}{6} + \dfrac{1}{8}\right)$

91. $\sqrt{29 + (-4)} - 7$

92. $8 - \sqrt{98 + (-3)} + 5$

93. $|10 + (-3)| - |-12 + (-6)|$

94. $|6 - 8| + |12 - 5|$

95. $\dfrac{3 - 4 + 5}{4 + (-2)}$

96. $\dfrac{12 - 14 + 6}{6 + (-2)}$

For Exercises 97–104, evaluate the expressions for $a = -2$, $b = -6$, and $c = -1$.

97. $(a + b) - c$

98. $(a - b) + c$

99. $a - (b + c)$

100. $a + (b - c)$

101. $(a - b) - c$

102. $(a + b) + c$

103. $a - (b - c)$

104. $a + (b + c)$

Problem Recognition Exercises

Addition and Subtraction of Real Numbers

1. State the rule for adding two negative numbers.

2. State the rule for adding a negative number to a positive number.

For Exercises 3–40, perform the indicated operations.

3. $65 - 24$

4. $42 - 29$

5. $13 - (-18)$

6. $22 - (-24)$

7. $4.8 - 6.1$

8. $3.5 - 7.1$

9. $4 + (-20)$

10. $5 + (-12)$

11. $\dfrac{1}{3} - \dfrac{5}{12}$

12. $\dfrac{3}{8} - \dfrac{1}{12}$

13. $-32 - 4$

14. $-51 - 8$

15. $-6 + (-6)$

16. $-25 + (-25)$

17. $-4 - \left(-\dfrac{5}{6}\right)$

18. $-2 - \left(-\dfrac{2}{5}\right)$

19. $-60 + 55$

20. $-55 + 23$

21. $-18 - (-18)$

22. $-3 - (-3)$

23. $-3.5 - 4.2$

24. $-6.6 - 3.9$

25. $-90 + (-24)$

26. $-35 + (-21)$

27. $-14 + (-2) - 16$

28. $-25 + (-6) - 15$

29. $-4.2 + 1.2 + 3.0$

30. $-4.6 + 8.6 + (-4.0)$

31. $-10 - 8 - 6 - 4 - 2$

32. $-100 - 90 - 80 - 70 - 60$

33. $-8 - [4 - (-3)]^2$

34. $10 - (-3 + 9)^2$

35. $18 - |9^2 + (-100)|$

36. $14 - |5^2 + (-5)|$

37. $6 - [1 - (2 - 3)]$

38. $9 - [3 - (-2 - 5)]$

39. $\left[-\dfrac{2}{3} - \left(-\dfrac{5}{6}\right)\right]^2$

40. $\left[-\dfrac{3}{4} - \left(-\dfrac{7}{8}\right)\right]^2$

Section 1.5   Multiplication and Division of Real Numbers

Objectives

1. **Multiplication of Real Numbers**
2. **Exponential Expressions**
3. **Division of Real Numbers**
4. **Applying the Order of Operations**

1. Multiplication of Real Numbers

Multiplication of real numbers can be interpreted as repeated addition. For example:

$$3(4) = 4 + 4 + 4 = 12 \qquad \text{Add 3 groups of 4.}$$

$$3(-4) = -4 + (-4) + (-4) = -12 \qquad \text{Add 3 groups of } -4.$$

These results suggest that the product of a positive number and a negative number is *negative.* Consider the following pattern of products.

$$
\begin{array}{rcr}
4 \times & 3 = & 12 \\
4 \times & 2 = & 8 \\
4 \times & 1 = & 4 \\
4 \times & 0 = & 0 \\
4 \times & -1 = & -4 \\
4 \times & -2 = & -8 \\
4 \times & -3 = & -12 \\
\end{array}
$$

The pattern decreases by 4 with each row.

Thus, the product of a positive number and a negative number must be *negative* for the pattern to continue.

Now suppose we have a product of two negative numbers. To determine the sign, consider the following pattern of products.

$$
\begin{array}{rl}
-4 \times 3 = & -12 \\
-4 \times 2 = & -8 \\
-4 \times 1 = & -4 \\
-4 \times 0 = & 0 \\
-4 \times -1 = & 4 \\
-4 \times -2 = & 8 \\
-4 \times -3 = & 12
\end{array}
$$

The pattern increases by 4 with each row.

Thus, the product of two negative numbers must be *positive* for the pattern to continue.

From the first four rows, we see that the product increases by 4 for each row. For the pattern to continue, it follows that the product of two negative numbers must be *positive*.

We now summarize the rules for multiplying real numbers.

PROCEDURE Multiplying Real Numbers

- The product of two real numbers with the *same* sign is positive.
    - Examples: $(5)(6) = 30$
    - $(-4)(-10) = 40$
- The product of two real numbers with *different* signs is negative.
    - Examples: $(-2)(5) = -10$
    - $(4)(-9) = -36$
- The product of any real number and zero is zero.
    - Examples: $(8)(0) = 0$
    - $(0)(-6) = 0$

Example 1 Multiplying Real Numbers

Multiply the real numbers.

a. $-8(-4)$ **b.** $-2.5(-1.7)$ **c.** $-7(10)$

d. $\dfrac{1}{2}(-8)$ **e.** $0(-8.3)$ **f.** $-\dfrac{2}{7}\left(-\dfrac{7}{2}\right)$

Solution:

a. $-8(-4) = 32$

b. $-2.5(-1.7) = 4.25$ $\Big\}$ *Same signs.* Product is positive.

c. $-7(10) = -70$

d. $\dfrac{1}{2}(-8) = -4$ $\Big\}$ *Different signs.* Product is negative.

e. $0(-8.3)$

  $= 0$ The product of any real number and zero is zero.

f. $-\dfrac{2}{7}\left(-\dfrac{7}{2}\right)$

  $= \dfrac{14}{14}$ *Same signs.* Product is positive.

  $= 1$ Reduce to lowest terms.

Skill Practice

Multiply.

1. $-9(-3)$ **2.** $-1.5(-1.5)$

3. $-6(4)$ **4.** $\dfrac{1}{3}(-15)$

5. $0(-4.1)$ **6.** $-\dfrac{5}{9}\left(-\dfrac{9}{5}\right)$

Answers

1. 27 **2.** 2.25 **3.** -24
4. -5 **5.** 0 **6.** 1

Observe the pattern for repeated multiplications.

$$(-1)(-1) \qquad \underbrace{(-1)(-1)(-1)} \qquad \underbrace{(-1)(-1)}(-1)(-1) \qquad \underbrace{(-1)(-1)}(-1)(-1)(-1)$$
$$= 1 \qquad = (1)(-1) \qquad = \underbrace{(1)(-1)}(-1) \qquad = \underbrace{(1)(-1)}(-1)(-1)$$
$$= -1 \qquad = (-1)(-1) \qquad = \underbrace{(-1)(-1)}(-1)$$
$$= 1 \qquad = (1)(-1)$$
$$= -1$$

The pattern demonstrated in these examples indicates that

Concept Connections

7. Without actually computing, determine if the product shown is positive or negative. Explain.

$(-526)(420)(-105)$

- The product of an even number of negative factors is positive.
- The product of an odd number of negative factors is negative.

2. Exponential Expressions

Recall that for any real number b and any positive integer, n:

$$b^n = \underbrace{b \cdot b \cdot b \cdot b \ldots \cdot b}_{n \text{ factors of } b}$$

Be particularly careful when evaluating exponential expressions involving negative numbers. An exponential expression with a negative base is written with parentheses around the base, such as $(-2)^4$.

To evaluate $(-2)^4$, the base -2 is multiplied four times:

$$(-2)^4 = (-2)(-2)(-2)(-2) = 16$$

If parentheses are *not* used, the expression -2^4 has a different meaning:

- The expression -2^4 has a base of 2 (not -2) and can be interpreted as $-1 \cdot 2^4$.

$$-2^4 = -1(2)(2)(2)(2) = -16$$

- The expression -2^4 can also be interpreted as the opposite of 2^4.

$$-2^4 = -(2 \cdot 2 \cdot 2 \cdot 2) = -16$$

Skill Practice

Simplify.

8. $(-7)^2$ **9.** -7^2

10. $\left(-\dfrac{2}{3}\right)^3$ **11.** -0.2^3

Example 2   Evaluating Exponential Expressions

Simplify.

a. $(-5)^2$ **b.** -5^2 **c.** $\left(-\dfrac{1}{2}\right)^3$ **d.** -0.4^3

Solution:

a. $(-5)^2 = (-5)(-5) = 25$ Multiply two factors of -5.

b. $-5^2 = -1(5)(5) = -25$ Multiply -1 by two factors of 5.

Avoiding Mistakes

The negative sign is not part of the base unless it is in parentheses with the base. Thus, in the expression -5^2, the exponent applies only to 5 and not to the negative sign.

Answers

7. The product is positive because there is an even number of negative factors.

8. 49   **9.** -49

10. $-\dfrac{8}{27}$ **11.** -0.008

c. $\left(-\dfrac{1}{2}\right)^3 = \left(-\dfrac{1}{2}\right)\left(-\dfrac{1}{2}\right)\left(-\dfrac{1}{2}\right) = -\dfrac{1}{8}$ Multiply three factors of $-\dfrac{1}{2}$.

d. $-0.4^3 = -1(0.4)(0.4)(0.4) = -0.064$ Multiply -1 by three factors of 0.4.

3. Division of Real Numbers

Two numbers are *reciprocals* if their product is 1. For example, $-\dfrac{2}{7}$ and $-\dfrac{7}{2}$ are reciprocals because $-\dfrac{2}{7}\left(-\dfrac{7}{2}\right) = 1$. Symbolically, if a is a nonzero real number, then the reciprocal of a is $\dfrac{1}{a}$ because $a \cdot \dfrac{1}{a} = 1$. This definition also implies that a number and its reciprocal have the same sign.

DEFINITION   The Reciprocal of a Real Number

Let a be a nonzero real number. Then, the **reciprocal** of a is $\dfrac{1}{a}$.

Recall that to subtract two real numbers, we add the opposite of the second number to the first number. In a similar way, division of real numbers is defined in terms of multiplication. To divide two real numbers, we multiply the first number by the reciprocal of the second number.

DEFINITION   Division of Real Numbers

Let a and b be real numbers such that $b \neq 0$. Then, $a \div b = a \cdot \dfrac{1}{b}$.

Consider the quotient $10 \div 5$. The reciprocal of 5 is $\dfrac{1}{5}$, so we have

multiply

$$10 \div 5 = 2 \qquad \text{or equivalently,} \qquad 10 \cdot \dfrac{1}{5} = 2$$

reciprocal

Because division of real numbers can be expressed in terms of multiplication, then the sign rules that apply to multiplication also apply to division.

PROCEDURE   Dividing Real Numbers

- The quotient of two real numbers with the *same* sign is positive.

    Examples: $24 \div 4 = 6$

    $-36 \div -9 = 4$

- The quotient of two real numbers with *different* signs is negative.

    Examples: $100 \div (-5) = -20$

    $-12 \div 4 = -3$

Multip
(provi

This r
divisio

  1. Th

  2. Th

    Fi
    ec
    T
  3. Th
    th
    0
    e

PRO

    Le

    1

4.

Ex

Simp

Sol

— Skill
Simplify.

12. −14

14. $\dfrac{-7.6}{-1.9}$

— **Skill Practice** —

Simplify.

23. $\dfrac{100 - 3[-1 + (2 - 6)^2]}{|20 - 25|}$

Example 6 **Applying the Order of Operations**

Simplify. $\dfrac{24 - 2[-3 + (5 - 8)]^2}{2|-12 + 3|}$

Solution:

$$\dfrac{24 - 2[-3 + (5 - 8)]^2}{2|-12 + 3|}$$ Simplify numerator and denominator separately.

$$= \dfrac{24 - 2[-3 + (-3)]^2}{2|-9|}$$ Simplify within the inner parentheses and absolute value.

$$= \dfrac{24 - 2[-6]^2}{2(9)}$$ Simplify within brackets, [ ]. Simplify the absolute value.

$$= \dfrac{24 - 2(36)}{2(9)}$$ Simplify exponents.

$$= \dfrac{24 - 72}{18}$$ Perform multiplication before subtraction.

$$= \dfrac{-48}{18} \text{ or } -\dfrac{8}{3}$$ Reduce to lowest terms.

TIP:
and den
fraction
then the
Therefor
simplifie

— **Skill** —
Simplify.

16. 12

17. $\dfrac{3}{4} \div$

— **Skill Practice** —

Given $a = -7$, evaluate the expressions.

24. a^2 **25.** $-a^2$

Example 7 **Evaluating an Algebraic Expression**

Given $y = -6$, evaluate the expressions.

a. y^2 **b.** $-y^2$

Solution:

a. y^2

$= (\ \ )^2$ When substituting a number for a variable, use parentheses.

$= (-6)^2$ Substitute $y = -6$.

$= 36$ Square -6, that is, $(-6)(-6) = 36$.

b. $-y^2$

$= -(\ \ )^2$ When substituting a number for a variable, use parentheses.

$= -(-6)^2$ Substitute $y = -6$.

$= -(36)$ Square -6.

$= -36$ Multiply by -1.

Answer

12. −2

15. $\dfrac{7}{3}$

Answers

23. 11   **24.** 49   **25.** −49

Calculator Connections

Topic: **Evaluating Exponential Expressions with Positive and Negative Bases**

Be particularly careful when raising a negative number to an even power on a calculator. For example, the expressions $(-4)^2$ and -4^2 have different values. That is, $(-4)^2 = 16$ and $-4^2 = -16$. Verify these expressions on a calculator.

Scientific Calculator:

To evaluate $(-4)^2$

Enter: **Result:** 16

To evaluate -4^2 on a scientific calculator, it is important to square 4 first and then take its opposite.

Enter: 4 $\boxed{x^2}$ $\boxed{+\!\circ\!-}$ **Result:** -16

Graphing Calculator:

```
(-4)²
             16
-4²
            -16
```

The graphing calculator allows for several methods of denoting the multiplication of two real numbers. For example, consider the product of -8 and 4.

```
-8*4
           -32
-8(4)
           -32
(-8)(4)
           -32
```

Calculator Exercises

Simplify the expression without the use of a calculator. Then use the calculator to verify your answer.

1. $-6(5)$ **2.** $\dfrac{-5.2}{2.6}$ **3.** $(-5)(-5)(-5)(-5)$ **4.** $(-5)^4$ **5.** -5^4

6. -2.4^2 **7.** $(-2.4)^2$ **8.** $(-1)(-1)(-1)$ **9.** $\dfrac{-8.4}{-2.1}$ **10.** $90 \div (-5)(2)$

Section 1.5 Practice Exercises

Study Skills Exercises

1. Look through the text, and write down a page number that contains:

   a. An Avoiding Mistakes box _____

   b. A Tip box _____

   c. A key term (shown in bold) _____

2. Define the key term **reciprocal of a real number**.

Review Exercises

For Exercises 3–6, determine if the expression is true or false.

3. $6 + (-2) > -5 + 6$

4. $|-6| + |-14| \le |-3| + |-17|$

5. $\sqrt{36} - |-6| > 0$

6. $\sqrt{9} + |-3| \le 0$

Objective 1: Multiplication of Real Numbers

For Exercises 7–18, multiply the real numbers. **(See Example 1.)**

7. $8(-7)$

8. $(-3) \cdot 4$

9. $(-6) \cdot 7$

10. $9(-5)$

11. $(-11)(-13)$

12. $(-5)(-26)$

13. $(-30)(-8)$

14. $(-16)(-8)$

15. $(-2.2)(5.8)$

16. $(9.1)(-4.5)$

17. $\left(-\dfrac{2}{3}\right)\left(-\dfrac{9}{8}\right)$

18. $\left(-\dfrac{5}{4}\right)\left(-\dfrac{12}{25}\right)$

Objective 2: Exponential Expressions

For Exercises 19–26, simplify the exponential expression. **(See Example 2.)**

19. $(-6)^2$

20. $(-10)^2$

21. -6^2

22. -10^2

23. $\left(-\dfrac{3}{5}\right)^3$

24. $\left(-\dfrac{5}{2}\right)^3$

25. $(-0.2)^4$

26. $(-0.1)^4$

Objective 3: Division of Real Numbers

For Exercises 27–34, divide the real numbers. **(See Examples 3–4.)**

27. $\dfrac{54}{-9}$

28. $\dfrac{-27}{3}$

29. $\dfrac{-100}{-10}$

30. $\dfrac{-120}{-40}$

31. $\dfrac{-14}{-7}$

32. $\dfrac{-21}{-3}$

33. $\dfrac{13}{-65}$

34. $\dfrac{7}{-77}$

For Exercises 35–42, show how multiplication can be used to check the division problems.

35. $\dfrac{14}{-2} = -7$

36. $\dfrac{-18}{-6} = 3$

37. $\dfrac{0}{-5} = 0$

38. $\dfrac{0}{-4} = 0$

39. $\dfrac{6}{0}$ is undefined

40. $\dfrac{-4}{0}$ is undefined

41. $-24 \div (-6) = 4$

42. $-18 \div 2 = -9$

Mixed Exercises

For Exercises 43–114, multiply or divide as indicated.

43. $2 \cdot 3$

44. $8 \cdot 6$

45. $2(-3)$

46. $8(-6)$

47. $(-2)3$

48. $(-8)6$

49. $(-2)(-3)$

50. $(-8)(-6)$

51. $24 \div 3$

52. $52 \div 2$

53. $24 \div (-3)$

54. $52 \div (-2)$

55. $(-24) \div 3$

56. $(-52) \div 2$

57. $(-24) \div (-3)$

58. $(-52) \div (-2)$

59. $-6 \cdot 0$

60. $-8 \cdot 0$

61. $-18 \div 0$

62. $-42 \div 0$

63. $0\left(-\dfrac{2}{5}\right)$

64. $0\left(-\dfrac{1}{8}\right)$

65. $0 \div \left(-\dfrac{1}{10}\right)$

66. $0 \div \left(\dfrac{4}{9}\right)$

67. $\dfrac{-9}{6}$

68. $\dfrac{-15}{10}$

69. $\dfrac{-30}{-100}$

70. $\dfrac{-250}{-1000}$

71. $\dfrac{26}{-13}$

72. $\dfrac{52}{-4}$

73. $1.72(-4.6)$

74. $361.3(-14.9)$

75. $-0.02(-4.6)$

76. $-0.06(-2.15)$

77. $\dfrac{14.4}{-2.4}$

78. $\dfrac{50.4}{-6.3}$

79. $\dfrac{-5.25}{-2.5}$

80. $\dfrac{-8.5}{-27.2}$

81. $(-3)^2$

82. $(-7)^2$

83. -3^2

84. -7^2

85. $\left(-\dfrac{4}{3}\right)^3$

86. $\left(-\dfrac{1}{5}\right)^3$

87. $(-0.2)^3$

88. $(-0.1)^6$

89. -0.2^4

90. -0.1^4

91. $87 \div (-3)$

92. $96 \div (-6)$

93. $-4(-12)$

94. $(-5)(-11)$

95. $2.8(-5.1)$

96. $(7.21)(-0.3)$

97. $(-6.8) \div (-0.02)$

98. $(-12.3) \div (-0.03)$

99. $\left(-\dfrac{2}{15}\right)\left(\dfrac{25}{3}\right)$

100. $\left(-\dfrac{5}{16}\right)\left(\dfrac{4}{9}\right)$

101. $\left(-\dfrac{7}{8}\right) \div \left(-\dfrac{9}{16}\right)$

102. $\left(-\dfrac{22}{23}\right) \div \left(-\dfrac{11}{3}\right)$

103. $(-2)(-5)(-3)$

104. $(-6)(-1)(-10)$

105. $(-8)(-4)(-1)(-3)$

106. $(-6)(-3)(-1)(-5)$

107. $100 \div (-10) \div (-5)$ **108.** $150 \div (-15) \div (-2)$

109. $-12 \div (-6) \div (-2)$

110. $-36 \div (-2) \div 6$

111. $\dfrac{2}{5} \cdot \dfrac{1}{3} \cdot \left(-\dfrac{10}{11}\right)$

112. $\left(-\dfrac{9}{8}\right) \cdot \left(-\dfrac{2}{3}\right) \cdot \left(1\dfrac{5}{12}\right)$ **113.** $\left(1\dfrac{1}{3}\right) \div 3 \div \left(-\dfrac{7}{9}\right)$

114. $-\dfrac{7}{8} \div \left(3\dfrac{1}{4}\right) \div (-2)$

Objective 4: Applying the Order of Operations

For Exercises 115–134, perform the indicated operations. **(See Examples 5–6.)**

115. $12 \div (-2)(4)$

116. $(-6) \cdot 7 \div (-2)$

117. $\left(-\dfrac{12}{5}\right) \div (-6) \cdot \left(-\dfrac{1}{8}\right)$

118. $10 \cdot \dfrac{1}{3} \div \dfrac{25}{6}$

119. $8 - 2^3 \cdot 5 + 3 - (-6)$

120. $-14 \div (-7) - 8 \cdot 2 + 3^3$

121. $-(2 - 8)^2 \div (-6) \cdot 2$

122. $-(3 - 5)^2 \cdot 6 \div (-4)$

123. $\dfrac{6(-4) - 2(5 - 8)}{-6 - 3 - 5}$

124. $\dfrac{3(-4) - 5(9 - 11)}{-9 - 2 - 3}$

125. $\dfrac{-4 + 5}{(-2) \cdot 5 + 10}$

126. $\dfrac{-3 + 10}{2(-4) + 8}$

127. $-4 - 3[2 - (-5 + 3)] - 8 \cdot 2^2$ **128.** $-6 - 5[-4 - (6 - 12)] + (-5)^2$ **129.** $-|-1| - |5|$

130. $-|-10| - |6|$ **131.** $\dfrac{|2 - 9| - |5 - 7|}{10 - 15}$ **132.** $\dfrac{|-2 + 6| - |3 - 5|}{13 - 11}$

133. $\dfrac{6 - 3[2 - (6 - 8)]^2}{-2|2 - 5|}$ **134.** $\dfrac{12 - 4[-6 - (5 - 8)]^2}{4|6 - 10|}$

For Exercises 135–140, evaluate the expression for $x = -2$, $y = -4$, and $z = 6$. **(See Example 7.)**

135. $x^2 - 2y$ **136.** $3y^2 - z$ **137.** $4(2x - z)$

138. $6(3x + y)$ **139.** $\dfrac{3x + 2y}{y}$ **140.** $\dfrac{2z - y}{x}$

141. Is the expression $\dfrac{10}{5x}$ equal to 10/5x? Explain. **142.** Is the expression 10/(5x) equal to $\dfrac{10}{5x}$? Explain.

For Exercises 143–150, translate the English phrase into an algebraic expression. Then evaluate the expression.

143. The product of -3.75 and 0.3 **144.** The product of -0.4 and -1.258

145. The quotient of $\frac{16}{5}$ and $\left(-\frac{8}{9}\right)$ **146.** The quotient of $\left(-\frac{3}{14}\right)$ and $\frac{1}{7}$

147. The number -0.4 plus the quantity 6 times -0.42 **148.** The number 0.5 plus the quantity -2 times 0.125

149. The number $-\frac{1}{4}$ minus the quantity 6 times $-\frac{1}{3}$ **150.** Negative five minus the quantity $\left(-\frac{5}{6}\right)$ times $\frac{3}{8}$

151. For 3 weeks, Jim pays \$2 a week for lottery tickets. Jim has one winning ticket for \$3. Write an expression that describes his net gain or loss. How much money has Jim won or lost?

152. Stephanie pays \$2 a week for 6 weeks for lottery tickets. Stephanie has one winning ticket for \$5. Write an expression that describes her net gain or loss. How much money has Stephanie won or lost?

153. Evaluate the expressions in parts (a) and (b).

   a. $-4 - 3 - 2 - 1$

   b. $-4(-3)(-2)(-1)$

   c. Explain the difference between the operations in parts (a) and (b).

154. Evaluate the expressions in parts (a) and (b).

   a. $-10 - 9 - 8 - 7$

   b. $-10(-9)(-8)(-7)$

   c. Explain the difference between the operations in parts (a) and (b).

Properties of Real Numbers and Simplifying Expressions	**Section 1.6**

1. Commutative Properties of Real Numbers

When getting dressed in the morning, it makes no difference whether you put on your left shoe first and then your right shoe, or vice versa. This example illustrates a process in which the order does not affect the outcome. Such a process or operation is said to be *commutative*.

In algebra, the operations of addition and multiplication are commutative because the order in which we add or multiply two real numbers does not affect the result. For example:

$$10 + 5 = 5 + 10 \qquad \text{and} \qquad 10 \cdot 5 = 5 \cdot 10$$

PROPERTY   Commutative Properties of Real Numbers

If a and b are real numbers, then

1. $a + b = b + a$ **commutative property of addition**

2. $ab = ba$ **commutative property of multiplication**

It is important to note that although the operations of addition and multiplication are commutative, subtraction and division are *not* commutative. For example:

$$\underbrace{10 - 5}_{5} \neq \underbrace{5 - 10}_{-5} \qquad \text{and} \qquad \underbrace{10 \div 5}_{2} \neq \underbrace{5 \div 10}_{\frac{1}{2}}$$

Example 1   Applying the Commutative Property of Addition

Use the commutative property of addition to rewrite each expression.

a. $-3 + (-7)$ **b.** $3x^3 + 5x^4$

Solution:

a. $-3 + (-7) = -7 + (-3)$

b. $3x^3 + 5x^4 = 5x^4 + 3x^3$

Recall that subtraction is not a commutative operation. However, if we rewrite $a - b$, as $a + (-b)$, we can apply the commutative property of addition. This is demonstrated in Example 2.

Objectives

1. **Commutative Properties of Real Numbers**
2. **Associative Properties of Real Numbers**
3. **Identity and Inverse Properties of Real Numbers**
4. **Distributive Property of Multiplication over Addition**
5. **Simplifying Algebraic Expressions**

Skill Practice

Use the commutative property of addition to rewrite each expression.

1. $-5 + 9$
2. $7y + x$

Answers

1. $9 + (-5)$ **2.** $x + 7y$

Example 2 Applying the Commutative Property of Addition

Rewrite the expression in terms of addition. Then apply the commutative property of addition.

a. $5a - 3b$ **b.** $z^2 - \frac{1}{4}$

Solution:

a. $5a - 3b$

$\quad = 5a + (-3b)$ Rewrite subtraction as addition of $-3b$.

$\quad = -3b + 5a$ Apply the commutative property of addition.

b. $z^2 - \frac{1}{4}$

$\quad = z^2 + \left(-\frac{1}{4}\right)$ Rewrite subtraction as addition of $-\frac{1}{4}$.

$\quad = -\frac{1}{4} + z^2$ Apply the commutative property of addition.

Example 3 Applying the Commutative Property of Multiplication

Use the commutative property of multiplication to rewrite each expression.

a. $12(-6)$ **b.** $x \cdot 4$

Solution:

a. $12(-6) = -6(12)$

b. $x \cdot 4 = 4 \cdot x$ (or simply $4x$)

2. Associative Properties of Real Numbers

The associative property of real numbers states that the manner in which three or more real numbers are grouped under addition or multiplication will not affect the outcome. For example:

$$(5 + 10) + 2 = 5 + (10 + 2) \quad \text{and} \quad (5 \cdot 10)2 = 5(10 \cdot 2)$$

$$15 + 2 = 5 + 12 \qquad\qquad (50)2 = 5(20)$$

$$17 = 17 \qquad\qquad\qquad 100 = 100$$

PROPERTY Associative Properties of Real Numbers

If a, b, and c represent real numbers, then

1. $(a + b) + c = a + (b + c)$ **associative property of addition**

2. $(ab)c = a(bc)$ **associative property of multiplication**

Answers

3. $8m + (-2n); \ -2n + 8m$

4. $\frac{1}{3}x + \left(-\frac{3}{4}\right); \ -\frac{3}{4} + \frac{1}{3}x$

5. $5(-2)$ **6.** $6y$

Example 4 Applying the Associative Property

Use the associative property of multiplication to rewrite each expression. Then simplify the expression if possible.

a. $(y + 5) + 6$ **b.** $4(5z)$ **c.** $-\dfrac{3}{2}\left(-\dfrac{2}{3}w\right)$

Solution:

a. $(y + 5) + 6$

$= y + (5 + 6)$ Apply the associative property of addition.

$= y + 11$ Simplify.

b. $4(5z)$

$= (4 \cdot 5)z$ Apply the associative property of multiplication.

$= 20z$ Simplify.

c. $-\dfrac{3}{2}\left(-\dfrac{2}{3}w\right)$

$= \left[-\dfrac{3}{2}\left(-\dfrac{2}{3}\right)\right]w$ Apply the associative property of multiplication.

$= 1w$ Simplify.

$= w$

Note: In most cases, a detailed application of the associative property will not be shown. Instead, the process will be written in one step, such as

$$(y + 5) + 6 = y + 11, \quad 4(5z) = 20z \quad \text{and} \quad -\dfrac{3}{2}\left(-\dfrac{2}{3}w\right) = w$$

Skill Practice

Use the associative property of addition or multiplication to rewrite each expression. Simplify if possible.

7. $(x + 4) + 3$

8. $-2(4x)$

9. $\dfrac{5}{4}\left(\dfrac{4}{5}t\right)$

3. Identity and Inverse Properties of Real Numbers

The number 0 has a special role under the operation of addition. Zero added to any real number does not change the number. Therefore, the number 0 is said to be the *additive identity* (also called the *identity element of addition*). For example:

$$-4 + 0 = -4 \qquad 0 + 5.7 = 5.7 \qquad 0 + \dfrac{3}{4} = \dfrac{3}{4}$$

The number 1 has a special role under the operation of multiplication. Any real number multiplied by 1 does not change the number. Therefore, the number 1 is said to be the *multiplicative identity* (also called the *identity element of multiplication*). For example:

$$(-8)1 = -8 \qquad 1(-2.85) = -2.85 \qquad 1\left(\dfrac{1}{5}\right) = \dfrac{1}{5}$$

PROPERTY **Identity Properties of Real Numbers**

If a is a real number, then

1. $a + 0 = 0 + a = a$ **identity property of addition**

2. $a \cdot 1 = 1 \cdot a = a$ **identity property of multiplication**

Answers

7. $x + (4 + 3)$; $x + 7$
8. $(-2 \cdot 4)x$; $-8x$
9. $\left(\dfrac{5}{4} \cdot \dfrac{4}{5}\right)t$; t

The sum of a number and its opposite equals 0. For example, $-12 + 12 = 0$. For any real number, a, the opposite of a (also called the *additive inverse* of a) is $-a$ and $a + (-a) = -a + a = 0$. The inverse property of addition states that the sum of any number and its additive inverse is the identity element of addition, 0. For example:

Number	Additive Inverse (Opposite)	Sum
9	-9	$9 + (-9) = 0$
-21.6	21.6	$-21.6 + 21.6 = 0$
$\dfrac{2}{7}$	$-\dfrac{2}{7}$	$\dfrac{2}{7} + \left(-\dfrac{2}{7}\right) = 0$

If b is a nonzero real number, then the reciprocal of b (also called the *multiplicative inverse* of b) is $\frac{1}{b}$. The inverse property of multiplication states that the product of b and its multiplicative inverse is the identity element of multiplication, 1. Symbolically, we have $b \cdot \frac{1}{b} = \frac{1}{b} \cdot b = 1$. For example:

Number	Multiplicative Inverse (Reciprocal)	Product
7	$\dfrac{1}{7}$	$7 \cdot \dfrac{1}{7} = 1$
3.14	$\dfrac{1}{3.14}$	$3.14\left(\dfrac{1}{3.14}\right) = 1$
$-\dfrac{3}{5}$	$-\dfrac{5}{3}$	$-\dfrac{3}{5}\left(-\dfrac{5}{3}\right) = 1$

Concept Connections

Fill in the blanks. State whether the property used is an inverse property or an identity property.

10. $5 + \square = 0$

11. $-8 \cdot \square = -8$

12. $\dfrac{1}{2} \cdot \square = 1$

13. $\dfrac{a}{b} + \square = \dfrac{a}{b}$

PROPERTY   Inverse Properties of Real Numbers

If a is a real number and b is a nonzero real number, then

1. $a + (-a) = -a + a = 0$ **inverse property of addition**

2. $b \cdot \dfrac{1}{b} = \dfrac{1}{b} \cdot b = 1$ **inverse property of multiplication**

4. Distributive Property of Multiplication over Addition

The operations of addition and multiplication are related by an important property called the **distributive property of multiplication over addition**. Consider the expression $6(2 + 3)$. The order of operations indicates that the sum $2 + 3$ is evaluated first, and then the result is multiplied by 6:

$$6(2 + 3)$$
$$= 6(5)$$
$$= 30$$

Answers

10. -5; inverse   **11.** 1; identity
12. 2; inverse   **13.** 0; identity

Notice that the same result is obtained if the factor of 6 is multiplied by each of the numbers 2 and 3, and then their products are added:

$6(2 + 3)$ The factor of 6 is distributed to the numbers 2 and 3.

$= 6(2) + 6(3)$

$=\ \ 12 + 18$

$=\ \ \ \ \ 30$

The distributive property of multiplication over addition states that this is true in general.

TIP: The mathematical definition of the distributive property is consistent with the everyday meaning of the word *distribute*. To distribute means to "spread out from one to many." In the mathematical context, the factor *a* is distributed to both *b* and *c* in the parentheses.

PROPERTY   Distributive Property of Multiplication over Addition

If a, b, and c are real numbers, then

$$a(b + c) = ab + ac \qquad \text{and} \qquad (b + c)a = ab + ac$$

Example 5   Applying the Distributive Property

Apply the distributive property: $2(a + 6b + 7)$

Solution:

$2(a + 6b + 7)$

$= 2(a + 6b + 7)$

$= 2(a) + 2(6b) + 2(7)$ Apply the distributive property.

$= 2a + 12b + 14$ Simplify.

Skill Practice

14. Apply the distributive property.

$7(x + 4y + z)$

TIP: Notice that the parentheses are removed after the distributive property is applied. Sometimes this is referred to as *clearing parentheses*.

Because the difference of two expressions $a - b$ can be written in terms of addition as $a + (-b)$, the distributive property can be applied when the operation of subtraction is present within the parentheses. For example:

$5(y - 7)$

$= 5[y + (-7)]$ Rewrite subtraction as addition of -7.

$= 5[y + (-7)]$ Apply the distributive property.

$= 5(y) + 5(-7)$

$= 5y + (-35), \text{ or } 5y - 35$ Simplify.

Answers

14. $7x + 28y + 7z$

TIP: Notice that a negative factor preceding the parentheses changes the signs of all the terms to which it is multiplied.

$-1(-3a + 2b + 5c)$
$= +3a - 2b - 5c$

Example 6 Applying the Distributive Property

Use the distributive property to rewrite each expression.

a. $-(-3a + 2b + 5c)$ **b.** $-6(2 - 4x)$

Solution:

a. $-(-3a + 2b + 5c)$

$= -1(-3a + 2b + 5c)$ The negative sign preceding the parentheses can be interpreted as taking the opposite of the quantity that follows or as $-1(-3a + 2b + 5c)$

$= -1(-3a + 2b + 5c)$

$= -1(-3a) + (-1)(2b) + (-1)(5c)$ Apply the distributive property.

$= 3a + (-2b) + (-5c)$ Simplify.

$= 3a - 2b - 5c$

b. $-6(2 - 4x)$

$= -6[2 + (-4x)]$ Change subtraction to addition of $-4x$.

$= -6[2 + (-4x)]$ Apply the distributive property. Notice that multiplying by -6 changes the signs of all terms to which it is applied.

$= -6(2) + (-6)(-4x)$

$= -12 + 24x$ Simplify.

Note: In most cases, the distributive property will be applied without as much detail as shown in Examples 5 and 6. Instead, the distributive property will be applied in one step.

$2(a + 6b + 7)$ $-(3a + 2b + 5c)$ $-6(2 - 4x)$
1 step $= 2a + 12b + 14$ 1 step $= -3a - 2b - 5c$ 1 step $= -12 + 24x$

5. Simplifying Algebraic Expressions

An algebraic expression is the sum of one or more terms. A term is a constant or the product or quotient of constants and variables. For example, the expression

$$-7x^2 + xy - 100 \quad \text{or} \quad -7x^2 + xy + (-100)$$

consists of the terms $-7x^2, xy$, and -100.

The terms $-7x^2$ and xy are **variable terms** and the term -100 is called a **constant term**. It is important to distinguish between a term and the factors within a term. For example, the quantity xy is one term, and the values x and y are factors within the term. The constant factor in a term is called the *numerical coefficient* (or simply **coefficient**) of the term. In the terms $-7x^2, xy$, and -100, the coefficients are $-7, 1$, and -100, respectively.

Terms are *like* terms if they each have the same variables and the corresponding variables are raised to the same powers. For example:

Like Terms	Unlike Terms	
$-3b$ and $5b$	$-5c$ and $7d$	(different variables)
$9p^2q^3$ and p^2q^3	$4p^2q^3$ and $8p^3q^2$	(different powers)
$5w$ and $2w$	$5w$ and 2	(different variables)

Example 7 Identifying Terms, Factors, Coefficients and *Like* Terms

a. List the terms of the expression $5x^2 - 3x + 2$.

b. Identify the coefficient of the term $6yz^3$.

c. Which of the pairs are *like* terms: $8b, 3b^2$ or $4c^2d, -6c^2d$?

Solution:

a. The terms of the expression $5x^2 - 3x + 2$ are $5x^2$, $-3x$, and 2.

b. The coefficient of $6yz^3$ is 6.

c. $4c^2d$ and $-6c^2d$ are *like* terms.

Skill Practice

17. List the terms in the expression.
$4xy - 9x^2 + 15$

18. Identify the coefficients of each term in the expression.
$2a - b + c - 80$

19. Which of the pairs are *like* terms?
$5x^3, 5x$ or $-7x^2, 11x^2$

Two terms can be added or subtracted only if they are *like* terms. To add or subtract *like* terms, we use the distributive property as shown in Example 8.

Example 8 Using the Distributive Property to Add and Subtract *Like* Terms

Add or subtract as indicated.

a. $7x + 2x$ **b.** $-2p + 3p - p$

Solution:

a. $7x + 2x$

$\qquad = (7 + 2)x$ Apply the distributive property.

$\qquad = 9x$ Simplify.

b. $-2p + 3p - p$

$\qquad = -2p + 3p - 1p$ Note that $-p$ equals $-1p$.

$\qquad = (-2 + 3 - 1)p$ Apply the distributive property.

$\qquad = (0)p$ Simplify.

$\qquad = 0$

Skill Practice

Simplify by adding *like* terms.
20. $8x + 3x$
21. $-6a + 4a + a$

Although the distributive property is used to add and subtract *like* terms, it is tedious to write each step. Observe that adding or subtracting *like* terms is a matter of adding or subtracting the coefficients and leaving the variable factors unchanged. This can be shown in one step, a shortcut that we will use throughout the text. For example:

$$7x + 2x = 9x \qquad -2p + 3p - 1p = 0p = 0 \qquad -3a - 6a = -9a$$

Answers
17. $4xy, -9x^2, 15$
18. $2, -1, 1, -80$
19. $-7x^2$ and $11x^2$ are *like* terms.
20. $11x$ **21.** $-a$

Example 9   Using the Distributive Property to Add and Subtract *Like* Terms

Simplify by combining *like* terms.

a. $3yz + 5 - 2yz + 9$ **b.** $1.2w^3 + 5.7w^3$

Solution:

a. $3yz + 5 - 2yz + 9$

$\quad = 3yz - 2yz + 5 + 9$ Arrange *like* terms together. Notice that constants such as 5 and 9 are *like* terms.

$\quad = 1yz + 14$ Combine *like* terms.

$\quad = yz + 14$

b. $1.2w^3 + 5.7w^3$

$\quad = 6.9w^3$ Combine *like* terms.

Examples 10 and 11 illustrate how the distributive property is used to clear parentheses.

Example 10   Clearing Parentheses and Combining *Like* Terms

Simplify by clearing parentheses and combining *like* terms. $5 - 2(3x + 7)$

Solution:

$5 - 2(3x + 7)$ The order of operations indicates that we must perform multiplication before subtraction.

It is important to understand that a factor of -2 (not 2) will be multiplied to all terms within the parentheses. To see why this is so, we may rewrite the subtraction in terms of addition.

$= 5 + (-2)(3x + 7)$ Change subtraction to addition.

$= 5 + (-2)(3x + 7)$ A factor of -2 is to be distributed to terms in the parentheses.

$= 5 + (-2)(3x) + (-2)(7)$ Apply the distributive property.

$= 5 + (-6x) + (-14)$ Simplify.

$= 5 + (-14) + (-6x)$ Arrange *like* terms together.

$= -9 + (-6x)$ Combine *like* terms.

$= -9 - 6x$ Simplify by changing addition of the opposite to subtraction.

Answers

22. $9pq - 15$

23. $13.4x^2$

24. $-10x + 44$

Example 11	**Clearing Parentheses and Combining *Like* Terms**

Simplify by clearing parentheses and combining *like* terms.

a. $10(5y + 2) - 6(y - 1)$

b. $\dfrac{1}{4}(4k + 2) - \dfrac{1}{2}(6k + 1)$

c. $-(4s - 6t) - (3t + 5s) - 2s$

Skill Practice

Clear the parentheses and combine *like* terms.

25. $5(2y + 3) - 2(3y + 1)$

26. $\dfrac{1}{2}(8x + 4) + \dfrac{1}{3}(3x - 9)$

27. $-4(x + 2y) - (2x - y) - 5x$

Solution:

a. $10(5y + 2) - 6(y - 1)$

$= 50y + 20 - 6y + 6$ Apply the distributive property. Notice that a factor of -6 is distributed through the second parentheses and changes the signs.

$= 50y - 6y + 20 + 6$ Arrange *like* terms together.

$= 44y + 26$ Combine *like* terms.

b. $\dfrac{1}{4}(4k + 2) - \dfrac{1}{2}(6k + 1)$

$= \dfrac{4}{4}k + \dfrac{2}{4} - \dfrac{6}{2}k - \dfrac{1}{2}$ Apply the distributive property. Notice that a factor of $-\frac{1}{2}$ is distributed through the second parentheses and changes the signs.

$= k + \dfrac{1}{2} - 3k - \dfrac{1}{2}$ Simplify fractions.

$= k - 3k + \dfrac{1}{2} - \dfrac{1}{2}$ Arrange *like* terms together.

$= -2k + 0$ Combine *like* terms.

$= -2k$

c. $-(4s - 6t) - (3t + 5s) - 2s$

$= -1(4s - 6t) - 1(3t + 5s) - 2s$ Notice that a factor of -1 is distributed through each parentheses.

$= -4s + 6t - 3t - 5s - 2s$ Apply the distributive property.

$= -4s - 5s - 2s + 6t - 3t$ Arrange *like* terms together.

$= -11s + 3t$ Combine *like* terms.

Answers

25. $4y + 13$ **26.** $5x - 1$

27. $-11x - 7y$

| **Example 12** | **Clearing Parentheses and Combining *Like* Terms** |

Simplify by clearing parentheses and combining *like* terms.

$$-7a - 4[3a - 2(a + 6)] - 4$$

Solution:

$-7a - 4[3a - 2(a + 6)] - 4$

$= -7a - 4[3a - 2a - 12] - 4$ Apply the distributive property to clear the innermost parentheses.

$= -7a - 4[a - 12] - 4$ Simplify within brackets by combining *like* terms.

$= -7a - 4a + 48 - 4$ Apply the distributive property to clear the brackets.

$= -11a + 44$ Combine *like* terms

Answer

28. $50y - 94$

Section 1.6   Practice Exercises

Study Skills Exercises

1. Write down the page number(s) for the Chapter Summary for this chapter. Describe one way in which you can use the Summary found at the end of each chapter.

2. Define the key term:

    a. commutative properties   **b. associative properties**   **c. identity properties**

    d. inverse properties   **e. distributive property of multiplication over addition**

    f. variable term   **g. constant term**   **h. coefficient**   **i. *like* terms**

Review Exercises

For Exercises 3–14, perform the indicated operations.

3. $(-6) + 14$ **4.** $(-2) + 9$ **5.** $-13 - (-5)$ **6.** $-1 - (-19)$

7. $18 \div (-4)$ **8.** $-27 \div 5$ **9.** $-3 \cdot 0$ **10.** $0(-15)$

11. $\dfrac{1}{2} + \dfrac{3}{8}$ **12.** $\dfrac{25}{21} - \dfrac{6}{7}$ **13.** $\left(-\dfrac{3}{5}\right)\left(\dfrac{4}{27}\right)$ **14.** $\left(-\dfrac{11}{12}\right) \div \left(-\dfrac{5}{4}\right)$

Objective 1: Commutative Properties of Real Numbers

For Exercises 15–22, rewrite each expression using the commutative property of addition or the commutative property of multiplication. **(See Examples 1 and 3.)**

15. $5 + (-8)$ **16.** $7 + (-2)$ **17.** $8 + x$ **18.** $p + 11$

19. $5(4)$ **20.** $10(8)$ **21.** $x(-12)$ **22.** $y(-23)$

For Exercises 23–26, rewrite each expression using addition. Then apply the commutative property of addition. **(See Example 2.)**

23. $x - 3$ **24.** $y - 7$ **25.** $4p - 9$ **26.** $3m - 12$

Objective 2: Associative Properties of Real Numbers

For Exercises 27–34, use the associative property of addition or multiplication to rewrite each expression. Then simplify the expression if possible. **(See Example 4.)**

27. $(x + 4) + 9$ **28.** $-3 + (5 + z)$ **29.** $-5(3x)$ **30.** $-12(4z)$

31. $\dfrac{6}{11}\left(\dfrac{11}{6}x\right)$ **32.** $\dfrac{3}{5}\left(\dfrac{5}{3}x\right)$ **33.** $-4\left(-\dfrac{1}{4}t\right)$ **34.** $-5\left(-\dfrac{1}{5}w\right)$

Objective 3: Identity and Inverse Properties of Real Numbers

35. What is another name for multiplicative inverse?   **36.** What is another name for additive inverse?

37. What is the additive identity?   **38.** What is the multiplicative identity?

Objective 4: Distributive Property of Multiplication over Addition

For Exercises 39–58, use the distributive property to clear parentheses. **(See Examples 5–6.)**

39. $6(5x + 1)$ **40.** $2(x + 7)$ **41.** $-2(a + 8)$ **42.** $-3(2z + 9)$

43. $3(5c - d)$ **44.** $4(w - 13z)$ **45.** $-7(y - 2)$ **46.** $-2(4x - 1)$

47. $-\dfrac{2}{3}(x - 6)$ **48.** $-\dfrac{1}{4}(2b - 8)$ **49.** $\dfrac{1}{3}(m - 3)$ **50.** $\dfrac{2}{5}(n - 5)$

51. $-(2p + 10)$ **52.** $-(7q + 1)$ **53.** $-2(-3w - 5z + 8)$ **54.** $-4(-7a - b - 3)$

55. $4(x + 2y - z)$ **56.** $-6(2a - b + c)$ **57.** $-(-6w + x - 3y)$ **58.** $-(-p - 5q - 10r)$

Mixed Exercises

For Exercises 59–62, use the associative property or distributive property to clear parentheses.

59. $2(3 + x)$ **60.** $5(4 + y)$ **61.** $4(6z)$ **62.** $8(2p)$

For Exercises 63–71, match the statement with the property that describes it.

63. $6 \cdot \dfrac{1}{6} = 1$

a. Commutative property of addition

64. $7(4 \cdot 9) = (7 \cdot 4)9$

b. Inverse property of multiplication

65. $2(3 + k) = 6 + 2k$

c. Commutative property of multiplication

66. $3 \cdot 7 = 7 \cdot 3$

d. Associative property of addition

67. $5 + (-5) = 0$

e. Identity property of multiplication

68. $18 \cdot 1 = 18$

f. Associative property of multiplication

69. $(3 + 7) + 19 = 3 + (7 + 19)$

g. Inverse property of addition

70. $23 + 6 = 6 + 23$

h. Identity property of addition

71. $3 + 0 = 3$

i. Distributive property of multiplication over addition

Objective 5: Simplifying Algebraic Expressions

For Exercises 72–75, for each expression list the terms and their coefficients. **(See Example 7.)**

72. $3xy - 6x^2 + y - 17$

Term	Coefficient

73. $2x - y + 18xy + 5$

Term	Coefficient

74. $x^4 - 10xy + 12 - y$

Term	Coefficient

75. $-x + 8y - 9x^2y - 3$

Term	Coefficient

76. Explain why $12x$ and $12x^2$ are not *like* terms.

77. Explain why $3x$ and $3xy$ are not *like* terms.

78. Explain why $7z$ and $\sqrt{13}z$ are *like* terms.

79. Explain why πx and $8x$ are *like* terms.

80. Write three different *like* terms.

81. Write three terms that are not *like*.

For Exercises 82–90, simplify by combining *like* terms. **(See Examples 8–9.)**

82. $5k - 10k - 12k + 16 + 7$

83. $-4p - 2p + 8p - 15 + 3$

84. $9x - 7x^2 + 12x + 14x^2$

85. $2y^2 - 8y + y - 5y^2 - 3y^2$

86. $4ab^2 + 2a^2b - 6ab^2 + 5 + 3a^2b - 2$

87. $8x^3y - 5xy + 3 - 7 + 6xy - x^3y$ **88.** $\dfrac{1}{4}a + b - \dfrac{3}{4}a - 5b$

89. $\dfrac{2}{5} + 2t - \dfrac{3}{5} + t - \dfrac{6}{5}$ **90.** $2.8z - 8.1z + 6 - 15.2$

For Exercises 91–118, simplify by clearing parentheses and combining *like* terms. **(See Examples 10–12.)**

91. $-3(2x - 4) + 10$ **92.** $-2(4a + 3) - 14$ **93.** $4(w + 3) - 12$

94. $5(2r + 6) - 30$ **95.** $5 - 3(x - 4)$ **96.** $4 - 2(3x + 8)$

97. $-3(2t + 4) + 8(2t - 4)$ **98.** $-5(5y + 9) + 3(3y + 6)$ **99.** $2(w - 5) - (2w + 8)$

100. $6(x + 3) - (6x - 5)$ **101.** $-\dfrac{1}{3}(6t + 9) + 10$ **102.** $-\dfrac{3}{4}(8 + 4q) + 7$

103. $10(5.1a - 3.1) + 4$ **104.** $100(-3.14p - 1.05) + 212$ **105.** $-4m + 2(m - 3) + 2m$

106. $-3b + 4(b + 2) - 8b$ **107.** $\dfrac{1}{2}(10q - 2) + \dfrac{1}{3}(2 - 3q)$ **108.** $\dfrac{1}{5}(15 - 4p) - \dfrac{1}{10}(10p + 5)$

109. $7n - 2(n - 3) - 6 + n$ **110.** $8k - 4(k - 1) + 7 - k$ **111.** $6(x + 3) - 12 - 4(x - 3)$

112. $5(y - 4) + 3 - 6(y - 7)$ **113.** $6.1(5.3z - 4.1) - 5.8$ **114.** $-3.6(1.7q - 4.2) + 14.6$

115. $6 + 2[-8 - 3(2x + 4)] + 10x$ **116.** $-3 + 5[-3 - 4(y + 2)] - 8y$

117. $1 - 3[2(z + 1) - 5(z - 2)]$ **118.** $1 - 6[3(2t + 2) - 8(t + 2)]$

Expanding Your Skills

For Exercises 119–126, determine if the expressions are equivalent. If two expressions are not equivalent, state why.

119. $3a + b, b + 3a$ **120.** $4y + 1, 1 + 4y$ **121.** $2c + 7, 9c$ **122.** $5z + 4, 9z$

123. $5x - 3, 3 - 5x$ **124.** $6d - 7, 7 - 6d$ **125.** $5x - 3, -3 + 5x$ **126.** $8 - 2x, -2x + 8$

127. Which grouping of terms is easier to compute, $(14\frac{2}{7} + 2\frac{1}{3}) + \frac{2}{3}$ or $14\frac{2}{7} + (2\frac{1}{3} + \frac{2}{3})$?

128. Which grouping of terms is easier to compute, $(5\frac{1}{8} + 18\frac{2}{5}) + 1\frac{3}{5}$ or $5\frac{1}{8} + (18\frac{2}{5} + 1\frac{3}{5})$?

129. As a small child in school, the great mathematician Karl Friedrich Gauss (1777–1855) was said to have found the sum of the integers from 1 to 100 mentally:

$$1 + 2 + 3 + 4 + \cdots + 99 + 100$$

Rather than adding the numbers sequentially, he added the numbers in pairs:

$$(1 + 99) + (2 + 98) + (3 + 97) + \cdots + 100$$

a. Use this technique to add the integers from 1 to 10.

$$1 + 2 + 3 + 4 + 5 + 6 + 7 + 8 + 9 + 10$$

b. Use this technique to add the integers from 1 to 20.

Group Activity

Evaluating Formulas Using a Calculator

Materials: A calculator

Estimated Time: 15 minutes

Group Size: 2

In this chapter, we learned one of the most important concepts in mathematics—the order of operations. The proper order of operations is required whenever we evaluate any mathematical expression. The following formulas are taken from applications from science, math, statistics, and business. These are just some samples of what you may encounter as you work your way through college.

For Exercises 1–8, substitute the given values into the formula. Then use a calculator and the proper order of operations to simplify the result. Round to three decimal places if necessary.

1. $F = \dfrac{9}{5}C + 32$ (biology) $C = 35$

2. $V = \dfrac{nRT}{P}$ (chemistry) $n = 1.00, R = 0.0821, T = 273.15, P = 1.0$

3. $R = k\left(\dfrac{L}{r^2}\right)$ (electronics) $k = 0.05, L = 200, r = 0.5$

4. $m = \dfrac{y_2 - y_1}{x_2 - x_1}$ (mathematics) $x_1 = -8.3, x_2 = 3.3, y_1 = 4.6, y_2 = -9.2$

5. $z = \dfrac{\bar{x} - \mu}{\dfrac{\sigma}{\sqrt{n}}}$ (statistics) $\bar{x} = 69, \mu = 55, \sigma = 20, n = 25$

6. $S = R\left[\dfrac{(1 + i)^n - 1}{i}\right]$ (finance) $R = 200, i = 0.08, n = 30$

7. $x = \dfrac{-b + \sqrt{b^2 - 4ac}}{2a}$ (mathematics) $a = 2, b = -7, c = -15$

8. $h = \dfrac{1}{2}gt^2 + v_0 t + h_0$ (physics) $g = -32, t = 2.4, v_0 = 192, h_0 = 288$

Chapter 1 Summary

Section 1.1   Sets of Numbers and the Real Number Line

Key Concepts

Natural numbers: $\{1, 2, 3, \ldots\}$

Whole numbers: $\{0, 1, 2, 3, \ldots\}$

Integers: $\{\ldots -3, -2, -1, 0, 1, 2, 3, \ldots\}$

Rational numbers: The set of numbers that can be expressed in the form $\frac{p}{q}$, where p and q are integers and q does not equal 0. In decimal form, rational numbers are terminating or repeating decimals.

Irrational numbers: A subset of the real numbers whose elements cannot be written as a ratio of two integers. In decimal form, irrational numbers are nonterminating, nonrepeating decimals.

Real numbers: The set of both the rational numbers and the irrational numbers.

Examples

Example 1

$-5, 0$, and 4 are integers.

$-\dfrac{5}{2}, -0.5$, and $0.\overline{3}$ are rational numbers.

$\sqrt{7}, -\sqrt{2}$, and π are irrational numbers.

Example 2

All real numbers can be located on the real number line.

Example 3

$a < b$	"a is less than b."
$a > b$	"a is greater than b."
$a \leq b$	"a is less than or equal to b."
$a \geq b$	"a is greater than or equal to b."

Two numbers that are the same distance from zero but on opposite sides of zero on the number line are called **opposites**. The opposite of a is denoted $-a$.

The **absolute value** of a real number, a, denoted $|a|$, is the distance between a and 0 on the number line.

If $a \geq 0$, $|a| = a$

If $a < 0$, $|a| = -a$

Example 3

$5 < 7$	"5 is less than 7."
$-2 > -10$	"-2 is greater than -10."
$y \leq 3.4$	"y is less than or equal to 3.4."
$x \geq \frac{1}{2}$	"x is greater than or equal to $\frac{1}{2}$."

Example 4

5 and -5 are opposites.

Example 5

$|7| = 7$

$|-7| = 7$

Section 1.2 — Exponents, Square Roots, and the Order of Operations

Key Concepts

A **variable** is a symbol or letter used to represent an unknown number.

A **constant** is a value that is not variable.

An algebraic **expression** is a collection of variables and constants under algebraic operations.

$$b^n = \underbrace{b \cdot b \cdot b \cdot b \ldots \cdot b}_{n \text{ factors of } b}$$ b is the **base**,
n is the **exponent**

$\sqrt{x}$ is the positive **square root** of x.

The Order of Operations

1. Simplify expressions within parentheses and other grouping symbols first.
2. Evaluate expressions involving exponents, radicals, and absolute values.
3. Perform multiplication or division in the order that they occur from left to right.
4. Perform addition or subtraction in the order that they occur from left to right.

Examples

Example 1

Variables:	x, y, z, a, b
Constants:	$2, -3, \pi$
Expressions:	$2x + 5, 3a + b^2$

Example 2

$5^3 = 5 \cdot 5 \cdot 5 = 125$

Example 3

$\sqrt{49} = 7$

Example 4

$10 + 5(3 - 1)^2 - \sqrt{5 - 1}$

$= 10 + 5(2)^2 - \sqrt{4}$ Work within grouping symbols.

$= 10 + 5(4) - 2$ Simplify exponents and radicals.

$= 10 + 20 - 2$ Perform multiplication.

$= 30 - 2$ Add and subtract, left to right

$= 28$

Section 1.3 — Addition of Real Numbers

Key Concepts

Addition of Two Real Numbers

Same Signs. Add the absolute values of the numbers and apply the common sign to the sum.

Different Signs. Subtract the smaller absolute value from the larger absolute value. Then apply the sign of the number having the larger absolute value.

Examples

Example 1

$-3 + (-4) = -7$

$-1.3 + (-9.1) = -10.4$

Example 2

$-5 + 7 = 2$

$\dfrac{2}{3} + \left(-\dfrac{7}{3}\right) = -\dfrac{5}{3}$

Section 1.4   Subtraction of Real Numbers

Key Concepts

Subtraction of Two Real Numbers

Add the opposite of the second number to the first number. That is,

$$a - b = a + (-b)$$

Examples

Example 1

$$7 - (-5) = 7 + (5) = 12$$

$$-3 - 5 = -3 + (-5) = -8$$

$$-11 - (-2) = -11 + (2) = -9$$

Section 1.5   Multiplication and Division of Real Numbers

Key Concepts

Multiplication and Division of Two Real Numbers

Same Signs.
Product is positive.
Quotient is positive.

Different Signs.
Product is negative.
Quotient is negative.

The **reciprocal** of a number a is $\frac{1}{a}$.

Multiplication and Division Involving Zero

The product of any real number and 0 is 0.

The quotient of 0 and any nonzero real number is 0.

The quotient of any nonzero real number and 0 is undefined.

Examples

Example 1

$$(-5)(-2) = 10 \qquad \frac{-20}{-4} = 5$$

Example 2

$$(-3)(7) = -21 \qquad \frac{-4}{8} = -\frac{1}{2}$$

Example 3

The reciprocal of -6 is $-\frac{1}{6}$.

Example 4

$$4 \cdot 0 = 0$$

$$0 \div 4 = 0$$

$4 \div 0$ is undefined.

Section 1.6	**Properties of Real Numbers and Simplifying Expressions**

Key Concepts

Properties of Real Numbers

Commutative Properties.

$a + b = b + a$

$ab = ba$

Associative Properties.

$(a + b) + c = a + (b + c)$

$(ab)c = a(bc)$

Identity Properties.

$0 + a = a$

$1 \cdot a = a$

Inverse Properties.

$a + (-a) = 0$

$b \cdot \dfrac{1}{b} = 1$ for $b \neq 0$

Distributive Property of Multiplication over Addition.

$a(b + c) = ab + ac$

A **term** is a constant or the product or quotient of constants and variables. The **coefficient** of a term is the numerical factor of the term.

Like **terms** have the same variables, and the corresponding variables have the same powers.

Terms can be added or subtracted if they are *like* terms. Sometimes it is necessary to clear parentheses before adding or subtracting *like* terms.

Examples

Example 1

$(-5) + (-7) = (-7) + (-5)$

$3 \cdot 8 = 8 \cdot 3$

Example 2

$(2 + 3) + 10 = 2 + (3 + 10)$

$(2 \cdot 4) \cdot 5 = 2 \cdot (4 \cdot 5)$

Example 3

$0 + (-5) = -5$

$1(-8) = -8$

Example 4

$1.5 + (-1.5) = 0$

$6 \cdot \dfrac{1}{6} = 1$

Example 5

$-2(x - 3y) = (-2)x + (-2)(-3y)$

$\qquad\qquad = -2x + 6y$

Example 6

$-2x$ is a term with coefficient -2.
yz^2 is a term with coefficient 1.

$3x$ and $-5x$ are *like* terms.
$4a^2b$ and $4ab$ are not *like* terms.

Example 7

$-2w - 4(w - 2) + 3$

$= -2w - 4w + 8 + 3$ Clear parentheses.

$= -6w + 11$ Combine *like* terms.

Chapter 1   Review Exercises

Section 1.1

1. Given the set $\{7, \frac{1}{3}, -4, 0, -\sqrt{3}, -0.\overline{2}, \pi, 1\}$,

   a. List the natural numbers.

   b. List the integers.

   c. List the whole numbers.

   d. List the rational numbers.

   e. List the irrational numbers.

   f. List the real numbers.

For Exercises 2–5, determine the absolute value.

2. $\left|\dfrac{1}{2}\right|$ **3.** $|-6|$ **4.** $|-\sqrt{7}|$ **5.** $|0|$

For Exercises 6–13, identify whether the inequality is true or false.

6. $-6 > -1$ **7.** $0 < -5$ **8.** $-10 \leq 0$

9. $5 \neq -5$ **10.** $7 \geq 7$ **11.** $7 \geq -7$

12. $0 \leq -3$ **13.** $-\dfrac{2}{3} \leq -\dfrac{2}{3}$

Section 1.2

For Exercises 14–19, translate the English phrases into algebraic expressions.

14. The product of x and $\dfrac{2}{3}$

15. The quotient of 7 and y

16. The sum of 2 and $3b$

17. The difference of a and 5

18. Two more than $5k$

19. Seven less than $13z$

For Exercises 20–23, evaluate the expressions for $x = 8$, $y = 4$, and $z = 1$.

20. $x - 2y$ **21.** $x^2 - y$

22. $\sqrt{x + z}$ **23.** $\sqrt{x + 2y}$

For Exercises 24–29, simplify the expressions.

24. 6^3 **25.** 15^2 **26.** $\sqrt{36}$

27. $\dfrac{1}{\sqrt{100}}$ **28.** $\left(\dfrac{1}{4}\right)^2$ **29.** $\left(\dfrac{3}{2}\right)^3$

For Exercises 30–33, perform the indicated operations.

30. $15 - 7 \cdot 2 + 12$ **31.** $|-11| + |5| - (7 - 2)$

32. $4^2 - (5 - 2)^2$ **33.** $22 - 3(8 \div 4)^2$

Section 1.3

For Exercises 34–46, add.

34. $-6 + 8$ **35.** $14 + (-10)$

36. $21 + (-6)$ **37.** $-12 + (-5)$

38. $\dfrac{2}{7} + \left(-\dfrac{1}{9}\right)$ **39.** $\left(-\dfrac{8}{11}\right) + \left(\dfrac{1}{2}\right)$

40. $\left(-\dfrac{1}{10}\right) + \left(-\dfrac{5}{6}\right)$ **41.** $\left(-\dfrac{5}{2}\right) + \left(-\dfrac{1}{5}\right)$

42. $-8.17 + 6.02$ **43.** $2.9 + (-7.18)$

44. $13 + (-2) + (-8)$ **45.** $-5 + (-7) + 20$

46. $2 + 5 + (-8) + (-7) + 0 + 13 + (-1)$

47. Under what conditions will the expression $a + b$ be negative?

48. The high temperatures (in degrees Celsius) for the province of Alberta, Canada, during a week in January were $-8, -1, -4, -3, -4, 0$, and 7. What was the average high temperature for that week? Round to the nearest tenth of a degree.

Section 1.4

For Exercises 49–61, subtract.

49. $13 - 25$ **50.** $31 - (-2)$

51. $-8 - (-7)$ **52.** $-2 - 15$

53. $\left(-\dfrac{7}{9}\right) - \dfrac{5}{6}$ **54.** $\dfrac{1}{3} - \dfrac{9}{8}$

55. $7 - 8.2$

56. $-1.05 - 3.2$

57. $-16.1 - (-5.9)$

58. $7.09 - (-5)$

59. $\dfrac{11}{2} - \left(-\dfrac{1}{6}\right) - \dfrac{7}{3}$

60. $-\dfrac{4}{5} - \dfrac{7}{10} - \left(-\dfrac{13}{20}\right)$

61. $6 - 14 - (-1) - 10 - (-21) - 5$

62. Under what conditions will the expression $a - b$ be negative?

For Exercises 63–67, write an algebraic expression and simplify.

63. -18 subtracted from -7

64. The difference of -6 and 41

65. Seven decreased by 13

66. Five subtracted from the difference of 20 and -7

67. The sum of 6 and -12, decreased by 21

68. In Nevada, the highest temperature ever recorded was $125°F$ and the lowest was $-50°F$. Find the difference between the highest and lowest temperatures. (*Source: Information Please Almanac*)

Section 1.5

For Exercises 69–86, multiply or divide as indicated.

69. $10(-17)$

70. $(-7)13$

71. $(-52) \div 26$

72. $(-48) \div (-16)$

73. $\dfrac{7}{4} \div \left(-\dfrac{21}{2}\right)$

74. $\dfrac{2}{3}\left(-\dfrac{12}{11}\right)$

75. $-\dfrac{21}{5} \cdot 0$

76. $\dfrac{3}{4} \div 0$

77. $0 \div (-14)$

78. $(-0.45)(-5)$

79. $\dfrac{-21}{14}$

80. $\dfrac{-13}{-52}$

81. $(5)(-2)(3)$

82. $(-6)(-5)(15)$

83. $\left(-\dfrac{1}{2}\right)\left(\dfrac{7}{8}\right)\left(-\dfrac{4}{7}\right)$

84. $\left(\dfrac{12}{13}\right)\left(-\dfrac{1}{6}\right)\left(\dfrac{13}{14}\right)$

85. $40 \div 4 \div (-5)$

86. $\dfrac{10}{11} \div \dfrac{7}{11} \div \dfrac{5}{9}$

For Exercises 87–92, perform the indicated operations.

87. $9 - 4[-2(4 - 8) - 5(3 - 1)]$

88. $\dfrac{8(-3) - 6}{-7 - (-2)}$

89. $\dfrac{2}{3} - \left(\dfrac{3}{8} + \dfrac{5}{6}\right) \div \dfrac{5}{3}$

90. $5.4 - (0.3)^2 \div 0.09$

91. $\dfrac{5 - [3 - (-4)^2]}{36 \div (-2)(3)}$

92. $|-8 + 5| - \sqrt{5^2 - 3^2}$

For Exercises 93–96, evaluate the expression with the given substitution.

93. $3(x + 2) \div y$ for $x = 4$ and $y = -9$

94. $a^2 - bc$ for $a = -6$, $b = 5$, and $c = 2$

95. $w + xy - \sqrt{z}$ for $w = 12$, $x = 6$, $y = -5$, and $z = 25$

96. $(u - v)^2 + (u^2 - v^2)$ for $u = 5$ and $v = -3$

 97. In statistics, the formula $x = \mu + z\sigma$ is used to find cutoff values for data that follow a bell-shaped curve. Find x if $\mu = 100$, $z = -1.96$, and $\sigma = 15$.

For Exercises 98–104, answer true or false. If a statement is false, explain why.

98. If n is positive, then $-n$ is negative.

99. If m is negative, then m^4 is negative.

100. If m is negative, then m^3 is negative.

101. If $m > 0$ and $n > 0$, then $mn > 0$.

102. If $p < 0$ and $q < 0$, then $pq < 0$.

103. A number and its reciprocal have the same signs.

104. A nonzero number and its opposite have different signs.

Section 1.6

For Exercises 105–112, answers may vary.

105. Give an example of the commutative property of addition.

106. Give an example of the associative property of addition.

107. Give an example of the inverse property of addition.

108. Give an example of the identity property of addition.

109. Give an example of the commutative property of multiplication.

110. Give an example of the associative property of multiplication.

111. Give an example of the inverse property of multiplication.

112. Give an example of the identity property of multiplication.

113. Explain why $5x - 2y$ is the same as $-2y + 5x$.

114. Explain why $3a - 9y$ is the same as $-9y + 3a$.

115. List the terms of the expression:
$$3y + 10x - 12 + xy$$

116. Identify the coefficients for the terms listed in Exercise 115.

For Exercises 117–118, simplify by combining *like* terms.

117. $3a + 3b - 4b + 5a - 10$

118. $-6p + 2q + 9 - 13q - p + 7$

For Exercises 119–120, use the distributive property to clear the parentheses.

119. $-2(4z + 9)$ **120.** $5(4w - 8y + 1)$

For Exercises 121–126, simplify the expression.

121. $2p - (p + 5) + 3$

122. $6(h + 3) - 7h - 4$

123. $\dfrac{1}{2}(-6q) + q - 4\left(3q + \dfrac{1}{4}\right)$

124. $0.3b + 12(0.2 - 0.5b)$

125. $-4[2(x + 1) - (3x + 8)]$

126. $5[(7y - 3) + 3(y + 8)]$

Chapter 1   Test

1. Is $0.\overline{315}$ a rational number or an irrational number? Explain your reasoning.

2. Plot the points on a number line: $|3|, 0, -2, 0.5,$ $\left|-\frac{3}{2}\right|, \sqrt{16}$.

3. Use the number line in Exercise 2 to identify whether the statements are true or false.

  a. $|3| < -2$ **b.** $0 \le \left|-\dfrac{3}{2}\right|$

  c. $-2 < 0.5$ **d.** $|3| \ge \left|-\dfrac{3}{2}\right|$

4. Use the definition of exponents to expand the expressions:

  a. $(4x)^3$ **b.** $4x^3$

5. **a.** Translate the expression into an English phrase: $2(a - b)$. (Answers may vary.)

    b. Translate the expression into an English phrase: $2a - b$. (Answers may vary.)

6. Translate the phrase into an algebraic expression: "The quotient of the square root of c and the square of d."

For Exercises 7–23, perform the indicated operations.

7. $18 + (-12)$

8. $-10 + (-9)$

9. $-15 - (-3)$

10. $21 - (-7)$

11. $-\dfrac{1}{8} + \left(-\dfrac{3}{4}\right)$

12. $-10.06 - (-14.72)$

13. $-14 + (-2) - 16$

14. $-84 \div 7$

15. $38 \div 0$

16. $7(-4)$

17. $-22 \cdot 0$

18. $(-16)(-2)(-1)(-3)$

19. $\dfrac{2}{5} \div \left(-\dfrac{7}{10}\right) \cdot \left(-\dfrac{7}{6}\right)$

20. $(8 - 10) \cdot \dfrac{3}{2} + (-5)$

21. $8 - [(2 - 4) - (8 - 9)]$

22. $\dfrac{\sqrt{5^2 - 4^2}}{|-12 + 3|}$

23. $\dfrac{|4 - 10|}{2 - 3(5 - 1)}$

24. Identify the property that justifies each statement.

a. $6(-8) = (-8)6$

b. $5 + 0 = 5$

c. $(2 + 3) + 4 = 2 + (3 + 4)$

d. $\dfrac{1}{7} \cdot 7 = 1$

e. $8[7(-3)] = (8 \cdot 7)(-3)$

For Exercises 25–29, simplify the expression.

25. $-5x - 4y + 3 - 7x + 6y - 7$

26. $-3(4m + 8p - 7)$

27. $3k - 20 + (-9k) + 12$

28. $4(p - 5) - (8p + 3)$

29. $\dfrac{1}{2}(12p - 4) + \dfrac{1}{3}(2 - 6p)$

For Exercises 30–33, evaluate the expression given the values $x = 4$ and $y = -3$ and $z = -7$.

30. $y^2 - x$

31. $3x - 2y$

32. $y(x - 2)$

33. $-y^2 - 4x + z$

For Exercises 34–35, translate the English statement to an algebraic expression. Then simplify the expression.

34. Subtract -4 from 12

35. Find the difference of 6 and 8

Linear Equations and Inequalities

2

CHAPTER OUTLINE

Chapter 2

In this chapter, we learn how to solve linear equations and inequalities in one variable. This important category of equations and inequalities is used in a variety of applications.

Use your skills at solving equations and simplifying expressions to complete this puzzle.

Across

2. Solve. $\dfrac{z}{-6} = -81$

4. Simplify. $15x - 5(x - 49) - 10x + 2000$

5. Smallest positive integer that can be used to clear decimals.
   $0.06x + 1.8 = 2 - 2.251$

Down

1. Solve. $2a - 3 = 45$

3. $-10^2 \cdot (-2)^3 \cdot 11$

4. Solve. $4(2x - 1) = 5x + 56$

5. Smallest positive integer that can be used to clear fractions.
   $\dfrac{1}{2}y - \dfrac{3}{4} = \dfrac{5}{3}y + 1$

Section 2.1   Addition, Subtraction, Multiplication, and Division Properties of Equality

Objectives

1. **Definition of a Linear Equation in One Variable**
2. **Addition and Subtraction Properties of Equality**
3. **Multiplication and Division Properties of Equality**
4. **Translations**

1. Definition of a Linear Equation in One Variable

An *equation* is a statement that indicates that two quantities are equal. The following are equations.

$$x = 5 \qquad y + 2 = 12 \qquad -4z = 28$$

All equations have an equal sign. Furthermore, notice that the equal sign separates the equation into two parts, the left-hand side and the right-hand side. **A solution to an equation** is a value of the variable that makes the equation a true statement. Substituting a solution into an equation for the variable makes the right-hand side equal to the left-hand side.

Equation	Solution	Check	
$x = 5$	5	$x = 5$ $\downarrow$ $5 = 5$ ✔	Substitute 5 for x. Right-hand side equals left-hand side.
$y + 2 = 12$	10	$y + 2 = 12$ $\downarrow$ $10 + 2 = 12$ ✔	Substitute 10 for y. Right-hand side equals left-hand side.
$-4z = 28$	-7	$-4z = 28$ $\downarrow$ $-4(-7) = 28$ ✔	Substitute -7 for z. Right-hand side equals left-hand side.

Avoiding Mistakes

Be sure to notice the difference between solving an equation versus simplifying an expression. For example, $2x + 1 = 7$ is an equation, whose solution is 3, while $2x + 1 + 7$ is an expression that simplifies to $2x + 8$.

Skill Practice

Determine if the number given is a solution to the equation.

1. $4x - 1 = 7$; 3
2. $-2y + 5 = 9$; -2

Example 1   Determining Whether a Number Is a Solution to an Equation

Determine whether the given number is a solution to the equation.

a. $4x + 7 = 5$; $-\frac{1}{2}$ 　　　　**b.** $-6w + 14 = 4$; 3

Solution:

a.　　$4x + 7 = 5$

$\qquad 4\left(-\frac{1}{2}\right) + 7 \overset{?}{=} 5$ 　　Substitute $-\frac{1}{2}$ for x.

$\qquad\qquad -2 + 7 \overset{?}{=} 5$ 　　Simplify.

$\qquad\qquad\qquad 5 \overset{?}{=} 5$ ✔ 　　Right-hand side equals the left-hand side.
 Thus, $-\frac{1}{2}$ *is a solution* to the equation $4x + 7 = 5$.

b.　$-6w + 14 = 4$

$\qquad -6(3) + 14 \overset{?}{=} 4$ 　　Substitute 3 for w.

$\qquad\qquad -18 + 14 \overset{?}{=} 4$ 　　Simplify.

$\qquad\qquad\qquad -4 \neq 4$ 　　Right-hand side does not equal left-hand side.
 Thus, 3 is *not a solution* to the equation $-6w + 14 = 4$.

Answers

1. No 　**2.** Yes

In the study of algebra, you will encounter a variety of equations. In this chapter, we will focus on a specific type of equation called a linear equation in one variable.

DEFINITION Linear Equation in One Variable

Let a and b be real numbers such that $a \neq 0$. A **linear equation in one variable** is an equation that can be written in the form

$$ax + b = 0$$

Notice that a linear equation in one variable has only one variable. Furthermore, because the variable has an implied exponent of 1, a linear equation is sometimes called a *first-degree equation*.

Linear Equation in One Variable	***Not* a Linear Equation in One Variable**
$2x + 3 = 0$	$4x^2 + 8 = 0$ (exponent for x is not 1)
$\frac{1}{5}a + \frac{2}{7} = 0$	$\frac{3}{4}a + \frac{5}{8}b = 0$ (more than one variable)

2. Addition and Subtraction Properties of Equality

If two equations have the same solution then the equations are equivalent. For example, the following equations are equivalent because the solution for each equation is 6.

Equivalent Equations	**Check the Solution 6**
$2x - 5 = 7$	$\longrightarrow$ $2(6) - 5 \overset{?}{=} 7 \Rightarrow 12 - 5 \overset{?}{=} 7$ ✔
$2x = 12$	$\longrightarrow$ $2(6) \overset{?}{=} 12 \Rightarrow \quad 12 \overset{?}{=} 12$ ✔
$x = 6$	$\longrightarrow$ $6 \overset{?}{=} 6 \Rightarrow \quad 6 \overset{?}{=} 6$ ✔

To solve a linear equation, $ax + b = 0$, the goal is to find *all* values of x that make the equation true. One general strategy for solving an equation is to rewrite it as an equivalent but simpler equation. This process is repeated until the equation can be written in the form $x = $ number. The addition and subtraction properties of equality help us do this.

PROPERTY Addition and Subtraction Properties of Equality

Let a, b, and c represent algebraic expressions.

1. **Addition property of equality:**   If $a = b$,
                                                    then $a + c = b + c$

2. ***Subtraction property of equality:**   If $a = b$,
                                                    then $a - c = b - c$

*The subtraction property of equality follows directly from the addition property, because subtraction is defined in terms of addition.

If $a + (-c) = b + (-c)$

then, $a - c = b - c$

Concept Connections

State whether the equation is a linear equation in one variable.

3. $5y + 1 = 0$

4. $2a^2 + 4 = 0$

5. $6 - 3x = 0$

6. $6t + 8y = 0$

Answers

3. Yes **4.** No **5.** Yes **6.** No

The addition and subtraction properties of equality indicate that adding or subtracting the same quantity on each side of an equation results in an equivalent equation. This is true because if two quantities are increased or decreased by the same amount, then the resulting quantities will also be equal (Figure 2-1).

Figure 2-1

Skill Practice

Solve the equations.

7. $v - 7 = 2$

8. $x + 4 = 4$

Example 2 Applying the Addition and Subtraction Properties of Equality

Solve the equations.

a. $p - 4 = 11$ **b.** $w + 5 = -2$

Solution:

In each equation, the goal is to isolate the variable on one side of the equation. To accomplish this, we use the fact that the sum of a number and its opposite is zero and the difference of a number and itself is zero.

a.
$$p - 4 = 11$$
$$p - 4 + 4 = 11 + 4 \qquad \text{To isolate } p, \text{ add } 4 \text{ to both sides } (-4 + 4 = 0).$$
$$p + 0 = 15 \qquad \text{Simplify.}$$
$$p = 15 \qquad \text{Check by substituting } p = 15 \text{ into the original equation.}$$

Check: $p - 4 = 11$

$15 - 4 \overset{?}{=} 11$

The solution is 15. $11 \overset{?}{=} 11$ ✔ True

b.
$$w + 5 = -2$$
$$w + 5 - 5 = -2 - 5 \qquad \text{To isolate } w, \text{ subtract } 5 \text{ from both sides.}$$
$$(5 - 5 = 0).$$
$$w + 0 = -7 \qquad \text{Simplify.}$$
$$w = -7 \qquad \text{Check by substituting } w = -7 \text{ into the original equation.}$$

Check: $w + 5 = -2$

$-7 + 5 \overset{?}{=} -2$

The solution is -7. $-2 \overset{?}{=} -2$ ✔ True

Answers

7. 9   **8.** 0

In some situations, rather than writing the solution to an equation, we may write the solution *set* to an equation. The **solution set** is the set of all solutions to an equation. For instance, in Example 2(b), the solution is -7, but the *solution set* would be written as $\{-7\}$. Check to determine how your instructor prefers that you write the answer.

Example 3 Applying the Addition and Subtraction Properties of Equality

Solve the equations.

a. $\dfrac{9}{4} = q - \dfrac{3}{4}$ **b.** $-1.2 + z = 4.6$

Skill Practice

Solve the equations.

9. $\dfrac{1}{4} = a - \dfrac{2}{3}$

10. $-8.1 + w = 11.5$

Solution:

a. $\dfrac{9}{4} = q - \dfrac{3}{4}$

$\dfrac{9}{4} + \dfrac{3}{4} = q - \dfrac{3}{4} + \dfrac{3}{4}$ To isolate q, add $\frac{3}{4}$ to both sides $\left(-\frac{3}{4} + \frac{3}{4} = 0\right)$.

$\dfrac{12}{4} = q + 0$ Simplify.

TIP: The variable may be isolated on either side of the equation.

$3 = q$ or equivalently, $q = 3$

Check: $\dfrac{9}{4} = q - \dfrac{3}{4}$

$\dfrac{9}{4} \overset{?}{=} 3 - \dfrac{3}{4}$ Substitute $q = 3$.

$\dfrac{9}{4} \overset{?}{=} \dfrac{12}{4} - \dfrac{3}{4}$ Common denominator

The solution is 3.

$\dfrac{9}{4} \overset{?}{=} \dfrac{9}{4}$ ✔ True

b. $-1.2 + z = 4.6$

$-1.2 + 1.2 + z = 4.6 + 1.2$ To isolate z, add 1.2 to both sides.

$0 + z = 5.8$

$z = 5.8$ Check: $-1.2 + z = 4.6$

$-1.2 + 5.8 \overset{?}{=} 4.6$ Substitute $z = 5.8$.

The solution is 5.8.

$4.6 \overset{?}{=} 4.6$ ✔ True

3. Multiplication and Division Properties of Equality

Adding or subtracting the same quantity to both sides of an equation results in an equivalent equation. In a similar way, multiplying or dividing both sides of an equation by the same nonzero quantity also results in an equivalent equation. This is stated formally as the multiplication and division properties of equality.

Answers

9. $\dfrac{11}{12}$ **10.** 19.6

11. The division property of equality states that if

$a = b$, then $\dfrac{a}{c} = \dfrac{b}{c}$

provided that $c \neq 0$. Why is the condition $c \neq 0$ necessary?

PROPERTY   Multiplication and Division Properties of Equality

Let a, b, and c represent algebraic expressions.

1. **Multiplication property of equality:**   If $a = b$,

    then $ac = bc$

2. ***Division property of equality:**   If $a = b$

    then $\dfrac{a}{c} = \dfrac{b}{c}$ (provided $c \neq 0$)

*The division property of equality follows directly from the multiplication property because division is defined as multiplication by the reciprocal.

If $a \cdot \dfrac{1}{c} = b \cdot \dfrac{1}{c}$ $(c \neq 0)$

then, $\dfrac{a}{c} = \dfrac{b}{c}$

To understand the multiplication property of equality, suppose we start with a true equation such as $10 = 10$. If both sides of the equation are multiplied by a constant such as 3, the result is also a true statement (Figure 2-2).

$10 = 10$

$3 \cdot 10 = 3 \cdot 10$

$30 = 30$

Figure 2-2

Similarly, if both sides of the equation are divided by a nonzero real number such as 2, the result is also a true statement (Figure 2-3).

$10 = 10$

$\dfrac{10}{2} = \dfrac{10}{2}$

$5 = 5$

Figure 2-3

TIP: The product of a number and its reciprocal is always 1. For example:

$\dfrac{1}{5}(5) = 1$

$-\dfrac{7}{2}\left(-\dfrac{2}{7}\right) = 1$

To solve an equation in the variable x, the goal is to write the equation in the form $x = $ number. In particular, notice that we desire the coefficient of x to be 1. That is, we want to write the equation as $1x = $ number. Therefore, to solve an equation such as $5x = 15$, we can multiply both sides of the equation by the reciprocal of the x-term coefficient. In this case, multiply both sides by the reciprocal of 5, which is $\frac{1}{5}$.

$5x = 15$

$\dfrac{1}{5}(5x) = \dfrac{1}{5}(15)$ Multiply by $\dfrac{1}{5}$.

$1x = 3$ The coefficient of the x-term is now 1.

$x = 3$

Answer

11. $c \neq 0$ because division by zero is undefined.

The division property of equality can also be used to solve the equation $5x = 15$ by dividing both sides by the coefficient of the x-term. In this case, divide both sides by 5 to make the coefficient of x equal to 1.

$$5x = 15$$

$$\frac{5x}{5} = \frac{15}{5} \qquad \text{Divide by 5.}$$

$$1x = 3 \qquad \text{The coefficient on the } x\text{-term is now 1.}$$

$$x = 3$$

TIP: The quotient of a nonzero real number and itself is always 1. For example:

$$\frac{5}{5} = 1$$

$$\frac{-3.5}{-3.5} = 1$$

Example 4 **Applying the Multiplication and Division Properties of Equality**

Solve the equations using the multiplication or division property of equality.

a. $12x = 60$ **b.** $48 = -8w$ **c.** $-x = 8$

Skill Practice

Solve the equations.
12. $4x = -20$
13. $100 = -4p$
14. $-y = -11$

Solution:

a. $12x = 60$

$$\frac{12x}{12} = \frac{60}{12} \qquad \text{To obtain a coefficient of 1 for the } x\text{-term,}$$
$$\text{divide both sides by 12.}$$

$$1x = 5 \qquad \text{Simplify.}$$

$$x = 5 \qquad \underline{\text{Check:}} \quad 12x = 60$$

$$12(5) \stackrel{?}{=} 60$$

The solution is 5. $60 \stackrel{?}{=} 60$ ✔ True

b. $48 = -8w$

$$\frac{48}{-8} = \frac{-8w}{-8} \qquad \text{To obtain a coefficient of 1 for the } w\text{-term, divide}$$
$$\text{both sides by } -8.$$

$$-6 = 1w \qquad \text{Simplify.}$$

$$-6 = w \qquad \underline{\text{Check:}} \ 48w = -8w$$

$$48 \stackrel{?}{=} -8(-6)$$

The solution is -6. $48 \stackrel{?}{=} 48$ ✔ True

c. $-x = 8$ Note that $-x$ is equivalent to $-1 \cdot x$.

$$-1x = 8$$

$$\frac{-1x}{-1} = \frac{8}{-1} \qquad \text{To obtain a coefficient of 1 for the } x\text{-term, divide}$$
$$\text{by } -1.$$

$$x = -8 \qquad \underline{\text{Check:}} \ -x = 8$$

$$-(-8) \stackrel{?}{=} 8$$

The solution is -8. $8 \stackrel{?}{=} 8$ ✔ True

TIP: In Example 4(c), we could also have *multiplied* both sides by -1 to create a coefficient of 1 on the x-term.

$$-x = 8$$

$$(-1)(-x) = (-1)8$$

$$x = -8$$

Answers
12. -5 **13.** -25 **14.** 11

Example 5 Applying the Multiplication and Division Properties of Equality

Solve the equations by using the multiplication or division property of equality.

a. $-\dfrac{2}{9}q = \dfrac{1}{3}$ **b.** $-3.43 = -0.7z$ **c.** $\dfrac{d}{6} = -4$

Solution:

a. $\qquad -\dfrac{2}{9}q = \dfrac{1}{3}$

$\left(-\dfrac{9}{2}\right)\left(-\dfrac{2}{9}q\right) = \dfrac{1}{3}\left(-\dfrac{9}{2}\right)$ To obtain a coefficient of 1 for the q-term, multiply by the reciprocal of $-\dfrac{2}{9}$, which is $-\dfrac{9}{2}$.

$1q = -\dfrac{3}{2}$ Simplify. The product of a number and its reciprocal is 1.

$q = -\dfrac{3}{2}$ Check: $\quad -\dfrac{2}{9}q = \dfrac{1}{3}$

$\qquad\qquad\qquad\qquad -\dfrac{2}{9}\left(-\dfrac{3}{2}\right) \overset{?}{=} \dfrac{1}{3}$

The solution is $-\dfrac{3}{2}$. $\dfrac{1}{3} \overset{?}{=} \dfrac{1}{3}$ ✔ True

> **TIP:** When applying the multiplication or division property of equality to obtain a coefficient of 1 for the variable term, we will generally use the following convention:
> - If the coefficient of the variable term is expressed as a fraction, we will usually multiply both sides by its reciprocal, as in Example 5(a).
> - If the coefficient of the variable term is an integer or decimal, we will divide both sides by the coefficient itself, as in Example 5(b).

b. $\qquad -3.43 = -0.7z$

$\dfrac{-3.43}{-0.7} = \dfrac{-0.7z}{-0.7}$ To obtain a coefficient of 1 for the z-term, divide by -0.7.

$4.9 = 1z$ Simplify.

$4.9 = z$

$z = 4.9$ Check: $-3.43 = -0.7z$

$\qquad\qquad\qquad\qquad -3.43 \overset{?}{=} -0.7(4.9)$

The solution is 4.9. $-3.43 \overset{?}{=} -3.43$ ✔ True

c. $\qquad \dfrac{d}{6} = -4$

$\dfrac{1}{6}d = -4$ $\dfrac{d}{6}$ is equivalent to $\dfrac{1}{6}d$.

$\dfrac{6}{1} \cdot \dfrac{1}{6}d = -4 \cdot \dfrac{6}{1}$ To obtain a coefficient of 1 for the d-term, multiply by the reciprocal of $\dfrac{1}{6}$, which is $\dfrac{6}{1}$.

$1d = -24$ Simplify.

$d = -24$ Check: $\dfrac{d}{6} = -4$

$\dfrac{-24}{6} \overset{?}{=} -4$

The solution is -24. $-4 \overset{?}{=} -4$ ✔ True

It is important to distinguish between cases where the addition or subtraction properties of equality should be used to isolate a variable versus those in which the multiplication or division property of equality should be used. Remember the goal is to isolate the variable term and obtain a coefficient of 1. Compare the equations:

$$5 + x = 20 \quad \text{and} \quad 5x = 20$$

In the first equation, the relationship between 5 and x is addition. Therefore, we want to reverse the process by subtracting 5 from both sides. In the second equation, the relationship between 5 and x is multiplication. To isolate x, we reverse the process by dividing by 5 or equivalently, multiplying by the reciprocal, $\frac{1}{5}$.

$$5 + x = 20 \qquad \text{and} \qquad 5x = 20$$

$$5 - 5 + x = 20 - 5 \qquad\qquad \dfrac{5x}{5} = \dfrac{20}{5}$$

$$x = 15 \qquad\qquad\qquad x = 4$$

4. Translations

Example 6 **Translating to a Linear Equation**

Write an algebraic equation to represent each English sentence. Then solve the equation.

a. The quotient of a number and 4 is 6.

b. The product of a number and 4 is 6.

c. Negative twelve is equal to the sum of -5 and a number.

d. The value 1.4 subtracted from a number is 5.7.

Solution:

For each case we will let x represent the unknown number.

a. The quotient of a number and 4 is 6.

$$\dfrac{x}{4} = 6$$

$$4 \cdot \dfrac{x}{4} = 4 \cdot 6 \qquad \text{Multiply both sides by 4.}$$

$$\dfrac{4}{1} \cdot \dfrac{x}{4} = 4 \cdot 6$$

$$x = 24 \qquad \text{Check: } \dfrac{24}{4} \overset{?}{=} 6 \text{ ✔ True}$$

Answers

18. $\dfrac{x}{-2} = 8$; The number is -16.

19. $-3x = -24$; The number is 8.
20. $y + 6 = -20$; The number is -26.
21. $13 = x - 5$; The number is 18.

b. The product of a number and 4 is 6.

$$4x = 6$$

$$\frac{4x}{4} = \frac{6}{4} \qquad \text{Divide both sides by 4.}$$

$$x = \frac{3}{2} \qquad \underline{\text{Check:}} \quad 4\left(\frac{3}{2}\right) \stackrel{?}{=} 6 \checkmark \quad \text{True}$$

c. Negative twelve is equal to the sum of -5 and a number.

$$-12 = -5 + x$$

$$-12 + 5 = -5 + 5 + x \qquad \text{Add 5 to both sides.}$$

$$-7 = x \qquad \underline{\text{Check:}} \quad -12 \stackrel{?}{=} -5 + (-7) \checkmark \quad \text{True}$$

d. The value 1.4 subtracted from a number is 5.7.

$$x - 1.4 = 5.7$$

$$x - 1.4 + 1.4 = 5.7 + 1.4 \qquad \text{Add 1.4 to both sides.}$$

$$x = 7.1 \qquad \underline{\text{Check:}} \quad 7.1 - 1.4 \stackrel{?}{=} 5.7 \checkmark \quad \text{True}$$

Section 2.1   Practice Exercises

Study Skills Exercises

1. After getting a test back, it is a good idea to correct the test so that you do not make the same errors again. One recommended approach is to use a clean sheet of paper, and divide the paper down the middle vertically as shown. For each problem that you missed on the test, rework the problem correctly on the left-hand side of the paper. Then give a written explanation on the right-hand side of the paper. To reinforce the correct procedure, do four more problems of that type.

Take the time this week to make corrections from your last test.

Perform the correct math here.	Explain the process here.
$\downarrow$	$\downarrow$
$2 + 4(5)$ $= 2 + 20$ $= 22$	Do multiplication before addition.

2. Define the key terms:

   a. linear equation in one variable   **b.** solution to an equation

   c. addition property of equality   **d.** subtraction property of equality

   e. multiplication property of equality   **f.** division property of equality   **g.** solution set

Objective 1: Definition of a Linear Equation in One Variable

For Exercises 3–6, identify the following as either an expression or an equation.

   3. $x - 4 + 5x$ **4.** $8x + 2 = 7$ **5.** $9 = 2x - 4$ **6.** $3x^2 + x = -3$

7. Explain how to determine if a number is a solution to an equation.

8. Explain why the equations $6x = 12$ and $x = 2$ are *equivalent equations*.

For Exercises 9–14, determine whether the given number is a solution to the equation. **(See Example 1.)**

9. $x - 1 = 5$; 4

10. $x - 2 = 1$; -1

11. $5x = -10$; -2

12. $3x = 21$; 7

13. $3x + 9 = 3$; -2

14. $2x - 1 = -3$; -1

Objective 2: Addition and Subtraction Properties of Equality

For Exercises 15–34, solve the equations using the addition or subtraction property of equality. Be sure to check your answers. **(See Examples 2–3.)**

15. $x + 6 = 5$

16. $x - 2 = 10$

17. $q - 14 = 6$

18. $w + 3 = -5$

19. $2 + m = -15$

20. $-6 + n = 10$

21. $-23 = y - 7$

22. $-9 = -21 + b$

23. $4 + c = 4$

24. $-13 + b = -13$

25. $4.1 = 2.8 + a$

26. $5.1 = -2.5 + y$

27. $5 = z - \dfrac{1}{2}$

28. $-7 = p + \dfrac{2}{3}$

29. $x + \dfrac{5}{2} = \dfrac{1}{2}$

30. $x - \dfrac{2}{3} = \dfrac{7}{3}$

31. $-6.02 + c = -8.15$

32. $p + 0.035 = -1.12$

33. $3.245 + t = -0.0225$

34. $-1.004 + k = 3.0589$

Objective 3: Multiplication and Division Properties of Equality

For Exercises 35–54, solve the equations using the multiplication or division property of equality. Be sure to check your answers. **(See Examples 4–5.)**

35. $6x = 54$

36. $2w = 8$

37. $12 = -3p$

38. $6 = -2q$

39. $-5y = 0$

40. $-3k = 0$

41. $-\dfrac{y}{5} = 3$

42. $-\dfrac{z}{7} = 1$

43. $\dfrac{4}{5} = -t$

44. $-\dfrac{3}{7} = -h$

45. $\dfrac{2}{5}a = -4$

46. $\dfrac{3}{8}b = -9$

47. $-\dfrac{1}{5}b = -\dfrac{4}{5}$

48. $-\dfrac{3}{10}w = \dfrac{2}{5}$

49. $-41 = -x$

50. $32 = -y$

51. $3.81 = -0.03p$

52. $2.75 = -0.5q$

53. $5.82y = -15.132$

54. $-32.3x = -0.4522$

Objective 4: Translations

For Exercises 55–64, write an algebraic equation to represent each English sentence. (Let x represent the unknown number.) Then solve the equation. **(See Example 6.)**

55. The sum of negative eight and a number is forty-two.

56. The sum of thirty-one and a number is thirteen.

57. The difference of a number and negative six is eighteen.

58. The sum of negative twelve and a number is negative fifteen.

59. The product of a number and seven is the same as negative sixty-three.

60. The product of negative three and a number is the same as twenty-four.

61. The quotient of a number and twelve is one-third.

62. Eighteen is equal to the quotient of a number and two.

63. The sum of a number and $\frac{5}{8}$ is $\frac{13}{8}$.

64. The difference of a number and $\frac{2}{3}$ is $\frac{1}{3}$.

Mixed Exercises

For Exercises 65–92, solve the equation using the appropriate property of equality.

65. $a - 9 = 1$

66. $b - 2 = -4$

67. $-9x = 1$

68. $-2k = -4$

69. $-\frac{2}{3}h = 8$

70. $\frac{3}{4}p = 15$

71. $\frac{2}{3} + t = 8$

72. $\frac{3}{4} + y = 15$

73. $\frac{r}{3} = -12$

74. $\frac{d}{-4} = 5$

75. $k + 16 = 32$

76. $-18 = -9 + t$

77. $16k = 32$

78. $-18 = -9t$

79. $7 = -4q$

80. $-3s = 10$

81. $-4 + q = 7$

82. $s - 3 = 10$

83. $-\frac{1}{3}d = 12$

84. $-\frac{2}{5}m = 10$

85. $4 = \frac{1}{2} + z$

86. $3 = \frac{1}{4} + p$

87. $1.2y = 4.8$

88. $4.3w = 8.6$

89. $4.8 = 1.2 + y$

90. $8.6 = w - 4.3$

91. $0.0034 = y - 0.405$

92. $-0.98 = m + 1.0034$

For Exercises 93–100, determine if the equation is a linear equation in one variable. Answer yes or no.

93. $4p + 5 = 0$

94. $3x - 5y = 0$

95. $4 + 2a^2 = 5$

96. $-8t = 7$

97. $x - 4 = 9$

98. $2x^3 + y = 0$

99. $19b = -3$

100. $13 + x = 19$

Expanding Your Skills

For Exercises 101–106, construct an equation with the given solution. Answers will vary.

101. $y = 6$

102. $x = 2$

103. $p = -4$

104. $t = -10$

105. $a = 0$

106. $k = 1$

For Exercises 107–110, simplify by collecting the *like* terms. Then solve the equation.

107. $5x - 4x + 7 = 8 - 2$

108. $2 + 3 = 2y + 1 - y$

109. $6p - 3p = 15 + 6$

110. $12 - 20 = 2t + 2t$

Solving Linear Equations

1. Solving Linear Equations Involving Multiple Steps

In Section 2.1, we studied a one-step process to solve linear equations by using the addition, subtraction, multiplication, and division properties of equality. In Example 1, we solve the equation $-2w - 7 = 11$. Solving this equation will require multiple steps. To understand the proper steps, always remember the ultimate goal—to isolate the variable. Therefore, we will first isolate the *term* containing the variable before dividing both sides by -2.

1. Solving Linear Equations Involving Multiple Steps
2. Procedure for Solving a Linear Equation in One Variable
3. Conditional Equations, Identities, and Contradictions

Example 1 Solving a Linear Equation

Solve the equation. $-2w - 7 = 11$

Solution:

$$-2w - 7 = 11$$

$$-2w - 7 + 7 = 11 + 7 \qquad \text{Add 7 to both sides of the equation. This isolates the } w\text{-term.}$$

$$-2w = 18$$

$$\frac{-2w}{-2} = \frac{18}{-2} \qquad \text{Next, apply the division property of equality to obtain a coefficient of 1 for } w. \text{ Divide by } -2 \text{ on both sides.}$$

$$w = -9$$

Check:

$$-2w - 7 = 11$$

$$-2(-9) - 7 \stackrel{?}{=} 11 \qquad \text{Substitute } w = -9 \text{ in the original equation.}$$

$$18 - 7 \stackrel{?}{=} 11$$

$$11 \stackrel{?}{=} 11 \checkmark \qquad \text{True. The solution is } -9.$$

Skill Practice

Solve the equation.

1. $-5y - 5 = 10$

Example 2 Solving a Linear Equation

Solve the equation. $2 = \frac{1}{5}x + 3$

Solution:

$$2 = \frac{1}{5}x + 3$$

$$2 - 3 = \frac{1}{5}x + 3 - 3 \qquad \text{Subtract 3 from both sides. This isolates the } x\text{-term.}$$

$$-1 = \frac{1}{5}x \qquad \text{Simplify.}$$

$$5(-1) = 5 \cdot \left(\frac{1}{5}x\right) \qquad \text{Next, apply the multiplication property of equality to obtain a coefficient of 1 for } x.$$

$$-5 = 1x$$

$$-5 = x \qquad \text{Simplify. The answer checks in the original equation.}$$

The solution is -5.

Skill Practice

Solve the equation.

2. $2 = \frac{1}{2}a - 7$

Answers

1. -3 **2.** 18

In Example 3, the variable x appears on both sides of the equation. In this case, apply the addition or subtraction property of equality to collect the variable terms on one side of the equation and the constant terms on the other side. Then use the multiplication or division property of equality to get a coefficient equal to 1.

Skill Practice

Solve the equation.

3. $10x - 3 = 4x - 2$

Example 3 **Solving a Linear Equation**

Solve the equation. $6x - 4 = 2x - 8$

Solution:

$$6x - 4 = 2x - 8$$

$$6x - 2x - 4 = 2x - 2x - 8 \qquad \text{Subtract } 2x \text{ from both sides leaving } 0x \text{ on the right-hand side.}$$

$$4x - 4 = 0x - 8 \qquad \text{Simplify.}$$

$$4x - 4 = -8 \qquad \text{The } x\text{-terms have now been combined on one side of the equation.}$$

$$4x - 4 + 4 = -8 + 4 \qquad \text{Add 4 to both sides of the equation. This combines the constant terms on the } \textit{other}$$

$$4x = -4 \qquad \text{side of the equation.}$$

$$\frac{4x}{4} = \frac{-4}{4} \qquad \text{To obtain a coefficient of 1 for } x, \text{ divide both sides of the equation by 4.}$$

$$x = -1$$

The solution is -1.

Check:

$$6x - 4 = 2x - 8$$

$$6(-1) - 4 \overset{?}{=} 2(-1) - 8$$

$$-6 - 4 \overset{?}{=} -2 - 8$$

$$-10 \overset{?}{=} -10 \ ✔ \quad \text{True}$$

TIP: It is important to note that the variable may be isolated on either side of the equation. We will solve the equation from Example 3 again, this time isolating the variable on the right-hand side.

$$6x - 4 = 2x - 8$$

$$6x - 6x - 4 = 2x - 6x - 8 \qquad \text{Subtract } 6x \text{ on both sides.}$$

$$0x - 4 = -4x - 8$$

$$-4 = -4x - 8$$

$$-4 + 8 = -4x - 8 + 8 \qquad \text{Add 8 to both sides.}$$

$$4 = -4x$$

$$\frac{4}{-4} = \frac{-4x}{-4} \qquad \text{Divide both sides by } -4.$$

$$-1 = x \quad \text{or equivalently } x = -1$$

Answer

3. $\dfrac{1}{6}$

2. Procedure for Solving a Linear Equation in One Variable

In some cases, it is necessary to simplify both sides of a linear equation before applying the properties of equality. Therefore, we offer the following steps to solve a linear equation in one variable.

PROCEDURE **Solving a Linear Equation in One Variable**

Step 1 Simplify both sides of the equation.
- Clear parentheses
- Combine *like* terms

Step 2 Use the addition or subtraction property of equality to collect the variable terms on one side of the equation.

Step 3 Use the addition or subtraction property of equality to collect the constant terms on the other side of the equation.

Step 4 Use the multiplication or division property of equality to make the coefficient of the variable term equal to 1.

Step 5 Check your answer.

Example 4 **Solving a Linear Equation**

Solve the equation. $7 + 3 = 2(p - 3)$

Solution:

$7 + 3 = 2(p - 3)$

$10 = 2p - 6$ **Step 1:** Simplify both sides of the equation by clearing parentheses and combining *like* terms.

   Step 2: The variable terms are already on one side.

$10 + 6 = 2p - 6 + 6$ **Step 3:** Add 6 to both sides to collect the constant terms on the other side.

$16 = 2p$

$\dfrac{16}{2} = \dfrac{2p}{2}$ **Step 4:** Divide both sides by 2 to obtain a coefficient of 1 for p.

$8 = p$ **Step 5:** Check:

$7 + 3 = 2(p - 3)$

$10 \stackrel{?}{=} 2(8 - 3)$

$10 \stackrel{?}{=} 2(5)$

The solution is 8. $10 \stackrel{?}{=} 10$ ✔ True

Skill Practice

Solve the equation.
4. $12 + 2 = 7(3 - y)$

Answer

4. 1

Skill Practice

Solve the equation.

5. $1.5t + 2.3 = 3.5t - 1.9$

Example 5   Solving a Linear Equation

Solve the equation. $2.2y - 8.3 = 6.2y + 12.1$

Solution:

$2.2y - 8.3 = 6.2y + 12.1$	**Step 1:** The right- and left-hand sides are already simplified.
$2.2y - 2.2y - 8.3 = 6.2y - 2.2y + 12.1$ $-8.3 = 4.0y + 12.1$	**Step 2:** Subtract $2.2y$ from both sides to collect the variable terms on one side of the equation.
$-8.3 - 12.1 = 4.0y + 12.1 - 12.1$ $-20.4 = 4.0y$	**Step 3:** Subtract 12.1 from both sides to collect the constant terms on the other side.
$\dfrac{-20.4}{4.0} = \dfrac{4.0y}{4.0}$ $-5.1 = y$	**Step 4:** To obtain a coefficient of 1 for the y-term, divide both sides of the equation by 4.0.
$y = -5.1$	**Step 5:** Check:

$$2.2y - 8.3 = 6.2y + 12.1$$
$$2.2(-5.1) - 8.3 \overset{?}{=} 6.2(-5.1) + 12.1$$
$$-11.22 - 8.3 \overset{?}{=} -31.62 + 12.1$$

The solution is -5.1.

$$-19.52 \overset{?}{=} -19.52 \ ✔ \ \text{True}$$

Skill Practice

Solve the equation.

6. $4(2y - 1) + y = 6y + 3 - y$

Example 6   Solving a Linear Equation

Solve the equation. $2 + 7x - 5 = 6(x + 3) + 2x$

Solution:

$2 + 7x - 5 = 6(x + 3) + 2x$	
$-3 + 7x = 6x + 18 + 2x$	**Step 1:** Add *like* terms on the left. Clear parentheses on the right.
$-3 + 7x = 8x + 18$	Combine *like* terms.
$-3 + 7x - 7x = 8x - 7x + 18$	**Step 2:** Subtract $7x$ from both sides.
$-3 = x + 18$	Simplify.
$-3 - 18 = x + 18 - 18$	**Step 3:** Subtract 18 from both sides.
$-21 = x$	
$x = -21$	**Step 4:** Because the coefficient of the x term is already 1, there is no need to apply the multiplication or division property of equality.

The solution is -21.

Step 5: The check is left to the reader.

Answers

5. 2.1   **6.** $\dfrac{7}{4}$

Example 7	**Solving a Linear Equation**

Solve the equation. $9 - (z - 3) + 4z = 4z - 5(z + 2) - 6$

Solution:

$$9 - (z - 3) + 4z = 4z - 5(z + 2) - 6$$

$9 - z + 3 + 4z = 4z - 5z - 10 - 6$	**Step 1:** Clear parentheses.
$12 + 3z = -z - 16$	Combine *like* terms.
$12 + 3z + z = -z + z - 16$	**Step 2:** Add z to both sides.
$12 + 4z = -16$	
$12 - 12 + 4z = -16 - 12$	**Step 3:** Subtract 12 from both sides.
$4z = -28$	
$\dfrac{4z}{4} = \dfrac{-28}{4}$	**Step 4:** Divide both sides by 4.
$z = -7$	**Step 5:** The check is left for the reader.

The solution is -7.

3. Conditional Equations, Identities, and Contradictions

The solutions to a linear equation are the values of x that make the equation a true statement. A linear equation has one unique solution. Some types of equations, however, have no solution while others have infinitely many solutions.

I. Conditional Equations

An equation that is true for some values of the variable but false for other values is called a **conditional equation**. The equation $x + 4 = 6$, for example, is true on the condition that $x = 2$. For other values of x, the statement $x + 4 = 6$ is false.

II. Contradictions

Some equations have no solution, such as $x + 1 = x + 2$. There is no value of x, that when increased by 1 will equal the same value increased by 2. If we tried to solve the equation by subtracting x from both sides, we get the contradiction $1 = 2$. This indicates that the equation has no solution. An equation that has no solution is called a **contradiction**.

$$x + 1 = x + 2$$
$$x - x + 1 = x - x + 2$$
$$1 = 2 \quad \text{(contradiction)} \qquad \text{No solution.}$$

III. Identities

An equation that has all real numbers as its solution set is called an **identity**. For example, consider the equation, $x + 4 = x + 4$. Because the left- and right-hand sides are equal, any real number substituted for x will result in equal quantities on both sides. If we subtract x from both sides of the equation, we get the identity $4 = 4$. In such a case, the solution is the set of all real numbers.

$$x + 4 = x + 4$$

$$x - x + 4 = x - x + 4$$

$$4 = 4 \quad \text{(identity)} \qquad \text{The solution is all real numbers.}$$

Skill Practice

Solve the equation. Identify the equation as a conditional equation, a contradiction, or an identity.

11. $4(2t + 1) - 1 = 8t + 3$

12. $3x - 5 = 4x + 1 - x$

13. $6(v - 2) = 2v - 4$

Example 8 Identifying Conditional Equations, Contradictions, and Identities

Solve the equation. Identify each equation as a conditional equation, a contradiction, or an identity.

a. $4k - 5 = 2(2k - 3) + 1$ **b.** $2(b - 4) = 2b - 7$ **c.** $3x + 7 = 2x - 5$

Solution:

a.

$4k - 5 = 2(2k - 3) + 1$	
$4k - 5 = 4k - 6 + 1$	Clear parentheses.
$4k - 5 = 4k - 5$	Combine *like* terms.
$4k - 4k - 5 = 4k - 4k - 5$	Subtract $4k$ from both sides.
$-5 = -5 \quad \text{(Identity)}$	

This is an identity. The solution is all real numbers.

b.

$2(b - 4) = 2b - 7$	
$2b - 8 = 2b - 7$	Clear parentheses.
$2b - 2b - 8 = 2b - 2b - 7$	Subtract $2b$ from both sides.
$-8 = -7 \quad \text{(Contradiction)}$	

This is a contradiction. There is no solution.

c.

$3x + 7 = 2x - 5$	
$3x - 2x + 7 = 2x - 2x - 5$	Subtract $2x$ from both sides.
$x + 7 = -5$	Simplify.
$x + 7 - 7 = -5 - 7$	Subtract 7 from both sides.
$x = -12 \quad \text{(Conditional equation)}$	

This is a conditional equation. The solution is -12. (The equation is true only on the condition that $x = -12$.)

Answers

11. All real numbers; the equation is an identity.

12. No solution; the equation is a contradiction.

13. 2; the equation is a conditional equation.

Section 2.2   Practice Exercises

Study Skills Exercises

1. Several strategies are given here about taking notes. Which would you do first to help make the most of note-taking? Put them in order of importance to you by labeling them with the numbers 1–6.

_____ Read your notes after class and complete any abbreviations or incomplete sentences.

_____ Highlight important terms and definitions.

_____ Review your notes from the previous class.

_____ Bring pencils (more than one) and paper to class.

_____ Sit in class where you can clearly read the board and hear your instructor.

_____ Keep your focus on the instructor looking for phrases such as, "The most important point is . . ." and "Here is where the problem usually occurs."

2. Define the key terms:

a. conditional equation   **b. contradiction**   **c. identity**

Review Exercises

For Exercises 3–6, simplify the expressions by clearing parentheses and combining *like* terms.

3. $5z + 2 - 7z - 3z$

4. $10 - 4w + 7w - 2 + w$

5. $-(-7p + 9) + (3p - 1)$

6. $8y - (2y + 3) - 19$

7. Explain the difference between simplifying an expression and solving an equation.

For Exercises 8–12, solve the equations using the addition, subtraction, multiplication, or division property of equality.

8. $5w = -30$

9. $-7y = 21$

10. $x + 8 = -15$

11. $z - 23 = -28$

12. $-\dfrac{9}{8} = -\dfrac{3}{4}k$

Objective 1: Solving Linear Equations Involving Multiple Steps

For Exercises 13–36, solve the equations using the steps outlined in the text. **(See Examples 1–3.)**

13. $6z + 1 = 13$

14. $5x + 2 = -13$

15. $3y - 4 = 14$

16. $-7w - 5 = -19$

17. $-2p + 8 = 3$

18. $2b - \dfrac{1}{4} = 5$

19. $0.2x + 3.1 = -5.3$

20. $-1.8 + 2.4a = -6.6$

21. $\dfrac{5}{8} = \dfrac{1}{4} - \dfrac{1}{2}p$

22. $\dfrac{6}{7} = \dfrac{1}{7} + \dfrac{5}{3}r$

23. $7w - 6w + 1 = 10 - 4$

24. $5v - 3 - 4v = 13$

25. $11h - 8 - 9h = -16$

26. $6u - 5 - 8u = -7$

27. $3a + 7 = 2a - 19$

28. $6b - 20 = 14 + 5b$

29. $-4r - 28 = -58 - r$

30. $-6x - 7 = -3 - 8x$

31. $-2z - 8 = -z$

32. $-7t + 4 = -6t$

33. $\dfrac{5}{6}x + \dfrac{2}{3} = -\dfrac{1}{6}x - \dfrac{5}{3}$

34. $\dfrac{3}{7}x - \dfrac{1}{4} = -\dfrac{4}{7}x - \dfrac{5}{4}$

35. $3y - 2 = 5y - 2$

36. $4 + 10t = -8t + 4$

Objective 2: Procedure for Solving a Linear Equation in One Variable

For Exercises 37–58, solve the equations using the steps outlined in the text. **(See Examples 4–7.)**

37. $4q + 14 = 2$

38. $6 = 7m - 1$

39. $-9 = 4n - 1$

40. $-\dfrac{1}{2} - 4x = 8$

41. $3(2p - 4) = 15$

42. $4(t + 15) = 20$

43. $6(3x + 2) - 10 = -4$

44. $4(2k + 1) - 1 = 5$

45. $3.4x - 2.5 = 2.8x + 3.5$

46. $5.8w + 1.1 = 6.3w + 5.6$

47. $17(s + 3) = 4(s - 10) + 13$

48. $5(4 + p) = 3(3p - 1) - 9$

49. $6(3t - 4) + 10 = 5(t - 2) - (3t + 4)$

50. $-5y + 2(2y + 1) = 2(5y - 1) - 7$

51. $5 - 3(x + 2) = 5$

52. $1 - 6(2 - h) = 7$

53. $3(2z - 6) - 4(3z + 1) = 5 - 2(z + 1)$

54. $-2(4a + 3) - 5(2 - a) = 3(2a + 3) - 7$

55. $-2[(4p + 1) - (3p - 1)] = 5(3 - p) - 9$

56. $5 - (6k + 1) = 2[(5k - 3) - (k - 2)]$

57. $3(-0.9n + 0.5) = -3.5n + 1.3$

58. $7(0.4m - 0.1) = 5.2m + 0.86$

Objective 3: Conditional Equations, Identities, and Contradictions

For Exercises 59–64, identify the equation as a conditional equation, a contradiction, or an identity. Then describe the solution. **(See Example 8.)**

59. $2(k - 7) = 2k - 13$

60. $5h + 4 = 5(h + 1) - 1$

61. $7x + 3 = 6(x - 2)$

62. $3y - 1 = 1 + 3y$

63. $3 - 5.2p = -5.2p + 3$

64. $2(q + 3) = 4q + q - 9$

65. A conditional linear equation has (choose one): One solution, no solution, or infinitely many solutions.

66. An equation that is a contradiction has (choose one): One solution, no solution, or infinitely many solutions.

67. An equation that is an identity has (choose one): One solution, no solution, or infinitely many solutions.

68. If the only solution to a linear equation is 5, then is the equation a conditional equation, an identity, or a contradiction?

Mixed Exercises

For Exercises 69–92, find the solution, if possible.

69. $4p - 6 = 8 + 2p$

70. $\dfrac{1}{2}t - 2 = 3$

71. $2k - 9 = -8$

72. $3(y - 2) + 5 = 5$

73. $7(w - 2) = -14 - 3w$

74. $0.24 = 0.4m$

75. $2(x + 2) - 3 = 2x + 1$

76. $n + \dfrac{1}{4} = -\dfrac{1}{2}$

77. $0.5b = -23$

78. $3(2r + 1) = 6(r + 2) - 6$

79. $8 - 2q = 4$

80. $\dfrac{x}{7} - 3 = 1$

81. $2 - 4(y - 5) = -4$

82. $4 - 3(4p - 1) = -8$

83. $0.4(a + 20) = 6$

84. $2.2r - 12 = 3.4$

85. $10(2n + 1) - 6 = 20(n - 1) + 12$

86. $\dfrac{2}{5}y + 5 = -3$

87. $c + 0.123 = 2.328$

88. $4(2z + 3) = 8(z - 3) + 36$

89. $\dfrac{4}{5}t - 1 = \dfrac{1}{5}t + 5$

90. $6g - 8 = 4 - 3g$

91. $8 - (3q + 4) = 6 - q$

92. $6w - (8 + 2w) = 2(w - 4)$

Expanding Your Skills

93. Suppose -5 is a solution to the equation $x + a = 10$. Find the value of a.

94. Suppose 6 is a solution to the equation $x + a = -12$. Find the value of a.

95. Suppose 3 is a solution to the equation $ax = 12$. Find the value of a.

96. Suppose 11 is a solution to the equation $ax = 49.5$. Find the value of a.

97. Write an equation that is an identity. Answers may vary.

98. Write an equation that is a contradiction. Answers may vary.

Linear Equations: Clearing Fractions and Decimals

Section 2.3

1. Solving Linear Equations with Fractions

Linear equations that contain fractions can be solved in different ways. The first procedure, illustrated here, uses the method outlined in Section 2.2.

Objectives

1. **Solving Linear Equations with Fractions**
2. **Solving Linear Equations with Decimals**

$$\frac{5}{6}x - \frac{3}{4} = \frac{1}{3}$$

$$\frac{5}{6}x - \frac{3}{4} + \frac{3}{4} = \frac{1}{3} + \frac{3}{4} \qquad \text{To isolate the variable term, add } \frac{3}{4} \text{ to both sides.}$$

$$\frac{5}{6}x = \frac{4}{12} + \frac{9}{12} \qquad \text{Find the common denominator on the right-hand side.}$$

$$\frac{5}{6}x = \frac{13}{12} \qquad \text{Simplify.}$$

$$\frac{6}{5}\left(\frac{5}{6}x\right) = \frac{\overset{1}{6}}{5}\left(\frac{13}{\underset{2}{12}}\right) \qquad \text{Multiply by the reciprocal of } \frac{5}{6}, \text{ which is } \frac{6}{5}.$$

$$x = \frac{13}{10} \qquad \text{The solution is } \frac{13}{10}.$$

Sometimes it is simpler to solve an equation with fractions by eliminating the fractions first using a process called **clearing fractions**. To clear fractions in the equation $\frac{5}{6}x - \frac{3}{4} = \frac{1}{3}$, we can multiply both sides of the equation by the least common denominator (LCD) of all terms in the equation. In this case, the LCD of $\frac{5}{6}x$, $-\frac{3}{4}$, and $\frac{1}{3}$ is 12. Because each denominator in the equation is a factor of 12, we can simplify common factors to leave integer coefficients for each term.

Skill Practice

Solve the equation by clearing fractions.

1. $\frac{2}{5}y + \frac{1}{2} = -\frac{7}{10}$

TIP: Recall that the multiplication property of equality indicates that multiplying both sides of an equation by a nonzero constant results in an equivalent equation.

Example 1   Solving a Linear Equation by Clearing Fractions

Solve the equation by clearing fractions first. $\frac{5}{6}x - \frac{3}{4} = \frac{1}{3}$

Solution:

$$\frac{5}{6}x - \frac{3}{4} = \frac{1}{3}$$

$$12\left(\frac{5}{6}x - \frac{3}{4}\right) = 12\left(\frac{1}{3}\right)$$ Multiply both sides of the equation by the LCD, 12.

$$\overset{2}{\cancel{12}}\left(\frac{5}{6}x\right) - \overset{3}{\cancel{12}}\left(\frac{3}{4}\right) = \overset{4}{\cancel{12}}\left(\frac{1}{3}\right)$$ Apply the distributive property (recall that $12 = \frac{12}{1}$).

$$2(5x) - 3(3) = 4(1)$$ Simplify common factors to clear the fractions.

$$10x - 9 = 4$$

$$10x - 9 + 9 = 4 + 9$$ Add 9 to both sides.

$$10x = 13$$

$$\frac{10x}{10} = \frac{13}{10}$$ Divide both sides by 10.

$$x = \frac{13}{10}$$ The solution is $\frac{13}{10}$.

TIP: The fractions in this equation can be eliminated by multiplying both sides of the equation by *any* common multiple of the denominators. For example, try multiplying both sides of the equation by 24:

$$24\left(\frac{5}{6}x - \frac{3}{4}\right) = 24\left(\frac{1}{3}\right)$$

$$\overset{4}{\cancel{24}}\left(\frac{5}{6}x\right) - \overset{6}{\cancel{24}}\left(\frac{3}{4}\right) = \overset{8}{\cancel{24}}\left(\frac{1}{3}\right)$$

$$20x - 18 = 8$$

$$20x = 26$$

$$\frac{20x}{20} = \frac{26}{20}$$

$$x = \frac{13}{10}$$

Answer

1. -3

In this section, we combine the process for clearing fractions and decimals with the general strategies for solving linear equations. To solve a linear equation, it is important to follow the steps listed below.

PROCEDURE Solving a Linear Equation in One Variable

Step 1 Simplify both sides of the equation.
- Clear parentheses
- Consider clearing fractions and decimals (if any are present) by multiplying both sides of the equation by a common denominator of all terms.
- Combine *like* terms

Step 2 Use the addition or subtraction property of equality to collect the variable terms on one side of the equation.

Step 3 Use the addition or subtraction property of equality to collect the constant terms on the other side of the equation.

Step 4 Use the multiplication or division property of equality to make the coefficient of the variable term equal to 1.

Step 5 Check your answer.

Example 2 Solving a Linear Equation with Fractions

Solve the equation. $\dfrac{1}{6}x - \dfrac{2}{3} = \dfrac{1}{5}x - 1$

Solution:

$$\frac{1}{6}x - \frac{2}{3} = \frac{1}{5}x - 1 \qquad \text{The LCD of } \tfrac{1}{6}x, \ -\tfrac{2}{3}, \text{ and } \tfrac{1}{5}x \text{ is } 30.$$

$$30\left(\frac{1}{6}x - \frac{2}{3}\right) = 30\left(\frac{1}{5}x - 1\right) \qquad \text{Multiply by the LCD, } 30.$$

$$\frac{\overset{5}{\cancel{30}}}{1} \cdot \frac{1}{\cancel{6}}x - \frac{\overset{10}{\cancel{30}}}{1} \cdot \frac{2}{\cancel{3}} = \frac{\overset{6}{\cancel{30}}}{1} \cdot \frac{1}{\cancel{5}}x - 30(1) \qquad \begin{array}{l}\text{Apply the distributive property} \\ \text{(recall } 30 = \tfrac{30}{1}\text{).}\end{array}$$

$$5x - 20 = 6x - 30 \qquad \text{Clear fractions.}$$

$$5x - 6x - 20 = 6x - 6x - 30 \qquad \text{Subtract } 6x \text{ from both sides.}$$

$$-x - 20 = -30$$

$$-x - 20 + 20 = -30 + 20 \qquad \text{Add } 20 \text{ to both sides.}$$

$$-x = -10$$

$$\frac{-x}{-1} = \frac{-10}{-1} \qquad \text{Divide both sides by } -1.$$

$$x = 10 \qquad \text{The solution is } 10.$$

Skill Practice

Solve the equation.

2. $\dfrac{2}{5}x - \dfrac{1}{2} = \dfrac{7}{4} + \dfrac{3}{10}x$

Answer

2. $\dfrac{45}{2}$

Skill Practice

Solve the equation.

3. $\frac{1}{5}(z + 1) + \frac{1}{4}(z + 3) = 2$

Example 3 **Solving a Linear Equation with Fractions**

Solve the equation. $\frac{1}{3}(x + 7) - \frac{1}{2}(x + 1) = 4$

Solution:

$$\frac{1}{3}(x + 7) - \frac{1}{2}(x + 1) = 4$$

$$\frac{1}{3}x + \frac{7}{3} - \frac{1}{2}x - \frac{1}{2} = 4 \qquad \text{Clear parentheses.}$$

$$6\left(\frac{1}{3}x + \frac{7}{3} - \frac{1}{2}x - \frac{1}{2}\right) = 6(4) \qquad \begin{array}{l}\text{The LCD of} \\ \frac{1}{3}x, \frac{7}{3}, -\frac{1}{2}x, \text{ and } -\frac{1}{2} \text{ is } 6.\end{array}$$

$$\frac{\overset{2}{\cancel{6}}}{1} \cdot \frac{1}{\cancel{3}}x + \frac{\overset{2}{\cancel{6}}}{1} \cdot \frac{7}{\cancel{3}} + \frac{\overset{3}{\cancel{6}}}{1}\left(-\frac{1}{\cancel{2}}x\right) + \frac{\overset{3}{\cancel{6}}}{1}\left(-\frac{1}{\cancel{2}}\right) = 6(4) \qquad \begin{array}{l}\text{Apply the distributive} \\ \text{property.}\end{array}$$

$$2x + 14 - 3x - 3 = 24$$

$$-x + 11 = 24 \qquad \text{Combine } like \text{ terms.}$$

$$-x + 11 - 11 = 24 - 11 \qquad \text{Subtract } 11.$$

$$-x = 13$$

$$\frac{-x}{-1} = \frac{13}{-1} \qquad \text{Divide by } -1.$$

$$x = -13 \qquad \begin{array}{l}\text{The check is left to the} \\ \text{reader.}\end{array}$$

TIP: In Example 3 both parentheses and fractions are present within the equation. In such a case, we recommend that you clear parentheses first. Then clear the fractions.

Skill Practice

Solve the equation.

4. $\frac{x + 1}{4} + \frac{x + 2}{6} = 1$

Example 4 **Solving a Linear Equation with Fractions**

Solve. $\frac{x - 2}{5} - \frac{x - 4}{2} = 2$

Solution:

$$\frac{x - 2}{5} - \frac{x - 4}{2} = \frac{2}{1} \qquad \text{The LCD of } \frac{x-2}{5}, \frac{x-4}{2}, \text{ and } \frac{2}{1} \text{ is } 10.$$

$$10\left(\frac{x - 2}{5} - \frac{x - 4}{2}\right) = 10\left(\frac{2}{1}\right) \qquad \text{Multiply both sides by } 10.$$

$$\frac{\overset{2}{\cancel{10}}}{1} \cdot \left(\frac{x - 2}{\cancel{5}}\right) - \frac{\overset{5}{\cancel{10}}}{1} \cdot \left(\frac{x - 4}{\cancel{2}}\right) = \frac{10}{1} \cdot \left(\frac{2}{1}\right) \qquad \text{Apply the distributive property.}$$

$$2(x - 2) - 5(x - 4) = 20 \qquad \text{Clear fractions.}$$

$$2x - 4 - 5x + 20 = 20 \qquad \text{Apply the distributive property.}$$

$$-3x + 16 = 20 \qquad \text{Simplify both sides of the equation.}$$

$$-3x + 16 - 16 = 20 - 16 \qquad \text{Subtract } 16 \text{ from both sides.}$$

$$-3x = 4$$

$$\frac{-3x}{-3} = \frac{4}{-3} \qquad \text{Divide both sides by } -3.$$

$$x = -\frac{4}{3} \qquad \text{The check is left to the reader.}$$

Avoiding Mistakes

In Example 4, several of the fractions in the equation have two terms in the numerator. It is important to enclose these fractions in parentheses when clearing fractions. In this way, we will remember to use the distributive property to multiply the factors shown in blue with both terms from the numerator of the fractions.

Answers

3. $\frac{7}{3}$ **4.** 1

2. Solving Linear Equations with Decimals

The same procedure used to clear fractions in an equation can be used to **clear decimals**. For example, consider the equation

$$2.5x + 3 = 1.7x - 6.6$$

Recall that any terminating decimal can be written as a fraction. Therefore, the equation can be interpreted as

$$\frac{25}{10}x + 3 = \frac{17}{10}x - \frac{66}{10}$$

A convenient common denominator of all terms is 10. Therefore, we can multiply the original equation by 10 to clear decimals. The result is

$$25x + 30 = 17x - 66$$

Multiplying by the appropriate power of 10 moves the decimal points so that all coefficients become integers.

Example 5 **Solving a Linear Equation Containing Decimals**

Solve the equation by clearing decimals. $2.5x + 3 = 1.7x - 6.6$

Solution:

$$2.5x + 3 = 1.7x - 6.6$$

$10(2.5x + 3) = 10(1.7x - 6.6)$	Multiply both sides of the equation by 10.
$25x + 30 = 17x - 66$	Apply the distributive property.
$25x - 17x + 30 = 17x - 17x - 66$	Subtract $17x$ from both sides.
$8x + 30 = -66$	
$8x + 30 - 30 = -66 - 30$	Subtract 30 from both sides.
$8x = -96$	
$\dfrac{8x}{8} = \dfrac{-96}{8}$	Divide both sides by 8.
$x = -12$	

Skill Practice

Solve.

5. $1.2w + 3.5 = 2.1 + w$

TIP: Notice that multiplying a decimal number by 10 has the effect of moving the decimal point one place to the right. Similarly, multiplying by 100 moves the decimal point two places to the right, and so on.

Answer

5. -7

Skill Practice

Solve.

6. $0.25(x + 2) - 0.15(x + 3) = 4$

TIP: The terms with the most digits following the decimal point are $-0.45x$ and -4.05. Each of these is written to the hundredths place. Therefore, we multiply both sides by 100.

Answer

6. 39.5

Example 6 **Solving a Linear Equation Containing Decimals**

Solve the equation by clearing decimals. $0.2(x + 4) - 0.45(x + 9) = 12$

Solution:

$$0.2(x + 4) - 0.45(x + 9) = 12$$

$$0.2x + 0.8 - 0.45x - 4.05 = 12 \qquad \text{Clear parentheses first.}$$

$$100(0.2x + 0.8 - 0.45x - 4.05) = 100(12) \qquad \text{Multiply both sides by 100.}$$

$$20x + 80 - 45x - 405 = 1200 \qquad \text{Apply the distributive property.}$$

$$-25x - 325 = 1200 \qquad \text{Simplify both sides.}$$

$$-25x - 325 + 325 = 1200 + 325 \qquad \text{Add 325 to both sides.}$$

$$-25x = 1525$$

$$\frac{-25x}{-25} = \frac{1525}{-25} \qquad \text{Divide both sides by } -25.$$

$$x = -61 \qquad \text{The check is left to the reader.}$$

Section 2.3 Practice Exercises

Study Skills Exercises

1. Instructors vary in what they emphasize on tests. For example, test material may come from the textbook, notes, handouts, or homework. What does your instructor emphasize?

2. Define the key terms:

    a. clearing fractions   **b. clearing decimals**

Review Exercises

For Exercises 3–6, solve the equation.

3. $5(x + 2) - 3 = 4x + 5$ **4.** $-2(2x - 4x) = 6 + 18$

5. $3(2y + 3) - 4(-y + 1) = 7y - 10$ **6.** $-(3w + 4) + 5(w - 2) - 3(6w - 8) = 10$

7. Solve the equation and describe the solution set. $7x + 2 = 7(x - 12)$

8. Solve the equation and describe the solution set. $2(3x - 6) = 3(2x - 4)$

Objective 1: Solving Linear Equations with Fractions

For Exercises 9–14, determine which of the values could be used to clear fractions or decimals in the given equation.

9. $\dfrac{2}{3}x - \dfrac{1}{6} = \dfrac{x}{9}$

    Values: 6, 9, 12, 18, 24, 36

10. $\dfrac{1}{4}x - \dfrac{2}{7} = \dfrac{1}{2}x + 2$

    Values: 4, 7, 14, 21, 28, 42

11. $0.02x + 0.5 = 0.35x + 1.2$

    Values: 10; 100; 1000; 10,000

12. $0.003 - 0.002x = 0.1x$

Values: 10; 100; 1000; 10,000

13. $\frac{1}{6}x + \frac{7}{10} = x$

Values: 3, 6, 10, 30, 60

14. $2x - \frac{5}{2} = \frac{x}{3} - \frac{1}{4}$

Values: 2, 3, 4, 6, 12, 24

For Exercises 15–36, solve the equation. **(See Examples 1–4.)**

15. $\frac{1}{2}x + 3 = 5$

16. $\frac{1}{3}y - 4 = 9$

17. $\frac{1}{6}y + 2 = \frac{5}{12}$

18. $\frac{2}{15}z + 3 = \frac{7}{5}$

19. $\frac{1}{3}q + \frac{3}{5} = \frac{1}{15}q - \frac{2}{5}$

20. $\frac{3}{7}x - 5 = \frac{24}{7}x + 7$

21. $\frac{12}{5}w + 7 = 31 - \frac{3}{5}w$

22. $-\frac{1}{9}p - \frac{5}{18} = -\frac{1}{6}p + \frac{1}{3}$

23. $\frac{1}{4}(3m - 4) - \frac{1}{5} = \frac{1}{4}m + \frac{3}{10}$

24. $\frac{1}{25}(20 - t) = \frac{4}{25}t - \frac{3}{5}$

25. $\frac{1}{6}(5s + 3) = \frac{1}{2}(s + 11)$

26. $\frac{1}{12}(4n - 3) = \frac{1}{4}(2n + 1)$

27. $\frac{2}{3}x + 4 = \frac{2}{3}x - 6$

28. $-\frac{1}{9}a + \frac{2}{9} = \frac{1}{3} - \frac{1}{9}a$

29. $\frac{1}{6}(2c - 1) = \frac{1}{3}c - \frac{1}{6}$

30. $\frac{3}{2}b - 1 = \frac{1}{8}(12b - 8)$

31. $\frac{2x + 1}{3} + \frac{x - 1}{3} = 5$

32. $\frac{4y - 2}{5} - \frac{y + 4}{5} = -3$

33. $\frac{3w - 2}{6} = 1 - \frac{w - 1}{3}$

34. $\frac{z - 7}{4} = \frac{6z - 1}{8} - 2$

35. $\frac{x + 3}{3} - \frac{x - 1}{2} = 4$

36. $\frac{5y - 1}{2} - \frac{y + 4}{5} = 1$

Objective 2: Solving Linear Equations with Decimals

For Exercises 37–54, solve the equation. **(See Examples 5–6.)**

37. $9.2y - 4.3 = 50.9$

38. $-6.3x + 1.5 = -4.8$

39. $21.1w + 4.6 = 10.9w + 35.2$

40. $0.05z + 0.2 = 0.15z - 10.5$

41. $0.2p - 1.4 = 0.2(p - 7)$

42. $0.5(3q + 87) = 1.5q + 43.5$

43. $0.20x + 53.60 = x$

44. $z + 0.06z = 3816$

45. $0.15(90) + 0.05p = 0.10(90 + p)$

46. $0.25(60) + 0.10x = 0.15(60 + x)$

47. $0.40(y + 10) - 0.60(y + 2) = 2$

48. $0.75(x - 2) + 0.25(x + 4) = 0.5$

49. $0.4x + 0.2 = -3.6 - 0.6x$

50. $0.12x + 3 - 0.8x = 0.22x - 0.6$

51. $0.06(x - 0.5) = 0.06x + 0.01$

52. $0.125x = 0.025(5x + 1)$

53. $-3.5x + 1.3 = -0.3(9x - 5)$

54. $x + 4 = 2(0.4x + 1.3)$

Mixed Exercises

For Exercises 55–64, solve the equation.

55. $0.2x - 1.8 = -3$

56. $9.8h + 2 = 3.8h + 20$

57. $\frac{1}{4}(x + 4) = \frac{1}{5}(2x + 3)$

58. $\frac{2}{3}(y - 1) = \frac{3}{4}(3y - 2)$

59. $0.3(x + 6) - 0.7(x + 2) = 4$

60. $0.05(2t - 1) - 0.03(4t - 1) = 0.2$

61. $\dfrac{2k + 5}{4} = 2 - \dfrac{k + 2}{3}$

62. $\dfrac{3d - 4}{6} + 1 = \dfrac{d + 1}{8}$

63. $\dfrac{1}{8}v + \dfrac{2}{3} = \dfrac{1}{6}v + \dfrac{3}{4}$

64. $\dfrac{2}{5}z - \dfrac{1}{4} = \dfrac{3}{10}z + \dfrac{1}{2}$

Expanding Your Skills

For Exercises 65–68, solve the equation.

65. $\dfrac{1}{2}a + 0.4 = -0.7 - \dfrac{3}{5}a$

66. $\dfrac{3}{4}c - 0.11 = 0.23(c - 5)$

67. $0.8 + \dfrac{7}{10}b = \dfrac{3}{2}b - 0.8$

68. $0.78 - \dfrac{1}{25}h = \dfrac{3}{5}h - 0.5$

Problem Recognition Exercises

Equations and Expressions

For Exercises 1–30, identify each exercise as an expression or an equation. Then simplify the expression or solve the equation.

1. $2b + 23 - 6b - 5$

2. $10p - 9 + 2p - 3 + 8p - 18$

3. $\dfrac{y}{4} = -2$

4. $-\dfrac{x}{2} = 7$

5. $3(4h - 2) - (5h - 8) = 8 - (2h + 3)$

6. $7y - 3(2y + 5) = 7 - (10 - 10y)$

7. $3(8z - 1) + 10 - 6(5 + 3z)$

8. $-5(1 - x) - 3(2x + 3) + 5$

9. $6c + 3(c + 1) = 10$

10. $-9 + 5(2y + 3) = -7$

11. $0.5(2a - 3) - 0.1 = 0.4(6 + 2a)$

12. $0.07(2v - 4) = 0.1(v - 4)$

13. $-\dfrac{5}{9}w + \dfrac{11}{12} = \dfrac{23}{36}$

14. $\dfrac{3}{8}t - \dfrac{5}{8} = \dfrac{1}{2}t + \dfrac{1}{8}$

15. $\dfrac{3}{4}x + \dfrac{1}{2} - \dfrac{1}{8}x + \dfrac{5}{4}$

16. $\dfrac{7}{3}(6 - 12t) + \dfrac{1}{2}(4t + 8)$

17. $2z - 7 = 2(z - 13)$

18. $-6x + 2(x + 1) = -2(2x + 3)$

19. $\dfrac{2x - 1}{4} + \dfrac{3x + 2}{6} = 2$

20. $\dfrac{w - 4}{6} - \dfrac{3w - 1}{2} = -1$

21. $4b - 8 - b = -3b + 2(3b - 4)$

22. $-k - 41 - 2 - k = -2(20 + k) - 3$

23. $\dfrac{4}{3}(6y - 3) = 0$

24. $\dfrac{1}{2}(2c - 4) + 3 = \dfrac{1}{3}(6c + 3)$

25. $3(x + 6) - 7(x + 2) - 4(1 - x)$

26. $-10(2k + 1) - 4(4 - 5k) + 25$

27. $3 - 2[4a - 5(a + 1)]$

28. $-9 - 4[3 - 2(q + 3)]$

29. $4 + 2[8 - (6 + x)] = -2(x - 1) - 4 + x$

30. $-1 - 5[2 + 3(w - 2)] = 5(w + 4)$

Applications of Linear Equations: Introduction to Problem Solving

1. Problem-Solving Strategies

Linear equations can be used to solve many real-world applications. However, with "word problems," students often do not know where to start. To help organize the problem-solving process, we offer the following guidelines:

Problem-Solving Flowchart for Word Problems

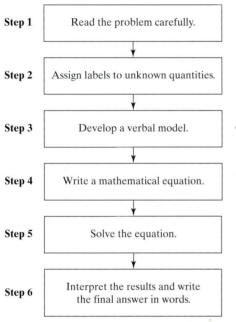

Step 1 Read the problem carefully.
• Familiarize yourself with the problem. Identify the unknown, and if possible, estimate the answer.

Step 2 Assign labels to unknown quantities.
• Identify the unknown quantity or quantities. Let x or another variable represent one of the unknowns. Draw a picture and write down relevant formulas.

Step 3 Develop a verbal model.
• Write an equation in *words*.

Step 4 Write a mathematical equation.
• Replace the verbal model with a mathematical equation using x or another variable.

Step 5 Solve the equation.
• Solve for the variable using the steps for solving linear equations.

Step 6 Interpret the results and write the final answer in words.
• Once you have obtained a numerical value for the variable, recall what it represents in the context of the problem. Can this value be used to determine other unknowns in the problem? Write an answer to the word problem in *words*.

Avoiding Mistakes

Once you have reached a solution to a word problem, verify that it is reasonable in the context of the problem.

2. Translations Involving Linear Equations

We have already practiced translating an English sentence to a mathematical equation. Recall from Section 1.2 that several key words translate to the algebraic operations of addition, subtraction, multiplication, and division.

Example 1   Translating to a Linear Equation

The sum of a number and negative eleven is negative fifteen. Find the number.

Solution:

Step 1: Read the problem.

Let x represent the unknown number.   **Step 2:** Label the unknown.

the sum of   is

$$(\text{a number}) + (-11) = (-15) \qquad \textbf{Step 3:} \text{ Develop a verbal model.}$$

$$x + (-11) = -15 \qquad \textbf{Step 4:} \text{ Write an equation.}$$

$$x + (-11) + 11 = -15 + 11 \qquad \textbf{Step 5:} \text{ Solve the equation.}$$

$$x = -4$$

The number is -4.   **Step 6:** Write the final answer in words.

Skill Practice

1. The sum of a number and negative seven is 12. Find the number.

Answer

1. The number is 19.

Example 2 Translating to a Linear Equation

Forty less than five times a number is fifty-two less than the number. Find the number.

Solution:

Let x represent the unknown number.

$$\underset{5x}{\left(\begin{array}{c}5\text{ times}\\\text{a number}\end{array}\right)} \overset{\text{less}}{\underset{-\;40}{-}} \overset{\text{is}}{\underset{=}{=}} \underset{x}{\left(\begin{array}{c}\text{the}\\\text{number}\end{array}\right)} \overset{\text{less}}{\underset{-\;52}{-}} (52)$$

$$5x - 40 = x - 52$$

$$5x - x - 40 = x - x - 52$$

$$4x - 40 = -52$$

$$4x - 40 + 40 = -52 + 40$$

$$4x = -12$$

$$\frac{4x}{4} = \frac{-12}{4}$$

$$x = -3$$

The number is -3.

Step 1: Read the problem.

Step 2: Label the unknown.

Step 3: Develop a verbal model.

Step 4: Write an equation.

Step 5: Solve the equation.

Step 6: Write the final answer in words.

Example 3 Translating to a Linear Equation

Twice the sum of a number and six is two more than three times the number. Find the number.

Solution:

Let x represent the unknown number.

$$\overset{\text{twice}}{\underset{2}{\downarrow}}\;\overset{\text{the sum}}{\underset{(x+6)}{\downarrow}}\;\overset{\text{is}}{\underset{=}{\downarrow}}\;\overset{\text{2 more than}}{\underset{3x+2}{\downarrow}}$$

$$\underset{\underset{\text{a number}}{\text{three times}}}{\uparrow}$$

Step 1: Read the problem.

Step 2: Label the unknown.

Step 3: Develop a verbal model.

Step 4: Write an equation.

Answers

2. The number is -8.
3. The number is -10.

$2(x + 6) = 3x + 2$	**Step 5:** Solve the equation.
$2x + 12 = 3x + 2$	
$2x - 2x + 12 = 3x - 2x + 2$	
$12 = x + 2$	
$12 - 2 = x + 2 - 2$	
$10 = x$	

The number is 10.   **Step 6:**   Write the final answer in
                                                         words.

> **Avoiding Mistakes**
>
> It is important to enclose "the sum of a number and six" within parentheses so that the entire quantity is multiplied by 2. Forgetting the parentheses would imply that only the x-term is multiplied by 2.
>
> Correct: $2(x + 6)$

3. Consecutive Integer Problems

The word *consecutive* means "following one after the other in order without gaps." The numbers 6, 7, and 8 are examples of three **consecutive integers**. The numbers $-4, -2, 0,$ and 2 are examples of **consecutive even integers**. The numbers 23, 25, and 27 are examples of **consecutive odd integers**.

Notice that any two consecutive integers differ by 1. Therefore, if x represents an integer, then $(x + 1)$ represents the next larger consecutive integer (Figure 2-4).

Consecutive integers differ by 1 unit.

Figure 2-4

Any two consecutive even integers differ by 2. Therefore, if x represents an even integer, then $(x + 2)$ represents the next consecutive larger even integer (Figure 2-5).

Consecutive even integers differ by 2 units.

Figure 2-5

Likewise, any two consecutive odd integers differ by 2. If x represents an odd integer, then $(x + 2)$ is the next larger odd integer (Figure 2-6).

Consecutive odd integers differ by 2 units.

Figure 2-6

Skill Practice

4. The sum of two consecutive even integers is 66. Find the integers.

Example 4 Solving an Application Involving Consecutive Integers

The sum of two consecutive odd integers is -188. Find the integers.

Solution:

In this example we have two unknown integers. We can let x represent either of the unknowns.

	Step 1: Read the problem.

Suppose x represents the first odd integer. **Step 2:** Label the variables.

Then $(x + 2)$ represents the second odd integer.

$$\left(\begin{array}{c}\text{First}\\\text{integer}\end{array}\right) + \left(\begin{array}{c}\text{second}\\\text{integer}\end{array}\right) = (\text{total})$$

Step 3: Write an equation in words.

$$x + (x + 2) = -188$$

Step 4: Write a mathematical equation.

$$x + (x + 2) = -188$$

$$2x + 2 = -188$$

Step 5: Solve for x.

$$2x + 2 - 2 = -188 - 2$$

$$2x = -190$$

$$\frac{2x}{2} = \frac{-190}{2}$$

$$x = -95$$

The first integer is $x = -95$.

Step 6: Interpret the results and write the answer in words.

The second integer is $x + 2 = -95 + 2 = -93$.

The two integers are -95 and -93.

TIP: With word problems, it is advisable to check that the answer is reasonable.

The numbers -95 and -93 are consecutive odd integers. Furthermore, their sum is -188 as desired.

Skill Practice

5. Five times the smallest of three consecutive integers is 17 less than twice the sum of the integers. Find the integers.

Example 5 Solving an Application Involving Consecutive Integers

Ten times the smallest of three consecutive integers is twenty-two more than three times the sum of the integers. Find the integers.

Solution:

	Step 1: Read the problem.

Let x represent the first integer. **Step 2:** Label the variables.

$x + 1$ represents the second consecutive integer.

$x + 2$ represents the third consecutive integer.

Answers

4. The integers are 32 and 34.
5. The integers are 11, 12, and 13.

$$\begin{pmatrix} 10 \text{ times} \\ \text{the first} \\ \text{integer} \end{pmatrix} = \begin{pmatrix} 3 \text{ times} \\ \text{the sum of} \\ \text{the integers} \end{pmatrix} + 22$$

Step 3: Write an equation in words.

$\underset{\substack{\text{10 times}\\\text{the first}\\\text{integer}}}{\downarrow} \quad \underset{\text{is}}{\downarrow} \quad \underset{\substack{\text{3 times}}}{\downarrow} \qquad\qquad \underset{\substack{\text{22 more}\\\text{than}}}{\downarrow}$

$10x = \quad 3\underbrace{[(x) + (x + 1) + (x + 2)]}_{\text{the sum of the integers}} + 22$

Step 4: Write a mathematical equation.

$10x = 3(x + x + 1 + x + 2) + 22$

Step 5: Solve the equation.

$10x = 3(3x + 3) + 22$

Clear parentheses.

$10x = 9x + 9 + 22$

Combine *like* terms.

$10x = 9x + 31$

$10x - 9x = 9x - 9x + 31$

$x = 31$

Isolate the *x*-terms on one side.

The first integer is $x = 31$.

Step 6: Interpret the results and write the answer in words.

The second integer is $x + 1 = 31 + 1 = 32$.

The third integer is $x + 2 = 31 + 2 = 33$.

The three integers are 31, 32, and 33.

4. Applications of Linear Equations

Example 6   **Using a Linear Equation in an Application**

A carpenter cuts a 6-ft board in two pieces. One piece must be three times as long as the other. Find the length of each piece.

Solution:

In this problem, one piece must be three times as long as the other. Thus, if x represents the length of one piece, then $3x$ can represent the length of the other.

Step 1: Read the problem completely.

x represents the length of the smaller piece.
$3x$ represents the length of the longer piece.

Step 2: Label the unknowns. Draw a figure.

Skill Practice

6. A plumber cuts a 96-in. piece of pipe into two pieces. One piece is five times longer than the other piece. How long is each piece?

Answer

6. One piece is 80 in. and the other is 16 in.

$$\begin{pmatrix}\text{Length of}\\\text{one piece}\end{pmatrix} + \begin{pmatrix}\text{length of}\\\text{other piece}\end{pmatrix} = \begin{pmatrix}\text{total length}\\\text{of the board}\end{pmatrix}$$

Step 3: Set up a verbal equation.

$$x \quad + \quad 3x \quad = \quad 6$$

Step 4: Write an equation.

$$4x = 6$$

Step 5: Solve the equation.

$$\frac{4x}{4} = \frac{6}{4}$$

$$x = 1.5$$

The smaller piece is $x = 1.5$ ft.

Step 6: Interpret the results.

The longer piece is $3x$ or $3(1.5 \text{ ft}) = 4.5$ ft.

> **TIP:** The variable can represent either unknown. In Example 6, if we let x represent the length of the longer piece of pipe, then $\frac{1}{3}x$ would represent the length of the smaller piece. The equation would become $x + \frac{1}{3}x = 6$. Try solving this equation and interpreting the result.

Skill Practice

7. There are 40 students in an algebra class. There are 4 more women than men. How many women and how many men are in the class?

Example 7   Using a Linear Equation in an Application

The hit movies *Spider-Man 3* and *Pirates of the Caribbean: At World's End* together brought in $265.8 million during their opening weekends. *Spider-Man 3* earned $78.3 million less than twice what *Pirates of the Caribbean* earned. How much revenue did each movie bring in during its opening weekend?

Solution:

In this example, we have two unknowns. The variable, x, can represent *either* quantity. However, the revenue from *Spider-Man 3* is given in terms of the revenue for *Pirates of the Caribbean: At World's End*.

Step 1: Read the problem.

Let x represent the revenue for *Pirates of the Caribbean*.

Step 2: Label the variables.

Then $2x - 78.3$ represents the revenue for *Spider-Man 3*.

Answer

7. There are 22 women and 18 men.

$$\begin{pmatrix}\text{Revenue from}\\\textit{Pirates}\end{pmatrix} + \begin{pmatrix}\text{revenue from}\\\textit{Spider-Man}\end{pmatrix} = \begin{pmatrix}\text{total}\\\text{revenue}\end{pmatrix}$$

Step 3: Write an equation in words.

$$x + (2x - 78.3) = 265.8$$

Step 4: Write an equation.

$$3x - 78.3 = 265.8$$

Step 5: Solve the equation.

$$3x - 78.3 + 78.3 = 265.8 + 78.3$$

$$3x = 344.1$$

$$\frac{3x}{3} = \frac{344.1}{3}$$

$$x = 114.7$$

- Revenue from *Pirates of the Caribbean:* $x = 114.7$
- Revenue from *Spider-Man 3:* $2x - 78.3 = 2(114.7) - 78.3 = 151.1$

The revenue from *Pirates of the Caribbean: At World's End* was \$114.7 million for its opening weekend. The revenue for *Spider-Man 3* was \$151.1 million.

Section 2.4 Practice Exercises

Study Skills Exercises

1. After doing a section of homework, check the odd-numbered answers in the back of the text. Choose a method to identify the exercises that gave you trouble (i.e., circle the number or put a star by the number). List some reasons why it is important to label these problems.

2. Define the key terms:

   a. consecutive integers **b. consecutive even integers** **c. consecutive odd integers**

Objective 2: Translations Involving Linear Equations

For Exercises 3–8, write an expression representing the unknown quantity.

3. In a math class, the number of students who received an "A" in the class was 5 more than the number of students who received a "B." If x represents the number of "B" students, write an expression for the number of "A" students.

4. At a recent motorcycle rally, the number of men exceeded the number of women by 216. If x represents the number of women, write an expression for the number of men.

5. Anna is three times as old as Jake. If x represents Jake's age, write an expression for Anna's age.

6. Rebecca downloaded twice as many songs as Nigel. If x represents the number of songs downloaded by Nigel, write an expression for the number downloaded by Rebecca.

7. Sidney made $20 more than three times Casey's weekly salary. If x represents Casey's weekly salary, write an expression for Sidney's weekly salary.

8. David scored 26 points less than twice the number of points Rich scored in a video game. If x represents the number of points scored by Rich, write an expression representing the number of points scored by David.

For Exercises 9–18, use the problem-solving flowchart on page 143. (**See Examples 1–3.**)

9. Six less than a number is –10. Find the number.

10. Fifteen less than a number is 41. Find the number.

11. Twice the sum of a number and seven is eight. Find the number.

12. Twice the sum of a number and negative two is sixteen. Find the number.

13. A number added to five is the same as twice the number. Find the number.

14. Three times a number is the same as the difference of twice the number and seven. Find the number.

15. The sum of six times a number and ten is equal to the difference of the number and fifteen. Find the number.

16. The difference of fourteen and three times a number is the same as the sum of the number and negative ten. Find the number.

17. If the difference of a number and four is tripled, the result is six more than the number. Find the number.

18. Twice the sum of a number and eleven is twenty-two less than three times the number. Find the number.

Objective 3: Consecutive Integer Problems

19. **a.** If x represents the smallest of three consecutive integers, write an expression to represent each of the next two consecutive integers.

    b. If x represents the largest of three consecutive integers, write an expression to represent each of the previous two consecutive integers.

20. **a.** If x represents the smallest of three consecutive odd integers, write an expression to represent each of the next two consecutive odd integers.

    b. If x represents the largest of three consecutive odd integers, write an expression to represent each of the previous two consecutive odd integers.

For Exercises 21–30, use the problem-solving flowchart from page 143. (**See Examples 4–5.**)

21. The sum of two consecutive integers is −67. Find the integers.

22. The sum of two consecutive odd integers is 52. Find the integers.

23. The sum of two consecutive odd integers is 28. Find the integers.

24. The sum of three consecutive even integers is 66. Find the integers.

25. The perimeter of a pentagon (a five-sided polygon) is 80 in. The five sides are represented by consecutive integers. Find the measures of the sides.

26. The perimeter of a triangle is 96 in. The lengths of the sides are represented by consecutive integers. Find the measures of the sides.

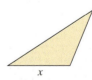

27. The sum of three consecutive even integers is 48 more than twice the smallest of the three integers. Find the integers.

28. The sum of three consecutive odd integers is 89 more than twice the largest integer. Find the integers.

29. Eight times the sum of three consecutive odd integers is 210 more than 10 times the middle integer. Find the integers.

30. Five times the sum of three consecutive even integers is 140 more than ten times the smallest. Find the integers.

Objective 4: Applications of Linear Equations

For Exercises 31–42, use the problem-solving flowchart (page 143) to solve the problems.

31. A board is 86 cm in length and must be cut so that one piece is 20 cm longer than the other piece. Find the length of each piece. (**See Example 6.**)

x $x + 20$

32. A rope is 54 in. in length and must be cut into two pieces. If one piece must be twice as long as the other, find the length of each piece.

33. Karen's age is 12 years more than Clarann's age. The sum of their ages is 58. Find their ages.

34. Maria's age is 15 years less than Orlando's age. The sum of their ages is 29. Find their ages.

35. For a recent year, 31 more Democrats than Republicans were in the U.S. House of Representatives. If the total number of representatives in the House from these two parties was 433, find the number of representatives from each party.

36. For a recent year, the number of men in the U.S. Senate totaled 4 more than 5 times the number of women. Find the number of men and the number of women in the Senate given that the Senate has 100 members.

37. Approximately 5.816 million people watch *The Oprah Winfrey Show*. This is 1.118 million more than watch *The Dr. Phil Show*. How many watch *The Dr. Phil Show*? (*Source: Neilson Media Research*) (**See Example 7.**)

38. Two of the largest Internet retailers are e-Bay and Amazon.com. Recently, the estimated U.S. sales of e-Bay were $0.1 billion less than twice the sales of Amazon.com. Given the total sales of $5.6 billion, determine the sales of e-Bay and Amazon.com.

39. The longest river in Africa is the Nile. It is 2455 km longer than the Congo River, also in Africa. The sum of the lengths of these rivers is 11,195 km. What is the length of each river?

40. The average depth of the Gulf of Mexico is three times the depth of the Red Sea. The difference between the average depths is 1078 m. What is the average depth of the Gulf of Mexico and the average depth of the Red Sea?

41. Asia and Africa are the two largest continents in the world. The land area of Asia is approximately 14,514,000 km² larger than the land area of Africa. Together their total area is 74,644,000 km². Find the land area of Asia and the land area of Africa.

42. Mt. Everest, the highest mountain in the world, is 2654 m higher than Mt. McKinley, the highest mountain in the United States. If the sum of their heights is 15,042 m, find the height of each mountain.

Mixed Exercises

43. A group of hikers walked from Hawk Mt. Shelter to Blood Mt. Shelter along the Appalachian Trail, a total distance of 20.5 mi. It took 2 days for the walk. The second day the hikers walked 4.1 mi less than they did on the first day. How far did they walk each day?

44. $120 is to be divided among three restaurant servers. Angie made $10 more than Marie. Gwen, who went home sick, made $25 less than Marie. How much money should each server get?

45. A 4-ft piece of PVC pipe is cut into three pieces. The longest piece is 12 in. longer than twice the smallest piece. The middle piece is 6 in. less than the longest piece. How long is each piece?

46. A 6-ft piece of copper wire must be cut into three pieces. The middle piece is 16 in. longer than the shortest piece. The longest piece is twice as long as the middle piece. How long is each piece?

47. Three consecutive integers are such that three times the largest exceeds the sum of the two smaller integers by 47. Find the integers.

48. Four times the smallest of three consecutive odd integers is 236 more than the sum of the other two integers. Find the integers.

49. In a recent year, the estimated earnings for Jennifer Lopez was $2.5 million more than half of the earnings for the band U2. If the total earnings were $106 million, what were the earnings for Jennifer Lopez and U2? (*Source: Forbes*)

50. Two of the longest-running TV series are *Gunsmoke* and *The Simpsons*. *Gunsmoke* ran 97 fewer episodes than twice the number of *The Simpsons*. If the total number of episodes is 998, how many of each show was produced?

51. Five times the difference of a number and three is four less than four times the number. Find the number.

52. Three times the difference of a number and seven is one less than twice the number. Find the number.

53. The sum of the page numbers on two facing pages in a book is 941. What are the page numbers?

54. Three raffle tickets are represented by three consecutive integers. If the sum of the three integers is 2,666,031, find the numbers.

55. If three is added to five times a number, the result is forty-three more than the number. Find the number.

56. If seven is added to three times a number, the result is thirty-one more than the number.

57. The deepest point in the Pacific Ocean is 676 m more than twice the deepest point in the Arctic Ocean. If the deepest point in the Pacific is 10,920 m, how many meters is the deepest point in the Arctic Ocean?

58. The area of Greenland is 201,900 km^2 less than three times the area of New Guinea. What is the area of New Guinea if the area of Greenland is 2,175,600 km^2?

59. The sum of twice a number and $\frac{3}{4}$ is the same as the difference of four times the number and $\frac{1}{8}$. Find the number.

60. The difference of a number and $-\frac{11}{12}$ is the same as the difference of three times the number and $\frac{1}{6}$. Find the number.

61. The product of a number and 3.86 is equal to 7.15 more than the number. Find the number.

62. The product of a number and 4.6 is 33.12 less than the number. Find the number.

Applications Involving Percents

1. Solving Basic Percent Equations

Recall from Section R.3 that the word *percent* means "per hundred." For example:

Percent	Interpretation
63% of homes have a computer	63 out of 100 homes have a computer.
5% sales tax	5¢ in tax is charged for every 100¢ in merchandise.
15% commission	$15 is earned in commission for every $100 sold.

Percents come up in a variety of applications in day-to-day life. Many such applications follow the basic percent equation:

$$\text{Amount} = (\text{percent})(\text{base}) \qquad \text{Basic percent equation}$$

In Example 1, we apply the basic percent equation to compute **sales tax**.

Example 1 Computing Sales Tax

A new digital camera costs $429.95.

a. Compute the sales tax if the tax rate is 4%.

b. Determine the total cost, including tax.

Solution:

	Step 1: Read the problem.
a. Let x represent the amount of tax.	**Step 2:** Label the variable.
Apply the percent equation to compute sales tax.	**Step 3:** Write a verbal equation.

$$\text{Amount} = (\text{percent}) \cdot (\text{base})$$
$$\downarrow \qquad\qquad \downarrow \qquad\qquad \downarrow$$
$$\text{Sales tax} = (\text{tax rate})(\text{price of merchandise})$$

$x = (0.04)(\$429.95)$	**Step 4:** Write a mathematical equation.
$x = \$17.198$	**Step 5:** Solve the equation.
$x = \$17.20$	Round to the nearest cent.
The tax on the merchandise is $17.20.	**Step 6:** Interpret the results.

b. The total cost is found by:

total cost = cost of merchandise + amount of tax

Therefore the total cost is $429.95 + $17.20 = $447.15.

In Example 2, we solve a problem in which the percent is unknown.

Example 2 Finding an Unknown Percent

A group of 240 college men were asked what intramural sport they most enjoyed playing. The results are in the graph. What percent of the men surveyed preferred tennis?

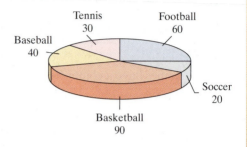

Solution:

	Step 1:	Read the problem.
Let x represent the unknown percent.	**Step 2:**	Label the variable.

The problem can be rephrased as:

$$30 \text{ is what percent of } 240?$$

	Step 3:	Write an equation in words.
$30 = x \cdot 240$	**Step 4:**	Write a mathematical equation.
$30 = 240x$	**Step 5:**	Solve the equation.
$\dfrac{30}{240} = \dfrac{240x}{240}$		Divide both sides by 240.
$0.125 = x$		
$0.125 \times 100\% = 12.5\%$	**Step 6:**	Interpret the results. Change the value of x to a percent by multiplying by 100%.

In this survey, 12.5% of men prefer tennis.

Example 3 Solving a Percent Equation with an Unknown Base

Andrea spends 20% of her monthly paycheck on rent each month. If her rent payment is $750, what is her monthly paycheck?

Solution:

	Step 1:	Read the problem.
Let x represent the amount of Andrea's monthly paycheck.	**Step 2:**	Label the variables.

The problem can be rephrased as:

$$\$750 \text{ is } 20\% \text{ of what number?}$$

	Step 3:	Write an equation in words.
$750 = 0.20 \cdot x$	**Step 4:**	Write a mathematical equation.

$$750 = 0.20x$$

Step 5: Solve the equation.

$$\frac{750}{0.20} = \frac{0.20x}{0.20}$$

Divide both sides by 0.20.

$$3750 = x$$

Andrea's monthly paycheck is $3750. **Step 6:** Interpret the results.

2. Applications Involving Simple Interest

One important application of percents is in computing simple interest on a loan or on an investment.

Simple interest is interest that is earned on principal (the original amount of money invested in an account). The following formula is used to compute simple interest:

$$\begin{pmatrix} \text{Simple} \\ \text{interest} \end{pmatrix} = \begin{pmatrix} \text{principal} \\ \text{invested} \end{pmatrix} \begin{pmatrix} \text{annual} \\ \text{interest rate} \end{pmatrix} \begin{pmatrix} \text{time} \\ \text{in years} \end{pmatrix}$$

This formula is often written symbolically as $I = Prt$.

For example, to find the simple interest earned on $2000 invested at 7.5% interest for 3 years, we have

$$I = Prt$$
$$\text{Interest} = (\$2000)(0.075)(3)$$
$$= \$450$$

Example 4 Applying Simple Interest

Jorge wants to save money for his daughter's college education. If Jorge needs to have $4340 at the end of 4 years, how much money would he need to invest at a 6% simple interest rate?

Solution:

Step 1: Read the problem.

Let P represent the original amount invested.   **Step 2:** Label the variables.

$$\begin{pmatrix} \text{Original} \\ \text{principal} \end{pmatrix} + (\text{interest}) = (\text{total})$$

Step 3: Write an equation in words.

$$\quad (P) \quad + \quad (Prt) \quad = (\text{total})$$

$$\quad P \quad + \; P(0.06)(4) = 4340$$

Step 4: Write a mathematical equation.

$$P + 0.24P = 4340$$

$$1.24P = 4340$$

Step 5: Solve the equation.

$$\frac{1.24P}{1.24} = \frac{4340}{1.24}$$

$$P = 3500$$

The original investment should be $3500. **Step 6:** Interpret the results and write the answer in words.

5. Cassandra invested some money in her bank account, and after 10 years at 4% simple interest, it has grown to $7700. What was the initial amount invested?

Avoiding Mistakes

The interest is computed on the original principal, P, not on the total amount $4340. That is, the interest is $P(0.06)(4)$, not $(\$4340)(0.06)(4)$.

Answer

5. The initial investment was $5500.

3. Applications Involving Discount and Markup

Applications involving percent increase and percent decrease are abundant in many real-world settings. Sales tax for example is essentially a markup by a state or local government. It is important to understand that percent increase or decrease is always computed on the original amount given.

In Example 5, we illustrate an example of percent decrease in an application where merchandise is discounted.

Skill Practice

6. An iPod is on sale for $151.20. This is after a 20% discount. What was the original cost of the iPod?

Example 5 Applying Percents to a Discount Problem

After a 38% discount, a used treadmill costs $868 on e-Bay. What was the original cost of the treadmill?

Solution:

	Step 1: Read the problem.
Let x be the original cost of the treadmill.	**Step 2:** Label the variables.
$\left(\begin{array}{c}\text{Original} \\ \text{cost}\end{array}\right) - (\text{discount}) = \left(\begin{array}{c}\text{sale} \\ \text{price}\end{array}\right)$	**Step 3:** Write an equation in words.
$x \qquad - 0.38(x) \qquad = 868$	**Step 4:** Write a mathematical equation. The discount is a percent of the *original* amount.
$x - 0.38x = 868$	**Step 5:** Solve the equation.
$0.62x = 868$	Combine *like* terms.
$\dfrac{0.62x}{0.62} = \dfrac{868}{0.62}$	Divide by 0.62.
$x = 1400$	
The original cost of the treadmill was $1400.	**Step 6:** Interpret the result.

Answer

6. The iPod originally cost $189.

Section 2.5 Practice Exercises

Study Skills Exercise

1. Define the key terms:

   a. sales tax   **b. simple interest**

Review Exercises

2. List the six steps to solve an application.

For Exercises 3–4, use the steps for problem solving to solve these applications.

3. Find two consecutive integers such that three times the larger is the same as 45 more than the smaller.

4. The height of the Great Pyramid of Giza is 17 m more than twice the height of the pyramid found in Saqqara. If the difference in their heights is 77 m, find the height of each pyramid.

Objective 1: Solving Basic Percent Equations

For Exercises 5–16, find the missing values.

5. 45 is what percent of 360?

6. 338 is what percent of 520?

7. 544 is what percent of 640?

8. 576 is what percent of 800?

9. What is 0.5% of 150?

10. What is 9.5% of 616?

11. What is 142% of 740?

12. What is 156% of 280?

13. 177 is 20% of what number?

14. 126 is 15% of what number?

15. 275 is 12.5% of what number?

16. 594 is 45% of what number?

17. A Craftsman drill is on sale for $99.99. If the sales tax rate is 7%, how much will Molly have to pay for the drill?
(See Example 1.)

18. Patrick purchased four new tires that were regularly priced at $94.99 each, but are on sale for $20 off per tire. If the sales tax rate is 6%, how much will be charged to Patrick's VISA card?

For Exercises 19–20, use the graph showing the distribution for leading forms of cancer in men. (*Source:* Centers for Disease Control)

19. If there are 700,000 cases of cancer in men in the United States, approximately how many are prostate cancer?

20. Approximately how many cases of lung cancer would be expected in 700,000 cancer cases among men in the United States?

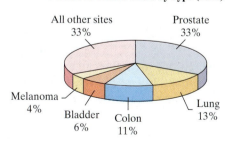

Percent of Cancer Cases by Type (Men)

21. There were 14,000 cases of cancer of the pancreas diagnosed out of 700,000 cancer cases. What percent is this? **(See Example 2.)**

22. There were 21,000 cases of leukemia diagnosed out of 700,000 cancer cases. What percent is this?

23. Javon is in a 28% tax bracket for his federal income tax. If the amount of money that he paid for federal income tax was $23,520, what was his taxable income? **(See Example 3.)**

24. In a recent survey of college-educated adults, 155 indicated that they regularly work more than 50 hr a week. If this represents 31% of those surveyed, how many people were in the survey?

Objective 2: Applications Involving Simple Interest

For Exercises 25–32, solve these equations involving simple interest.

25. How much interest will Pam earn in 4 years if she invests $3000 in an account that pays 3.5% simple interest?

26. How much interest will Roxanne have to pay if she borrows $2000 for 2 yr at a simple interest rate of 4%?

27. Bob borrowed some money for 1 yr at 5% simple interest. If he had to pay back a total of $1260, how much did he originally borrow? **(See Example 4.)**

28. Mike borrowed some money for 2 yr at 6% simple interest. If he had to pay back a total of $3640, how much did he originally borrow?

29. If $1500 grows to $1950 after 5 yr, find the simple interest rate.

30. If $9000 grows to $10,440 in 2 yr, find the simple interest rate.

31. Perry is planning a vacation to Europe in 2 yr. How much should he invest in a certificate of deposit that pays 3% simple interest to get the $3500 that he needs for the trip? Round to the nearest dollar.

32. Sherica invested in a mutual fund and at the end of 20 yr she has $14,300 in her account. If the mutual fund returned an average yield of 8%, how much did she originally invest?

Objective 3: Applications Involving Discount and Markup

33. A Pioneer car CD/MP3 player costs $170. Circuit City has it on sale for 12% off with free installation.

    a. What is the discount on the CD/MP3 player?

    b. What is the sale price?

34. A laptop computer, originally selling for $899.00, is on sale for 10% off.

    a. What is the discount on the laptop?

    b. What is the sale price?

35. A Sony digital camera is on sale for $400.00. This price is 15% off the original price. What was the original price? Round to the nearest cent. **(See Example 5.)**

36. The *Star Wars: Episode III* DVD is on sale for $18. If this represents an 18% discount rate, what was the original price of the DVD?

37. The original price of an Audio Jukebox was $250. It is on sale for $220. What percent discount does this represent?

38. During the holiday season, the Xbox 360 sold for $425.00 in stores. This product was in such demand that it sold for $800 online. What percent markup does this represent? (Round to the nearest whole percent.)

39. In one area, the cable company marked up the monthly cost by 6%. The new cost is $63.60 per month. What was the cost before the increase?

40. A doctor ordered a dosage of medicine for a patient. After 2 days, she increased the dosage by 20% and the new dosage came to 18 cc. What was the original dosage?

Mixed Exercises

41. Sun Lei bought a laptop computer for $1800. The total cost, including tax, came to $1890. What is the tax rate?

42. Jamie purchased a compact disk and paid $18.26. If the disk price is $16.99, what is the sales tax rate (round to the nearest tenth of a percent)?

43. For a recent year, admission to Walt Disney World cost $74.37, including taxes of 11%. What was the original ticket price before taxes?

44. A hotel room rented for 5 nights costs $706.25 including 13% in taxes. Find the original price of the room (before tax) for the 5 nights. Then find the price per night.

45. Deon purchased a house and sold it for a 24% profit. If he sold the house for $260,400, what was the original purchase price?

46. To meet the rising cost of energy, the yearly membership at a YMCA had to be increased by 12.5% from the past year. The yearly membership fee is currently $450. What was the cost of membership last year?

47. Alina earns $1600 per month plus a 12% commission on pharmaceutical sales. If she sold $25,000 in pharmaceuticals one month, what was her salary that month?

48. Dan sold a beachfront home for $650,000. If his commission rate is 4%, what did he earn in commission?

49. Diane sells women's sportswear at a department store. She earns a regular salary and, as a bonus, she receives a commission of 4% on all sales over $200. If Diane earned an extra $25.80 last week in commission, how much merchandise did she sell over $200?

50. For selling software, Tom received a bonus commission based on sales over $500. If he received $180 in commission for selling a total of $2300 worth of software, what is his commission rate?

Formulas and Applications of Geometry

Section 2.6

1. Formulas and Literal Equations

Objectives

Literal equations are equations that contain several variables. A formula is a literal equation with a specific application. For example, the perimeter of a triangle (distance around the triangle) can be found by the formula $P = a + b + c$, where a, b, and c are the lengths of the sides (Figure 2-7).

1. **Formulas and Literal Equations**
2. **Geometry Applications**

$$P = a + b + c$$
$$= 5 \text{ ft} + 7 \text{ ft} + 8 \text{ ft}$$
$$= 20 \text{ ft}$$

Figure 2-7

In this section, we will learn how to rewrite formulas to solve for a different variable within the formula. Suppose, for example, that the perimeter of a triangle is known and two of the sides are known (say, sides a and b). Then the third side, c, can be found by subtracting the lengths of the known sides from the perimeter (Figure 2-8).

If the perimeter is 20 ft, then
$$c = P - a - b$$
$$= 20 \text{ ft} - 5 \text{ ft} - 7 \text{ ft}$$
$$= 8 \text{ ft}$$

Figure 2-8

To solve a formula for a different variable, we use the same properties of equality outlined in the earlier sections of this chapter. For example, consider the two equations $2x + 3 = 11$ and $wx + y = z$. Suppose we want to solve for x in each case:

$2x + 3 = 11$			$wx + y = z$	
$2x + 3 - 3 = 11 - 3$	Subtract 3.		$wx + y - y = z - y$	Subtract y.
$2x = 8$			$wx = z - y$	
$\dfrac{2x}{2} = \dfrac{8}{2}$	Divide by 2.		$\dfrac{wx}{w} = \dfrac{z - y}{w}$	Divide by w.
$x = 4$			$x = \dfrac{z - y}{w}$	

The equation on the left has only one variable and we are able to simplify the equation to find a numerical value for x. The equation on the right has multiple variables. Because we do not know the values of w, y, and z, we are not able to simplify further. The value of x is left as a formula in terms of w, y, and z.

Example 1   Solving for an Indicated Variable

Solve for the indicated variables.

a. $d = rt$ for t **b.** $5x + 2y = 12$ for y

Solution:

a. $d = rt$ for t The goal is to isolate the variable t.

$\dfrac{d}{r} = \dfrac{rt}{r}$ Because the relationship between r and t is multiplication, we reverse the process by dividing both sides by r.

$\dfrac{d}{r} = t$, or equivalently $t = \dfrac{d}{r}$

b. $5x + 2y = 12$ for y The goal is to solve for y.

$5x - 5x + 2y = 12 - 5x$ Subtract $5x$ from both sides to isolate the y-term.

$2y = -5x + 12$ $-5x + 12$ is the same as $12 - 5x$.

$\dfrac{2y}{2} = \dfrac{-5x + 12}{2}$ Divide both sides by 2 to isolate y.

$y = \dfrac{-5x + 12}{2}$

TIP: In the expression $\dfrac{-5x + 12}{2}$ do not try to divide the 2 into the 12. The divisor of 2 is dividing the entire quantity, $-5x + 12$ (not just the 12).

We may, however, apply the divisor to each term individually in the numerator. That is, $\dfrac{-5x + 12}{2}$ can be written in several different forms. Each is correct.

$$y = \dfrac{-5x + 12}{2} \quad \text{or} \quad y = \dfrac{-5x}{2} + \dfrac{12}{2} \Rightarrow y = -\dfrac{5x}{2} + 6$$

Example 2	Solving Formulas for an Indicated Variable

The formula $C = \frac{5}{9}(F - 32)$ is used to find the temperature, C, in degrees Celsius for a given temperature expressed in degrees Fahrenheit, F. Solve the formula $C = \frac{5}{9}(F - 32)$ for F.

Solution:

$$C = \frac{5}{9}(F - 32)$$

$$C = \frac{5}{9}F - \frac{5}{9} \cdot 32 \qquad \text{Clear parentheses.}$$

$$C = \frac{5}{9}F - \frac{160}{9} \qquad \text{Multiply: } \frac{5}{9} \cdot \frac{32}{1} = \frac{160}{9}.$$

$$9(C) = 9\left(\frac{5}{9}F - \frac{160}{9}\right) \qquad \text{Multiply by the LCD to clear fractions.}$$

$$9C = \frac{9}{1} \cdot \frac{5}{9}F - \frac{9}{1} \cdot \frac{160}{9} \qquad \text{Apply the distributive property.}$$

$$9C = 5F - 160 \qquad \text{Simplify.}$$

$$9C + 160 = 5F - 160 + 160 \qquad \text{Add } 160 \text{ to both sides.}$$

$$9C + 160 = 5F$$

$$\frac{9C + 160}{5} = \frac{5F}{5} \qquad \text{Divide both sides by } 5.$$

$$\frac{9C + 160}{5} = F$$

The answer may be written in several forms:

$$F = \frac{9C + 160}{5} \qquad \text{or} \qquad F = \frac{9C}{5} + \frac{160}{5} \quad \Rightarrow \quad F = \frac{9}{5}C + 32$$

Skill Practice

3. Solve for the indicated variable.

$$y = \frac{1}{3}(x - 7) \text{ for } x.$$

2. Geometry Applications

In Section R.4, we presented numerous facts and formulas related to geometry. Sometimes these are needed to solve applications in geometry.

Example 3	Solving a Geometry Application Involving Perimeter

The length of a rectangular lot is 1 m less than twice the width. If the perimeter is 190 m, find the length and width.

Solution:

	Step 1: Read the problem.

Let x represent the width of the rectangle.   **Step 2:** Label the variables.

Then $2x - 1$ represents the length.

x

$2x - 1$

Skill Practice

4. The length of a rectangle is 10′ less than twice the width. If the perimeter is 178′ find the length and width.

$$P = 2l + 2w$$ **Step 3:** Perimeter formula

$$190 = 2(2x - 1) + 2(x)$$ **Step 4:** Write an equation in terms of x.

$$190 = 4x - 2 + 2x$$ **Step 5:** Solve for x.

$$190 = 6x - 2$$

$$192 = 6x$$

$$\frac{192}{6} = \frac{6x}{6}$$

$$32 = x$$

The width is $x = 32$.

The length is $2x - 1 = 2(32) - 1 = 63$. **Step 6:** Interpret the results and write the answer in words.

The width of the rectangular lot is 32 m and the length is 63 m.

Recall some facts about angles.

- Two angles are complementary if the sum of their measures is 90°.
- Two angles are supplementary if the sum of their measures is 180°.
- The sum of the measures of the angles within a triangle is 180°.

Skill Practice

5. Two complementary angles are constructed so that one measures 1° less than six times the other. Find the measures of the angles.

Example 4 Solving a Geometry Application Involving Complementary Angles

Two complementary angles are drawn such that one angle is 4° more than seven times the other angle. Find the measure of each angle.

Solution:

Step 1: Read the problem.

Let x represent the measure of one angle. **Step 2:** Label the variables.

Then $7x + 4$ represents the measure of the other angle.

The angles are complementary, so their sum must be 90°.

$$\left(\begin{array}{c}\text{Measure of}\\\text{first angle}\end{array}\right) + \left(\begin{array}{c}\text{measure of}\\\text{second angle}\end{array}\right) = 90°$$ **Step 3:** Create a verbal equation.

$$x + 7x + 4 = 90$$ **Step 4:** Write a mathematical equation.

$$8x + 4 = 90$$ **Step 5:** Solve for x.

$$8x = 86$$

$$\frac{8x}{8} = \frac{86}{8}$$

$$x = 10.75$$

Answer

5. 13° and 77°

One angle is $x = 10.75$.

The other angle is $7x + 4 = 7(10.75) + 4 = 79.25$.

The angles are $10.75°$ and $79.25°$.

Step 6: Interpret the results and write the answer in words.

Example 5 Solving a Geometry Application

One angle in a triangle is twice as large as the smallest angle. The third angle is 10° more than seven times the smallest angle. Find the measure of each angle.

Solution:

Step 1: Read the problem.

Let x represent the measure of the smallest angle.

Step 2: Label the variables.

Then $2x$ and $7x + 10$ represent the measures of the other two angles.

The sum of the angles must be 180°.

$$\left(\begin{array}{c}\text{Measure of}\\\text{first angle}\end{array}\right) + \left(\begin{array}{c}\text{measure of}\\\text{second angle}\end{array}\right) + \left(\begin{array}{c}\text{measure of}\\\text{third angle}\end{array}\right) = 180°$$

Step 3: Create a verbal equation.

$$x \quad + \quad 2x \quad + \quad (7x + 10) = 180$$

Step 4: Write a mathematical equation.

$$x + 2x + 7x + 10 = 180$$

Step 5: Solve for x.

$$10x + 10 = 180$$
$$10x = 170$$
$$x = 17$$

Step 6: Interpret the results and write the answer in words.

The smallest angle is $x = 17$.

The other angles are $2x = 2(17) = 34$

$$7x + 10 = 7(17) + 10 = 129$$

The angles are $17°$, $34°$, and $129°$.

Skill Practice

6. In a triangle, the measure of the first angle is 80° greater than the measure of the second angle. The measure of the third angle is twice that of the second. Find the measures of the angles.

Answer

6. $25°$, $50°$, and $105°$

Example 6 Solving a Geometry Application Involving Circumference

The distance around a circular garden is 188.4 ft. Find the radius to the nearest tenth of a foot (Figure 2-9). Use 3.14 for π.

$C = 188.4$ ft

Figure 2-9

Solution:

$$C = 2\pi r \qquad \text{Use the formula for the circumference of a circle.}$$

$$188.4 = 2\pi r \qquad \text{Substitute 188.4 for } C.$$

$$\frac{188.4}{2\pi} = \frac{2\pi r}{2\pi} \qquad \text{Divide both sides by } 2\pi.$$

$$\frac{188.4}{2\pi} = r$$

$$r \approx \frac{188.4}{2(3.14)}$$

$$= 30.0$$

The radius is approximately 30.0 ft.

Answer

7. 2.0 cm

Calculator Connections

Topic: Using the π Key on a Calculator

In Example 6 we could have obtained a more accurate result if we had used the π key on the calculator.

Note that parentheses are required to divide 188.4 by the quantity 2π. This guarantees that the calculator follows the implied order of operations. Without parentheses, the calculator would divide 188.4 by 2 and then multiply the result by π.

Scientific Calculator

Enter: 188.4 ÷ (2 × π) =   Result: 29.98479128 correct

Enter: 188.4 ÷ 2 × π =   Result: 295.938028 incorrect

Graphing Calculator

```
188.4/(2π)
        29.98479128  ←── Correct
188.4/2π
        295.938028   ←── Incorrect
```

Calculator Exercises

Approximate the expressions with a calculator. Round to three decimal places if necessary.

1. $\dfrac{880}{2\pi}$ **2.** $\dfrac{1600}{\pi(4)^2}$ **3.** $\dfrac{20}{5\pi}$ **4.** $\dfrac{10}{7\pi}$

Section 2.6 Practice Exercises

Study Skills Exercises

1. A good technique for studying for a test is to choose four problems from each section of the chapter and write the problems along with the directions on a 3×5 card. On the back of the card, put the page number where you found that problem. Then shuffle the cards and test yourself on the procedure to solve each problem. If you find one that you do not know how to solve, look at the page number and do several of that type. Write four problems you would choose for this section.

2. Define the key term: **literal equation**

Review Exercises

For Exercises 3–8, solve the equation.

3. $3(2y + 3) - 4(-y + 1) = 7y - 10$

4. $-(3w + 4) + 5(w - 2) - 3(6w - 8) = 10$

5. $\frac{1}{2}(x - 3) + \frac{3}{4} = 3x - \frac{3}{4}$

6. $\frac{5}{6}x + \frac{1}{2} = \frac{1}{4}(x - 4)$

7. $0.5(y + 2) - 0.3 = 0.4y + 0.5$

8. $0.25(500 - x) + 0.15x = 75$

Objective 1: Formulas and Literal Equations

For Exercises 9–40, solve for the indicated variable. **(See Examples 1–2.)**

9. $P = a + b + c$ for a

10. $P = a + b + c$ for b

11. $x = y - z$ for y

12. $c + d = e$ for d

13. $p = 250 + q$ for q

14. $y = 35 + x$ for x

15. $A = bh$ for b

16. $d = rt$ for r

17. $PV = nrt$ for t

18. $P_1V_1 = P_2V_2$ for V_1

19. $x - y = 5$ for x

20. $x + y = -2$ for y

21. $3x + y = -19$ for y

22. $x - 6y = -10$ for x

23. $2x + 3y = 6$ for y

24. $5x + 2y = 10$ for y

25. $-2x - y = 9$ for x

26. $3x - y = -13$ for x

27. $4x - 3y = 12$ for y

28. $6x - 3y = 4$ for y

29. $ax + by = c$ for y

30. $ax + by = c$ for x

31. $A = P(1 + rt)$ for t

32. $P = 2(L + w)$ for L

33. $a = 2(b + c)$ for c

34. $3(x + y) = z$ for x

35. $Q = \frac{x + y}{2}$ for y

36. $Q = \frac{a - b}{2}$ for a

37. $M = \frac{a}{S}$ for a

38. $A = \frac{1}{3}(a + b + c)$ for c

39. $P = I^2R$ for R

40. $F = \frac{GMm}{d^2}$ for m

Objective 2: Geometry Applications

For Exercises 41–62, use the problem-solving flowchart (page 143) from Section 2.4.

41. The perimeter of a rectangular garden is 24 ft. The length is 2 ft more than the width. Find the length and the width of the garden. **(See Example 3.)**

42. In a small rectangular wallet photo, the width is 7 cm less than the length. If the border (perimeter) of the photo is 34 cm, find the length and width.

43. The length of a rectangular parking area is four times the width. The perimeter is 300 yd. Find the length and width of the parking area.

44. The width of Jason's workbench is $\frac{1}{2}$ the length. The perimeter is 240 in. Find the length and the width of the workbench.

45. A builder buys a rectangular lot of land such that the length is 5 m less than two times the width. If the perimeter is 590 m, find the length and the width.

46. The perimeter of a rectangular pool is 140 yd. If the length is 20 yd less than twice the width, find the length and the width.

$2w - 5$

w

$2w - 20$

47. A triangular parking lot has two sides that are the same length and the third side is 5 m longer. If the perimeter is 71 m, find the lengths of the sides.

48. The perimeter of a triangle is 16 ft. One side is 3 ft longer than the shortest side. The third side is 1 ft longer than the shortest side. Find the lengths of all the sides.

49. Sometimes memory devices are helpful for remembering mathematical facts. Recall that the sum of two complementary angles is 90°. That is, two complementary angles when added together form a right angle or "corner." The words *Complementary* and *Corner* both start with the letter "*C*." Derive your own memory device for remembering that the sum of two supplementary angles is 180°.

$x° + y° = 90°$

Complementary angles form a "Corner"

$x° + y° = 180°$

Supplementary angles . . .

50. What do you know about the measures of two vertical angles?

51. Two angles are complementary. One angle is 20° less than the other angle. Find the measures of the angles. **(See Example 4.)**

52. Two angles are complementary. One angle is 4° less than three times the other angle. Find the measures of the angles.

53. Two angles are supplementary. One angle is three times as large as the other angle. Find the measures of the angles.

54. Two angles are supplementary. One angle is 6° more than four times the other. Find the measures of the two angles.

55. Find the measures of the vertical angles labeled in the figure by first solving for x.

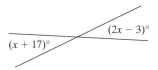

56. Find the measures of the vertical angles labeled in the figure by first solving for y.

57. The largest angle in a triangle is three times the smallest angle. The middle angle is two times the smallest angle. Given that the sum of the angles in a triangle is 180°, find the measure of each angle. **(See Example 5.)**

58. The smallest angle in a triangle measures 90° less than the largest angle. The middle angle measures 60° less than the largest angle. Find the measure of each angle.

59. The smallest angle in a triangle is half the largest angle. The middle angle measures 30° less than the largest angle. Find the measure of each angle.

60. The largest angle of a triangle is three times the middle angle. The smallest angle measures 10° less than the middle angle. Find the measure of each angle.

61. Find the value of x and the measure of each angle labeled in the figure.

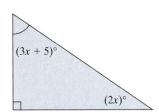

62. Find the value of y and the measure of each angle labeled in the figure.

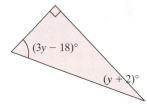

63. a. A rectangle has length l and width w. Write a formula for the area.

b. Solve the formula for the width, w.

c. The area of a rectangular volleyball court is 1740.5 ft² and the length is 59 ft. Find the width.

64. a. A parallelogram has height h and base b. Write a formula for the area.

b. Solve the formula for the base, b.

c. Find the base of the parallelogram pictured if the area is 40 m².

65. a. A rectangle has length *l* and width *w*. Write a formula for the perimeter.

    b. Solve the formula for the length, *l*.

    c. The perimeter of the soccer field at Giants Stadium is 338 m. If the width is 66 m, find the length.

Perimeter = 338 m

66 m

67. a. A circle has a radius of *r*. Write a formula for the circumference. **(See Example 6.)**

    b. Solve the formula for the radius, *r*.

    c. The circumference of the circular Buckingham Fountain in Chicago is approximately 880 ft. Find the radius. Round to the nearest foot.

66. a. A triangle has height *h* and base *b*. Write a formula for the area.

    b. Solve the formula for the height, *h*.

    c. Find the height of the triangle pictured if the area is 12 km².

$h = ?$

$b = 6$ km

68. a. The length of each side of a square is *s*. Write a formula for the perimeter of the square.

    b. Solve the formula for the length of a side, *s*.

    c. The Pyramid of Khufu (known as the Great Pyramid) at Giza has a square base. If the distance around the bottom is 921.6 m, find the length of the sides at the bottom of the pyramid.

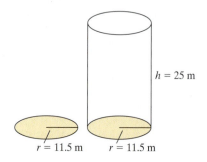

s

s

s

s

Expanding Your Skills

For Exercises 69–70, find the indicated area or volume. Be sure to include the proper units and round each answer to two decimal places if necessary.

69. a. Find the area of a circle with radius 11.5 m.

    b. Find the volume of a right circular cylinder with radius 11.5 m and height 25 m.

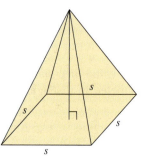

$h = 25$ m

$r = 11.5$ m $r = 11.5$ m

70. a. Find the area of a parallelogram with base 30 in. and height 12 in.

    b. Find the area of a triangle with base 30 in. and height 12 in.

    c. Compare the areas found in parts (a) and (b).

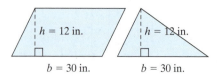

$h = 12$ in. $h = 12$ in.

$b = 30$ in. $b = 30$ in.

Mixture Applications and Uniform Motion

1. Applications Involving Cost

In Examples 1 and 2, we will look at different kinds of mixture problems. The first example "mixes" two types of movie tickets, adult tickets that sell for $8 and children's tickets that sell for $6. Furthermore, there were 300 tickets sold for a total revenue of $2040. Before attempting the problem, we should try to gain some familiarity. Let's try a few combinations to see how many of each type of ticket might have been sold.

Suppose 100 adult tickets were sold and 200 children's tickets were sold (a total of 300 tickets).

- 100 adult tickets at $8 each gives $100(\$8) = \$800$
- 200 children's tickets at $6 each gives $200(\$6) = \$1200$

Total revenue: $2000 (not enough)

Suppose 150 adult tickets were sold and 150 children's tickets were sold (a total of 300 tickets).

- 150 adult tickets at $8 each gives $150(\$8) = \$1200$
- 150 children's tickets at $6 each gives $150(\$6) = \$900$

Total revenue: $2100 (too much)

As you can see, the trial-and-error process can be tedious and time-consuming. Therefore we will use algebra to determine the correct combination of each type of ticket.

Suppose we let x represent the number of adult tickets, then the number of children's tickets is the *total minus x*. That is

$$\begin{pmatrix} \text{Number of} \\ \text{children's tickets} \end{pmatrix} = \begin{pmatrix} \text{total number} \\ \text{of tickets} \end{pmatrix} - \begin{pmatrix} \text{number of} \\ \text{adult tickets, } x \end{pmatrix}$$

Number of children's tickets = $300 - x$.

Notice that the number of tickets sold times the price per ticket gives the revenue.

- x adult tickets at $8 each gives a revenue of: $x(\$8)$ or simply $8x$.
- $300 - x$ children's tickets at $6 each gives: $(300 - x)(\$6)$ or $6(300 - x)$

This will help us set up an equation in Example 1.

Example 1   Solving a Mixture Problem Involving Ticket Sales

At one showing of *Shrek 3*, there were 300 tickets sold. Adult tickets cost $8 and tickets for children cost $6. If the total revenue from ticket sales was $2040, determine the number of each type of ticket sold.

Solution:

Let x represent the number of adult tickets sold.

$300 - x$ is the number of children's tickets.

Step 1:   Read the problem.

Step 2:   Label the variables.

	$8 Tickets	**$6 Tickets**	**Total**
Number of tickets	x	$300 - x$	300
Revenue	$8x$	$6(300 - x)$	2040

$$\left(\begin{array}{c}\text{Revenue from} \\ \text{adult tickets}\end{array}\right) + \left(\begin{array}{c}\text{revenue from} \\ \text{children's tickets}\end{array}\right) = \left(\begin{array}{c}\text{total} \\ \text{revenue}\end{array}\right)$$

Step 3: Write an equation in words.

$$8x \quad + \quad 6(300 - x) \quad = \quad 2040$$

Step 4: Write a mathematical equation.

$$8x + 6(300 - x) = 2040$$

Step 5: Solve the equation.

$$8x + 1800 - 6x = 2040$$
$$2x + 1800 = 2040$$
$$2x = 240$$
$$x = 120$$

Step 6: Interpret the results.

There were 120 adult tickets sold.
The number of children's tickets is $300 - x$ which is 180.

2. Applications Involving Mixtures

Example 2 Solving a Mixture Application

How many liters (L) of a 60% antifreeze solution must be added to 8 L of a 10% antifreeze solution to produce a 20% antifreeze solution?

Solution:

The information can be organized in a table. Notice that an algebraic equation is derived from the second row of the table. This relates the number of liters of pure antifreeze in each container.

Step 1: Read the problem.

	60% Antifreeze	10% Antifreeze	Final Mixture: 20% Antifreeze
Number of liters of solution	x	8	$(8 + x)$
Number of liters of pure antifreeze	$0.60x$	$0.10(8)$	$0.20(8 + x)$

Step 2: Label the variables.

The amount of pure antifreeze in the final solution equals the sum of the amounts of antifreeze in the first two solutions.

$$\left(\begin{array}{c}\text{Pure antifreeze} \\ \text{from solution 1}\end{array}\right) + \left(\begin{array}{c}\text{pure antifreeze} \\ \text{from solution 2}\end{array}\right) = \left(\begin{array}{c}\text{pure antifreeze} \\ \text{in the final solution}\end{array}\right)$$

Step 3: Write an equation in words.

$$0.60x \quad + \quad 0.10(8) \quad = \quad 0.20(8 + x)$$
$$0.60x + 0.10(8) = 0.20(8 + x)$$

Step 4: Write a mathematical equation.

Answer

3. 5 gal is needed.

$$0.6x + 0.8 = 1.6 + 0.2x \qquad \text{Step 5:} \quad \text{Solve the equation.}$$

$$0.6x - 0.2x + 0.8 = 1.6 + 0.2x - 0.2x \qquad \text{Subtract } 0.2x.$$

$$0.4x + 0.8 = 1.6$$

$$0.4x + 0.8 - 0.8 = 1.6 - 0.8 \qquad \text{Subtract } 0.8.$$

$$0.4x = 0.8$$

$$\frac{0.4x}{0.4} = \frac{0.8}{0.4} \qquad \text{Divide by } 0.4.$$

$$x = 2 \qquad \text{Step 6:} \quad \text{Interpret the result.}$$

Therefore, 2 L of 60% antifreeze solution is necessary to make a final solution that is 20% antifreeze.

3. Applications Involving Uniform Motion

The formula (distance) = (rate)(time) or simply, $d = rt$, relates the distance traveled to the rate of travel and the time of travel.

For example, if a car travels at 60 mph for 3 hr, then

$$d = (60 \text{ mph})(3 \text{ hours})$$

$$= 180 \text{ miles}$$

If a car travels at 60 mph for x hr, then

$$d = (60 \text{ mph})(x \text{ hours})$$

$$= 60x \text{ miles}$$

Example 3 **Solving an Application Involving Distance, Rate, and Time**

One bicyclist rides 4 mph faster than another bicyclist. The faster rider takes 3 hr to complete a race, while the slower rider takes 4 hr. Find the speed for each rider.

Solution:

Step 1: Read the problem.

The problem is asking us to find the speed of each rider.

Let x represent the speed of the slower rider. Then $(x + 4)$ is the speed of the faster rider.

Step 2: Label the variables and organize the information given in the problem. A distance-rate-time chart may be helpful.

	Distance	**Rate**	**Time**
Faster rider	$3(x + 4)$	$x + 4$	3
Slower rider	$4(x)$	x	4

To complete the first column, we can use the relationship, $d = rt$.

Skill Practice

4. An express train travels 25 mph faster than a cargo train. It takes the express train 6 hr to travel a route, and it takes 9 hr for the cargo train to travel the same route. Find the speed of each train.

Answer

4. The express train travels 75 mph, and the cargo train travels 50 mph.

Because the riders are riding in the same race, their distances are equal.

$$\left(\begin{array}{c} \text{Distance} \\ \text{by faster rider} \end{array}\right) = \left(\begin{array}{c} \text{distance} \\ \text{by slower rider} \end{array}\right)$$

Step 3: Set up a verbal model.

$$3(x + 4) = 4(x)$$

Step 4: Write a mathematical equation.

$$3x + 12 = 4x$$

Step 5: Solve the equation.

$$12 = x$$

Subtract $3x$ from both sides.

The variable x represents the slower rider's rate. The quantity $x + 4$ is the faster rider's rate. Thus, if $x = 12$, then $x + 4 = 16$.

The slower rider travels 12 mph and the faster rider travels 16 mph.

Avoiding Mistakes

Check that the answer is reasonable. If the slower rider rides at 12 mph for 4 hr, he travels 48 mi. If the faster rider rides at 16 mph for 3 hr, he also travels 48 mi as expected.

Skill Practice

5. A Piper Cub airplane has an average air speed that is 10 mph faster than a Cessna 150 airplane. If the combined distance traveled by these two small planes is 690 miles after 3 hr, what is the average speed of each plane?

Example 4   Solving an Application Involving Distance, Rate, and Time

Two families that live 270 mi apart plan to meet for an afternoon picnic at a park that is located between their two homes. Both families leave at 9.00 A.M., but one family averages 12 mph faster than the other family. If the families meet at the designated spot $2\frac{1}{2}$ hr later, determine

a. The average rate of speed for each family.

b. The distance each family traveled to the picnic.

Solution:

For simplicity, we will call the two families, Family A and Family B. Let Family A be the family that travels at the slower rate (Figure 2-10).

Step 1: Read the problem and draw a sketch.

Figure 2-10

Let x represent the rate of Family A.

Then $(x + 12)$ is the rate of Family B.

Step 2: Label the variables.

	Distance	Rate	Time
Family A	2.5x	x	2.5
Family B	2.5(x + 12)	x + 12	2.5

To complete the first column, we can use the relationship $d = rt$.

Answer

5. The Cessna's speed is 110 mph, and the Piper Cub's speed is 120 mph.

To set up an equation, recall that the total distance between the two families is given as 270 miles.

$$\begin{pmatrix} \text{Distance} \\ \text{traveled by} \\ \text{Family A} \end{pmatrix} + \begin{pmatrix} \text{distance} \\ \text{traveled by} \\ \text{Family B} \end{pmatrix} = \begin{pmatrix} \text{total} \\ \text{distance} \end{pmatrix}$$

Step 3: Create a verbal equation.

$$2.5x + 2.5(x + 12) = 270$$

Step 4: Write a mathematical equation.

$$2.5x + 2.5(x + 12) = 270$$
$$2.5x + 2.5x + 30 = 270$$
$$5.0x + 30 = 270$$
$$5x = 240$$
$$x = 48$$

Step 5: Solve for x.

a. Family A traveled 48 (mph).

Family B traveled $x + 12 = 48 + 12 = 60$ (mph).

Step 6: Interpret the results and write the answer in words.

b. To compute the distance each family traveled, use $d = rt$.

Family A traveled: $(48 \text{ mph})(2.5 \text{ hr}) = 120 \text{ mi}$

Family B traveled: $(60 \text{ mph})(2.5 \text{ hr}) = 150 \text{ mi}$

Section 2.7 Practice Exercises

Study Skills

1. The following is a list of steps to help you solve word problems. Check those that you follow on a regular basis when solving a word problem. Place an asterisk next to the steps that you need to improve.

_____ Read through the entire problem before writing anything down.

_____ Write down exactly what you are being asked to find.

_____ Write down what is known and assign variables to what is unknown.

_____ Draw a figure or diagram if it will help you understand the problem.

_____ Highlight key words like total, sum, difference, etc.

_____ Translate the word problem to a mathematical problem.

_____ After solving, check that your answer makes sense.

Review Exercises

For Exercises 2–3, solve for the indicated variable.

2. $ax - by = c$ for x

3. $cd = r$ for c

For Exercises 4–6, solve the equation.

4. $-2d + 11 = 4 - d$ **5.** $3(2y + 5) - 8(y - 1) = 3y + 3$ **6.** $0.02x + 0.04(10 - x) = 1.26$

Objective 1: Applications Involving Cost

For Exercises 7–12, write an algebraic expression as indicated.

7. Two numbers total 200. Let t represent one of the numbers. Write an algebraic expression for the other number.

8. The total of two numbers is 43. Let s represent one of the numbers. Write an algebraic expression for the other number.

9. Olivia needs to bring 100 cookies to her friend's party. She has already baked x cookies. Write an algebraic expression for the number of cookies Olivia still needs to bake.

10. Rachel needs a mixture of 55 pounds (lb) of nuts consisting of peanuts and cashews. Let p represent the number of pounds of peanuts in the mixture. Write an algebraic expression for the number of pounds of cashews that she needs to add.

11. Max has a total of $3000 in two bank accounts. Let y represent the amount in one account. Write an algebraic expression for the amount in the other account.

12. Roberto has a total of $7500 in two savings accounts. Let z represent the amount in one account. Write an algebraic expression for the amount in the other account.

13. A church had an ice cream social and sold tickets for $3 and $2. When the social was over, 81 tickets had been sold totaling $215. How many of each type of ticket did the church sell? **(See Example 1.)**

	$3 Tickets	$2 Tickets	Total
Number of tickets			
Cost of tickets			

14. Anna is a teacher at an elementary school. She purchased 72 tickets to take the first-grade children and some parents on a field trip to the zoo. She purchased children's tickets for $10 each and adults' tickets for $18 each. She spent a total of $856. How many of each ticket did she buy?

	Adults	Children	Total
Number of tickets			
Cost of tickets			

15. Josh downloaded 25 tunes from an online site for his MP3 player. Some songs cost $0.90 each while others were $1.50 each. He spent a total of $27.30. How many of each type of song did he download?

16. During the past year, Mrs. Singh purchased 32 books at a wholesale club store. She purchased softcover books for $4.50 each and hardcover books for $13.50 each. The total cost of the books was $243. How many of each type of book did she purchase?

17. Angelina purchased 45 bottled drinks for her graduation party. She purchased soda for $1.60 each and bottles of flavored water for $2.00 each. The total cost of the drinks was $80. How many of each type did she buy?

18. Mr. Garvey purchased 58 food items at a fast food restaurant for his Little League team. He purchased hamburgers for $2.50 each and french fries for $1.50 each. He spent a total of $127. How many hamburgers and how many french fries did he purchase?

Objective 2: Applications Involving Mixtures

19. A container holds 7 ounces (oz) of liquid. Let x represent the number of ounces of liquid in another container. Write an expression for the total amount of liquid.

20. A bucket contains 2.5 L of a bleach solution. Let n represent the number of liters of bleach solution in a second bucket. Write an expression for the total amount of bleach solution.

21. If Miguel invests $2000 in a certificate of deposit and d dollars in a stock, write an expression for the total amount he invested.

22. James has $5000 in one savings account. Let y represent the amount he has in another savings account. Write an expression for the total amount of money in both accounts.

23. How many ounces of a 50% antifreeze solution must be mixed with 10 oz of an 80% antifreeze solution to produce a 60% antifreeze solution? **(See Example 2.)**

	50% Antifreeze	80% Antifreeze	Final Mixture: 60% Antifreeze
Number of ounces of solution			
Number of ounces of pure antifreeze			

24. How many liters of a 10% alcohol solution must be mixed with 12 L of a 5% alcohol solution to produce an 8% alcohol solution?

	10% Alcohol	5% Alcohol	Final Mixture: 8% Alcohol
Number of liters of solution			
Number of liters of pure alcohol			

25. A pharmacist needs to mix a 1% saline (salt) solution with 24 milliliters (mL) of a 16% saline solution to obtain a 9% saline solution. How many milliliters of the 1% solution must she use?

26. A landscaper needs to mix a 75% pesticide solution with 30 gal of a 25% pesticide solution to obtain a 60% pesticide solution. How many gallons of the 75% solution must he use?

27. To clean a concrete driveway, a contractor needs a solution that is 30% acid. How many ounces of a 50% acid solution must be mixed with 15 ounces of a 21% solution to obtain a 30% acid solution?

28. A veterinarian needs a mixture that contains 12% of a certain medication to treat an injured bird. How many milliliters of a 16% solution should be mixed with 6 mL of a 7% solution to obtain a solution that is 12% medication?

Objective 3: Applications Involving Uniform Motion

29. a. If a car travels 60 mph for 5 hr, find the distance traveled.

b. If a car travels at x mi per hour for 5 hr, write an expression that represents the distance traveled.

c. If a car travels at $x + 12$ mph for 5 hr, write an expression that represents the distance traveled.

30. a. If a plane travels 550 mph for 2.5 hr, find the distance traveled.

b. If a plane travels at x mi per hour for 2.5 hr, write an expression that represents the distance traveled.

c. If a plane travels at $x - 100$ mph for 2.5 hr, write an expression that represents the distance traveled.

31. A woman can walk 2 mph faster down a trail to Cochita Lake than she can on the return trip uphill. It takes her 2 hr to get to the lake and 4 hr to return. What is her speed walking down to the lake? **(See Example 3.)**

	Distance	Rate	Time
Downhill to the lake			
Uphill from the lake			

32. A car travels 20 mph slower in a bad rain storm than in sunny weather. The car travels the same distance in 2 hr in sunny weather as it does in 3 hr in rainy weather. Find the speed of the car in sunny weather.

	Distance	Rate	Time
Rain storm			
Sunny weather			

33. Bryan hiked up to the top of City Creek in 3 hr and then returned down the canyon to the trailhead in another 2 hr. His speed downhill was 1 mph faster than his speed uphill. How far up the canyon did he hike?

34. Kevin hiked up Lamb's Canyon in 2 hr and then ran back down in 1 hr. His speed running downhill was 2.5 mph greater than his speed hiking uphill. How far up the canyon did he hike?

35. Hazel and Emilie fly from Atlanta to San Diego. The flight from Atlanta to San Diego is against the wind and takes 4 hr. The return flight with the wind takes 3.5 hr. If the wind speed is 40 mph, find the speed of the plane in still air.

36. A boat on the Potomac River travels the same distance downstream in $\frac{2}{3}$ hr as it does going upstream in 1 hr. If the speed of the current is 3 mph, find the speed of the boat in still water.

37. Two cars are 200 mi apart and traveling toward each other on the same road. They meet in 2 hr. One car is traveling 4 mph faster than the other. What is the speed of each car? **(See Example 4.)**

38. Two cars are 238 mi apart and traveling toward each other along the same road. They meet in 2 hr. One car is traveling 5 mph slower than the other. What is the speed of each car?

39. After Hurricane Katrina, a rescue vehicle leaves a station at noon and heads for New Orleans. An hour later a second vehicle traveling 10 mph faster leaves the same station. By 4:00 P.M., the first vehicle reaches its destination, and the second is still 10 mi away. How fast is each vehicle?

40. A truck leaves a truck stop at 9:00 A.M. and travels toward Sturgis, Wyoming. At 10:00 A.M., a motorcycle leaves the same truck stop and travels the same route. The motorcycle travels 15 mph faster than the truck. By noon, the truck has traveled 20 mi further than the motorcycle. How fast is each vehicle?

41. Two boats traveling the same direction leave a harbor at noon. After 2 hr, they are 40 mi apart. If one boat travels twice as fast as the other, find the rate of each boat.

42. Two canoes travel down a river, starting at 9:00 A.M. One canoe travels twice as fast as the other. After 3.5 hr, the canoes are 5.25 mi apart. Find the speed of each canoe.

Mixed Exercises

43. A certain granola mixture is 10% peanuts.

  a. If a container has 20 lb of granola, how many pounds of peanuts are there?

  b. If a container has x pounds of granola, write an expression that represents the number of pounds of peanuts in the granola.

  c. If a container has $x + 3$ lb of granola, write an expression that represents the number of pounds of peanuts.

44. A certain blend of coffee sells for $9.00 per pound.

  a. If a container has 20 lb of coffee, how much will it cost.

  b. If a container has x lb of coffee, write an expression that represents the cost.

  c. If a container has $40 - x$ lb of this coffee, write an expression that represents the cost.

45. The Coffee Company mixes coffee worth $12 per pound with coffee worth $8 per pound to produce 50 lb of coffee worth $8.80 per pound. How many pounds of the $12 coffee and how many pounds of the $8 coffee must be used?

	$12 Coffee	$8 Coffee	Total
Number of pounds			
Value of coffee			

46. The Nut House sells pecans worth $4 per pound and cashews worth $6 per pound. How many pounds of pecans and how many pounds of cashews must be mixed to form 16 lb of a nut mixture worth $4.50 per pound?

	$4 Pecans	$6 Cashews	Total
Number of pounds			
Value of nuts			

47. A boat in distress, 21 nautical miles from a marina, travels toward the marina at 3 knots (nautical miles per hour). A coast guard cruiser leaves the marina and travels toward the boat at 25 knots. How long will it take for the boats to reach each other?

48. An air traffic controller observes a plane heading from New York to San Francisco traveling at 450 mph. At the same time, another plane leaves San Francisco and travels 500 mph to New York. If the distance between the airports is 2850 mi, how long will it take for the planes to pass each other?

49. Surfer Sam purchased a total of 21 items at the surf shop. He bought wax for $3.00 per package and sunscreen for $8.00 per bottle. He spent a total amount of $88.00. How many of each item did he purchase?

50. Tonya Toast loves jam. She purchased 30 jars of gourmet jam for $178.50. She bought raspberry jam for $6.25 per jar and strawberry jam for $5.50 per jar. How many jars of each did she purchase?

51. How many quarts of 85% chlorine solution must be mixed with 5 quarts of 25% chlorine solution to obtain a 45% chlorine solution?

52. How many liters of a 58% sugar solution must be added to 14 L of a 40% sugar solution to obtain a 50% sugar solution?

Expanding Your Skills

53. How much pure water must be mixed with 12 L of a 40% alcohol solution to obtain a 15% alcohol solution? (*Hint:* Pure water is 0% alcohol.)

54. How much pure water must be mixed with 10 oz of a 60% alcohol solution to obtain a 25% alcohol solution?

55. Amtrak Acela Express is a high speed train that runs in the United States between Washington, D.C., and Boston. In Japan, a bullet train along the Sanyo line operates at an average speed of 60 km/hr faster than the Amtrak Acela Express. It takes the Japanese bullet train 2.7 hr to travel the same distance as the Acela Express can travel in 3.375 hr. Find the speed of each train.

56. Amtrak Acela Express is a high-speed train along the northeast corridor between Washington, D.C., and Boston. Since its debut, it cuts the travel time from 4 hr 10 min to 3 hr 20 min. On average, if the Acela Express is 30 mph faster than the old train, find the speed of the Acela Express. (*Hint:* 4 hr 10 min = $4\frac{1}{6}$ hr.)

Section 2.8 Linear Inequalities

1. Graphing Linear Inequalities

Consider the following two statements.

$$2x + 7 = 11 \quad \text{and} \quad 2x + 7 < 11$$

The first statement is an equation (it has an = sign). The second statement is an inequality (it has an inequality symbol, <). In this section, we will learn how to solve linear *inequalities*, such as $2x + 7 < 11$.

> **DEFINITION** **A Linear Inequality in One Variable**
>
> A **linear inequality in one variable**, x, is defined as any relationship of the form:
>
> $ax + b < 0$, $ax + b \leq 0$, $ax + b > 0$, or $ax + b \geq 0$, where $a \neq 0$.

The following inequalities are linear equalities in one variable.

$$2x - 3 < 0 \qquad -4z - 3 > 0 \qquad a \leq 4 \qquad 5.2y \geq 10.4$$

The number line is a useful tool to visualize the solution set of an equation or inequality. For example, the solution set to the equation $x = 2$ is {2} and may be graphed as a single point on the number line.

$x = 2$ −6 −5 −4 −3 −2 −1   0   1   2   3   4   5   6

The solution set to an inequality is the set of real numbers that make the inequality a true statement. For example, the solution set to the inequality $x \geq 2$ is all real numbers 2 or greater. Because the solution set has an infinite number of values, we cannot list all of the individual solutions. However, we can graph the solution set on the number line.

$x \geq 2$ −6 −5 −4 −3 −2 −1   0   1   2   3   4   5   6

The square bracket symbol, [, is used on the graph to indicate that the point $x = 2$ is included in the solution set. By convention, square brackets, either [or], are used to *include* a point on a number line. Parentheses, (or), are used to *exclude* a point on a number line.

The solution set of the inequality $x > 2$ includes the real numbers greater than 2 but not including 2. Therefore, a (symbol is used on the graph to indicate that $x = 2$ is not included.

$x > 2$ −6 −5 −4 −3 −2 −1   0   1   2   3   4   5   6

In Example 1, we demonstrate how to graph linear inequalities. To graph an inequality means that we graph its *solution set*. That is, we graph all of the values on the number line that make the inequality true.

Graph the solution sets.

a. $x > -1$ **b.** $c \le \dfrac{7}{3}$ **c.** $3 > y$

Skill Practice

Graph the solution sets.

1. $y < 0$

2. $x \ge -\dfrac{5}{4}$

3. $5 \ge a$

Solution:

a. $x > -1$

The solution set is the set of all real numbers strictly greater than -1. Therefore, we graph the region on the number line to the right of -1. Because $x = -1$ is not included in the solution set, we use the (symbol at $x = -1$.

b. $c \le \frac{7}{3}$ is equivalent to $c \le 2\frac{1}{3}$.

The solution set is the set of all real numbers less than or equal to $2\frac{1}{3}$. Therefore, graph the region on the number line to the left of and including $2\frac{1}{3}$. Use the symbol] to indicate that $c = 2\frac{1}{3}$ is included in the solution set.

c. $3 > y$ This inequality reads "3 is greater than y." This is equivalent to saying, "y is less than 3." The inequality $3 > y$ can also be written as $y < 3$.

$y < 3$

The solution set is the set of real numbers less than 3. Therefore, graph the region on the number line to the left of 3. Use the symbol) to denote that the endpoint, 3, is not included in the solution.

TIP: Some textbooks use a closed circle or an open circle (● or ○) rather than a bracket or parenthesis to denote inclusion or exclusion of a value on the real number line. For example, the solution sets for the inequalities $x > -1$ and $c \le \frac{7}{3}$ are graphed here.

$x > -1$

$c \le \frac{7}{3}$

A statement that involves more than one inequality is called a **compound inequality**. One type of compound inequality is used to indicate that one number is between two others. For example, the inequality $-2 < x < 5$ means that $-2 < x$ and $x < 5$. In words, this is easiest to understand if we read the variable first: x is greater than -2 and x is less than 5. The numbers satisfied by these two conditions are those between -2 and 5.

Answers

Skill Practice

Graph the solution set.

4. $0 \le y \le 8.5$

Example 2   **Graphing a Compound Inequality**

Graph the solution set of the inequality: $-4.1 < y \le -1.7$

Solution:

$-4.1 < y \le -1.7$ means that

$-4.1 < y$ and $y \le -1.7$

-4.1 -1.7

Shade the region of the number line greater than -4.1 and less than or equal to -1.7.

Concept Connections

5. Translate the phrase below to set-builder notation.

The set of all b such that b is greater than or equal to -20.

6. Translate the phrase below to interval notation.

The set of all x such that x is less than -30.

---

2. Set-Builder Notation and Interval Notation

Graphing the solution set to an inequality is one way to define the set. Two other methods are to use **set-builder notation** or **interval notation**.

Set-Builder Notation

The solution to the inequality $x \ge 2$ can be expressed in set-builder notation as follows:

the set of all x such that x is greater than or equal to 2

Interval Notation

To understand interval notation, first think of a number line extending infinitely far to the right and infinitely far to the left. Sometimes we use the infinity symbol, ∞, or negative infinity symbol, $-\infty$, to label the far right and far left ends of the number line (Figure 2-11).

$-\infty$ —————————————— ∞

0

Figure 2-11

To express the solution set of an inequality in interval notation, sketch the graph first. Then use the endpoints to define the interval.

Inequality	Graph	Interval Notation
$x \ge 2$		$[2, \infty)$

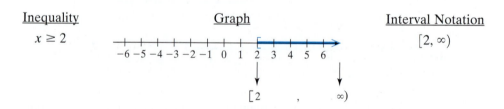

$[2$, $\infty)$

The graph of the solution set $x \ge 2$ begins at 2 and extends infinitely far to the right. The corresponding interval notation begins at 2 and extends to ∞. Notice that a square bracket [is used at 2 for both the graph and the interval notation. A parenthesis is always used at ∞ and for $-\infty$, because there is no endpoint.

Answers

4.

0   8.5

5. $\{b \mid b \ge -20\}$

6. $(-\infty, -30)$

> **PROCEDURE** Using Interval Notation
> - The endpoints used in interval notation are always written from left to right. That is, the smaller number is written first, followed by a comma, followed by the larger number.
> - A parenthesis, (or), indicates that an endpoint is excluded from the set.
> - A square bracket, [or], indicates that an endpoint is included in the set.
> - Parentheses, (and), are always used with $-\infty$ and ∞, respectively.

In Table 2-1, we present examples of eight different scenarios for interval notation and the corresponding graph.

Table 2-1

Interval Notation	Graph	Interval Notation	Graph
(a, ∞)		$[a, \infty)$	
$(-\infty, a)$		$(-\infty, a]$	
(a, b)		$[a, b]$	
$(a, b]$		$[a, b)$	

Example 3 Using Set-Builder Notation and Interval Notation

Complete the chart.

Set-Builder Notation	Graph	Interval Notation
		$[-\frac{1}{2}, \infty)$
$\{y \mid -2 \leq y < 4\}$		

Solution:

Set-Builder Notation	Graph	Interval Notation
$\{x \mid x < -3\}$		$(-\infty, -3)$
$\{x \mid x \geq -\frac{1}{2}\}$		$[-\frac{1}{2}, \infty)$
$\{y \mid -2 \leq y < 4\}$		$[-2, 4)$

Skill Practice

Express each of the following in set-builder notation and interval notation.

7.

8. $x < \dfrac{3}{2}$

9.

Answers

7. $\{x \mid x \geq -2\}; [-2, \infty)$

8. $\left\{x \mid x < \dfrac{3}{2}\right\}; \left(-\infty, \dfrac{3}{2}\right)$

9. $\{x \mid -3 < x \leq 1\}; (-3, 1]$

3. Addition and Subtraction Properties of Inequality

The process to solve a linear inequality is very similar to the method used to solve linear equations. Recall that adding or subtracting the same quantity to both sides of an equation results in an equivalent equation. The addition and subtraction properties of inequality state that the same is true for an inequality.

> **PROPERTY** **Addition and Subtraction Properties of Inequality**
>
> Let a, b, and c represent real numbers.
>
> **1.** *Addition Property of Inequality: If $\qquad a < b$,
>   then $\quad a + c < b + c$
>
> **2.** *Subtraction Property of Inequality: If $\qquad a < b$,
>   then $\quad a - c < b - c$
>
> *These properties may also be stated for $a \le b$, $a > b$, and $a \ge b$.

To illustrate the addition and subtraction properties of inequality, consider the inequality $5 > 3$. If we add or subtract a real number such as 4 to both sides, the left-hand side will still be greater than the right-hand side. (See Figure 2-12.)

$5 > 3$ $5 + 4 > 3 + 4$

Figure 2-12

Skill Practice

Solve the inequality. Graph the solution set and express in interval notation.

10. $2y - 5 < y - 11$

Example 4   **Solving a Linear Inequality**

Solve the inequality and graph the solution set. Express the solution set in set-builder notation and in interval notation.

$$-2p + 5 < -3p + 6$$

Solution:

$$-2p + 5 < -3p + 6$$

$$-2p + 3p + 5 < -3p + 3p + 6 \qquad \text{Addition property of inequality (add } 3p \text{ to both sides).}$$

$$p + 5 < 6 \qquad \text{Simplify.}$$

$$p + 5 - 5 < 6 - 5 \qquad \text{Subtraction property of inequality.}$$

$$p < 1$$

Graph:

Set-builder notation: $\{p \mid p < 1\}$

Interval notation: $(-\infty, 1)$

Answer

10.

$(-\infty, -6)$

TIP: The solution to an inequality gives a set of values that make the original inequality true. Therefore, you can test your final answer by using *test points.* That is, pick a value in the proposed solution set and verify that it makes the original inequality true. Furthermore, any test point picked outside the solution set should make the original inequality false. For example,

Pick $p = -4$ as an arbitrary test point within the proposed solution set.

$$-2p + 5 < -3p + 6$$
$$-2(-4) + 5 \overset{?}{<} -3(-4) + 6$$
$$8 + 5 \overset{?}{<} 12 + 6$$
$$13 < 18 \checkmark \qquad \text{True}$$

Pick $p = 3$ as an arbitrary test point outside the proposed solution set.

$$-2p + 5 < -3p + 6$$
$$-2(3) + 5 \overset{?}{<} -3(3) + 6$$
$$-6 + 5 \overset{?}{<} -9 + 6$$
$$-1 \overset{?}{<} -3 \qquad \text{False}$$

4. Multiplication and Division Properties of Inequality

Multiplying both sides of an equation by the same quantity results in an equivalent equation. However, the same is not always true for an inequality. If you multiply or divide an inequality by a negative quantity, the direction of the inequality symbol must be reversed.

For example, consider multiplying or dividing the inequality, $4 < 5$ by -1.

Figure 2-13

The number 4 lies to the left of 5 on the number line. However, -4 lies to the right of -5 (Figure 2-13). Changing the sign of two numbers changes their relative position on the number line. This is stated formally in the multiplication and division properties of inequality.

PROPERTY Multiplication and Division Properties of Inequality

Let a, b, and c represent real numbers.

*If c is positive and $a < b$, then $ac < bc$ and $\dfrac{a}{c} < \dfrac{b}{c}$

*If c is negative and $a < b$, then $ac > bc$ and $\dfrac{a}{c} > \dfrac{b}{c}$

The second statement indicates that if both sides of an inequality are multiplied or divided by a negative quantity, the inequality sign must be reversed.

*These properties may also be stated for $a \le b$, $a > b$, and $a \ge b$.

Example 5 Solving a Linear Inequality

Solve the inequality $-5x - 3 \leq 12$. Graph the solution set and write the answer in interval notation.

Solution:

$$-5x - 3 \leq 12$$

$$-5x - 3 + 3 \leq 12 + 3 \qquad \text{Add } 3 \text{ to both sides.}$$

$$-5x \leq 15$$

$$\frac{-5x}{-5} \geq \frac{15}{-5} \qquad \text{Divide by } -5. \text{ Reverse the direction of the inequality sign.}$$

$$x \geq -3$$

Interval notation: $[-3, \infty)$

TIP: The inequality $-5x - 3 \leq 12$ could have been solved by isolating x on the right-hand side of the inequality. This would create a positive coefficient on the variable term and eliminate the need to divide by a negative number.

$$-5x - 3 \leq 12$$

$$-3 \leq 5x + 12$$

$$-15 \leq 5x \qquad \text{Notice that the coefficient of } x \text{ is positive.}$$

$$\frac{-15}{5} \leq \frac{5x}{5} \qquad \text{Do not reverse the inequality sign because we are dividing by a positive number.}$$

$$-3 \leq x, \text{ or equivalently, } x \geq -3$$

Example 6 Solving a Linear Inequality

Solve the inequality. Graph the solution set and write the answer in interval notation.

$$1.4x + 4.5 < 0.2x - 0.3$$

Solution:

$$1.4x + 4.5 < 0.2x - 0.3$$

$$1.4x - 0.2x + 4.5 < 0.2x - 0.2x - 0.3 \qquad \text{Subtract } 0.2x \text{ from both sides.}$$

$$1.2x + 4.5 < -0.3 \qquad \text{Simplify.}$$

$$1.2x + 4.5 - 4.5 < -0.3 - 4.5 \qquad \text{Subtract } 4.5 \text{ from both sides.}$$

$$1.2x < -4.8 \qquad \text{Simplify.}$$

$$\frac{1.2x}{1.2} < \frac{-4.8}{1.2} \qquad \text{Divide by } 1.2. \text{ The direction of the inequality sign is } \textit{not} \text{ reversed because we divided by a positive number.}$$

$$x < -4$$

Interval notation: $(-\infty, -4)$

Answers

11.

$(-\infty, -4)$

12.

$(5, \infty)$

Example 7 Solving Linear Inequalities

Solve the inequality $-\frac{1}{4}k + \frac{1}{6} \le 2 + \frac{2}{3}k$. Graph the solution set and write the answer in interval notation.

Solution:

$$-\frac{1}{4}k + \frac{1}{6} \le 2 + \frac{2}{3}k$$

$$12\left(-\frac{1}{4}k + \frac{1}{6}\right) \le 12\left(2 + \frac{2}{3}k\right)$$

Multiply both sides by 12 to clear fractions. (Because we multiplied by a positive number, the inequality sign is not reversed.)

$$\frac{12}{1}\left(-\frac{1}{4}k\right) + \frac{12}{1}\left(\frac{1}{6}\right) \le 12(2) + \frac{12}{1}\left(\frac{2}{3}k\right)$$

Apply the distributive property.

$$-3k + 2 \le 24 + 8k$$

Simplify.

$$-3k - 8k + 2 \le 24 + 8k - 8k$$

Subtract $8k$ from both sides.

$$-11k + 2 \le 24$$

$$-11k + 2 - 2 \le 24 - 2$$

Subtract 2 from both sides.

$$-11k \le 22$$

$$\frac{-11k}{-11} \ge \frac{22}{-11}$$

Divide both sides by -11. Reverse the inequality sign.

$$k \ge -2$$

Graph:

$$\xrightarrow{\hspace{0.3cm}\underset{-4\;-3\;-2\;-1\;\;0\;\;\;1\;\;\;2\;\;\;3\;\;\;4}{\hspace{1.5cm}[\!\!\!\!\!\!\!\!\rule{0pt}{0pt}}\hspace{0.3cm}}$$

Interval notation: $[-2, \infty)$

5. Solving Inequalities of the Form $a < x < b$

To solve a compound inequality of the form $a < x < b$ we can work with the inequality as a three-part inequality and isolate the variable, x, as demonstrated in Example 8.

Example 8 Solving a Compound Inequality of the Form $a < x < b$

Solve the inequality: $-3 \le 2x + 1 < 7$. Graph the solution and write the answer in interval notation.

Solution:

To solve the compound inequality $-3 \le 2x + 1 < 7$ isolate the variable x in the middle. The operations performed on the middle portion of the inequality must also be performed on the left-hand side and right-hand side.

$$-3 \leq 2x + 1 < 7$$

$$-3 - 1 \leq 2x + 1 - 1 < 7 - 1 \qquad \text{Subtract } 1 \text{ from all three parts of the inequality.}$$

$$-4 \leq 2x < 6 \qquad \text{Simplify.}$$

$$\frac{-4}{2} \leq \frac{2x}{2} < \frac{6}{2} \qquad \text{Divide by } 2 \text{ in all three parts of the inequality.}$$

$$-2 \leq x < 3$$

Graph:
$$\begin{array}{c}-6\ -5\ -4\ -3\ -2\ -1\ \ 0\ \ 1\ \ 2\ \ 3\ \ 4\ \ 5\ \ 6\end{array}$$

Interval notation: $[-2, 3)$

6. Applications of Linear Inequalities

Table 2-2 provides several commonly used translations to express inequalities.

Table 2-2

English Phrase	Mathematical Inequality
a is less than b	$a < b$
a is greater than b a exceeds b	$a > b$
a is less than or equal to b a is at most b a is no more than b	$a \leq b$
a is greater than or equal to b a is at least b a is no less than b	$a \geq b$

Skill Practice

Translate the English phrase into a mathematical inequality.

15. Bill needs a score of at least 92 on the final exam. Let x represent Bill's score.

16. Fewer than 19 cars are in the parking lot. Let c represent the number of cars.

17. The heights, h, of women who wear petite size clothing are typically between 58 in. and 63 in., inclusive.

Example 9 Translating Expressions Involving Inequalities

Translate the English phrases into mathematical inequalities.

a. Claude's annual salary, s, is no more than $40,000.

b. A citizen must be at least 18 years old to vote. (Let a represent a citizen's age.)

c. An amusement park ride has a height requirement between 48 in. and 70 in. (Let h represent height in inches.)

Solution:

a. $s \leq 40,000$ Claude's annual salary, s, is no more than $40,000.

b. $a \geq 18$ A citizen must be at least 18 years old to vote.

c. $48 < h < 70$ An amusement park ride has a height requirement between 48 in. and 70 in.

Linear inequalities are found in a variety of applications. Example 10 can help you determine the minimum grade you need on an exam to get an A in your math course.

Answers

15. $x \geq 92$

16. $c < 19$

17. $58 \leq h \leq 63$

Example 10 Solving an Application with Linear Inequalities

To earn an A in a math class, Alsha must average at least 90 on all of her tests. Suppose Alsha has scored 79, 86, 93, 90, and 95 on her first five math tests. Determine the minimum score she needs on her sixth test to get an A in the class.

Solution:

Let x represent the score on the sixth exam.   Label the variable.

$$\left(\begin{array}{c}\text{Average of}\\\text{all tests}\end{array}\right) \geq 90$$ Create a verbal model.

$$\frac{79 + 86 + 93 + 90 + 95 + x}{6} \geq 90$$ The average score is found by taking the sum of the test scores and dividing by the number of scores.

$$\frac{443 + x}{6} \geq 90$$ Simplify.

$$6\left(\frac{443 + x}{6}\right) \geq (90)6$$ Multiply both sides by 6 to clear fractions.

$$443 + x \geq 540$$ Solve the inequality.

$$x \geq 540 - 443$$ Subtract 443 from both sides.

$$x \geq 97$$ Interpret the results.

Alsha must score at least 97 on her sixth exam to receive an A in the course.

Answer

18. Simon needs at least 85.

Section 2.8 **Practice Exercises**

Boost your GRADE at ALEKS.com!

- Practice Problems
- Self-Tests
- NetTutor
- e-Professors
- Videos

Study Skills Exercises

1. Find the page numbers for the Chapter Review Exercises, the Chapter Test, and the Cumulative Review Exercises for this chapter.

Chapter Review Exercises _____   Chapter Test _____

Cumulative Review Exercises _____

Compare these features and state the advantages of each.

2. Define the key terms:

　a. linear inequality in one variable　　**b. compound inequality**

　c. set-builder notation　　**d. interval notation**

Review Problems

3. Solve the equation.　$3(x + 2) - (2x - 7) = -(5x - 1) - 2(x + 6)$

4. Solve the equation.　$6 - 8(x + 3) + 5x = 5x - (2x - 5) + 13$

Objectives 1–2: Graphing Linear Inequalities; Set-Builder Notation and Interval Notation

For Exercises 5–10, graph each inequality and write the solution set in interval notation. **(See Examples 1–3.)**

Set-Builder Notation	Graph	Interval Notation
5. $\{x \mid x \geq 6\}$		
6. $\left\{x \mid \frac{1}{2} < x \leq 4\right\}$		
7. $\{x \mid x \leq 2.1\}$		
8. $\left\{x \mid x > \frac{7}{3}\right\}$		
9. $\{x \mid -2 < x \leq 7\}$		
10. $\{x \mid x < -5\}$		

For Exercises 11–16, write each set in set-builder notation and in interval notation. **(See Examples 1–3.)**

Set-Builder Notation	Graph	Interval Notation
11.		
12.		
13.		
14.		
15.		
16.		

For Exercises 17–22, graph each set and write the set in set-builder notation. **(See Examples 1–3.)**

Set-Builder Notation	Graph	Interval Notation
17.		$[18, \infty)$
18.		$[-10, -2]$
19.		$(-\infty, -0.6)$
20.		$\left(-\infty, \frac{5}{3}\right)$
21.		$[-3.5, 7.1)$
22.		$[-10, \infty)$

Objectives 3–4: Properties of Inequality

For Exercises 23–30, solve the equation in part (a). For part (b), solve the inequality and graph the solution set. **(See Examples 4–7.)**

23. a. $x + 3 = 6$

    b. $x + 3 > 6$

24. a. $y - 6 = 12$

    b. $y - 6 \geq 12$

25. a. $p - 4 = 9$

    b. $p - 4 \leq 9$

26. a. $k + 8 = 10$

    b. $k + 8 < 10$

27. a. $4c = -12$ **28. a.** $5d = -35$ **29. a.** $-10z = 15$ **30. a.** $-2w = 14$

  b. $4c < -12$ **b.** $5d > -35$ **b.** $-10z \le 15$ **b.** $-2w < 14$

Objective 5: Solving Inequalities of the Form $a < x < b$

For Exercises 31–36, graph the solution. **(See Example 8.)**

31. $-1 < y \le 4$ **32.** $2.5 \le t < 5.7$ **33.** $0 < x + 3 < 8$

34. $-2 \le x - 4 \le 3$ **35.** $8 \le 4x \le 24$ **36.** $-9 < 3x < 12$

Mixed Exercises

For Exercises 37–84, solve the inequality. Graph the solution set and write the set in interval notation.

37. $x + 5 \le 6$ **38.** $y - 7 < 6$ **39.** $3q - 7 > 2q + 3$

40. $5r + 4 \ge 4r - 1$ **41.** $4 < 1 + z$ **42.** $3 > z - 6$

43. $2 \ge a - 6$ **44.** $7 \le b + 12$ **45.** $3c > 6$

46. $4d \le 12$ **47.** $-3c > 6$ **48.** $-4d \le 12$

49. $-h \le -14$ **50.** $-q > -7$ **51.** $12 \ge -\dfrac{x}{2}$

52. $6 < -\dfrac{m}{3}$ **53.** $-2 \le p + 1 < 4$ **54.** $0 < k + 7 < 6$

55. $-3 < 6h - 3 < 12$ **56.** $-6 \le 4a - 2 \le 12$ **57.** $5 < \dfrac{1}{2}x < 6$

58. $-6 \le 3x \le 12$ **59.** $-5 \le 4x - 1 < 15$ **60.** $-2 < \dfrac{1}{3}x - 2 \le 2$

61. $54 \le 0.6z$ **62.** $28 < -0.7w$ **63.** $-\dfrac{2}{3}y < 6$

64. $\dfrac{3}{4}x \le -12$ **65.** $-2x - 4 \le 11$ **66.** $-3x + 1 > 0$

67. $-12 > 7x + 9$ **68.** $8 < 2x - 10$ **69.** $-7b - 3 \le 2b$

70. $3t \geq 7t - 35$

71. $4n + 2 < 6n + 8$

72. $2w - 1 \leq 5w + 8$

73. $8 - 6(x - 3) > -4x + 12$

74. $3 - 4(h - 2) > -5h + 6$

75. $3(x + 1) - 2 \leq \frac{1}{2}(4x - 8)$

76. $8 - (2x - 5) \geq \frac{1}{3}(9x - 6)$

77. $\frac{7}{6}p + \frac{4}{3} \geq \frac{11}{6}p - \frac{7}{6}$

78. $\frac{1}{3}w - \frac{1}{2} \leq \frac{5}{6}w + \frac{1}{2}$

79. $\frac{y - 6}{3} > y + 4$

80. $\frac{5t + 7}{2} < t - 4$

81. $-1.2a - 0.4 < -0.4a + 2$

82. $-0.4c + 1.2 > -2c - 0.4$

83. $-2x + 5 \geq -x + 5$

84. $4x - 6 < 5x - 6$

For Exercises 85–88, determine whether the given number is a solution to the inequality.

85. $-2x + 5 < 4$; $x = -2$

86. $-3y - 7 > 5$; $y = 6$

87. $4(p + 7) - 1 > 2 + p$; $p = 1$

88. $3 - k < 2(-1 + k)$; $k = 4$

Objective 6: Applications of Linear Inequalities

89. Let x represent a student's average in a math class. The grading scale is given here.

A	$93 \leq x \leq 100$
B+	$89 \leq x < 93$
B	$84 \leq x < 89$
C+	$80 \leq x < 84$
C	$75 \leq x < 80$
F	$0 \leq x < 75$

   a. Write the range of scores corresponding to each letter grade in interval notation.

   b. If Stephan's average is 84.01, what grade will he receive?

   c. If Estella's average is 79.89, what grade will she receive?

90. Let x represent a student's average in a science class. The grading scale is given here.

A	$90 \leq x \leq 100$
B+	$86 \leq x < 90$
B	$80 \leq x < 86$
C+	$76 \leq x < 80$
C	$70 \leq x < 76$
D+	$66 \leq x < 70$
D	$60 \leq x < 66$
F	$0 \leq x < 60$

   a. Write the range of scores corresponding to each letter grade in interval notation.

   b. If Jacque's average is 89.99, what is her grade?

   c. If Marc's average is 66.01, what is his grade?

For Exercises 91–100, translate the English phrase into a mathematical inequality. **(See Example 9.)**

91. The length of a fish, L, was at least 10 in.

92. Tasha's average test score, t, exceeded 90.

93. The wind speed, w, exceeded 75 mph.

94. The height of a cave, h, was no more than 2 ft.

95. The temperature of the water in Blue Spring, t, is no more than 72°F.

96. The temperature on the tennis court, t, was no less than 100°F.

97. The length of the hike, L, was no less than 8 km.

98. The depth, d, of a certain pool was at most 10 ft.

99. The snowfall, h, in Monroe County is between 2 inches and 5 inches.

100. The cost, c, of carpeting a room is between $300 and $400.

101. The average summer rainfall for Miami, Florida, for June, July, and August is 7.4 in. per month. If Miami receives 5.9 in. of rain in June and 6.1 in. in July, how much rain is required in August to exceed the 3-month summer average? **(See Example 10.)**

102. The average winter snowfall for Burlington, Vermont, for December, January, and February is 18.7 in. per month. If Burlington receives 22 in. of snow in December and 24 in. in January, how much snow is required in February to exceed the 3-month winter average?

103. An artist paints wooden birdhouses. She buys the birdhouses for $9 each. However, for large orders, the price per birdhouse is discounted by a percentage off the original price. Let x represent the number of birdhouses ordered. The corresponding discount is given in the table.

Size of Order	Discount
$x \le 49$	0%
$50 \le x \le 99$	5%
$100 \le x \le 199$	10%
$x \ge 200$	20%

a. If the artist places an order for 190 birdhouses, compute the total cost.

b. Which costs more: 190 birdhouses or 200 birdhouses? Explain your answer.

104. A wholesaler sells T-shirts to a surf shop at $8 per shirt. However, for large orders, the price per shirt is discounted by a percentage off the original price. Let x represent the number of shirts ordered. The corresponding discount is given in the table.

Number of Shirts Ordered	Discount
$x \le 24$	0%
$25 \le x \le 49$	2%
$50 \le x \le 99$	4%
$100 \le x \le 149$	6%
$x \ge 150$	8%

a. If the surf shop orders 50 shirts, compute the total cost.

b. Which costs more: 148 shirts or 150 shirts? Explain your answer.

105. Maggie sells lemonade at an art show. She has a fixed cost of $75 to cover the registration fee for the art show. In addition, her cost to produce each lemonade is $0.17. If x represents the number of lemonades, then the total cost to produce x lemonades is given by:

$$\text{Cost} = 75 + 0.17x$$

If Maggie sells each lemonade for $2, then her revenue (the amount she brings in) for selling x lemonades is given by:

$$\text{Revenue} = 2.00x$$

a. Write an inequality that expresses the number of lemonades, x, that Maggie must sell to make a profit. Profit is realized when the revenue is greater than the cost (Revenue > Cost).

b. Solve the inequality in part (a).

106. Two rental car companies rent subcompact cars at a discount. Company A rents for $14.95 per day plus 22 cents per mile. Company B rents for $18.95 a day plus 18 cents per mile. Let x represent the number of miles driven in one day.

The cost to rent a subcompact car for one day from Company A is: $\text{Cost}_A = 14.95 + 0.22x$

The cost to rent a subcompact car for one day from Company B is: $\text{Cost}_B = 18.95 + 0.18x$

a. Write an inequality that expresses the number of miles, x, for which the daily cost to rent from Company A is less than the daily cost to rent from Company B.

b. Solve the inequality in part (a).

Expanding Your Skills

For Exercises 107–112, solve the inequality. Graph the solution set and write the set in interval notation.

107. $3(x + 2) - (2x - 7) \leq (5x - 1) - 2(x + 6)$

108. $6 - 8(y + 3) + 5y > 5y - (2y - 5) + 13$

109. $-2 - \dfrac{w}{4} \leq \dfrac{1 + w}{3}$

110. $\dfrac{z - 3}{4} - 1 > \dfrac{z}{2}$

111. $-0.703 < 0.122p - 2.472$

112. $3.88 - 1.335t \geq 5.66$

Group Activity

Computing Body Mass Index (BMI)

Materials: Calculator

Estimated Time: 10 minutes

Group Size: 2

Body mass index is a statistical measure of an individual's weight in relation to the person's height. It is computed by

$$\text{BMI} = \frac{703W}{h^2}$$
where W is a person's weight in *pounds.*
h is the person's height in *inches.*

The NIH categorizes body mass indices as follows:

1. Compute the body mass index for a person 5′4″ tall weighing 160 lb. Is this person's weight considered ideal?

Body Mass Index (BMI)	Weight Status
$18.5 \leq \text{BMI} \leq 24.9$	considered ideal
$25.0 \leq \text{BMI} \leq 29.9$	considered overweight
$\text{BMI} \geq 30.0$	considered obese

2. At the time that basketball legend Michael Jordan played for the Chicago Bulls, he was 210 lb and stood 6′6″ tall. What was Michael Jordan's body mass index?

3. For a fixed height, body mass index is a function of a person's weight only. For example, for a person 72 in. tall (6 ft), solve the following inequality to determine the person's ideal weight range.

$$18.5 \leq \frac{703W}{(72)^2} \leq 24.9$$

4. At the time that professional bodybuilder, Jay Cutler, won the Mr. Olympia contest he was 260 lb and stood 5′10″ tall.

   a. What was Jay Cutler's body mass index?

   b. As a body builder, Jay Cutler has an extraordinarily small percentage of body fat. Yet, according to the chart, would he be considered overweight or obese? Why do you think that the formula is not an accurate measurement of Mr. Cutler's weight status?

Chapter 2   Summary

| Section 2.1 | Addition, Subtraction, Multiplication, and Division Properties of Equality |

Key Concepts

An equation is an algebraic statement that indicates two expressions are equal. A **solution to an equation** is a value of the variable that makes the equation a true statement. The set of all solutions to an equation is the solution set of the equation.

A **linear equation in one variable** can be written in the form $ax + b = 0$, where $a \neq 0$.

Addition Property of Equality:

If $a = b$, then $a + c = b + c$

Subtraction Property of Equality:

If $a = b$, then $a - c = b - c$

Multiplication Property of Equality:

If $a = b$, then $ac = bc$

Division Property of Equality:

If $a = b$, then $\dfrac{a}{c} = \dfrac{b}{c}$ $(c \neq 0)$

Examples

Example 1

$2x + 1 = 9$ is an equation with solution 4.

Check: $2(4) + 1 \overset{?}{=} 9$

$8 + 1 \overset{?}{=} 9$

$9 \overset{?}{=} 9$ ✔   True

Example 2

$x - 5 = 12$

$x - 5 + 5 = 12 + 5$

$x = 17$ The solution is 17.

Example 3

$z + 1.44 = 2.33$

$z + 1.44 - 1.44 = 2.33 - 1.44$

$z = 0.89$ The solution is 0.89.

Example 4

$\dfrac{3}{4}x = 12$

$\dfrac{4}{3} \cdot \dfrac{3}{4}x = 12 \cdot \dfrac{4}{3}$

$x = 16$ The solution is 16.

Example 5

$16 = 8y$

$\dfrac{16}{8} = \dfrac{8y}{8}$

$2 = y$ The solution is 2.

| **Section 2.4** | **Applications of Linear Equations: Introduction to Problem Solving** |

Key Concepts

Problem-Solving Steps for Word Problems:

1. Read the problem carefully.
2. Assign labels to unknown quantities.
3. Develop a verbal model.
4. Write a mathematical equation.
5. Solve the equation.
6. Interpret the results and write the answer in words.

Examples

Example 1

The perimeter of a triangle is 54 m. The lengths of the sides are represented by three consecutive even integers. Find the lengths of the three sides.

1. Read the problem.

2. Let x represent one side, $x + 2$ represent the second side, and $x + 4$ represent the third side.

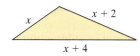

3. (First side) + (second side) + (third side) = perimeter

4. $x + (x + 2) + (x + 4) = 54$

5. $3x + 6 = 54$
   $3x = 48$
   $x = 16$

6. $x = 16$ represents the length of the shortest side. The lengths of the other sides are given by $x + 2 = 18$ and $x + 4 = 20$.

   The lengths of the three sides are 16 m, 18 m, and 20 m.

Section 2.5   Applications Involving Percents

Key Concepts

The following formula will help you solve basic percent problems.

$$\text{Amount} = (\text{percent})(\text{base})$$

One common use of percents is in computing **sales tax**.

Another use of percent is in computing **simple interest** using the formula:

$$\begin{pmatrix}\text{Simple} \\ \text{interest}\end{pmatrix} = (\text{principal})\begin{pmatrix}\text{annual} \\ \text{interest} \\ \text{rate}\end{pmatrix}\begin{pmatrix}\text{time in} \\ \text{years}\end{pmatrix}$$

or $I = Prt.$

Examples

Example 1

A dinette set costs $1260.00 after a 5% sales tax is included. What was the price before tax?

$$\begin{pmatrix}\text{Price} \\ \text{before tax}\end{pmatrix} + (\text{tax}) = \begin{pmatrix}\text{total} \\ \text{price}\end{pmatrix}$$

$$x \qquad + 0.05x = 1260$$

$$1.05x = 1260$$

$$x = 1200$$

The dinette set cost $1200 before tax.

Example 2

John Li invests $5400 at 2.5% simple interest. How much interest does he earn after 5 years?

$$I = Prt$$

$$I = (\$5400)(0.025)(5)$$

$$I = \$675$$

Section 2.6   Formulas and Applications of Geometry

Key Concepts

A **literal equation** is an equation that has more than one variable. Often such an equation can be manipulated to solve for different variables.

Formulas from Section R.4 can be used in applications involving geometry.

Examples

Example 1

$$P = 2a + b, \text{ solve for } a.$$

$$P - b = 2a + b - b$$

$$P - b = 2a$$

$$\frac{P - b}{2} = \frac{2a}{2}$$

$$\frac{P - b}{2} = a \quad \text{or} \quad a = \frac{P - b}{2}$$

Example 2

Find the length of a side of a square whose perimeter is 28 ft.

Use the formula $P = 4s$. Substitute 28 for P and solve:

$$P = 4s$$

$$28 = 4s$$

$$7 = s$$

The length of a side of the square is 7 ft.

Section 2.7   Mixture Applications and Uniform Motion

Key Concepts

Example 1 illustrates a mixture problem.

Example 1

How much 80% disinfectant solution should be mixed with 8 L of a 30% disinfectant solution to make a 40% solution?

	80% Solution	30% Solution	40% Solution
Amount of Solution	x	8	$x + 8$
Amount of Pure Disinfectant	$0.80x$	$0.30(8)$	$0.40(x + 8)$

$0.80x + 0.30(8) = 0.40(x + 8)$

$\quad 0.80x + 2.4 = 0.40x + 3.2$

$\quad 0.40x + 2.4 = 3.2 \qquad$ Subtract $0.40x$.

$\qquad 0.40x = 0.80 \qquad$ Subtract 2.4.

$\qquad\quad x = 2 \qquad$ Divide by 0.40.

2 L of 80% solution is needed.

Examples

Example 2 illustrates a uniform motion problem.

Example 2

Jack and Diane participate in a bicycle race. Jack rides the first half of the race in 1.5 hr. Diane rides the second half at a rate 5 mph slower than Jack and completes her portion in 2 hr. How fast does each person ride?

	Distance	Rate	Time
Jack	$1.5x$	x	1.5
Diane	$2(x - 5)$	$x - 5$	2

$\begin{pmatrix} \text{Distance} \\ \text{Jack rides} \end{pmatrix} = \begin{pmatrix} \text{Distance} \\ \text{Diane rides} \end{pmatrix}$

$\quad 1.5x \quad = \quad 2(x - 5)$

$\qquad 1.5x = 2x - 10$

$\qquad -0.5x = -10 \qquad$ Subtract $2x$.

$\qquad\quad x = 20 \qquad$ Divide by -0.5.

Jack's speed is x. Jack rides 20 mph. Diane's speed is $x - 5$, which is 15 mph.

Section 2.8   Linear Inequalities

Key Concepts

A **linear inequality in one variable**, x, is any relationship in the form: $ax + b < 0$, $ax + b > 0$, $ax + b \le 0$, or $ax + b \ge 0$, where $a \ne 0$.

The solution set to an inequality can be expressed as a graph or in **set-builder notation** or in **interval notation**.

When graphing an inequality or when writing interval notation, a parenthesis, (or), is used to denote that an endpoint is *not included* in a solution set. A square bracket, [or], is used to show that an endpoint *is included* in a solution set. Parenthesis (or) are always used with $-\infty$ and ∞, respectively.

The inequality $a < x < b$ is used to show that x is greater than a and less than b. That is, x is *between a* and b.

Multiplying or dividing an inequality by a negative quantity requires the direction of the inequality sign to be reversed.

Examples

Example 1

$\qquad -2x + 6 \ge 14$

$-2x + 6 - 6 \ge 14 - 6 \qquad$ Subtract 6.

$\qquad -2x \ge 8 \qquad$ Simplify.

$\qquad \dfrac{-2x}{-2} \le \dfrac{8}{-2} \qquad$ Divide by -2. Reverse the inequality sign.

$\qquad x \le -4$

Set-builder notation: $\{x \mid x \le -4\}$

Graph:

Interval notation: $(-\infty, -4]$

Chapter 2   Review Exercises

Section 2.1

1. Label the following as either an expression or an equation:

   a. $3x + y = 10$ **b.** $9x + 10y - 2xy$

   c. $4(x + 3) = 12$ **d.** $-5x = 7$

2. Explain how to determine whether an equation is linear in one variable.

3. Identify which equations are linear.

   a. $4x^2 + 8 = -10$ **b.** $x + 18 = 72$

   c. $-3 + 2y^2 = 0$ **d.** $-4p - 5 = 6p$

4. For the equation, $4y + 9 = -3$, determine if the given numbers are solutions.

   a. $y = 3$ **b.** $y = 0$

   c. $y = -3$ **d.** $y = -2$

For Exercises 5–12, solve the equation using the addition property, subtraction property, multiplication property, or division property of equality.

5. $a + 6 = -2$ 6. $6 = z - 9$

7. $-\dfrac{3}{4} + k = \dfrac{9}{2}$ 8. $0.1r = 7$

9. $-5x = 21$ 10. $\dfrac{t}{3} = -20$

11. $-\dfrac{2}{5}k = \dfrac{4}{7}$ 12. $-m = -27$

13. The quotient of a number and negative six is equal to negative ten. Find the number.

14. The difference of a number and $-\frac{1}{8}$ is $\frac{5}{12}$. Find the number.

15. Four subtracted from a number is negative twelve. Find the number.

16. Six subtracted from a number is negative eight. Find the number.

Section 2.2

For Exercises 17–28, solve the equation.

17. $4d + 2 = 6$ 18. $5c - 6 = -9$

19. $-7c = -3c - 9$ 20. $-28 = 5w + 2$

21. $\dfrac{b}{3} + 1 = 0$ 22. $\dfrac{2}{3}h - 5 = 7$

23. $-3p + 7 = 5p + 1$ 24. $4t - 6 = -12t + 16$

25. $4a - 9 = 3(a - 3)$ 26. $3(2c + 5) = -2(c - 8)$

27. $7b + 3(b - 1) + 16 = 2(b + 8)$

28. $2 + (17 - x) + 2(x - 1) = 4(x + 2) - 8$

29. Explain the difference between an equation that is a contradiction and an equation that is an identity.

30. Label each equation as a conditional equation, a contradiction, or an identity.

   a. $x + 3 = 3 + x$ **b.** $3x - 19 = 2x + 1$

   c. $5x + 6 = 5x - 28$ **d.** $2x - 8 = 2(x - 4)$

   e. $-8x - 9 = -8(x - 9)$

Section 2.3

For Exercises 31–48, solve the equation.

31. $\dfrac{x}{8} - \dfrac{1}{4} = \dfrac{1}{2}$ 32. $\dfrac{y}{15} - \dfrac{2}{3} = \dfrac{4}{5}$

33. $\dfrac{x + 5}{2} - \dfrac{2x + 10}{9} = 5$

34. $\dfrac{x - 6}{3} - \dfrac{2x + 8}{2} = 12$

35. $\dfrac{1}{10}p - 3 = \dfrac{2}{5}p$ 36. $\dfrac{1}{4}y - \dfrac{3}{4} = \dfrac{1}{2}y + 1$

37. $-\dfrac{1}{4}(2 - 3t) = \dfrac{3}{4}$ 38. $\dfrac{2}{7}(w + 4) = \dfrac{1}{2}$

39. $17.3 - 2.7q = 10.55$

40. $4.9z + 4.6 = 3.2z - 2.2$

41. $5.74a + 9.28 = 2.24a - 5.42$

42. $62.84t - 123.66 = 4(2.36 + 2.4t)$

43. $0.05x + 0.10(24 - x) = 0.75(24)$

44. $0.20(x + 4) + 0.65x = 0.20(854)$

45. $100 - (t - 6) = -(t - 1)$

46. $3 - (x + 4) + 5 = 3x + 10 - 4x$

47. $5t - (2t + 14) = 3t - 14$

48. $9 - 6(2z + 1) = -3(4z - 1)$

Section 2.4

49. Twelve added to the sum of a number and two is forty-four. Find the number.

50. Twenty added to the sum of a number and six is thirty-seven. Find the number.

51. Three times a number is the same as the difference of twice the number and seven. Find the number.

52. Eight less than five times a number is forty-eight less than the number. Find the number.

53. Three times the largest of three consecutive even integers is 76 more than the sum of the other two integers. Find the integers.

54. Ten times the smallest of three consecutive integers is 213 more than the sum of the other two integers. Find the integers.

55. The perimeter of a triangle is 78 in. The lengths of the sides are represented by three consecutive integers. Find the lengths of the sides of the triangle.

56. The perimeter of a pentagon (a five-sided polygon) is 190 cm. The five sides are represented by consecutive integers. Find the lengths of the sides.

57. The minimum salary for a major league baseball player in 1985 was $60,000. This was twice the minimum salary in 1980. What was the minimum salary in 1980?

58. The state of Indiana has approximately 2.1 million more people than Kentucky. Together their population totals 10.3 million. Approximately how many people are in each state?

Section 2.5

 For Exercises 59–68, solve the problems involving percents.

59. What is 35% of 68?   **60.** What is 4% of 720?

61. 53.5 is what percent of 428?

62. 68.4 is what percent of 72?

63. 24 is 15% of what number?

64. 8.75 is 0.5% of what number?

65. A novel is discounted 30%. The sale price is $20.65. What was the original price?

66. A couple spent a total of $50.40 for dinner. This included a 20% tip and 6% sales tax on the price of the meal. What was the price of the dinner before tax and tip?

67. Anna Tsao invested $3000 in an account paying 8% simple interest.

    a. How much interest will she earn in $3\frac{1}{2}$ years?

    b. What will her balance be at that time?

68. Eduardo invested money in an account earning 4% simple interest. At the end of 5 years, he had a total of $14,400. How much money did he originally invest?

Section 2.6

For Exercises 69–76, solve for the indicated variable.

69. $C = K - 273$ for K

70. $K = C + 273$ for C

71. $P = 4s$ for s

72. $P = 3s$ for s

73. $y = mx + b$ for x

74. $a + bx = c$ for x

75. $2x + 5y = -2$ for y

76. $4(a + b) = Q$ for b

For Exercises 77–83, use the appropriate geometry formula to solve the problem.

77. The volume of a cone is given by the formula

$$V = \frac{1}{3}\pi r^2 h.$$

    a. Solve the formula for h.

    b. Find the height of a right circular cone whose volume is 47.8 in.^{3} and whose radius is 3 in. Round to the nearest tenth of an inch.

78. Find the height of a parallelogram whose area is 42 m^2 and whose base is 6 m.

79. The smallest angle of a triangle is 2° more than $\frac{1}{4}$ of the largest angle. The middle angle is 2° less than the largest angle. Find the measure of each angle.

80. One angle is 6° less than twice a second angle. If the two angles are complementary, what are their measures?

81. A rectangular window has width 1 ft less than its length. The perimeter is 18 ft. Find the length and the width of the window.

82. Find the measure of the vertical angles by first solving for x.

$(2x + 25)°$ $(4x - 15)°$

83. Find the measure of angle y.

37°

y

Section 2.7

84. In stormy conditions, a delivery truck can travel a route in 14 hr. In good weather, the same trip can be made in 10.5 hr because the truck travels 15 km/hr faster. Find the speed of the truck in stormy weather and the speed in good weather.

85. Winston and Gus ride their bicycles in a relay. Each rider rides the same distance. Winston rides 3 mph faster than Gus and finishes the course in 2.5 hr. Gus finishes in 3 hr. How fast does each person ride?

86. Two cars leave a rest stop on Interstate I-10 at the same time. One heads east and the other heads west. One car travels 55 mph and the other 62 mph. How long will it take for them to be 327.6 mi apart?

87. Two hikers begin at the same time at opposite ends of a 9-mi trail and walk toward each other. One hiker walks 2.5 mph and the other walks 1.5 mph. How long will it be before they meet?

88. How much ground beef with 24% fat should be mixed with 8 lb of ground sirloin that is 6% fat to make a mixture that is 9.6% fat?

89. A soldering compound with 40% lead (the rest is tin) must be combined with 80 lb of solder that is 75% lead to make a compound that is 68% lead? How much solder with 40% lead should be used?

Section 2.8

For Exercises 90–92, graph the inequalities and write the set in interval notation.

90. $\{x \mid x > -2\}$

91. $\left\{x \mid x \leq \dfrac{1}{2}\right\}$

92. $\{x \mid -1 < x \leq 4\}$

 93. A landscaper buys potted geraniums from a nursery at a price of $5 per plant. However, for large orders, the price per plant is discounted by a percentage off the original price. Let x represent the number of potted plants ordered. The corresponding discount is given in the following table.

Number of Plants	Discount
$x \leq 99$	0%
$100 \leq x \leq 199$	2%
$200 \leq x \leq 299$	4%
$x \geq 300$	6%

a. Find the cost to purchase 130 plants.

b. Which costs more, 300 plants or 295 plants? Explain your answer.

For Exercises 94–103, solve the inequality. Graph the solution set and express the answer in interval notation.

94. $c + 6 < 23$ $\longrightarrow$

95. $3w - 4 > -5$ $\longrightarrow$

96. $-2x - 7 \geq 5$ $\longrightarrow$

97. $5(y + 2) \leq -4$ $\longrightarrow$

98. $-\dfrac{3}{7}a \leq -21$ $\longrightarrow$

99. $1.3 > 0.4t - 12.5$ $\longrightarrow$

100. $4k + 23 < 7k - 31$ $\longrightarrow$

101. $\dfrac{6}{5}h - \dfrac{1}{5} \leq \dfrac{3}{10} + h$ $\longrightarrow$

102. $-6 < 2b \leq 14$ $\longrightarrow$

103. $-2 \leq z + 4 \leq 9$ $\longrightarrow$

104. The summer average rainfall for Bermuda for June, July, and August is 5.3 in. per month. If Bermuda receives 6.3 in. of rain in June and 7.1 in. in July, how much rain is required in August to exceed the 3-month summer average?

105. Reggie sells hot dogs at a ballpark. He has a fixed cost of $33 to use the concession stand at the park. In addition, the cost for each hot dog is $0.40. If x represents the number of hot dogs sold, then the total cost is given by

$$\text{Cost} = 33 + 0.40x$$

If Reggie sells each hot dog for $1.50, then his revenue (the amount he brings in) for selling x hot dogs is given by

$$\text{Revenue} = 1.50x$$

a. Write an inequality that expresses the number of hot dogs, x, that Reggie must sell to make a profit. Profit is realized when the revenue is greater than the cost (revenue > cost).

b. Solve the inequality in part (a).

Chapter 2   Test

1. Which of the equations have $x = -3$ as a solution?

a. $4x + 1 = 10$ **b.** $6(x - 1) = x - 21$

c. $5x - 2 = 2x + 1$ **d.** $\dfrac{1}{3}x + 1 = 0$

2. a. Simplify: $3x - 1 + 2x + 8$

b. Solve: $3x - 1 = 2x + 8$

For Exercises 3–13, solve the equation.

3. $t + 3 = -13$ **4.** $8 = p - 4$

5. $\dfrac{t}{8} = -\dfrac{2}{9}$ **6.** $-3x + 5 = -2$

7. $2(p - 4) = p + 7$

8. $2 + d = 2 - 3(d - 5) - 2$

9. $\dfrac{3}{7} + \dfrac{2}{5}x = -\dfrac{1}{5}x + 1$ **10.** $3h + 1 = 3(h + 1)$

11. $\dfrac{3x + 1}{2} - \dfrac{4x - 3}{3} = 1$

12. $0.5c - 1.9 = 2.8 + 0.6c$

13. $-5(x + 2) + 8x = -2 + 3x - 8$

14. Solve the equation for y: $3x + y = -4$

15. Solve $C = 2\pi r$ for r.

16. 13% of what is 11.7?

17. One number is four plus one-half of another. The sum of the numbers is 31. Find the numbers.

18. The perimeter of a pentagon (a five-sided polygon) is 315 in. The five sides are represented by consecutive integers. Find the measures of the sides.

19. The total bill for a pair of basketball shoes (including sales tax) is $87.74. If the tax rate is 7%, find the cost of the shoes before tax.

20. A couple purchased two hockey tickets and two basketball tickets for $153.92. A hockey ticket cost $4.32 more than a basketball ticket. What were the prices of the individual tickets?

21. Clarita borrowed money at a 6% simple interest rate. If she paid back a total of $8000 at the end of 10 yr, how much did she originally borrow?

22. The length of a soccer field for international matches is 40 m less than twice its width. If the perimeter is 370 m, what are the dimensions of the field?

23. Given the triangle, find the measures of each angle by first solving for y.

24. Paula mixes macadamia nuts that cost $9.00 per pound with 50 lb of peanuts that cost $5.00 per pound. How many pounds of macadamia nuts should she mix to make a nut mixture that costs $6.50 per pound?

25. Two families leave their homes at the same time to meet for lunch. The families live 210 mi apart, and one family drives 5 mph slower than the other. If it takes them 2 hr to meet at a point between their homes, how fast does each family travel?

26. Two angles are complementary. One angle is 26° more than the other angle. What are the measures of the angles?

27. Graph the inequalities and write the sets in interval notation.
    a. $\{x \mid x < 0\}$
    b. $\{x \mid -2 \le x < 5\}$

For Exercises 28–31, solve the inequality. Graph the solution and write the solution set in interval notation.

28. $5x + 14 > -2x$

29. $2(3 - x) \ge 14$

30. $3(2y - 4) + 1 > 2(2y - 3) - 8$

31. $-13 \le 3p + 2 \le 5$

32. The average winter snowfall for Syracuse, New York, for December, January, and February is 27.5 in. per month. If Syracuse receives 24 in. of snow in December and 32 in. in January, how much snow is required in February to exceed the 3-month average?

Chapters 1–2   Cumulative Review Exercises

For Exercises 1–5, perform the indicated operations.

1. $\left| -\dfrac{1}{5} + \dfrac{7}{10} \right|$

2. $5 - 2[3 - (4 - 7)]$

3. $-\dfrac{2}{3} + \left(\dfrac{1}{2}\right)^2$

4. $-3^2 + (-5)^2$

5. $\sqrt{5 - (-20) - 3^2}$

For Exercises 6–7, translate the mathematical expressions and simplify the results.

6. The square root of the difference of five squared and nine

7. The sum of -14 and 12

8. List the terms of the expression: $-7x^2y + 4xy - 6$

9. Simplify: $-4[2x - 3(x + 4)] + 5(x - 7)$

For Exercises 10–15, solve the equations.

10. $8t - 8 = 24$

11. $-2.5x - 5.2 = 12.8$

12. $-5(p - 3) + 2p = 3(5 - p)$

13. $\dfrac{x + 3}{5} - \dfrac{x + 2}{2} = 2$

14. $\dfrac{2}{9}x - \dfrac{1}{3} = x + \dfrac{1}{9}$

15. $-0.6w = 48$

16. The sum of two consecutive odd integers is 156. Find the integers.

17. The total bill for a man's three-piece suit (including sales tax) is $374.50. If the tax rate is 7%, find the cost of the suit before tax.

18. The area of a triangle is 41 cm². Find the height of the triangle if the base is 12 cm.

12 cm

For Exercises 19–20, solve the inequality. Graph the solution set on a number line and express the solution in interval notation.

19. $-3x - 3(x + 1) < 9$ ⟶

20. $-6 \le 2x - 4 \le 14$ ⟶

Graphing Linear Equations in Two Variables

3

Chapter 3

In this chapter, we study graphing and focus on the graphs of lines. As you work through this chapter, make note of the key terms. Then try to find them in this word search puzzle.

origin   slope
axis   point
line   quadrant
parallel   ordered pair
intercept

Section 3.1	Rectangular Coordinate System

Objectives

1. **Interpreting Graphs**
2. **Plotting Points in a Rectangular Coordinate System**
3. **Applications of Plotting and Identifying Points**

1. Interpreting Graphs

Mathematics is a powerful tool used by scientists and has directly contributed to the highly technical world in which we live. Applications of mathematics have led to advances in the sciences, business, computer technology, and medicine.

One fundamental application of mathematics is the graphical representation of numerical information (or **data**). For example, Table 3-1 represents the number of clients admitted to a drug and alcohol rehabilitation program over a 12-month period.

Table 3-1

	Month	Number of Clients
Jan.	1	55
Feb.	2	62
March	3	64
April	4	60
May	5	70
June	6	73
July	7	77
Aug.	8	80
Sept.	9	80
Oct.	10	74
Nov.	11	85
Dec.	12	90

In table form, the information is difficult to picture and interpret. It appears that on a monthly basis, the number of clients fluctuates. However, when the data are represented in a graph, an upward trend is clear (Figure 3-1).

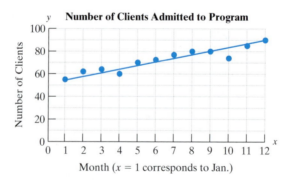

Figure 3-1

From the increase in clients shown in this graph, management for the rehabilitation center might make plans for the future. If the trend continues, management might consider expanding its facilities and increasing its staff to accommodate the expected increase in clients.

Example 1 Interpreting a Graph

Refer to Figure 3-1 and Table 3-1.

a. For which month was the number of clients the greatest?

b. How many clients were served in the first month (January)?

c. Which month corresponds to 60 clients served?

d. Between which two months did the number of clients decrease?

e. Between which two months did the number of clients remain the same?

Solution:

a. Month 12 (December) corresponds to the highest point on the graph, which represents the most clients.

b. In month 1 (January), there were 55 clients served.

c. Month 4 (April).

d. The number of clients decreased between months 3 and 4 and between months 9 and 10.

e. The number of clients remained the same between months 8 and 9.

2. Plotting Points in a Rectangular Coordinate System

In Example 1, two variables are represented, time and the number of clients. To picture two variables, we use a graph with two number lines drawn at right angles to each other (Figure 3-2). This forms a **rectangular coordinate system**. The horizontal line is called the **x-axis**, and the vertical line is called the **y-axis**. The point where the lines intersect is called the **origin**. On the x-axis, the numbers to the right of the origin are positive and the numbers to the left are negative. On the y-axis, the numbers above the origin are positive and the numbers below are negative. The x- and y-axes divide the graphing area into four regions called **quadrants**.

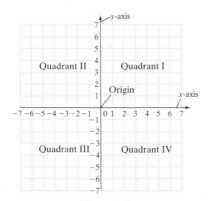

Figure 3-2

Points graphed in a rectangular coordinate system are defined by two numbers as an **ordered pair**, (x, y). The first number (called the **x-coordinate**, or the abscissa) is the horizontal position from the origin. The second number (called the **y-coordinate**, or the ordinate) is the vertical position from the origin. Example 2 shows how points are plotted in a rectangular coordinate system.

Answers

1. Month 10 (October)
2. 30
3. Month 2 (February)
4. Months 1 and 6 (January and June)

5. Plot the points.

$A(3, 4)$ $B(-2, 2)$

$C(4, 0)$ $D\left(\dfrac{5}{2}, -\dfrac{1}{2}\right)$

$E(-5, -2)$

Example 2   **Plotting Points in a Rectangular Coordinate System**

Plot the points.

a. $(4, 5)$ **b.** $(-4, -5)$ **c.** $(-1, 3)$ **d.** $(3, -1)$

e. $\left(\dfrac{1}{2}, -\dfrac{7}{3}\right)$ **f.** $(-2, 0)$ **g.** $(0, 0)$ **h.** $(\pi, 1.1)$

Solution:

See Figure 3-3.

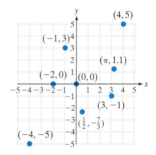

a. The ordered pair $(4, 5)$ indicates that $x = 4$ and $y = 5$. Beginning at the origin, move 4 units in the positive x-direction (4 units to the right), and from there move 5 units in the positive y-direction (5 units up). Then plot the point. The point is in Quadrant I.

b. The ordered pair $(-4, -5)$ indicates that $x = -4$ and $y = -5$. Move 4 units in the negative x-direction (4 units to the left), and from there move 5 units in the negative y-direction (5 units down). Then plot the point. The point is in Quadrant III.

Figure 3-3

c. The ordered pair $(-1, 3)$ indicates that $x = -1$ and $y = 3$. Move 1 unit to the left and 3 units up. The point is in Quadrant II.

d. The ordered pair $(3, -1)$ indicates that $x = 3$ and $y = -1$. Move 3 units to the right and 1 unit down. The point is in Quadrant IV.

TIP: Notice that changing the order of the x- and y-coordinates changes the location of the point. The point $(-1, 3)$ for example is in Quadrant II, whereas $(3, -1)$ is in Quadrant IV (Figure 3-3). This is why points are represented by *ordered* pairs. The order of the coordinates is important.

e. The improper fraction $-\dfrac{7}{3}$ can be written as the mixed number $-2\dfrac{1}{3}$. Therefore, to plot the point $\left(\dfrac{1}{2}, -\dfrac{7}{3}\right)$ move to the right $\dfrac{1}{2}$ unit, and down $2\dfrac{1}{3}$ units. The point is in Quadrant IV.

f. The point $(-2, 0)$ indicates $y = 0$. Therefore, the point is on the x-axis.

g. The point $(0, 0)$ is at the origin.

h. The irrational number, π, can be approximated as 3.14. Thus, the point $(\pi, 1.1)$ is located approximately 3.14 units to the right and 1.1 units up. The point is in Quadrant I.

Answer

5.

3. Applications of Plotting and Identifying Points

The effective use of graphs for mathematical models requires skill in identifying points and interpreting graphs.

Example 3 Determining Points from a Graph

A map of a national park is drawn so that the origin is placed at the ranger station (Figure 3-4). Four fire observation towers are located at points $A, B, C,$ and D. Estimate the coordinates of the fire towers relative to the ranger station (all distances are in miles).

Figure 3-4

Solution:

Point A: $(-1, -3)$

Point B: $(-2, 3)$

Point C: $(3\frac{1}{2}, 1\frac{1}{2})$ or $(\frac{7}{2}, \frac{3}{2})$ or $(3.5, 1.5)$

Point D: $(1\frac{1}{2}, -2)$ or $(\frac{3}{2}, -2)$ or $(1.5, -2)$

Skill Practice

6. Towers are located at points $A, B, C,$ and D. Estimate the coordinates of the towers.

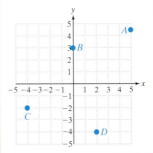

Example 4 Plotting Points in an Application

The daily low temperatures (in degrees Fahrenheit) for one week in January for Sudbury, Ontario, Canada, are given in Table 3-2.

a. Write an ordered pair for each row in the table using the day number as the x-coordinate and the temperature as the y-coordinate.

b. Plot the ordered pairs from part (a) on a rectangular coordinate system.

Table 3-2

Day Number, x	Temperature (°F), y
1	-3
2	-5
3	1
4	6
5	5
6	0
7	-4

Skill Practice

7. The table shows the number of homes sold in one town for a 6-month period. Plot the ordered pairs.

Month, x	Number Sold, y
1	20
2	25
3	28
4	40
5	45
6	30

Solution:

a. Each ordered pair represents the day number and the corresponding low temperature for that day.

$(1, -3)$ $(2, -5)$ $(3, 1)$ $(4, 6)$ $(5, 5)$ $(6, 0)$ $(7, -4)$

b.

TIP: The graph in Example 4(b) shows only Quadrants I and IV because all x-coordinates are positive.

Answers

6. $A(5, 4\frac{1}{2})$
   $B(0, 3)$
   $C(-4, -2)$
   $D(2, -4)$

7.

Section 3.1 | Practice Exercises

Study Skills Exercises

1. Before you proceed further in Chapter 3, make your test corrections for the Chapter 2 test. See Exercise 1 of Section 2.1 for instructions.

2. Define the key terms:

   a. data   b. ordered pair   c. origin   d. quadrant   e. rectangular coordinate system

   f. *x*-axis   g. *y*-axis   h. *x*-coordinate   i. *y*-coordinate

Objective 1: Interpreting Graphs

For Exercises 3–6, refer to the graphs to answer the questions.
(See Example 1.)

 3. The number of patients served by a certain hospice care center for the first 12 months after it opened is shown in the graph.

   a. For which month was the number of patients greatest?

   b. How many patients did the center serve in the first month?

   c. Between which months did the number of patients decrease?

   d. Between which two months did the number of patients remain the same?

   e. Which month corresponds to 40 patients served?

   f. Approximately how many patients were served during the 10th month?

Number of Patients Served by Month

4. The number of housing permits (in thousands) issued by a county in Texas between 2003 and 2009 is shown in the graph.

   a. For which year was the number of permits greatest?

   b. How many permits did the county issue in 2003?

   c. Between which years did the number of permits decrease?

   d. Between which two years did the number of permits remain the same?

   e. Which year corresponds to 7000 permits issued?

Number of Housing Permits

5. The price per share of a stock (in dollars) over a period of 5 days is shown in the graph.

   a. Interpret the meaning of the ordered pair (1, 89.25).

   b. What was the increase in price between day 3 and day 4?

   c. What was the decrease in price between day 4 and day 5?

Price per Share ($)

6. The price per share of a stock (in dollars) over a period of 5 days is shown in the graph.

    a. Interpret the meaning of the ordered pair $(1, 10.125)$.

    b. What was the decrease between day 4 and day 5?

Price per Share ($)

Objective 2: Plotting Points in a Rectangular Coordinate System

7. Plot the points on a rectangular coordinate system. (See Example 2.)

    a. $(2, 6)$ **b.** $(6, 2)$ **c.** $(-7, 3)$

    d. $(-7, -3)$ **e.** $(0, -3)$ **f.** $(-3, 0)$

    g. $(6, -4)$ **h.** $(0, 5)$

8. Plot the points on a rectangular coordinate system.

    a. $(4, 5)$ **b.** $(-4, 5)$ **c.** $(-6, 0)$

    d. $(6, 0)$ **e.** $(4, -5)$ **f.** $(-4, -5)$

    g. $(0, -2)$ **h.** $(0, 0)$

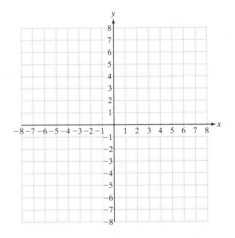

9. Plot the points on a rectangular coordinate system.

    a. $(-1, 5)$ **b.** $(0, 4)$ **c.** $\left(-2, -\dfrac{3}{2}\right)$

    d. $(2, -0.75)$ **e.** $(4, 2)$ **f.** $(-6, 0)$

10. Plot the points on a rectangular coordinate system.

    a. $(7, 0)$ **b.** $(-3, -1)$ **c.** $(6\frac{3}{5}, 1)$

    d. $(0, 1.5)$ **e.** $\left(\dfrac{1}{4}, -4\right)$ **f.** $\left(-\dfrac{1}{4}, 4\right)$

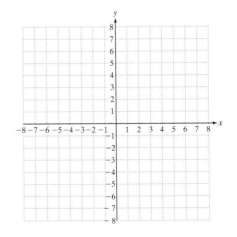

For Exercises 11–18, identify the quadrant in which the given point is found.

11. $(13, -2)$ **12.** $(25, 16)$ **13.** $(-8, 14)$ **14.** $(-82, -71)$

15. $(-5, -19)$ **16.** $(-31, 6)$ **17.** $\left(\dfrac{5}{2}, \dfrac{7}{4}\right)$ **18.** $(9, -40)$

19. Explain why the point $(0, -5)$ is *not* located in Quadrant IV.

20. Explain why the point $(-1, 0)$ is *not* located in Quadrant II.

21. Where is the point $\left(\dfrac{7}{8}, 0\right)$ located?

22. Where is the point $\left(0, \dfrac{6}{5}\right)$ located?

Objective 3: Applications of Plotting and Identifying Points

For Exercises 23–24, refer to the graph. **(See Example 3.)**

23. Estimate the coordinates of the points A, B, C, D, E, and F.

24. Estimate the coordinates of the points G, H, I, J, K, and L.

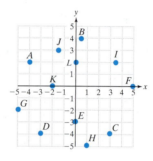

25. A map of a park is laid out with the visitor center located at the origin. Five visitors are in the park located at points A, B, C, D, and E. All distances are in meters.

     a. Estimate the coordinates of each hiker. **(See Example 3.)**

     b. How far apart are visitors C and D?

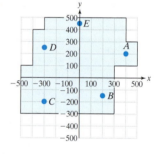

26. A townhouse has a sprinkler system in the backyard. With the water source at the origin, the sprinkler heads are located at points A, B, C, D, and E. All distances are in feet.

     a. Estimate the coordinates of each sprinkler head.

     b. How far is the distance from sprinkler head B to C?

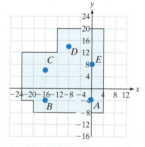

27. A movie theater has kept records of popcorn sales versus movie attendance.

     a. Use the table shown on page 213 to write the corresponding ordered pairs using the movie attendance as the x-variable and sales of popcorn as the y-variable. Interpret the meaning of the first ordered pair. **(See Example 4.)**

b. Plot the data points on a rectangular coordinate system.

Movie Attendance (Number of People)	Sales of Popcorn ($)
250	225
175	193
315	330
220	209
450	570
400	480
190	185

28. The age and systolic blood pressure (in millimeters of mercury, mm Hg) for eight different women are given in the table.

   a. Write the corresponding ordered pairs using the woman's age as the *x*-variable and the systolic blood pressure as the *y*-variable. Interpret the meaning of the first ordered pair.

   b. Plot the data points on a rectangular coordinate system.

Age (Years)	Systolic Blood Pressure (mm Hg)
57	149
41	120
71	158
36	115
64	151
25	110
40	118
77	165

29. The income level defining the poverty line for an individual is given for selected years between 1980 and 2005. Let *x* represent the number of years since 1980. Let *y* represent the income defining the poverty level. (*Source: U.S. Department of the Census*)

   (0, 4300)   (5, 5600)   (10, 6800)

   (15, 7900)   (20, 9000)   (25, 10,500)

   a. Interpret the meaning of the ordered pair (10, 6800).

   b. Plot the points on a rectangular coordinate system.

30. The following ordered pairs give the population of the U.S. colonies from 1700 to 1770. Let *x* represent the year, where *x* = 0 corresponds to 1700, *x* = 10 corresponds to 1710, and so on. Let *y* represent the population of the colonies.
   (*Source: Information Please Almanac*)

   (0, 251000)   (10, 332000)   (20, 466000)

   (30, 629000)   (40, 906000)   (50, 1171000)

   (60, 1594000)   (70, 2148000)

   a. Interpret the meaning of the ordered pair (10, 332000).

   b. Plot the points on a rectangular coordinate system.

31. The following table shows the average temperature in degrees Celsius for Montreal, Quebec, Canada, by month.

a. Write the corresponding ordered pairs, letting $x = 1$ correspond to the month of January.

b. Plot the ordered pairs on a rectangular coordinate system.

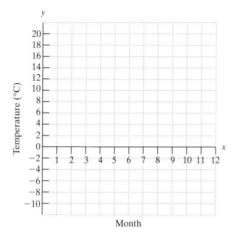

Month, x		Temperature (°C), y
Jan.	1	−10.2
Feb.	2	−9.0
March	3	−2.5
April	4	5.7
May	5	13.0
June	6	18.3
July	7	20.9
Aug.	8	19.6
Sept.	9	14.8
Oct.	10	8.7
Nov.	11	2.0
Dec.	12	−6.9

32. The table shows the average temperature in degrees Fahrenheit for Fairbanks, Alaska, by month.

a. Write the corresponding ordered pairs, letting $x = 1$ correspond to the month of January.

b. Plot the ordered pairs on a rectangular coordinate system.

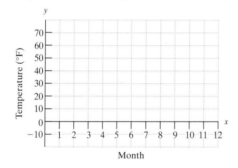

Month, x		Temperature (°F), y
Jan.	1	−12.8
Feb.	2	−4.0
March	3	8.4
April	4	30.2
May	5	48.2
June	6	59.4
July	7	61.5
Aug.	8	56.7
Sept.	9	45.0
Oct.	10	25.0
Nov.	11	6.1
Dec.	12	−10.1

Expanding Your Skills

33. The data in the table give the percent of males and females who have completed 4 or more years of college education for selected years. Let x represent the number of years since 1960. Let y represent the percent of men and the percent of women that completed 4 or more years of college.

Year	x	Percent, y Men	Percent, y Women
1960	0	9.7	5.8
1970	10	13.5	8.1
1980	20	20.1	12.8
1990	30	24.4	18.4
2000	40	27.8	23.6
2005	45	28.9	26.5

a. Plot the data points for men and for women on the same graph.

b. Is the percentage of men with 4 or more years of college increasing or decreasing?

c. Is the percentage of women with 4 or more years of college increasing or decreasing?

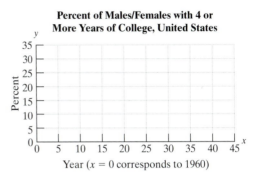

Percent of Males/Females with 4 or More Years of College, United States

34. Use the data and graph from Exercise 33 to answer the questions.

a. In which year was the difference in percentages between men and women with 4 or more years of college the greatest?

b. In which year was the difference in percentages between men and women the least?

c. If the trend continues beyond the data in the graph, does it seem possible that in the future, the percentage of women with 4 or more years of college will be greater than or equal to the percentage of men?

Linear Equations in Two Variables

1. Definition of a Linear Equation in Two Variables

Recall that an equation in the form $ax + b = 0$, where $a \neq 0$, is called a linear equation in one variable. A solution to such an equation is a value of x that makes the equation a true statement. For example, $3x + 6 = 0$ has a solution of -2.

In this section, we will look at linear equations in *two* variables.

Objectives

1. **Definition of a Linear Equation in Two Variables**
2. **Graphing Linear Equations in Two Variables by Plotting Points**
3. ***x*- and *y*-Intercepts**
4. **Horizontal and Vertical Lines**

> **DEFINITION   Linear Equation in Two Variables**
>
> Let A, B, and C be real numbers such that A and B are not both zero. Then, an equation that can be written in the form:
>
> $$Ax + By = C$$
>
> is called a **linear equation in two variables**.

The equation $x + y = 4$ is a linear equation in two variables. A solution to such an equation is an ordered pair (x, y) that makes the equation a true statement. Several solutions to the equation $x + y = 4$ are listed here:

Solution:	Check:
(x, y)	$x + y = 4$
$(2, 2)$	$(2) + (2) = 4$ ✔
$(1, 3)$	$(1) + (3) = 4$ ✔
$(4, 0)$	$(4) + (0) = 4$ ✔
$(-1, 5)$	$(-1) + (5) = 4$ ✔

By graphing these ordered pairs, we see that the solution points line up (Figure 3-5).

Notice that there are infinitely many solutions to the equation $x + y = 4$ so they cannot all be listed. Therefore, to visualize all solutions to the equation $x + y = 4$, we draw the line through the points in the graph. Every point on the line represents an ordered pair solution to the equation $x + y = 4$, and the line represents the set of *all* solutions to the equation.

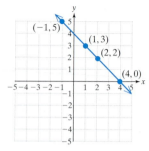

Figure 3-5

Skill Practice

Given the equation
$3x - 2y = -12$
determine whether the given ordered pair is a solution.

1. $(4, 0)$
2. $(-2, 3)$
3. $\left(1, \dfrac{15}{2}\right)$

Example 1 **Determining Solutions to a Linear Equation**

For the linear equation, $4x - 5y = 8$, determine whether the given ordered pair is a solution.

a. $(2, 0)$ **b.** $(3, 1)$ **c.** $\left(1, -\dfrac{4}{5}\right)$

Solution:

a. $4x - 5y = 8$

$4(2) - 5(0) \overset{?}{=} 8$ Substitute $x = 2$ and $y = 0$.

$8 - 0 \overset{?}{=} 8$ ✔ True The ordered pair $(2, 0)$ is a solution.

b. $4x - 5y = 8$

$4(3) - 5(1) \overset{?}{=} 8$ Substitute $x = 3$ and $y = 1$.

$12 - 5 \neq 8$ The ordered pair $(3, 1)$ is *not* a solution.

c. $4x - 5y = 8$

$4(1) - 5\left(-\dfrac{4}{5}\right) \overset{?}{=} 8$ Substitute $x = 1$ and $y = -\dfrac{4}{5}$.

$4 + 4 \overset{?}{=} 8$ ✔ True The ordered pair $\left(1, -\dfrac{4}{5}\right)$ is a solution.

---

2. Graphing Linear Equations in Two Variables by Plotting Points

In this section, we will graph linear equations in two variables.

> **DEFINITION** **The Graph of an Equation in Two Variables**
>
> The graph of an equation in two variables is the graph of all ordered pair solutions to the equation.

The word *linear* means "relating to or resembling a line." It is not surprising then that the solution set for any linear equation in two variables forms a line in a rectangular coordinate system. Because two points determine a line, to graph a linear equation it is sufficient to find two solution points and draw the line between them. We will find three solution points and use the third point as a check point. This process is demonstrated in Example 2.

Answers

1. No 2. Yes 3. Yes

Example 2   **Graphing a Linear Equation**

Graph the equation $x - 2y = 8$.

Solution:

We will find three ordered pairs that are solutions to $x - 2y = 8$. To find the ordered pairs, choose arbitrary values of x or y, such as those shown in the table. Then complete the table to find the corresponding ordered pairs.

x	y
2	
	-1
0	

→ (2,)
→ (, -1)
→ (0,)

TIP: Usually we try to choose arbitrary values that will be convenient to graph.

From the first row, substitute $x = 2$:

$x - 2y = 8$

$(2) - 2y = 8$

$-2y = 8 - 2$

$-2y = 6$

$y = -3$

From the second row, substitute $y = -1$:

$x - 2y = 8$

$x - 2(-1) = 8$

$x + 2 = 8$

$x = 8 - 2$

$x = 6$

From the third row, substitute $x = 0$:

$x - 2y = 8$

$(0) - 2y = 8$

$-2y = 8$

$y = -4$

The completed table is shown below with the corresponding ordered pairs.

x	y
2	-3
6	-1
0	-4

→ (2, -3)
→ (6, -1)
→ (0, -4)

To graph the equation, plot the three solutions and draw the line through the points (Figure 3-6).

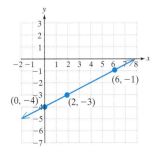

Figure 3-6

TIP: Only two points are needed to graph a line. However, in Example 2, we found a third ordered pair, $(0, -4)$. Notice that this point "lines up" with the other two points. If the three points do not line up, then we know that a mistake was made in solving for at least one of the ordered pairs.

In Example 2, the original values for x and y given in the table were picked arbitrarily by the authors. It is important to note, however, that once you pick an arbitrary value for x, the corresponding y-value is determined by the equation. Similarly, once you pick an arbitrary value for y, the x-value is determined by the equation.

Skill Practice

4. Graph the equation.
$2x + y = 6$

Answer

4.

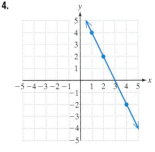

5. Graph the equation.
   $2x + 3y = 12$

Example 3   Graphing a Linear Equation

Graph the equation $4x + 3y = 15$.

Solution:

We will find three ordered pairs that are solutions to the equation $4x + 3y = 15$. In the table, we have selected arbitrary values for x and y and must complete the ordered pairs. Notice that in this case, we are choosing zero for x and zero for y to illustrate that the resulting equation is often easy to solve.

x	y	
0		→ $(0, \ )$
	0	→ $(\ , 0)$
3		→ $(3, \ )$

From the first row, substitute $x = 0$:

$$4x + 3y = 15$$
$$4(0) + 3y = 15$$
$$3y = 15$$
$$y = 5$$

From the second row, substitute $y = 0$:

$$4x + 3y = 15$$
$$4x + 3(0) = 15$$
$$4x = 15$$
$$x = \frac{15}{4} \text{ or } 3\frac{3}{4}$$

From the third row, substitute $x = 3$:

$$4x + 3y = 15$$
$$4(3) + 3y = 15$$
$$12 + 3y = 15$$
$$3y = 3$$
$$y = 1$$

The completed table is shown with the corresponding ordered pairs.

x	y	
0	5	→ $(0, 5)$
$3\frac{3}{4}$	0	→ $(3\frac{3}{4}, 0)$
3	1	→ $(3, 1)$

To graph the equation, plot the three solutions and draw the line through the points (Figure 3-7).

Figure 3-7

Answer

5.

Example 4   Graphing a Linear Equation in Two Variables

Graph the equation $y = -\dfrac{1}{3}x + 1$.

Solution:

Because the y-variable is isolated in the equation, it is easy to substitute a value for x and simplify the right-hand side to find y. Since any number for x can be picked, choose numbers that are multiples of 3. These will simplify easily when multiplied by $-\frac{1}{3}$.

x	y
3	
0	
−3	

$y = -\dfrac{1}{3}x + 1$

<u>Let $x = 3$:</u>

$y = -\dfrac{1}{3}(3) + 1$

$y = -1 + 1$

$y = 0$

<u>Let $x = 0$:</u>

$y = -\dfrac{1}{3}(0) + 1$

$y = 0 + 1$

$y = 1$

<u>Let $x = -3$:</u>

$y = -\dfrac{1}{3}(-3) + 1$

$y = 1 + 1$

$y = 2$

x	y
3	0
0	1
−3	2

The line through the three ordered pairs $(3, 0), (0, 1)$, and $(-3, 2)$ is shown in Figure 3-8. The line represents the set of all solutions to the equation $y = -\frac{1}{3}x + 1$.

Figure 3-8

3. x- and y-Intercepts

The x- and y-intercepts are the points where the graph intersects the x- and y-axes, respectively. From Example 4, we see that the x-intercept is at the point $(3, 0)$ and the y-intercept is at the point $(0, 1)$. See Figure 3-8. Notice that a y-intercept is a point on the y-axis and must have an x-coordinate of 0. Likewise, an x-intercept is a point on the x-axis and must have a y-coordinate of 0.

DEFINITION *x- and y-Intercepts*

An **x-intercept** of a graph is a point $(a, 0)$ where the graph intersects the x-axis.

A **y-intercept** of a graph is a point $(0, b)$ where the graph intersects the y-axis.

In some applications, an x-intercept is defined as the x-coordinate of a point of intersection that a graph makes with the x-axis. For example, if an x-intercept is at the point $(3, 0)$, it is sometimes stated simply as 3 (the y-coordinate is assumed to be 0). Similarly, a y-intercept is sometimes defined as the y-coordinate of a point of intersection that a graph makes with the y-axis. For example, if a y-intercept is at the point $(0, 7)$, it may be stated simply as 7 (the x-coordinate is assumed to be 0).

Although any two points may be used to graph a line, in some cases it is convenient to use the x- and y-intercepts of the line. To find the x- and y-intercepts of any two-variable equation in x and y, follow these steps:

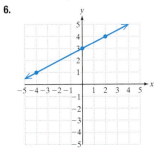

PROCEDURE *Finding x- and y-Intercepts*

Step 1 Find the x-intercept(s) by substituting $y = 0$ into the equation and solving for x.

Step 2 Find the y-intercept(s) by substituting $x = 0$ into the equation and solving for y.

Given the equation
$x - 3y = -4$,

7. Find the x-intercept.
8. Find the y-intercept.
9. Graph the equation.

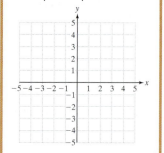

Example 5 Finding the x- and y-Intercepts of a Line

Given the equation $-3x + 2y = 8$,

a. Find the x-intercept.

b. Find the y-intercept.

c. Graph the equation.

Solution:

a. To find the x-intercept, substitute $y = 0$.

$$-3x + 2y = 8$$
$$-3x + 2(0) = 8$$
$$-3x \qquad = 8$$
$$\frac{-3x}{-3} = \frac{8}{-3}$$
$$x = -\frac{8}{3}$$

The x-intercept is $\left(-\frac{8}{3}, 0\right)$.

b. To find the y-intercept, substitute $x = 0$.

$$-3x + 2y = 8$$
$$-3(0) + 2y = 8$$
$$2y = 8$$
$$y = 4$$

The y-intercept is $(0, 4)$.

c. The line through the ordered pairs $\left(-\frac{8}{3}, 0\right)$ and $(0, 4)$ is shown in Figure 3-9. Note that the point $\left(-\frac{8}{3}, 0\right)$ can be written as $\left(-2\frac{2}{3}, 0\right)$.

The line represents the set of all solutions to the equation $-3x + 2y = 8$.

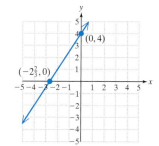

Figure 3-9

Example 6 Finding the x- and y-Intercepts of a Line

Given the equation $4x + 5y = 0$,

a. Find the x-intercept.

b. Find the y-intercept.

c. Graph the equation.

Solution:

a. To find the x-intercept, substitute $y = 0$.

$$4x + 5y = 0$$
$$4x + 5(0) = 0$$
$$4x \qquad = 0$$
$$x = 0$$

The x-intercept is $(0, 0)$.

b. To find the y-intercept, substitute $x = 0$.

$$4x + 5y = 0$$
$$4(0) + 5y = 0$$
$$5y = 0$$
$$y = 0$$

The y-intercept is $(0, 0)$.

Answers

7. $(-4, 0)$ **8.** $\left(0, \frac{4}{3}\right)$

9.

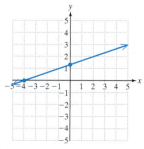

c. Because the *x*-intercept and the *y*-intercept are the same point (the origin), one or more additional points are needed to graph the line. In the table, we have arbitrarily selected additional values for *x* and *y* to find two more points on the line.

x	y
−5	
	2

Let $x = -5$:
$$4x + 5y = 0$$
$$4(-5) + 5y = 0$$
$$-20 + 5y = 0$$
$$5y = 20$$
$$y = 4$$

Let $y = 2$:
$$4x + 5y = 0$$
$$4x + 5(2) = 0$$
$$4x + 10 = 0$$
$$4x = -10$$
$$x = -\frac{10}{4}$$
$$x = -\frac{5}{2}$$

The line through the ordered pairs $(0, 0)$, $(-5, 4)$, and $\left(-\frac{5}{2}, 2\right)$ is shown in Figure 3-10. Note that the point $\left(-\frac{5}{2}, 2\right)$ can be written as $\left(-2\frac{1}{2}, 2\right)$.

The line represents the set of all solutions to the equation $4x + 5y = 0$.

x	y
−5	4
$-\frac{5}{2}$	2
0	0

Figure 3-10

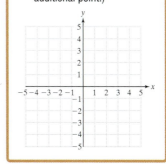
4. Horizontal and Vertical Lines

Recall that a linear equation can be written in the form of $Ax + By = C$, where *A* and *B* are not both zero. However, if *A* or *B* is 0, then the line is either parallel to the *x*-axis (horizontal) or parallel to the *y*-axis (vertical), respectively.

DEFINITION Equations of Vertical and Horizontal Lines

1. A **vertical line** can be represented by an equation of the form, $x = k$, where *k* is a constant.

2. A **horizontal line** can be represented by an equation of the form, $y = k$, where *k* is a constant.

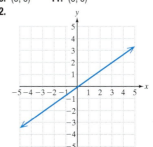

13. Graph the equation.
$y = -2$

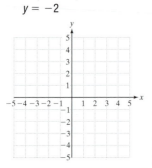

14. Graph the equation.
$3x = 12$

Answers

13.

14.

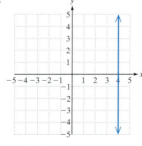

Example 7 **Graphing a Horizontal Line**

Graph the equation $y = 3$.

Solution:

Because this equation is in the form $y = k$, the line is horizontal and must cross the y-axis at $y = 3$ (Figure 3-11).

Figure 3-11

Alternative Solution:

Create a table of values for the equation $y = 3$. The choice for the y-coordinate must be 3, but x can be any real number.

x	y
0	3
1	3
2	3

x can be any number. y must be 3.

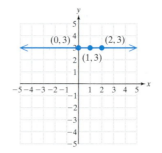

TIP: Notice that a horizontal line has a y-intercept, but does not have an x-intercept (unless the horizontal line is the x-axis itself).

Example 8 **Graphing a Vertical Line**

Graph the equation $4x = -8$.

Solution:

Because the equation does not have a y-variable, we can solve the equation for x.

$4x = -8$ is equivalent to $x = -2$

This equation is in the form $x = k$, indicating that the line is vertical and must cross the x-axis at $x = -2$ (Figure 3-12).

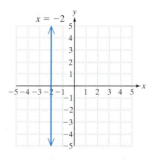

Figure 3-12

Alternative Solution:

Create a table of values for the equation $x = -2$. The choice for the x-coordinate must be -2, but y can be any real number.

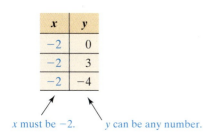

x	y
-2	0
-2	3
-2	-4

x must be -2. y can be any number.

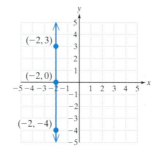

TIP: Notice that a vertical line has an x-intercept but does not have a y-intercept (unless the vertical line is the y-axis itself).

Calculator Connections

Topic: Graphing Linear Equations on an Appropriate Viewing Window

A viewing window of a graphing calculator shows a portion of a rectangular coordinate system. The standard viewing window for many calculators shows the x-axis between -10 and 10 and the y-axis between -10 and 10 (Figure 3-13). Furthermore, the scale defined by the "tick" marks on both the x- and y-axes is usually set to 1.

The "Standard Viewing Window"

Figure 3-13

To graph an equation in x and y on a graphing calculator, the equation must be written with the y-variable isolated. For example, to graph the equation $x + 3y = 3$, we solve for y by applying the steps for solving a literal equation. The result, $y = -\frac{1}{3}x + 1$, can now be entered into a graphing calculator. To enter the equation $y = -\frac{1}{3}x + 1$ use parentheses around the fraction $\frac{1}{3}$. The *Graph* option displays the graph of the line.

Sometimes the standard viewing window does not provide an adequate display for the graph of an equation. For example, the graph of $y = -x + 15$ is visible only in a small portion of the upper right corner of the standard viewing window.

To see where this line crosses the x- and y-axes, we can change the viewing window to accommodate larger values of x and y. Most calculators have a *Range* feature or *Window* feature that allows the user to change the minimum and maximum x- and y-values.

To get a better picture of the equation $y = -x + 15$, change the minimum x-value to -10 and the maximum x-value to 20. Similarly, use a minimum y-value of -10 and a maximum y-value of 20.

Calculator Exercises

For Exercises 1–8, graph the equations on the standard viewing window.

1. $y = -2x + 5$ **2.** $y = 3x - 1$

3. $y = \dfrac{1}{2}x - \dfrac{7}{2}$

4. $y = -\dfrac{3}{4}x + \dfrac{5}{3}$

5. $4x - 7y = 21$

6. $2x + 3y = 12$

7. $-3x - 4y = 6$

8. $-5x + 4y = 10$

For Exercises 9–12, graph the equations on the given viewing window.

9. $y = 3x + 15$

Window: $-10 \le x \le 10$
$\qquad\quad -5 \le y \le 20$

10. $y = -2x - 25$

Window: $-30 \le x \le 30$
$\qquad\quad -30 \le y \le 30$

Xscl = 3 (sets the x-axis tick marks to increments of 3)

Yscl = 3 (sets the y-axis tick marks to increments of 3)

11. $y = -0.2x + 0.04$

Window: $-0.1 \le x \le 0.3$
$\qquad\quad -0.1 \le y \le 0.1$

Xscl = 0.01 (sets the x-axis tick marks to increments of 0.01)

Yscl = 0.01 (sets the y-axis tick marks to increments of 0.01)

12. $y = 0.3x - 0.5$

Window: $-1 \le x \le 3$
$\qquad\quad -1 \le y \le 1$

Xscl = 0.1 (sets the x-axis tick marks to increments of 0.1)

Yscl = 0.1 (sets the y-axis tick marks to increments of 0.1)

Section 3.2 Practice Exercises

Study Skills Exercises

1. Check your progress by answering these questions.

Yes _____ No _____ Did you have sufficient time to study for the test on Chapter 2? If not, what could you have done to create more time for studying?

Yes _____ No _____ Did you work all of the assigned homework problems in Chapter 2?

Yes _____ No _____ If you encountered difficulty, did you see your instructor or tutor for help?

Yes _____ No _____ Have you taken advantage of the textbook supplements such as the *Student Solutions Manual*?

2. Define the key terms:

a. horizontal line   **b. linear equation in two variables**   **c. vertical line**

d. x-intercept   **e. y-intercept**

Review Exercises

For Exercises 3–8, refer to the figure to give the coordinates of the labeled points, and state the quadrant or axis where the point is located.

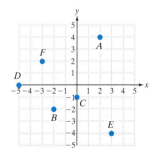

3. A **4.** B

5. C **6.** D

7. E **8.** F

Objective 1: Definition of a Linear Equation in Two Variables

For Exercises 9–17, determine if the given ordered pair is a solution to the equation. **(See Example 1.)**

9. $x - y = 6$; $(8, 2)$

10. $y = 3x - 2$; $(1, 1)$

 11. $y = -\dfrac{1}{3}x + 3$; $(-3, 4)$

12. $y = -\dfrac{5}{2}x + 5$; $(-2, 0)$

13. $4x + 5y = 20$; $(-5, -4)$

14. $y = 7$; $(0, 7)$

15. $y = -2$; $(-2, 6)$

16. $x = 1$; $(0, 1)$

17. $x = -5$; $(-5, 6)$

Objective 2: Graphing Linear Equations in Two Variables by Plotting Points

For Exercises 18–31, complete each table, and graph the corresponding ordered pairs. Draw the line defined by the points to represent all solutions to the equation. **(See Examples 2–4.)**

18. $x + y = 3$

x	y
2	
	3
-1	
	0

19. $x + y = -2$

x	y
1	
	0
-3	
	2

20. $y = 5x + 1$

x	y
1	
	1
-1	

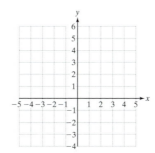

21. $y = -3x - 3$

x	y
-2	
	0
-4	

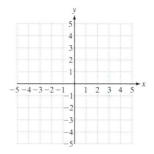

22. $2x - 3y = 6$

x	y
0	
	0
2	

23. $4x + 2y = 8$

x	y
0	
	0
3	

24. $y = \dfrac{2}{7}x - 5$

x	y
7	
-7	
0	

25. $y = -\dfrac{3}{5}x - 2$

x	y
0	
5	
10	

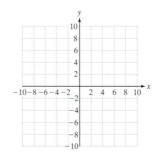

26. $y = 3$

x	y
2	
0	
−1	

27. $y = -2$

x	y
0	
−3	
5	

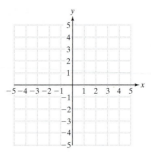

28. $x = -4$

x	y
	1
	−2
	4

29. $x = \frac{3}{2}$

x	y
	−1
	2
	−3

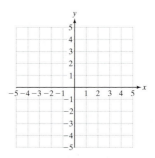

30. $y = -3.4x + 5.8$

x	y
0	
1	
2	

31. $y = -1.2x + 4.6$

x	y
0	
1	
2	

For Exercises 32–43, graph the lines by making a table of at least three ordered pairs and plotting the points.

32. $x = y + 2$

33. $x - y = 4$

34. $-3x + y = -6$

35. $2x - 5y = 10$

36. $y = 4x$

37. $y = -2x$

38. $y = -\dfrac{1}{2}x + 3$

39. $y = \dfrac{1}{4}x - 2$

40. $x + y = 0$

41. $-x + y = 0$

42. $50x - 40y = 200$

43. $-30x - 20y = 60$

Objective 3: *x*- and *y*-Intercepts

44. The *x*-intercept is on which axis?

45. The *y*-intercept is on which axis?

For Exercises 46–49, estimate the coordinates of the *x*- and *y*-intercepts.

46.

47.

48.

49.

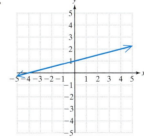

For Exercises 50–61, find the x- and y-intercepts (if they exist), and graph the line. **(See Examples 5–6.)**

50. $5x + 2y = 5$

51. $4x - 3y = -9$

52. $y = \dfrac{2}{3}x - 1$

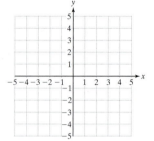

53. $y = -\dfrac{3}{4}x + 2$

54. $x - 3 = y$

55. $2x + 8 = y$

56. $-3x + y = 0$

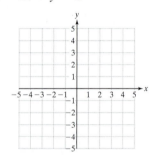

57. $2x - 2y = 0$

58. $25y = 10x + 100$

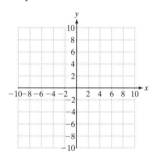

59. $20x = -40y + 200$

60. $x = 2y$

61. $x = -5y$

Objective 4: Horizontal and Vertical Lines

For Exercises 62–65, answer true or false. If the statement is false, rewrite it to be true.

62. The line $x = 3$ is horizontal.

63. The line $y = -4$ is horizontal.

64. A line parallel to the y-axis is vertical.

65. A line perpendicular to the x-axis is vertical.

For Exercises 66–74,

a. Identify the equation as representing a horizontal or vertical line.

b. Graph the line.

c. Identify the x- and y-intercepts if they exist. **(See Examples 7–8.)**

66. $x = 3$

67. $y = -1$

68. $-2y = 8$

69. $5x = 20$

70. $x + 3 = 7$

71. $y - 8 = -13$

72. $3y = 0$

73. $5x = 0$

74. $2x + 7 = 10$

75. Explain why not every line has both an x- and a y-intercept.

76. Which of the lines has an x-intercept?

    a. $2x - 3y = 6$ **b.** $x = 5$ **c.** $2y = 8$ **d.** $-x + y = 0$

77. Which of the lines has a y-intercept?

    a. $y = 2$ **b.** $x + y = 0$ **c.** $2x - 10 = 2$ **d.** $x + 4y = 8$

Expanding Your Skills

78. The store "CDs R US" sells all compact disks for $13.99. The following equation represents the revenue, y, (in dollars) generated by selling x CDs.

$$y = 13.99x \quad (x \geq 0)$$

a. Find y when $x = 13$.

b. Find x when $y = 279.80$.

c. Write the ordered pairs from parts (a) and (b), and interpret their meaning in the context of the problem.

d. Graph the ordered pairs and the line defined by the points.

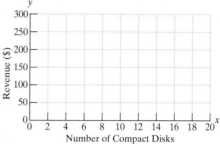

79. The value of a car depreciates once it is driven off of the dealer's lot. For a Hyundai Accent, the value of the car is given by the equation $y = -1531x + 11{,}599$ $(x \geq 0)$ where y is the value of the car in dollars x years after its purchase. (*Source: Kelly Blue Book*)

a. Find y when $x = 1$.

b. Find x when $y = 7006$.

c. Write the ordered pairs from parts (a) and (b), and interpret their meaning in the context of the problem.

| **Section 3.3** | **Slope of a Line and Rate of Change** |

Objectives

1. **Introduction to Slope**
2. **Slope Formula**
3. **Parallel and Perpendicular Lines**
4. **Applications of Slope: Rate of Change**

1. Introduction to Slope

The x- and y-intercepts represent the points where a line crosses the x- and y-axes. Another important feature of a line is its slope. Geometrically, the slope of a line measures the "steepness" of the line. For example, two ski runs are depicted by the lines in Figure 3-14.

Beginner's Hill   Daredevil Hill

Figure 3-14

By visual inspection, Daredevil Hill is "steeper" than Beginner's Hill. To measure the slope of a line quantitatively, consider two points on the line. The **slope** of the line is the ratio of the vertical change (change in y) between the two points and the horizontal change (change in x). As a memory device, we might think of the slope of a line as "rise over run." See Figure 3-15.

$$\text{Slope} = \frac{\text{change in } y}{\text{change in } x} = \frac{\text{rise}}{\text{run}}$$

Change in y (rise)   Change in x (run)

Figure 3-15

To move from point A to point B on Beginner's Hill, rise 2 ft and move to the right 6 ft (Figure 3-16).

To move from point A to point B on Daredevil Hill, rise 12 ft and move to the right 6 ft (Figure 3-17).

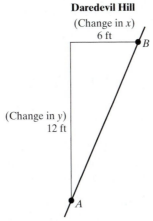

Daredevil Hill

(Change in x)
6 ft

B

(Change in y)
12 ft

A

Figure 3-17

Beginner's Hill

(Change in x)
6 ft

(Change in y)
2 ft

B

A

Figure 3-16

$$\text{Slope} = \frac{\text{change in } y}{\text{change in } x} = \frac{2 \text{ ft}}{6 \text{ ft}} = \frac{1}{3}$$

$$\text{Slope} = \frac{\text{change in } y}{\text{change in } x} = \frac{12 \text{ ft}}{6 \text{ ft}} = \frac{2}{1} = 2$$

The slope of Daredevil Hill is greater than the slope of Beginner's Hill, confirming the observation that Daredevil Hill is steeper. On Daredevil Hill there is a 12-ft change in elevation for every 6 ft of horizontal distance (a 2:1 ratio). On Beginner's Hill there is only a 2-ft change in elevation for every 6 ft of horizontal distance (a 1:3 ratio).

Example 1   **Finding Slope in an Application**

Determine the slope of the ramp up the stairs.

Solution:

$$\text{Slope} = \frac{\text{change in } y}{\text{change in } x} = \frac{8 \text{ ft}}{16 \text{ ft}}$$

$$\frac{8}{16} = \frac{1}{2} \quad \text{Write the ratio for the slope and simplify.}$$

The slope is $\frac{1}{2}$.

8 ft

16 ft

Skill Practice

1. Determine the slope of the aircraft's takeoff path.

500 ft

6000 ft

2. Slope Formula

The slope of a line may be found using any two points on the line—call these points (x_1, y_1) and (x_2, y_2). The change in y between the points can be found by taking the difference of the y-values: $y_2 - y_1$. The change in x can be found by taking the difference of the x-values in the same order: $x_2 - x_1$ (Figure 3-18).

The slope of a line is often symbolized by the letter m and is given by the following formula.

(x_2, y_2)

Change in y
$y_2 - y_1$

(x_1, y_1)

$x_2 - x_1$
Change in x

Figure 3-18

Answer

1. $\dfrac{500}{6000} = \dfrac{1}{12}$

> **FORMULA** **Slope Formula**
>
> The **slope** of a line passing through the distinct points (x_1, y_1) and (x_2, y_2) is
>
> $$m = \frac{y_2 - y_1}{x_2 - x_1} \quad \text{provided } x_2 - x_1 \neq 0$$

Example 2 **Finding the Slope of a Line Given Two Points**

Find the slope of the line through the points $(-1, 3)$ and $(-4, -2)$.

Solution:

To use the slope formula, first label the coordinates of each point and then substitute the coordinates into the slope formula.

$$\underset{(x_1, y_1)}{(-1, 3)} \quad \text{and} \quad \underset{(x_2, y_2)}{(-4, -2)} \qquad \text{Label the points.}$$

$$m = \frac{y_2 - y_1}{x_2 - x_1} = \frac{(-2) - (3)}{(-4) - (-1)} \qquad \text{Apply the slope formula.}$$

$$= \frac{-5}{-3}$$

$$= \frac{5}{3} \qquad \text{Simplify to lowest terms.}$$

The slope of the line can be verified from the graph (Figure 3-19).

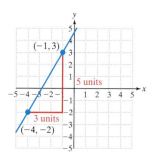

Figure 3-19

> **TIP:** The slope formula is not dependent on which point is labeled (x_1, y_1) and which point is labeled (x_2, y_2). In Example 2, reversing the order in which the points are labeled results in the same slope.
>
> $$\underset{(x_2, y_2)}{(-1, 3)} \quad \text{and} \quad \underset{(x_1, y_1)}{(-4, -2)} \qquad \text{Label the points.}$$
>
> $$m = \frac{(3) - (-2)}{(-1) - (-4)} = \frac{5}{3} \qquad \text{Apply the slope formula.}$$

Answer

2. $\frac{1}{6}$

When you apply the slope formula, you will see that the slope of a line may be positive, negative, zero, or undefined.

- Lines that increase, or rise, from left to right have a positive slope.
- Lines that decrease, or fall, from left to right have a negative slope.
- Horizontal lines have a slope of zero.
- Vertical lines have an undefined slope.

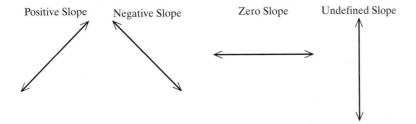

Positive Slope   Negative Slope   Zero Slope   Undefined Slope

---

Concept Connections

3. Label each line as having a slope that is positive, negative, zero, or undefined.

a.

b.

c.

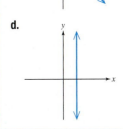

d.

Example 3 | **Finding the Slope of a Line Given Two Points**

Find the slope of the line passing through the points $(-5, 0)$ and $(2, -3)$.

Solution:

$$\begin{array}{ccc} (-5, 0) & \text{and} & (2, -3) \\ (x_1, y_1) & & (x_2, y_2) \end{array}$$ Label the points.

$$m = \frac{y_2 - y_1}{x_2 - x_1} = \frac{(-3) - (0)}{(2) - (-5)}$$ Apply the slope formula.

$$= \frac{-3}{7} \quad \text{or} \quad -\frac{3}{7}$$ Simplify.

By graphing the points $(-5, 0)$ and $(2, -3)$, we can verify that the slope is $-\frac{3}{7}$ (Figure 3-20). Notice that the line slopes downward from left to right.

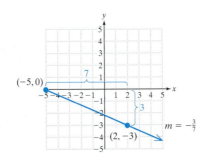

Figure 3-20

Skill Practice

Find the slope of the line through the given points.

4. $(0, -8)$ and $(-2, -2)$

Answers

3. **a.** Zero   **b.** Positive
   c. Negative   **d.** Undefined
4. -3

Example 4 Determining the Slope of a Vertical Line

Find the slope of the line passing through the points $(2, -1)$ and $(2, 4)$.

Solution:

$$\underset{(x_1, y_1)}{(2, -1)} \quad \text{and} \quad \underset{(x_2, y_2)}{(2, 4)} \qquad \text{Label the points.}$$

$$m = \frac{y_2 - y_1}{x_2 - x_1} = \frac{(4) - (-1)}{(2) - (2)} \qquad \text{Apply the slope formula.}$$

$$m = \frac{5}{0} \quad \text{Undefined}$$

Because the slope, m, is undefined, we expect the points to form a vertical line as shown in Figure 3-21.

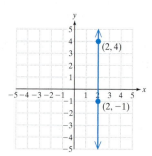

Figure 3-21

Example 5 Determine the Slope of a Horizontal Line

Find the slope of the line passing through the points $(3, -2)$ and $(-4, -2)$.

Solution:

$$\underset{(x_1, y_1)}{(3, -2)} \quad \text{and} \quad \underset{(x_2, y_2)}{(-4, -2)} \qquad \text{Label the points.}$$

$$m = \frac{y_2 - y_1}{x_2 - x_1} = \frac{(-2) - (-2)}{(-4) - (3)} \qquad \text{Apply the slope formula.}$$

$$m = \frac{-2 + 2}{-4 - 3} = \frac{0}{-7} = 0$$

Because the slope is 0, we expect the points to form a horizontal line, as shown in Figure 3-22.

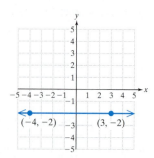

Figure 3-22

3. Parallel and Perpendicular Lines

Lines in the same plane that do not intersect are called **parallel lines**. Parallel lines have the same slope and different *y*-intercepts (Figure 3-23).

Lines that intersect at a right angle are **perpendicular lines**. If two lines are perpendicular then the slope of one line is the opposite of the reciprocal of the slope of the other line (provided neither line is vertical) (Figure 3-24).

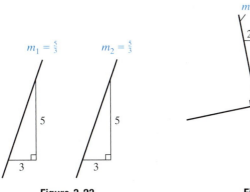

Figure 3-23   **Figure 3-24**

PROPERTY Slopes of Parallel Lines

If m_1 and m_2 represent the slopes of two parallel (nonvertical) lines, then

$$m_1 = m_2.$$

See Figure 3-23.

PROPERTY Slopes of Perpendicular Lines

If $m_1 \neq 0$ and $m_2 \neq 0$ represent the slopes of two perpendicular lines, then

$$m_1 = -\frac{1}{m_2} \text{ or equivalently, } m_1 m_2 = -1. \text{ See Figure 3-24.}$$

Example 6   **Determining the Slope of Parallel and Perpendicular Lines**

Suppose a given line has a slope of $-\frac{1}{4}$.

a. Find the slope of a line parallel to the given line.

b. Find the slope of a line perpendicular to the given line.

Solution:

a. Parallel lines must have the same slope. The slope of a line parallel to the given line is: $m = -\dfrac{1}{4}$.

b. Perpendicular lines must have opposite and reciprocal slopes. The slope of a line perpendicular to the given line is: $m = +\dfrac{4}{1}$ or simply, $m = 4$.

If the slopes of two lines are known, then we can compare the slopes to determine if the lines are parallel, perpendicular, or neither.

Skill Practice

Determine if lines l_1 and l_2 are parallel, perpendicular, or neither.

9. l_1: $(-2, -3)$ and $(4, -1)$
   l_2: $(0, 2)$ and $(-3, 1)$

Example 7 **Determining If Lines Are Parallel, Perpendicular, or Neither**

Lines l_1 and l_2 pass through the given points. Determine if l_1 and l_2 are parallel, perpendicular, or neither.

$$l_1: \quad (2, -7) \text{ and } (4, 1) \qquad l_2: \quad (-3, 1) \text{ and } (1, 0)$$

Solution:

Find the slope of each line.

l_1: $(2, -7)$ and $(4, 1)$
$\quad\;(x_1, y_1) \qquad (x_2, y_2)$

$$m_1 = \frac{1 - (-7)}{4 - 2}$$

$$m_1 = \frac{8}{2}$$

$$m_1 = 4$$

l_2: $(-3, 1)$ and $(1, 0)$
$\quad\;(x_1, y_1) \qquad (x_2, y_2)$

$$m_2 = \frac{0 - 1}{1 - (-3)}$$

$$m_2 = \frac{-1}{4}$$

$$m_2 = -\frac{1}{4}$$

Avoiding Mistakes

You can check that two lines are perpendicular by checking that the product of their slopes is -1.

$$4\left(-\frac{1}{4}\right) = -1$$

One slope is the opposite of the reciprocal of the other slope. Therefore, the lines are perpendicular.

4. Applications of Slope: Rate of Change

In many applications, the interpretation of slope refers to the *rate of change* of the y-variable to the x-variable.

Skill Practice

10. In the year 2000, the population of Alaska was approximately 630,000. By 2005, it had grown to 670,000. Use the ordered pairs (2000, 630,000) and (2005, 670,000) to determine the slope of the line through the points. Then interpret the meaning in the context of this problem.

Example 8 **Interpreting Slope in an Application**

The annual median income for males in the United States for selected years is shown in Figure 3-25. The trend is approximately linear. Find the slope of the line and interpret the meaning of the slope in the context of the problem.

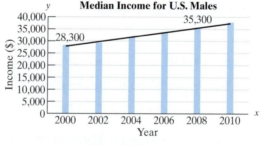

Figure 3-25

Source: U.S. Department of the Census

Answers

9. Parallel
10. $m = 8000$; The population of Alaska increased at a rate of 8000 people per year.

Solution:

To determine the slope we need to know two points on the line. From the graph, the median income for males in the year 2000 was approximately \$28,300. This gives us the ordered pair (2000, 28,300). In the year 2008, the income was $35,300. This gives the ordered pair (2008, 35,300).

$$\underset{(x_1,\,y_1)}{(2000,\,28{,}300)} \quad \text{and} \quad \underset{(x_2,\,y_2)}{(2008,\,35{,}300)} \qquad \text{Label the points.}$$

$$m = \frac{y_2 - y_1}{x_2 - x_1} = \frac{35{,}300 - 28{,}300}{2008 - 2000} \qquad \text{Apply the slope formula.}$$

$$= \frac{7000}{8}$$

$$= 875 \qquad \text{Simplify.}$$

The slope is 875. This tells us the rate of change of the y-variable (income) to the x-variable (years). This means that men's median income in the United States increased at a rate of $875 per year during this time period.

Section 3.3 Practice Exercises

Boost *your* GRADE at ALEKS.com!

- Practice Problems
- Self-Tests
- NetTutor
- e-Professors
- Videos

Study Skills Exercises

1. Make up a practice test for yourself. Use examples or exercises from the text. Be sure to cover each concept that was presented.

2. Define the key terms: **a. parallel lines**   **b. perpendicular lines**   **c. slope**

Review Exercises

For Exercises 3–8, find the x- and y-intercepts (if they exist). Then graph the line.

3. $x - 3y = 6$

4. $x - 5 = 2$

5. $y = \dfrac{2}{3}x$

6. $2y - 3 = 0$

7. $4x + y = 8$

8. $2x = 4y$

Objective 1: Introduction to Slope

9. Determine the pitch (slope) of the roof.
(See Example 1.)

10. Determine the slope of the stairs.

11. Calculate the slope of the handrail.

12. Determine the slope of the treadmill.

Objective 2: Slope Formula

For Exercises 13–16, fill in the blank with the appropriate term: zero, negative, positive, or undefined.

13. The slope of a line parallel to the y-axis is _____.

14. The slope of a horizontal line is ____.

15. The slope of a line that rises from left to right is _____.

16. The slope of a line that falls from left to right is _____.

For Exercises 17–25, determine if the slope is positive, negative, zero, or undefined.

17.

18.

19.

20.

21.

22.

23.

24.

25.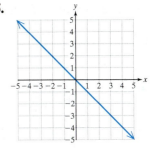

For Exercises 26–34, determine the slope by using the slope formula and any two points on the line. Check your answer by drawing a right triangle and labeling the "rise" and "run."

26.

27.

28.

29.

30.

31.

32.

33.

34.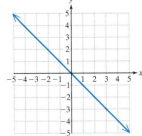

For Exercises 35–52, find the slope of the line that passes through the two points. **(See Examples 2–5.)**

35. $(2, 4)$ and $(-1, 3)$

36. $(0, 4)$ and $(3, 0)$

37. $(-2, 3)$ and $(-1, 0)$

38. $(-3, -4)$ and $(1, -5)$

39. $(1, 5)$ and $(-4, 2)$

40. $(-6, -1)$ and $(-2, -3)$

41. $(5, 3)$ and $(-2, 3)$

42. $(0, -1)$ and $(-4, -1)$

43. $(2, -7)$ and $(2, 5)$

44. $(-4, 3)$ and $(-4, -4)$

45. $\left(\dfrac{1}{2}, \dfrac{3}{5}\right)$ and $\left(\dfrac{1}{4}, -\dfrac{4}{5}\right)$

46. $\left(-\dfrac{2}{7}, \dfrac{1}{3}\right)$ and $\left(\dfrac{8}{7}, -\dfrac{5}{6}\right)$

47. $(3, -1)$ and $(-5, 6)$ **48.** $(-6, 5)$ and $(-10, 4)$ **49.** $(6.8, -3.4)$ and $(-3.2, 1.1)$

50. $(-3.15, 8.25)$ and $(6.85, -4.25)$ **51.** $(1994, 3.5)$ and $(2000, 2.6)$ **52.** $(1988, 4.65)$ and $(1998, 9.25)$

Objective 3: Parallel and Perpendicular Lines

For Exercises 53–60, the slope of a line is given. **(See Example 6.)**

a. Determine the slope of a line parallel to the given line.

b. Determine the slope of a line perpendicular to the given line.

53. $m = -2$ **54.** $m = \dfrac{2}{3}$

55. $m = 0$ **56.** The slope is undefined.

57. $m = \dfrac{4}{5}$ **58.** $m = -4$

59. The slope is undefined.   **60.** $m = 0$

For Exercises 61–66, let m_1 and m_2 represent the slopes of two lines. Determine if the lines are parallel, perpendicular, or neither. **(See Example 6.)**

61. $m_1 = -2, m_2 = \dfrac{1}{2}$ **62.** $m_1 = \dfrac{2}{3}, m_2 = \dfrac{3}{2}$ **63.** $m_1 = 1, m_2 = \dfrac{4}{4}$

64. $m_1 = \dfrac{3}{4}, m_2 = -\dfrac{8}{6}$ **65.** $m_1 = \dfrac{2}{7}, m_2 = -\dfrac{2}{7}$ **66.** $m_1 = 5, m_2 = 5$

For Exercises 67–72, find the slopes of the lines l_1 and l_2 defined by the two given points. Then determine whether l_1 and l_2 are parallel, perpendicular, or neither. **(See Example 7.)**

67. l_1: $(2, 4)$ and $(-1, -2)$
l_2: $(1, 7)$ and $(0, 5)$

68. l_1: $(0, 0)$ and $(-2, 4)$
l_2: $(1, -5)$ and $(-1, -1)$

69. l_1: $(1, 9)$ and $(0, 4)$
l_2: $(5, 2)$ and $(10, 1)$

70. l_1: $(3, -4)$ and $(-1, -8)$
l_2: $(5, -5)$ and $(-2, 2)$

71. l_1: $(4, 4)$ and $(0, 3)$
l_2: $(1, 7)$ and $(-1, -1)$

72. l_1: $(3, 5)$ and $(-2, -5)$
l_2: $(2, 0)$ and $(-4, -3)$

Objective 4: Applications of Slope: Rate of Change

73. In 1980, there were 304 thousand male inmates in federal and state prisons. By 2005, the number increased to 1479 thousand. Let x represent the year, and let y represent the number of prisoners (in thousands). **(See Example 8.)**

a. Using the ordered pairs $(1980, 304)$ and $(2005, 1479)$, find the slope of the line.

b. Interpret the slope in the context of this problem.

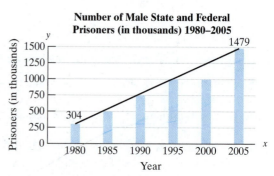

Number of Male State and Federal Prisoners (in thousands) 1980–2005

(*Source:* U.S. Bureau of Justice Statistics)

74. In the year 1980, there were 12 thousand female inmates in federal and state prisons. By 2005, the number increased to 102 thousand.

  Let x represent the year, and let y represent the number of prisoners (in thousands).

  a. Using the ordered pairs (1980, 12) and (2005, 102), find the slope of the line.

  b. Interpret the slope in the context of this problem.

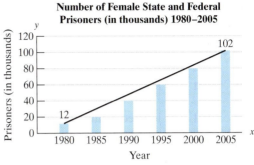

Number of Female State and Federal Prisoners (in thousands) 1980–2005

(*Source:* U.S. Bureau of Justice Statistics)

75. In 1985, the U.S. Postal Service charged $0.22 for first class letters and cards up to 1 oz. By 2007, the price had increased to $0.41. In the graph, x represents the year, and y represents the cost for 1 oz of first class postage.

  a. Determine the slope of the line. Round to three decimal places.

  b. Interpret the meaning of the slope in the context of this problem.

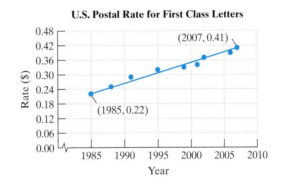

U.S. Postal Rate for First Class Letters

76. The distance, d (in miles), between a lightning strike and an observer is given by the equation $d = 0.2t$, where t is the time (in seconds) between seeing lightning and hearing thunder.

Distance Between Lightning Strike and an Observer

$d = 0.2t$

  a. If an observer counts 5 sec between seeing lightning and hearing thunder, how far away was the lightning strike?

  b. If an observer counts 10 sec between seeing lightning and hearing thunder, how far away was the lightning strike?

  c. If an observer counts 15 sec between seeing lightning and hearing thunder, how far away was the lightning strike?

  d. What is the slope of the line? Interpret the meaning of the slope in the context of this problem.

> **DEFINITION Slope-Intercept Form of a Line**
>
> $y = mx + b$ is the slope-intercept form of a line.
>
> m is the slope and the point $(0, b)$ is the y-intercept.

Example 1 Identifying the Slope and y-Intercept of a Line

For each equation, identify the slope and y-intercept.

a. $y = 3x - 1$ **b.** $y = -2.7x + 5$ **c.** $y = 4x$

Solution:

Each equation is written in slope-intercept form, $y = mx + b$. The slope is the coefficient of x, and the y-intercept is determined by the constant term.

a. $y = 3x - 1$ The slope is 3.   The y-intercept is $(0, -1)$.

b. $y = -2.7x + 5$ The slope is -2.7.   The y-intercept is $(0, 5)$.

c. $y = 4x$ can be written as $y = 4x + 0$.   The slope is 4.
    The y-intercept is $(0, 0)$.

Given the equation of a line, we can write the equation in slope-intercept form by solving the equation for the y-variable. This is demonstrated in the next example.

Example 2 Identifying the Slope and y-Intercept of a Line

Given the equation of the line $-5x - 2y = 6$,

a. Write the equation in slope-intercept form.

b. Identify the slope and y-intercept.

Solution:

a. Write the equation in slope-intercept form, $y = mx + b$, by solving for y.

$$-5x - 2y = 6$$

$$-2y = 5x + 6 \qquad \text{Add } 5x \text{ to both sides.}$$

$$\frac{-2y}{-2} = \frac{5x + 6}{-2} \qquad \text{Divide both sides by } -2.$$

$$y = \frac{5x}{-2} + \frac{6}{-2} \qquad \text{Divide each term by } -2 \text{ and simplify.}$$

$$y = -\frac{5}{2}x - 3 \qquad \text{Slope-intercept form.}$$

b. The slope is $-\frac{5}{2}$, and the y-intercept is $(0, -3)$.

2. Graphing a Line from Its Slope and y-Intercept

Slope-intercept form is a useful tool to graph a line. The y-intercept is a known point on the line. The slope indicates the direction of the line and can be used to find a second point. Using slope-intercept form to graph a line is demonstrated in the next example.

Example 3 Graphing a Line Using the Slope and *y*-Intercept

Graph the equation of the line $y = -\frac{5}{2}x - 3$ by using the slope and *y*-intercept.

Solution:

First plot the *y*-intercept, $(0, -3)$.

The slope, $m = -\frac{5}{2}$, can be written as

$$m = \frac{-5}{2} \longleftarrow \text{The change in } y \text{ is } -5.$$
$$\longleftarrow \text{The change in } x \text{ is } 2.$$

To find a second point on the line, start at the *y*-intercept and move down 5 units and to the right 2 units. Then draw the line through the two points (Figure 3-26).

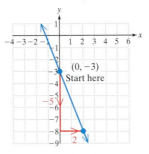

Figure 3-26

Similarly, the slope can be written as

$$m = \frac{5}{-2} \longleftarrow \text{The change in } y \text{ is } 5.$$
$$\longleftarrow \text{The change in } x \text{ is } -2.$$

To find a second point, start at the *y*-intercept and move up 5 units and to the left 2 units. Then draw the line through the two points (Figure 3-27).

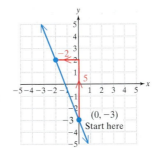

Figure 3-27

Example 4 Graphing a Line from Its Slope and *y*-Intercept

Graph the equation of the line $y = 4x$ by using the slope and *y*-intercept.

Solution:

The line can be written as $y = 4x + 0$. Therefore, we can plot the *y*-intercept at $(0, 0)$. The slope $m = 4$ can be written as

$$m = \frac{4}{1} \quad \begin{array}{l} \nwarrow \text{The change in } y \text{ is } 4. \\ \swarrow \text{The change in } x \text{ is } 1. \end{array}$$

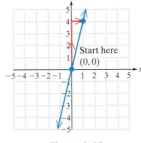

Figure 3-28

To find a second point on the line, start at the *y*-intercept and move up 4 units and to the right 1 unit. Then draw the line through the two points (Figure 3-28).

3. Determining Whether Two Lines Are Parallel, Perpendicular, or Neither

The slope-intercept form provides a means to find the slope of a line by inspection. Recall that if the slopes of two lines are known, then we can compare the slopes to determine if the lines are parallel, perpendicular, or neither parallel nor perpendicular. (Two distinct nonvertical lines are parallel if their slopes are equal. Two lines are perpendicular if the slope of one line is the opposite of the reciprocal of the slope of the other line.)

Skill Practice

6. Graph the equation by using the slope and the *y*-intercept.
$$y = 2x - 3$$

Skill Practice

7. Graph the equation by using the slope and the *y*-intercept.
$$y = -\frac{1}{4}x$$

Answers

6.

7.

Example 5 **Determining If Two Lines Are Parallel, Perpendicular, or Neither**

For each pair of lines, determine if they are parallel, perpendicular, or neither.

a. l_1: $y = 3x - 5$ **b.** l_1: $x - 3y = -9$

l_2: $y = 3x + 1$ l_2: $3x = -y + 4$

c. l_1: $y = \frac{3}{2}x + 2$ **d.** l_1: $x = 2$

l_2: $y = \frac{2}{3}x + 1$ l_2: $2y = 8$

Solution:

a. l_1: $y = 3x - 5$ The slope of l_1 is 3.

l_2: $y = 3x + 1$ The slope of l_2 is 3.

Because the slopes are the same, the lines are parallel.

b. First write the equation of each line in slope-intercept form.

l_1: $x - 3y = -9$ l_2: $3x = -y + 4$

$\qquad -3y = -x - 9$ $\qquad 3x + y = 4$

$\qquad \dfrac{-3y}{-3} = \dfrac{-x}{-3} - \dfrac{9}{-3}$ $\qquad y = -3x + 4$

$\qquad\qquad y = \dfrac{1}{3}x + 3$

l_1: $y = \frac{1}{3}x + 3$ The slope of l_1 is $\frac{1}{3}$.

l_2: $y = -3x + 4$ The slope of l_2 is -3.

The slope of $\frac{1}{3}$ is the opposite of the reciprocal of -3. Therefore, the lines are perpendicular.

c. l_1: $y = \frac{3}{2}x + 2$ The slope of l_1 is $\frac{3}{2}$.

l_2: $y = \frac{2}{3}x + 1$ The slope of l_2 is $\frac{2}{3}$.

The slopes are not the same. Therefore, the lines are not parallel. The values of the slopes are reciprocals, but they are not opposite in sign. Therefore, the lines are not perpendicular. The lines are neither parallel nor perpendicular.

d. The equation $x = 2$ represents a vertical line because the equation is in the form $x = k$.

The equation $2y = 8$ can be simplified to $y = 4$, which represents a horizontal line.

In this example, we do not need to analyze the slopes because vertical lines and horizontal lines are perpendicular.

4. Writing an Equation of a Line Using Slope-Intercept Form

The slope-intercept form of a line can be used to write an equation of a line when the slope is known and the y-intercept is known.

Example 6 Writing an Equation of a Line Using Slope-Intercept Form

Write an equation of the line whose slope is $\frac{2}{3}$ and whose y-intercept is $(0, 8)$.

Solution:

The slope is given as $m = \frac{2}{3}$, and the y-intercept $(0, b)$ is given as $(0, 8)$. Substitute the values $m = \frac{2}{3}$ and $b = 8$ into the slope-intercept form of a line.

$$y = mx + b$$
$$y = \frac{2}{3}x + 8$$

Skill Practice

12. Write an equation of the line whose slope is -4 and y-intercept is $(0, -10)$.

Example 7 Writing an Equation of a Line Using Slope-Intercept Form

Write an equation of the line that passes through the origin and whose slope is -2.

Solution:

The slope is given as $m = -2$. Furthermore, the line passes through the origin. Therefore, the y-intercept is $(0, 0)$ and the corresponding value of b is 0. The slope-intercept form of the line becomes:

$$y = mx + b$$

$$y = -2x + 0 \text{ or equivalently } y = -2x.$$

Skill Practice

13. Write an equation of the line that passes through the origin and has slope $\frac{6}{5}$.

Answers

12. $y = -4x - 10$

13. $y = \frac{6}{5}x$

Calculator Connections

Topic: Using the *ZSquare* Option in Zoom

In Example 5(b) we found that the equations $y = \frac{1}{3}x + 3$ and $y = -3x + 4$ represent perpendicular lines. We can verify our results by graphing the lines on a graphing calculator.

Notice that the lines do not appear perpendicular in the calculator display. That is, they do not appear to form a right angle at the point of intersection. Because many calculators have a rectangular screen, the standard viewing window is elongated in the horizontal direction. To eliminate this distortion, try using a *ZSquare* option, which is located under the Zoom menu. This feature will set the viewing window so that equal distances on the display denote an equal number of units on the graph.

Calculator Exercises

For each pair of lines, determine if the lines are parallel, perpendicular, or neither. Then use a square viewing window to graph the lines on a graphing calculator to verify your results.

1. $x + y = 1$
    $x - y = -3$

2. $3x + y = -2$
    $6x + 2y = 6$

3. $2x - y = 4$
    $3x + 2y = 4$

4. Graph the lines defined by $y = x + 1$ and $y = 0.99x + 3$. Are these lines parallel? Explain.

5. Graph the lines defined by $y = -2x - 1$ and $y = -2x - 0.99$. Are these lines the same? Explain.

6. Graph the line defined by $y = 0.001x + 3$. Is this line horizontal? Explain.

Section 3.4   Practice Exercises

Study Skills Exercises

1. When taking a test, go through the test and do all the problems that you know first. Then go back and work on the problems that were more difficult. Give yourself a time limit for how much time you spend on each problem (maybe 3 to 5 min the first time through). Circle the importance of each statement.

	not important	somewhat important	very important
a. Read through the entire test first.	1	2	3
b. If time allows, go back and check each problem.	1	2	3
c. Write out all steps instead of doing the work in your head.	1	2	3

2. Define the key terms:

    a. slope-intercept form of a line   **b. standard form of a line**

Review Exercises

For Exercises 3–10, determine the x- and y-intercepts, if they exist.

3. $x - 5y = 10$ **4.** $3x + y = -12$ **5.** $3y = -9$ **6.** $2 + y = 5$

7. $-4x = 6y$ **8.** $-x + 3 = 8$ **9.** $5x = 20$ **10.** $y = \dfrac{1}{2}x$

Objective 1: Slope-Intercept Form of a Line

For Exercises 11–30, identify the slope and y-intercept, if they exist. **(See Examples 1–2.)**

11. $y = -2x + 3$ **12.** $y = \dfrac{2}{3}x + 5$ **13.** $y = x - 2$

14. $y = -x + 6$ **15.** $y = -x$ **16.** $y = -5x$

17. $y = \dfrac{3}{4}x - 1$ **18.** $y = x - \dfrac{5}{3}$ **19.** $2x - 5y = 4$

20. $3x + 2y = 9$ **21.** $3x - y = 5$ **22.** $7x - 3y = -6$

23. $x + y = 6$ **24.** $x - y = 1$ **25.** $x + 6 = 8$

26. $-4 + x = 1$ **27.** $-8y = 2$ **28.** $1 - y = 9$

29. $3y - 2x = 0$ **30.** $5x = 6y$

Objective 2: Graphing a Line from Its Slope and *y*-Intercept

For Exercises 31–34, graph the line using the slope and *y*-intercept. **(See Examples 3–4.)**

31. Graph the line through the point $(0, 2)$, having a slope of -4.

32. Graph the line through the point $(0, -1)$, having a slope of -3.

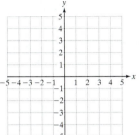

33. Graph the line through the point $(0, -5)$, having a slope of $\frac{3}{2}$.

34. Graph the line through the point $(0, 3)$, having a slope of $-\frac{1}{4}$.

For Exercises 35–40, match the equation with the graph (a–f) by identifying if the slope is positive or negative and if the *y*-intercept is positive, negative, or zero.

35. $y = 2x + 3$

36. $y = -3x - 2$

37. $y = -\frac{1}{3}x + 3$

38. $y = \frac{1}{2}x - 2$

39. $y = x$

40. $y = -2x$

a.

b.

c.

d.

e.

f.

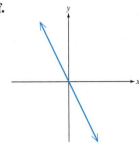

For Exercises 41–52, write each equation in slope-intercept form (if possible) and graph the line. (See Examples 3–4.)

41. $x - 2y = 6$

42. $5x - 2y = 2$

43. $2x + y = 9$

44. $-6x + y = 8$

45. $2x = -4y + 6$

46. $3x = y - 7$

47. $x + y = 0$

48. $x - y = 0$

49. $5y = 4x$

50. $-2x = 5y$

51. $3y + 2 = 0$

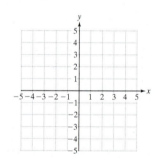

52. $1 + 5y = 6$

Objective 3: Determining Whether Two Lines Are Parallel, Perpendicular, or Neither

For Exercises 53–68, determine if the equations represent parallel lines, perpendicular lines, or neither. (See Example 5.)

53. l_1: $y = -2x - 3$

l_2: $y = \dfrac{1}{2}x + 4$

54. l_1: $y = \dfrac{4}{3}x - 2$

l_2: $y = -\dfrac{3}{4}x + 6$

55. l_1: $y = \dfrac{4}{5}x - \dfrac{1}{2}$

l_2: $y = \dfrac{5}{4}x - \dfrac{2}{3}$

56. l_1: $y = \dfrac{1}{5}x + 1$

l_2: $y = 5x - 3$

57. l_1: $y = -9x + 6$

l_2: $y = -9x - 1$

58. l_1: $y = 4x - 1$

l_2: $y = 4x + \dfrac{1}{2}$

59. l_1: $x = 3$

l_2: $y = \dfrac{7}{4}$

60. l_1: $y = \dfrac{2}{3}$

l_2: $x = 6$

61. l_1: $2x = 4$

l_2: $6 = x$

62. l_1: $2y = 7$

l_2: $y = 4$

63. l_1: $2x + 3y = 6$

l_2: $3x - 2y = 12$

64. l_1: $4x + 5y = 20$

l_2: $5x - 4y = 60$

65. l_1: $4x + 2y = 6$ ● **66.** l_1: $3x + y = 5$ **67.** l_1: $y = \dfrac{1}{5}x - 3$ **68.** l_1: $y = \dfrac{1}{3}x + 2$

   l_2: $4x + 8y = 16$ l_2: $x + 3y = 18$ l_2: $2x - 10y = 20$ l_2: $-x + 3y = 12$

Objective 4: Writing an Equation of a Line Using Slope-Intercept Form

For Exercises 69–78, write an equation of the line given the following information. Write the answer in slope-intercept form if possible. **(See Examples 6–7.)**

69. The slope is $-\frac{1}{3}$, and the y-intercept is $(0, 2)$.

70. The slope is $\frac{2}{3}$, and the y-intercept is $(0, -1)$.

71. The slope is 10, and the y-intercept is $(0, -19)$.   ● **72.** The slope is -14, and the y-intercept is $(0, 2)$.

73. The slope is 0, and the y-intercept is -11.

74. The slope is 0, and the y-intercept is $\dfrac{6}{7}$.

75. The slope is 5, and the line passes through the origin.

76. The slope is -3, and the line passes through the origin.

77. The slope is 6, and the line passes through the point $(0, -2)$.

78. The slope is -4, and the line passes through the point $(0, -3)$.

Expanding Your Skills

79. The cost for a rental car is $49.95 per day plus a flat fee of $31.95 for insurance. The equation, $C = 49.95x + 31.95$ represents the total cost, C (in dollars), to rent the car for x days.

   a. Identify the slope. Interpret the meaning of the slope in the context of this problem.

   b. Identify the C-intercept. Interpret the meaning of the C-intercept in the context of this problem.

   c. Use the equation to determine how much it would cost to rent the car for 1 week.

80. A phone bill is determined each month by a $16.95 flat fee plus $0.10/min of long distance. The equation $C = 0.10x + 16.95$ represents the total monthly cost, C, for x minutes of long distance.

   a. Identify the slope. Interpret the meaning of the slope in the context of this problem.

   b. Identify the C-intercept. Interpret the meaning of the C-intercept in the context of this problem.

   c. Use the equation to determine the total cost of 234 min of long distance.

81. A linear equation is written in standard form if it can be written as $Ax + By = C$, where A and B are not both zero. Write the equation $Ax + By = C$ in slope-intercept form to show that the slope is given by the ratio, $-\frac{A}{B}$. ($B \neq 0$.)

For Exercises 82–85, use the result of Exercise 81 to find the slope of the line.

82. $2x + 5y = 8$ **83.** $6x + 7y = -9$ **84.** $4x - 3y = -5$ **85.** $11x - 8y = 4$

Problem Recognition Exercises

Linear Equations in Two Variables

For Exercises 1–20, choose the equation(s) from the column on the right whose graph satisfies the condition described. Give all possible answers.

1. Line whose slope is positive.

2. Line whose slope is negative.

3. Line that passes through the origin.

4. Line that contains the point $(3, -2)$.

5. Line whose y-intercept is $(0, 4)$.

6. Line whose y-intercept is $(0, -5)$.

7. Line whose slope is $\dfrac{1}{2}$.

8. Line whose slope is -2.

9. Line whose slope is 0.

10. Line whose slope is undefined.

11. Line that is parallel to the line with equation $y = -\dfrac{2}{3}x + 4$.

12. Line perpendicular to the line with equation $y = 2x + 9$.

13. Line that is vertical.

14. Line that is horizontal.

15. Line whose x-intercept is $(10, 0)$.

16. Line whose x-intercept is $(6, 0)$.

17. Line that is parallel to the x-axis.

18. Line that is perpendicular to the y-axis.

19. Line with a negative slope and positive y-intercept.

20. Line with a positive slope and negative y-intercept.

a. $y = 5x$

b. $2x + 3y = 12$

c. $y = \dfrac{1}{2}x - 5$

d. $3x - 6y = 10$

e. $2y = -8$

f. $y = -2x + 4$

g. $3x = 1$

h. $x + 2y = 6$

Point-Slope Formula

1. Writing an Equation of a Line Using the Point-Slope Formula

In Section 3.4, the slope-intercept form of a line was used as a tool to construct an equation of a line. Another useful tool to determine an equation of a line is the point-slope formula. The point-slope formula can be derived from the slope formula as follows:

Suppose a line passes through a given point (x_1, y_1) and has slope m. If (x, y) is any other point on the line, then:

$$m = \frac{y - y_1}{x - x_1} \qquad \text{Slope formula}$$

$$m(x - x_1) = \frac{y - y_1}{x - x_1}(x - x_1) \qquad \text{Clear fractions.}$$

$$m(x - x_1) = y - y_1$$

$$y - y_1 = m(x - x_1) \qquad \text{Point-slope formula}$$

> **FORMULA Point-Slope Formula**
>
> The **point-slope formula** is given by
>
> $$y - y_1 = m(x - x_1)$$
>
> where m is the slope of the line and (x_1, y_1) is a known point on the line.

Example 1 demonstrates how to use the point-slope formula to find an equation of a line when a point on the line and slope are given.

Example 1 Writing an Equation of a Line Using the Point-Slope Formula

Use the point-slope formula to write an equation of the line having a slope of 3 and passing through the point $(-2, -4)$. Write the answer in slope-intercept form.

Solution:

The slope of the line is given: $m = 3$.

A point on the line is given: $(x_1, y_1) = (-2, -4)$.

Objectives

1. Writing an Equation of a Line Using the Point-Slope Formula
2. Writing an Equation of a Line Given Two Points
3. Writing an Equation of a Line Parallel or Perpendicular to Another Line
4. Different Forms of Linear Equations: A Summary

Skill Practice

1. Use the point-slope formula to write an equation of the line having a slope of -4 and passing through $(-1, 5)$. Write the answer in slope-intercept form.

Answer

1. $y = -4x + 1$

The point-slope formula:

$$y - y_1 = m(x - x_1)$$

$y - (-4) = 3[x - (-2)]$ Substitute $m = 3$, $x_1 = -2$, and $y_1 = -4$.

$y + 4 = 3(x + 2)$ Simplify. Because the final answer is required in slope-intercept form, simplify the equation and solve for y.

$y + 4 = 3x + 6$ Apply the distributive property.

$y = 3x + 2$ Slope-intercept form

The equation $y = 3x + 2$ from Example 1 is graphed in Figure 3-29. Notice that the line does indeed pass through the point $(-2, -4)$.

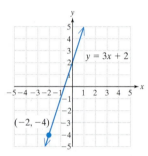

Figure 3-29

2. Writing an Equation of a Line Given Two Points

Example 2 is similar to Example 1; however, the slope must first be found from two given points.

┌─ **Skill Practice** ─┐

2. Use the point-slope formula to write an equation of the line passing through the points $(1, -1)$ and $(-1, -5)$.

Example 2   Writing an Equation of a Line Given Two Points

Use the point-slope formula to find an equation of the line passing through the points $(-2, 5)$ and $(4, -1)$. Write the final answer in slope-intercept form.

Solution:

Given two points on a line, the slope can be found with the slope formula.

$$\begin{array}{ccc} (-2, 5) & \text{and} & (4, -1) \\ (x_1, y_1) & & (x_2, y_2) \end{array}$$ Label the points.

$$m = \frac{y_2 - y_1}{x_2 - x_1} = \frac{(-1) - (5)}{(4) - (-2)} = \frac{-6}{6} = -1$$

To apply the point-slope formula, use the slope, $m = -1$ and either given point. We will choose the point $(-2, 5)$ as (x_1, y_1).

TIP: The point-slope formula can be applied using either given point for (x_1, y_1). In Example 2, using the point $(4, -1)$ for (x_1, y_1) produces the same result.

$$y - y_1 = m(x - x_1)$$
$$y - (-1) = -1(x - 4)$$
$$y + 1 = -x + 4$$
$$y = -x + 3$$

$$y - y_1 = m(x - x_1)$$

$y - 5 = -1[x - (-2)]$ Substitute $m = -1$, $x_1 = -2$, and $y_1 = 5$.

$y - 5 = -1(x + 2)$ Simplify.

$$y - 5 = -x - 2$$

$$y = -x + 3$$

Answer

2. $y = 2x - 3$

The solution to Example 2 can be checked by graphing the line $y = -x + 3$ using the slope and y-intercept. Notice that the line passes through the points $(-2, 5)$ and $(4, -1)$ as expected. See Figure 3-30.

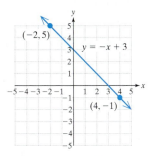

Figure 3-30

3. Writing an Equation of a Line Parallel or Perpendicular to Another Line

| **Example 3** | Writing an Equation of a Line Parallel to Another Line |

Use the point-slope formula to find an equation of the line passing through the point $(-1, 0)$ and parallel to the line $y = -4x + 3$. Write the final answer in slope-intercept form.

Solution:

Figure 3-31 shows the line $y = -4x + 3$ (pictured in black) and a line parallel to it (pictured in blue) that passes through the point $(-1, 0)$. The equation of the given line, $y = -4x + 3$, is written in slope-intercept form, and its slope is easily identified as -4. The line parallel to the given line must also have a slope of -4.

Apply the point-slope formula using $m = -4$ and the point $(x_1, y_1) = (-1, 0)$.

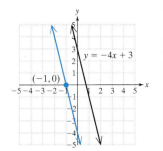

Figure 3-31

$$y - y_1 = m(x - x_1)$$
$$y - 0 = -4[x - (-1)]$$
$$y = -4(x + 1)$$
$$y = -4x - 4$$

TIP: When writing an equation of a line, slope-intercept form or standard form is usually preferred. For instance, the solution to Example 3 can be written as follows.

Slope-intercept form:
$y = -4x - 4$

Standard form:
$4x + y = -4$

| **Example 4** | Writing an Equation of a Line Perpendicular to Another Line |

Use the point-slope formula to find an equation of the line passing through the point $(-3, 1)$ and perpendicular to the line $3x + y = -2$. Write the final answer in slope-intercept form.

Solution:

The given line can be written in slope-intercept form as $y = -3x - 2$. The slope of this line is -3. Therefore, the slope of a line perpendicular to the given line is $\frac{1}{3}$.

Answers

3. $y = \frac{3}{4}x - 4$ 4. $y = 2x - 16$

Apply the point-slope formula with $m = \frac{1}{3}$, and $(x_1, y_1) = (-3, 1)$.

$y - y_1 = m(x - x_1)$	Point-slope formula
$y - (1) = \frac{1}{3}[x - (-3)]$	Substitute $m = \frac{1}{3}$, $x_1 = -3$, and $y_1 = 1$.
$y - 1 = \frac{1}{3}(x + 3)$	To write the final answer in slope-intercept form, simplify the equation and solve for y.
$y - 1 = \frac{1}{3}x + 1$	Apply the distributive property.
$y = \frac{1}{3}x + 2$	Add 1 to both sides.

A sketch of the perpendicular lines $y = \frac{1}{3}x + 2$ and $y = -3x - 2$ is shown in Figure 3-32. Notice that the line $y = \frac{1}{3}x + 2$ passes through the point $(-3, 1)$.

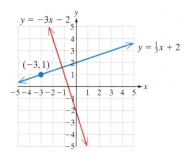

Figure 3-32

4. Different Forms of Linear Equations: A Summary

A linear equation can be written in several different forms, as summarized in Table 3-3.

Table 3-3

Form	Example	Comments
Standard Form $Ax + By = C$	$4x + 2y = 8$	A and B must not both be zero.
Horizontal Line $y = k$ (k is constant)	$y = 4$	The slope is zero, and the y-intercept is $(0, k)$.
Vertical Line $x = k$ (k is constant)	$x = -1$	The slope is undefined, and the x-intercept is $(k, 0)$.
Slope-Intercept Form $y = mx + b$ the slope is m y-intercept is $(0, b)$	$y = -3x + 7$ Slope $= -3$ y-intercept is $(0, 7)$	Solving a linear equation for y results in slope-intercept form. The coefficient of the x-term is the slope, and the constant defines the location of the y-intercept.
Point-Slope Formula $y - y_1 = m(x - x_1)$	$m = -3$ $(x_1, y_1) = (4, 2)$ $y - 2 = -3(x - 4)$	This formula is typically used to build an equation of a line when a point on the line is known and the slope of the line is known.

Concept Connections

5. Explain why the point-slope formula cannot be used to write an equation for a vertical line.
6. Explain how to write an equation for a vertical line.

Answers

5. The point-slope formula requires a value for the slope and a vertical line has an undefined slope.
6. The form $x = k$ must be used. Then the appropriate value is substituted for k.

Although standard form and slope-intercept form can be used to express an equation of a line, often the slope-intercept form is used to give a *unique* representation of the line. For example, the following linear equations are all written in standard form, yet they each define the same line.

$$2x + 5y = 10$$

$$-4x - 10y = -20$$

$$6x + 15y = 30$$

$$\frac{2}{5}x + y = 2$$

The line can be written uniquely in slope-intercept form as: $y = -\frac{2}{5}x + 2$.

Although it is important to understand and apply slope-intercept form and the point-slope formula, they are not necessarily applicable to all problems, particularly when dealing with a horizontal or vertical line.

| Example 5 | Writing an Equation of a Line |

Find an equation of the line passing through the point $(2, -4)$ and parallel to the x-axis.

Solution:

Because the line is parallel to the x-axis, the line must be horizontal. Recall that all horizontal lines can be written in the form $y = k$, where k is a constant. A quick sketch can help find the value of the constant. See Figure 3-33.

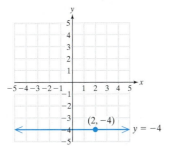

Figure 3-33

Because the line must pass through a point whose y-coordinate is -4, then the equation of the line must be $y = -4$.

Skill Practice

7. Write an equation for the vertical line that passes through the point $(-7, 2)$.

Answer

7. $x = -7$

Section 3.5 Practice Exercises

Study Skills Exercises

1. Prepare a one-page summary sheet with the most important information that you need for the test. On the day of the test, look at this sheet several times to refresh your memory instead of trying to memorize new information.

2. Define the key term: **point-slope formula**

Review Exercises

For Exercises 3–6, graph the equations.

3. $2x - 3y = -3$

4. $y = -2x$

5. $3 - y = 9$

6. $y = \dfrac{4}{5}x$

For Exercises 7–10, find the slope of the line that passes through the given points.

7. $(1, -3)$ and $(2, 6)$

8. $(2, -4)$ and $(-2, 4)$

9. $(-2, 5)$ and $(5, 5)$

10. $(6.1, 2.5)$ and $(6.1, -1.5)$

Objective 1: Writing an Equation of a Line Using the Point-Slope Formula

For Exercises 11–22, use the point-slope formula (if possible) to write an equation of the line given the following information. **(See Example 1.)**

11. The slope is 3, and the line passes through the point $(-2, 1)$.

12. The slope is -2, and the line passes through the point $(1, -5)$.

13. The slope is -4, and the line passes through the point $(-3, -2)$.

14. The slope is 5, and the line passes through the point $(-1, -3)$.

15. The slope is $-\frac{1}{2}$, and the line passes through $(-1, 0)$.

16. The slope is $-\frac{3}{4}$, and the line passes through $(2, 0)$.

17. The slope is $\frac{1}{4}$, and the line passes through the point $(-8, 6)$.

18. The slope is $\frac{2}{5}$, and the line passes through the point $(-5, 4)$.

19. The slope is 4.5, and the line passes through the point $(5.2, -2.2)$.

20. The slope is -3.6, and the line passes through the point $(10.0, 8.2)$.

21. The slope is 0, and the line passes through the point $(3, -2)$.

22. The slope is 0, and the line passes through the point $(0, 5)$.

Objective 2: Writing an Equation of a Line Given Two Points

For Exercises 23–28, use the point-slope formula to write an equation of the line given the following information. **(See Example 2.)**

23. The line passes through the points $(-2, -6)$ and $(1, 0)$.

24. The line passes through the points $(-2, 5)$ and $(0, 1)$.

25. The line passes through the points $(0, -4)$ and $(-1, -3)$.

26. The line passes through the points $(1, -3)$ and $(-7, 2)$.

27. The line passes through the points $(2.2, -3.3)$ and $(12.2, -5.3)$.

28. The line passes through the points $(4.7, -2.2)$ and $(-0.3, 6.8)$.

For Exercises 29–34, find an equation of the line through the given points. Write the final answer in slope-intercept form.

29.

30.

31.

32.

33.

34.

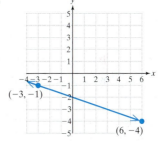

Objective 3: Writing an Equation of a Line Parallel or Perpendicular to Another Line

For Exercises 35–44, use the point-slope formula to write an equation of the line given the following information. **(See Examples 3–4.)**

35. The line passes through the point $(-3, 1)$ and is parallel to the line $y = 4x + 3$.

36. The line passes through the point $(4, -1)$ and is parallel to the line $y = 3x + 1$.

37. The line passes through the point $(4, 0)$ and is parallel to the line $3x + 2y = 8$.

38. The line passes through the point $(2, 0)$ and is parallel to the line $5x + 3y = 6$.

39. The line passes through the point $(-5, 2)$ and is perpendicular to the line $y = \frac{1}{2}x + 3$.

40. The line passes through the point $(-2, -2)$ and is perpendicular to the line $y = \frac{1}{3}x - 5$.

41. The line passes through the point $(0, -6)$ and is perpendicular to the line $-5x + y = 4$.

42. The line passes through the point $(0, -8)$ and is perpendicular to the line $2x - y = 5$.

43. The line passes through the point $(4, 4)$ and is parallel to the line $3x - y = 6$.

44. The line passes through the point $(-1, -7)$ and is parallel to the line $5x + y = -5$.

Objective 4: Different Forms of Linear Equations: A Summary

For Exercises 45–50, match the form or formula on the left with its name on the right.

45. $x = k$ i. Standard form

46. $y = mx + b$ ii. Point-slope formula

47. $m = \dfrac{y_2 - y_1}{x_2 - x_1}$ iii. Horizontal line

48. $y - y_1 = m(x - x_1)$ iv. Vertical line

49. $y = k$ v. Slope-intercept form

50. $Ax + By = C$ vi. Slope formula

For Exercises 51–60, find an equation for the line given the following information. **(See Example 5.)**

51. The line passes through the point $(3, 1)$ and is parallel to the line $y = -4$. See the figure.

52. The line passes through the point $(-1, 1)$ and is parallel to the line $y = 2$. See the figure.

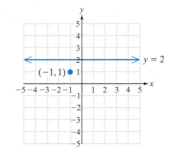

53. The line passes through the point $(2, 6)$ and is perpendicular to the line $y = 1$. (*Hint:* Sketch the line first.)

54. The line passes through the point $(0, 3)$ and is perpendicular to the line $y = -5$. (*Hint:* Sketch the line first.)

55. The line passes through the point $(2, 2)$ and is perpendicular to the line $x = 0$.

56. The line passes through the point $(5, -2)$ and is perpendicular to the line $x = 0$.

57. The slope is undefined, and the line passes through the point $(-6, -3)$.

58. The slope is undefined, and the line passes through the point $(2, -1)$.

59. The line passes through the points $(-4, 0)$ and $(-4, 3)$.

60. The line passes through the points $(1, 3)$ and $(1, -4)$.

Applications of Linear Equations and Modeling | Section 3.6

1. Interpreting a Linear Equation in Two Variables

Linear equations can often be used to describe (or model) the relationship between two variables in a real-world event.

Example 1   Interpreting a Linear Equation

The number of tigers in India decreased from 1900 to 2005. This decrease can be approximated by the equation $y = -350x + 42{,}000$. The variable y represents the number of tigers left in India, and x represents the number of years since 1900.

a. Use the equation to predict the number of tigers in 1960.

b. Use the equation to predict the number of tigers in 2010.

c. Determine the slope of the line. Interpret the meaning of the slope in terms of the number of tigers and the year.

d. Determine the x-intercept. Interpret the meaning of the x-intercept in terms of the number of tigers.

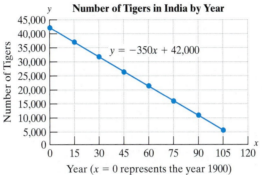

Number of Tigers in India by Year

$y = -350x + 42{,}000$

Number of Tigers

Year ($x = 0$ represents the year 1900)

Solution:

a. The year 1960 is 60 yr since 1900. Substitute $x = 60$ into the equation.

$$y = -350x + 42{,}000$$

$$y = -350(60) + 42{,}000$$

$$= 21{,}000$$ There were approximately 21,000 tigers in India in 1960.

b. The year 2010 is 110 yr since 1900. Substitute $x = 110$.

$$y = -350(110) + 42{,}000$$

$$= 3500$$ There will be approximately 3500 tigers in India in 2010.

c. The slope is -350.   The slope means that the tiger population is decreasing by 350 tigers per year.

d. To find the x-intercept, substitute $y = 0$.

$$y = -350x + 42{,}000$$

$$0 = -350x + 42{,}000$$ Substitute 0 for y.

$$-42{,}000 = -350x$$

$$120 = x$$

The x-intercept is $(120, 0)$. This means that 120 yr after the year 1900, the tiger population would be expected to reach zero. That is, in the year 2020, there will be no tigers left in India if this linear trend continues.

2. Writing a Linear Equation Using Observed Data Points

Example 2   **Writing a Linear Equation from Observed Data Points**

The monthly sales of hybrid cars sold in the United States are given for a recent year. The sales for the first 8 months of the year are shown in Figure 3-34. The value $x = 0$ represents January, $x = 1$ represents February, and so on.

Figure 3-34

a. Use the data points from Figure 3-34 to find a linear equation that represents the monthly sales of hybrid cars in the United States. Let x represent the month number and let y represent the number of vehicles sold.

b. Use the linear equation in part (a) to estimate the number of hybrid vehicles sold in month 7 (August).

Solution:

a. The ordered pairs $(0, 14{,}400)$ and $(5, 23{,}400)$ are given in the graph. Use these points to find the slope.

$$\underset{(x_1,\, y_1)}{(0,\ 14{,}400)} \quad \text{and} \quad \underset{(x_2,\, y_2)}{(5,\ 23{,}400)} \qquad \text{Label the points.}$$

$$m = \frac{y_2 - y_1}{x_2 - x_1} = \frac{23{,}400 - 14{,}400}{5 - 0}$$

$$= \frac{9000}{5}$$

$$= 1800 \qquad \text{The slope is 1800. This indicates that sales increased by approximately 1800 per month during this time period.}$$

With $m = 1800$, and the y-intercept given as $(0, 14{,}400)$, we have the following linear equation in slope-intercept form.

$$y = 1800x + 14{,}400$$

b. To approximate the sales in month number 7, substitute $x = 7$ into the equation from part (a).

$$y = 1800(7) + 14{,}400 \qquad \text{Substitute } x = 7.$$

$$= 27{,}000$$

The monthly sales for August (month 7) would be 27,000 vehicles.

3. Writing a Linear Equation Given a Fixed Value and a Rate of Change

Another way to look at the equation $y = mx + b$ is to identify the term mx as the variable term and the term b as the constant term. The value of the term mx will change with the value of x (this is why the slope, m, is called a *rate of change*). However, the term b will remain constant regardless of the value of x. With these ideas in mind, we can write a linear equation if the rate of change and the constant are known.

Example 3   Finding a Linear Equation

A stack of posters to advertise a school play costs $19.95 plus $1.50 per poster at the printer.

  a. Write a linear equation to compute the cost, c, of buying x posters.

  b. Use the equation to compute the cost of 125 posters.

Solution:

  a. The constant cost is $19.95. The variable cost is $1.50 per poster. If m is replaced with 1.50 and b is replaced with 19.95, the equation is

$$c = 1.50x + 19.95 \qquad \text{where } c \text{ is the cost (in dollars) of buying } x \text{ posters.}$$

  b. Because x represents the number of posters, substitute $x = 125$.

$$c = 1.50(125) + 19.95$$
$$= 187.5 + 19.95$$
$$= 207.45$$

The total cost of buying 125 posters is $207.45.

Calculator Connections

Topic: Using the Evaluate Feature on a Graphing Calculator

In Example 3, the equation $c = 1.50x + 19.95$ was used to represent the cost, c, to buy x posters. To graph this equation on a graphing calculator, first replace the variable c by y.

$$y = 1.50x + 19.95$$

We enter the equation into the calculator and set the viewing window.

To evaluate the equation for a user-defined value of x, use the *Value* feature in the CALC menu.

In this case, we entered $x = 125$, and the calculator returned $y = 207.45$.

Calculator Exercises

Use a graphing calculator to graph the lines on an appropriate viewing window. Evaluate the equation at the given values of x.

  1. $y = -4.6x + 27.1$ at $x = 3$

  2. $y = -3.6x - 42.3$ at $x = 0$

  3. $y = 40x + 105$ at $x = 6$

  4. $y = 20x - 65$ at $x = 8$

Section 3.6 Practice Exercises

Study Skills Exercise

1. On test day, take a look at any formulas or important points that you had to memorize before you enter the classroom. Then when you sit down to take your test, write these formulas on the test or on scrap paper. This is called a memory dump. Write down the formulas from Chapter 3.

Review Exercises

2. Determine the slope of the line defined by $2x - 8y = 15$.

For Exercises 3–8, find the x- and y-intercepts of the lines, if possible.

3. $5x + 6y = 30$

4. $3x + 4y = 1$

5. $y = -2x - 4$

6. $y = 5x$

7. $y = -9$

8. $x = 2$

Objective 1: Interpreting a Linear Equation in Two Variables

9. The minimum hourly wage, y (in dollars per hour), in the United States can be approximated by the equation $y = 0.14x + 1.60$. In this equation, x represents the number of years since 1970 ($x = 0$ represents 1970, $x = 5$ represents 1975, and so on). **(See Example 1.)**

    a. Use the equation to approximate the minimum wage in the year 1980.

    b. Use the equation to predict the minimum wage in 2010.

    c. Determine the y-intercept. Interpret the meaning of the y-intercept in the context of this problem.

    d. Determine the slope. Interpret the meaning of the slope in the context of this problem.

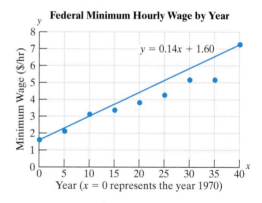

10. The graph depicts the rise in the number of jail inmates in the United States since 1995. Two linear equations are given: one to describe the number of female inmates and one to describe the number of male inmates by year.

    Let y represent the number of inmates (in thousands). Let x represent the number of years since 1995.

    a. What is the slope of the line representing the number of female inmates? Interpret the meaning of the slope in the context of this problem.

    b. What is the slope of the line representing the number of male inmates? Interpret the meaning of the slope in the context of this problem.

    c. Which group, males or females, has the larger slope? What does this imply about the rise in the number of male and female prisoners?

    d. Assuming this trend continues, use the equation to predict the number of female inmates in 2015.

(*Source:* U.S. Bureau of Justice Statistics)

11. The average daily temperature in January for cities along the eastern seaboard of the United States and Canada generally decreases for cities farther north. A city's latitude in the northern hemisphere is a measure of how far north it is on the globe.

The average temperature, y (measured in degrees Fahrenheit), can be described by the equation

$y = -2.333x + 124.0$ where x is the latitude of the city.

City	x Latitude (°N)	y Average Daily Temperature (°F)
Jacksonville, FL	30.3	52.4
Miami, FL	25.8	67.2
Atlanta, GA	33.8	41.0
Baltimore, MD	39.3	31.8
Boston, MA	42.3	28.6
Atlantic City, NJ	39.4	30.9
New York, NY	40.7	31.5
Portland, ME	43.7	20.8
Charlotte, NC	35.2	39.3
Norfolk, VA	36.9	39.1

(*Source:* U.S. National Oceanic and Atmospheric Administration)

a. Use the equation to predict the average daily temperature in January for Philadelphia, Pennsylvania, whose latitude is 40.0°N. Round to one decimal place.

b. Use the equation to predict the average daily temperature in January for Edmundston, New Brunswick, Canada, whose latitude is 47.4°N. Round to one decimal place.

c. What is the slope of the line? Interpret the meaning of the slope in terms of latitude and temperature.

d. From the equation, determine the value of the x-intercept. Round to one decimal place. Interpret the meaning of the x-intercept in terms of latitude and temperature.

12. The graph shows the number of points scored by Shaquille O'Neal and by Allen Iverson according to the number of minutes played for several games. Two linear equations are given: one to describe the number of points scored by O'Neal and one to describe the number of points scored by Iverson. In both equations, y represents the number of points scored and x represents the number of minutes played.

a. What is the slope of the line representing the number of points scored by O'Neal? Interpret the meaning of the slope in the context of this problem.

b. What is the slope of the line representing the number of points scored by Iverson? Interpret the meaning of the slope in the context of this problem.

c. According to these linear equations, approximately how many points would each player expect to score if he played for 36 min? Round to the nearest point.

13. The electric bill charge for a certain utility company is $0.095 per kilowatt-hour. The total cost, y, depends on the number of kilowatt-hours, x, according to the equation $y = 0.095x$, $x \geq 0$.

   a. Determine the cost of using 1000 kilowatt-hours.

   b. Determine the cost of using 2000 kilowatt-hours.

   c. Determine the y-intercept. Interpret the meaning of the y-intercept in the context of this problem.

   d. Determine the slope. Interpret the meaning of the slope in the context of this problem.

14. For a recent year, children's admission to the Minnesota State Fair was $8. Ride tickets were $0.75 each. The equation $y = 0.75x + 8$ represented the cost, y, in dollars to be admitted to the fair and to purchase x ride tickets.

   a. Determine the slope of the line represented by $y = 0.75x + 8$. Interpret the meaning of the slope in the context of this problem.

   b. Determine the y-intercept. Interpret its meaning in the context of this problem.

   c. Use the equation to determine how much money a child needed for admission and to ride 10 rides.

Objective 2: Writing a Linear Equation Using Observed Data Points

15. The average length of stay for community hospitals decreased in the United States from 1980 to 2005. Let x represent the number of years since 1980. Let y represent the average length of a hospital stay in days. **(See Example 2.)**

   a. Find a linear equation that relates the average length of hospital stays versus the year.

   b. Use the linear equation found in part (a) to predict the average length of stay in community hospitals in the year 2010. Round to the nearest day.

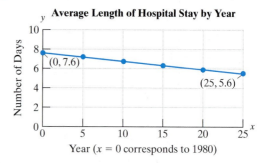

Average Length of Hospital Stay by Year

$(0, 7.6)$ $(25, 5.6)$

Number of Days

Year ($x = 0$ corresponds to 1980)

16. The figure depicts a relationship between a person's height, y (in inches), and the length of the person's arm, x (measured in inches from shoulder to wrist).

   a. Use the points $(17, 57.75)$ and $(24, 82.25)$ to find a linear equation relating height to arm length.

   b. What is the slope of the line? Interpret the slope in the context of this problem.

   c. Use the equation from part (a) to estimate the height of a person whose arm length is 21.5 in.

Height Versus Arm Length

$(24, 82.25)$

$(17, 57.75)$

Height (inches)

Arm Length (inches)

17. The graph shows the average height for boys based on age. Let x represent a boy's age, and let y represent his height (in inches).

   a. Find a linear equation that represents the height of a boy versus his age.

   b. Use the linear equation found in part (a) to predict the average height of a 5-year-old boy.

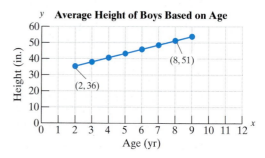

Average Height of Boys Based on Age

$(8, 51)$

$(2, 36)$

Height (in.)

Age (yr)

(*Source:* National Parenting Council)

18. Wind energy is one type of renewable energy that does not produce dangerous greenhouse gases as a by-product. The graph shows the consumption of wind energy in the United States for selected years. The variable y represents the amount of wind energy in trillions of Btu, and the variable x represents the number of years since 2000.

    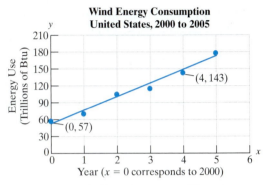

    Wind Energy Consumption United States, 2000 to 2005

    (*Source:* United States Department of Energy)

    a. Use the points $(0, 57)$ and $(4, 143)$ to determine the slope of the line.

    b. Interpret the slope in the context of this problem?

    c. Use the points $(0, 57)$ and $(4, 143)$ to find a linear equation relating the consumption of wind energy, y, to the number of years, x, since 2000.

    d. If this linear trend continues beyond the observed data values, use the equation in part (c) to predict the consumption of wind energy in the year 2010.

Objective 3: Writing a Linear Equation Given a Fixed Value and a Rate of Change

19. The cost to rent a car, y, for 1 day is $20 plus $0.25 per mile. **(See Example 3.)**

    a. Write a linear equation to compute the cost, y, of driving a car x miles for 1 day.

    b. Use the equation to compute the cost of driving 258 miles in the rental car for 1 day.

20. A phone bill is determined each month by a $18.95 flat fee plus $0.08 per minute of long distance.

    a. Write a linear equation to compute the monthly cost of a phone bill, y, if x minutes of long distance are used.

    b. Use the equation to compute the phone bill for a month in which 1 hr and 27 min of long distance was used.

21. The cost to rent a 10 ft by 10 ft storage space is $90 per month plus a nonrefundable deposit of $105.

    a. Write a linear equation to compute the cost, y, of renting a 10 ft by 10 ft space for x months.

    b. What is the cost of renting such a storage space for 1 year (12 months)?

22. An air-conditioning and heating company has a fixed monthly cost of $5000. Furthermore, each service call costs the company $25.

    a. Write a linear equation to compute the total cost, y, for 1 month if x service calls are made.

    b. Use the equation to compute the cost for 1 month if 150 service calls are made.

23. A bakery that specializes in bread rents a booth at a flea market. The daily cost to rent the booth is $100. Each loaf of bread costs the bakery $0.80 to produce.

    a. Write a linear equation to compute the total cost, y, for 1 day if x loaves of bread are produced.

    b. Use the equation to compute the cost for 1 day if 200 loaves of bread are produced.

24. A beverage company rents a booth at an art show to sell lemonade. The daily cost to rent a booth is $35. Each lemonade costs $0.50 to produce.

    a. Write a linear equation to compute the total cost, y, for 1 day if x lemonades are produced.

    b. Use the equation to compute the cost for 1 day if 350 lemonades are produced.

Objectives

1. **Definition of a Relation**
2. **Definition of a Function**
3. **Vertical Line Test**
4. **Function Notation**
5. **Domain and Range of a Function**
6. **Applications of Functions**

1. Definition of a Relation

The number of points scored by LeBron James during the first six games of a recent basketball season is shown in Table 3-4.

Table 3-4

Game, x	Number of Points, y		Ordered Pair
1	26	→	$(1, 26)$
2	35	→	$(2, 35)$
3	16	→	$(3, 16)$
4	34	→	$(4, 34)$
5	19	→	$(5, 19)$
6	38	→	$(6, 38)$

Each ordered pair from Table 3-4 shows a correspondence, or relationship, between the game number and the number of points scored by LeBron James. The set of ordered pairs: $\{(1, 26), (2, 35), (3, 16), (4, 34), (5, 19), (6, 38)\}$ defines a relation between the game number and the number of points scored.

> **DEFINITION Relation in x and y**
>
> Any set of ordered pairs, (x, y), is called a **relation** in x and y. Furthermore:
> - The set of first components in the ordered pairs is called the **domain** of the relation.
> - The set of second components in the ordered pairs is called the **range** of the relation.

Skill Practice

1. Find the domain and range of the relation. $\{(0, 1), (4, 5), (-6, 8), (4, 13), (-8, 8)\}$

Example 1 Finding the Domain and Range of a Relation

Find the domain and range of the relation linking the game number to the number of points scored by James in the first six games of the season:

$$\{(1, 26), (2, 35), (3, 16), (4, 34), (5, 19), (6, 38)\}$$

Solution:

Domain: $\{1, 2, 3, 4, 5, 6\}$ (Set of first coordinates)

Range: $\{26, 35, 16, 34, 19, 38\}$ (Set of second coordinates)

The domain consists of the game numbers for the first six games of the season. The range represents the corresponding number of points.

Answer

1. Domain: $\{0, 4, -6, -8\}$;
   Range: $\{1, 5, 8, 13\}$

| Example 2 | Finding the Domain and Range of a Relation |

The three women represented in Figure 3-35 each have children. Molly has one child, Peggy has two children, and Joanne has three children.

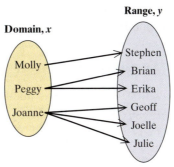

Figure 3-35

a. If the set of mothers is given as the domain and the set of children is the range, write a set of ordered pairs defining the relation given in Figure 3-35.

b. Write the domain and range of the relation.

Solution:

a. {(Molly, Stephen), (Peggy, Brian), (Peggy, Erika), (Joanne, Geoff), (Joanne, Joelle), (Joanne, Julie)}

b. Domain: {Molly, Peggy, Joanne}
   Range: {Stephen, Brian, Erika, Geoff, Joelle, Julie}

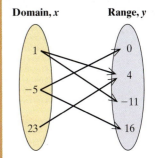
2. Definition of a Function

In mathematics, a special type of relation, called a function, is used extensively.

> **DEFINITION** Function
>
> Given a relation in x and y, we say "y is a **function** of x" if for each element x in the domain, there is exactly one value of y in the range.
>
> *Note:* This means that no two ordered pairs may have the same first coordinate and different second coordinates.

To understand the difference between a relation that is a function and one that is not a function, consider Example 3.

| Example 3 | Determining Whether a Relation Is a Function |

Determine whether the following relations are functions:

a. {(2, −3), (4, 1), (3, −1), (2, 4)}   b. {(−3, 1), (0, 2), (4, −3), (1, 5), (−2, 1)}

Solution:

a. This relation is defined by the set of ordered pairs.

same x-values

{(2, −3), (4, 1), (3, −1), (2, 4)}

different y-values

When $x = 2$, there are two possibilities for y: $y = -3$ and $y = 4$.

This relation is *not* a function because for $x = 2$, there is more than one corresponding element in the range.

b. This relation is defined by the set of ordered pairs: {(−3, 1), (0, 2), (4, −3), (1, 5), (−2, 1)}. Notice that no two ordered pairs have the same value of x but different values of y. Therefore, this relation *is* a function.

In Example 2, the relation linking the set of mothers with their respective children is *not* a function. The domain elements, "Peggy" and "Joanne," each have more than one child. Because these *x*-values in the domain have more than one corresponding *y*-value in the range, the relation is not a function.

3. Vertical Line Test

A relation that is not a function has at least one domain element, *x*, paired with more than one range element, *y*. For example, the ordered pairs (2, 1) and (2, 4) do not make a function. On a graph, these two points are aligned vertically in the *xy*-plane, and a vertical line drawn through one point also intersects the other point (Figure 3-36). Thus, if a vertical line drawn through a graph of a relation intersects the graph in more than one point, the relation cannot be a function. This idea is stated formally as the **vertical line test**.

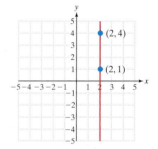

Figure 3-36

> **PROCEDURE** **Using the Vertical Line Test**
>
> Consider a relation defined by a set of points (*x*, *y*) on a rectangular coordinate system. Then the graph defines *y* as a function of *x* if no vertical line intersects the graph in more than one point.

The vertical line test also implies that if any vertical line drawn through the graph of a relation intersects the relation in more than one point, then the relation does *not* define *y* as a function of *x*.

The vertical line test can be demonstrated by graphing the ordered pairs from the relations in Example 3 (Figure 3-37 and Figure 3-38).

$$\{(2, -3), (4, 1), (3, -1), (2, 4)\} \qquad \{(-3, 1), (0, 2), (4, -3), (1, 5), (-2, 1)\}$$

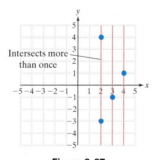

Figure 3-37

Not a Function

A vertical line intersects in
more than one point.

Figure 3-38

Function

No vertical line intersects
more than once.

The relations in Examples 1, 2, and 3 consist of a finite number of ordered pairs. A relation may, however, consist of an *infinite* number of points defined by an equation or by a graph. For example, the equation $y = x + 1$ defines infinitely many ordered pairs whose *y*-coordinate is one more than its *x*-coordinate. These ordered pairs cannot all be listed but can be depicted in a graph.

The vertical line test is especially helpful in determining whether a relation is a function based on its graph.

Example 4 **Using the Vertical Line Test**

Use the vertical line test to determine whether the following relations are functions.

a.

b.

Solution:

a.

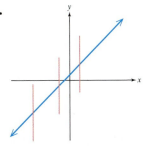

Function
No vertical line intersects more than once.

b.

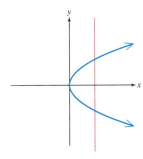

Not a Function
A vertical line intersects in more than one point.

4. Function Notation

A function is defined as a relation with the added restriction that each value of the domain corresponds to only one value in the range. In mathematics, functions are often given by rules or equations to define the relationship between two or more variables. For example, the equation, $y = x + 1$ defines the set of ordered pairs such that the y-value is one more than the x-value.

When a function is defined by an equation, we often use **function notation**. For example, the equation $y = x + 1$ may be written in function notation as

$$f(x) = x + 1$$

where f is the name of the function, x is an input value from the domain of the function, and $f(x)$ is the function value (or y-value) corresponding to x.

The notation $f(x)$ is read as "f of x" or "the value of the function, f, at x."

A function may be evaluated at different values of x by substituting values of x from the domain into the function. For example, for the function defined by $f(x) = x + 1$ we can evaluate f at $x = 3$ by using substitution.

$f(x) = x + 1$

$f(3) = (3) + 1$

$f(3) = 4$ This is read as "f of 3 equals 4."

Thus, when $x = 3$, the corresponding function value is 4. This can also be interpreted as an ordered pair: $(3, 4)$

The names of functions are often given by either lowercase letters or uppercase letters such as f, g, h, p, k, M, and so on.

Avoiding Mistakes

The notation $f(x)$ is read as "f of x" and does *not* imply multiplication.

Answers

6. Function   **7.** Not a function

Example 5   **Evaluating a Function**

Given the function defined by $h(x) = x^2 - 2$, find the function values.

a. $h(0)$ **b.** $h(1)$ **c.** $h(2)$ **d.** $h(-1)$ **e.** $h(-2)$

Solution:

a. $h(x) = x^2 - 2$

$h(0) = (0)^2 - 2$ Substitute $x = 0$ into the function.

$= 0 - 2$

$= -2$ $h(0) = -2$ means that when $x = 0$, $y = -2$, yielding the ordered pair $(0, -2)$.

b. $h(x) = x^2 - 2$

$h(1) = (1)^2 - 2$ Substitute $x = 1$ into the function.

$= 1 - 2$

$= -1$ $h(1) = -1$ means that when $x = 1$, $y = -1$, yielding the ordered pair $(1, -1)$.

c. $h(x) = x^2 - 2$

$h(2) = (2)^2 - 2$ Substitute $x = 2$ into the function.

$= 4 - 2$

$= 2$ $h(2) = 2$ means that when $x = 2$, $y = 2$, yielding the ordered pair $(2, 2)$.

d. $h(x) = x^2 - 2$

$h(-1) = (-1)^2 - 2$ Substitute $x = -1$ into the function.

$= 1 - 2$

$= -1$ $h(-1) = -1$ means that when $x = -1$, $y = -1$, yielding the ordered pair $(-1, -1)$.

e. $h(x) = x^2 - 2$

$h(-2) = (-2)^2 - 2$ Substitute $x = -2$ into the function.

$= 4 - 2$

$= 2$ $h(-2) = 2$ means that when $x = -2$, $y = 2$, yielding the ordered pair $(-2, 2)$.

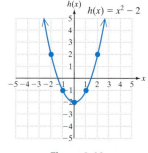

Figure 3-39

The rule $h(x) = x^2 - 2$ is equivalent to the equation $y = x^2 - 2$. The function values $h(0), h(1), h(2), h(-1)$, and $h(-2)$ correspond to the y-values in the ordered pairs $(0, -2), (1, -1), (2, 2), (-1, -1)$, and $(-2, 2)$, respectively. These points can be used to sketch a graph of the function (Figure 3-39).

5. Domain and Range of a Function

A function is a relation, and it is often necessary to determine its domain and range. Consider a function defined by the equation $y = f(x)$. The domain of f is the set of all x-values within the ordered pairs that make up the function. The range is the set of y-values.

Example 6 **Finding the Domain and Range of a Function**

Find the domain and range of the functions based on the graph of the function. Express the answers in interval notation.

a.

b.

Solution:

a.
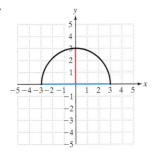

The horizontal "span" of the graph is determined by the x-values of the points. This is the domain. In this graph, the x-values in the domain are bounded between -3 and 3. (Shown in blue.)

Domain: $[-3, 3]$

The vertical "span" of the graph is determined by the y-values of the points. This is the range.

The y-values in the range are bounded between 0 and 3. (Shown in red.)

Range: $[0, 3]$

b.
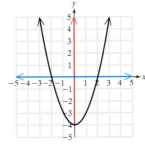

The function extends infinitely far to the left and right. The domain is shown in blue.

Domain: $(-\infty, \infty)$

The y-values extend infinitely far in the positive direction, but are bounded below at $y = -4$. (Shown in red.)

Range: $[-4, \infty)$

Skill Practice

Find the domain and range of the functions based on the graph of the function.

13.

14.

Answers

13. Domain: $[0, 5]$
     Range: $[2, 4]$
14. Domain: $(-\infty, \infty)$
     Range: $\{3\}$

6. Applications of Functions

Skill Practice

The function defined by

$$S(x) = 6x^2 \quad (x \ge 0)$$

indicates the surface area of the cube whose side is length x (in inches).

15. Evaluate $S(5)$.

16. Interpret the function value, $S(5)$.

Example 7   Using a Function in an Application

The score a student receives on an exam is a function of the number of hours the student spends studying. The function defined by

$$P(x) = \frac{100x^2}{40 + x^2} \quad (x \ge 0)$$

indicates that a student will achieve a score of $P\%$ after studying for x hours.

a. Evaluate $P(0)$, $P(10)$, and $P(20)$.

b. Interpret the function values from part (a) in the context of this problem.

Solution:

a. $P(x) = \dfrac{100x^2}{40 + x^2}$

$$P(0) = \frac{100(0)^2}{40 + (0)^2} \qquad P(10) = \frac{100(10)^2}{40 + (10)^2} \qquad P(20) = \frac{100(20)^2}{40 + (20)^2}$$

$$P(0) = \frac{0}{40} \qquad P(10) = \frac{10{,}000}{140} \qquad P(20) = \frac{40{,}000}{440}$$

$$P(0) = 0 \qquad P(10) = \frac{500}{7} \approx 71.4 \qquad P(20) = \frac{1000}{11} \approx 90.9$$

b. $P(0) = 0$ means that for 0 hr spent studying, the student will receive 0% on the exam.

$P(10) \approx 71.4$ means that for 10 hr spent studying, the student will receive approximately 71.4% on the exam.

$P(20) \approx 90.9$ means that for 20 hr spent studying, the student will receive approximately 90.9% on the exam.

The graph of $P(x) = \dfrac{100x^2}{40 + x^2}$ is shown in Figure 3-40.

Student Score (Percent) as a Function of Study Time

Figure 3-40

Answers

15. 150

16. For a cube 5 in. on a side, the surface area is 150 in.²

Calculator Connections

Topic: Graphing Functions

A graphing calculator can be used to graph a function. We replace $f(x)$ by y and enter the defining expression into the calculator. For example,

$$f(x) = \frac{1}{4}x^3 - x^2 - x + 4 \quad \text{becomes} \quad y = \frac{1}{4}x^3 - x^2 - x + 4.$$

Calculator Exercises

Use a graphing calculator to graph the following functions.

1. $f(x) = x^2 - 5x + 2$

2. $g(x) = -x^2 + 4x + 5$

3. $m(x) = \frac{1}{3}x^3 + x^2 - 3x - 1$

4. $n(x) = x^3 - 9x$

Section 3.7 Practice Exercises

Study Skills Exercises

1. List three ways that your friends or family can help support you in your academic pursuits.

2. Define the key terms:

    a. domain
    b. function
    c. function notation

    d. range
    e. relation
    f. vertical line test

Review Exercises

3. Given the equation $y = 3x - 5$:

    a. Find the x- and y-intercepts.

    b. Find the slope of the line.

    c. Graph the line.

4. Given the equation $2x - 5y = 12$

    a. Find the x- and y-intercepts.

    b. Find the slope of the line.

    c. Graph the line.

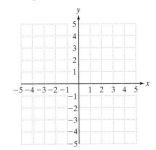

Objective 1: Definition of a Relation

For Exercises 5–14, determine the domain and range of the relation. **(See Examples 1–2.)**

5. $\{(4, 2), (3, 7), (4, 1), (0, 6)\}$

6. $\{(-3, -1), (-2, 6), (1, 3), (1, -2)\}$

7. $\{(\frac{1}{2}, 3), (0, 3), (1, 3)\}$

8. $\{(9, 6), (4, 6), (-\frac{1}{3}, 6)\}$

9. $\{(0, 0), (5, 0), (-8, 2), (8, 5)\}$

10. $\{(\frac{1}{2}, -\frac{1}{2}), (-4, 0), (0, -\frac{1}{2}), (\frac{1}{2}, 0)\}$

11.

12.

13.

14.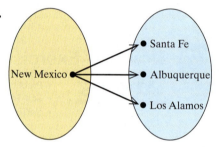

Objective 2: Definition of a Function

15. How can you tell if a set of ordered pairs represents a function?

16. Refer back to Exercises 6, 8, 10, 12, and 14. Identify which relations are functions.

17. Refer back to Exercises 5, 7, 9, 11, and 13. Identify which relations are functions. **(See Example 3.)**

Objective 3: Vertical Line Test

18. How can you tell from the graph of a relation if the relation is a function?

For Exercises 19–27, determine if the relation defines y as a function of x. **(See Example 4.)**

19.

20.

21.

22.

23.

24.

25.

26.

27.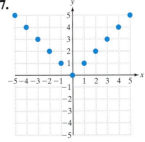

Objective 4: Function Notation

28. Explain how you would evaluate $f(x) = 3x^2$ at $x = -1$.

For Exercises 29–36, evaluate the given functions. **(See Example 5.)**

29. Let $f(x) = 2x - 5$.

Find: $f(0)$
$f(2)$
$f(-3)$

30. Let $g(x) = x^2 + 1$.

Find: $g(0)$
$g(-1)$
$g(3)$

31. Let $h(x) = \dfrac{1}{x + 4}$.

Find: $h(1)$
$h(0)$
$h(-2)$

32. Let $p(x) = \sqrt{x + 4}$.

Find: $p(0)$
$p(-4)$
$p(5)$

33. Let $m(x) = |5x - 7|$.

Find: $m(0)$
$m(1)$
$m(2)$

34. Let $w(x) = |2x - 3|$.

Find: $w(0)$
$w(1)$
$w(2)$

35. Let $n(x) = \sqrt{x - 2}$.

Find: $n(2)$
$n(3)$
$n(6)$

36. Let $t(x) = \dfrac{1}{x - 3}$.

Find: $t(1)$
$t(-1)$
$t(2)$

Objective 5: Domain and Range of a Function

For Exercises 37–40, match the domain and range given with a possible graph. **(See Example 6.)**

37. Domain: $(-\infty, \infty)$

Range: $[1, \infty)$

38. Domain: $[-4, 4]$

Range: $[-2, 2]$

39. Domain: $[-2, \infty)$

Range: $(-\infty, \infty)$

40. Domain: $(-\infty, \infty)$

Range: $(-\infty, \infty)$

a.

b.

c.

d.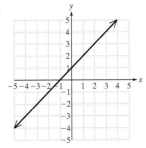

For Exercises 41–44, translate the expressions into English phrases.

41. $f(6) = 2$ **42.** $f(-2) = -14$ **43.** $g\left(\dfrac{1}{2}\right) = \dfrac{1}{4}$ **44.** $h(k) = k^2$

45. Consider a function defined by $y = f(x)$. The function value $f(2) = 7$ corresponds to what ordered pair?

46. Consider a function defined by $y = f(x)$. The function value $f(-3) = -4$ corresponds to what ordered pair?

47. Consider a function defined by $y = f(x)$. The function value $f(0) = 8$ corresponds to what ordered pair?

48. Consider a function defined by $y = f(x)$. The function value $f(4) = 0$ corresponds to what ordered pair?

Objective 6: Applications of Functions

49. In the absence of air resistance, the speed, s (in feet per second: ft/sec), of an object in free fall is a function of the number of seconds, t, after it was dropped: **(See Example 7.)**

$$s(t) = 32t$$

a. Find $s(1)$, and interpret the meaning of this function value in terms of speed and time.

b. Find $s(2)$, and interpret the meaning in terms of speed and time.

c. Find $s(10)$, and interpret the meaning in terms of speed and time.

d. A ball dropped from the top of the Sears Tower in Chicago falls for approximately 9.2 sec. How fast was the ball going the instant before it hit the ground?

50. The number of people diagnosed with skin cancer, $N(x)$, can be approximated by $N(x) = 45{,}625(1 + 0.029x)$. For this function, x represents the number of years since 2003. (*Source:* Centers for Disease Control)

a. Evaluate $N(0)$ and interpret its meaning in the context of this problem.

b. Evaluate $N(7)$ and interpret its meaning in the context of this problem. Round to the nearest whole number.

51. A punter kicks a football straight up with an initial velocity of 64 ft/sec. The height of the ball, h (in feet), is a function of the number of seconds, t, after the ball is kicked:

$$h(t) = -16t^2 + 64t + 3$$

a. Find $h(0)$, and interpret the meaning of the function value in terms of time and height.

b. Find $h(1)$, and interpret the meaning in terms of time and height.

c. Find $h(2)$, and interpret the meaning in terms of time and height.

d. Find $h(4)$, and interpret the meaning in terms of time and height.

52. For people 16 years old and older, the maximum recommended heart rate, *M* (in beats per minute: beats/min), is a function of a person's age, *x* (in years).

$$M(x) = 220 - x \text{ for } x \geq 16$$

a. Find $M(16)$, and interpret the meaning in terms of maximum recommended heart rate and age.

b. Find $M(30)$, and interpret the meaning in terms of maximum recommended heart rate and age.

c. Find $M(60)$, and interpret the meaning in terms of maximum recommended heart rate and age.

d. Find your own maximum recommended heart rate.

Group Activity

Modeling a Linear Equation

Materials: Yardstick or other device for making linear measurements

Estimated Time: 15–20 minutes

Group Size: 3

1. The members of each group should measure the length of their arms (in inches) from elbow to wrist. Record this measurement as *x* and the person's height (in inches) as *y*. Write these values as ordered pairs for each member of the group. Then write the ordered pairs on the board.

2. Next, copy the ordered pairs collected from all groups in the class and plot the ordered pairs. (This is called a "scatter diagram.")

3. Select two ordered pairs that seem to follow the upward trend of the data. Using these data points, determine the slope of the line.

Slope: _____

4. Using the data points and slope from question 3, find an equation of the line through the two points. Write the equation in slope-intercept form, $y = mx + b$.

Equation: _____

5. Using the equation from question 4, estimate the height of a person whose arm length from elbow to wrist is 8.5 in.

6. Suppose a crime scene investigator uncovers a partial skeleton and identifies a bone as a human ulna (the ulna is one of two bones in the forearm and extends from elbow to wrist). If the length of the bone is 12 in., estimate the height of the person before death. Would you expect this person to be male or female?

Chapter 3   Summary

Section 3.1   Rectangular Coordinate System

Key Concepts

Graphical representation of numerical **data** is often helpful to study problems in real-world applications.

A **rectangular coordinate system** is made up of a horizontal line called the **x-axis** and a vertical line called the **y-axis**. The point where the lines meet is the **origin**. The four regions of the plane are called **quadrants**.

The point (x, y) is an **ordered pair**. The first element in the ordered pair is the point's horizontal position from the origin. The second element in the ordered pair is the point's vertical position from the origin.

Example

Example 1

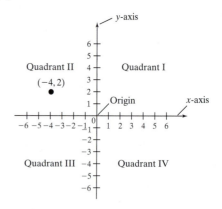

Section 3.2   Linear Equations in Two Variables

Key Concepts

An equation written in the form $Ax + By = C$ (where A and B are not both zero) is a **linear equation in two variables**.

A solution to a linear equation in x and y is an ordered pair (x, y) that makes the equation a true statement. The graph of the set of all solutions of a linear equation in two variables is a line in a rectangular coordinate system.

A linear equation can be graphed by finding at least two solutions and graphing the line through the points.

Examples

Example 1

Graph the equation $2x + y = 2$.

Select arbitrary values of x or y such as those shown in the table. Then complete the table to find the corresponding ordered pairs.

x	y	
0	2	$\longrightarrow (0, 2)$
−1	4	$\longrightarrow (-1, 4)$
1	0	$\longrightarrow (1, 0)$

Example 2

For the line $2x + y = 2$, the x-intercept is $(1, 0)$ and the y-intercept is $(0, 2)$.

An **x-intercept** of a graph is a point $(a, 0)$ where the graph intersects the x-axis.
A **y-intercept** of a graph is a point $(0, b)$ where the graph intersects the y-axis.

A **vertical line** can be represented by an equation of the form $x = k$.
A **horizontal line** can be represented by an equation of the form $y = k$.

Example 3

$x = 3$ represents a vertical line

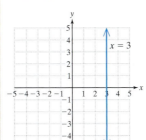

$y = 3$ represents a horizontal line

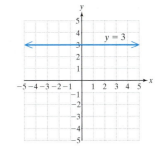

Section 3.3   Slope of a Line and Rate of Change

Key Concepts

The **slope**, m, of a line between two points (x_1, y_1) and (x_2, y_2) is given by

$$m = \frac{y_2 - y_1}{x_2 - x_1} \quad \text{or} \quad \frac{\text{change in } y}{\text{change in } x}$$

The slope of a line may be positive, negative, zero, or undefined.

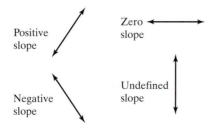

If m_1 and m_2 represent the slopes of two **parallel lines** (nonvertical), then $m_1 = m_2$.

If $m_1 \neq 0$ and $m_2 \neq 0$ represent the slopes of two nonvertical **perpendicular lines**, then

$$m_1 = -\frac{1}{m_2} \text{ or equivalently, } m_1 m_2 = -1.$$

Examples

Example 1

Find the slope of the line between $(1, -5)$ and $(-3, 7)$.

$$m = \frac{7 - (-5)}{-3 - 1} = \frac{12}{-4} = -3$$

Example 2

The slope of the line $y = -2$ is 0 because the line is horizontal.

Example 3

The slope of the line $x = 4$ is undefined because the line is vertical.

Example 4

The slopes of two distinct lines are given. Determine whether the lines are parallel, perpendicular, or neither.

a. $m_1 = -7$ and $m_2 = -7$ Parallel

b. $m_1 = -\dfrac{1}{5}$ and $m_2 = 5$ Perpendicular

c. $m_1 = -\dfrac{3}{2}$ and $m_2 = -\dfrac{2}{3}$ Neither

Section 3.4   Slope-Intercept Form of a Line

Key Concepts

The **slope-intercept form** of a line is

$$y = mx + b$$

where m is the slope of the line and $(0, b)$ is the y-intercept.

Slope-intercept form is used to identify the slope and y-intercept of a line when the equation is given.

Slope-intercept form can also be used to graph a line.

Examples

Example 1

Find the slope and y-intercept.

$7x - 2y = 4$

$$-2y = -7x + 4 \qquad \text{Solve for } y.$$

$$\frac{-2y}{-2} = \frac{-7x}{-2} + \frac{4}{-2}$$

$$y = \frac{7}{2}x - 2$$

The slope is $\frac{7}{2}$. The y-intercept is $(0, -2)$.

Example 2

Graph the line.

$$y = \frac{7}{2}x - 2$$

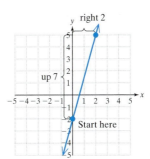

Section 3.5   Point-Slope Formula

Key Concepts

The **point-slope formula** is used primarily to construct an equation of a line given a point and the slope.

Equations of Lines—A Summary:

Standard form: $Ax + By = C$
Horizontal line: $y = k$
Vertical line: $x = k$
Slope-intercept form: $y = mx + b$
Point-slope formula: $y - y_1 = m(x - x_1)$

Examples

Example 1

Find an equation of the line passing through the point $(6, -4)$ and having a slope of $-\frac{1}{2}$.

Label the given information:
$m = -\frac{1}{2}$ and $(x_1, y_1) = (6, -4)$

$$y - y_1 = m(x - x_1)$$

$$y - (-4) = -\frac{1}{2}(x - 6)$$

$$y + 4 = -\frac{1}{2}x + 3$$

$$y = -\frac{1}{2}x - 1$$

Section 3.6   Applications of Linear Equations and Modeling

Key Concepts

Linear equations can often be used to describe or model the relationship between variables in a real-world event. In such applications, the slope may be interpreted as a rate of change.

Examples

Example 1

The number of drug-related arrests for a small city has been growing approximately linearly since 1980.

    Let y represent the number of drug arrests, and let x represent the number of years after 1980.

Number of Drug Arrests

Year ($x = 0$ represents 1980)

a. Use the ordered pairs $(0, 1890)$ and $(20, 3210)$ to find an equation of the line shown in the graph.

$$m = \frac{y_2 - y_1}{x_2 - x_1} = \frac{3210 - 1890}{20 - 0}$$

$$= \frac{1320}{20} = 66$$

The slope is 66, indicating that the number of drug arrests is increasing at a rate of 66 per year. $m = 66$, and the y-intercept is $(0, 1890)$. Hence:

$$y = mx + b \implies y = 66x + 1890$$

b. Use the equation in part (a) to predict the number of drug-related arrests in the year 2010. (The year 2010 is 30 years after 1980. Hence, $x = 30$.)

$$y = 66(30) + 1890$$

$$y = 3870$$

The number of drug arrests is predicted to be 3870 by the year 2010.

Section 3.7   Introduction to Functions

Key Concepts

Any set of ordered pairs, (x, y), is called a **relation** in x and y.

The **domain** of a relation is the set of first components in the ordered pairs in the relation. The **range** of a relation is the set of second components in the ordered pairs.

Given a relation in x and y, we say "y is a **function** of x" if for each element x in the domain, there is exactly one value y in the range.

Vertical Line Test for Functions

Consider any relation defined by a set of points (x, y) on a rectangular coordinate system. Then the graph defines y as a function of x if no vertical line intersects the graph in more than one point.

Function Notation

$f(x)$ is the value of the function, f, at x.

Examples

Example 1

Find the domain and range of the relation.

$\{(0, 0), (1, 1), (2, 4), (3, 9), (-1, 1), (-2, 4), (-3, 9)\}$

Domain: $\{0, 1, 2, 3, -1, -2, -3\}$

Range: $\{0, 1, 4, 9\}$

Example 2

Function: $\{(1, 3), (2, 5), (6, 3)\}$

Not a function: $\{(1, 3), (2, 5), (1, -2)\}$

different y-values for the same x-value

Example 3

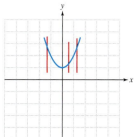

Not a Function

Vertical line intersects more than once.

Function

No vertical line intersects more than once.

Example 4

Given $f(x) = -3x^2 + 5x$, find $f(-2)$.

$f(-2) = -3(-2)^2 + 5(-2)$

$\qquad = -12 - 10$

$\qquad = -22$

Chapter 3   Review Exercises

Section 3.1

1. Graph the points on a rectangular coordinate system.

   a. $\left(\dfrac{1}{2}, 5\right)$ **b.** $(-1, 4)$ **c.** $(2, -1)$

   d. $(0, 3)$ **e.** $(0, 0)$ **f.** $\left(-\dfrac{8}{5}, 0\right)$

   g. $(-2, -5)$ **h.** $(3, 1)$

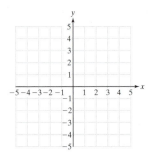

2. Estimate the coordinates of the points A, B, C, D, E, and F.

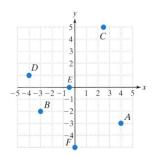

For Exercises 3–8, determine the quadrant in which the given point is found.

   3. $(-2, -10)$ **4.** $(-4, 6)$

   5. $(3, -5)$ **6.** $\left(\dfrac{1}{2}, \dfrac{7}{5}\right)$

   7. $(\pi, -2.7)$ **8.** $(-1.2, -6.8)$

   9. On which axis is the point $(2, 0)$ found?

   10. On which axis is the point $(0, -3)$ found?

11. The price per share of a stock (in dollars) over a period of 5 days is shown in the graph.

   a. Interpret the meaning of the ordered pair $(1, 26.25)$.

   b. On which day was the price the highest?

   c. What was the increase in price between day 1 and day 2?

12. The number of space shuttle launches for selected years is given by the ordered pairs. Let x represent the number of years since 1995. Let y represent the number of launches.

   $(1, 7)$ $(2, 8)$ $(3, 5)$ $(4, 3)$

   $(5, 5)$ $(6, 6)$ $(7, 5)$ $(8, 1)$

   a. Interpret the meaning of the ordered pair $(8, 1)$.

   b. Plot the points on a rectangular coordinate system.

Section 3.2

For Exercises 13–16, determine if the given ordered pair is a solution to the equation.

13. $5x - 3y = 12;$ $(0, 4)$

14. $2x - 4y = -6;$ $(3, 0)$

15. $y = \dfrac{1}{3}x - 2;$ $(9, 1)$

16. $y = -\dfrac{2}{5}x + 1;$ $(-10, 5)$

For Exercises 17–20, complete the table and graph the corresponding ordered pairs. Graph the line through the points to represent all solutions to the equation.

17. $3x - y = 5$

x	y
2	
	4
1	

18. $\dfrac{1}{2}x + 3y = 6$

x	y
	2
-2	
	3

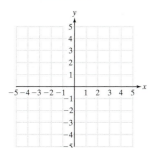

19. $y = \dfrac{2}{3}x - 1$

x	y
0	
3	
-6	

20. $y = -2x - 3$

x	y
0	
-3	
1	

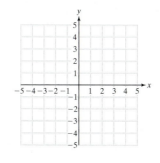

For Exercises 21–24, graph the equation.

21. $x + 2y = 4$

22. $x - y = 5$

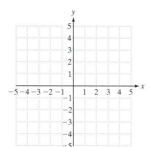

23. $y = 3x$

24. $y = \dfrac{1}{4}x$

For Exercises 25–28, identify the line as horizontal or vertical. Then graph the equation.

25. $3x - 2 = 10$

26. $2x + 1 = -2$

27. $6y + 1 = 13$

28. $5y - 1 = 14$

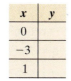

For Exercises 29–36, find the *x*- and *y*-intercepts if they exist.

29. $-4x + 8y = 12$

30. $2x + y = 6$

31. $y = 8x$

32. $5x - y = 0$

33. $6y = -24$

34. $2y - 3 = 1$

35. $2x + 5 = 0$

36. $-3x + 1 = 0$

Section 3.3

37. What is the slope of the ladder leaning up against the wall?

12 ft

5 ft

38. Point *A* is located 4 units down and 2 units to the right of point *B*. What is the slope of the line through points *A* and *B*?

39. Determine the slope of the line that passes through the points $(7, -9)$ and $(-5, -1)$.

40. Determine the slope of the line that has *x*- and *y*-intercepts of $(-1, 0)$ and $(0, 8)$.

41. Determine the slope of the line that passes through the points $(3, 0)$ and $(3, -7)$.

42. Determine the slope of the horizontal line given by $y = -1$.

43. A given line has a slope of -5.

 a. What is the slope of a line parallel to the given line?

 b. What is the slope of a line perpendicular to the given line?

44. A given line has a slope of 0.

 a. What is the slope of a line parallel to the given line?

 b. What is the slope of a line perpendicular to the given line?

For Exercises 45–48, find the slopes of the lines l_1 and l_2 from the two given points. Then determine whether l_1 and l_2 are parallel, perpendicular, or neither.

45. l_1: $(3, 7)$ and $(0, 5)$

 l_2: $(6, 3)$ and $(-3, -3)$

46. l_1: $(-2, 1)$ and $(-1, 9)$

 l_2: $(0, -6)$ and $(2, 10)$

47. l_1: $(0, \frac{5}{6})$ and $(2, 0)$

 l_2: $(0, \frac{6}{5})$ and $(-\frac{1}{2}, 0)$

48. l_1: $(1, 1)$ and $(1, -8)$

 l_2: $(4, -5)$ and $(7, -5)$

Section 3.4

For Exercises 49–54, write each equation in slope-intercept form. Identify the slope and the *y*-intercept.

49. $5x - 2y = 10$

50. $3x + 4y = 12$

51. $x - 3y = 0$

52. $5y - 8 = 4$

53. $2y = -5$

54. $y - x = 0$

For Exercises 55–59, determine whether the equations represent parallel lines, perpendicular lines, or neither.

55. l_1: $y = \dfrac{3}{5}x + 3$

 l_2: $y = \dfrac{5}{3}x + 1$

56. l_1: $2x - 5y = 10$

 l_2: $5x + 2y = 20$

57. l_1: $3x + 2y = 6$

 l_2: $-6x - 4y = 4$

58. l_1: $y = \dfrac{1}{4}x - 3$

 l_2: $-x + 4y = 8$

59. l_1: $2x = 4$

 l_2: $y = 6$

60. Write an equation of the line whose slope is $-\frac{4}{3}$ and whose *y*-intercept is $(0, -1)$.

61. Write an equation of the line that passes through the origin and has a slope of 5.

Section 3.5

62. Write a linear equation in two variables in slope-intercept form. (Answers may vary.)

63. Write a linear equation in two variables in standard form. (Answers may vary.)

64. Write the slope formula to find the slope of the line between the points (x_1, y_1) and (x_2, y_2).

65. Write the point-slope formula.

66. Write an equation of a vertical line (answers may vary).

67. Write an equation of a horizontal line (answers may vary).

For Exercises 68–73, write an equation of a line given the following information.

68. The slope is -6, and the line passes through the point $(-1, 8)$.

69. The slope is $\frac{2}{3}$, and the line passes through the point $(5, 5)$.

70. The line passes through the points $(0, -4)$ and $(8, -2)$.

71. The line passes through the points $(2, -5)$ and $(8, -5)$.

72. The line passes through the point $(5, 12)$ and is perpendicular to the line $y = -\frac{5}{6}x - 3$.

73. The line passes through the point $(-6, 7)$ and is parallel to the line $4x - y = 0$.

Section 3.6

74. The graph shows the average height for girls based on age (*Source:* National Parenting Council). Let x represent a girl's age, and let y represent her height (in inches).

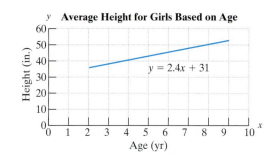

a. Use the equation to estimate the average height of a 7-year-old girl.

b. What is the slope of the line? Interpret the meaning of the slope in the context of the problem.

75. The number of drug prescriptions has increased between 1995 and 2007 (see graph). Let x represent the number of years since 1995. Let y represent the number of prescriptions (in millions).

a. Using the ordered pairs $(4, 2688)$ and $(12, 3784)$ find the slope of the line.

b. Interpret the meaning of the slope in the context of this problem.

c. Find a linear equation that represents the number of prescriptions, y, versus the year, x.

d. Predict the number of prescriptions for the year 2010.

76. A water purification company charges $20 per month and a $55 installation fee.

a. Write a linear equation to compute the total cost, y, of renting this system for x months.

b. Use the equation from part (a) to determine the total cost to rent the system for 9 months.

77. A small cleaning company has a fixed monthly cost of $700 and a variable cost of $8 per service call.

a. Write a linear equation to compute the total cost, y, of making x service calls in one month.

b. Use the equation from part (a) to determine the total cost of making 80 service calls.

Section 3.7

For Exercises 78–83, state the domain and range of each relation. Then determine whether the relation is a function.

78. $\{(6, 3), (10, 3), (-1, 3), (0, 3)\}$

79. $\{(2, 0), (2, 1), (2, -5), (2, 2)\}$

80.

81.

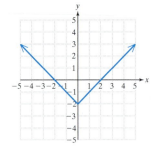

82. $\{(4, 23), (3, -2), (-6, 5), (4, 6)\}$

83. $\{(3, 0), (-4, \frac{1}{2}), (0, 3), (2, -12)\}$

84. Given the function defined by $f(x) = x^3$ find:

　a. $f(0)$　　**b.** $f(2)$　　**c.** $f(-3)$

　d. $f(-1)$　　**e.** $f(4)$

85. Given the function defined by $g(x) = \dfrac{x}{5 - x}$. Find:

　a. $g(0)$　　**b.** $g(4)$　　**c.** $g(-1)$

　d. $g(3)$　　**e.** $g(-5)$

86. The landing distance that a certain plane will travel on a runway is determined by the initial landing speed at the instant the plane touches down. The following function relates landing distance, $D(x)$, to initial landing speed, x, where $x \geq 15$.

$$D(x) = \frac{1}{10}x^2 - 3x + 22$$

where D is in feet and x is in feet per second.

Distance Plane Travels on Runway Versus Speed of Plane

a. Find $D(90)$, and interpret the meaning of the function value in terms of landing speed and length of the runway.

b. Find $D(110)$, and interpret the meaning in terms of landing speed and length of the runway.

Chapter 3　Test

1. In which quadrant is the given point found?

　a. $\left(-\dfrac{7}{2}, 4\right)$　　**b.** $(4.6, -2)$　　**c.** $(-37, -45)$

2. What is the y-coordinate for a point on the x-axis?

3. What is the x-coordinate for a point on the y-axis?

4. The following table depicts a boy's height versus his age. Let x represent the boy's age and y represent his height.

Age (years), x	Height (inches), y
5	46
7	50
9	55
11	60

　a. Write the data as ordered pairs and interpret the meaning of the first ordered pair.

　b. Graph the ordered pairs on a rectangular coordinate system.

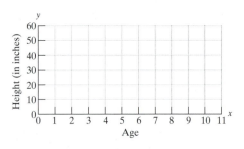

　c. From the graph and table estimate the boy's height at age 10.

　d. The data appear to follow an upward trend up to the boy's teenage years. Do you think this trend will continue? Would it be reasonable to use these data to predict the boy's height at age 25?

5. Determine whether the ordered pair is a solution to the equation $2x - y = 6$.

a. $(0, 6)$

b. $(4, 2)$

c. $(3, 0)$

d. $\left(\dfrac{9}{2}, 3\right)$

6. Given the equation $y = \frac{1}{4}x - 2$, complete the table. Plot the ordered pairs and graph the line through the points to represent the set of all solutions to the equation.

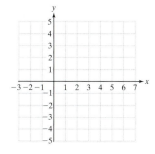

x	y
0	
4	
6	

7. If x represents an adult's age, then the person's maximum recommended heart rate, y, during exercise is approximated by the equation

$$y = 220 - x \quad (x \geq 18)$$

a. Use the equation to find the maximum recommended heart rate for a person who is 18 years old.

b. Use the equation to complete the following ordered pairs: $(20, \quad)$, $(30, \quad)$, $(40, \quad)$, $(50, \quad)$, $(60, \quad)$.

For Exercises 8–9, determine whether the equation represents a horizontal or vertical line. Then graph the line.

8. $-6y = 18$

9. $5x + 1 = 8$

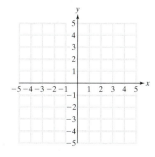

For Exercises 10–13, determine the x- and y-intercepts if they exist.

10. $-4x + 3y = 6$

11. $2y = 6x$

12. $x = 4$

13. $y - 3 = 0$

14. What is the slope of the hill?

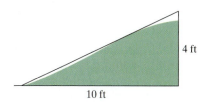

4 ft

10 ft

15. a. Find the slope of the line that passes through the points $(-2, 0)$ and $(-5, -1)$.

b. Find the slope of the line $4x - 3y = 9$.

16. a. What is the slope of a line parallel to the line $x + 4y = -16$?

b. What is the slope of a line perpendicular to the line $x + 4y = -16$?

17. a. What is the slope of the line $x = 5$?

b. What is the slope of the line $y = -3$?

For Exercises 18–19, graph the equations.

18. $y = 8x + 2$

19. $2x + 9y = 0$

20. Determine whether the equations represent parallel lines, perpendicular lines, or neither.

l_1: $2y = 3x - 3$ l_2: $4x = -6y + 1$

21. Write an equation of the line that has y-intercept $\left(0, \frac{1}{2}\right)$ and slope $\frac{1}{4}$.

22. Write an equation of the line that passes through the points $(2, 8)$, and $(4, 1)$.

23. Write an equation of the line that passes through the point $(2, -6)$ and is parallel to the x-axis.

24. Write an equation of the line that passes through the point $(3, 0)$ and is parallel to the line $2x + 6y = -5$.

25. Write an equation of the line that passes through the point $(-3, -1)$ and is perpendicular to the line $x + 3y = 9$.

26. Hurricane Floyd dumped rain at an average rate of $\frac{3}{4}$ in./hr on Southport, North Carolina. Further inland, in Lumberton, North Carolina, the storm dropped $\frac{1}{2}$ in. of rain per hour. The following graph depicts the total amount of rainfall (in inches) versus the time (in hours) for both locations in North Carolina.

a. What is the slope of the line representing the rainfall for Southport?

b. What is the slope of the line representing the rainfall for Lumberton?

27. To attend a state fair, the cost is $10 per person to cover exhibits and musical entertainment. There is an additional cost of $1.50 per ride.

a. Write an equation that gives the total cost, y, of visiting the state fair and going on x rides.

b. Use the equation from part (a) to determine the cost of going to the state fair and going on 10 rides.

28. The number of medical doctors for selected years is shown in the graph. Let x represent the number of years since 1980, and let y represent the number of medical doctors (in thousands) in the United States.

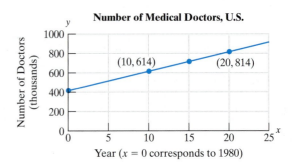

a. Find the slope of the line shown in the graph. Interpret the meaning of the slope in the context of this problem.

b. Find an equation of the line.

c. Use the equation from part (b) to predict the number of medical doctors in the United States for the year 2010.

29. Given the relation $\{(0, -1), (2, 3), (-15, -8), (4, 4), (9, -1)\}$:

a. State the domain.

b. State the range.

c. Determine whether the relation is a function.

30. Given the function defined by $f(x) = x^2 - x$, find

a. $f(3)$ **b.** $f(-3)$

Chapters 1–3 Cumulative Review Exercises

1. Identify the numbers as rational or irrational.

   a. -3 b. $\dfrac{5}{4}$ c. $\sqrt{10}$ d. 0

2. Write the opposite and the absolute value for each number.

   a. $-\dfrac{2}{3}$ b. 5.3

3. Simplify the expression using the order of operations: $32 \div 2 \cdot 4 + 5$

4. Add: $3 + (-8) + 2 + (-10)$

5. Subtract: $16 - 5 - (-7)$

For Exercises 6–7, translate the English phrase into an algebraic expression. Then evaluate the expression.

6. The quotient of $\dfrac{3}{4}$ and $-\dfrac{7}{8}$.

7. The product of -2.1 and -6.

8. Name the property that is illustrated by the following statement. $6 + (8 + 2) = (6 + 8) + 2$

For Exercises 9–12, solve the equation.

9. $6x - 10 = 14$ 10. $3(m + 2) - 3 = 2m + 8$

11. $\dfrac{2}{3}y - \dfrac{1}{6} = y + \dfrac{4}{3}$ 12. $1.7z + 2 = -2(0.3z + 1.3)$

13. The area of Texas is $267{,}277$ mi^2. If this is 712 mi^2 less than 29 times the area of Maine, find the area of Maine.

14. For the formula $3a + b = c$, solve for a.

15. Graph the equation $-6x + 2y = 0$.

16. Find the x- and y-intercepts of $-2x + 4y = 4$.

17. Write the equation in slope-intercept form. Then identify the slope and the y-intercept. $3x + 2y = -12$

18. Explain why the line $2x + 3 = 5$ has only one intercept.

19. Find an equation of a line passing through $(2, -5)$ with slope -3.

20. Find an equation of the line passing through $(0, 6)$ and $(-3, 4)$.

Solve the system by graphing.

5. $y = -3x$
 $x = -1$

Solve the system by graphing.

6. $y = 2x - 3$
 $6x + 2y = 4$

Answers

5.

6.

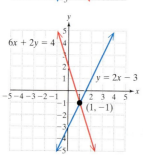

Example 2 **Solving a System of Linear Equations by Graphing**

Solve the system by the graphing method. $y = 2x$
 $y = 2$

Solution:

The equation $y = 2x$ is written in slope-intercept form as $y = 2x + 0$. The line passes through the origin, with a slope of 2.

The line $y = 2$ is a horizontal line and has a slope of 0.

Because the lines have different slopes, the lines must be different and nonparallel. From this, we know that the lines must intersect at exactly one point. Graph the lines to find the point of intersection (Figure 4-2).

The point $(1, 2)$ appears to be the point of intersection. This can be confirmed by substituting $x = 1$ and $y = 2$ into both original equations.

Figure 4-2

$$y = 2x \qquad (2) \stackrel{?}{=} 2(1) ✔ \quad \text{True}$$
$$y = 2 \qquad (2) \stackrel{?}{=} 2 ✔ \quad \text{True}$$

The solution is $(1, 2)$.

Example 3 **Solving a System of Linear Equations by Graphing**

Solve the system by the graphing method.

$$x - 2y = -2$$
$$-3x + 2y = 6$$

Solution:

To graph each equation, write the equation in slope-intercept form: $y = mx + b$.

Equation 1

$$x - 2y = -2$$
$$-2y = -x - 2$$
$$\frac{-2y}{-2} = \frac{-x}{-2} - \frac{2}{-2}$$
$$y = \frac{1}{2}x + 1$$

Equation 2

$$-3x + 2y = 6$$
$$2y = 3x + 6$$
$$\frac{2y}{2} = \frac{3x}{2} + \frac{6}{2}$$
$$y = \frac{3}{2}x + 3$$

Figure 4-3

From their slope-intercept forms, we see that the lines have different slopes, indicating that the lines are different and nonparallel. Therefore, the lines must intersect at exactly one point. Graph the lines to find that point (Figure 4-3).

The point $(-2, 0)$ appears to be the point of intersection. This can be confirmed by substituting $x = -2$ and $y = 0$ into both equations.

$$x - 2y = -2 \longrightarrow (-2) - 2(0) \overset{?}{=} -2 \ \checkmark \quad \text{True}$$

$$-3x + 2y = 6 \longrightarrow -3(-2) + 2(0) \overset{?}{=} 6 \ \checkmark \quad \text{True}$$

The solution is $(-2, 0)$.

TIP: In Examples 2 and 3, the lines could also have been graphed by using the x- and y-intercepts or by using a table of points. However, the advantage of writing the equations in slope-intercept form is that we can compare the slopes and y-intercepts of each line.

1. If the slopes differ, the lines are different and nonparallel and must cross in exactly one point.

2. If the slopes are the same and the y-intercepts are different, the lines are parallel and will not intersect.

3. If the slopes are the same and the y-intercepts are the same, the two equations represent the same line.

Example 4 Graphing an Inconsistent System

Solve the system by graphing.

$$-x + 3y = -6$$

$$6y = 2x + 6$$

Solution:

To graph the lines, write each equation in slope-intercept form.

Equation 1	**Equation 2**
$-x + 3y = -6$	$6y = 2x + 6$
$3y = x - 6$	
$\dfrac{3y}{3} = \dfrac{x}{3} - \dfrac{6}{3}$	$\dfrac{6y}{6} = \dfrac{2x}{6} + \dfrac{6}{6}$
$y = \dfrac{1}{3}x - 2$	$y = \dfrac{1}{3}x + 1$

Because the lines have the same slope but different y-intercepts, they are parallel (Figure 4-4). Two parallel lines do not intersect, which implies that the system has no solution. The system is inconsistent.

Figure 4-4

Example 5   Graphing a Dependent System

Solve the system by graphing.

$$x + 4y = 8$$

$$y = -\frac{1}{4}x + 2$$

Solution:

Write the first equation in slope-intercept form. The second equation is already in slope-intercept form.

Equation 1	**Equation 2**

$$x + 4y = 8 \qquad\qquad y = -\frac{1}{4}x + 2$$

$$4y = -x + 8$$

$$\frac{4y}{4} = \frac{-x}{4} + \frac{8}{4}$$

$$y = -\frac{1}{4}x + 2$$

Figure 4-5

Answer

8.

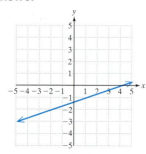

$$\left\{ (x, y) \mid y = \frac{1}{3}x - \frac{4}{3} \right\}$$

The system is dependent.

Notice that the slope-intercept forms of the two lines are identical. Therefore, the equations represent the same line (Figure 4-5). The system is dependent, and the solution to the system of equations is the set of all points on the line.

Because the ordered pairs in the solution set cannot all be listed, we can write the solution in set-builder notation: $\{(x, y) \mid y = -\frac{1}{4}x + 2\}$. This can be read as "the set of all ordered pairs" (x, y) such that the ordered pairs satisfy the equation $y = -\frac{1}{4}x + 2$.

In summary:

- There are infinitely many solutions to the system of equations.
- The solution set is $\{(x, y) \mid y = -\frac{1}{4}x + 2\}$.
- The system is dependent.

Calculator Connections

Topic: Graphing Systems of Linear Equations in Two Variables

The solution to a system of equations can be found by using either a *Trace* feature or an *Intersect* feature on a graphing calculator to find the point of intersection between two graphs.

For example, consider the system:

$$-2x + y = 6$$

$$5x + y = -1$$

First graph the equations together on the same viewing window. Recall that to enter the equations into the calculator, the equations must be written with the *y*-variable isolated.

Isolate y.

$$-2x + y = 6 \longrightarrow y = 2x + 6$$

$$5x + y = -1 \longrightarrow y = -5x - 1$$

By inspection of the graph, it appears that the solution is $(-1, 4)$. The *Trace* option on the calculator may come close to $(-1, 4)$ but may not show the exact solution (Figure 4-6). However, an *Intersect* feature on a graphing calculator may provide the exact solution (Figure 4-7). See your user's manual for further details.

Using *Trace*

Figure 4-6

Using *Intersect*

Figure 4-7

Calculator Exercises

Use a graphing calculator to graph each linear equation on the same viewing window. Use a *Trace* or *Intersect* feature to find the point(s) of intersection.

1. $y = 2x - 3$
$y = -4x + 9$

2. $y = -\dfrac{1}{2}x + 2$
$y = \dfrac{1}{3}x - 3$

3. $x + y = 4$ (Example 1)
$-2x + y = -5$

4. $x - 2y = -2$ (Example 3)
$-3x + 2y = 6$

5. $-x + 3y = -6$ (Example 4)
$6y = 2x + 6$

6. $x + 4y = 8$ (Example 5)
$y = -\dfrac{1}{4}x + 2$

Section 4.1   Practice Exercises

Study Skills Exercises

1. Figure out your grade at this point. Are you earning the grade that you want? If not, maybe organizing a study group would help.

In a study group, check the activities that you might try to help you learn and understand the material.

_____ Quiz each other by asking each other questions.

_____ Practice teaching each other.

_____ Share and compare class notes.

_____ Support and encourage each other.

_____ Work together on exercises and sample problems.

2. Define the key terms:

    a. system of linear equations   **b. solution to a system of linear equations**   **c. consistent system**

    d. inconsistent system   **e. dependent system**   **f. independent system**

Objective 1: Determining Solutions to a System of Linear Equations

For Exercises 3–10, determine if the given point is a solution to the system. **(See Example 1.)**

3. $3x - y = 7$ $(2, -1)$

    $x - 2y = 4$

4. $x - y = 3$ $(4, 1)$

    $x + y = 5$

5. $4y = -3x + 12$ $(0, 4)$

    $y = \dfrac{2}{3}x - 4$

6. $y = -\dfrac{1}{3}x + 2$ $(9, -1)$

    $x = 2y + 6$

7. $3x - 6y = 9$ $\left(4, \dfrac{1}{2}\right)$

    $x - 2y = 3$

8. $x - y = 4$ $(6, 2)$

    $3x - 3y = 12$

9. $\dfrac{1}{3}x = \dfrac{2}{5}y - \dfrac{4}{5}$ $(0, 2)$

    $\dfrac{3}{4}x + \dfrac{1}{2}y = 2$

10. $\dfrac{1}{4}x + \dfrac{1}{2}y = \dfrac{3}{2}$ $(4, 1)$

    $y = \dfrac{3}{2}x - 6$

Objective 2: Dependent and Inconsistent Systems of Linear Equations

For Exercises 11–14, match the graph of the system of equations with the appropriate description of the solution.

11.

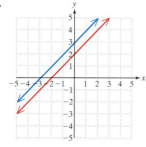

    a. The solution is $(1, 3)$.

    b. No solution.

    c. There are infinitely many solutions.

    d. The solution is $(0, 0)$.

12.

13.

14.

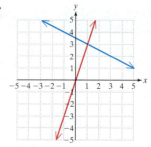

15. Graph each system of equations. **(See Examples 2–5.)**

   a. $y = 2x - 3$ **b.** $y = 2x + 1$ **c.** $y = 3x - 5$

   $\quad y = 2x + 5$ $\quad y = 4x - 5$ $\quad y = 3x - 5$

For Exercises 16–26, determine which system of equations (a, b, or c) makes the statement true. (*Hint:* Refer to the graphs from Exercise 15.)

a. $y = 2x - 3$ **b.** $y = 2x + 1$ **c.** $y = 3x - 5$

$\quad y = 2x + 5$ $\quad y = 4x - 5$ $\quad y = 3x - 5$

16. The lines are parallel.   **17.** The lines coincide.

18. The lines intersect at exactly one point. **19.** The system is inconsistent.

20. The system is dependent.   **21.** The lines have the same slope but different
                                                         y-intercepts.

22. The lines have the same slope and same   **23.** The lines have different slopes.
      y-intercept.

24. The system has exactly one solution.   **25.** The system has infinitely many solutions.

26. The system has no solution.

Objective 3: Solving Systems of Linear Equations by Graphing

For Exercises 27–52, solve the systems by graphing. If a system does not have a unique solution, identify the system as inconsistent or dependent. **(See Examples 2–5.)**

27. $y = -x + 4$ **28.** $y = 3x + 2$ **29.** $2x + y = 0$

$\quad y = x - 2$ $\quad y = 2x$ $\quad 3x + y = 1$

    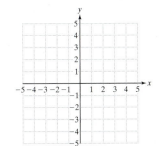

30. $x + y = -1$
$2x - y = -5$

31. $2x + y = 6$
$x = 1$

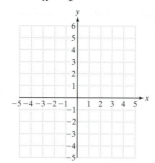

32. $4x + 3y = 9$
$x = 3$

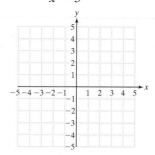

33. $-6x - 3y = 0$
$4x + 2y = 4$

34. $2x - 6y = 12$
$-3x + 9y = 12$

35. $-2x + y = 3$
$6x - 3y = -9$

36. $x + 3y = 0$
$-2x - 6y = 0$

37. $y = 6$
$2x + 3y = 12$

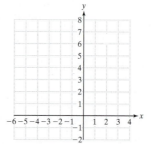

38. $y = -2$
$x - 2y = 10$

39. $-5x + 3y = -9$
$y = \dfrac{5}{3}x - 3$

40. $4x + 2y = 6$
$y = -2x + 3$

41. $x = 4 + y$
$3y = -3x$

42. $3y = 4x$

$x - y = -1$

43. $-x + y = 3$

$4y = 4x + 6$

44. $x - y = 4$

$3y = 3x + 6$

45. $x = 4$

$2y = 4$

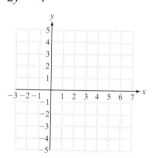

46. $-3x = 6$

$y = 2$

47. $4x + 4y = 8$

$5x + 5y = 5$

48. $2x + 3y = 8$

$-4x - 6y = 6$

49. $2x + y = 4$

$4x - 2y = -4$

50. $6x + 6y = 3$

$2x - y = 4$

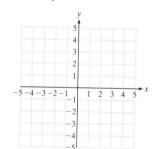

51. $y = 0.5x + 2$

$-x + 2y = 4$

52. $3x - 4y = 6$

$-6x + 8y = -12$

2. Solutions to Systems of Linear Equations: A Summary

The following summary reviews the three different geometric relationships between two lines and the solutions to the corresponding systems of equations.

> **PROCEDURE** Interpreting Solutions to a System of Two Linear Equations
> - The lines may intersect at one point (yielding one unique solution).
> - The lines may be parallel and have no point of intersection (yielding no solution). This is detected algebraically when a contradiction (false statement) is obtained (for example, $0 = -3$ and $12 = 6$).
> - The lines may be the same and intersect at all points on the line (yielding an infinite number of solutions). This is detected algebraically when an identity is obtained (for example, $0 = 0$ and $12 = 12$).

3. Applications of the Substitution Method

In Chapter 2, we solved word problems using one linear equation and one variable. In this chapter, we investigate application problems with two unknowns. In such a case, we can use two variables to represent the unknown quantities. However, if two variables are used, we must write a system of *two* distinct equations.

Skill Practice

6. One number is 16 more than another. Their sum is 92. Use a system of equations to find the numbers.

Example 6 Applying the Substitution Method

One number is 3 more than 4 times another. Their sum is 133. Find the numbers.

Solution:

We can use two variables to represent the two unknown numbers.

Let x represent one number.
Let y represent the other number.   Label the variables.

We must now write two equations. Each of the first two sentences gives a relationship between x and y:

One number is 3 more than 4 times another. $\longrightarrow x = 4y + 3$ (first equation)

Their sum is 133. $\longrightarrow x + y = 133$ (second equation)

Step 1: Notice that x is already isolated in the first equation.

$(4y + 3) + y = 133$

Step 2: Substitute $x = 4y + 3$ into the second equation, $x + y = 133$.

$5y + 3 = 133$

Step 3: Solve the resulting equation.

$5y = 130$

$y = 26$

Answer

6. One number is 38, and the other number is 54.

$x = 4y + 3$

$x = 4(26) + 3$ **Step 4:** To solve for x, substitute $y = 26$ into the equation $x = 4y + 3$.

$x = 104 + 3$

$x = 107$

One number is 26, and the other is 107.

TIP: Check that the numbers 26 and 107 meet the conditions of Example 6.
- 4 times 26 is 104. Three more than 104 is 107. ✔
- The sum of the numbers should be 133: 26 + 107 = 133. ✔

Example 7 **Using the Substitution Method in a Geometry Application**

Two angles are supplementary. The measure of one angle is 15° more than twice the measure of the other angle. Find the measures of the two angles.

Solution:

Let x represent the measure of one angle.
Let y represent the measure of the other angle.

The sum of the measures of supplementary angles is $180°$ ⟶ $x + y = 180$

The measure of one angle is 15° more than twice the other angle ⟶ $x = 2y + 15$

$x + y = 180$

$x = 2y + 15$ **Step 1:** The x-variable in the second equation is already isolated.

$(2y + 15) + y = 180$ **Step 2:** Substitute $2y + 15$ into the first equation for x.

$2y + 15 + y = 180$ **Step 3:** Solve the resulting equation.

$3y + 15 = 180$

$3y = 165$

$y = 55$

$x = 2y + 15$ **Step 4:** Substitute $y = 55$ into the equation $x = 2y + 15$.

$x = 2(55) + 15$

$x = 110 + 15$

$x = 125$

One angle is 55°, and the other is 125°.

Skill Practice

7. The measure of one angle is 2° less than 3 times the measure of another angle. The angles are complementary. Use a system of equations to find the measures of the two angles.

TIP: Check that the angles 55° and 125° meet the conditions of Example 7.
- Because 55° + 125° = 180°, the angles are supplementary. ✔
- The angle 125° is 15° more than twice 55°: 125° = 2(55°) + 15°. ✔

Answer

7. The measures of the angles are 23° and 67°.

Section 4.2 Practice Exercises

Review Exercises

For Exercises 1–6, write each pair of lines in slope-intercept form. Then identify whether the lines intersect in exactly one point or if the lines are parallel or coinciding.

1. $2x - y = 4$
$-2y = -4x + 8$

2. $x - 2y = 5$
$3x = 6y + 15$

3. $2x + 3y = 6$
$x - y = 5$

4. $x - y = -1$
$x + 2y = 4$

5. $2x = \frac{1}{2}y + 2$
$4x - y = 13$

6. $4y = 3x$
$3x - 4y = 15$

Objective 1: Solving Systems of Linear Equations by the Substitution Method

For Exercises 7–10, solve each system using the substitution method. **(See Example 1.)**

7. $3x + 2y = -3$
$y = 2x - 12$

8. $4x - 3y = -19$
$y = -2x + 13$

9. $x = -4y + 16$
$3x + 5y = 20$

10. $x = -y + 3$
$-2x + y = 6$

11. Given the system: $4x - 2y = -6$
$\qquad\qquad\qquad 3x + y = 8$

a. Which variable from which equation is easiest to isolate and why?

b. Solve the system using the substitution method.

12. Given the system: $x - 5y = 2$
$\qquad\qquad\qquad 11x + 13y = 22$

a. Which variable from which equation is easiest to isolate and why?

b. Solve the system using the substitution method.

For Exercises 13–48, solve each system using the substitution method. **(See Examples 1–5.)**

13. $x = 3y - 1$
$2x - 4y = 2$

14. $2y = x + 9$
$y = -3x + 1$

15. $-2x + 5y = 5$
$x = 4y - 10$

16. $y = -2x + 27$
$3x - 7y = -2$

17. $4x - y = -1$
$2x + 4y = 13$

18. $5x - 3y = -2$
$10x - y = 1$

19. $4x - 3y = 11$
$x = 5$

20. $y = -3x - 9$
$y = 12$

21. $4x = 8y + 4$
$5x - 3y = 5$

22. $3y = 6x - 6$
$-3x + y = -4$

23. $x - 3y = -11$
$6x - y = 2$

24. $-2x - y = 9$
$x + 7y = 15$

25. $3x + 2y = -1$
$\frac{3}{2}x + y = 4$

26. $5x - 2y = 6$
$-\frac{5}{2}x + y = 5$

27. $10x - 30y = -10$
$2x - 6y = -2$

28. $3x + 6y = 6$
$-6x - 12y = -12$

29. $2x + y = 3$
$y = -7$

30. $-3x = 2y + 23$
$x = -1$

31. $x + 2y = -2$
$4x = -2y - 17$

32. $x + y = 1$
$2x - y = -2$

33. $y = -\frac{1}{2}x - 4$
$y = 4x - 13$

34. $y = \frac{2}{3}x - 3$
$y = 6x - 19$

35. $y = 6$
$y - 4 = -2x - 6$

36. $x = 9$
$x - 3 = 6y + 12$

37. $3x + 2y = 4$

$2x - 3y = -6$

38. $4x + 3y = 4$

$-2x + 5y = -2$

39. $y = 0.25x + 1$

$-x + 4y = 4$

40. $y = 0.75x - 3$

$-3x + 4y = -12$

41. $11x + 6y = 17$

$5x - 4y = 1$

42. $3x - 8y = 7$

$10x - 5y = 45$

43. $x + 2y = 4$

$4y = -2x - 8$

44. $-y = x - 6$

$2x + 2y = 4$

45. $2x = 3 - y$

$x + y = 4$

46. $2x = 4 + 2y$

$3x + y = 10$

47. $\dfrac{x}{3} + \dfrac{y}{2} = -4$

$x - 3y = 6$

48. $x - 2y = -5$

$\dfrac{2x}{3} + \dfrac{y}{3} = 0$

Objective 3: Applications of the Substitution Method

For Exercises 49–58, set up a system of linear equations and solve for the indicated quantities. **(See Examples 6–7.)**

49. Two numbers have a sum of 106. One number is 10 less than the other. Find the numbers.

 50. Two positive numbers have a difference of 8. The larger number is 2 less than 3 times the smaller number. Find the numbers.

51. The difference between two positive numbers is 26. The larger number is three times the smaller. Find the numbers.

52. The sum of two numbers is 956. One number is 94 less than 6 times the other. Find the numbers.

53. Two angles are supplementary. One angle is 15° more than 10 times the other angle. Find the measure of each angle.

54. Two angles are complementary. One angle is 1° less than 6 times the other angle. Find the measure of each angle.

 55. Two angles are complementary. One angle is 10° more than 3 times the other angle. Find the measure of each angle.

56. Two angles are supplementary. One angle is 5° less than twice the other angle. Find the measure of each angle.

57. In a right triangle, one of the acute angles is 6° less than the other acute angle. Find the measure of each acute angle.

58. In a right triangle, one of the acute angles is 9° less than twice the other acute angle. Find the measure of each acute angle.

Expanding Your Skills

59. The following system of equations is dependent and has infinitely many solutions. Find three ordered pairs that are solutions to the system of equations.

$$y = 2x + 3$$
$$-4x + 2y = 6$$

60. The following system of equations is dependent and has infinitely many solutions. Find three ordered pairs that are solutions to the system of equations.

$$y = -x + 1$$
$$2x + 2y = 2$$

Solving Systems of Equations by the Addition Method

Objectives

1. Solving a System of Linear Equations by the Addition Method
2. Summary of Methods for Solving Linear Equations in Two Variables

1. Solving a System of Linear Equations by the Addition Method

Thus far in Chapter 4 we have used the graphing method and the substitution method to solve a system of linear equations in two variables. In this section, we present another algebraic method to solve a system of linear equations, called the *addition method* (sometimes called the *elimination method*). The purpose of the addition method is to eliminate one variable.

Skill Practice

Solve the system using the addition method.

1. $x + y = 13$
   $2x - y = 2$

Example 1 Using the Addition Method

Solve the system using the addition method.

$$x + y = -2$$
$$x - y = -6$$

Solution:

Notice that the coefficients of the y-variables are opposites:

Coefficient is 1.
$$x + 1y = -2$$
$$x - 1y = -6$$
Coefficient is -1.

Because the coefficients of the y-variables are opposites, we can add the two equations to eliminate the y-variable.

$$x + y = -2$$
$$\underline{x - y = -6}$$
$$2x \quad = -8 \quad \leftarrow \text{ After adding the equations, we have one equation and one variable.}$$

$$2x = -8 \quad \text{Solve the resulting equation.}$$
$$x = -4$$

To find the value of y, substitute $x = -4$ into *either* of the original equations.

$$x + y = -2 \quad \text{First equation}$$
$$(-4) + y = -2$$
$$y = -2 + 4$$
$$y = 2$$

The solution is $(-4, 2)$.

TIP: Notice that the value $x = -4$ could have been substituted into the second equation to obtain the same value for y.

$$x - y = -6$$
$$(-4) - y = -6$$
$$-y = -6 + 4$$
$$-y = -2$$
$$y = 2$$

Check:

$$x + y = -2 \longrightarrow (-4) + (2) \stackrel{?}{=} -2 \longrightarrow -2 \stackrel{?}{=} -2 ✔ \text{ True}$$
$$x - y = -6 \longrightarrow (-4) - (2) \stackrel{?}{=} -6 \longrightarrow -6 \stackrel{?}{=} -6 ✔ \text{ True}$$

Answer

1. $(5, 8)$

It is important to note that the addition method works on the premise that the two equations have *opposite* values for the coefficients of one of the variables. Sometimes it is necessary to manipulate the original equations to create two coefficients that are opposites. This is accomplished by multiplying one or both equations by an appropriate constant. The process is outlined as follows.

> **PROCEDURE** **Solving a System of Equations by the Addition Method**
>
> **Step 1** Write both equations in standard form: $Ax + By = C$.
> **Step 2** Clear fractions or decimals (optional).
> **Step 3** Multiply one or both equations by nonzero constants to create opposite coefficients for one of the variables.
> **Step 4** Add the equations from step 3 to eliminate one variable.
> **Step 5** Solve for the remaining variable.
> **Step 6** Substitute the known value from step 5 into one of the original equations to solve for the other variable.
> **Step 7** Check the solution in both equations.

Example 2   **Solving a System of Linear Equations Using the Addition Method**

Solve the system using the addition method.

$$3x + 5y = 17$$
$$2x - y = -6$$

Solution:

$3x + 5y = 17$ **Step 1:** Both equations are already written in standard form.

$2x - y = -6$ **Step 2:** There are no fractions or decimals.

Notice that neither the coefficients of x nor the coefficients of y are opposites. However, multiplying the second equation by 5 creates the term $-5y$ in the second equation. This is the opposite of the term $+5y$ in the first equation.

$$3x + 5y = 17 \qquad\qquad 3x + 5y = 17$$
$$2x - y = -6 \xrightarrow{\text{Multiply by 5.}} \underline{10x - 5y = -30}$$
$$13x = -13$$

Step 3: Multiply the second equation by 5.

Step 4: Add the equations.

$$13x = -13 \qquad \textbf{Step 5:} \quad \text{Solve the equation.}$$

$$x = -1$$

$$3x + 5y = 17 \qquad \text{First equation}$$
$$3(-1) + 5y = 17$$
$$-3 + 5y = 17$$
$$5y = 20$$
$$y = 4$$

Step 6: Substitute $x = -1$ into one of the original equations.

The solution is $(-1, 4)$.

Step 7: Check the solution in both original equations.

Check:

$$3x + 5y = 17 \longrightarrow 3(-1) + 5(4) \stackrel{?}{=} 17 \longrightarrow -3 + 20 \stackrel{?}{=} 17 \checkmark \text{ True}$$
$$2x - y = -6 \longrightarrow 2(-1) - (4) \stackrel{?}{=} -6 \longrightarrow -2 - 4 \stackrel{?}{=} -6 \checkmark \text{ True}$$

Skill Practice

Solve the system using the addition method.
2. $4x + 3y = 3$
$\ x - 2y = 9$

Answer
2. $(3, -3)$

In Example 3, the system of equations uses the variables a and b instead of x and y. In such a case, we will write the solution as an ordered pair with the variables written in alphabetical order, such as (a, b).

Skill Practice

Solve the system using the addition method.

3. $8n = 4 - 5m$
$7m + 6n = -10$

Example 3 Solve a System of Linear Equations Using the Addition Method

Solve the system using the addition method.

$$5b = 7a + 8$$
$$-4a - 2b = -10$$

Solution:

Step 1: Write the equations in standard form.

The first equation becomes: $\quad\quad 5b = 7a + 8 \longrightarrow -7a + 5b = 8$

The system becomes: $\quad\quad -7a + 5b = 8$
$$-4a - 2b = -10$$

Step 2: There are no fractions or decimals.

Step 3: We need to obtain opposite coefficients on either the a or b term.

Notice that neither the coefficients of a nor the coefficients of b are opposites. However, it is possible to change the coefficients of b to 10 and -10 (this is because the LCM of 5 and 2 is 10). This is accomplished by multiplying the first equation by 2 and the second equation by 5.

$\begin{array}{l} -7a + 5b = 8 \\ -4a - 2b = -10 \end{array} \quad \begin{array}{l} \text{Multiply by 2.} \\ \longrightarrow \\ \longrightarrow \\ \text{Multiply by 5.} \end{array} \quad \begin{array}{l} -14a + 10b = 16 \\ \underline{-20a - 10b = -50} \\ -34a = -34 \end{array}$ **Step 4:** Add the equations.

$$-34a = -34$$ **Step 5:** Solve the resulting equation.

$$\frac{-34a}{-34} = \frac{-34}{-34}$$

$$a = 1$$

$5b = 7a + 8 \quad$ First equation **Step 6:** Substitute $a = 1$ into one of the original equations.

$$5b = 7(1) + 8$$

$$5b = 15$$

$$b = 3$$

The solution is $(1, 3)$. **Step 7:** Check the solution in the original equations.

Check:

$5b = 7a + 8 \longrightarrow 5(3) \overset{?}{=} 7(1) + 8 \longrightarrow 15 \overset{?}{=} 7 + 8$ ✔ True

$-4a - 2b = -10 \longrightarrow -4(1) - 2(3) \overset{?}{=} -10 \longrightarrow -4 - 6 \overset{?}{=} -10$ ✔ True

Answer

3. $(-4, 3)$

Example 4 **Solving a System of Linear Equations Using the Addition Method**

Solve the system using the addition method.

$$34x - 22y = 4$$

$$17x - 88y = -19$$

Skill Practice

Solve the system using the addition method.

4. $15x - 16y = 1$

$45x + 4y = 16$

Solution:

The equations are already in standard form. There are no fractions or decimals to clear.

$$34x - 22y = 4 \longrightarrow 34x - 22y = 4$$

$$17x - 88y = -19 \xrightarrow{\text{Multiply by } -2.} \underline{-34x + 176y = 38}$$

$$154y = 42$$

Solve for y. $\quad 154y = 42$

$$\frac{154y}{154} = \frac{42}{154}$$

Simplify. $\quad y = \dfrac{3}{11}$

To find the value of x, we normally substitute y into one of the original equations and solve for x. In this example, we will show an alternative method for finding x. By repeating the addition method, this time eliminating y, we can solve for x. This approach allows us to avoid substitution of the fractional value for y.

$$34x - 22y = 4 \xrightarrow{\text{Multiply by } -4.} -136x + 88y = -16$$

$$17x - 88y = -19 \longrightarrow \underline{17x - 88y = -19}$$

$$-119x = -35$$

Solve for x. $\quad -119x = -35$

$$\frac{-119x}{-119} = \frac{-35}{-119}$$

Simplify. $\quad x = \dfrac{5}{17}$

The solution is $\left(\frac{5}{17}, \frac{3}{11}\right)$. These values can be checked in the original equations:

$$34x - 22y = 4 \qquad\qquad 17x - 88y = -19$$

$$34\left(\frac{5}{17}\right) - 22\left(\frac{3}{11}\right) \overset{?}{=} 4 \qquad\qquad 17\left(\frac{5}{17}\right) - 88\left(\frac{3}{11}\right) \overset{?}{=} -19$$

$$10 - 6 \overset{?}{=} 4 \text{ ✔ True} \qquad\qquad 5 - 24 \overset{?}{=} -19 \text{ ✔ True}$$

Answer

4. $\left(\dfrac{1}{3}, \dfrac{1}{4}\right)$

Example 5 **Solving a System of Linear Equations**

Solve the system using the addition method.

$$2x - 5y = 10$$

$$\frac{1}{2}x - \frac{5}{4}y = 1$$

Solution:

$2x - 5y = 10$

$\dfrac{1}{2}x - \dfrac{5}{4}y = 1$ **Step 1:** The equations are in standard form.

Step 2: Multiply both sides of the second equation by 4 to clear fractions.

$$\frac{1}{2}x - \frac{5}{4}y = 1 \longrightarrow 4\left(\frac{1}{2}x - \frac{5}{4}y\right) = 4(1) \longrightarrow 2x - 5y = 4$$

Now the system becomes $\quad 2x - 5y = 10$

$$2x - 5y = 4$$

To make either the x-coefficients or y-coefficients opposites, multiply either equation by -1.

$2x - 5y = 10 \xrightarrow{\text{Multiply by } -1.} -2x + 5y = -10$ **Step 3:** Create opposite coefficients.

$2x - 5y = 4 \longrightarrow \underline{\quad 2x - 5y = \quad\; 4}$

$\qquad\qquad\qquad\qquad\qquad\qquad 0 = \;-6$ **Step 4:** Add the equations.

Because the result is a contradiction, there is no solution, and the system of equations is inconsistent. Writing each line in slope-intercept form verifies that the lines are parallel (Figure 4-8).

$$2x - 5y = 10 \xrightarrow{\text{slope-intercept form}} y = \frac{2}{5}x - 2$$

$$\frac{1}{2} - \frac{5}{4}y = 1 \xrightarrow{\text{slope-intercept form}} y = \frac{2}{5}x - \frac{4}{5}$$

There is no solution.

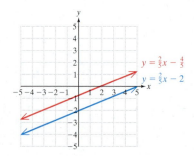

Figure 4-8

Example 6   **Solving a System of Linear Equations**

Solve the system by the addition method.

$$3x - y = 4$$

$$2y = 6x - 8$$

Solution:

$3x - y = 4 \longrightarrow 3x - y = 4$ **Step 1:** Write the equations in standard form.

$2y = 6x - 8 \longrightarrow -6x + 2y = -8$ **Step 2:** There are no fractions or decimals.

Notice that the equations differ exactly by a factor of -2, which indicates that these two equations represent the same line. Multiply the first equation by 2 to create opposite coefficients for the variables.

$$\begin{array}{ll} 3x - y = 4 \xrightarrow{\text{Multiply by 2.}} & 6x - 2y = 8 \\ -6x + 2y = -8 & \underline{-6x + 2y = -8} \\ & 0 = 0 \end{array}$$

Step 3: Create opposite coefficients.

Step 4: Add the equations.

Because the resulting equation is an identity, the original equations represent the same line. This can be confirmed by writing each equation in slope-intercept form.

$$3x - y = 4 \longrightarrow -y = -3x + 4 \longrightarrow y = 3x - 4$$

$$-6x + 2y = -8 \longrightarrow 2y = 6x - 8 \longrightarrow y = 3x - 4$$

The solution is the set of all points on the line, or equivalently, $\{(x, y) | y = 3x - 4\}$.

2. Summary of Methods for Solving Linear Equations in Two Variables

If no method of solving a system of linear equations is specified, you may use the method of your choice. However, we recommend the following guidelines:

1. If one of the equations is written with a variable isolated, the substitution method is a good choice. For example:

$$2x + 5y = 2 \qquad \text{or} \qquad y = \frac{1}{3}x - 2$$

$$x = y - 6 \qquad\qquad\qquad x - 6y = 9$$

2. If both equations are written in standard form, $Ax + By = C$, where none of the variables has coefficients of 1 or -1, then the addition method is a good choice.

$$4x + 5y = 12$$

$$5x + 3y = 15$$

3. If both equations are written in standard form, $Ax + By = C$, and at least one variable has a coefficient of 1 or -1, then either the substitution method or the addition method is a good choice.

Section 4.3 Practice Exercises

Study Skills Exercise

1. Now that you have learned three methods of solving a system of linear equations with two variables, choose a system and solve it all three ways. There are two advantages to this. One is to check your answer (you should get the same answer using all three methods). The second advantage is to show you which method is the easiest for you to use.

Solve the system by using the graphing method, the substitution method, and the addition method.

$$2x + y = -7$$
$$x - 10 = 4y$$

Review Exercises

For Exercises 2–5, check whether the given ordered pair is a solution to the system.

2. $x + y = 8$ $(5, 3)$
  $y = x - 2$

3. $x = y + 1$ $(3, 2)$
  $-x + 2y = 0$

4. $3x + 2y = 14$ $(5, -2)$
  $5x - 2y = 29$

5. $x = 2y - 11$ $(-3, 4)$
  $-x + 5y = 23$

Objective 1: Solving a System of Linear Equations by the Addition Method

For Exercises 6–7, answer as true or false.

6. Given the system $5x - 4y = 1$
                          $7x - 2y = 5$

a. To eliminate the y-variable using the addition method, multiply the second equation by 2.

b. To eliminate the x-variable, multiply the first equation by 7 and the second equation by -5.

7. Given the system $3x + 5y = -1$
                          $9x - 8y = -26$

a. To eliminate the x-variable using the addition method, multiply the first equation by -3.

b. To eliminate the y-variable, multiply the first equation by 8 and the second equation by -5.

8. Given the system $3x - 4y = 2$
                          $17x + \ y = 35$

a. Which variable, x or y, is easier to eliminate using the addition method?

b. Solve the system using the addition method.

9. Given the system $-2x + 5y = -15$
                          $6x - 7y = 21$

a. Which variable, x or y, is easier to eliminate using the addition method?

b. Solve the system using the addition method.

For Exercises 10–25, solve the systems using the addition method. **(See Examples 1–4.)**

10. $x + 2y = 8$
  $5x - 2y = 4$

11. $2x - 3y = 11$
  $-4x + 3y = -19$

12. $a + b = 3$
  $3a + b = 13$

13. $-2u + 6v = 10$
  $-2u + \ v = -5$

14. $-3x + \ y = 1$
  $-6x - 2y = -2$

15. $5m - 2n = 4$
  $3m + \ n = 9$

16. $3x - 5y = 13$
$x - 2y = 5$

17. $7a + 2b = -1$
$3a - 4b = 19$

18. $6c - 2d = -2$
$5c = -3d + 17$

19. $2s + 3t = -1$
$5s = 2t + 7$

20. $6y - 4z = -2$
$4y + 6z = 42$

21. $4k - 2r = -4$
$3k - 5r = 18$

22. $2x + 3y = 6$
$x - y = 5$

23. $6x + 6y = 8$
$9x - 18y = -3$

24. $2x - 5y = 4$
$3x - 3y = 4$

25. $6x - 5y = 7$
$4x - 6y = 7$

26. In solving a system of equations, suppose you get the statement $0 = 5$. How many solutions will the system have? What can you say about the graphs of these equations?

27. In solving a system of equations, suppose you get the statement $0 = 0$. How many solutions will the system have? What can you say about the graphs of these equations?

28. In solving a system of equations, suppose you get the statement $3 = 3$. How many solutions will the system have? What can you say about the graphs of these equations?

29. In solving a system of equations, suppose you get the statement $2 = -5$. How many solutions will the system have? What can you say about the graphs of these equations?

30. Suppose in solving a system of linear equations, you get the statement $x = 0$. How many solutions will the system have? What can you say about the graphs of these equations?

31. Suppose in solving a system of linear equations, you get the statement $y = 0$. How many solutions will the system have? What can you say about the graphs of these equations?

For Exercises 32–43, solve the system of equations using the addition method. **(See Examples 5–6.)**

32. $-2x + y = -5$
$8x - 4y = 12$

33. $x - 3y = 2$
$-5x + 15y = 10$

34. $x + 2y = 2$
$-3x - 6y = -6$

35. $4x - 3y = 6$
$-12x + 9y = -18$

36. $3a + 2b = 11$
$7a - 3b = -5$

37. $4y + 5z = -2$
$5y - 3z = 16$

38. $3x - 5y = 7$
$5x - 2y = -1$

39. $4s + 3t = 9$
$3s + 4t = 12$

40. $2x + 2 = -3y + 9$
$3x - 10 = -4y$

41. $-3x + 6 + 7y = 5$
$5y = 2x$

42. $4x - 5y = 0$
$8(x - 1) = 10y$

43. $y = 2x + 1$
$-3(2x - y) = 0$

Objective 2: Summary of Methods for Solving Linear Equations in Two Variables

For Exercises 44–63, solve the system of equations by either the addition method or the substitution method.

44. $5x - 2y = 4$
$y = -3x + 9$

45. $-x = 8y + 5$
$4x - 3y = -20$

46. $0.1x + 0.1y = 0.6$
$0.1x - 0.1y = 0.1$

47. $0.1x + 0.1y = 0.2$
$0.1x - 0.1y = 0.3$

48. $3x = 5y - 9$
$2y = 3x + 3$

49. $10x - 5 = 3y$
$4x + 5y = 2$

50. $y = -5x - 5$
$6x - 3 = -3y$

51. $4x + 5y = -2$
$3x = -2y - 5$

52. $x = -\dfrac{1}{2}$
$6x - 5y = -8$

53. $4x - 2y = 1$
       $y = 3$

54. $0.02x + 0.04y = 0.12$
       $0.03x - 0.05y = -0.15$

55. $-0.04x + 0.03y = 0.03$
       $-0.06x - 0.02y = -0.02$

56. $8x - 16y = 24$
       $2x - 4y = 0$

57. $y = -\dfrac{1}{2}x - 5$
       $2x + 4y = -8$

58. $\dfrac{m}{2} + \dfrac{n}{5} = \dfrac{13}{10}$
       $3m - 3n = m - 10$

59. $\dfrac{a}{4} - \dfrac{3b}{2} = \dfrac{15}{2}$
       $a + 2b = -10$

60. $2m - 6n = m + 4$
       $3m + 8 = 5m - n$

61. $m - 3n = 10$
       $3m + 12n = -12$

62. $9a - 2b = 8$
       $18a + 6 = 4b + 22$

63. $a = 5 + 2b$
       $3a - 6b = 15$

For Exercises 64–67, set up a system of linear equations, and solve for the indicated quantities.

64. The sum of two positive numbers is 26. Their difference is 14. Find the numbers.

65. The difference of two positive numbers is 2. The sum of the numbers is 36. Find the numbers.

66. Eight times the smaller of two numbers plus 2 times the larger number is 44. Three times the smaller number minus 2 times the larger number is zero. Find the numbers.

67. Six times the smaller of two numbers minus the larger number is −9. Ten times the smaller number plus five times the larger number is 5. Find the numbers.

For Exercises 68–70, solve the system by using each of the three methods: (a) the graphing method, (b) the substitution method, and (c) the addition method.

68. $2x + y = 1$
       $-4x - 2y = -2$

69. $3x + y = 6$
       $-2x + 2y = 4$

70. $2x - 2y = 6$
       $5y = 5x + 5$

Expanding Your Skills

71. Explain why a system of linear equations cannot have exactly two solutions.

72. The solution to the system of linear equations is $(1, 2)$. Find A and B.
$$Ax + 3y = 8$$
$$x + By = -7$$

73. The solution to the system of linear equations is $(-3, 4)$. Find A and B.
$$4x + Ay = -32$$
$$Bx + 6y = 18$$

Problem Recognition Exercises

Systems of Equations

For Exercises 1–6 determine the number of solutions to the system without solving the system. Explain your answers.

1. $y = -4x + 2$
$y = -4x + 2$

2. $y = -4x + 6$
$y = -4x + 1$

3. $y = 4x - 3$
$y = -4x + 5$

4. $y = 7$
$2x + 3y = 1$

5. $2x + 3y = 1$
$2x + 3y = 8$

6. $8x - 2y = 6$
$12x - 3y = 9$

For Exercises 7–26, solve the system using the method of your choice.

7. $x = -2y + 5$
$2x - 4y = 10$

8. $y = -3x - 4$
$2x - y = 9$

9. $3x - 2y = 22$
$5x + 2y = 10$

10. $-4x + 2y = -2$
$4x - 5y = -7$

11. $\dfrac{1}{3}x + \dfrac{1}{2}y = \dfrac{2}{3}$

$-\dfrac{2}{3}x + y = -\dfrac{4}{3}$

12. $\dfrac{1}{4}x + \dfrac{2}{5}y = 6$

$\dfrac{1}{2}x - \dfrac{1}{10}y = 3$

13. $2c + 7d = -1$
$c = 2$

14. $-3w + 5z = -6$
$z = -4$

15. $y = 0.4x - 0.3$
$-4x + 10y = 20$

16. $x = -0.5y + 0.1$
$-10x - 5y = 2$

17. $3a + 7b = -3$
$-11a + 3b = 11$

18. $2v - 5w = 10$
$9v + 7w = 45$

19. $y = 2x - 14$
$4x - 2y = 28$

20. $x = 5y - 9$
$-2x + 10y = 18$

21. $x + y = 3200$
$0.06x + 0.04y = 172$

22. $x + y = 4500$
$0.07x + 0.05y = 291$

23. $3x + y - 7 = x - 4$
$3x - 4y + 4 = -6y + 5$

24. $7y - 8y - 3 = -3x + 4$
$10x - 5y - 12 = 13$

25. $3x - 6y = -1$
$9x + 4y = 8$

26. $8x - 2y = 5$
$12x + 4y = -3$

Section 4.4 Applications of Linear Equations in Two Variables

Skill Practice

1. Lynn went to a fast-food restaurant and spent $9.00. She purchased 4 hamburgers and 5 orders of fries. The next day, Ricardo went to the same restaurant and purchased 10 hamburgers and 7 orders of fries. He spent $18.10. Use a system of equations to determine the cost of a burger and the cost of an order of fries.

1. Applications Involving Cost

In Sections 2.4–2.7, we solved several applied problems by setting up a linear equation in one variable. When solving an application that involves two unknowns, sometimes it is convenient to use a system of linear equations in two variables.

> **Example 1** Using a System of Linear Equations Involving Cost
>
> At a movie theater a couple buys one large popcorn and two drinks for $5.75. A group of teenagers buys two large popcorns and five drinks for $13.00. Find the cost of one large popcorn and the cost of one drink.

Solution:

In this application we have two unknowns, which we can represent by x and y.

Let x represent the cost of one large popcorn.
Let y represent the cost of one drink.

We must now write two equations. Each of the first two sentences in the problem gives a relationship between x and y:

$$\left(\begin{matrix}\text{Cost of 1} \\ \text{large popcorn}\end{matrix}\right) + \left(\begin{matrix}\text{cost of 2} \\ \text{drinks}\end{matrix}\right) = \left(\begin{matrix}\text{total} \\ \text{cost}\end{matrix}\right) \longrightarrow x + 2y = 5.75$$

$$\left(\begin{matrix}\text{Cost of 2} \\ \text{large popcorns}\end{matrix}\right) + \left(\begin{matrix}\text{cost of 5} \\ \text{drinks}\end{matrix}\right) = \left(\begin{matrix}\text{total} \\ \text{cost}\end{matrix}\right) \longrightarrow 2x + 5y = 13.00$$

To solve this system, we may either use the substitution method or the addition method. We will use the substitution method by solving for x in the first equation.

$$x + 2y = 5.75 \longrightarrow x = -2y + 5.75 \qquad \text{Isolate } x \text{ in the first equation.}$$

$$2x + 5y = 13.00$$

$$2(-2y + 5.75) + 5y = 13.00 \qquad \text{Substitute } x = -2y + 5.75 \text{ into the other equation.}$$

$$-4y + 11.50 + 5y = 13.00$$

$$y + 11.50 = 13.00 \qquad \text{Solve for } y.$$

$$y = 1.50$$

$$x = -2y + 5.75$$

$$x = -2(1.50) + 5.75 \qquad \text{Substitute } y = 1.50 \text{ into the equation}$$

$$x = -3.00 + 5.75 \qquad x = -2y + 5.75.$$

$$x = 2.75$$

The cost of one large popcorn is $2.75 and the cost of one drink is $1.50.

Check by verifying that the solutions meet the specified conditions.

$$1 \text{ popcorn} + 2 \text{ drinks} = 1(\$2.75) + 2(\$1.50) = \$5.75 ✔ \quad \text{True}$$

$$2 \text{ popcorns} + 5 \text{ drinks} = 2(\$2.75) + 5(\$1.50) = \$13.00 ✔ \quad \text{True}$$

Answer

1. The cost of a burger is $1.25 and the cost of an order of fries is $0.80.

2. Applications Involving Principal and Interest

In Section 2.4, we learned that simple interest is interest computed on the principal amount of money invested (or borrowed). Simple interest, I, is found by using the formula

$$I = Prt \quad \text{where } P \text{ is the principal,}$$
$$r \text{ is the annual interest rate, and}$$
$$t \text{ is the time in years.}$$

If the amount of time is taken to be 1 year, we have: $I = Pr(1)$ or simply $I = Pr$.

In Example 2, we apply the concept of simple interest to two accounts to produce a desired amount of interest after 1 yr.

| Example 2 | Using a System of Linear Equations Involving Investments |

Joanne has a total of $6000 to deposit in two accounts. One account earns 3.5% simple interest and the other earns 2.5% simple interest. If the total amount of interest at the end of 1 yr is $195, find the amount she deposited in each account.

Solution:

Let x represent the principal deposited in the 2.5% account.
Let y represent the principal deposited in the 3.5% account.

	2.5% Account	3.5% Account	Total
Principal	x	y	6000
Interest $(I = Pr)$	$0.025x$	$0.035y$	195

Each row of the table yields an equation in x and y:

$$\begin{pmatrix} \text{Principal} \\ \text{invested} \\ \text{at } 2.5\% \end{pmatrix} + \begin{pmatrix} \text{principal} \\ \text{invested} \\ \text{at } 3.5\% \end{pmatrix} = \begin{pmatrix} \text{total} \\ \text{principal} \end{pmatrix} \longrightarrow x + y = 6000$$

$$\begin{pmatrix} \text{Interest} \\ \text{earned} \\ \text{at } 2.5\% \end{pmatrix} + \begin{pmatrix} \text{interest} \\ \text{earned} \\ \text{at } 3.5\% \end{pmatrix} = \begin{pmatrix} \text{total} \\ \text{interest} \end{pmatrix} \longrightarrow 0.025x + 0.035y = 195$$

We will choose the addition method to solve the system of equations. First multiply the second equation by 1000 to clear decimals.

Multiply by -25.

$$x + y = 6000 \longrightarrow \quad x + y = 6000 \quad \longrightarrow -25x - 25y = -150{,}000$$
$$0.025x + 0.035y = 195 \longrightarrow 25x + 35y = 195{,}000 \longrightarrow \underline{\quad 25x + 35y = \quad 195{,}000}$$

Multiply by 1000.

$$10y = \quad 45{,}000$$

$$10y = 45{,}000 \qquad \text{After eliminating the } x\text{-variable, solve for } y.$$

$$\frac{10y}{10} = \frac{45{,}000}{10}$$

$$y = 4500 \qquad \text{The amount invested in the 3.5\% account is \$4500.}$$

$x + y = 6000$ Substitute $y = 4500$ into the equation $x + y = 6000$.

$x + 4500 = 6000$

$x = 1500$ The amount invested in the 2.5% account is $1500.

Joanne deposited $1500 in the 2.5% account and $4500 in the 3.5% account.

To check, verify that the conditions of the problem have been met.

1. The sum of $1500 and $4500 is $6000 as desired. ✔ True

2. The interest earned on $1500 at 2.5% is: $0.025(\$1500) = \$37.50$
    The interest earned on $4500 at 3.5% is: $0.035(\$4500) = \$157.50$

    Total interest: $195.00 ✔ True

3. Applications Involving Mixtures

Skill Practice

3. How many ounces of 20% and 35% acid solution should be mixed together to obtain 15 oz of 30% acid solution.

Example 3 **Using a System of Linear Equations in a Mixture Application**

A 10% alcohol solution is mixed with a 40% alcohol solution to produce 30 L of a 20% alcohol solution. Find the number of liters of 10% solution and the number of liters of 40% solution required for this mixture.

Solution:

Each solution contains a percentage of alcohol plus some other mixing agent such as water. Before we set up a system of equations to model this situation, it is helpful to have background understanding of the problem. In Figure 4-9, the liquid depicted in blue is pure alcohol and the liquid shown in gray is the mixing agent (such as water). Together these liquids form a solution. (Realistically the mixture may not separate as shown, but this image may be helpful for your understanding.)

Let x represent the number of liters of 10% solution.
Let y represent the number of liters of 40% solution.

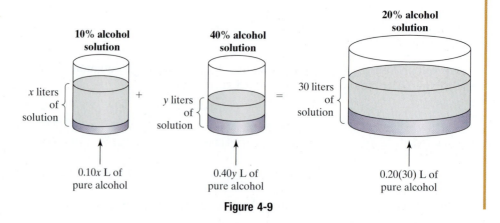

Figure 4-9

Answer

3. 10 oz of the 35% solution, and 5 oz of the 20% solution.

The information given in the statement of the problem can be organized in a chart.

	10% Alcohol	40% Alcohol	20% Alcohol
Number of liters of solution	x	y	30
Number of liters of pure alcohol	$0.10x$	$0.40y$	$0.20(30) = 6$

From the first row, we have

$$\left(\begin{array}{c} \text{Amount of} \\ 10\% \text{ solution} \end{array}\right) + \left(\begin{array}{c} \text{amount of} \\ 40\% \text{ solution} \end{array}\right) = \left(\begin{array}{c} \text{total amount} \\ \text{of } 20\% \text{ solution} \end{array}\right) \longrightarrow x + y = 30$$

From the second row, we have

$$\left(\begin{array}{c} \text{Amount of} \\ \text{alcohol in} \\ 10\% \text{ solution} \end{array}\right) + \left(\begin{array}{c} \text{amount of} \\ \text{alcohol in} \\ 40\% \text{ solution} \end{array}\right) = \left(\begin{array}{c} \text{total amount of} \\ \text{alcohol in} \\ 20\% \text{ solution} \end{array}\right) \longrightarrow 0.10x + 0.40y = 6$$

We will solve the system with the addition method by first clearing decimals.

$$
\begin{array}{lll}
& & \text{Multiply by } -1. \\
x + y = 30 \longrightarrow & x + y = 30 \longrightarrow & -x - y = -30 \\
0.10x + 0.40y = 6 \longrightarrow & x + 4y = 60 \longrightarrow & \underline{x + 4y = 60} \\
& \text{Multiply by 10.} & 3y = 30
\end{array}
$$

$3y = 30$ After eliminating the x-variable, solve for y.

$y = 10$ 10 L of 40% solution is needed.

$x + y = 30$ Substitute $y = 10$ into either of the original equations.

$x + (10) = 30$

$x = 20$ 20 L of 10% solution is needed.

10 L of 40% solution must be mixed with 20 L of 10% solution.

4. Applications Involving Distance, Rate, and Time

The following formula relates the distance traveled to the rate and time of travel.

$$d = rt \quad \text{distance} = \text{rate} \cdot \text{time}$$

For example, if a car travels at 60 mph for 3 hr, then

$$d = (60 \text{ mph})(3 \text{ hr})$$

$$= 180 \text{ mi}$$

If a car travels at 60 mph for x hr, then

$$d = (60 \text{ mph})(x \text{ hr})$$

$$= 60x \text{ mi}$$

The relationship $d = rt$ is used in Example 4.

Example 4 **Using a System of Linear Equations in a Distance, Rate, and Time Application**

A plane travels with a tail wind from Kansas City, Missouri, to Denver, Colorado, a distance of 600 mi in 2 hr. The return trip against a head wind takes 3 hr. Find the speed of the plane in still air, and find the speed of the wind.

Solution:

Let p represent the speed of the plane in still air.
Let w represent the speed of the wind.

Notice that when the plane travels with the wind, the net speed is $p + w$. When the plane travels against the wind, the net speed is $p - w$.

The information given in the problem can be organized in a chart.

	Distance	Rate	Time
With a tail wind	600	$p + w$	2
Against a head wind	600	$p - w$	3

To set up two equations in p and w, recall that $d = rt$.

From the first row, we have

$$\binom{\text{Distance}}{\text{with the wind}} = \binom{\text{rate with}}{\text{the wind}}\binom{\text{time traveled}}{\text{with the wind}} \longrightarrow 600 = (p + w) \cdot 2$$

From the second row, we have

$$\binom{\text{Distance}}{\text{against the wind}} = \binom{\text{rate against}}{\text{the wind}}\binom{\text{time traveled}}{\text{against the wind}} \longrightarrow 600 = (p - w) \cdot 3$$

Using the distributive property to clear parentheses produces the following system:

$$2p + 2w = 600$$
$$3p - 3w = 600$$

The coefficients of the w-variable can be changed to 6 and -6 by multiplying the first equation by 3 and the second equation by 2.

$$2p + 2w = 600 \xrightarrow{\text{Multiply by 3.}} 6p + 6w = 1800$$
$$3p - 3w = 600 \xrightarrow{\text{Multiply by 2.}} \underline{6p - 6w = 1200}$$
$$12p \quad\quad = 3000$$

$$12p = 3000$$
$$\frac{12p}{12} = \frac{3000}{12}$$
$$p = 250 \quad \text{The speed of the plane in still air is 250 mph.}$$

TIP: To create opposite coefficients on the w-variables, we could have divided the first equation by 2 and divided the second equation by 3:

Divide by 2.

$2p + 2w = 600 \longrightarrow p + w = 300$

$3p - 3w = 600 \longrightarrow p - w = 200$

Divide by 3.

$2p \quad\quad = 500$

$p = 250$

$2p + 2w = 600$ Substitute $p = 250$ into the first equation.

$2(250) + 2w = 600$

$500 + 2w = 600$

$2w = 100$

$w = 50$ The speed of the wind is 50 mph.

The speed of the plane in still air is 250 mph. The speed of the wind is 50 mph.

Section 4.4 Practice Exercises

Review Exercises

For Exercises 1–4, solve each system of equations by three different methods:

   a. Graphing method   **b.** Substitution method   **c.** Addition method

1. $-2x + y = 6$

   $2x + y = 2$

2. $x - y = 2$

   $x + y = 6$

3. $y = -2x + 6$

   $4x - 2y = 8$

4. $2x = y + 4$

   $4x = 2y + 8$

For Exercises 5–8, set up a system of linear equations in two variables to solve for the unknown quantities.

5. One number is eight more than twice another. Their sum is 20. Find the numbers.

6. The difference of two positive numbers is 264. The larger number is three times the smaller number. Find the numbers.

7. Two angles are complementary. The measure of one angle is 10° less than nine times the measure of the other. Find the measure of each angle.

8. Two angles are supplementary. The measure of one angle is 9° more than twice the measure of the other angle. Find the measure of each angle.

Objective 1: Applications Involving Cost

9. Kent bought three DVDs and two CDs for $62.50. Demond bought one DVD and four CDs for $72.50. Find the cost of one DVD and the cost of one CD. (See Example 1.)

10. Tanya bought three adult tickets and one child's ticket to a movie for $23.00. Li bought two adult tickets and five children's tickets for $30.50. Find the cost of one adult ticket and the cost of one children's ticket.

11. Linda bought 100 shares of a technology stock and 200 shares of a mutual fund for $3800. Her sister, Sandy, bought 300 shares of technology stock and 50 shares of a mutual fund for $5350. Find the cost per share of the technology stock, and the cost per share of the mutual fund.

12. Two video games and three DVDs can be rented for $19.15. Four video games and one DVD can be rented for $17.35. Find the cost to rent one video game and the cost to rent one DVD.

13. Patricia buys a combination of 42¢ stamps and 59¢ stamps at the Post Office. If she spends exactly $22.70 on 50 stamps, how many of each type did she buy?

14. Bennett purchased some beef and some chicken for a family barbeque. The beef cost $6.00 per pound and the chicken cost $4.50 per pound. He bought a total of 18 lb of meat and spent $96. How much of each type of meat did he purchase?

Objective 2: Applications Involving Principal and Interest

15. Shanelle invested $10,000, and at the end of 1 yr, she received $805 in interest. She invested part of the money in an account earning 10% simple interest and the remaining money in an account earning 7% simple interest. How much did she invest in each account? (See Example 2.)

	10% Account	7% Account	Total
Principal invested			
Interest earned			

16. $12,000 was invested in two accounts, one earning 12% simple interest and the other earning 8% simple interest. If the total interest at the end of 1 yr was $1240, how much was invested in each account?

	12% Account	8% Account	Total
Principal invested			
Interest earned			

17. Troy borrowed a total of $12,000 in two different loans to help pay for his new Chevy Silverado. One loan charges 9% simple interest, and the other charges 6% simple interest. If he is charged $810 in interest after 1 yr, find the amount borrowed at each rate.

18. Blake has a total of $4000 to invest in two accounts. One account earns 2% simple interest, and the other earns 5% simple interest. How much should be invested in both accounts to earn exactly $155 at the end of 1 yr?

19. Suppose a rich uncle dies and leaves you an inheritance of $30,000. You decide to invest part of the money in a relatively safe bond fund that returns 8%. You invest the rest of the money in a riskier stock fund that you hope will return 12% at the end of 1 yr. If you need $3120 at the end of 1 yr to make a down payment on a car, how much should you invest at each rate?

20. As part of his retirement strategy, John plans to invest $200,000 in two different funds. He projects that the moderately high risk investments should return, over time, about 9% per year, while the low risk investments should return about 4% per year. If he wants a supplemental income of $12,000 a year, how should he divide his investments?

Objective 3: Applications Involving Mixtures

21. How much 50% disinfectant solution must be mixed with a 40% disinfectant solution to produce 25 gal of a 46% disinfectant solution? **(See Example 3.)**

	50% Mixture	40% Mixture	46% Mixture
Amount of solution			
Amount of disinfectant			

22. How many gallons of 20% antifreeze solution and a 10% antifreeze solution must be mixed to obtain 40 gal of a 16% antifreeze solution?

	20% Mixture	10% Mixture	16% Mixture
Amount of solution			
Amount of antifreeze			

23 How much 45% disinfectant solution must be mixed with a 30% disinfectant solution to produce 20 gal of a 39% disinfectant solution?

24. How many gallons of a 25% antifreeze solution and a 15% antifreeze solution must be mixed to obtain 15 gal of a 23% antifreeze solution?

25. A nurse needs 50 mL of a 16% salt solution for a patient. She can only find a 13% salt solution and an 18% salt solution in the supply room. How many milliliters of the 13% solution should be mixed with the 18% solution to produce the desired amount of the 16% solution?

26. Meadowsilver Dairy keeps two kinds of milk on hand, skim milk that has 0.3% butterfat and whole milk that contains 3.3% butterfat. How many gallons of each type of milk does the company need to produce 300 gallons of 1% milk for the P&A grocery store?

Objective 4: Applications Involving Distance, Rate, and Time

27. It takes a boat 2 hr to go 16 mi downstream with the current and 4 hr to return against the current. Find the speed of the boat in still water and the speed of the current. **(See Example 4.)**

	Distance	Rate	Time
Downstream			
Return			

28. A boat takes 1.5 hr to go 12 mi upstream against the current. It can go 24 mi downstream with the current in the same amount of time. Find the speed of the current and the speed of the boat in still water.

	Distance	Rate	Time
Upstream			
Downstream			

29. A plane can fly 960 mi with the wind in 3 hr. It takes the same amount of time to fly 840 mi against the wind. What is the speed of the plane in still air and the speed of the wind?

30. A plane flies 720 mi with the wind in 3 hr. The return trip takes 4 hr. What is the speed of the wind and the speed of the plane in still air?

31. Tony Markins flew from JFK Airport to London. It took him 6 hr to fly with the wind, and 8 hr on the return flight against the wind. If the distance is approximately 3600 mi, determine the speed of the plane in still air and the speed of the wind.

32. A riverboat cruise upstream on the Mississippi River from New Orleans, Louisiana, to Natchez, Mississippi, takes 10 hr and covers 140 mi. The return trip downstream with the current takes only 7 hr. Find the speed of the riverboat in still water and the speed of the current.

Mixed Exercises

33. Debi has $2.80 in a collection of dimes and nickels. The number of nickels is five more than the number of dimes. Find the number of each type of coin.

34. A child is collecting state quarters and new $1 coins. If she has a total of 25 coins, and the number of quarters is nine more than the number of dollar coins, how many of each type of coin does she have?

35. In the 1961–1962 NBA basketball season, Wilt Chamberlain of the Philadelphia Warriors made 2432 baskets. Some of the baskets were free throws (worth 1 point each) and some were field goals (worth 2 points each). The number of field goals was 762 more than the number of free throws.

   a. How many field goals did he make and how many free throws did he make?

   b. What was the total number of points scored?

   c. If Wilt Chamberlain played 80 games during this season, what was the average number of points per game?

36. In the 1971–1972 NBA basketball season, Kareem Abdul-Jabbar of the Milwaukee Bucks made 1663 baskets. Some of the baskets were free throws (worth 1 point each) and some were field goals (worth 2 points each). The number of field goals he scored was 151 more than twice the number of free throws.

   a. How many field goals did he make and how many free throws did he make?

   b. What was the total number of points scored?

   c. If Kareem Abdul-Jabbar played 81 games during this season, what was the average number of points per game?

37. A small plane can fly 350 mi with a tailwind in $1\frac{3}{4}$ hr. In the same amount of time, the same plane can travel only 210 mi with a headwind. What is the speed of the plane in still air and the speed of the wind?

38. A plane takes 2 hr to travel 1000 mi with the wind. It can travel only 880 mi against the wind in the same amount of time. Find the speed of the wind and the speed of the plane in still air.

39. At the holidays, Erica likes to sell a candy/nut mixture to her neighbors. She wants to combine candy that costs $1.80 per pound with nuts that cost $1.20 per pound. If Erica needs 20 lb of mixture that will sell for $1.56 per pound, how many pounds of candy and how many pounds of nuts should she use?

40. Mary Lee's natural food store sells a combination of teas. The most popular is a mixture of a tea that sells for $3.00 per pound with one that sells for $4.00 per pound. If she needs 40 lb of tea that will sell for $3.65 per pound, how many pounds of each tea should she use?

41. A total of $60,000 is invested in two accounts, one that earns 5.5% simple interest, and one that earns 6.5% simple interest. If the total interest at the end of 1 yr is $3750, find the amount invested in each account.

42. Jacques borrows a total of $15,000. Part of the money is borrowed from a bank that charges 12% simple interest per year. Jacques borrows the remaining part of the money from his sister and promises to pay her 7% simple interest per year. If Jacques' total interest for the year is $1475, find the amount he borrowed from each source.

43. In the 1994 Super Bowl, the Dallas Cowboys scored four more points than twice the number of points scored by the Buffalo Bills. If the total number of points scored by both teams was 43, find the number of points scored by each team.

44. In the 1973 Super Bowl, the Miami Dolphins scored twice as many points as the Washington Redskins. If the total number of points scored by both teams was 21, find the number of points scored by each team.

Expanding Your Skills

45. In a survey conducted among 500 college students, 340 said that the campus lacked adequate lighting. If $\frac{4}{5}$ of the women and $\frac{1}{2}$ of the men said that they thought the campus lacked adequate lighting, how many men and how many women were in the survey?

46. During a 1-hr television program, there were 22 commercials. Some commercials were 15 sec and some were 30 sec long. Find the number of 15-sec commercials and the number of 30-sec commercials if the total playing time for commercials was 9.5 min.

Linear Inequalities and Systems of Inequalities in Two Variables

Section 4.5

1. Graphing Linear Inequalities in Two Variables

A **linear inequality in two variables** x and y is an inequality that can be written in one of the following forms: $ax + by < c$, $ax + by > c$, $ax + by \leq c$, or $ax + by \geq c$.

A solution to a linear inequality in two variables is an ordered pair that makes the inequality true. For example, solutions to the inequality $x + y < 3$ are ordered pairs (x, y) such that the sum of the x- and y-coordinates is less than 3. Several such examples are $(0, 0), (-2, -2), (3, -2),$ and $(-4, 1)$. There are actually infinitely many solutions to this inequality, and therefore it is convenient to express the solution set as a graph. The shaded area in Figure 4-10 represents all solutions (x, y), whose coordinates total less than 3.

Objectives

1. **Graphing Linear Inequalities in Two Variables**
2. **Graphing Systems of Linear Inequalities in Two Variables**

Concept Connections

Determine which of the ordered pairs are solutions to the inequality $y < 4x - 2$.

1. $(3, -1)$ **2.** $(0, 0)$

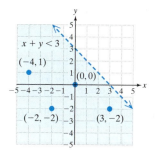

Figure 4-10

To graph a linear inequality in two variables we will use a process called the **test point method**. To use the test point method, first graph the related equation. In this case, the related equation represents a line in the xy-plane. Then choose a test point *not* on the line to determine which side of the line to shade. This process is demonstrated in Example 1.

Answers

1. Solution
2. Not a solution

Example 5   Graphing a System of Linear Inequalities

Graph the solution set. $y > \dfrac{1}{2}x - 2$

$$x + y \leq 1$$

Solution:

Sketch each inequality.

$y > \dfrac{1}{2}x - 2 \xrightarrow{\quad \text{Related equation} \quad} y = \dfrac{1}{2}x - 2$ $x + y \leq 1 \xrightarrow{\quad \text{Related equation} \quad} x + y = 1$

The line $y = \dfrac{1}{2}x - 2$ is drawn in red in Figure 4-20. Substituting the test point $(0, 0)$ into the inequality results in a true statement. Therefore, we shade above the line.

The line $x + y = 1$ is drawn in blue in Figure 4-21. Substituting the test point $(0, 0)$ into the inequality results in a true statement. Therefore, we shade below the line.

Figure 4-20

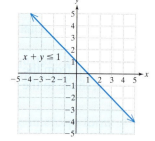

Figure 4-21

Next, we draw these regions on the same graph. The intersection ("overlap") is shown in purple (Figure 4-22).

In Figure 4-23, we show the solution to the system of inequalities. Notice that the portions of the lines not bounding the solution are dashed.

Figure 4-22

Figure 4-23

Answer

7.

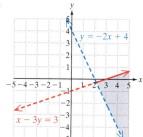

Section 4.5 Practice Exercises

Boost *your* GRADE at
ALEKS.com!

• Practice Problems
• Self-Tests
• NetTutor

• e-Professors
• Videos

Study Skills Exercise

1. Define the key terms:

   a. linear inequality in two variables

   b. test point method

Review Exercises

For Exercises 2–4, graph the equations.

2. $x = -3$

3. $y = \dfrac{3}{5}x + 2$

4. $y = -\dfrac{4}{3}x$

Objective 1: Graphing Linear Inequalities in Two Variables

5. When is a solid line used in the graph of a linear inequality in two variables?

6. When is a dashed line used in the graph of a linear inequality in two variables?

7. What does the shaded region represent in the graph of a linear inequality in two variables?

8. When graphing a linear inequality in two variables, how do you determine which side of the boundary line to shade?

9. Which is the graph of $-2x - y \le 2$?

   a.

   b.

   c.

10. Which is the graph of $-3x + y > -1$?

a.

b.

c.
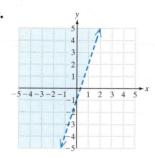

For Exercises 11–16, graph the solution set. Then write three ordered pairs that are solutions to the inequality. **(See Examples 1–4.)**

11. $y \geq -x + 5$

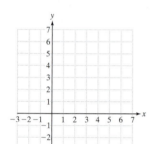

12. $y \leq 2x - 1$

13. $y < 4x$

14. $y > -5x$

15. $3x + 7y \leq 14$

16. $5x - 6y \geq 18$

For Exercises 17–34, graph the solution set. **(See Examples 1–4.)**

17. $x - y > 6$

18. $x + y < 5$

19. $x \geq -1$

20. $x \leq 6$

21. $y < 3$

22. $y > -3$

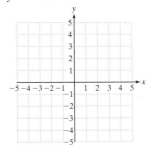

23. $y \leq -\dfrac{3}{4}x + 2$

24. $y \geq \dfrac{2}{3}x + 1$

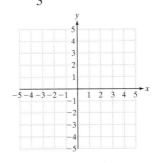

25. $y - 2x > 0$

26. $y + 3x < 0$

27. $x \leq 0$

28. $y \leq 0$

29. $y \geq 0$

30. $x \geq 0$

31. $-x \leq \dfrac{1}{2}y - 2$

32. $-3 + 2x \leq -y$

33. $2x > 3y$

34. $-4x > 5y$

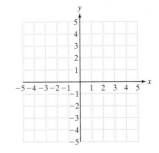

35. a. Describe the graph of the inequality $x + y > 4$. Find three solutions to the inequality (answers will vary).

b. Describe the graph of the equation $x + y = 4$. Find three solutions to the equation (answers will vary).

c. Describe the graph of the inequality $x + y < 4$. Find three solutions to the inequality (answers will vary).

36. a. Describe the graph of the inequality $x + y < 3$. Find three solutions to the inequality (answers will vary).

b. Describe the graph of the equation $x + y = 3$. Find three solutions to the equation (answers will vary).

c. Describe the graph of the inequality $x + y > 3$. Find three solutions to the inequality (answers will vary).

Objective 2: Graphing Systems of Linear Inequalities in Two Variables

For Exercises 37–54, graph the solution set. **(See Example 5.)**

37. $2x + y < 3$

$y \ge x + 3$

38. $x + y < 3$

$y - x \ge 0$

39. $x + y \ge -3$

$x - 2y \ge 6$

40. $y \ge -3x + 4$

$x + y \le 4$

41. $2x + 3y < 6$

$3x + y > -5$

42. $-2x - y < 5$

$x + 2y \ge 2$

43. $y > 2x$

$y > -4x$

44. $2y \ge 6x$

$y \le x$

45. $y < \dfrac{1}{2}x - 1$

$5x + y \le -12$

46. $y \geq \dfrac{1}{3}x + 2$

$4x + y < -2$

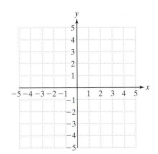

47. $y < 4$

$4x + 3y \geq 12$

48. $x \geq -3$

$2x + 4y < 4$

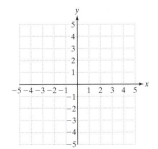

49. $x > -4$

$y \leq 3$

50. $x \leq 3$

$y > 1$

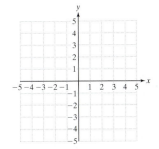

51. $2x \geq 5$

$6 > 3y$

52. $4y \geq 6$

$8 > 2x$

53. $x \geq -4$

$x \leq 1$

54. $y \geq -2$

$y \leq 3$

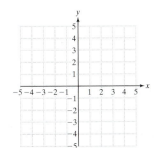

Group Activity

Creating Linear Models from Data

Materials: Two pieces of rope for each group. The ropes should be of different thicknesses. The piece of thicker rope should be between 4 and 5 ft long. The thinner piece of rope should be 8 to 12 in. shorter than the thicker rope. You will also need a yardstick or other device for making linear measurements.

Estimated Time: 30–35 minutes

Group Size: 4 (2 pairs)

1. Each group of 4 should divide into two pairs, and each pair will be given a piece of rope. Each pair will measure the initial length of rope. Then students will tie a series of knots in the rope and measure the new length after each knot is tied. (*Hint:* Try to tie the knots with an equal amount of force each time. Also, as the ropes are straightened for measurement, try to use the same amount of tension in the rope.) The results should be recorded in the table.

Thick Rope		Thin Rope	
Number of Knots, x	Length (in.), y	Number of Knots, x	Length (in.), y
0		0	
1		1	
2		2	
3		3	
4		4	

2. Graph each set of data points. Use a different color pen or pencil for each set of points. Does it appear that each set of data follows a linear trend? Draw a line through each set of points.

3. Each time a knot is tied, the rope decreases in length. Using the results from question 1, compute the average amount of length lost per knot tied.

For the thick rope, the length decreases by _____ inches per knot tied.

For the thin rope, the length decreases by _____ inches per knot tied.

4. For each set of data points, find an equation of the line through the points. Write the equation in slope-intercept form, $y = mx + b$.

[*Hint:* The slope of the line will be negative and will be represented by the amount of length lost per knot (see question 3). The value of b will be the original length of the rope.]

Equation for the thick rope: _____

Equation for the thin rope: _____

5. Next, you will try to predict the number of knots that you need to tie in each rope so that the ropes will be equal in length. To do this, solve the system of equations in question 4.

Solution to the system of equations: (_____, _____)

               number of knots, x length, y

Interpret the meaning of the ordered pair in terms of the number of knots tied and the length of the ropes.

6. Check your answer from question 5 by actually tying the required number of knots in each rope. After doing this, are the ropes the same length? What is the length of each rope? Does this match the length predicted from question 5?

Chapter 4   Summary

Solving Systems of Equations by the Graphing Method

Key Concepts

A **system of two linear equations** can be solved by graphing.

A **solution to a system of linear equations** is an ordered pair that satisfies each equation in the system. Graphically, this represents a point of intersection of the lines.

There may be one solution, infinitely many solutions, or no solution.

One solution	Infinitely many solutions	No solution
Consistent	Consistent	Inconsistent
Independent	Dependent	Independent

A system of equations is **consistent** if there is at least one solution. A system is **inconsistent** if there is no solution.

A linear system in x and y is **dependent** if two equations represent the same line. The solution set is the set of all points on the line.

If two linear equations represent different lines, then the system of equations is **independent**.

Examples

Example 1

Solve by graphing.

$$x + y = 3$$
$$2x - y = 0$$

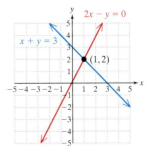

The solution is $(1, 2)$.

Example 2

Solve by graphing.

$$3x - 2y = 2$$
$$-6x + 4y = 4$$

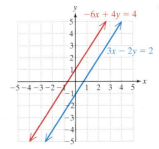

There is no solution. The system is inconsistent.

Example 3

Solve by graphing.

$$x + 2y = 2$$
$$-3x - 6y = -6$$

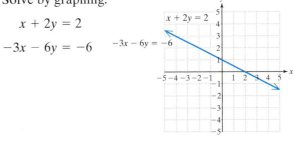

The system is dependent, and the solution set consists of all points on the line, given by
$\{(x, y) \mid x + 2y = 2\}$.

Section 4.2   Solving Systems of Equations by the Substitution Method

Key Concepts

Steps to Solve a System of Equations by the

Substitution Method:

1. Isolate one of the variables from one equation.
2. Substitute the quantity found in step 1 into the other equation.
3. Solve the resulting equation.
4. Substitute the value found in step 3 back into the equation in step 1 to find the remaining variable.
5. Check the solution in both original equations and write the answer as an ordered pair.

An inconsistent system has no solution and is detected algebraically by a contradiction (such as $0 = 3$).

If two linear equations represent the same line, the system is dependent. This is detected algebraically by an identity (such as $0 = 0$).

Examples

Example 1

Solve by the substitution method.

$$x + 4y = -11$$

$$3x - 2y = -5$$

Isolate x in the first equation: $x = -4y - 11$
Substitute into the second equation.

$$3(-4y - 11) - 2y = -5 \qquad \text{Solve the equation.}$$

$$-12y - 33 - 2y = -5$$

$$-14y = 28$$

$$y = -2$$

$$\begin{aligned} && \text{Substitute} \\ x &= -4y - 11 & y = -2. \\ x &= -4(-2) - 11 & \text{Solve for } x. \\ x &= -3 \end{aligned}$$

The solution is $(-3, -2)$ and checks in both original equations.

Example 2

Solve by the substitution method.

$$3x + y = 4$$

$$-6x - 2y = 2$$

Isolate y in the first equation: $y = -3x + 4$.
Substitute into the second equation.

$$-6x - 2(-3x + 4) = 2$$

$$-6x + 6x - 8 = 2$$

$$-8 = 2 \qquad \text{Contradiction}$$

The system is inconsistent and has no solution.

Example 3

Solve by the substitution method.

$$y = x + 2 \qquad y \text{ is already isolated.}$$

$$x - y = -2$$

$$x - (x + 2) = -2 \qquad \text{Substitute } y = x + 2 \text{ into the}$$

$$x - x - 2 = -2 \qquad \text{second equation.}$$

$$-2 = -2 \qquad \text{Identity}$$

The system is dependent. The solution set is all points on the line $y = x + 2$ or $\{(x, y) \mid y = x + 2\}$.

Section 4.3 Solving Systems of Equations by the Addition Method

Key Concepts

Solving a System of Linear Equations by the Addition Method:

1. Write both equations in standard form: $Ax + By = C$.
2. Clear fractions or decimals (optional).
3. Multiply one or both equations by a nonzero constant to create opposite coefficients for one of the variables.
4. Add the equations to eliminate one variable.
5. Solve for the remaining variable.
6. Substitute the known value into one of the original equations to solve for the other variable.
7. Check the solution in both equations.

Examples

Example 1

Solve by using the addition method.

$$5x = -4y - 7 \qquad \text{Write the first equation in}$$
$$6x - 3y = 15 \qquad \text{standard form.}$$

Multiply by 3.
$$5x + 4y = -7 \longrightarrow 15x + 12y = -21$$
$$6x - 3y = 15 \longrightarrow 24x - 12y = 60$$
Multiply by 4.
$$\overline{ 39x = 39}$$
$$x = 1$$

$$5x = -4y - 7$$
$$5(1) = -4y - 7$$
$$5 = -4y - 7$$
$$12 = -4y$$
$$-3 = y \qquad \text{The solution is } (1, -3) \text{ and checks in both original equations.}$$

Section 4.4 | Applications of Linear Equations in Two Variables

Examples

Example 1

A riverboat travels 36 mi with the current in 2 hr. The return trip takes 3 hr against the current. Find the speed of the current and the speed of the boat in still water.

Let x represent the speed of the boat in still water.
Let y represent the speed of the current.

	Distance	Rate	Time
Against current	36	$x - y$	3
With current	36	$x + y$	2

Distance $= $ (rate)(time)

$36 = (x - y) \cdot 3 \longrightarrow 36 = 3x - 3y$

$36 = (x + y) \cdot 2 \longrightarrow 36 = 2x + 2y$

$36 = 3x - 3y \xrightarrow{\text{Multiply by 2.}} 72 = 6x - 6y$

$36 = 2x + 2y \xrightarrow{\text{Multiply by 3.}} \underline{108 = 6x + 6y}$

$ \quad 180 = 12x$

$ 15 = x$

$36 = 2(15) + 2y$

$36 = 30 + 2y$

$6 = 2y$

$3 = y$

The speed of the boat in still water is 15 mph, and the speed of the current is 3 mph.

Example 2

Diane borrows a total of $15,000. Part of the money is borrowed from a lender that charges 8% simple interest. She borrows the rest of the money from her mother and will pay back the money at 5% interest. If the total interest after one year is $900, how much did she borrow from each source?

	8%	5%	Total
Principal	x	y	15,000
Interest	$0.08x$	$0.05y$	900

$x + y = 15{,}000$

$0.08x + 0.05y = 900$

Substitute $x = 15{,}000 - y$ into the second equation.

$0.08(15{,}000 - y) + 0.05y = 900$

$1200 - 0.08y + 0.05y = 900$

$1200 - 0.03y = 900$

$-0.03y = -300$

$y = 10{,}000$

$x = 15{,}000 - 10{,}000$

$= 5{,}000$

The amount borrowed at 8% is $5,000.
The amount borrowed from her mother is $10,000.

Section 4.5   Linear Inequalities and Systems of Inequalities in Two Variables

Key Concepts

A **linear inequality in two variables** can be written in one of the forms: $ax + by < c$, $ax + by > c$, $ax + by \leq c$, or $ax + by \geq c$.

Steps for Using the Test Point Method to Solve a Linear Inequality in Two Variables:

1. Set up the related *equation*.
2. Graph the related equation. This will be a line in the *xy*-plane.
   - If the original inequality is a strict inequality, $<$ or $>$, then the line is *not* part of the solution set. Therefore, graph the boundary as a dashed line.

   - If the original inequality is not strict, $\leq$ or $\geq$, then the line *is* part of the solution set. Therefore, graph the boundary as a solid line.

3. Choose a point not on the line and substitute its coordinates into the original inequality.
   - If the test point makes the inequality true, then the region it represents is part of the solution set. Shade that region.
   - If the test point makes the inequality false, then the other region is part of the solution set and should be shaded.

Example

Example 1

Graph the inequality. $2x - y < 4$

1. The related equation is $2x - y = 4$.
2. Graph the equation $2x - y = 4$ (dashed line).

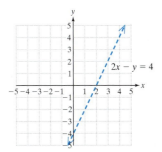

3. Choose an arbitrary test point not on the line such as $(0, 0)$.

$$2x - y < 4$$
$$2(0) - (0) \overset{?}{<} 4$$
$$0 \overset{?}{<} 4 \; ✔ \quad \text{True}$$

Shade the region represented by the test point (in this case, above the line).

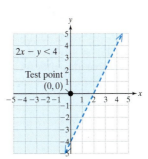

Chapter 4 Review Exercises

Section 4.1

For Exercises 1–4, determine if the ordered pair is a solution to the system.

1. $x - 4y = -4$ $(4, 2)$

    $x + 2y = 8$

2. $x - 6y = 6$ $(12, 1)$

    $-x + y = 4$

3. $3x + y = 9$ $(1, 3)$

    $y = 3$

4. $2x - y = 8$ $(2, -4)$

    $x = 2$

For Exercises 5–10, identify whether the system represents intersecting lines, parallel lines, or coinciding lines by comparing their slopes and y-intercepts.

5. $y = -\dfrac{1}{2}x + 4$

    $y = x - 1$

6. $y = -3x + 4$

    $y = 3x + 4$

7. $y = -\dfrac{4}{7}x + 3$

    $y = -\dfrac{4}{7}x - 5$

8. $y = 5x - 3$

    $y = \dfrac{1}{5}x - 3$

9. $y = 9x - 2$

    $9x - y = 2$

10. $x = -5$

    $y = 2$

For Exercises 11–18, solve the systems by graphing. If a system does not have a unique solution, identify the system as inconsistent or dependent.

11. $y = -\dfrac{2}{3}x - 2$

    $-x + 3y = -6$

12. $y = -2x - 1$

    $2x + 3y = 5$

13. $4x = -2y + 10$

    $2x + y = 5$

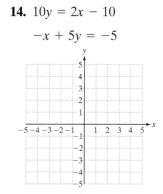

14. $10y = 2x - 10$

    $-x + 5y = -5$

15. $6x - 3y = 9$

    $y = -1$

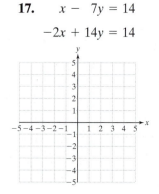

16. $5x + y = -11$

    $x = -1$

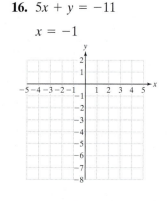

17. $x - 7y = 14$

    $-2x + 14y = 14$

18. $y = -5x + 6$

    $10x + 2y = 6$

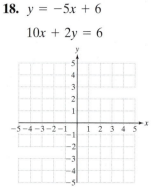

19. A rental car company rents a compact car for $20 a day plus $0.25 per mile. A midsize car rents for $30 a day plus $0.20 per mile.

The cost, y_c, to rent a compact car for one day is given by the equation:

$y_c = 20 + 0.25x$ where x is the number of miles driven

The cost, y_m, to rent a midsize car for one day is given by the equation:

$y_m = 30 + 0.20x$ where x is the number of miles driven

Find the number of miles at which the cost to rent either car for a day would be the same. Confirm your answer with the graph.

Cost to Rent a Car Versus Mileage

Section 4.2

For Exercises 20–23, solve the systems using the substitution method.

20. $6x + y = 2$
$y = 3x - 4$

21. $2x + 3y = -5$
$x = y - 5$

22. $2x + 6y = 10$
$x = -3y + 6$

23. $4x + 2y = 4$
$y = -2x + 2$

24. Given the system:

$$x + 2y = 11$$
$$5x + 4y = 40$$

a. Which variable from which equation is easiest to isolate and why?

b. Solve the system using the substitution method.

25. Given the system:

$$4x - 3y = 9$$
$$2x + y = 12$$

a. Which variable from which equation is easiest to isolate and why?

b. Solve the system using the substitution method.

For Exercises 26–29, solve the systems using the substitution method.

26. $3x - 2y = 23$
$x + 5y = -15$

27. $x + 5y = 20$
$3x + 2y = 8$

28. $x - 3y = 9$
$5x - 15y = 45$

29. $-3x + y = 15$
$6x - 2y = 12$

30. The difference of two positive numbers is 42. The larger number is 2 more than 6 times the smaller number. Find the numbers.

31. In a right triangle, one of the acute angles is 6° less than the other acute angle. Find the measure of each acute angle.

32. Two angles are supplementary. One angle measures 14° less than two times the other angle. Find the measure of each angle.

Section 4.3

33. Explain the process for solving a system of two equations using the addition method.

34. Given the system:

$$3x - 5y = 1$$
$$2x - y = -4$$

a. Which variable, x or y, is easier to eliminate using the addition method? (Answers may vary.)

b. Solve the system using the addition method.

35. Given the system:

$$9x - 2y = 14$$
$$4x + 3y = 14$$

a. Which variable, x or y, is easier to eliminate using the addition method? (Answers may vary.)

b. Solve the system using the addition method.

For Exercises 36–43, solve the systems using the addition method.

36. $2x + 3y = 1$

$\quad\quad x - 2y = 4$

37. $\quad x + 3y = 0$

$\quad\quad -3x - 10y = -2$

38. $8x + 8 = -6y + 6$

$\quad\quad 10x = 9y - 8$

39. $12x = 5y + 5$

$\quad\quad 5y = -1 - 4x$

40. $-4x - 6y = -2$

$\quad\quad 6x + 9y = 3$

41. $-8x - 4y = 16$

$\quad\quad 10x + 5y = 5$

42. $\dfrac{1}{2}x - \dfrac{3}{4}y = -\dfrac{1}{2}$

$\quad\quad \dfrac{1}{3}x + y = -\dfrac{10}{3}$

43. $0.5x - 0.2y = 0.5$

$\quad\quad 0.4x + 0.7y = 0.4$

44. Given the system:

$$4x + 9y = -7$$
$$y = 2x - 13$$

   a. Which method would you choose to solve the system, the substitution method or the addition method? Explain your choice. (Answers may vary.)

   b. Solve the system.

45. Given the system:

$$5x - 8y = -2$$
$$3x - y = -5$$

   a. Which method would you choose to solve the system, the substitution method or the addition method? Explain your choice. (Answers may vary.)

   b. Solve the system.

Section 4.4

46. Miami Metrozoo charges $11.50 for adult admission and $6.75 for children under 12. The total bill before tax for a school group of 60 people is $443. How many adults and how many children were admitted?

47. As part of his retirement strategy Winston plans to invest $600,000 in two different funds. He projects that the high-risk investments should return, over time, about 12% per year, while the low-risk investments should return about 4% per year. If he wants a supplemental income of $30,000 a year, how should he divide his investments?

48. Suppose that whole milk with 4% fat is mixed with 1% low fat milk to make a 2% reduced fat milk. How much of the whole milk should be mixed with the low fat milk to make 60 gal of 2% reduced fat milk?

49. A boat travels 80 mi downstream with the current in 4 hr and 80 mi upstream against the current in 5 hr. Find the speed of the current and the speed of the boat in still water.

50. A plane travels 870 miles against a headwind in 3 hr. Traveling with a tailwind, the plane travels 700 mi in 2 hr. Find the speed of the plane in still air and the speed of the wind.

51. At Conseco Fieldhouse, home of the Indiana Pacers, the total cost of a soft drink and a hot dog is $8.00. The price of the hot dog is $1.00 more than the cost of the soft drink. Find the cost of a soft drink and the cost of a hot dog.

52. In a recent election 5700 votes were cast and 3675 voters voted for the winning candidate. If $\frac{5}{8}$ of the women and $\frac{2}{3}$ of the men voted for the winning candidate, how many men and how many women voted?

53. Ray played two rounds of golf at Pebble Beach for a total score of 154. If his score in the second round is 10 more than his score in the first round, find the scores for each round.

Section 4.5

For Exercises 54–57, graph the inequalities. Then write three ordered pairs that are in the solution set (answers may vary).

54. $y < 3x - 1$

55. $y > -2x + 6$

56. $-2x - 3y \geq 8$

57. $4x - 2y \leq 10$

For Exercises 58–63, graph the solution set.

58. $x - 5y \geq 0$

59. $7x - y \leq 0$

60. $x > 5$

61. $y < -4$

62. $y \geq 0$

63. $x \geq 0$

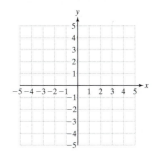

For Exercises 64–67, graph the solution set.

64. $2x - y \geq 8$

    $x + y \leq 3$

65. $y \leq x - 1$

    $x + 2y \geq 4$

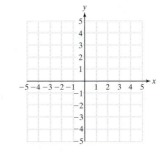

66. $y \leq 2x$

    $-2x - y > -3$

67. $y \leq 4$

    $2x - y < 1$

Chapter 4   Test

1. Write each line in slope-intercept form. Then determine if the lines represent intersecting lines, parallel lines, or coinciding lines.

$$5x + 2y = -6$$

$$-\frac{5}{2}x - y = -3$$

For Exercises 2–3 solve the system by graphing.

2. $y = 2x - 4$

$-2x + 3y = 0$

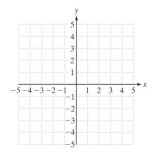

3. $2x + 4y = 6$

$2y - 3 = -x$

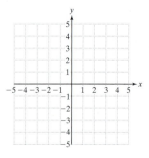

4. Solve the system using the substitution method.

$$x = 5y - 2$$

$$2x + y = -4$$

5. In the 2005 WNBA (basketball) season, the league's leading scorer was Sheryl Swoopes from the Houston Comets. Swoopes scored 17 points more than the second leading scorer, Lauren Jackson from the Seattle Storm. Together they scored a total of 1211 points. How many points did each player score?

6. Solve the system using the addition method.

$$3x - 6y = 8$$

$$2x + 3y = 3$$

7. How many milliliters of a 50% acid solution and how many milliliters of a 20% acid solution must be mixed to produce 36 mL of a 30% acid solution?

8. a. How many solutions does a system of two linear equations have if the equations represent parallel lines?

 b. How many solutions does a system of two linear equations have if the equations represent coinciding lines?

 c. How many solutions does a system of two linear equations have if the equations represent intersecting lines?

For Exercises 9–14, solve the systems using any method.

9. $\frac{1}{3}x + y = \frac{7}{3}$

$x = \frac{3}{2}y - 11$

10. $2(x - 6) = y$

$2x - \frac{1}{2}y = x + 5$

11. $3x - 4y = 29$

$2x + 5y = -19$

12. $2x = 6y - 14$

$2y = 3 - x$

13. $-0.25x - 0.05y = 0.2$

$10x + 2y = -8$

14. $3x + 3y = -2y - 7$

$-3y = 10 - 4x$

15. At Best Buy, Latrell buys four CDs and two DVDs for $54 from the sale rack. Kendra buys two CDs and three DVDs from the same rack for $49. What is the price per CD and the price per DVD?

16. The cost to ride the trolley one-way in San Diego is $2.25. Kelly and Hazel had to buy eight tickets for their group.

 a. What was the total amount of money required?

 b. Kelly and Hazel had only quarters and $1 bills. They also determined that they used twice as many quarters as $1 bills. How many quarters and how many $1 bills did they use?

17. Suppose a total of $5000 is borrowed from two different loans. One loan charges 10% simple interest, and the other charges 8% simple interest. How much was borrowed at each rate if $424 in interest is charged at the end of 1 yr?

18. During the first 13 yr of his football career, Jerry Rice scored a total of 166 touchdowns. One touchdown was scored on a kickoff return, and the remaining 165 were scored rushing or receiving. The number of receiving touchdowns he scored was 5 more than 15 times the number of rushing touchdowns he scored. How many receiving touchdowns and how many rushing touchdowns did he score?

19. A plane travels 910 mi in 2 hr against the wind and 1090 mi in 2 hr with the same wind. Find the speed of the plane in still air and the speed of the wind.

20. The number of calories in a piece of cake is 20 less than 3 times the number of calories in a scoop of ice cream. Together, the cake and ice cream have 460 calories. How many calories are in each?

21. How much 10% acid solution should be mixed with a 25% acid solution to create 100 mL of a 16% acid solution?

22. Graph the solution set. $5x - y \geq -6$.

23. Graph the solution set.

$$2x + y > 1$$
$$x + y < 2$$

Chapters 1–4   Cumulative Review Exercises

1. Simplify.

$$\frac{|2 - 5| + 10 \div 2 + 3}{\sqrt{10^2 - 8^2}}$$

2. Solve for x. $\frac{1}{3}x - \frac{3}{4} = \frac{1}{2}(x + 2)$

3. Solve for a. $-4(a + 3) + 2 = -5(a + 1) + a$

4. Solve for y. $3x - 2y = 6$

5. Solve for z. Graph the solution set on a number line and write the solution in interval notation:

$$-2(3z + 1) \leq 5(z - 3) + 10$$

6. The largest angle in a triangle is 110°. Of the remaining two angles, one is 4° less than the other angle. Find the measure of the three angles.

7. Two hikers start at opposite ends of an 18-mi trail and walk toward each other. One hiker walks predominately down hill and averages 2 mph faster than the other hiker. Find the average rate of each hiker if they meet in 3 hr.

8. Jesse Ventura became the 38th governor of Minnesota by receiving 37% of the votes. If approximately 2,060,000 votes were cast, how many did Mr. Ventura get?

9. The YMCA wants to raise $2500 for its summer program for disadvantaged children. If the YMCA has already raised $900, what percent of its goal has been achieved?

10. Two angles are complementary. One angle measures 17° more than the other angle. Find the measure of each angle.

11. Solve for x. $z = \dfrac{x - m}{5}$

12. Solve for y. $2x - 3y = 6$

13. The slope of a given line is $-\frac{2}{3}$.

    a. What is the slope of a line parallel to the given line?

    b. What is the slope of a line perpendicular to the given line?

14. Find an equation of the line passing through the point $(2, -3)$ and having a slope of -3. Write the final answer in slope-intercept form.

15. Sketch the following equations on the same graph.

    a. $2x + 5y = 10$

    b. $2y = 4$

    c. Find the point of intersection and check the solution in each equation.

    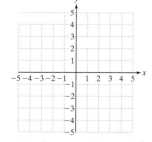

16. Solve the system of equations by using the substitution method.

    $$2x + 5y = 10$$
    $$2y = 4$$

17. a. Graph the line $2x + y = 3$.

    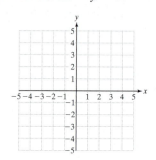

b. Graph the inequality $2x + y < 3$.

c. Explain the difference between the graphs in parts (a) and (b).

18. How many gallons of a 15% antifreeze solution should be mixed with a 60% antifreeze solution to produce 60 gal of a 45% antifreeze solution?

19. Use a system of linear equations to solve for x and y.

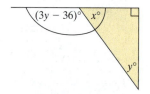

20. In 1920, the average speed for the winner of the Indianapolis 500 car race was 88.6 mph. By 1990, the average speed of the winner was 186.0 mph.

    a. Find the slope of the line shown in the figure. Round to one decimal place.

    b. Interpret the meaning of the slope in the context of this problem.

Polynomials and Properties of Exponents

5

CHAPTER OUTLINE

Chapter 5

In Chapter 5, we begin by covering the properties and applications of expressions involving exponents, including scientific notation. Then we introduce the concept of a polynomial and learn how to add, subtract, multiply, and divide polynomials.

The following is a Sudoku puzzle. Use the clues given to fill in the boxes labeled A–E. Then fill in the remaining part of the grid so that every row, every column, and every 2 × 3 box contains the digits 1 through 6.

A. The number of terms in a trinomial

B. The constant term in the polynomial: $4x^2 - 3x + 5$

C. The value of the missing exponent: $x \cdot x^5 \cdot x^{-4} = x^?$

D. The degree of the polynomial: $-2y^6 - 3y^4 + 5y^2 + 9$

E. Simplified form of $(-2)^2 \cdot 3^0$

| Section 5.1 | **Exponents: Multiplying and Dividing Common Bases** |

Objectives

1. Review of Exponential Notation
2. Evaluating Expressions with Exponents
3. Multiplying and Dividing Common Bases
4. Simplifying Expressions with Exponents
5. Applications of Exponents

1. Review of Exponential Notation

Recall that an **exponent** is used to show repeated multiplication of the **base**.

DEFINITION Definition of b^n

Let b represent any real number and n represent a positive integer. Then,

$$b^n = \underbrace{b \cdot b \cdot b \cdot b \cdot \ldots b}_{n \text{ factors of } b}$$

Skill Practice

For each expression, identify the base and exponent.

1. 8^3
2. $\left(-\dfrac{1}{4}\right)^2$
3. 0.2^4

Example 1 Evaluating Expressions with Exponents

For each expression, identify the exponent and base. Then evaluate the expression.

a. 6^2 **b.** $\left(-\dfrac{1}{2}\right)^3$ **c.** 0.8^4

Solution:

Expression	Base	Exponent	Result
a. 6^2	6	2	$(6)(6) = 36$
b. $\left(-\dfrac{1}{2}\right)^3$	$-\dfrac{1}{2}$	3	$\left(-\dfrac{1}{2}\right)\left(-\dfrac{1}{2}\right)\left(-\dfrac{1}{2}\right) = -\dfrac{1}{8}$
c. 0.8^4	0.8	4	$(0.8)(0.8)(0.8)(0.8) = 0.4096$

Note that if no exponent is explicitly written for an expression, then the expression has an implied exponent of 1. For example, $x = x^1$.

Consider an expression such as $3y^6$. The factor 3 has an exponent of 1, and the factor y has an exponent of 6. That is, the expression $3y^6$ is interpreted as $3^1 y^6$.

2. Evaluating Expressions with Exponents

Recall from Section 1.2 that particular care must be taken when evaluating exponential expressions involving negative numbers. An exponential expression with a negative base is written with parentheses around the base, such as $(-3)^2$.

To evaluate $(-3)^2$, we have: $(-3)^2 = (-3)(-3) = 9$

If no parentheses are present, the expression -3^2, is the *opposite* of 3^2, or equivalently, $-1 \cdot 3^2$.

$$-3^2 = -1(3^2) = -1(3)(3) = -9$$

Answers

1. Base 8; exponent 3
2. Base $-\dfrac{1}{4}$; exponent 2
3. Base 0.2; exponent 4

Example 2 **Evaluating Expressions with Exponents**

Evaluate each expression.

a. -5^4 **b.** $(-5)^4$ **c.** $(-0.2)^3$ **d.** -0.2^3

Skill Practice
Evaluate.
4. -2^4 **5.** $(-2)^4$
6. $(-0.1)^3$ **7.** -0.1^3

Solution:

a. -5^4

$\quad = -1 \cdot 5^4$ 5 is the base with exponent 4.

$\quad = -1 \cdot 5 \cdot 5 \cdot 5 \cdot 5$ Multiply -1 times four factors of 5.

$\quad = -625$

b. $(-5)^4$

$\quad = (-5)(-5)(-5)(-5)$ Parentheses indicate that -5 is the base with exponent 4.

$\quad = 625$ Multiply four factors of -5.

c. $(-0.2)^3$ Parentheses indicate that -0.2 is the base with exponent 3.

$\quad = (-0.2)(-0.2)(-0.2)$ Multiply three factors of -0.2.

$\quad = -0.008$

d. -0.2^3

$\quad = -1 \cdot 0.2^3$ 0.2 is the base with exponent 3.

$\quad = -1 \cdot 0.2 \cdot 0.2 \cdot 0.2$ Multiply -1 times three factors of 0.2.

$\quad = -0.008$

Concept Connections

8. Will the value of a negative base with an odd exponent be negative or positive?
9. Will the value of a negative base with an even exponent be negative or positive?

Example 3 **Evaluating Expressions with Exponents**

Evaluate each expression for $a = 2$ and $b = -3$.

a. $5a^2$ **b.** $(5a)^2$ **c.** $5ab^2$ **d.** $(b + a)^2$

Skill Practice
Evaluate each expression for $x = 2$ and $y = -5$.
10. $6x^2$ **11.** $(6x)^2$
12. $2xy^2$ **13.** $(y - x)^2$

Solution:

a. $5a^2$

$\quad = 5(\quad)^2$ Use parentheses to substitute a number for a variable.

$\quad = 5(2)^2$ Substitute $a = 2$.

$\quad = 5(4)$ Simplify.

$\quad = 20$

b. $(5a)^2$

$\quad = [5(\quad)]^2$ Use parentheses to substitute a number for a variable. The original parentheses are replaced with brackets.

$\quad = [5(2)]^2$ Substitute $a = 2$.

$\quad = (10)^2$ Simplify inside the parentheses first.

$\quad = 100$

Answers
4. -16 **5.** 16   **6.** -0.001
7. -0.001 **8.** Negative
9. Positive   **10.** 24   **11.** 144
12. 100   **13.** 49

c. $5ab^2$

$= 5(2)(-3)^2$ Substitute $a = 2, b = -3$.

$= 5(2)(9)$ Simplify.

$= 90$ Multiply.

d. $(b + a)^2$

$= [(-3) + (2)]^2$ Substitute $b = -3$ and $a = 2$.

$= (-1)^2$ Simplify within the parentheses first.

$= 1$

3. Multiplying and Dividing Common Bases

In this section, we investigate the effect of multiplying or dividing two quantities with the same base. For example, consider the expressions: x^5x^2 and $\frac{x^5}{x^2}$. Simplifying each expression, we have:

$$x^5x^2 = (x \cdot x \cdot x \cdot x \cdot x)(x \cdot x) = \overbrace{x \cdot x \cdot x \cdot x \cdot x \cdot x \cdot x}^{7 \text{ factors of } x} = x^7$$

$$\frac{x^5}{x^2} = \frac{x \cdot x \cdot x \cdot \cancel{x} \cdot \cancel{x}}{\cancel{x} \cdot \cancel{x}} = \frac{x \cdot x \cdot x}{1} = x^3$$

These examples suggest that to multiply two quantities with the same base, we add the exponents. To divide two quantities with the same base, we subtract the exponent in the denominator from the exponent in the numerator. These rules are stated formally in the following two properties.

PROPERTY Multiplication of Like Bases

Assume that b is a real number and that m and n represent positive integers. Then,

$$b^m b^n = b^{m+n}$$

PROPERTY Division of Like Bases

Assume that $b \neq 0$ is a real number and that m and n represent positive integers such that $m > n$. Then,

$$\frac{b^m}{b^n} = b^{m-n}$$

| **Example 4**   Simplifying Expressions with Exponents |

Simplify the expressions.

a. $w^3 w^4$ **b.** $2^3 \cdot 2^4$

Solution:

a. $w^3 w^4$ $(w \cdot w \cdot w)(w \cdot w \cdot w \cdot w)$

  $= w^{3+4}$ To multiply like bases, add the exponents.

  $= w^7$

b. $2^3 \cdot 2^4$ $(2 \cdot 2 \cdot 2)(2 \cdot 2 \cdot 2 \cdot 2)$

  $= 2^{3+4}$ To multiply like bases, add the exponents (the base is unchanged).

  $= 2^7$ or 128

| **Example 5**   Simplifying Expressions with Exponents |

Simplify the expressions.

a. $\dfrac{t^6}{t^4}$ **b.** $\dfrac{5^6}{5^4}$

Solution:

a. $\dfrac{t^6}{t^4}$ $\dfrac{t \cdot t \cdot t \cdot t \cdot t \cdot t}{t \cdot t \cdot t \cdot t}$

  $= t^{6-4}$ To divide like bases subtract the exponents.

  $= t^2$

b. $\dfrac{5^6}{5^4}$ $\dfrac{5 \cdot 5 \cdot 5 \cdot 5 \cdot 5 \cdot 5}{5 \cdot 5 \cdot 5 \cdot 5}$

  $= 5^{6-4}$ To divide like bases subtract the exponents (the base is unchanged).

  $= 5^2$ or 25

Example 6 Simplifying Expressions with Exponents

Simplify the expressions.

a. $\dfrac{z^4 z^5}{z^3}$ **b.** $\dfrac{10^7}{10^2 \cdot 10}$

Solution:

a. $\dfrac{z^4 z^5}{z^3}$

$= \dfrac{z^{4+5}}{z^3}$ Add the exponents in the numerator (the base is unchanged).

$= \dfrac{z^9}{z^3}$

$= z^{9-3}$ Subtract the exponents.

$= z^6$

b. $\dfrac{10^7}{10^2 \cdot 10}$

$= \dfrac{10^7}{10^2 \cdot 10^1}$ Note that 10 is equivalent to 10^1.

$= \dfrac{10^7}{10^{2+1}}$ Add the exponents in the denominator (the base is unchanged).

$= \dfrac{10^7}{10^3}$

$= 10^{7-3}$ Subtract the exponents.

$= 10^4$ or 10,000 Simplify.

4. Simplifying Expressions with Exponents

Example 7 Simplifying Expressions with Exponents

Use the commutative and associative properties of real numbers and the properties of exponents to simplify the expressions.

a. $(3p^2 q^4)(2pq^5)$ **b.** $\dfrac{16w^9 z^3}{4w^8 z}$

Solution:

a. $(3p^2 q^4)(2pq^5)$

$= (3 \cdot 2)(p^2 p)(q^4 q^5)$ Apply the associative and commutative properties of multiplication to group coefficients and like bases.

$= (3 \cdot 2)p^{2+1} q^{4+5}$ Add the exponents when multiplying like bases.

$= 6p^3 q^9$ Simplify.

b. $\dfrac{16w^9 z^3}{4w^8 z}$

$= \left(\dfrac{16}{4}\right)\left(\dfrac{w^9}{w^8}\right)\left(\dfrac{z^3}{z}\right)$ Group coefficients and like bases.

$= 4w^{9-8}z^{3-1}$ Subtract the exponents when dividing like bases.

$= 4wz^2$ Simplify.

5. Applications of Exponents

Simple interest on an investment or loan is computed by the formula $I = Prt$, where P is the amount of principal, r is the annual interest rate, and t is the time in years. Simple interest is based only on the original principal. However, in most day-to-day applications, the interest computed on money invested or borrowed is compound interest. **Compound interest** is computed on the original principal and on the interest already accrued.

Suppose $1000 is invested at 8% interest for 3 yr. Compare the total amount in the account if the money earns simple interest versus if the interest is compounded annually.

Simple Interest

The simple interest earned is given by $I = Prt$

$$= (1000)(0.08)(3)$$

$$= \$240$$

The total amount in the account after 3 yr is $1240.

Compound Interest (Annual)

To compute interest compounded annually over a period of 3 yr, compute the interest earned in the first year. Then add the interest earned in the first year to the principal. This value then becomes the principal on which to base the interest earned in the second year. We repeat this process, finding the interest for the second and third years based on the principal and interest earned in the preceding years. This process is outlined using a table.

Year	Interest Earned $I = Prt$	Total Amount in the Account
First year	$I = (\$1000)(0.08)(1) = \$80$	$\$1000 + \$80 = \$1080$
Second year	$I = (\$1080)(0.08)(1) = \$86.40$	$\$1080 + \$86.40 = \$1166.40$
Third year	$I = (\$1166.40)(0.08)(1) \approx \$93.31$	$\$1166.40 + 93.31 = \mathbf{\$1259.71}$

The total amount in the account found by compounding interest annually is $1259.71. The difference in the account balance for interest compounded annually versus for simple interest is $1259.71 - \$1240 = \$19.71$.

The total amount, A, in an account earning compound annual interest may be computed quickly using the following formula:

$A = P(1 + r)^t$ where P is the amount of principal, r is the annual interest rate (expressed in decimal form), and t is the number of years.

22. Identify the values for the variables P, r, and t if $19,000 is invested for 10 years at 7% interest compounded annually.

For example, for $1000 invested at 8% interest compounded annually for 3 yr, we have $P = 1000$, $r = 0.08$, and $t = 3$.

$$A = P(1 + r)^t$$
$$A = 1000(1 + 0.08)^3$$
$$= 1000(1.08)^3$$
$$= 1000(1.259712)$$
$$= 1259.712$$

Rounding to the nearest cent, we have $A = \$1259.71$, as expected.

23. Find the amount in an account after 3 yr if the initial investment is $4000 invested at 5% interest compounded annually.

Example 8   **Using Exponents in an Application**

Find the amount in an account after 8 yr if the initial investment is $7000, invested at 2.25% interest compounded annually.

Solution:

Identify the values for each variable.

$$P = 7000$$

$r = 0.0225$ Note that the decimal form of a percent is used for calculations.

$$t = 8$$

$A = P(1 + r)^t$

$= 7000(1 + 0.0225)^8$ Substitute.

$= 7000(1.0225)^8$ Simplify inside the parentheses.

$= 7000(1.194831142)$ Approximate $(1.0225)^8$.

$= 8363.82$ Multiply (round to the nearest cent).

The amount in the account after 8 yr is $8363.82.

Answers

22. $P = 19000$; $r = 0.07$; $t = 10$
23. $4630.50

Calculator Connections

In Example 8, it was necessary to evaluate the expression $(1.0225)^8$. Recall that the $\boxed{\wedge}$ or $\boxed{y^x}$ key may be used to enter expressions with exponents.

Scientific Calculator

Enter: 1.0225 $\boxed{y^x}$ 8 $\boxed{=}$ **Result:** $\boxed{1.194831142}$

Graphing Calculator

```
1.0225^8
        1.194831142
```

Calculator Exercises

Use a calculator to evaluate the expressions.

1. $(1.06)^5$

2. $(1.02)^{40}$

3. $5000(1.06)^5$

4. $2000(1.02)^{40}$

5. $3000(1 + 0.06)^2$

6. $1000(1 + 0.05)^3$

Section 5.1   Practice Exercises

For this exercise set, assume all variables represent nonzero real numbers.

Study Skills Exercise

1. Define the key terms:

   a. exponent   b. base   c. simple interest   d. compound interest

Objective 1: Review of Exponential Notation

For Exercises 2–9, identify the base and the exponent. **(See Example 1.)**

2. c^3

3. x^4

4. 5^2

5. 3^5

6. $(-4)^8$

7. $(-1)^4$

8. x

9. q

10. What base corresponds to the exponent 5 in the expression $x^3y^5z^2$?

11. What base corresponds to the exponent 2 in the expression w^3v^2?

12. What is the exponent for the factor of 2 in the expression $2x^3$?

13. What is the exponent for the factor of p in the expression pq^7?

For Exercises 14–22, write the expression using exponents.

14. $(4n)(4n)(4n)$

15. $(-6b)(-6b)$

16. $4 \cdot n \cdot n \cdot n$

17. $-6 \cdot b \cdot b$

18. $(x - 5)(x - 5)(x - 5)$

19. $(y + 2)(y + 2)(y + 2)(y + 2)$

20. $\dfrac{4}{x \cdot x \cdot x \cdot x \cdot x}$

21. $\dfrac{-2}{t \cdot t \cdot t}$

22. $\dfrac{5 \cdot x \cdot x \cdot x}{(y - 7)(y - 7)}$

Objective 2: Evaluating Expressions with Exponents

For Exercises 23–30, evaluate the two expressions and compare the answers. Do the expressions have the same value? **(See Example 2.)**

23. -5^2 and $(-5)^2$

24. -3^4 and $(-3)^4$

25. -2^5 and $(-2)^5$

26. -5^3 and $(-5)^3$

27. $\left(\dfrac{1}{2}\right)^3$ and $\dfrac{1}{2^3}$

28. $\left(\dfrac{1}{5}\right)^2$ and $\dfrac{1}{5^2}$

29. $\left(\dfrac{3}{10}\right)^2$ and $(0.3)^2$

30. $\left(\dfrac{7}{10}\right)^3$ and $(0.7)^3$

For Exercises 31–38, evaluate the expressions.

31. 16^1

32. 20^1

33. $(-1)^{21}$

34. $(-1)^{30}$

35. $\left(-\dfrac{1}{3}\right)^2$

36. $\left(-\dfrac{1}{4}\right)^3$

37. $-\left(\dfrac{2}{5}\right)^2$

38. $-\left(\dfrac{3}{5}\right)^2$

For Exercises 39–46, simplify using the order of operations.

39. $3 \cdot 2^4$
40. $2 \cdot 0^5$
41. $-4(-1)^7$
42. $-3(-1)^4$

43. $6^2 - 3^3$
44. $4^3 + 2^3$
45. $2 \cdot 3^2 + 4 \cdot 2^3$
46. $6^2 - 3 \cdot 1^3$

For Exercises 47–58, evaluate each expression for $a = -4$ and $b = 5$. **(See Example 3.)**

47. $-4b^2$
48. $3a^2$
49. $(-4b)^2$
50. $(3a)^2$

51. $(a + b)^2$
52. $(a - b)^2$
53. $a^2 + 2ab + b^2$
54. $a^2 - 2ab + b^2$

55. $-10ab^2$
56. $-6a^3b$
57. $-10a^2b$
58. $-a^2b$

Objective 3: Multiplying and Dividing Common Bases

59. Expand the following expressions first. Then simplify using exponents.

   a. $x^4 \cdot x^3$ **b.** $5^4 \cdot 5^3$

60. Expand the following expressions first. Then simplify using exponents.

   a. $y^2 \cdot y^4$ **b.** $3^2 \cdot 3^4$

For Exercises 61–72, simplify the expressions. Write the answers in exponent form. **(See Example 4.)**

61. $z^5 z^3$
62. $w^4 w^7$
63. $a \cdot a^8$
64. $p^4 p$

65. $4^5 \cdot 4^9$
66. $6^7 \cdot 6^5$
67. $\left(\dfrac{2}{3}\right)^3 \left(\dfrac{2}{3}\right)$
68. $\left(\dfrac{1}{x}\right)\left(\dfrac{1}{x}\right)^2$

69. $c^5 c^2 c^7$
70. $b^7 b^2 b^8$
71. $x \cdot x^4 \cdot x^{10} \cdot x^3$
72. $z^7 \cdot z^{11} \cdot z^{60} \cdot z$

73. Expand the expressions. Then simplify.

   a. $\dfrac{p^8}{p^3}$ **b.** $\dfrac{8^8}{8^3}$

74. Expand the expressions. Then simplify.

   a. $\dfrac{w^5}{w^2}$ **b.** $\dfrac{4^5}{4^2}$

For Exercises 75–90, simplify the expressions. Write the answers in exponent form. **(See Examples 5–6.)**

75. $\dfrac{x^8}{x^6}$
76. $\dfrac{z^5}{z^4}$
77. $\dfrac{a^{10}}{a}$
78. $\dfrac{b^{12}}{b}$

79. $\dfrac{7^{13}}{7^6}$
80. $\dfrac{2^6}{2^4}$
81. $\dfrac{5^8}{5}$
82. $\dfrac{3^5}{3}$

83. $\dfrac{y^{13}}{y^{12}}$
84. $\dfrac{w^7}{w^6}$
85. $\dfrac{h^3 h^8}{h^7}$
86. $\dfrac{n^5 n^4}{n^2}$

87. $\dfrac{7^2 \cdot 7^6}{7}$
88. $\dfrac{5^3 \cdot 5^8}{5}$
89. $\dfrac{10^{20}}{10^3 \cdot 10^8}$
90. $\dfrac{3^{15}}{3^2 \cdot 3^{10}}$

Objective 4: Simplifying Expressions with Exponents (Mixed Exercises)

For Exercises 91–112, use the commutative and associative properties of real numbers and the properties of exponents to simplify the expressions. **(See Example 7.)**

91. $(2x^3)(3x^4)$
92. $(10y)(2y^3)$
93. $(5a^2b)(8a^3b^4)$
94. $(10xy^3)(3x^4y)$

95. $(r^6s^4)(13r^2s)$
96. $(6p^2q^8)(7p^5q^3)$
97. $s^3 \cdot t^5 \cdot t \cdot t^{10} \cdot s^6$
98. $c \cdot c^4 \cdot d^2 \cdot c^3 \cdot d^3$

99. $(-2v^2)(3v)(5v^5)$ **100.** $(10q^5)(-3q^8)(q)$ **101.** $\left(\frac{2}{3}m^{13}n^8\right)(24m^7n^2)$ **102.** $\left(\frac{1}{4}c^6d^6\right)(28c^2d^7)$

103. $\dfrac{14c^4d^5}{7c^3d}$ **104.** $\dfrac{36h^5k^2}{9h^3k}$ **105.** $\dfrac{z^3z^{11}}{z^4z^6}$ **106.** $\dfrac{w^{12}w^2}{w^4w^5}$

107. $\dfrac{25h^3jk^5}{12h^2k}$ **108.** $\dfrac{15m^5np^{12}}{4mp^9}$ **109.** $(-4p^6q^8r^4)(2pqr^2)$ **110.** $(-5a^4bc)(-10a^2b)$

111. $\dfrac{-12s^2tu^3}{4su^2}$ **112.** $\dfrac{15w^5x^{10}y^3}{-15w^4x}$

Objective 5: Applications of Exponents

 Use the formula $A = P(1 + r)^t$ for Exercises 113–116. **(See Example 8.)**

113. Find the amount in an account after 2 yr if the initial investment is $5000, invested at 7% interest compounded annually.

114. Find the amount in an account after 5 yr if the initial investment is $2000, invested at 4% interest compounded annually.

115. Find the amount in an account after 3 yr if the initial investment is $4000, invested at 6% interest compounded annually.

116. Find the amount in an account after 4 yr if the initial investment is $10,000, invested at 5% interest compounded annually.

 For Exercises 117–120, use the geometry formulas found in Section R.4.

117. Find the area of the pizza shown in the figure. Round to the nearest square inch.

118. Find the volume of the sphere shown in the figure. Round to the nearest cubic centimeter.

16 in.

$r = 3$ cm

119. Find the volume of a spherical balloon that is 8 in. in diameter. Round to the nearest cubic inch.

120. Find the area of a circular pool 50 ft in diameter. Round to the nearest square foot.

Expanding Your Skills

For Exercises 121–128, simplify the expressions using the addition or subtraction rules of exponents. Assume that a, b, m, and n represent positive integers.

121. x^nx^{n+1} **122.** y^ay^{2a} **123.** $p^{3m+5}p^{-m-2}$ **124.** $q^{4b-3}q^{-4b+4}$

125. $\dfrac{z^{b+1}}{z^b}$ **126.** $\dfrac{w^{5n+3}}{w^{2n}}$ **127.** $\dfrac{r^{3a+3}}{r^{3a}}$ **128.** $\dfrac{t^{3+2m}}{t^{2m}}$

Section 5.2 **More Properties of Exponents**

Objectives

1. **Power Rule for Exponents**
2. **The Properties**
   $(ab)^m = a^m b^m$ and
   $\left(\dfrac{a}{b}\right)^m = \dfrac{a^m}{b^m}$

1. Power Rule for Exponents

The expression $(x^2)^3$ indicates that the quantity x^2 is cubed.

$$(x^2)^3 = (x^2)(x^2)(x^2) = (x \cdot x)(x \cdot x)(x \cdot x) = x^6$$

From this example, it appears that to raise a base to successive powers, we multiply the exponents and leave the base unchanged. This is stated formally as the power rule for exponents.

> **PROPERTY   Power Rule for Exponents**
>
> Assume that b is a real number and that m and n represent positive integers. Then,
>
> $$(b^m)^n = b^{m \cdot n}$$

Skill Practice

Simplify the expressions.

1. $(y^3)^5$
2. $(2^8)^{10}$
3. $(q^5 q^4)^3$

Example 1   **Simplifying Expressions with Exponents**

Simplify the expressions.

a. $(s^4)^2$ **b.** $(3^4)^2$ **c.** $(x^2 x^5)^4$

Solution:

a. $(s^4)^2$

   $= s^{4 \cdot 2}$ Multiply exponents (the base is unchanged).

   $= s^8$

b. $(3^4)^2$

   $= 3^{4 \cdot 2}$ Multiply exponents (the base is unchanged).

   $= 3^8$ or 6561

c. $(x^2 x^5)^4$

   $= (x^7)^4$ Simplify inside the parentheses by adding exponents.

   $= x^{7 \cdot 4}$ Multiply exponents (the base is unchanged).

   $= x^{28}$

2. The Properties $(ab)^m = a^m b^m$ and $\left(\dfrac{a}{b}\right)^m = \dfrac{a^m}{b^m}$

Consider the following expressions and their simplified forms:

$$(xy)^3 = (xy)(xy)(xy) = (x \cdot x \cdot x)(y \cdot y \cdot y) = x^3 y^3$$

$$\left(\frac{x}{y}\right)^3 = \left(\frac{x}{y}\right)\left(\frac{x}{y}\right)\left(\frac{x}{y}\right) = \left(\frac{x \cdot x \cdot x}{y \cdot y \cdot y}\right) = \frac{x^3}{y^3}$$

The expressions were simplified using the commutative and associative properties of multiplication. The simplified forms for each expression could have been reached in one step by applying the exponent to each factor inside the parentheses.

Answers

1. y^{15} **2.** 2^{80} **3.** q^{27}

PROPERTY **Power of a Product and Power of a Quotient**

Assume that a and b are real numbers. Let m represent a positive integer. Then,

$$(ab)^m = a^m b^m$$

$$\left(\frac{a}{b}\right)^m = \frac{a^m}{b^m}, \quad b \neq 0$$

Applying these properties of exponents, we have

$$(xy)^3 = x^3 y^3 \quad \text{and} \quad \left(\frac{x}{y}\right)^3 = \frac{x^3}{y^3}$$

Example 2 **Simplifying Expressions with Exponents**

Simplify the expressions.

a. $(-2xyz)^4$ **b.** $(5x^2 y^7)^3$ **c.** $\left(\frac{2}{5}\right)^3$ **d.** $\left(\frac{1}{3xy^4}\right)^2$

Skill Practice

Simplify the expressions.

4. $(3abc)^5$ **5.** $(-2t^2 w^4)^3$

6. $\left(\frac{3}{4}\right)^3$ **7.** $\left(\frac{2x^3}{y^5}\right)^2$

Solution:

a. $(-2xyz)^4$

$= (-2)^4 x^4 y^4 z^4$ — Raise each factor within parentheses to the fourth power.

$= 16 x^4 y^4 z^4$

b. $(5x^2 y^7)^3$

$= 5^3 (x^2)^3 (y^7)^3$ — Raise each factor within parentheses to the third power.

$= 125 x^6 y^{21}$ — Multiply exponents and simplify.

c. $\left(\frac{2}{5}\right)^3$

$= \frac{(2)^3}{(5)^3}$ — Apply the exponent to each factor in parentheses.

$= \frac{8}{125}$ — Simplify.

d. $\left(\frac{1}{3xy^4}\right)^2$

$= \frac{1^2}{3^2 x^2 (y^4)^2}$ — Square each factor within parentheses.

$= \frac{1}{9 x^2 y^8}$ — Multiply exponents and simplify.

Answers

4. $3^5 a^5 b^5 c^5$ or $243 a^5 b^5 c^5$

5. $-8 t^6 w^{12}$ **6.** $\frac{27}{64}$ **7.** $\frac{4x^6}{y^{10}}$

The properties of exponents can be used along with the properties of real numbers to simplify complicated expressions.

Example 3 Simplifying Expressions with Exponents

Simplify the expression. $\dfrac{(x^2)^6(x^3)}{(x^7)^2}$

Solution:

$\dfrac{(x^2)^6(x^3)}{(x^7)^2}$ Clear parentheses by applying the power rule.

$= \dfrac{x^{2\cdot 6}x^3}{x^{7\cdot 2}}$ Multiply exponents.

$= \dfrac{x^{12}x^3}{x^{14}}$

$= \dfrac{x^{12+3}}{x^{14}}$ Add exponents in the numerator.

$= \dfrac{x^{15}}{x^{14}}$

$= x^{15-14}$ Subtract exponents.

$= x$ Simplify.

Example 4 Simplifying Expressions with Exponents

Simplify the expression. $(3cd^2)(2cd^3)^3$

Solution:

$(3cd^2)(2cd^3)^3$ Clear parentheses by applying the power rule.

$= 3cd^2 \cdot 2^3 c^3 d^9$ Raise each factor in the second parentheses to the third power.

$= 3 \cdot 2^3 cc^3 d^2 d^9$ Group like factors.

$= 3 \cdot 8c^{1+3} d^{2+9}$ Add exponents on like bases.

$= 24c^4 d^{11}$ Simplify.

Example 5   **Simplifying Expressions with Exponents**

Simplify the expression. $\left(\dfrac{x^7yz^4}{8xz^3}\right)^2$

Solution:

$\left(\dfrac{x^7yz^4}{8xz^3}\right)^2$

$= \left(\dfrac{x^{7-1}yz^{4-3}}{8}\right)^2$ First simplify inside the parentheses by subtracting exponents on like bases.

$= \left(\dfrac{x^6yz}{8}\right)^2$

$= \dfrac{(x^6)^2y^2z^2}{8^2}$ Apply the power rule of exponents.

$= \dfrac{x^{12}y^2z^2}{64}$

Section 5.2   Practice Exercises

Boost your GRADE at ALEKS.com!

ALEKS version 3.0

- Practice Problems
- Self-Tests
- NetTutor
- e-Professors
- Videos

For this exercise set assume all variables represent nonzero real numbers.

Review Exercises

For Exercises 1–8, simplify.

1. $4^2 \cdot 4^7$ **2.** $5^8 \cdot 5^3 \cdot 5$ **3.** $a^{13} \cdot a \cdot a^6$ **4.** $y^{14}y^3$

5. $\dfrac{d^{13}d}{d^5}$ **6.** $\dfrac{3^8 \cdot 3}{3^2}$ **7.** $\dfrac{7^{11}}{7^5}$ **8.** $\dfrac{z^4}{z^3}$

9. Explain when to add exponents versus when to subtract exponents.

10. Explain when to add exponents versus when to multiply exponents.

Objective 1: Power Rule for Exponents

For Exercises 11–22, simplify and write answers in exponent form. **(See Example 1.)**

11. $(5^3)^4$ **12.** $(2^8)^7$ **13.** $(12^3)^2$ **14.** $(6^4)^4$

15. $(y^7)^2$ **16.** $(z^6)^4$ **17.** $(w^5)^5$ **18.** $(t^3)^6$

 19. $(a^2a^4)^6$ **20.** $(z \cdot z^3)^2$ **21.** $(y^3y^4)^2$ **22.** $(w^5w)^4$

23. Evaluate the two expressions and compare the answers: $(2^2)^3$ and $(2^3)^2$.

24. Evaluate the two expressions and compare the answers: $(4^4)^2$ and $(4^2)^4$.

25. Evaluate the two expressions and compare the answers. Which expression is greater? Why?

$$2^{(2^4)} \quad \text{and} \quad (2^2)^4$$

26. Evaluate the two expressions and compare the answers. Which expression is greater? Why?

$$3^{(2^4)} \quad \text{and} \quad (3^2)^4$$

Objective 2: The Properties $(ab)^m = a^m b^m$ and $\left(\dfrac{a}{b}\right)^m = \dfrac{a^m}{b^m}$

For Exercises 27–42, use the appropriate property to clear the parentheses. **(See Example 2.)**

27. $(5w)^2$

28. $(4y)^3$

29. $(srt)^4$

30. $(wxy)^6$

31. $\left(\dfrac{2}{r}\right)^4$

32. $\left(\dfrac{1}{t}\right)^8$

33. $\left(\dfrac{x}{y}\right)^5$

34. $\left(\dfrac{w}{z}\right)^7$

35. $(-3a)^4$

36. $(2x)^5$

37. $(-3abc)^3$

38. $(-5xyz)^2$

39. $\left(-\dfrac{4}{x}\right)^3$

40. $\left(-\dfrac{1}{w}\right)^4$

41. $\left(-\dfrac{a}{b}\right)^2$

42. $\left(-\dfrac{r}{s}\right)^3$

Mixed Exercises

For Exercises 43–74, simplify the expressions. **(See Examples 3–5.)**

43. $(6u^2v^4)^3$

44. $(3a^5b^2)^6$

45. $5(x^2y)^4$

46. $18(u^3v^4)^2$

47. $(-h^4)^7$

48. $(-k^6)^3$

49. $(-m^2)^6$

50. $(-n^3)^8$

51. $\left(\dfrac{4}{rs^4}\right)^5$

52. $\left(\dfrac{2}{h^7k}\right)^3$

53. $\left(\dfrac{3p}{q^3}\right)^5$

54. $\left(\dfrac{5x^2}{y^3}\right)^4$

55. $\dfrac{y^8(y^3)^4}{(y^2)^3}$

56. $\dfrac{(w^3)^2(w^4)^5}{(w^4)^2}$

57. $(x^2)^5(x^3)^7$

58. $(y^3)^4(y^2)^5$

59. $(2a^2b)^3(5a^4b^3)^2$

60. $(4c^3d^5)^2(3cd^3)^2$

61. $(-2p^2q^4)^4$

62. $(-7x^4y^5)^2$

63. $(-m^7n^3)^5$

64. $(-a^3b^6)^7$

65. $\dfrac{(5a^3b)^4(a^2b)^4}{(5ab)^2}$

66. $\dfrac{(6s^3)^2(s^4t^5)^2}{(3s^4t^2)^2}$

67. $\left(\dfrac{2c^3d^4}{3c^2d}\right)^2$

68. $\left(\dfrac{x^3y^5z}{5xy^2}\right)^2$

69. $(2c^3d^2)^5\left(\dfrac{c^6d^8}{4c^2d}\right)^3$

70. $\left(\dfrac{s^5t^6}{2s^2t}\right)^2(10s^3t^3)^2$

71. $\left(\dfrac{-3a^3b}{c^2}\right)^3$

72. $\left(\dfrac{-4x^2}{y^4z}\right)^3$

73. $\dfrac{(-8b^6)^2(b^3)^5}{4b}$

74. $\dfrac{(-6a^2)^2(a^3)^4}{9a}$

Expanding Your Skills

For Exercises 75–82, simplify the expressions using the addition or subtraction properties of exponents. Assume that a, b, m, and n represent positive integers.

75. $(x^m)^2$

76. $(y^3)^n$

77. $(5a^{2n})^3$

78. $(3b^4)^m$

79. $\left(\dfrac{m^2}{n^3}\right)^b$

80. $\left(\dfrac{x^5}{y^3}\right)^m$

81. $\left(\dfrac{3a^3}{5b^4}\right)^n$

82. $\left(\dfrac{4m^6}{3n^2}\right)^b$

Definitions of b^0 and b^{-n}

In Sections 5.1 and 5.2, we learned several rules that enable us to manipulate expressions containing *positive* integer exponents. In this section, we present definitions that can be used to simplify expressions with negative exponents or with an exponent of zero.

Objectives

1. Definition of b^0
2. Definition of b^{-n}
3. Properties of Integer Exponents: A Summary

1. Definition of b^0

To begin, consider the following pattern.

$3^3 = 27$ Divide by 3.

$3^2 = 9$ Divide by 3.

$3^1 = 3$ Divide by 3.

$3^0 = 1$

As the exponents decrease by 1, the resulting expressions are divided by 3.

For the pattern to continue, we define $3^0 = 1$.

This pattern suggests that we should define an expression with a zero exponent as follows.

DEFINITION Definition of b^0

Let b be a nonzero real number. Then, $b^0 = 1$.

Avoiding Mistakes

$b^0 = 1$ provided that b is not zero. Therefore, the expression 0^0 cannot be simplified by this rule.

Example 1 Simplifying Expressions with a Zero Exponent

Simplify (assume $z \neq 0$).

a. 4^0 **b.** $(-4)^0$ **c.** -4^0

d. z^0 **e.** $-4z^0$ **f.** $(4z)^0$

Skill Practice

Evaluate the expressions. Assume all variables represent nonzero real numbers.

1. 7^0 2. $(-7)^0$
3. -5^0 4. y^0
5. $-2x^0$ 6. $(2x)^0$

Solution:

a. $4^0 = 1$ By definition

b. $(-4)^0 = 1$ By definition

c. $-4^0 = -1 \cdot 4^0 = -1 \cdot 1 = -1$ The exponent 0 applies only to 4.

d. $z^0 = 1$ By definition

e. $-4z^0 = -4 \cdot z^0 = -4 \cdot 1 = -4$ The exponent 0 applies only to z.

f. $(4z)^0 = 1$ The parentheses indicate that the exponent, 0, applies to both factors 4 and z.

The definition of b^0 is consistent with the other properties of exponents learned thus far. For example, we know that $1 = \frac{5^3}{5^3}$. If we subtract exponents, the result is 5^0.

Subtract exponents.

$$1 = \frac{5^3}{5^3} = 5^{3-3} = 5^0. \qquad \text{Therefore, } 5^0 \text{ must be defined as 1.}$$

Answers

1. 1   2. 1   3. −1
4. 1   5. −2   6. 1

2. Definition of b^{-n}

To understand the concept of a *negative* exponent, consider the following pattern.

$3^3 = 27$

     Divide by 3.

$3^2 = 9$

     Divide by 3.   As the exponents decrease by

$3^1 = 3$

     Divide by 3.   1, the resulting expressions are

$3^0 = 1$

     Divide by 3.   divided by 3.

     Divide by 3.

$3^{-1} = \dfrac{1}{3}$ ←   For the pattern to continue, we define $3^{-1} = \dfrac{1}{3^1} = \dfrac{1}{3}$.

$3^{-2} = \dfrac{1}{9}$ ←   For the pattern to continue, we define $3^{-2} = \dfrac{1}{3^2} = \dfrac{1}{9}$.

$3^{-3} = \dfrac{1}{27}$ ←   For the pattern to continue, we define $3^{-3} = \dfrac{1}{3^3} = \dfrac{1}{27}$.

This pattern suggests that $3^{-n} = \frac{1}{3^n}$ for all integers, n. In general, we have the following definition involving negative exponents.

> **DEFINITION**   **Definition of b^{-n}**
>
> Let n be an integer and b be a nonzero real number. Then,
>
> $$b^{-n} = \left(\dfrac{1}{b}\right)^n \quad \text{or} \quad \dfrac{1}{b^n}$$

The definition of b^{-n} implies that to evaluate b^{-n}, take the reciprocal of the base and change the sign of the exponent.

Change the sign of the exponent.

Change the sign of the exponent.

$$4^{-2} = \left(\dfrac{1}{4}\right)^2 \quad \text{or} \quad \dfrac{1}{4^2} \qquad\qquad \left(\dfrac{a}{b}\right)^{-n} = \left(\dfrac{b}{a}\right)^n$$

Reciprocal of the base   Reciprocal of the base

Skill Practice

Simplify.

7. p^{-4} **8.** 3^{-3}

9. $(-5)^{-2}$

Avoiding Mistakes

A negative exponent does *not* affect the sign of the base.

Answers

7. $\dfrac{1}{p^4}$ **8.** $\dfrac{1}{3^3}$ or $\dfrac{1}{27}$

9. $\dfrac{1}{(-5)^2}$ or $\dfrac{1}{25}$

Example 2   **Simplifying Expressions with Negative Exponents**

Simplify. Assume that $c \neq 0$.   **a.** c^{-3} **b.** 5^{-1} **c.** $(-3)^{-4}$

Solution:

a. $c^{-3} = \dfrac{1}{c^3}$ By definition

b. $5^{-1} = \dfrac{1}{5^1}$ By definition

      $= \dfrac{1}{5}$ Simplify.

c. $(-3)^{-4} = \dfrac{1}{(-3)^4}$ The base is -3 and must be enclosed in parentheses.

       $= \dfrac{1}{81}$ Simplify. Note that $(-3)^4 = (-3)(-3)(-3)(-3) = 81$.

Example 3 **Simplifying Expressions with Negative Exponents**

Simplify. Assume that $y \neq 0$. **a.** $\left(\dfrac{1}{4}\right)^{-2}$ **b.** $\left(-\dfrac{3}{5}\right)^{-3}$ **c.** $\dfrac{1}{y^{-5}}$

Solution:

a. $\left(\dfrac{1}{4}\right)^{-2} = 4^2$ Take the reciprocal of the base, and change the sign of the exponent.

$= 16$ Simplify.

b. $\left(-\dfrac{3}{5}\right)^{-3} = \left(-\dfrac{5}{3}\right)^{3}$ Take the reciprocal of the base, and change the sign of the exponent.

$= -\dfrac{125}{27}$ Simplify.

c. $\dfrac{1}{y^{-5}} = \left(\dfrac{1}{y}\right)^{-5}$ Apply the power of a quotient rule from Section 5.2.

$= (y)^5$ Take the reciprocal of the base, and change the sign of the exponent.

$= y^5$

Example 4 **Simplifying Expressions with Negative Exponents**

Simplify. Assume that $x \neq 0$. **a.** $(5x)^{-3}$ **b.** $5x^{-3}$ **c.** $-5x^{-3}$

Solution:

a. $(5x)^{-3} = \left(\dfrac{1}{5x}\right)^{3}$ Take the reciprocal of the base, and change the sign of the exponent.

$= \dfrac{(1)^3}{(5x)^3}$ Apply the exponent of 3 to each factor within parentheses.

$= \dfrac{1}{125x^3}$ Simplify.

b. $5x^{-3} = 5 \cdot x^{-3}$ Note that the exponent, -3, applies only to x.

$= 5 \cdot \dfrac{1}{x^3}$ Rewrite x^{-3} as $\dfrac{1}{x^3}$.

$= \dfrac{5}{x^3}$ Multiply.

c. $-5x^{-3} = -5 \cdot x^{-3}$ Note that the exponent, -3, applies only to x, and that -5 is a coefficient.

$= -5 \cdot \dfrac{1}{x^3}$ Rewrite x^{-3} as $\dfrac{1}{x^3}$.

$= -\dfrac{5}{x^3}$ Multiply.

It is important to note that the definition of b^{-n} is consistent with the other properties of exponents learned thus far. For example, consider the expression

$$\frac{x^4}{x^7} = \frac{\cancel{x} \cdot \cancel{x} \cdot \cancel{x} \cdot \cancel{x}}{\cancel{x} \cdot \cancel{x} \cdot \cancel{x} \cdot \cancel{x} \cdot x \cdot x \cdot x} = \frac{1}{x^3}$$

Subtract exponents.

By subtracting exponents, we have $\quad \dfrac{x^4}{x^7} = x^{4-7} = x^{-3}$

Hence, $x^{-3} = \dfrac{1}{x^3}$

3. Properties of Integer Exponents: A Summary

The definitions of b^0 and b^{-n} allow us to extend the properties of exponents learned in Sections 5.1 and 5.2 to include integer exponents. These are summarized in Table 5-1.

Table 5-1

Properties of Integer Exponents		
Assume that a and b are real numbers ($b \neq 0$) and that m and n represent integers.		
Property	**Example**	**Details/Notes**
Multiplication of Like Bases $$b^m b^n = b^{m+n}$$	$$b^2 b^4 = b^{2+4} = b^6$$	$$b^2 b^4 = (b \cdot b)(b \cdot b \cdot b \cdot b) = b^6$$
Division of Like Bases $$\frac{b^m}{b^n} = b^{m-n}$$	$$\frac{b^5}{b^2} = b^{5-2} = b^3$$	$$\frac{b^5}{b^2} = \frac{\cancel{b} \cdot \cancel{b} \cdot b \cdot b \cdot b}{\cancel{b} \cdot \cancel{b}} = b^3$$
The Power Rule $$(b^m)^n = b^{m \cdot n}$$	$$(b^4)^2 = b^{4 \cdot 2} = b^8$$	$$(b^4)^2 = (b \cdot b \cdot b \cdot b)(b \cdot b \cdot b \cdot b) = b^8$$
Power of a Product $$(ab)^m = a^m b^m$$	$$(ab)^3 = a^3 b^3$$	$$(ab)^3 = (ab)(ab)(ab)$$ $$= (a \cdot a \cdot a)(b \cdot b \cdot b) = a^3 b^3$$
Power of a Quotient $$\left(\frac{a}{b}\right)^m = \frac{a^m}{b^m}$$	$$\left(\frac{a}{b}\right)^3 = \frac{a^3}{b^3}$$	$$\left(\frac{a}{b}\right)^3 = \left(\frac{a}{b}\right)\left(\frac{a}{b}\right)\left(\frac{a}{b}\right) = \frac{a \cdot a \cdot a}{b \cdot b \cdot b} = \frac{a^3}{b^3}$$
Definitions		
Assume that b is a real number ($b \neq 0$) and that n represents an integer.		
Definition	**Example**	**Details/Notes**
$$b^0 = 1$$	$$(4)^0 = 1$$	Any nonzero quantity raised to the zero power equals 1.
$$b^{-n} = \left(\frac{1}{b}\right)^n = \frac{1}{b^n}$$	$$b^{-5} = \left(\frac{1}{b}\right)^5 = \frac{1}{b^5}$$	To simplify a negative exponent, take the reciprocal of the base and make the exponent positive.

Example 5 **Simplifying Expressions with Exponents**

Simplify the expressions. Write the answers with positive exponents only. Assume all variables are nonzero.

a. $\dfrac{a^3 b^{-2}}{c^{-5}}$ **b.** $\dfrac{x^2 x^{-7}}{x^3}$ **c.** $\dfrac{z^2}{w^{-4} w^4 z^{-8}}$

Solution:

a. $\dfrac{a^3 b^{-2}}{c^{-5}}$

$= \dfrac{a^3}{1} \cdot \dfrac{b^{-2}}{1} \cdot \dfrac{1}{c^{-5}}$

$= \dfrac{a^3}{1} \cdot \dfrac{1}{b^2} \cdot \dfrac{c^5}{1}$ Simplify negative exponents.

$= \dfrac{a^3 c^5}{b^2}$ Multiply.

b. $\dfrac{x^2 x^{-7}}{x^3}$

$= \dfrac{x^{2+(-7)}}{x^3}$ Add the exponents in the numerator.

$= \dfrac{x^{-5}}{x^3}$ Simplify.

$= x^{-5-3}$ Subtract the exponents.

$= x^{-8}$

$= \dfrac{1}{x^8}$ Simplify the negative exponent.

c. $\dfrac{z^2}{w^{-4} w^4 z^{-8}}$

$= \dfrac{z^2}{w^{-4+4} z^{-8}}$ Add the exponents in the denominator.

$= \dfrac{z^2}{w^0 z^{-8}}$

$= \dfrac{z^2}{(1) z^{-8}}$ Recall that $w^0 = 1$.

$= z^{2-(-8)}$ Subtract the exponents.

$= z^{10}$ Simplify.

Answers

16. $\dfrac{z^8}{y^4 x^6}$ **17.** $\dfrac{1}{x^9}$ **18.** $p^3 z^2$

Example 6 **Simplifying Expressions with Exponents**

Simplify the expressions. Write the answers with positive exponents only. Assume that all variables are nonzero.

a. $(-4ab^{-2})^{-3}$ **b.** $\left(\dfrac{2p^{-4}q^3}{5p^2q}\right)^{-2}$

Solution:

a. $(-4ab^{-2})^{-3}$

$= (-4)^{-3}a^{-3}(b^{-2})^{-3}$ Apply the power rule of exponents.

$= (-4)^{-3}a^{-3}b^6$

$= \dfrac{1}{(-4)^3} \cdot \dfrac{1}{a^3} \cdot b^6$ Simplify the negative exponents.

$= \dfrac{1}{-64} \cdot \dfrac{1}{a^3} \cdot b^6$ Simplify.

$= -\dfrac{b^6}{64a^3}$ Multiply fractions.

b. $\left(\dfrac{2p^{-4}q^3}{5p^2q}\right)^{-2}$ First simplify within the parentheses.

$= \left(\dfrac{2p^{-4-2}q^{3-1}}{5}\right)^{-2}$ Divide like bases by subtracting exponents.

$= \left(\dfrac{2p^{-6}q^2}{5}\right)^{-2}$ Simplify.

$= \dfrac{(2p^{-6}q^2)^{-2}}{(5)^{-2}}$ Apply the power rule of a quotient.

$= \dfrac{2^{-2}(p^{-6})^{-2}(q^2)^{-2}}{5^{-2}}$ Apply the power rule of a product.

$= \dfrac{2^{-2}p^{12}q^{-4}}{5^{-2}}$ Simplify.

$= \dfrac{5^2p^{12}}{2^2q^4}$ Simplify the negative exponents.

$= \dfrac{25p^{12}}{4q^4}$ Simplify.

Answers

19. $\dfrac{x^4}{25y^6}$ **20.** $\dfrac{16x^8}{9y^2}$

Example 7 Simplifying an Expression with Exponents

Simplify the expression $2^{-1} + 3^{-1} + 5^0$. Write the answer with positive exponents only.

Solution:

$2^{-1} + 3^{-1} + 5^0$

$= \dfrac{1}{2} + \dfrac{1}{3} + 1$ Simplify negative exponents. Simplify $5^0 = 1$.

$= \dfrac{3}{6} + \dfrac{2}{6} + \dfrac{6}{6}$ The least common denominator is 6.

$= \dfrac{11}{6}$ Simplify.

Section 5.3   Practice Exercises

Boost your GRADE at ALEKS.com!

 ALEKS version 3.0

- Practice Problems
- Self-Tests
- NetTutor
- e-Professors
- Videos

For this set of exercises, assume all variables represent nonzero real numbers.

Study Skills Exercise

1. To help you remember the properties of exponents, write them on 3×5 cards. On each card, write a property on one side and an example using that property on the other side. Keep these cards with you, and when you have a spare moment (such as waiting at the doctor's office), pull out these cards and go over the properties.

Review Exercises

For Exercises 2–9, simplify the expressions.

2. $b^3 b^8$ **3.** $c^7 c^2$ **4.** $\dfrac{x^6}{x^2}$ **5.** $\dfrac{y^9}{y^8}$

6. $\dfrac{9^4 \cdot 9^8}{9}$ **7.** $\dfrac{3^{14}}{3^3 \cdot 3^5}$ **8.** $(6ab^3 c^2)^5$ **9.** $(7w^7 z^2)^4$

Objective 1: Definition of b^0

10. Simplify.   **11.** Simplify.   **12.** Simplify.

   a. 8^0 **b.** $\dfrac{8^4}{8^4}$ **a.** d^0 **b.** $\dfrac{d^3}{d^3}$ **a.** m^0 **b.** $\dfrac{m^5}{m^5}$

For Exercises 13–24, simplify the expression. **(See Example 1.)**

13. p^0 **14.** k^0 **15.** 5^0 **16.** 2^0

17. -4^0 **18.** -1^0 **19.** $(-6)^0$ **20.** $(-2)^0$

21. $(8x)^0$ **22.** $(-3y^3)^0$ **23.** $-7x^0$ **24.** $6y^0$

Objective 2: Definition of b^{-n}

25. Simplify and write the answers with positive exponents.

    a. t^{-5} **b.** $\dfrac{t^3}{t^8}$

26. Simplify and write the answers with positive exponents.

    a. 4^{-3} **b.** $\dfrac{4^2}{4^5}$

For Exercises 27–46, simplify. **(See Examples 2–4.)**

27. $\left(\dfrac{2}{7}\right)^{-3}$ **28.** $\left(\dfrac{5}{4}\right)^{-1}$ **29.** $\left(-\dfrac{1}{5}\right)^{-2}$ **30.** $\left(-\dfrac{1}{3}\right)^{-3}$

31. a^{-3} **32.** c^{-5} **33.** 12^{-1} **34.** 4^{-2}

35. $(4b)^{-2}$ **36.** $(3z)^{-1}$ **37.** $6x^{-2}$ **38.** $7y^{-1}$

39. $(-8)^{-2}$ **40.** -8^{-2} **41.** $-3y^{-4}$ **42.** $-6a^{-2}$

43. $(-t)^{-3}$ **44.** $(-r)^{-5}$ **45.** $\dfrac{1}{a^{-5}}$ **46.** $\dfrac{1}{b^{-6}}$

Objective 3: Properties of Integer Exponents: A Summary

47. Explain what is wrong with the following logic. $\dfrac{x^4}{x^{-6}} = x^{4-6} = x^{-2}$

48. Explain what is wrong with the following logic. $\dfrac{y^5}{y^{-3}} = y^{5-3} = y^2$

49. Explain what is wrong with the following logic. $2a^{-3} = \dfrac{1}{2a^3}$

50. Explain what is wrong with the following logic. $5b^{-2} = \dfrac{1}{5b^2}$

Mixed Exercises

For Exercises 51–94, simplify the expression. Write the answer with positive exponents only. **(See Examples 5–6.)**

51. $x^{-8}x^4$ **52.** s^5s^{-6} **53.** $a^{-8}a^8$ **54.** q^3q^{-3}

55. $y^{17}y^{-13}$ **56.** $b^{20}b^{-14}$ **57.** $(m^{-6}n^9)^3$ **58.** $(c^4d^{-5})^{-2}$

59. $(-3j^{-5}k^6)^4$ **60.** $(6xy^{-11})^{-3}$ **61.** $\dfrac{p^3}{p^9}$ **62.** $\dfrac{q^2}{q^{10}}$

63. $\dfrac{r^{-5}}{r^{-2}}$ **64.** $\dfrac{u^{-2}}{u^{-6}}$ **65.** $\dfrac{a^2}{a^{-6}}$ **66.** $\dfrac{p^3}{p^{-5}}$

67. $\dfrac{y^{-2}}{y^6}$ **68.** $\dfrac{s^{-4}}{s^3}$ **69.** $\dfrac{7^3}{7^2 \cdot 7^8}$ **70.** $\dfrac{3^4 \cdot 3}{3^7}$

71. $\dfrac{a^2a}{a^3}$ **72.** $\dfrac{t^5}{t^2t^3}$ **73.** $\dfrac{a^{-1}b^2}{a^3b^8}$ **74.** $\dfrac{k^{-4}h^{-1}}{k^6h}$

75. $\dfrac{w^{-8}(w^2)^{-5}}{w^3}$ **76.** $\dfrac{p^2p^{-7}}{(p^2)^3}$ **77.** $\dfrac{3^{-2}}{3}$ **78.** $\dfrac{5^{-1}}{5}$

79. $\left(\dfrac{p^{-1}q^5}{p^{-6}}\right)^0$ **80.** $\left(\dfrac{ab^{-4}}{a^{-5}}\right)^0$ **81.** $(8x^3y^0)^{-2}$ **82.** $(3u^2v^0)^{-3}$

83. $(-8y^{-12})(2y^{16}z^{-2})$

84. $(5p^{-2}q^5)(-2p^{-4}q^{-1})$

85. $\dfrac{-18a^{10}b^6}{108a^{-2}b^6}$

86. $\dfrac{-35x^{-4}y^{-3}}{-21x^2y^{-3}}$

87. $\dfrac{(-4c^{12}d^7)^2}{(5c^{-3}d^{10})^{-1}}$

88. $\dfrac{(s^3t^{-2})^4}{(3s^{-4}t^6)^{-2}}$

89. $\left(\dfrac{2}{p^6p^3}\right)^{-3}$

90. $\left(\dfrac{5x}{x^7}\right)^{-2}$

91. $\left(\dfrac{5cd^{-3}}{10d^5}\right)^{-2}$

92. $\left(\dfrac{4m^{10}n^4}{2m^{12}n^{-2}}\right)^{-1}$

93. $(2xy^3)\left(\dfrac{9xy}{4x^3y^2}\right)$

94. $(-3a^3)\left(\dfrac{ab}{27a^4b^2}\right)$

For Exercises 95–102, simplify the expression. **(See Example 7.)**

95. $5^{-1} + 2^{-2}$

96. $4^{-2} + 8^{-1}$

97. $10^0 - 10^{-1}$

98. $3^0 - 3^{-2}$

99. $2^{-2} + 1^{-2}$

100. $4^{-1} + 8^{-1}$

101. $4 \cdot 5^0 - 2 \cdot 3^{-1}$

102. $2 \cdot 4^0 - 3 \cdot 4^{-1}$

Scientific Notation

Section 5.4

Objectives

1. **Writing Numbers in Scientific Notation**
2. **Writing Numbers in Standard Form**
3. **Multiplying and Dividing Numbers in Scientific Notation**

1. Writing Numbers in Scientific Notation

In many applications in mathematics, it is necessary to work with very large or very small numbers. For example, the number of movie tickets sold in the United States recently is estimated to be 1,500,000,000. The weight of a flea is approximately 0.00066 lb. To avoid writing numerous zeros in very large or small numbers, scientific notation was devised as a shortcut.

The principle behind scientific notation is to use a power of 10 to express the magnitude of the number. For example, the numbers 4000 and 0.07 can be written as:

$$4000 = 4 \times 1000 = 4 \times 10^3$$

$$0.07 = 7.0 \times 0.01 = 7.0 \times 10^{-2} \qquad \text{Note that } 10^{-2} = \frac{1}{100} = 0.01$$

> **DEFINITION Scientific Notation**
>
> A positive number expressed in the form: $a \times 10^n$, where $1 \le a < 10$ and n is an integer is said to be written in **scientific notation**.

To write a positive number in scientific notation, we apply the following guidelines:

1. Move the decimal point so that its new location is to the right of the first nonzero digit. The number should now be greater than or equal to 1 but less than 10. Count the number of places that the decimal point is moved.

2. If the original number is *large* (greater than or equal to 10), use the number of places the decimal point was moved as a *positive* power of 10.

$$450,000 \qquad = 4.5 \times 100,000 = 4.5 \times 10^5$$
5 places

3. If the original number is *small* (between 0 and 1), use the number of places the decimal point was moved as a *negative* power of 10.

$$0.0002 \qquad = 2.0 \times 0.0001 = 2.0 \times 10^{-4}$$
4 places

4. If the original number is greater than or equal to 1 but less than 10, use 0 as the power of 10.

$7.592 = 7.592 \times 10^0$ *Note:* A number between 1 and 10 is seldom written in scientific notation.

5. If the original number is negative, then $-10 < a \le -1$.

$$-450,000 = -4.5 \times 100,000 = -4.5 \times 10^5$$

5 places

Skill Practice

Write the numbers in scientific notation.

1. 175,000,000
2. 0.000005

Example 1 Writing Numbers in Scientific Notation

Write the numbers in scientific notation.

a. 53,000   **b.** 0.00053

Solution:

a. $53,000. = 5.3 \times 10^4$ To write 53,000 in scientific notation, the decimal point must be moved four places to the left. Because 53,000 is larger than 10, a *positive* power of 10 is used.

b. $0.00053 = 5.3 \times 10^{-4}$ To write 0.00053 in scientific notation, the decimal point must be moved four places to the right. Because 0.00053 is less than 1, a *negative* power of 10 is used.

Skill Practice

Write the numbers in scientific notation.

3. The population of the Earth is approximately 6,360,000,000.
4. The weight of a grain of salt is approximately 0.000002 ounces.

Example 2 Writing Numbers in Scientific Notation

Write the numerical values in scientific notation.

a. The number of movie tickets sold in the United States recently is estimated to be 1,500,000,000.

b. The weight of a flea is approximately 0.00066 lb.

c. The temperature on a January day in Fargo dropped to $-43°F$.

d. A bench is 8.2 ft long.

Solution:

a. $1,500,000,000 = 1.5 \times 10^9$ **b.** $0.00066 \text{ lb} = 6.6 \times 10^{-4} \text{ lb}$

c. $-43°F = -4.3 \times 10^1 \, °F$ **d.** $8.2 \text{ ft} = 8.2 \times 10^0 \text{ ft}$

---

2. Writing Numbers in Standard Form

Example 3 Writing Numbers in Standard Form

Write the numerical values in standard form.

a. The mass of a proton is approximately 1.67×10^{-24} g.

b. The "nearby" star Vega is approximately 1.552×10^{14} miles from Earth.

Answers
1. 1.75×10^8 **2.** 5.0×10^{-6}
3. 6.36×10^9 **4.** 2.0×10^{-6} oz

Solution:

a. 1.67×10^{-24} g = 0.000 000 000 000 000 000 000 001 67 g

Because the power of 10 is negative, the value of 1.67×10^{-24} is a decimal number between 0 and 1. Move the decimal point 24 places to the *left*.

b. 1.552×10^{14} miles = 155,200,000,000,000 miles

Because the power of 10 is a positive integer, the value of 1.552×10^{14} is a large number greater than 10. Move the decimal point 14 places to the *right*.

3. Multiplying and Dividing Numbers in Scientific Notation

To multiply or divide two numbers in scientific notation, use the commutative and associative properties of multiplication to group the powers of 10. For example:

$$400 \times 2000 = (4 \times 10^2)(2 \times 10^3) = (4 \cdot 2) \times (10^2 \cdot 10^3) = 8 \times 10^5$$

$$\frac{0.00054}{150} = \frac{5.4 \times 10^{-4}}{1.5 \times 10^2} = \left(\frac{5.4}{1.5}\right) \times \left(\frac{10^{-4}}{10^2}\right) = 3.6 \times 10^{-6}$$

Example 4 **Multiplying and Dividing Numbers in Scientific Notation**

Multiply or divide as indicated.

a. $(8.7 \times 10^4)(2.5 \times 10^{-12})$

b. $\dfrac{4.25 \times 10^{13}}{8.5 \times 10^{-2}}$

Solution:

a. $(8.7 \times 10^4)(2.5 \times 10^{-12})$

$= (8.7 \cdot 2.5) \times (10^4 \cdot 10^{-12})$ Commutative and associative properties of multiplication

$= 21.75 \times 10^{-8}$ The number 21.75 is not in proper scientific notation because 21.75 is not between 1 and 10.

$= (2.175 \times 10^1) \times 10^{-8}$ Rewrite 21.75 as 2.175×10^1.

$= 2.175 \times (10^1 \times 10^{-8})$ Associative property of multiplication

$= 2.175 \times 10^{-7}$ Simplify.

b. $\dfrac{4.25 \times 10^{13}}{8.5 \times 10^{-2}}$

$= \left(\dfrac{4.25}{8.5}\right) \times \left(\dfrac{10^{13}}{10^{-2}}\right)$ Commutative and associative properties

$= 0.5 \times 10^{15}$ The number 0.5×10^{15} is not in proper scientific notation because 0.5 is not between 1 and 10.

$= (5.0 \times 10^{-1}) \times 10^{15}$ Rewrite 0.5 as 5.0×10^{-1}.

$= 5.0 \times (10^{-1} \times 10^{15})$ Associative property of multiplication

$= 5.0 \times 10^{14}$ Simplify.

Skill Practice

Write the numerical values in standard form.

5. The probability of winning the California Super Lotto Jackpot is 5.5×10^{-8}.

6. The Sun's mass is 2×10^{30} kilograms.

Skill Practice

Multiply or divide as indicated

7. $(7 \times 10^5)(5 \times 10^3)$

8. $\dfrac{1 \times 10^{-2}}{4 \times 10^{-7}}$

Answers

5. 0.000 000 055

6. 2,000,000,000,000,000,000,000,000,000,000

7. 3.5×10^9

8. 2.5×10^4

Calculator Connections

Topic: Using Scientific Notation

Both scientific and graphing calculators can perform calculations involving numbers written in scientific notation. Most calculators use an $\boxed{\text{EE}}$ key or an $\boxed{\text{EXP}}$ key to enter the power of 10.

Scientific Calculator

Enter: 2.7 $\boxed{\text{EE}}$ 5 $\boxed{=}$ or   2.7 $\boxed{\text{EXP}}$ 5 $\boxed{=}$ **Result:** | 270000 |

Enter: 7.1 $\boxed{\text{EE}}$ 3 $\boxed{+\circ-}$ $\boxed{=}$ or   7.1 $\boxed{\text{EXP}}$ 3 $\boxed{+\circ-}$ $\boxed{=}$ **Result:** | 0.0071 |

Graphing Calculator

```
2.7E5
           270000
7.1E-3
             .0071
```

We recommend that you use parentheses to enclose each number written in scientific notation when performing calculations. Try using your calculator to perform the calculations from Example 4.

a. $(8.7 \times 10^4)(2.5 \times 10^{-12})$ **b.** $\dfrac{4.25 \times 10^{13}}{8.5 \times 10^{-2}}$

Scientific Calculator

Enter: $\boxed{(}$ 8.7 $\boxed{\text{EE}}$ 4 $\boxed{)}$ $\boxed{\times}$ $\boxed{(}$ 2.5 $\boxed{\text{EE}}$ 12 $\boxed{+\circ-}$ $\boxed{)}$ $\boxed{=}$ **Result:** | 0.000000218 |

Enter: $\boxed{(}$ 4.25 $\boxed{\text{EE}}$ 13 $\boxed{)}$ $\boxed{\div}$ $\boxed{(}$ 8.5 $\boxed{\text{EE}}$ 2 $\boxed{+\circ-}$ $\boxed{)}$ $\boxed{=}$ **Result:** | 5E14 |

Notice that the answer to part (b) is shown on the calculator in scientific notation. The calculator does not have enough room to display 14 zeros. Also notice that the calculator rounds the answer to part (a). The exact answer is 2.175×10^{-7} or 0.0000002175.

Graphing Calculator

```
(8.7E4)*(2.5E-12
)
         2.175E-7
(4.25E13)/(8.5E-
2)
             5E14
```

Avoiding Mistakes

A display of 5E14 on a calculator does not mean 5^{14}. It is scientific notation and means 5×10^{14}.

Calculator Exercises

Use a calculator to perform the indicated operations:

1. $(5.2 \times 10^6)(4.6 \times 10^{-3})$

2. $(2.19 \times 10^{-8})(7.84 \times 10^{-4})$

3. $\dfrac{4.76 \times 10^{-5}}{2.38 \times 10^9}$

4. $\dfrac{8.5 \times 10^4}{4.0 \times 10^{-1}}$

5. $\dfrac{(9.6 \times 10^7)(4.0 \times 10^{-3})}{2.0 \times 10^{-2}}$

6. $\dfrac{(5.0 \times 10^{-12})(6.4 \times 10^{-5})}{(1.6 \times 10^{-8})(4.0 \times 10^2)}$

Section 5.4 Practice Exercises

Study Skills Exercise

1. Define the key term: **scientific notation**

Review Exercises

For Exercises 2–13, simplify the expression. Assume all variables represent nonzero real numbers.

2. $a^3 a^{-4}$

3. $b^5 b^8$

4. $10^3 \cdot 10^{-4}$

5. $10^5 \cdot 10^8$

6. $\dfrac{x^3}{x^6}$

7. $\dfrac{y^2}{y^7}$

8. $(c^4 d^2)^3$

9. $(x^5 y^{-3})^4$

10. $\dfrac{z^9 z^4}{z^3}$

11. $\dfrac{w^{-2} w^5}{w^{-1}}$

12. $\dfrac{10^9 \cdot 10^4}{10^3}$

13. $\dfrac{10^{-2} \cdot 10^5}{10^{-1}}$

Objective 1: Writing Numbers in Scientific Notation

14. Explain how scientific notation might be valuable in studying astronomy. Answers may vary.

15. Explain how you would write the number 0.000 000 000 23 in scientific notation.

16. Explain how you would write the number 23,000,000,000,000 in scientific notation.

For Exercises 17–28, write the number in scientific notation. **(See Example 1.)**

17. 50,000

18. 900,000

19. 208,000

20. 420,000,000

21. 6,010,000

22. 75,000

23. 0.000008

24. 0.003

25. 0.000125

26. 0.00000025

27. 0.006708

28. 0.02004

For Exercises 29–34, write the numbers in scientific notation. **(See Example 2.)**

29. The mass of a proton is approximately 0.000 000 000 000 000 000 000 0017 g.

30. The total combined salaries of the president, vice president, senators, and representatives of the United States federal government is approximately $85,000,000.

31. The Bill Gates Foundation has over $27,000,000,000 from which it makes contributions to global charities.

32. One gram is equivalent to 0.0035 oz.

33. In the world's largest tanker disaster, *Amoco Cadiz* spilled 68,000,000 gal of oil off Portsall, France, causing widespread environmental damage over 100 miles of Brittany coast.

34. The human heart pumps about 1400 L of blood per day. That means that it pumps approximately 10,000,000 L per year.

Objective 2: Writing Numbers in Standard Form

35. Explain how you would write the number 3.1×10^{-9} in standard form.

36. Explain how you would write the number 3.1×10^{9} in standard form.

For Exercises 37–52, write the numbers in standard form. **(See Example 3.)**

37. 5×10^{-5}

38. 2×10^{-7}

39. 2.8×10^{3}

40. 9.1×10^{6}

41. 6.03×10^{-4}

42. 7.01×10^{-3}

43. 2.4×10^{6}

44. 3.1×10^{4}

45. 1.9×10^{-2}

46. 2.8×10^{-6}

47. 7.032×10^{3}

48. 8.205×10^{2}

49. One picogram (pg) is equal to 1×10^{-12} g.

50. A nanometer (nm) is approximately 3.94×10^{-8} in.

51. A normal diet contains between 1.6×10^{3} Cal and 2.8×10^{3} Cal per day.

52. The total land area of Texas is approximately 2.62×10^{5} square miles.

Objective 3: Multiplying and Dividing Numbers in Scientific Notation

For Exercises 53–72, multiply or divide as indicated. Write the answers in scientific notation. **(See Example 4.)**

53. $(2.5 \times 10^{6})(2.0 \times 10^{-2})$

54. $(2.0 \times 10^{-7})(3.0 \times 10^{13})$

55. $(1.2 \times 10^{4})(3 \times 10^{7})$

56. $(3.2 \times 10^{-3})(2.5 \times 10^{8})$

57. $\dfrac{7.7 \times 10^{6}}{3.5 \times 10^{2}}$

58. $\dfrac{9.5 \times 10^{11}}{1.9 \times 10^{3}}$

59. $\dfrac{9.0 \times 10^{-6}}{4.0 \times 10^{7}}$

60. $\dfrac{7.0 \times 10^{-2}}{5.0 \times 10^{9}}$

61. $(8.0 \times 10^{10})(4.0 \times 10^{3})$

62. $(6.0 \times 10^{-4})(3.0 \times 10^{-2})$

63. $(3.2 \times 10^{-4})(7.6 \times 10^{-7})$

64. $(5.9 \times 10^{12})(3.6 \times 10^{9})$

65. $\dfrac{2.1 \times 10^{11}}{7.0 \times 10^{-3}}$

66. $\dfrac{1.6 \times 10^{14}}{8.0 \times 10^{-5}}$

67. $\dfrac{5.7 \times 10^{-2}}{9.5 \times 10^{-8}}$

68. $\dfrac{2.72 \times 10^{-6}}{6.8 \times 10^{-4}}$

69. $6{,}000{,}000{,}000 \times 0.0000000023$

70. $0.000055 \times 40{,}000$

71. $\dfrac{0.0000000003}{6000}$

72. $\dfrac{420{,}000}{0.0000021}$

Mixed Exercises

73. If a piece of paper is 3.0×10^{-3} in. thick, how thick is a stack of 1.25×10^{3} pieces of paper?

74. A box of staples contains 5.0×10^{3} staples and weighs 15 oz. How much does one staple weigh? Write your answer in scientific notation.

75. Bill Gates owned approximately 1,100,000,000 shares of Microsoft stock. If the stock price was $27 per share, how much was Bill Gates' stock worth?

76. A state lottery had a jackpot of 5.2×10^7. This week the winner was a group of office employees that included 13 people. How much would each person receive?

77. Dinosaurs became extinct about 65 million years ago.

   a. Write the number 65 million in scientific notation.

   b. How many days is 65 million years?

   c. How many hours is 65 million years?

   d. How many seconds is 65 million years?

78. The Earth is 111,600,000 km from the Sun.

   a. Write the number 111,600,000 in scientific notation.

   b. If there are 1000 m in a kilometer, how many meters is the Earth from the Sun?

   c. If there are 100 cm in a meter, how many centimeters is the Earth from the Sun?

Problem Recognition Exercises

Properties of Exponents

Simplify completely. Assume that all variables represent nonzero real numbers.

1. t^3t^5

2. $2^3 2^5$

3. $\dfrac{y^7}{y^2}$

4. $\dfrac{p^9}{p^3}$

5. $(r^2 s^4)^2$

6. $(ab^3 c^2)^3$

7. $\dfrac{w^4}{w^{-2}}$

8. $\dfrac{m^{-14}}{m^2}$

9. $\dfrac{y^{-7} x^4}{z^{-3}}$

10. $\dfrac{a^3 b^{-6}}{c^{-8}}$

11. $(2.5 \times 10^{-3})(5.0 \times 10^5)$

12. $(3.1 \times 10^6)(4.0 \times 10^{-2})$

13. $\dfrac{4.8 \times 10^7}{6.0 \times 10^{-2}}$

14. $\dfrac{5.4 \times 10^{-2}}{9.0 \times 10^6}$

15. $\dfrac{1}{p^{-6} p^{-8} p^{-1}}$

16. $p^6 p^8 p$

17. $\dfrac{v^9}{v^{11}}$

18. $(c^5 d^4)^{10}$

19. $\left(\dfrac{1}{2}\right)^{-1} + \left(\dfrac{1}{3}\right)^0$

20. $\left(\dfrac{1}{4}\right)^0 - \left(\dfrac{1}{5}\right)^{-1}$

21. $(2^5 b^{-3})^{-3}$

22. $(3^{-2} y^3)^{-2}$

23. $\left(\dfrac{3x}{2y}\right)^{-4}$

24. $\left(\dfrac{6c}{5d^3}\right)^{-2}$

25. $(3ab^2)(a^2 b)^3$

26. $(4x^2 y^3)^3 (xy^2)$

27. $\left(\dfrac{xy^2}{x^3 y}\right)^4$

28. $\left(\dfrac{a^3 b}{a^5 b^3}\right)^5$

29. $\dfrac{(t^{-2})^3}{t^{-4}}$

30. $\dfrac{(p^3)^{-4}}{p^{-5}}$

31. $\left(\dfrac{2w^2 x^3}{3y^0}\right)^3$

32. $\left(\dfrac{5a^0 b^4}{4c^3}\right)^2$

33. $\dfrac{q^3 r^{-2}}{s^{-1} t^5}$

34. $\dfrac{n^{-3} m^2}{p^{-3} q^{-1}}$

35. $\dfrac{(y^{-3})^2 (y^5)}{(y^{-3})^{-4}}$

36. $\dfrac{(w^2)^{-4} (w^{-2})}{(w^5)^{-4}}$

37. $\left(\dfrac{-2a^2 b^{-3}}{a^{-4} b^{-5}}\right)^{-3}$

38. $\left(\dfrac{-3x^{-4} y^3}{2x^5 y^{-2}}\right)^{-2}$

39. $(5h^{-2} k^0)^3 (5k^{-2})^{-4}$

40. $(6m^3 n^{-5})^{-4} (6m^0 n^{-2})^5$

8. Explain the difference between 3.0×10^7 and 3^7.

9. Explain the difference between 4.0×10^{-2} and 4^{-2}.

Objective 1: Introduction to Polynomials

10. Write the polynomial in descending order. $10 - 8a - a^3 + 2a^2 + a^5$

11. Write the polynomial in descending order:

$$6 + 7x^2 - 7x^4 + 9x$$

12. Write the polynomial in descending order:

$$\frac{1}{2}y + y^2 - 12y^4 + y^3 - 6$$

For Exercises 13–24, categorize the expression as a monomial, a binomial, or a trinomial. Then identify the coefficient and degree of the leading term. **(See Example 1.)**

13. $10a^2 + 5a$

14. $7z + 13z^2 - 15$

15. $6x^2$

16. 9

17. $2t - t^4$

18. $7x + 2$

19. $12y^4 - 3y + 1$

20. $5bc^2$

21. 23

22. $4 - 2c$

23. $-32xyz$

24. $w^4 - w^2$

Objective 2: Addition of Polynomials

25. Explain why the terms $3x$ and $3x^2$ are not *like* terms.

26. Explain why the terms $4w^3$ and $4z^3$ are not *like* terms.

For Exercises 27–42, add the polynomials. **(See Examples 2–3.)**

27. $23x^2y + 12x^2y$

28. $-5ab^3 + 17ab^3$

29. $(6y + 3x) + (4y - 3x)$

30. $(2z - 5h) + (-3z + h)$

31. $3b^5d^2 + (5b^5d^2 - 9d)$

32. $4c^2d^3 + (3cd - 10c^2d^3)$

33. $(7y^2 + 2y - 9) + (-3y^2 - y)$

34. $(-3w^2 + 4w - 6) + (5w^2 + 2)$

35. $\quad 6a + 2b - 5c$
$+ \underline{\ -2a - 2b - 3c}$

36. $\quad -13x + 5y + 10z$
$+ \underline{\ -3x - 3y + \ 2z}$

37. $\left(\dfrac{2}{5}a + \dfrac{1}{4}b - \dfrac{5}{6}\right) + \left(\dfrac{3}{5}a - \dfrac{3}{4}b - \dfrac{7}{6}\right)$

38. $\left(\dfrac{5}{9}x + \dfrac{1}{10}y\right) + \left(-\dfrac{4}{9}x + \dfrac{3}{10}y\right)$

39. $\left(z - \dfrac{8}{3}\right) + \left(\dfrac{4}{3}z^2 - z + 1\right)$

40. $\left(-\dfrac{7}{5}r + 1\right) + \left(-\dfrac{3}{5}r^2 + \dfrac{7}{5}r + 1\right)$

41. $\quad 7.9t^3 \qquad + 2.6t - 1.1$
$+ \underline{\qquad - 3.4t^2 + 3.4t - 3.1}$

42. $\quad 0.34y^2 \qquad + 1.23$
$+ \underline{\qquad\quad 3.42y - 7.56}$

Objective 3: Subtraction of Polynomials

For Exercises 43–48, find the opposite of each polynomial. **(See Example 4.)**

43. $4h - 5$

44. $5k - 12$

45. $-2m^2 + 3m - 15$

46. $-n^2 - 6n + 9$

47. $3v^3 + 5v^2 + 10v + 22$

48. $7u^4 + 3v^2 + 17$

For Exercises 49–68, subtract the polynomials. **(See Examples 5–6.)**

49. $4a^3b^2 - 12a^3b^2$

50. $5yz^4 - 14yz^4$

51. $-32x^3 - 21x^3$

52. $-23c^5 - 12c^5$

53. $(7a - 7) - (12a - 4)$

54. $(4x + 3v) - (-3x + v)$

55. $(4k + 3) - (-12k - 6)$

56. $(3h - 15) - (8h + 13)$

57. $25s - (23s + 14)$

58. $3x^2 - (-x^2 - 12)$

59. $(5t^2 - 3t - 2) - (2t^2 + t + 1)$

60. $(k^2 + 2k + 1) - (3k^2 - 6k + 2)$

61. $\begin{array}{r} 10r - 6s + 2t \\ -\ (12r - 3s -\ t) \\ \hline \end{array}$

62. $\begin{array}{r} a - 14b + 7c \\ -\ (-3a -\ 8b + 2c) \\ \hline \end{array}$

63. $\left(\dfrac{7}{8}x + \dfrac{2}{3}y - \dfrac{3}{10}\right) - \left(\dfrac{1}{8}x + \dfrac{1}{3}y\right)$

64. $\left(r - \dfrac{1}{12}s\right) - \left(\dfrac{1}{2}r - \dfrac{5}{12}s - \dfrac{4}{11}\right)$

65. $\left(\dfrac{2}{3}h^2 - \dfrac{1}{5}h - \dfrac{3}{4}\right) - \left(\dfrac{4}{3}h^2 - \dfrac{4}{5}h + \dfrac{7}{4}\right)$

66. $\left(\dfrac{3}{8}p^3 - \dfrac{5}{7}p^2 - \dfrac{2}{5}\right) - \left(\dfrac{5}{8}p^3 - \dfrac{2}{7}p^2 + \dfrac{7}{5}\right)$

67. $\begin{array}{r} 4.5x^4 - 3.1x^2\qquad\ \ - 6.7 \\ -\ (2.1x^4 \qquad\qquad + 4.4x\qquad\ ) \\ \hline \end{array}$

68. $\begin{array}{r} 1.3c^3 \qquad\qquad\quad + 4.8 \\ -\ (\qquad 4.3c^2 - 2c - 2.2) \\ \hline \end{array}$

69. Find the difference of $(4b^3 + 6b - 7)$ and $(-12b^2 + 11b + 5)$.

70. Find the difference of $(-5y^2 + 3y - 21)$ and $(-4y^2 - 5y + 23)$.

71. Subtract $(3x^3 - 5x + 10)$ from $(-2x^2 + 6x - 21)$. **(See Example 7.)**

72. Subtract $(7a^5 - 2a^3 - 5a)$ from $(3a^5 - 9a^2 + 3a - 8)$.

Objective 4: Polynomials and Applications to Geometry

73. Find a polynomial that represents the perimeter of the figure.

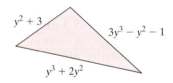

74. Find a polynomial that represents the perimeter of the figure.

75. If the perimeter of the figure can be represented by the polynomial $5a^2 - 2a + 1$, find a polynomial that represents the length of the missing side. **(See Example 8.)**

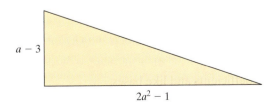

76. If the perimeter of the figure can be represented by the polynomial $6w^3 - 2w - 3$, find a polynomial that represents the length of the missing side.

Skill Practice

Divide the polynomials using the appropriate method of division.

6. $\dfrac{6x^3 - x^2 + 3x - 5}{2x + 3}$

7. $\dfrac{9w^3 - 18w^2 + 6w + 12}{3w}$

Example 5 **Determining Whether Long Division Is Necessary**

Determine whether long division is necessary for each division of polynomials.

a. $\dfrac{2p^5 - 8p^4 + 4p - 16}{p^2 - 2p + 1}$

b. $\dfrac{2p^5 - 8p^4 + 4p - 16}{2p^2}$

c. $(3z^3 - 5z^2 + 10) \div (15z^3)$

d. $(3z^3 - 5z^2 + 10) \div (3z + 1)$

Solution:

a. $\dfrac{2p^5 - 8p^4 + 4p - 16}{p^2 - 2p + 1}$ The divisor has three terms. Use long division.

b. $\dfrac{2p^5 - 8p^4 + 4p - 16}{2p^2}$ The divisor has one term. No long division.

c. $(3z^3 - 5z^2 + 10) \div (15z^3)$ The divisor has one term. No long division.

d. $(3z^3 - 5z^2 + 10) \div (3z + 1)$ The divisor has two terms. Use long division.

TIP: Recall that
- Long division is used when the divisor has *two or more terms*.
- If the divisor has *one term*, then divide each term in the dividend by the monomial divisor.

Answers

6. $3x^2 - 5x + 9 + \dfrac{-32}{2x + 3}$

7. $3w^2 - 6w + 2 + \dfrac{4}{w}$

Section 5.7   Practice Exercises

Review Exercises

For Exercises 1–10, perform the indicated operations.

1. $(6z^5 - 2z^3 + z - 6) - (10z^4 + 2z^3 + z^2 + z)$

2. $(7a^2 + a - 6) + (2a^2 + 5a + 11)$

3. $(10x + y)(x - 3y)$

4. $8b^2(2b^2 - 5b + 12)$

5. $(10x + y) + (x - 3y)$

6. $(2w^3 + 5)^2$

7. $\left(\dfrac{4}{3}y^2 - \dfrac{1}{2}y + \dfrac{3}{8}\right) - \left(\dfrac{1}{3}y^2 + \dfrac{1}{4}y - \dfrac{1}{8}\right)$

8. $\left(\dfrac{7}{8}w - 1\right)\left(\dfrac{7}{8}w + 1\right)$

9. $(a + 3)(a^2 - 3a + 9)$

10. $(2x + 1)(5x - 3)$

Objective 1: Division by a Monomial

11. There are two methods for dividing polynomials. Explain when long division is used.

12. Explain how to check a polynomial division problem.

13. a. Divide $\dfrac{15t^3 + 18t^2}{3t}$

    b. Check by multiplying the quotient by the divisor.

14. a. Divide $(-9y^4 + 6y^2 - y) \div (3y)$

    b. Check by multiplying the quotient by the divisor.

For Exercises 15–30, divide the polynomials. **(See Example 1.)**

15. $(6a^2 + 4a - 14) \div (2)$

16. $\dfrac{4b^2 + 16b - 12}{4}$

17. $\dfrac{-5x^2 - 20x + 5}{-5}$

18. $\dfrac{-3y^3 + 12y - 6}{-3}$

19. $\dfrac{3p^3 - p^2}{p}$

20. $(7q^4 + 5q^2) \div q$

21. $(4m^2 + 8m) \div 4m^2$

22. $\dfrac{n^2 - 8}{n}$

23. $\dfrac{14y^4 - 7y^3 + 21y^2}{-7y^2}$

24. $(25a^5 - 5a^4 + 15a^3 - 5a) \div (-5a)$

25. $(4x^3 - 24x^2 - x + 8) \div (4x)$

26. $\dfrac{20w^3 + 15w^2 - w + 5}{10w}$

27. $\dfrac{-a^3b^2 + a^2b^2 - ab^3}{-a^2b^2}$

28. $(3x^4y^3 - x^2y^2 - xy^3) \div (-x^2y^2)$

29. $(6t^4 - 2t^3 + 3t^2 - t + 4) \div (2t^3)$

30. $\dfrac{2y^3 - 2y^2 + 3y - 9}{2y^2}$

Objective 2: Long Division

31. a. Divide $(z^2 + 7z + 11) \div (z + 5)$

    b. Check by multiplying the quotient by the divisor and adding the remainder.

32. a. Divide $\dfrac{2w^2 - 7w + 3}{w - 4}$

    b. Check by multiplying the quotient by the divisor and adding the remainder.

For Exercises 33–56, divide the polynomials. **(See Examples 2–4.)**

33. $\dfrac{t^2 + 4t + 5}{t + 1}$

34. $(3x^2 + 8x + 5) \div (x + 2)$

35. $(7b^2 - 3b - 4) \div (b - 1)$

36. $\dfrac{w^2 - w - 2}{w - 2}$

37. $\dfrac{5k^2 - 29k - 6}{5k + 1}$

38. $(4y^2 + 25y - 21) \div (4y - 3)$

39. $(4p^3 + 12p^2 + p - 12) \div (2p + 3)$

40. $\dfrac{12a^3 - 2a^2 - 17a - 5}{3a + 1}$

Section 5.7   Division of Polynomials

Key Concepts

Division of Polynomials

1. Division by a monomial, use the properties:

$$\frac{a+b}{c} = \frac{a}{c} + \frac{b}{c} \quad \text{and} \quad \frac{a-b}{c} = \frac{a}{c} - \frac{b}{c}$$

2. If the divisor has more than one term, use long division.

Examples

Example 1

Divide: $\dfrac{-3x^2 - 6x + 9}{-3x}$

$$= \frac{-3x^2}{-3x} - \frac{6x}{-3x} + \frac{9}{-3x}$$

$$= x + 2 - \frac{3}{x}$$

Example 2

Divide: $(3x^2 - 5x + 1) \div (x + 2)$

$$
\begin{array}{r}
3x - 11 \\
x+2\overline{)3x^2 - 5x + 1} \\
-(3x^2 + 6x) \\
\hline
-11x + 1 \\
-(-11x - 22) \\
\hline
23
\end{array}
$$

$$3x - 11 + \frac{23}{x+2}$$

Chapter 5   Review Exercises

Section 5.1

For Exercises 1–4, identify the base and the exponent.

1. 5^3 **2.** x^4 **3.** $(-2)^0$ **4.** y

5. Evaluate the expressions.

   a. 6^2 **b.** $(-6)^2$ **c.** -6^2

6. Evaluate the expressions.

   a. 4^3 **b.** $(-4)^3$ **c.** -4^3

For Exercises 7–18, simplify and write the answers in exponent form. Assume that all variables represent nonzero real numbers.

7. $5^3 \cdot 5^{10}$ **8.** $a^7 a^4$

9. $x \cdot x^6 \cdot x^2$ **10.** $6^3 \cdot 6 \cdot 6^5$

11. $\dfrac{10^7}{10^4}$ **12.** $\dfrac{y^{14}}{y^8}$

13. $\dfrac{b^9}{b}$ **14.** $\dfrac{7^8}{7}$

15. $\dfrac{k^2 k^3}{k^4}$ **16.** $\dfrac{8^4 \cdot 8^7}{8^{11}}$

17. $\dfrac{2^8 \cdot 2^{10}}{2^3 \cdot 2^7}$ **18.** $\dfrac{q^3 q^{12}}{qq^8}$

19. Explain why $2^2 \cdot 4^4$ does *not* equal 8^6.

20. Explain why $\frac{10^5}{5^2}$ does *not* equal 2^3.

For Exercises 21–22, use the formula

$$A = P(1 + r)^t$$

21. Find the amount in an account after 3 yr if the initial investment is $6000, invested at 6% interest compounded annually.

22. Find the amount in an account after 2 yr if the initial investment is $20,000, invested at 5% interest compounded annually.

Section 5.2

For Exercises 23–40, simplify the expressions. Write the answers in exponent form. Assume all variables represent nonzero real numbers.

23. $(7^3)^4$

24. $(c^2)^6$

25. $(p^4 p^2)^3$

26. $(9^5 \cdot 9^2)^4$

27. $\left(\dfrac{a}{b}\right)^2$

28. $\left(\dfrac{1}{3}\right)^4$

29. $\left(\dfrac{5}{c^2 d^5}\right)^2$

30. $\left(-\dfrac{m^2}{4n^6}\right)^5$

31. $(2ab^2)^4$

32. $(-x^7 y)^2$

33. $\left(\dfrac{-3x^3}{5y^2 z}\right)^3$

34. $\left(\dfrac{r^3}{s^2 t^6}\right)^5$

35. $\dfrac{a^4 (a^2)^8}{(a^3)^3}$

36. $\dfrac{(8^3)^4 \cdot 8^{10}}{(8^4)^5}$

37. $\dfrac{(4h^2 k)^2 (h^3 k)^4}{(2hk^3)^2}$

38. $\dfrac{(p^3 q)^3 (2p^2 q^4)^4}{(8p)(pq^3)^2}$

39. $\left(\dfrac{2x^4 y^3}{4xy^2}\right)^2$

40. $\left(\dfrac{a^4 b^6}{ab^4}\right)^3$

Section 5.3

For Exercises 41–62, simplify the expressions. Assume all variables represent nonzero real numbers.

41. 8^0

42. $(-b)^0$

43. $-x^0$

44. 1^0

45. $2y^0$

46. $(2y)^0$

47. z^{-5}

48. 10^{-4}

49. $(6a)^{-2}$

50. $6a^{-2}$

51. $4^0 + 4^{-2}$

52. $9^{-1} + 9^0$

53. $t^{-6} t^{-2}$

54. $r^8 r^{-9}$

55. $\dfrac{12x^{-2} y^3}{6x^4 y^{-4}}$

56. $\dfrac{8ab^{-3} c^0}{10a^{-5} b^{-4} c^{-1}}$

57. $(-2m^2 n^{-4})^{-4}$

58. $(3u^{-5} v^2)^{-3}$

59. $\dfrac{(k^{-6})^{-2}(k^3)}{5k^{-6} k^0}$

60. $\dfrac{(3h)^{-2}(h^{-5})^{-3}}{h^{-4} h^8}$

61. $2 \cdot 3^{-1} - 6^{-1}$

62. $2^{-1} - 2^{-2} + 2^0$

Section 5.4

63. Write the numbers in scientific notation.

a. In a recent year there were 97,400,000 packages of M&Ms sold in the United States.

b. The thickness of a piece of paper is 0.0042 in.

64. Write the numbers in standard form.

a. A pH of 10 means the hydrogen ion concentration is 1×10^{-10} units.

b. A fund-raising event for neurospinal research raised 2.56×10^5.

For Exercises 65–68, perform the indicated operations. Write the answers in scientific notation.

65. $(4.1 \times 10^{-6})(2.3 \times 10^{11})$

66. $\dfrac{9.3 \times 10^3}{6.0 \times 10^{-7}}$

67. $\dfrac{2000}{0.000008}$

68. $(0.000078)(21,000,000)$

69. Use your calculator to evaluate 5^{20}. Why is scientific notation necessary on your calculator to express the answer?

70. Use your calculator to evaluate $(0.4)^{30}$. Why is scientific notation necessary on your calculator to express the answer?

71. The average distance between the Earth and Sun is 9.3×10^7 mi.

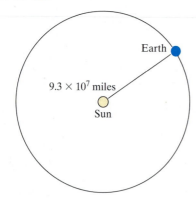

Earth

9.3×10^7 miles

Sun

a. If the Earth's orbit is approximated by a circle, find the total distance the Earth travels around the Sun in one orbit. (*Hint:* The circumference of a circle is given by $C = 2\pi r$.) Express the answer in scientific notation.

b. If the Earth makes one complete trip around the Sun in 1 yr (365 days = 8.76×10^3 hr), find the average speed that the Earth travels around the Sun in miles per hour. Express the answer in scientific notation.

72. The average distance between the planet Mercury and the Sun is 3.6×10^7 mi.

a. If Mercury's orbit is approximated by a circle, find the total distance Mercury travels around the Sun in one orbit. (*Hint:* The circumference of a circle is given by $C = 2\pi r$.) Express the answer in scientific notation.

b. If Mercury makes one complete trip around the Sun in 88 days (2.112×10^3 hr), find the average speed that Mercury travels around the Sun in miles per hour. Express the answer in scientific notation.

Section 5.5

73. For the polynomial $7x^4 - x + 6$

a. Classify as a monomial, a binomial, or a trinomial.

b. Identify the degree of the polynomial.

c. Identify the leading coefficient.

74. For the polynomial $2y^3 - 5y^7$

a. Classify as a monomial, a binomial, or a trinomial.

b. Identify the degree of the polynomial.

c. Identify the leading coefficient.

For Exercises 75–80, add or subtract as indicated.

75. $(4x + 2) + (3x - 5)$

76. $(7y^2 - 11y - 6) - (8y^2 + 3y - 4)$

77. $(9a^2 - 6) - (-5a^2 + 2a)$

78. $\left(5x^3 - \dfrac{1}{4}x^2 + \dfrac{5}{8}x + 2\right) + \left(\dfrac{5}{2}x^3 + \dfrac{1}{2}x^2 - \dfrac{1}{8}x\right)$

79. $8w^4 - 6w + 3$
$ + 2w^4 + 2w^3 - w + 1$

80. $-0.02b^5 + b^4 - 0.7b + 0.3$
$+ 0.03b^5 - 0.1b^3 + b + 0.03$

81. Subtract $(9x^2 + 4x + 6)$ from $(7x^2 - 5x)$.

82. Find the difference of $(x^2 - 5x - 3)$ and $(6x^2 + 4x + 9)$.

83. Write a trinomial of degree 2 with a leading coefficient of -5. (Answers may vary.)

84. Write a binomial of degree 5 with leading coefficient 6. (Answers may vary.)

85. Find a polynomial that represents the perimeter of the given rectangle.

$2w + 3$

w

Section 5.6

For Exercises 86–103, multiply the expressions.

86. $(25x^4y^3)(-3x^2y)$

87. $(9a^6)(2a^2b^4)$

88. $5c(3c^3 - 7c + 5)$

89. $(x^2 + 5x - 3)(-2x)$

90. $(5k - 4)(k + 1)$

91. $(4t - 1)(5t + 2)$

92. $(q + 8)(6q - 1)$

93. $(2a - 6)(a + 5)$

94. $\left(7a + \dfrac{1}{2}\right)^2$

95. $(b - 4)^2$

96. $(4p^2 + 6p + 9)(2p - 3)$

97. $(2w - 1)(-w^2 - 3w - 4)$

98. $2x^2 - 3x + 4$
$\times 2x - 1$

99.
$$\begin{array}{r} 4a^2 + a - 5 \\ \times \quad 3a + 2 \end{array}$$

100. $(b - 4)(b + 4)$

101. $\left(\dfrac{1}{3}r^4 - s^2\right)\left(\dfrac{1}{3}r^4 + s^2\right)$

102. $(-7z^2 + 6)^2$

103. $(2h + 3)(h^4 - h^3 + h^2 - h + 1)$

104. Find a polynomial that represents the area of the given rectangle

2x − 5

x + 4

Section 5.7

For Exercises 105–117, divide the polynomials.

105. $\dfrac{20y^3 - 10y^2}{5y}$

106. $(18a^3b^2 - 9a^2b - 27ab^2) \div 9ab$

107. $(12x^4 - 8x^3 + 4x^2) \div (-4x^2)$

108. $\dfrac{10z^7w^4 - 15z^3w^2 - 20zw}{-20z^2w}$

109. $\dfrac{x^2 + 7x + 10}{x + 5}$

110. $(2t^2 + t - 10) \div (t - 2)$

111. $(2p^2 + p - 16) \div (2p + 7)$

112. $\dfrac{5a^2 + 27a - 22}{5a - 3}$

113. $\dfrac{b^3 - 125}{b - 5}$

114. $(z^3 + 4z^2 + 5z + 20) \div (5 + z^2)$

115. $(y^4 - 4y^3 + 5y^2 - 3y + 2) \div (y^2 + 3)$

116. $(3t^4 - 8t^3 + t^2 - 4t - 5) \div (3t^2 + t + 1)$

117. $\dfrac{2w^4 + w^3 + 4w - 3}{2w^2 - w + 3}$

Chapter 5   Test

Assume all variables represent nonzero real numbers.

1. Expand the expression using the definition of exponents, then simplify: $\dfrac{3^4 \cdot 3^3}{3^6}$

For Exercises 2–11, simplify the expression. Write the answer with positive exponents only.

2. $9^5 \cdot 9$

3. $\dfrac{q^{10}}{q^2}$

4. $(3a^2b)^3$

5. $\left(\dfrac{2x}{y^3}\right)^4$

6. $(-7)^0$

7. c^{-3}

8. $\dfrac{14^3 \cdot 14^9}{14^{10} \cdot 14}$

9. $\dfrac{(s^2t)^3(7s^4t)^4}{(7s^2t^3)^2}$

10. $(2a^0b^{-6})^2$

11. $\left(\dfrac{6a^{-5}b}{8ab^{-2}}\right)^{-2}$

12. a. Write the number in scientific notation: 43,000,000,000

b. Write the number in standard form: 5.6×10^{-6}

13. The average amount of water flowing over Niagara Falls is 1.68×10^5 m³/min.

a. How many cubic meters of water flow over the falls in one day?

b. How many cubic meters of water flow over the falls in one year?

14. Write the polynomial in descending order:
$4x + 5x^3 - 7x^2 + 11$

    a. Identify the degree of the polynomial.

    b. Identify the leading coefficient of the polynomial.

15. Perform the indicated operations.

$(7w^2 - 11w - 6) + (8w^2 + 3w + 4) - (-9w^2 - 5w + 2)$

16. Subtract $(3x^2 - 5x^3 + 2x)$ from $(10x^3 - 4x^2 + 1)$.

For Exercises 17–23, multiply the polynomials.

17. $-2x^3(5x^2 + x - 15)$ **18.** $(4a - 3)(2a - 1)$

19. $(4y - 5)(y^2 - 5y + 3)$ **20.** $(2 + 3b)(2 - 3b)$

21. $(5z - 6)^2$ **22.** $(5x + 3)(3x - 2)$

23. $(y^2 - 5y + 2)(y - 6)$

24. Find the perimeter and the area of the rectangle shown in the figure.

For Exercises 25–28, divide:

25. $(-12x^8 + x^6 - 8x^3) \div (4x^2)$

26. $\dfrac{2y^2 - 13y + 21}{y - 3}$

27. $(-5w^2 + 2w^3 - 2w + 5) \div (2w + 3)$

28. $\dfrac{3x^4 + x^3 + 4x - 33}{x^2 + 4}$

Chapters 1–5 Cumulative Review Exercises

For Exercises 1–2, simplify completely.

1. $-5 - \dfrac{1}{2}[4 - 3(-7)]$ **2.** $|-3^2 + 5|$

3. Translate the phrase into a mathematical expression and simplify:

The difference of the square of five and the square root of four.

4. Solve for x: $\dfrac{1}{2}(x - 6) + \dfrac{2}{3} = \dfrac{1}{4}x$

5. Solve for y: $-2y - 3 = -5(y - 1) + 3y$

6. For a point in a rectangular coordinate system, in which quadrant are both the x- and y-coordinates negative?

7. For a point in a rectangular coordinate system, on which axis is the x-coordinate zero and the y-coordinate nonzero?

8. In a triangle, one angle measures 23° more than the smallest angle. The third angle measures 10° more than the sum of the other two angles. Find the measure of each angle.

9. A snow storm lasts for 9 hr and dumps snow at a rate of $1\frac{1}{2}$ in./hr. If there was already 6 in. of snow on the ground before the storm, the snow depth is given by the equation:

$y = \dfrac{3}{2}x + 6$ where y is the snow depth in inches and $x \geq 0$ is the time in hours.

    a. Find the snow depth after 4 hr.

    b. Find the snow depth at the end of the storm.

    c. How long had it snowed when the total depth of snow was $14\frac{1}{4}$ in.?

10. Solve the system of equations.

$$5x + 3y = -3$$
$$3x + 2y = -1$$

11. Solve the inequality. Graph the solution set on the real number line and express the solution in interval notation. $2 - 3(2x + 4) \leq -2x - (x - 5)$

$\longrightarrow$

For Exercises 12–15, perform the indicated operations.

12. $(2x^2 + 3x - 7) - (-3x^2 + 12x + 8)$

13. $(2y + 3z)(-y - 5z)$

14. $(4t - 3)^2$

15. $\left(\dfrac{2}{5}a + \dfrac{1}{3}\right)\left(\dfrac{2}{5}a - \dfrac{1}{3}\right)$

For Exercises 16–17, divide the polynomials.

16. $(12a^4b^3 - 6a^2b^2 + 3ab) \div (-3ab)$

17. $\dfrac{4m^3 - 5m + 2}{m - 2}$

For Exercises 18–19, use the properties of exponents to simplify the expressions. Write the answers with positive exponents only. Assume all variables represent nonzero real numbers.

18. $\left(\dfrac{2c^2d^4}{8cd^6}\right)^2$

19. $\dfrac{10a^{-2}b^{-3}}{5a^0b^{-6}}$

20. Perform the indicated operations, and write the final answer in scientific notation.

$$\dfrac{8.2 \times 10^{-2}}{2.0 \times 10^{-5}}$$

Objectives

1. **Identifying the Greatest Common Factor**
2. **Factoring out the Greatest Common Factor**
3. **Factoring out a Negative Factor**
4. **Factoring out a Binomial Factor**
5. **Factoring by Grouping**

1. Identifying the Greatest Common Factor

Chapter 6 is devoted to a mathematical operation called **factoring**. To factor an integer means to write the integer as a product of two or more integers. To factor a polynomial means to express the polynomial as a product of two or more polynomials.

In the product $2 \cdot 5 = 10$, for example, 2 and 5 are factors of 10.

In the product $(3x + 4)(2x - 1) = 6x^2 + 5x - 4$, the quantities $(3x + 4)$ and $(2x - 1)$ are factors of $6x^2 + 5x - 4$.

We begin our study of factoring by factoring integers. The number 20, for example, can be factored as $1 \cdot 20$ or $2 \cdot 10$ or $4 \cdot 5$ or $2 \cdot 2 \cdot 5$. The product $2 \cdot 2 \cdot 5$ (or equivalently $2^2 \cdot 5$) consists only of prime numbers and is called the **prime factorization**.

The **greatest common factor** (denoted **GCF**) of two or more integers is the greatest factor common to each integer. To find the greatest common factor of two or more integers, it is often helpful to express the numbers as a product of prime factors as shown in the next example.

Example 1 Identifying the GCF

Find the greatest common factor.

   a. 24 and 36   **b.** 105, 40, and 60

Solution:

First find the prime factorization of each number. Then find the product of common factors.

a.
$$\begin{array}{ll} 2\underline{|24} & 2\underline{|36} \\ 2\underline{|12} & 2\underline{|18} \\ 2\underline{|6} & 3\underline{|9} \\ 3 & 3 \end{array}$$

Factors of 24 = $2 \cdot 2 \cdot 2 \cdot 3$
Factors of 36 = $2 \cdot 2 \cdot 3 \cdot 3$

Common factors are circled.

The numbers 24 and 36 share two factors of 2 and one factor of 3. Therefore, the greatest common factor is $2 \cdot 2 \cdot 3 = 12$.

b.
$$\begin{array}{lll} 5\underline{|105} & 5\underline{|40} & 5\underline{|60} \\ 3\underline{|21} & 2\underline{|8} & 3\underline{|12} \\ 7 & 2\underline{|4} & 2\underline{|4} \\ & 2 & 2 \end{array}$$

Factors of 105 = $3 \cdot 7 \cdot 5$

Factors of 40 = $2 \cdot 2 \cdot 2 \cdot 5$

Factors of 60 = $2 \cdot 2 \cdot 3 \cdot 5$

The greatest common factor is 5.

In Example 2, we find the greatest common factor of two or more variable terms.

Example 2 Identifying the Greatest Common Factor

Find the GCF among each group of terms.

a. $7x^3, 14x^2, 21x^4$ **b.** $15a^4b, 25a^3b^2$ **c.** $8c^2d^7e, 6c^3d^4$

Solution:

List the factors of each term.

a. $7x^3 = 7 \cdot x \cdot x \cdot x$

$14x^2 = 2 \cdot 7 \cdot x \cdot x$ The GCF is $7x^2$.

$21x^4 = 3 \cdot 7 \cdot x \cdot x \cdot x \cdot x$

b. $15a^4b = 3 \cdot 5 \cdot a \cdot a \cdot a \cdot a \cdot b$

$25a^3b^2 = 5 \cdot 5 \cdot a \cdot a \cdot a \cdot b \cdot b$ The GCF is $5a^3b$.

TIP: Notice that the expressions $15a^4b$ and $25a^3b^2$ share factors of 5, a, and b. The GCF is the product of the common factors, where each factor is raised to the lowest power to which it occurs in all the original expressions.

$\left.\begin{array}{l} 15a^4b = 3 \cdot 5a^4b \\ 25a^3b^2 = 5^2a^3b^2 \end{array}\right\}$ $\left.\begin{array}{l} \text{Lowest power of 5 is 1:} \quad 5^1 \\ \text{Lowest power of } a \text{ is 3:} \quad a^3 \\ \text{Lowest power of } b \text{ is 1:} \quad b^1 \end{array}\right\}$ The GCF is $5a^3b$.

c. $\left.\begin{array}{l} 8c^2d^7e = 2^3c^2d^7e \\ 6c^3d^4 = 2 \cdot 3c^3d^4 \end{array}\right\}$ The common factors are 2, c, and d.

$\left.\begin{array}{l} \text{The lowest power of 2 is 1:} \quad 2^1 \\ \text{The lowest power of } c \text{ is 2:} \quad c^2 \\ \text{The lowest power of } d \text{ is 4:} \quad d^4 \end{array}\right\}$ The GCF is $2c^2d^4$.

Sometimes polynomials share a common binomial factor, as shown in Example 3.

Example 3 Finding the Greatest Common Binomial Factor

Find the greatest common factor between the terms: $3x(a + b)$ and $2y(a + b)$

Solution:

$\left.\begin{array}{l} 3x(a + b) \\ 2y(a + b) \end{array}\right\}$ The only common factor is the binomial $(a + b)$.

The GCF is $(a + b)$.

2. Factoring out the Greatest Common Factor

The process of factoring a polynomial is the reverse process of multiplying polynomials. Both operations use the distributive property: $ab + ac = a(b + c)$.

Multiply

$$5y(y^2 + 3y + 1) = 5y(y^2) + 5y(3y) + 5y(1)$$
$$= 5y^3 + 15y^2 + 5y$$

Factor

$$5y^3 + 15y^2 + 5y = 5y(y^2) + 5y(3y) + 5y(1)$$
$$= 5y(y^2 + 3y + 1)$$

PROCEDURE Factoring out the Greatest Common Factor

Step 1 Identify the GCF of all terms of the polynomial.

Step 2 Write each term as the product of the GCF and another factor.

Step 3 Use the distributive property to remove the GCF.

Note: To check the factorization, multiply the polynomials to remove parentheses.

Skill Practice

Factor out the GCF.

7. $6w + 18$

8. $21m^3 - 7m$

Example 4 Factoring out the Greatest Common Factor

Factor out the GCF.

a. $4x - 20$ **b.** $6w^2 + 3w$

Solution:

a. $4x - 20$ The GCF is 4.

$\quad = 4(x) - 4(5)$ Write each term as the product of the GCF and another factor.

$\quad = 4(x - 5)$ Use the distributive property to factor out the GCF.

TIP: Any factoring problem can be checked by multiplying the factors:

Check: $4(x - 5) = 4x - 20$ ✔

Avoiding Mistakes

In Example 4(b), the GCF, $3w$, is equal to one of the terms of the polynomial. In such a case, you must leave a 1 in place of that term after the GCF is factored out.

b. $6w^2 + 3w$ The GCF is $3w$.

$\quad = 3w(2w) + 3w(1)$ Write each term as the product of $3w$ and another factor.

$\quad = 3w(2w + 1)$ Use the distributive property to factor out the GCF.

Check: $3w(2w + 1) = 6w^2 + 3w$ ✔

Answers

7. $6(w + 3)$ **8.** $7m(3m^2 - 1)$

Example 5   Factoring out the Greatest Common Factor

Factor out the GCF.

  a. $15y^3 + 12y^4$ **b.** $9a^4b - 18a^5b + 27a^6b$

Skill Practice

Factor out the GCF.
  9. $9y^2 - 6y^5$
  10. $50s^3t - 40st^2 + 10st$

Solution:

a. $15y^3 + 12y^4$ The GCF is $3y^3$.

    $= 3y^3(5) + 3y^3(4y)$ Write each term as the product of $3y^3$ and another factor.

    $= 3y^3(5 + 4y)$ Use the distributive property to factor out the GCF.

         Check: $3y^3(5 + 4y) = 15y^3 + 12y^4$ ✔

b. $9a^4b - 18a^5b + 27a^6b$ The GCF is $9a^4b$.

    $= 9a^4b(1) - 9a^4b(2a) + 9a^4b(3a^2)$ Write each term as the product of $9a^4b$ and another factor.

    $= 9a^4b(1 - 2a + 3a^2)$ Use the distributive property to factor out the GCF.

        Check: $9a^4b(1 - 2a + 3a^2)$
                        $= 9a^4b - 18a^5b + 27a^6b$ ✔

The greatest common factor of the polynomial $2x + 5y$ is 1. If we factor out the GCF, we have $1(2x + 5y)$. A polynomial whose only factors are itself and 1 is called a **prime polynomial**. A prime polynomial cannot be factored further.

3. Factoring out a Negative Factor

Usually it is advantageous to factor out the *opposite* of the GCF when the leading coefficient of the polynomial is negative. This is demonstrated in the next example. Notice that this *changes the signs* of the remaining terms inside the parentheses.

Example 6   Factoring out a Negative Factor

Factor out -3 from the polynomial $-3x^2 + 6x - 33$.

Skill Practice

  11. Factor out -2 from the polynomial.
     $-2x^2 - 10x + 16$

Solution:

  $-3x^2 + 6x - 33$ The GCF is 3. However, in this case, we will factor out the *opposite* of the GCF, -3.

  $= -3(x^2) + (-3)(-2x) + (-3)(11)$ Write each term as the product of -3 and another factor.

  $= -3[x^2 + (-2x) + 11]$ Factor out -3.

  $= -3(x^2 - 2x + 11)$ Simplify. Notice that each sign within the trinomial has changed.

    Check: $-3(x^2 - 2x + 11) = -3x^2 + 6x - 33$ ✔

Answers
  9. $3y^2(3 - 2y^3)$
  10. $10st(5s^2 - 4t + 1)$
  11. $-2(x^2 + 5x - 8)$

Example 7 **Factoring out a Negative Factor**

Factor out the quantity $-4pq$ from the polynomial $-12p^3q - 8p^2q^2 + 4pq^3$.

Solution:

$-12p^3q - 8p^2q^2 + 4pq^3$ The GCF is $4pq$. However, in this case, we will factor out the *opposite* of the GCF, $-4pq$.

$= -4pq(3p^2) + (-4pq)(2pq) + (-4pq)(-q^2)$ Write each term as the product of $-4pq$ and another factor.

$= -4pq[3p^2 + 2pq + (-q^2)]$ Factor out $-4pq$. Notice that each sign within the trinomial has changed.

$= -4pq(3p^2 + 2pq - q^2)$ To verify that this is the correct factorization and that the signs are correct, multiply the factors.

Check: $-4pq(3p^2 + 2pq - q^2) = -12p^3q - 8p^2q^2 + 4pq^3$ ✔

4. Factoring out a Binomial Factor

The distributive property can also be used to factor out a common factor that consists of more than one term as shown in Example 8.

Example 8 **Factoring out a Binomial Factor**

Factor out the GCF: $2w(x + 3) - 5(x + 3)$

Solution:

$2w(x + 3) - 5(x + 3)$ The greatest common factor is the quantity $(x + 3)$.

$= (x + 3)(2w) - (x + 3)(5)$ Write each term as the product of $(x + 3)$ and another factor.

$= (x + 3)(2w - 5)$ Use the distributive property to factor out the GCF.

5. Factoring by Grouping

When two binomials are multiplied, the product before simplifying contains four terms. For example:

$$(x + 4)(3a + 2b) = (x + 4)(3a) + (x + 4)(2b)$$

$$= (x + 4)(3a) + (x + 4)(2b)$$

$$= 3ax + 12a + 2bx + 8b$$

Answers

12. $-5xy(2x - 1 + 3y)$
13. $(a + b)(8y + 9)$

In Example 9, we learn how to reverse this process. That is, given a four-term polynomial, we will factor it as a product of two binomials. The process is called *factoring by grouping*.

PROCEDURE   Factoring by Grouping

To factor a four-term polynomial by grouping:

Step 1   Identify and factor out the GCF from all four terms.
Step 2   Factor out the GCF from the first pair of terms. Factor out the GCF from the second pair of terms. (Sometimes it is necessary to factor out the opposite of the GCF.)
Step 3   If the two terms share a common binomial factor, factor out the binomial factor.

Example 9   **Factoring by Grouping**

Factor by grouping: $3ax + 12a + 2bx + 8b$

Skill Practice

Factor by grouping.
14. $5x + 10y + ax + 2ay$

Solution:

$3ax + 12a + 2bx + 8b$	**Step 1:** Identify and factor out the GCF from all four terms. In this case, the GCF is 1.
$= 3ax + 12a \mid + 2bx + 8b$	Group the first pair of terms and the second pair of terms.
$= 3a(x + 4) + 2b(x + 4)$	**Step 2:** Factor out the GCF from each pair of terms. *Note:* The two terms now share a common binomial factor of $(x + 4)$.
$= (x + 4)(3a + 2b)$	**Step 3:** Factor out the common binomial factor.

Check: $(x + 4)(3a + 2b) = 3ax + 2bx + 12a + 8b$ ✔

Note: Step 2 results in two terms with a common binomial factor. If the two binomials are different, step 3 cannot be performed. In such a case, the original polynomial may not be factorable by grouping, or different pairs of terms may need to be grouped and inspected.

TIP: One frequently asked question when factoring is whether the order can be switched between the factors. The answer is yes. Because multiplication is commutative, the order in which the factors are written does not matter.

$$(x + 4)(3a + 2b) = (3a + 2b)(x + 4)$$

Answer
14. $(x + 2y)(5 + a)$

Skill Practice

Factor by grouping.

15. $tu - tv - 2u + 2v$

Avoiding Mistakes

In step 2, the expression $a(x + y) - b(x + y)$ is not yet factored completely because it is a *difference*, not a product. To factor the expression, you must carry it one step further.

$$a(x + y) - b(x + y)$$
$$= (x + y)(a - b)$$

The factored form must be represented as a product.

Example 10   Factoring by Grouping

Factor by grouping. $ax + ay - bx - by$

Solution:

$$ax + ay - bx - by$$

Step 1: Identify and factor out the GCF from all four terms. In this case, the GCF is 1.

$$= ax + ay \mid - bx - by$$

Group the first pair of terms and the second pair of terms.

$$= a(x + y) - b(x + y)$$

Step 2: Factor out a from the first pair of terms.

Factor out $-b$ from the second pair of terms. (This causes sign changes within the second parentheses. The terms in parentheses now match.)

$$= (x + y)(a - b)$$

Step 3: Factor out the common binomial factor.

$$\underline{\text{Check:}} \ (x + y)(a - b) = x(a) + x(-b) + y(a) + y(-b)$$
$$= ax - bx + ay - by \ ✔$$

Skill Practice

Factor by grouping.

16. $3ab^2 + 6b^2 - 12ab - 24b$

Example 11   Factoring by Grouping

Factor by grouping. $16w^4 - 40w^3 - 12w^2 + 30w$

Solution:

$$16w^4 - 40w^3 - 12w^2 + 30w$$

Step 1: Identify and factor out the GCF from all four terms. In this case, the GCF is $2w$.

$$= 2w[8w^3 - 20w^2 - 6w + 15]$$

$$= 2w[8w^3 - 20w^2 \mid - 6w + 15]$$

Group the first pair of terms and the second pair of terms.

$$= 2w[4w^2(2w - 5) - 3(2w - 5)]$$

Step 2: Factor out $4w^2$ from the first pair of terms.

Factor out -3 from the second pair of terms. (This causes sign changes within the second parentheses. The terms in parentheses now match.)

$$= 2w[(2w - 5)(4w^2 - 3)]$$

Step 3: Factor out the common binomial factor.

$$= 2w(2w - 5)(4w^2 - 3)$$

Answers

15. $(u - v)(t - 2)$
16. $3b(a + 2)(b - 4)$

Section 6.1 Practice Exercises

Study Skills Exercises

1. The final exam is just around the corner. Your old tests and quizzes provide good material to study for the final exam. Use your old tests to make a list of the chapters on which you need to concentrate. Ask your professor for help if there are still concepts that you do not understand.

2. Define the key terms:

   a. **factoring**

   b. **greatest common factor (GCF)**

   c. **prime factorization**

   d. **prime polynomial**

Objective 1: Identifying the Greatest Common Factor

For Exercises 3–14, identify the greatest common factor. **(See Examples 1–3.)**

3. $28, 63$

4. $24, 40$

5. $42, 30, 60$

6. $20, 52, 32$

7. $3xy, 7y$

8. $10mn, 11n$

9. $12w^3z, 16w^2z$

10. $20cd, 15c^3d$

11. $8x^3y^4z^2, 12xy^5z^4, 6x^2y^8z^3$

12. $15r^2s^2t^5, 5r^3s^4t^3, 30r^4s^3t^2$

13. $7(x - y), 9(x - y)$

14. $(2a - b), 3(2a - b)$

Objective 2: Factoring out the Greatest Common Factor

15. a. Use the distributive property to multiply $3(x - 2y)$.

    b. Use the distributive property to factor $3x - 6y$.

16. a. Use the distributive property to multiply $a^2(5a + b)$.

    b. Use the distributive property to factor $5a^3 + a^2b$.

For Exercises 17–36, factor out the GCF. **(See Examples 4–5.)**

17. $4p + 12$

18. $3q - 15$

19. $5c^2 - 10c + 15$

20. $16d^3 + 24d^2 + 32d$

21. $x^5 + x^3$

22. $y^2 - y^3$

23. $t^4 - 4t + 8t^2$

24. $7r^3 - r^5 + r^4$

25. $2ab + 4a^3b$

26. $5u^3v^2 - 5uv$

27. $38x^2y - 19x^2y^4$

28. $100a^5b^3 + 16a^2b$

29. $6x^3y^5 - 18xy^9z$

30. $15mp^7q^4 + 12m^4q^3$

31. $5 + 7y^3$

32. $w^3 - 5u^3v^2$

33. $42p^3q^2 + 14pq^2 - 7p^4q^4$

34. $8m^2n^3 - 24m^2n^2 + 4m^3n$

35. $t^5 + 2rt^3 - 3t^4 + 4r^2t^2$

36. $u^2v + 5u^3v^2 - 2u^2 + 8uv$

Objective 3: Factoring out a Negative Factor

37. For the polynomial $-2x^3 - 4x^2 + 8x$

    a. Factor out $-2x$.   **b.** Factor out $2x$.

38. For the polynomial $-9y^5 + 3y^3 - 12y$

    a. Factor out $-3y$.   **b.** Factor out $3y$.

39. Factor out -1 from the polynomial $-8t^2 - 9t - 2$.

40. Factor out -1 from the polynomial $-6x^3 - 2x - 5$.

For Exercises 41–46, factor out the opposite of the greatest common factor. **(See Examples 6–7.)**

41. $-15p^3 - 30p^2$

42. $-24m^3 - 12m^4$

43. $-q^4 + 2q^2 - 9q$

44. $-r^3 + 9r^2 - 5r$

45. $-7x - 6y - 2z$

46. $-4a + 5b - c$

Objective 4: Factoring out a Binomial Factor

For Exercises 47–52, factor out the GCF. **(See Example 8.)**

47. $13(a + 6) - 4b(a + 6)$

48. $7(x^2 + 1) - y(x^2 + 1)$

49. $8v(w^2 - 2) + (w^2 - 2)$

50. $t(r + 2) + (r + 2)$

51. $21x(x + 3) + 7x^2(x + 3)$

52. $5y^3(y - 2) - 15y(y - 2)$

Objective 5: Factoring by Grouping

For Exercises 53–72, factor by grouping. **(See Examples 9–10.)**

53. $8a^2 - 4ab + 6ac - 3bc$

54. $4x^3 + 3x^2y + 4xy^2 + 3y^3$

55. $3q + 3p + qr + pr$

56. $xy - xz + 7y - 7z$

57. $6x^2 + 3x + 4x + 2$

58. $4y^2 + 8y + 7y + 14$

59. $2t^2 + 6t - 5t - 15$

60. $2p^2 - p - 6p + 3$

61. $6y^2 - 2y - 9y + 3$

62. $5a^2 + 30a - 2a - 12$

63. $b^4 + b^3 - 4b - 4$

64. $8w^5 + 12w^2 - 10w^3 - 15$

65. $3j^2k + 15k + j^2 + 5$

66. $2ab^2 - 6ac + b^2 - 3c$

67. $14w^6x^6 + 7w^6 - 2x^6 - 1$

68. $18p^4q - 9p^5 - 2q + p$

69. $ay + bx + by + ax$
    (*Hint*: Rearrange the terms.)

70. $2c + 3ay + ac + 6y$

71. $vw^2 - 3 + w - 3wv$

72. $2x^2 + 6m + 12 + x^2m$

Mixed Exercises

For Exercises 73–78, factor out the GCF first. Then factor by grouping. **(See Example 11.)**

73. $15x^4 + 15x^2y^2 + 10x^3y + 10xy^3$

74. $2a^3b - 4a^2b + 32ab - 64b$

75. $4abx - 4b^2x - 4ab + 4b^2$

76. $p^2q - pq^2 - rp^2q + rpq^2$

77. $6st^2 - 18st - 6t^4 + 18t^3$

78. $15j^3 - 10j^2k - 15j^2k^2 + 10jk^3$

79. The formula $P = 2l + 2w$ represents the perimeter, P, of a rectangle given the length, l, and the width, w. Factor out the GCF, and write an equivalent formula in factored form.

80. The formula $P = 2a + 2b$ represents the perimeter, P, of a parallelogram given the base, b, and an adjacent side, a. Factor out the GCF, and write an equivalent formula in factored form.

81. The formula $S = 2\pi r^2 + 2\pi rh$ represents the surface area, S, of a cylinder with radius, r, and height, h. Factor out the GCF, and write an equivalent formula in factored form.

82. The formula $A = P + Prt$ represents the total amount of money, A, in an account that earns simple interest at a rate, r, for t years. Factor out the GCF, and write an equivalent formula in factored form.

Expanding Your Skills

83. Factor out $\dfrac{1}{7}$ from $\dfrac{1}{7}x^2 + \dfrac{3}{7}x - \dfrac{5}{7}$.

84. Factor out $\dfrac{1}{5}$ from $\dfrac{6}{5}y^2 - \dfrac{4}{5}y + \dfrac{1}{5}$.

85. Factor out $\dfrac{1}{4}$ from $\dfrac{5}{4}w^2 + \dfrac{3}{4}w + \dfrac{9}{4}$.

86. Factor out $\dfrac{1}{6}$ from $\dfrac{1}{6}p^2 - \dfrac{3}{6}p + \dfrac{5}{6}$.

87. Write a polynomial that has a GCF of $3x$. (Answers may vary.)

88. Write a polynomial that has a GCF of $7y$. (Answers may vary.)

89. Write a polynomial that has a GCF of $4p^2q$. (Answers may vary.)

90. Write a polynomial that has a GCF of $2ab^2$. (Answers may vary.)

Factoring Trinomials of the Form $x^2 + bx + c$ | Section 6.2

1. Factoring Trinomials with a Leading Coefficient of 1

Objective

1. Factoring Trinomials with a Leading Coefficient of 1

In Section 5.6, we learned how to multiply two binomials. We also saw that such a product often results in a trinomial. For example,

$$\underset{\substack{\text{Product of}\\\text{first terms}}}{} \qquad \underset{\substack{\text{Product of}\\\text{last terms}}}{}$$

$$(x + 3)(x + 7) = x^2 + \underbrace{7x + 3x}_{\substack{\text{Sum of products of inner}\\\text{terms and outer terms}}} + 21 = x^2 + 10x + 21$$

In this section, we want to reverse the process. That is, given a trinomial, we want to *factor* it as a product of two binomials. In particular, we begin our study with the case in which a trinomial has a leading coefficient of 1.

Consider the quadratic trinomial $x^2 + bx + c$. To produce a leading term of x^2, we can construct binomials of the form $(x + \ )(x + \ )$. The remaining terms can be obtained from two integers, p and q, whose product is c and whose sum is b.

$$x^2 + bx + c = (x + p)(x + q) = x^2 + qx + px + pq$$
$$= x^2 + \underbrace{(q + p)}_{\text{Sum} = b}x + \underbrace{pq}_{\text{Product} = c}$$

This process is demonstrated in Example 1.

Skill Practice

Factor.

1. $x^2 - 5x - 14$

Example 1 Factoring a Trinomial of the Form $x^2 + bx + c$

Factor: $x^2 + 4x - 45$

Solution:

$x^2 + 4x - 45 = (x + \square)(x + \square)$ The product of the first terms in the binomials must equal the leading term of the trinomial $x \cdot x = x^2$.

We must fill in the blanks with two integers whose product is -45 and whose sum is 4. The factors must have opposite signs to produce a negative product. The possible factorizations of -45 are:

Product $= -45$	Sum
$-1 \cdot 45$	44
$-3 \cdot 15$	12
$-5 \cdot 9$	4
$-9 \cdot 5$	-4
$-15 \cdot 3$	-12
$-45 \cdot 1$	-44

$x^2 + 4x - 45 = (x + \square)(x + \square)$

$\qquad\qquad = (x + (-5))(x + 9)$ Fill in the blanks with -5 and 9,

$\qquad\qquad = (x - 5)(x + 9)$ Factored form

$\qquad\qquad$ Check:

$\qquad\qquad (x - 5)(x + 9) = x^2 + 9x - 5x - 45$

$\qquad\qquad\qquad\qquad\qquad = x^2 + 4x - 45 \checkmark$

One frequently asked question is whether the order of factors can be reversed. The answer is yes because multiplication of polynomials is a commutative operation. Therefore, in Example 1, we can express the factorization as $(x - 5)(x + 9)$ or as $(x + 9)(x - 5)$.

Skill Practice

Factor.

2. $z^2 - 16z + 48$

Example 2 Factoring a Trinomial of the Form $x^2 + bx + c$

Factor: $w^2 - 15w + 50$

Solution:

$w^2 - 15w + 50 = (w + \square)(w + \square)$ The product $w \cdot w = w^2$.

Find two integers whose product is 50 and whose sum is -15. To form a positive product, the factors must be either both positive or both negative. The sum must be negative, so we will choose negative factors of 50.

Answers

1. $(x - 7)(x + 2)$
2. $(z - 4)(z - 12)$

$$\underline{\text{Product} = 50} \qquad \underline{\text{Sum}}$$
$$(-1)(-50) \qquad -51$$
$$(-2)(-25) \qquad -27$$
$$(-5)(-10) \qquad -15$$

$$w^2 - 15w + 50 = (w + \square)(w + \square)$$
$$= (w + (-5))(w + (-10))$$
$$= (w - 5)(w - 10) \qquad \text{Factored form}$$

Check:

$$(w - 5)(w - 10) = w^2 - 10w - 5w + 50$$
$$= w^2 - 15w + 50 \checkmark$$

Practice will help you become proficient in factoring polynomials. As you do your homework, keep these important guidelines in mind:

- To factor a trinomial, write the trinomial in descending order such as $x^2 + bx + c$.
- For all factoring problems, always factor out the GCF from all terms first.

Furthermore, we offer the following rules for determining the signs within the binomial factors.

PROCEDURE **Sign Rules for Factoring Trinomials**

Given the trinomial $x^2 + bx + c$, the signs within the binomial factors are determined as follows:

Case 1 If c is *positive*, then the signs in the binomials must be the same (either both positive or both negative). The correct choice is determined by the middle term. If the middle term is positive, then both signs must be positive. If the middle term is negative, then both signs must be negative.

c is positive. c is positive.

$$x^2 + 6x + 8 \qquad\qquad x^2 - 6x + 8$$
$$(x + 2)(x + 4) \qquad\qquad (x - 2)(x - 4)$$

Same signs   Same signs

Case 2 If c is *negative*, then the signs in the binomials must be different.

c is negative. c is negative.

$$x^2 + 2x - 35 \qquad\qquad x^2 - 2x - 35$$
$$(x + 7)(x - 5) \qquad\qquad (x - 7)(x + 5)$$

Different signs   Different signs

Skill Practice

Factor.

3. $-5w + w^2 - 6$

4. $30y^3 + 2y^4 + 112y^2$

Example 3 Factoring Trinomials

Factor. **a.** $-8p - 48 + p^2$ **b.** $-40t - 30t^2 + 10t^3$

Solution:

a. $-8p - 48 + p^2$

$= p^2 - 8p - 48$ Write in descending order.

$= (p \quad \square)(p \quad \square)$ Find two integers whose product is -48 and whose sum is -8. The numbers are -12 and 4.

$= (p - 12)(p + 4)$ Factored form

b. $-40t - 30t^2 + 10t^3$

$= 10t^3 - 30t^2 - 40t$ Write in descending order.

$= 10t(t^2 - 3t - 4)$ Factor out the GCF.

$= 10t(t \quad \square)(t \quad \square)$ Find two integers whose product is -4 and whose sum is -3. The numbers are -4 and 1.

$= 10t(t - 4)(t + 1)$ Factored form

Skill Practice

Factor.

5. $-x^2 + x + 12$

6. $-3a^2 + 15ab - 12b^2$

Example 4 Factoring Trinomials

Factor. **a.** $-a^2 + 6a - 8$ **b.** $-2c^2 - 22cd - 60d^2$

Solution:

a. $-a^2 + 6a - 8$

$= -1(a^2 - 6a + 8)$ It is generally easier to factor a trinomial with a *positive* leading coefficient. Therefore, we will factor out -1 from all terms.

$= -1(a \quad \square)(a \quad \square)$ Find two integers whose product is 8 and whose sum is -6. The numbers are -4 and -2.

$= -1(a - 4)(a - 2)$

Avoiding Mistakes

Recall that factoring out -1 from a polynomial changes the signs of all terms within parentheses.

b. $-2c^2 - 22cd - 60d^2$

$= -2(c^2 + 11cd + 30d^2)$ Factor out -2.

$= -2(c \quad \square d)(c \quad \square d)$ Notice that the second pair of terms has a factor of d. This will produce a product of d^2.

$= -2(c + 5d)(c + 6d)$ Find two integers whose product is 30 and whose sum is 11. The numbers are 5 and 6.

To factor a trinomial of the form $x^2 + bx + c$, we must find two integers whose product is c and whose sum is b. If no such integers exist, then the trinomial is not factorable and is called a **prime polynomial**.

Answers

3. $(w - 6)(w + 1)$
4. $2y^2(y + 8)(y + 7)$
5. $-(x - 4)(x + 3)$
6. $-3(a - b)(a - 4b)$

Example 5	Factoring Trinomials

Factor: $x^2 - 13x + 8$

Solution:

$x^2 - 13x + 8$ The trinomial is in descending order. The GCF is 1.

$= (x \quad \square)(x \quad \square)$ Find two integers whose product is 8 and whose sum is -13. No such integers exist.

The trinomial $x^2 - 13x + 8$ is prime.

Skill Practice

Factor.

7. $x^2 - 7x + 28$

Answer

7. Prime

Section 6.2   Practice Exercises

Boost your GRADE at ALEKS.com!

ALEKS version 3.0

- Practice Problems
- Self-Tests
- NetTutor
- e-Professors
- Videos

Study Skills Exercises

1. Sometimes the problems on a test do not appear in the same order as the concepts appear in the text. In order to better prepare for a test, try to practice with problems taken from the book but placed in random order. Choose 30 problems from various chapters, randomize the order, and use the problems to review for the test. Repeat the process several times for additional practice.

2. Define the key term **prime polynomial**.

Review Exercises

For Exercises 3–6, factor completely.

3. $4x^3y^7 - 12x^4y^5 + 8xy^8$

4. $9a^6b^3 - 27a^3b^6 - 3a^2b^2$

5. $ax + 2bx - 5a - 10b$

6. $m^2 - mx - 3pm + 3px$

Objective 1: Factoring Trinomials with a Leading Coefficient of 1

For Exercises 7–20, factor completely. **(See Examples 1, 2, and 5.)**

7. $x^2 + 10x + 16$

8. $y^2 + 18y + 80$

9. $z^2 - 11z + 18$

10. $w^2 - 7w + 12$

11. $z^2 - 3z - 18$

12. $w^2 + 4w - 12$

13. $p^2 - 3p - 40$

14. $a^2 - 10a + 9$

 15. $t^2 + 6t - 40$

16. $m^2 - 12m + 11$

17. $x^2 - 3x + 20$

18. $y^2 + 6y + 18$

19. $n^2 + 8n + 16$

20. $v^2 + 10v + 25$

For Exercises 21–24, assume that b and c represent positive integers.

21. When factoring a polynomial of the form $x^2 + bx + c$, pick an appropriate combination of signs.

    a. $(\quad+\quad)(\quad+\quad)$ **b.** $(\quad-\quad)(\quad-\quad)$ **c.** $(\quad+\quad)(\quad-\quad)$

22. When factoring a polynomial of the form $x^2 + bx - c$, pick an appropriate combination of signs.

    a. $(\quad+\quad)(\quad+\quad)$ **b.** $(\quad-\quad)(\quad-\quad)$ **c.** $(\quad+\quad)(\quad-\quad)$

23. When factoring a polynomial of the form $x^2 - bx - c$, pick an appropriate combination of signs.

    a. $(\quad+\quad)(\quad+\quad)$ **b.** $(\quad-\quad)(\quad-\quad)$ **c.** $(\quad+\quad)(\quad-\quad)$

24. When factoring a polynomial of the form $x^2 - bx + c$, pick an appropriate combination of signs.

    a. $(\quad+\quad)(\quad+\quad)$ **b.** $(\quad-\quad)(\quad-\quad)$ **c.** $(\quad+\quad)(\quad-\quad)$

25. Which is the correct factorization of $y^2 - y - 12$? Explain. $(y - 4)(y + 3)$ or $(y + 3)(y - 4)$

26. Which is the correct factorization of $x^2 + 14x + 13$? Explain. $(x + 13)(x + 1)$ or $(x + 1)(x + 13)$

27. Which is the correct factorization of $w^2 + 2w + 1$? Explain. $(w + 1)(w + 1)$ or $(w + 1)^2$

28. Which is the correct factorization of $z^2 - 4z + 4$? Explain. $(z - 2)(z - 2)$ or $(z - 2)^2$

29. In what order should a trinomial be written before attempting to factor it?

30. Referring to page 439, write two important guidelines to follow when factoring trinomials.

For Exercises 31–48, factor completely. Be sure to factor out the GCF when necessary. **(See Examples 3–4.)**

31. $-13x + x^2 - 30$ **32.** $12y - 160 + y^2$ **33.** $-18w + 65 + w^2$

34. $17t + t^2 + 72$ **35.** $22t + t^2 + 72$ **36.** $10q - 1200 + q^2$

37. $3x^2 - 30x - 72$ **38.** $2z^2 + 4z - 198$ **39.** $8p^3 - 40p^2 + 32p$

40. $5w^4 - 35w^3 + 50w^2$ **41.** $y^4z^2 - 12y^3z^2 + 36y^2z^2$ **42.** $t^4u^2 + 6t^3u^2 + 9t^2u^2$

43. $-x^2 + 10x - 24$ **44.** $-y^2 - 12y - 35$ **45.** $-m^2 + m + 6$

46. $-n^2 + 5n + 6$ **47.** $-4 - 2c^2 - 6c$ **48.** $-40d - 30 - 10d^2$

Mixed Exercises

For Exercises 49–66, factor completely.

49. $x^3y^3 - 19x^2y^3 + 60xy^3$ **50.** $y^2z^5 + 17yz^5 + 60z^5$ **51.** $12p^2 - 96p + 84$

52. $5w^2 - 40w - 45$ **53.** $-2m^2 + 22m - 20$ **54.** $-3x^2 - 36x - 81$

55. $c^2 + 6cd + 5d^2$ **56.** $x^2 + 8xy + 12y^2$ **57.** $a^2 - 9ab + 14b^2$

58. $m^2 - 15mn + 44n^2$ **59.** $a^2 + 4a + 18$ **60.** $b^2 - 6a + 15$

61. $2q + q^2 - 63$ **62.** $-32 - 4t + t^2$ **63.** $x^2 + 20x + 100$

64. $z^2 - 24z + 144$ **65.** $t^2 + 18t - 40$ **66.** $d^2 + 2d - 99$

67. A student factored a trinomial as $(2x - 4)(x - 3)$. The instructor did not give full credit. Why?

68. A student factored a trinomial as $(y + 2)(5y - 15)$. The instructor did not give full credit. Why?

69. What polynomial factors as $(x - 4)(x + 13)$?

70. What polynomial factors as $(q - 7)(q + 10)$?

Expanding Your Skills

For Exercises 71–74, factor completely.

71. $x^4 + 10x^2 + 9$ **72.** $y^4 + 4y^2 - 21$ **73.** $w^4 + 2w^2 - 15$ **74.** $p^4 - 13p^2 + 40$

75. Find all integers, b, that make the trinomial $x^2 + bx + 6$ factorable.

76. Find all integers, b, that make the trinomial $x^2 + bx + 10$ factorable.

77. Find a value of c that makes the trinomial $x^2 + 6x + c$ factorable.

78. Find a value of c that makes the trinomial $x^2 + 8x + c$ factorable.

Factoring Trinomials: Trial-and-Error Method   Section 6.3

In Section 6.2, we learned how to factor trinomials of the form $x^2 + bx + c$. These trinomials have a leading coefficient of 1. In this section and the next, we will consider the more general case in which the leading coefficient may be *any* integer. That is, we will factor quadratic trinomials of the form $ax^2 + bx + c$ (where $a \neq 0$). The method presented in this section is called the trial-and-error method.

Objective

1. **Factoring Trinomials by the Trial-and-Error Method**

1. Factoring Trinomials by the Trial-and-Error Method

To understand the basis of factoring trinomials of the form $ax^2 + bx + c$, first consider the multiplication of two binomials:

$$(2x + 3)(1x + 2) = 2x^2 + \underbrace{4x + 3x}_{\text{Sum of products of inner terms and outer terms}} + 6 = 2x^2 + 7x + 6$$

To factor the trinomial, $2x^2 + 7x + 6$, this operation is reversed.

$$2x^2 + 7x + 6 = (\square x \quad \square)(\square x \quad \square)$$

We need to fill in the blanks so that the product of the first terms in the binomials is $2x^2$ and the product of the last terms in the binomials is 6. Furthermore, the factors of $2x^2$ and 6 must be chosen so that the sum of the products of the inner terms and outer terms equals $7x$.

To produce the product $2x^2$, we might try the factors $2x$ and x within the binomials:

$$(2x \quad \square)(x \quad \square)$$

To produce a product of 6, the remaining terms in the binomials must either both be positive or both be negative. To produce a positive middle term, we will try positive factors of 6 in the remaining blanks until the correct product is found. The possibilities are $1 \cdot 6, 2 \cdot 3, 3 \cdot 2,$ and $6 \cdot 1$.

$$(2x + 1)(x + 6) = 2x^2 + 12x + 1x + 6 = 2x^2 + 13x + 6 \qquad \text{Wrong middle term}$$

$$(2x + 2)(x + 3) = 2x^2 + 6x + 2x + 6 = 2x^2 + 8x + 6 \qquad \text{Wrong middle term}$$

$$(2x + 3)(x + 2) = 2x^2 + 4x + 3x + 6 = 2x^2 + 7x + 6 \qquad \text{Correct!}$$

$$(2x + 6)(x + 1) = 2x^2 + 2x + 6x + 6 = 2x^2 + 8x + 6 \qquad \text{Wrong middle term}$$

The correct factorization of $2x^2 + 7x + 6$ is $(2x + 3)(x + 2)$. ✔

As this example shows, we factor a trinomial of the form $ax^2 + bx + c$ by shuffling the factors of a and c within the binomials until the correct product is obtained. However, sometimes it is not necessary to test all the possible combinations of factors. In the previous example, the GCF of the original trinomial is 1. Therefore, any binomial factor whose terms share a common factor *greater than 1* does not need to be considered. In this case, the possibilities $(2x + 2)(x + 3)$ and $(2x + 6)(x + 1)$ cannot work.

$$(2x + 2)(x + 3) \qquad (2x + 6)(x + 1)$$

Common factor of 2   Common factor of 2

The steps to factor a trinomial by the trial-and-error method are outlined in the following box.

PROCEDURE   Trial-and-Error Method to Factor $ax^2 + bx + c$

Step 1 Factor out the GCF.

Step 2 List all pairs of positive factors of a and pairs of positive factors of c. Consider the reverse order for one of the lists of factors.

Step 3 Construct two binomials of the form:

Factors of a

$$(\square x \quad \square)(\square x \quad \square)$$

Factors of c

Step 4 Test each combination of factors and signs until the correct product is found.

Step 5 If no combination of factors produces the correct product, the trinomial cannot be factored further and is a **prime polynomial**.

Before we begin Example 1, keep these two important guidelines in mind:

- For any factoring problem you encounter, always factor out the GCF from all terms first.
- To factor a trinomial, write the trinomial in the form $ax^2 + bx + c$.

Example 1 **Factoring a Trinomial by the Trial-and-Error Method**

Factor the trinomial by the trial-and-error method: $10x^2 - 9x - 1$

Skill Practice

Factor using the trial-and-error method.

1. $3b^2 + 8b + 4$

Solution:

$10x^2 - 9x - 1$ **Step 1:** Factor out the GCF from all terms. In this case, the GCF is 1.

                                     The trinomial is written in the form $ax^2 + bx + c$.

To factor $10x^2 - 9x - 1$, two binomials must be constructed in the form:

$(\Box x \;\; \Box)(\Box x \;\; \Box)$ **Step 2:** To produce the product $10x^2$, we might try $5x$ and $2x$, or $10x$ and $1x$. To produce a product of -1, we will try the factors $(1)(-1)$ and $(-1)(1)$.

                       Step 3: Construct all possible binomial factors using different combinations of the factors of $10x^2$ and -1.

$(5x + 1)(2x - 1) = 10x^2 - 5x + 2x - 1 = 10x^2 - 3x - 1$ Wrong middle term

$(5x - 1)(2x + 1) = 10x^2 + 5x - 2x - 1 = 10x^2 + 3x - 1$ Wrong middle term

Because the numbers 1 and -1 did not produce the correct trinomial when coupled with $5x$ and $2x$, try using $10x$ and $1x$.

$(10x - 1)(1x + 1) = 10x^2 + 10x - 1x - 1 = 10x^2 + 9x - 1$ Wrong middle term

$(10x + 1)(1x - 1) = 10x^2 - 10x + 1x - 1 = 10x^2 - 9x - 1$ Correct!

Therefore, $10x^2 - 9x - 1 = (10x + 1)(x - 1)$.

In Example 1, the factors of -1 must have opposite signs to produce a negative product. Therefore, one binomial factor is a sum and one is a difference. Determining the correct signs is an important aspect of factoring trinomials. We suggest the following guidelines:

Answer

1. $(3b + 2)(b + 2)$

PROCEDURE Sign Rules for the Trial-and-Error Method

Given the trinomial $ax^2 + bx + c, (a > 0)$, the signs can be determined as follows:

- If *c is positive*, then the signs in the binomials must be the same (either both positive or both negative). The correct choice is determined by the middle term. If the middle term is positive, then both signs must be positive. If the middle term is negative, then both signs must be negative.

$$\overset{c \text{ is positive}}{\underset{\downarrow}{}}$$

$$20x^2 + 43x + 21 \qquad\qquad 20x^2 - 43x + 21$$

$$\underbrace{(4x + 3)(5x + 7)}_{\text{Same signs}} \qquad \underbrace{(4x - 3)(5x - 7)}_{\text{Same signs}}$$

- If *c is negative*, then the signs in the binomial must be different. The middle term in the trinomial determines which factor gets the positive sign and which gets the negative sign.

$$\overset{c \text{ is negative}}{\underset{\downarrow}{}}$$

$$x^2 + 3x - 28 \qquad\qquad x^2 - 3x - 28$$

$$\underbrace{(x + 7)(x - 4)}_{\text{Different signs}} \qquad \underbrace{(x - 7)(x + 4)}_{\text{Different signs}}$$

Skill Practice

Factor.

2. $-25w + 6w^2 + 4$

Answer

2. $(6w - 1)(w - 4)$

Example 2 Factoring a Trinomial

Factor the trinomial: $13y - 6 + 8y^2$

Solution:

$13y - 6 + 8y^2$

$= 8y^2 + 13y - 6$	Write the polynomial in descending order.
$(\square y \;\; \square)(\square y \;\; \square)$	**Step 1:** The GCF is 1.

Factors of 8	Factors of 6	
$1 \cdot 8$	$1 \cdot 6$	**Step 2:** List the positive factors of 8 and positive factors of 6. Consider the reverse order in one list of factors.
$2 \cdot 4$	$2 \cdot 3$	
	$\left.\begin{array}{c}3 \cdot 2 \\ 6 \cdot 1\end{array}\right\}$ (reverse order)	

Step 3: Construct all possible binomial factors using different combinations of the factors of 8 and 6.

$$\left.\begin{array}{l}(2y \;\; 1)(4y \;\; 6) \\ (2y \;\; 2)(4y \;\; 3) \\ (2y \;\; 3)(4y \;\; 2) \\ (2y \;\; 6)(4y \;\; 1) \\ (1y \;\; 1)(8y \;\; 6) \\ (1y \;\; 3)(8y \;\; 2)\end{array}\right\}$$

Without regard to signs, these factorizations cannot work because the terms in the binomials share a common factor greater than 1.

Test the remaining factorizations. Keep in mind that to produce a product of -6, the signs within the parentheses must be opposite (one positive and one negative). Also, the sum of the products of the inner terms and outer terms must be combined to form $13y$.

$(1y \quad 6)(8y \quad 1)$ *Incorrect.*   Wrong middle term.

Regardless of the signs, the product of inner terms, $48y$, and the product of outer terms, $1y$, cannot be combined to form the middle term $13y$.

$(1y \quad 2)(8y \quad 3)$ *Correct.*   The terms $16y$ and $3y$ can be combined to form the middle term $13y$, provided the signs are applied correctly. We require $+16y$ and $-3y$.

The correct factorization of $8y^2 + 13y - 6$ is $(y + 2)(8y - 3)$.

Remember that the first step in any factoring problem is to remove the GCF. By removing the GCF, the remaining terms of the trinomial will be simpler and may have smaller coefficients.

Example 3   Factoring a Trinomial by the Trial-and-Error Method

Factor the trinomial by the trial-and-error method: $40x^3 - 104x^2 + 10x$

Skill Practice

Factor.

3. $8t^3 + 38t^2 + 24t$

Solution:

$40x^3 - 104x^2 + 10x$

$= 2x(20x^2 - 52x + 5)$ **Step 1:** The GCF is $2x$.

$= 2x(\Box x \quad \Box)(\Box x \quad \Box)$ **Step 2:** List the factors of 20 and factors of 5. Consider the reverse order in one list of factors.

Factors of 20	Factors of 5
$1 \cdot 20$	$1 \cdot 5$
$2 \cdot 10$	$5 \cdot 1$
$4 \cdot 5$	

Step 3: Construct all possible binomial factors using different combinations of the factors of 20 and factors of 5. The signs in the parentheses must both be negative.

TIP: Notice that when the GCF, $2x$, is removed from the original trinomial, the new trinomial has smaller coefficients. This makes the factoring process simpler. It is easier to list the factors of 20 and 5 than the factors of 40 and 10.

$= 2x(1x - 1)(20x - 5)$
$= 2x(2x - 1)(10x - 5)$ *Incorrect.*   Once the GCF has been removed
$= 2x(4x - 1)(5x - 5)$ from the original polynomial, the binomial factors cannot contain a GCF greater than 1.

$= 2x(1x - 5)(20x - 1)$ *Incorrect.*   Wrong middle term.

$2x(x - 5)(20x - 1)$
$= 2x(20x^2 - 1x - 100x + 5)$
$= 2x(20x^2 - 101x + 5)$

$= 2x(4x - 5)(5x - 1)$ *Incorrect.*   Wrong middle term.

$2x(4x - 5)(5x - 1)$
$= 2x(20x^2 - 4x - 25x + 5)$
$= 2x(20x^2 - 29x + 5)$

$= 2x(2x - 5)(10x - 1)$ *Correct.* $2x(2x - 5)(10x - 1)$
$= 2x(20x^2 - 2x - 50x + 5)$
$= 2x(20x^2 - 52x + 5)$
$= 40x^3 - 104x^2 + 10x$

The correct factorization is $2x(2x - 5)(10x - 1)$.

Answer

3. $2t(4t + 3)(t + 4)$

Often it is easier to factor a trinomial when the leading coefficient is positive. If the leading coefficient is negative, consider factoring out the opposite of the GCF.

Example 4   Factoring a Trinomial by the Trial-and-Error Method

Factor: $-45x^2 - 3xy + 18y^2$

Solution:

$-45x^2 - 3xy + 18y^2$

$= -3(15x^2 + xy - 6y^2)$ **Step 1:** Factor out -3 to make the leading coefficient positive.

$= -3(\Box x \quad \Box y)(\Box x \quad \Box y)$ **Step 2:** List the factors of 15 and 6.

Factors of 15	Factors of 6
$1 \cdot 15$	$1 \cdot 6$
$3 \cdot 5$	$2 \cdot 3$
	$3 \cdot 2$
	$6 \cdot 1$

Step 3: We will construct all binomial combinations, without regard to signs first.

$\left.\begin{array}{l} -3(x \quad y)(15x \quad 6y) \\ -3(x \quad 2y)(15x \quad 3y) \\ -3(3x \quad 3y)(5x \quad 2y) \\ -3(3x \quad 6y)(5x \quad y) \end{array}\right\}$ *Incorrect.*   The binomials contain a common factor.

Test the remaining factorizations. The signs within parentheses must be opposite to produce a product of $-6y^2$. Also, the sum of the products of the inner terms and outer terms must be combined to form $1xy$.

$-3(x \quad 3y)(15x \quad 2y)$ *Incorrect.*   Regardless of signs, $45xy$ and $2xy$ cannot be combined to equal xy.

$-3(x \quad 6y)(15x \quad y)$ *Incorrect.*   Regardless of signs, $90xy$ and xy cannot be combined to equal xy.

$-3(3x \quad y)(5x \quad 6y)$ *Incorrect.*   Regardless of signs, $5xy$ and $18xy$ cannot be combined to equal xy.

$-3(3x \quad 2y)(5x \quad 3y)$ *Correct.*   The terms $10xy$ and $9xy$ can be combined to form xy provided that the signs are applied correctly. We require $10xy$ and $-9xy$.

$-3(3x + 2y)(5x - 3y)$ Factored form

Recall that a prime polynomial is a polynomial whose only factors are itself and 1. Not every trinomial is factorable by the methods presented in this text.

Example 5 **Factoring a Trinomial by the Trial-and-Error Method**

Factor the trinomial by the trial-and-error method: $2p^2 - 8p + 3$

Skill Practice

Factor.

5. $3a^2 + a + 4$

Solution:

$2p^2 - 8p + 3$ **Step 1:** The GCF is 1.

$= (1p \ \square)(2p \ \square)$ **Step 2:** List the factors of 2 and the factors of 3.

Factors of 2	Factors of 3
$1 \cdot 2$	$1 \cdot 3$
	$3 \cdot 1$

Step 3: Construct all possible binomial factors using different combinations of the factors of 2 and 3. Because the third term in the trinomial is positive, both signs in the binomial must be the same. Because the middle term coefficient is negative, both signs will be negative.

$(p - 1)(2p - 3) = 2p^2 - 3p - 2p + 3$

$= 2p^2 - 5p + 3$ *Incorrect.* Wrong middle term.

$(p - 3)(2p - 1) = 2p^2 - p - 6p + 3$

$= 2p^2 - 7p + 3$ *Incorrect.* Wrong middle term.

None of the combinations of factors results in the correct product. Therefore, the polynomial $2p^2 - 8p + 3$ is prime and cannot be factored further.

In Example 6, we use the trial-and-error method to factor a higher degree trinomial into two binomial factors.

Example 6 **Factoring a Higher Degree Trinomial**

Factor the trinomial: $3x^4 + 8x^2 + 5$

Skill Practice

Factor.

6. $2y^4 - y^2 - 15$

Solution:

$3x^4 + 8x^2 + 5$ **Step 1:** The GCF is 1.

$= (\square x^2 + \square)(\square x^2 + \square)$ **Step 2:** To produce the product $3x^4$, we must use $3x^2$, and $1x^2$. To produce a product of 5, we will try the factors $(1)(5)$ and $(5)(1)$.

Step 3: Construct all possible binomial factors using the combinations of factors of $3x^4$ and 5.

$(3x^2 + 1)(x^2 + 5) = 3x^4 + 15x^2 + 1x^2 + 5 = 3x^4 + 16x^2 + 5$ Wrong middle term

$(3x^2 + 5)(x^2 + 1) = 3x^4 + 3x^2 + 5x^2 + 5 = 3x^4 + 8x^2 + 5$ Correct!

Therefore, $3x^4 + 8x^2 + 5 = (3x^2 + 5)(x^2 + 1)$

Answers

5. Prime   **6.** $(y^2 - 3)(2y^2 + 5)$

Section 6.3 Practice Exercises

Study Skills Exercises

1. In addition to studying the material for a test, here are some other activities that people use when preparing for a test. Circle the importance of each statement.

	not important	somewhat important	very important
a. Get a good night's sleep the night before the test.	1	2	3
b. Eat a good meal before the test.	1	2	3
c. Wear comfortable clothes on the day of the test.	1	2	3
d. Arrive early to class on the day of the test.	1	2	3

2. Define the key term **prime polynomial**.

Review Exercises

For Exercises 3–8, factor completely.

3. $21a^2b^2 + 12ab^2 - 15a^2b$ **4.** $5uv^2 - 10u^2v + 25u^2v^2$ **5.** $mn - m - 2n + 2$

6. $5x - 10 - xy + 2y$ **7.** $6a^2 - 30a - 84$ **8.** $10b^2 + 20b - 240$

Objective 1: Factoring Trinomials by the Trial-and-Error Method

For Exercises 9–12, assume a, b, and c represent positive integers.

9. When factoring a polynomial of the form $ax^2 + bx + c$, pick an appropriate combination of signs.

   a. $(\ +\ )(\ +\ )$

   b. $(\ -\ )(\ -\ )$

   c. $(\ +\ )(\ -\ )$

10. When factoring a polynomial of the form $ax^2 - bx - c$, pick an appropriate combination of signs.

   a. $(\ +\ )(\ +\ )$

   b. $(\ -\ )(\ -\ )$

   c. $(\ +\ )(\ -\ )$

11. When factoring a polynomial of the form $ax^2 - bx + c$, pick an appropriate combination of signs.

   a. $(\ +\ )(\ +\ )$

   b. $(\ -\ )(\ -\ )$

   c. $(\ +\ )(\ -\ )$

12. When factoring a polynomial of the form $ax^2 + bx - c$, pick an appropriate combination of signs.

   a. $(\ +\ )(\ +\ )$

   b. $(\ -\ )(\ -\ )$

   c. $(\ +\ )(\ -\ )$

For Exercises 13–30, factor completely by using the trial-and-error method. **(See Examples 1, 2, and 5.)**

13. $2y^2 - 3y - 2$ **14.** $2w^2 + 5w - 3$ **15.** $3n^2 + 13n + 4$

16. $2a^2 + 7a + 6$ **17.** $5x^2 - 14x - 3$ **18.** $7y^2 + 9y - 10$

19. $12c^2 - 5c - 2$ **20.** $6z^2 + z - 12$ **21.** $-12 + 10w^2 + 37w$

22. $-10 + 10p^2 + 21p$ **23.** $-5q - 6 + 6q^2$ **24.** $17a - 2 + 3a^2$

25. $6b - 23 + 4b^2$ **26.** $8 + 7x^2 - 18x$ **27.** $-8 + 25m^2 - 10m$

28. $8q^2 + 31q - 4$ **29.** $6y^2 + 19xy - 20x^2$ **30.** $12y^2 - 73yz + 6z^2$

For Exercises 31–38, factor completely. Be sure to factor out the GCF first. **(See Examples 3–4.)**

31. $2m^2 - 12m - 80$ **32.** $3c^2 - 33c + 72$ **33.** $2y^5 + 13y^4 + 6y^3$

34. $3u^8 - 13u^7 + 4u^6$ **35.** $-a^2 - 15a + 34$ **36.** $-x^2 - 7x - 10$

37. $80m^2 - 100mp - 30p^2$ **38.** $60w^2 + 550wz - 500z^2$

For Exercises 39–44, factor the higher degree polynomial. **(See Example 6.)**

39. $x^4 + 10x^2 + 9$ **40.** $y^4 + 4y^2 - 21$ **41.** $w^4 + 2w^2 - 15$

42. $p^4 - 13p^2 + 40$ **43.** $2x^4 - 7x^2 - 15$ **44.** $5y^4 + 11y^2 + 2$

Mixed Exercises

For Exercises 45–92, factor the trinomial completely.

45. $20z - 18 - 2z^2$ **46.** $25t - 5t^2 - 30$ **47.** $42 - 13q + q^2$ **48.** $-5w - 24 + w^2$

49. $6t^2 + 7t - 3$ **50.** $4p^2 - 9p + 2$ **51.** $4m^2 - 20m + 25$ **52.** $16r^2 + 24r + 9$

53. $5c^2 - c + 2$ **54.** $7s^2 + 2s + 9$ **55.** $6x^2 - 19xy + 10y^2$ **56.** $15p^2 + pq - 2q^2$

57. $12m^2 + 11mn - 5n^2$ **58.** $4a^2 + 5ab - 6b^2$ **59.** $30r^2 + 5r - 10$ **60.** $36x^2 - 18x - 4$

61. $4s^2 - 8st + t^2$ **62.** $6u^2 - 10uv + 5v^2$ **63.** $10t^2 - 23t - 5$ **64.** $16n^2 + 14n + 3$

65. $14w^2 + 13w - 12$ **66.** $12x^2 - 16x + 5$ **67.** $x^2 + 7x - 18$ **68.** $y^2 - 6y - 40$

69. $a^2 - 10a - 24$ **70.** $b^2 + 6b - 7$ **71.** $r^2 + 5r - 24$ **72.** $t^2 + 20t + 100$

73. $x^2 + 9xy + 20y^2$ **74.** $p^2 - 13pq + 36q^2$ **75.** $v^2 + 2v + 15$ **76.** $x^2 - x - 1$

77. $a^2 + 21ab + 20b^2$ **78.** $x^2 - 17xy - 18y^2$ **79.** $t^2 - 10t + 21$ **80.** $z^2 - 15z + 36$

81. $5d^3 + 3d^2 - 10d$ **82.** $3y^3 - y^2 + 12y$ **83.** $4b^3 - 4b^2 - 80b$ **84.** $2w^2 + 20w + 42$

85. $x^2y^2 - 13xy^2 + 30y^2$ **86.** $p^2q^2 - 14pq^2 + 33q^2$ **87.** $-12u^3 - 22u^2 + 20u$ **88.** $-18z^4 + 15z^3 + 12z^2$

89. $8x^4 + 14x^2 + 3$ **90.** $6y^4 - 5y^2 - 4$ **91.** $10z^4 + 9z^2 - 9$ **92.** $6p^4 + 17p^2 + 10$

Expanding Your Skills

For Exercises 93–96, each pair of trinomials looks similar but differs by one sign. Factor each trinomial, and see how their factored forms differ.

93. **a.** $x^2 - 10x - 24$

    b. $x^2 - 10x + 24$

94. **a.** $x^2 - 13x - 30$

    b. $x^2 - 13x + 30$

95. **a.** $x^2 - 5x - 6$

    b. $x^2 - 5x + 6$

96. **a.** $x^2 - 10x + 9$

    b. $x^2 + 10x + 9$

Section 6.4   Factoring Trinomials: AC-Method

Objective

1. **Factoring Trinomials by the AC-Method**

In Section 6.2, we factored trinomials with a leading coefficient of 1. In Section 6.3, we learned the trial-and-error method to factor the more general case in which the leading coefficient is any integer. In this section, we provide an alternative method to factor trinomials, called the ac-method.

1. Factoring Trinomials by the AC-Method

The product of two binomials results in a four-term expression that can sometimes be simplified to a trinomial. To factor the trinomial, we want to reverse the process.

Multiply:

> Multiply the binomials.   Add the middle terms.

$$(2x + 3)(x + 2) = \longrightarrow 2x^2 + 4x + 3x + 6 = \longrightarrow 2x^2 + 7x + 6$$

Factor:

$$2x^2 + 7x + 6 = \longrightarrow 2x^2 + 4x + 3x + 6 = \longrightarrow (2x + 3)(x + 2)$$

> Rewrite the middle term as   Factor by grouping.
> a sum or difference of terms.

To factor a quadratic trinomial, $ax^2 + bx + c$, by the ac-method, we rewrite the middle term, bx, as a sum or difference of terms. The goal is to produce a four-term polynomial that can be factored by grouping. The process is outlined as follows.

> **PROCEDURE** **AC-Method: Factoring $ax^2 + bx + c$ ($a \neq 0$)**
>
> **Step 1** Factor out the GCF from all terms.
>
> **Step 2** Multiply the coefficients of the first and last terms (ac).
>
> **Step 3** Find two integers whose product is ac and whose sum is b. (If no pair of integers can be found, then the trinomial cannot be factored further and is a **prime polynomial**.)
>
> **Step 4** Rewrite the middle term, bx, as the sum of two terms whose coefficients are the integers found in step 3.
>
> **Step 5** Factor the polynomial by grouping.

The ac-method for factoring trinomials is illustrated in Example 1. However, before we begin, keep these two important guidelines in mind:

- For any factoring problem you encounter, always factor out the GCF from all terms first.
- To factor a trinomial, write the trinomial in the form $ax^2 + bx + c$.

Example 1   Factoring a Trinomial by the AC-Method

Factor the trinomial by the ac-method: $2x^2 + 7x + 6$

Skill Practice

Factor by the ac-method.
1. $2x^2 + 5x + 3$

Solution:

$2x^2 + 7x + 6$ **Step 1:** Factor out the GCF from all terms. In this case, the GCF is 1.

$2x^2 + 7x + 6$ **Step 2:** The trinomial is written in the form $ax^2 + bx + c$.

$a = 2, b = 7, c = 6$ Find the product $ac = (2)(6) = 12$.

$$\begin{array}{ll} \underline{12} & \underline{12} \\ 1 \cdot 12 & (-1)(-12) \\ 2 \cdot 6 & (-2)(-6) \\ 3 \cdot 4 & (-3)(-4) \end{array}$$

Step 3: List all factors of ac and search for the pair whose sum equals the value of b. That is, list the factors of 12 and find the pair whose sum equals 7.

The numbers 3 and 4 satisfy both conditions: $3 \cdot 4 = 12$ and $3 + 4 = 7$.

$2x^2 + 7x + 6$

$= 2x^2 + 3x + 4x + 6$ **Step 4:** Write the middle term of the trinomial as the sum of two terms whose coefficients are the selected pair of numbers: 3 and 4.

$= 2x^2 + 3x \mid + 4x + 6$ **Step 5:** Factor by grouping.

$= x(2x + 3) + 2(2x + 3)$

$= (2x + 3)(x + 2)$

Check: $(2x + 3)(x + 2) = 2x^2 + 4x + 3x + 6$

$\qquad\qquad\qquad = 2x^2 + 7x + 6$ ✔

TIP: One frequently asked question is whether the order matters when we rewrite the middle term of the trinomial as two terms (step 3). The answer is no. From the previous example, the two middle terms in step 3 could have been reversed to obtain the same result:

$$2x^2 + 7x + 6$$
$$= 2x^2 + 4x + 3x + 6$$
$$= 2x(x + 2) + 3(x + 2)$$
$$= (x + 2)(2x + 3)$$

This example also points out that the order in which two factors are written does not matter. The expression $(x + 2)(2x + 3)$ is equivalent to $(2x + 3)(x + 2)$ because multiplication is a commutative operation.

Answer

1. $(x + 1)(2x + 3)$

Example 5   Factoring a Trinomial by the AC-Method

Factor the trinomial by the ac-method: $2p^2 - 8p + 3$

Solution:

$2p^2 - 8p + 3$ **Step 1:** The GCF is 1.

                      Step 2: The product $ac = 6$.

6	6
$1 \cdot 6$	$(-1)(-6)$
$2 \cdot 3$	$(-2)(-3)$

Step 3: List the factors of 6. Notice that no pair of factors has a sum of -8. Therefore, the trinomial cannot be factored.

The trinomial $2p^2 - 8p + 3$ is a prime polynomial.

In Example 6, we use the ac-method to factor a higher degree trinomial.

Example 6   Factoring a Higher Degree Trinomial

Factor the trinomial. $2x^4 + 5x^2 + 2$

Solution:

$2x^4 + 5x^2 + 2$ **Step 1:** The GCF is 1.

$a = 2, b = 5, c = 2$ **Step 2:** Find the product $ac = (2)(2) = 4$.

                                   Step 3: The numbers 1 and 4 have a product of 4 and a sum of 5.

$2x^4 + x^2 + 4x^2 + 2$ **Step 4:** Rewrite the middle term, $5x^2$, as $x^2 + 4x^2$.

$2x^4 + x^2 + 4x^2 + 2$ **Step 5:** Factor by grouping.

$x^2(2x^2 + 1) + 2(2x^2 + 1)$

$(2x^2 + 1)(x^2 + 2)$ Factored form.

Answers

5. Prime

6. $(3y^2 - 4)(y^2 + 2)$

Section 6.4   Practice Exercises

Boost *your* GRADE at
ALEKS.com!

ALEKS®
version 3.0

• Practice Problems   • e-Professors
• Self-Tests   • Videos
• NetTutor

Study Skills Exercise

1. Define the key term **prime polynomial**.

Review Exercises

For Exercises 2–4, factor completely.

2. $5x(x - 2) - 2(x - 2)$ **3.** $8(y + 5) + 9y(y + 5)$ **4.** $6ab + 24b - 12a - 48$

Objective 1: Factoring Trinomials by the AC-Method

For Exercises 5–12, find the pair of integers whose product and sum are given.

5. Product: 12   Sum: 13   **6.** Product: 12   Sum: 7

7. Product: 8 Sum: −9

8. Product: −4 Sum: −3

9. Product: −20 Sum: 1

10. Product: −6 Sum: −1

11. Product: −18 Sum: 7

12. Product: −72 Sum: −6

For Exercises 13–42, factor the trinomials using the ac-method. **(See Examples 1–6.)**

13. $3x^2 + 13x + 4$

14. $2y^2 + 7y + 6$

15. $4w^2 - 9w + 2$

16. $2p^2 - 3p - 2$

17. $2m^2 + 5m - 3$

18. $6n^2 + 7n - 3$

19. $8k^2 - 6k - 9$

20. $9h^2 - 12h + 4$

21. $4k^2 - 20k + 25$

22. $16h^2 + 24h + 9$

23. $5x^2 + x + 7$

24. $4y^2 - y + 2$

25. $10 + 9z^2 - 21z$

26. $13x + 4x^2 - 12$

27. $50y + 24 + 14y^2$

28. $-24 + 10w + 4w^2$

29. $12y^2 + 8yz - 15z^2$

30. $20a^2 + 3ab - 9b^2$

31. $-15w^2 + 22w + 5$

32. $-16z^2 + 34z + 15$

33. $-12x^2 + 20xy - 8y^2$

34. $-6p^2 - 21pq - 9q^2$

35. $18y^3 + 60y^2 + 42y$

36. $8t^3 - 4t^2 - 40t$

37. $a^4 + 5a^2 + 6$

38. $y^4 - 2y^2 - 35$

39. $6x^4 - x^2 - 15$

40. $8t^4 + 2t^2 - 3$

41. $8p^4 + 37p^2 - 15$

42. $2a^4 + 11a^2 + 14$

Mixed Exercises

For Exercises 43–75, factor completely.

43. $20p^2 - 19p + 3$

44. $4p^2 + 5pq - 6q^2$

45. $6u^2 - 19uv + 10v^2$

46. $15m^2 + mn - 2n^2$

47. $12a^2 + 11ab - 5b^2$

48. $3r^2 - rs - 14s^2$

49. $3h^2 + 19hk - 14k^2$

50. $2x^2 - 13xy + y^2$

51. $3p^2 + 20pq - q^2$

52. $3 - 14z + 16z^2$

53. $10w + 1 + 16w^2$

54. $b^2 + 16 - 8b$

55. $1 + q^2 - 2q$

56. $25x - 5x^2 - 30$

57. $20a - 18 - 2a^2$

58. $-6 - t + t^2$

59. $-6 + m + m^2$

60. $72x^2 + 18x - 2$

61. $20y^2 - 78y - 8$

62. $p^3 - 6p^2 - 27p$

63. $w^5 - 11w^4 + 28w^3$

64. $3x^3 + 10x^2 + 7x$

65. $4r^3 + 3r^2 - 10r$

66. $2p^3 - 38p^2 + 120p$

67. $4q^3 - 4q^2 - 80q$

68. $x^2y^2 + 14x^2y + 33x^2$

69. $a^2b^2 + 13ab^2 + 30b^2$

70. $-k^2 - 7k - 10$

71. $-m^2 - 15m + 34$

72. $-3n^2 - 3n + 90$

73. $-2h^2 + 28h - 90$

74. $x^4 - 7x^2 + 10$

75. $m^4 + 10m^2 + 21$

76. Is the expression $(2x + 4)(x - 7)$ factored completely? Explain why or why not.

77. Is the expression $(3x + 1)(5x - 10)$ factored completely? Explain why or why not.

Objectives

1. **Factoring a Difference of Squares**
2. **Factoring Perfect Square Trinomials**

1. Factoring a Difference of Squares

Up to this point, we have learned several methods of factoring, including:

- Factoring out the greatest common factor from a polynomial
- Factoring a four-term polynomial by grouping
- Factoring trinomials by the ac-method or by the trial-and-error method

In this section, we begin by factoring a special binomial called a difference of squares. Recall from Section 5.6 that the product of two conjugates results in a **difference of squares**:

$$(a + b)(a - b) = a^2 - b^2$$

Therefore, to factor a difference of squares, the process is reversed. Identify a and b and construct the conjugate factors.

> **FORMULA** **Factored Form of a Difference of Squares**
>
> $$a^2 - b^2 = (a + b)(a - b)$$

We know that the numbers, 1, 4, 9, 16, 25, and so on are perfect squares. It is also important to recognize that a variable expression is a perfect square if its exponent is a multiple of 2. For example:

Perfect Squares

$$x^2 = (x)^2$$
$$x^4 = (x^2)^2$$
$$x^6 = (x^3)^2$$
$$x^8 = (x^4)^2$$
$$x^{10} = (x^5)^2$$

Skill Practice

Factor completely.
1. $a^2 - 64$
2. $25q^2 - 49w^2$
3. $98m^3n - 50mn$

Example 1 Factoring Differences of Squares

Factor the binomials.

a. $y^2 - 25$ **b.** $49s^2 - 4t^4$ **c.** $18w^2z - 2z$

Solution:

a. $y^2 - 25$ The binomial is a difference of squares.

$\quad = (y)^2 - (5)^2$ Write in the form: $a^2 - b^2$, where $a = y$, $b = 5$.

$\quad = (y + 5)(y - 5)$ Factor as $(a + b)(a - b)$.

b. $49s^2 - 4t^4$ The binomial is a difference of squares.

$\quad = (7s)^2 - (2t^2)^2$ Write in the form $a^2 - b^2$, where $a = 7s$ and $b = 2t^2$.

$\quad = (7s + 2t^2)(7s - 2t^2)$ Factor as $(a + b)(a - b)$.

Answers

1. $(a + 8)(a - 8)$
2. $(5q + 7w)(5q - 7w)$
3. $2mn(7m + 5)(7m - 5)$

c. $18w^2z - 2z$ The GCF is $2z$.

$= 2z(9w^2 - 1)$ $(9w^2 - 1)$ is a difference of squares.

$= 2z[(3w)^2 - (1)^2]$ Write in the form: $a^2 - b^2$, where $a = 3w, b = 1$.

$= 2z(3w + 1)(3w - 1)$ Factor as $(a + b)(a - b)$.

The difference of squares $a^2 - b^2$ factors as $(a - b)(a + b)$. However, the *sum* of squares is not factorable.

> **PROPERTY   Sum of Squares**
>
> Suppose a and b have no common factors. Then the **sum of squares** $a^2 + b^2$ is *not* factorable over the real numbers.
>
> That is, $a^2 + b^2$ is prime over the real numbers.

To see why $a^2 + b^2$ is not factorable, consider the product of binomials:

$(a + b)(a - b) = a^2 - b^2$ Wrong sign

$(a + b)(a + b) = a^2 + 2ab + b^2$ Wrong middle term

$(a - b)(a - b) = a^2 - 2ab + b^2$ Wrong middle term

After exhausting all possibilities, we see that if a and b share no common factors, then the sum of squares $a^2 + b^2$ is a prime polynomial.

Example 2   **Factoring Binomials**

Factor the binomials, if possible. **a.** $p^2 - 9$ **b.** $p^2 + 9$

Solution:

a. $p^2 - 9$ Difference of squares

$= (p - 3)(p + 3)$ Factor as $a^2 - b^2 = (a - b)(a + b)$.

b. $p^2 + 9$ Sum of squares

    Prime (cannot be factored)

Skill Practice

Factor the binomials, if possible.

4. $t^2 - 144$

5. $t^2 + 144$

Some factoring problems require several steps. Always be sure to factor completely.

Example 3   **Factoring the Difference of Squares**

Factor completely. $w^4 - 81$

Solution:

$w^4 - 81$ The GCF is 1. $w^4 - 81$ is a difference of squares.

$= (w^2)^2 - (9)^2$ Write in the form: $a^2 - b^2$, where $a = w^2, b = 9$.

$= (w^2 + 9)(w^2 - 9)$ Factor as $(a + b)(a - b)$.

$= (w^2 + 9)(w + 3)(w - 3)$ Note that $w^2 - 9$ can be factored further as a difference of squares. (The binomial $w^2 + 9$ is a sum of squares and cannot be factored further.)

Skill Practice

Factor completely.

6. $y^4 - 1$

Answers

4. $(t - 12)(t + 12)$
5. Prime
6. $(y + 1)(y - 1)(y^2 + 1)$

Example 4 **Factoring a Polynomial**

Factor completely. $y^3 - 5y^2 - 4y + 20$

Solution:

$y^3 - 5y^2 - 4y + 20$ The GCF is 1. The polynomial has four terms. Factor by grouping.

$= y^3 - 5y^2 \mid - 4y + 20$

$= y^2(y - 5) - 4(y - 5)$

$= (y - 5)(y^2 - 4)$ The expression $y^2 - 4$ is a difference of squares and can be factored further as $(y - 2)(y + 2)$.

$= (y - 5)(y - 2)(y + 2)$

Check: $(y - 5)(y - 2)(y + 2) = (y - 5)(y^2 - 2y + 2y - 4)$

$= (y - 5)(y^2 - 4)$

$= (y^3 - 4y - 5y^2 + 20)$

$= y^3 - 5y^2 - 4y + 20$ ✔

2. Factoring Perfect Square Trinomials

Recall from Section 5.6 that the square of a binomial always results in a **perfect square trinomial**.

$$(a + b)^2 = (a + b)(a + b) \xrightarrow{\text{Multiply.}} = a^2 + 2ab + b^2$$
$$(a - b)^2 = (a - b)(a - b) \xrightarrow{\text{Multiply.}} = a^2 - 2ab + b^2$$

For example, $(3x + 5)^2 = (3x)^2 + 2(3x)(5) + (5)^2$

$$= 9x^2 + 30x + 25 \text{ (perfect square trinomial)}$$

We now want to reverse this process by factoring a perfect square trinomial. The trial-and-error method or the ac-method can always be used; however, if we recognize the pattern for a perfect square trinomial, we can use one of the following formulas to reach a quick solution.

> **FORMULA** **Factored Form of a Perfect Square Trinomial**
> $$a^2 + 2ab + b^2 = (a + b)^2$$
> $$a^2 - 2ab + b^2 = (a - b)^2$$

For example, $9x^2 + 30x + 25$ is a perfect square trinomial with $a = 3x$ and $b = 5$. Therefore, it factors as

$$9x^2 + 30x + 25 = (3x)^2 + 2(3x)(5) + (5)^2 = (3x + 5)^2$$
$$a^2 \quad + 2 \ (a) \ (b) + (b)^2 = (a \ + \ b)^2$$

To apply the formula to factor a perfect square trinomial, we must first be sure that the trinomial is indeed a perfect square trinomial.

PROCEDURE Checking for a Perfect Square Trinomial

Step 1 Determine whether the first and third terms are both perfect squares and have positive coefficients.

Step 2 If this is the case, identify a and b, and determine if the middle term equals $2ab$ or $-2ab$.

Example 5 Factoring Perfect Square Trinomials

Factor the trinomials completely.

a. $x^2 + 14x + 49$ **b.** $25y^2 - 20y + 4$

Solution:

a. $x^2 + 14x + 49$ The GCF is 1.

- The first and third terms are positive.

- The first term is a perfect square: $x^2 = (x)^2$.

Perfect squares

- The third term is a perfect square: $49 = (7)^2$.

$x^2 + 14x + 49$

- The middle term is twice the product of x and 7: $14x = 2(x)(7)$

$= (x)^2 + 2(x)(7) + (7)^2$ The trinomial is in the form $a^2 + 2ab + b^2$, where $a = x$ and $b = 7$.

$= (x + 7)^2$ Factor as $(a + b)^2$.

b. $25y^2 - 20y + 4$ The GCF is 1.

Perfect squares

- The first and third terms are positive.

- The first term is a perfect square: $25y^2 = (5y)^2$.

$25y^2 - 20y + 4$

- The third term is a perfect square: $4 = (2)^2$.

$= (5y)^2 - 2(5y)(2) + (2)^2$ • In the middle: $20y = 2(5y)(2)$

$= (5y - 2)^2$ Factor as $(a - b)^2$.

TIP: The sign of the middle term in a perfect square trinomial determines the sign within the binomial of the factored form.

$a^2 + 2ab + b^2 = (a + b)^2$

$a^2 - 2ab + b^2 = (a - b)^2$

Answers

8. $(x - 3)^2$

9. $(9w + 4)^2$

Example 6   **Factoring Perfect Square Trinomials**

Factor the trinomials completely.

a. $18c^3 - 48c^2d + 32cd^2$ **b.** $5w^2 + 50w + 45$

Solution:

a. $18c^3 - 48c^2d + 32cd^2$

$= 2c(9c^2 - 24cd + 16d^2)$ The GCF is $2c$.

Perfect squares

$= 2c(9c^2 - 24cd + 16d^2)$

- The first and third terms are positive.

- The first term is a perfect square: $9c^2 = (3c)^2$.

- The third term is a perfect square: $16d^2 = (4d)^2$.

$= 2c[(3c)^2 - 2(3c)(4d) + (4d)^2]$ • In the middle: $24cd = 2(3c)(4d)$

$= 2c(3c - 4d)^2$ Factor as $(a - b)^2$.

b. $5w^2 + 50w + 45$

$= 5(w^2 + 10w + 9)$ The GCF is 5.

Perfect squares

The first and third terms are perfect squares.

$$w^2 = (w)^2 \quad \text{and} \quad 9 = (3)^2$$

$= 5(w^2 + 10w + 9)$

However, the middle term is not 2 times the product of w and 3. Therefore, this is not a perfect square trinomial.

$$10w \neq 2(w)(3)$$

$= 5(w + 9)(w + 1)$ To factor, use the trial-and-error method.

TIP: To help you identify a perfect square trinomial, we recommend that you familiarize yourself with the first several perfect squares.

$(1)^2 = 1$	$(6)^2 = 36$	$(11)^2 = 121$
$(2)^2 = 4$	$(7)^2 = 49$	$(12)^2 = 144$
$(3)^2 = 9$	$(8)^2 = 64$	$(13)^2 = 169$
$(4)^2 = 16$	$(9)^2 = 81$	$(14)^2 = 196$
$(5)^2 = 25$	$(10)^2 = 100$	$(15)^2 = 225$

If you do not recognize that a trinomial is a perfect square trinomial, you may still use the trial-and-error method or ac-method to factor it.

| Section 6.5 | **Practice Exercises** |

Study Skills Exercise

1. Define the key terms:

    a. difference of squares   **b. sum of squares**   **c. perfect square trinomial**

Review Exercises

For Exercises 2–10, factor completely.

2. $3x^2 + x - 10$

3. $6x^2 - 17x + 5$

4. $6a^2b + 3a^3b$

5. $15x^2y^5 - 10xy^6$

6. $5p^2q + 20p^2 - 3pq - 12p$

7. $ax + ab - 6x - 6b$

8. $-6x + 5 + x^2$

9. $6y - 40 + y^2$

10. $a^2 + 7a + 1$

Objective 1: Factoring a Difference of Squares

11. What binomial factors as $(x - 5)(x + 5)$?

12. What binomial factors as $(n - 3)(n + 3)$?

13. What binomial factors as $(2p - 3q)(2p + 3q)$?

14. What binomial factors as $(7x - 4y)(7x + 4y)$?

For Exercises 15–38, factor the binomials completely. **(See Examples 1–3.)**

15. $x^2 - 36$

16. $r^2 - 81$

17. $3w^2 - 300$

18. $t^3 - 49t$

19. $4a^2 - 121b^2$

20. $9x^2 - y^2$

 21. $49m^2 - 16n^2$

22. $100a^2 - 49b^2$

23. $9q^2 + 16$

24. $36 + s^2$

25. $y^2 - 4z^2$

26. $b^2 - 144c^2$

27. $a^2 - b^4$

28. $y^4 - x^2$

29. $25p^2q^2 - 1$

30. $81s^2t^2 - 1$

31. $c^6 - 25$

32. $z^6 - 4$

33. $50 - 32t^2$

34. $63 - 7h^2$

35. $z^4 - 16$

36. $y^4 - 625$

37. $a^4 - \dfrac{1}{81}$

38. $\dfrac{1}{16} - z^4$

For Exercises 39–46, factor the polynomials completely. **(See Example 4.)**

39. $x^3 + 5x^2 - 9x - 45$

40. $b^3 + 6b^2 - 4b - 24$

41. $c^3 - c^2 - 25c + 25$

42. $t^3 + 2t^2 - 16t - 32$

43. $2x^2 - 18 + x^2y - 9y$

44. $5a^2 - 5 + a^2b - b$

45. $x^2y^2 - 9x^2 - 4y^2 + 36$

46. $w^2z^2 - w^2 - 25z^2 + 25$

Objective 2: Factoring Perfect Square Trinomials

47. Multiply. $(3x + 5)^2$

48. Multiply. $(2y - 7)^2$

49. a. Which trinomial is a perfect square trinomial? $x^2 + 4x + 4$ or $x^2 + 5x + 4$

    b. Factor the trinomials from part (a).

50. a. Which trinomial is a perfect square trinomial? $x^2 + 13x + 36$ or $x^2 + 12x + 36$

    b. Factor the trinomials from part (a).

For Exercises 51–68, factor completely. (*Hint:* Look for the pattern of a perfect square trinomial.) **(See Examples 5–6.)**

51. $x^2 + 18x + 81$

52. $y^2 - 8y + 16$

53. $25z^2 - 20z + 4$

54. $36p^2 + 60p + 25$

55. $49a^2 + 42ab + 9b^2$

56. $25m^2 - 30mn + 9n^2$

57. $-2y + y^2 + 1$

58. $4 + w^2 - 4w$

59. $80z^2 + 120zw + 45w^2$

60. $36p^2 - 24pq + 4q^2$

61. $9y^2 + 78y + 25$

62. $4y^2 + 20y + 9$

63. $2a^2 - 20a + 50$

64. $3t^2 + 18t + 27$

65. $4x^2 + x + 9$

66. $c^2 - 4c + 16$

67. $4x^2 + 4xy + y^2$

68. $100y^2 + 20yz + z^2$

Expanding Your Skills

For Exercises 69–76, factor the difference of squares.

69. $(y - 3)^2 - 9$

70. $(x - 2)^2 - 4$

71. $(2p + 1)^2 - 36$

72. $(4q + 3)^2 - 25$

73. $16 - (t + 2)^2$

74. $81 - (a + 5)^2$

75. $100 - (2b - 5)^2$

76. $49 - (3k - 7)^2$

77. a. Write a polynomial that represents the area of the shaded region in the figure.

   b. Factor the expression from part (a).

78. a. Write a polynomial that represents the area of the shaded region in the figure.

   b. Factor the expression from part (a).

Section 6.6 | Sum and Difference of Cubes

Objectives

1. Factoring a Sum or Difference of Cubes
2. Factoring Binomials: A Summary

1. Factoring a Sum or Difference of Cubes

A binomial $a^2 - b^2$ is a difference of squares and can be factored as $(a - b)(a + b)$. Furthermore, if a and b share no common factors, then a sum of squares $a^2 + b^2$ is not factorable over the real numbers. In this section, we will learn that both a difference of cubes, $a^3 - b^3$, and a sum of cubes, $a^3 + b^3$, are factorable.

> **FORMULA Factored Form of a Sum or Difference of Cubes**
>
> **Sum of Cubes:** $a^3 + b^3 = (a + b)(a^2 - ab + b^2)$
>
> **Difference of Cubes:** $a^3 - b^3 = (a - b)(a^2 + ab + b^2)$

Multiplication can be used to confirm the formulas for factoring a sum or difference of cubes:

$$(a + b)(a^2 - ab + b^2) = a^3 - a^2b + ab^2 + a^2b - ab^2 + b^3 = a^3 + b^3 \checkmark$$

$$(a - b)(a^2 + ab + b^2) = a^3 + a^2b + ab^2 - a^2b - ab^2 - b^3 = a^3 - b^3 \checkmark$$

To help you remember the formulas for factoring a sum or difference of cubes, keep the following guidelines in mind:

- The factored form is the product of a binomial and a trinomial.
- The first and third terms in the trinomial are the squares of the terms within the binomial factor.
- Without regard to signs, the middle term in the trinomial is the product of terms in the binomial factor.

Square the first term of the binomial.   Product of terms in the binomial

$$x^3 + 8 = (x)^3 + (2)^3 = (x + 2)[(x)^2 - (x)(2) + (2)^2]$$

Square the last term of the binomial.

- The sign within the binomial factor is the same as the sign of the original binomial.
- The first and third terms in the trinomial are always positive.
- The sign of the middle term in the trinomial is opposite the sign within the binomial.

Same sign   Positive

$$x^3 + 8 = (x)^3 + (2)^3 = (x + 2)[(x)^2 - (x)(2) + (2)^2]$$

Opposite signs

To help you recognize a sum or difference of cubes, we recommend that you familiarize yourself with the first several perfect cubes:

Perfect Cubes	**Perfect Cubes**
$1 = (1)^3$	$216 = (6)^3$
$8 = (2)^3$	$343 = (7)^3$
$27 = (3)^3$	$512 = (8)^3$
$64 = (4)^3$	$729 = (9)^3$
$125 = (5)^3$	$1000 = (10)^3$

It is also helpful to recognize that a variable expression is a perfect cube if its exponent is a multiple of 3. For example:

Perfect Cubes

$$x^3 = (x)^3$$
$$x^6 = (x^2)^3$$
$$x^9 = (x^3)^3$$
$$x^{12} = (x^4)^3$$

Example 1 **Factoring a Sum of Cubes**

Factor: $w^3 + 64$

Solution:

$w^3 + 64$ $\qquad$ w^3 and 64 are perfect cubes.

$= (w)^3 + (4)^3$ $\qquad$ Write as $a^3 + b^3$, where $a = w$, $b = 4$.

$a^3 + b^3 = (a + b)(a^2 - ab + b^2)$ $\qquad$ Apply the formula for a sum of cubes.

$(w)^3 + (4)^3 = (w + 4)[(w)^2 - (w)(4) + (4)^2]$

$= (w + 4)(w^2 - 4w + 16)$ $\qquad$ Simplify.

Skill Practice

Factor.

1. $p^3 + 125$

Answer

1. $(p + 5)(p^2 - 5p + 25)$

Example 2 Factoring a Difference of Cubes

Factor: $27p^3 - 1000q^3$

Solution:

$27p^3 - 1000q^3$ $27p^3$ and $1000q^3$ are perfect cubes.

$(3p)^3 - (10q)^3$ Write as $a^3 - b^3$, where $a = 3p$, $b = 10q$.

$\quad a^3 - b^3 = (a - b)(a^2 + ab + b^2)$ Apply the formula for a difference of cubes.

$(3p)^3 - (10q)^3 = (3p - 10q)[(3p)^2 + (3p)(10q) + (10q)^2]$

$\qquad\qquad\qquad = (3p - 10q)(9p^2 + 30pq + 100q^2)$ Simplify.

2. Factoring Binomials: A Summary

After removing the GCF, the next step in any factoring problem is to recognize what type of pattern it follows. Exponents that are divisible by 2 are perfect squares and those divisible by 3 are perfect cubes. The formulas for factoring binomials are summarized in the following box:

FORMULA   **Factored Forms of Binomials**

Difference of Squares: $a^2 - b^2 = (a + b)(a - b)$

Difference of Cubes: $a^3 - b^3 = (a - b)(a^2 + ab + b^2)$

Sum of Cubes: $a^3 + b^3 = (a + b)(a^2 - ab + b^2)$

Example 3 Factoring Binomials

Factor completely.

a. $27y^3 + 1$ **b.** $m^2 - \dfrac{1}{4}$ **c.** $z^6 - 8w^3$

Solution:

a. $27y^3 + 1$ Sum of cubes: $27y^3 = (3y)^3$ and $1 = (1)^3$.

$\quad = (3y)^3 + (1)^3$ Write as $a^3 + b^3$, where $a = 3y$ and $b = 1$.

$\quad = (3y + 1)((3y)^2 - (3y)(1) + (1)^2)$ Apply the formula $a^3 + b^3 = (a + b)(a^2 - ab + b^2)$.

$\quad = (3y + 1)(9y^2 - 3y + 1)$ Simplify.

b. $m^2 - \dfrac{1}{4}$ Difference of squares

$\quad = (m)^2 - \left(\dfrac{1}{2}\right)^2$ Write as $a^2 - b^2$, where $a = m$ and $b = \frac{1}{2}$.

$\quad = \left(m + \dfrac{1}{2}\right)\left(m - \dfrac{1}{2}\right)$ Apply the formula $a^2 - b^2 = (a + b)(a - b)$.

c. $z^6 - 8w^3$ Difference of cubes: $z^6 = (z^2)^3$ and $8w^3 = (2w)^3$

$= (z^2)^3 - (2w)^3$ Write as $a^3 - b^3$, where $a = z^2$ and $b = 2w$.

$= (z^2 - 2w)[(z^2)^2 + (z^2)(2w) + (2w)^2]$ Apply the formula $a^3 - b^3 = (a - b)(a^2 + ab + b^2)$.

$= (z^2 - 2w)(z^4 + 2z^2w + 4w^2)$ Simplify.

Each of the factorizations in Example 3 can be checked by multiplying.

Some factoring problems require more than one method of factoring. In general, when factoring a polynomial, be sure to factor completely.

Example 4 **Factoring Polynomials**

Factor completely. $3y^4 - 48$

Solution:

$3y^4 - 48$

$= 3(y^4 - 16)$ Factor out the GCF. The binomial is a difference of squares.

$= 3[(y^2)^2 - (4)^2]$ Write as $a^2 - b^2$, where $a = y^2$ and $b = 4$.

$= 3(y^2 + 4)(y^2 - 4)$ Apply the formula $a^2 - b^2 = (a + b)(a - b)$.

$y^2 + 4$ is a sum of squares and cannot be factored.

$= 3(y^2 + 4)(y + 2)(y - 2)$ $y^2 - 4$ is a difference of squares and can be factored further.

Example 5 **Factoring Polynomials**

Factor completely. $4x^3 + 4x^2 - 25x - 25$

Solution:

$4x^3 + 4x^2 - 25x - 25$ The GCF is 1.

$= 4x^3 + 4x^2 - 25x - 25$ The polynomial has four terms. Factor by grouping.

$= 4x^2(x + 1) - 25(x + 1)$

$= (x + 1)(4x^2 - 25)$ $4x^2 - 25$ is a difference of squares.

$= (x + 1)(2x + 5)(2x - 5)$

Skill Practice

Factor completely.

8. $z^6 - 64$

Example 6 Factoring Binomials

Factor the binomial $x^6 - y^6$ as

a. A difference of cubes   **b.** A difference of squares

Solution:

a. $x^6 - y^6$

$$= \underbrace{(x^2)^3 - (y^2)^3}_{\text{Difference of cubes}}$$

Write as $a^3 - b^3$, where $a = x^2$ and $b = y^2$.

$$= (x^2 - y^2)[(x^2)^2 + (x^2)(y^2) + (y^2)^2]$$

Apply the formula $a^3 - b^3 = (a - b)(a^2 + ab + b^2)$.

$$= (x^2 - y^2)(x^4 + x^2y^2 + y^4)$$

Factor $x^2 - y^2$ as a difference of squares.

$$= (x + y)(x - y)(x^4 + x^2y^2 + y^4)$$

b. $x^6 - y^6$

$$= \underbrace{(x^3)^2 - (y^3)^2}_{\text{Difference of squares}}$$

Write as $a^2 - b^2$, where $a = x^3$ and $b = y^3$.

$$= (x^3 + y^3)(x^3 - y^3)$$

Apply the formula $a^2 - b^2 = (a + b)(a - b)$.

Factor $x^3 + y^3$ as a sum of cubes.

Factor $x^3 - y^3$ as a difference of cubes.

$$= (x + y)(x^2 - xy + y^2)(x - y)(x^2 + xy + y^2)$$

Notice that the expressions x^6 and y^6 are both perfect squares and perfect cubes because both exponents are multiples of 2 and of 3. Consequently, $x^6 - y^6$ can be factored initially as either the difference of squares or as the difference of cubes. In such a case, it is recommended that you factor the expression as a difference of squares first because it factors more completely into polynomials of lower degree.

$$x^6 - y^6 = (x + y)(x^2 - xy + y^2)(x - y)(x^2 + xy + y^2)$$

Answer

8. $(z + 2)(z - 2)(z^2 + 2z + 4)$
$(z^2 - 2z + 4)$

Section 6.6 Practice Exercises

Study Skills Exercise

1. Define the key terms.

    a. sum of cubes   **b. difference of cubes**

Review Exercises

For Exercises 2–10, factor completely.

2. $600 - 6x^2$

3. $20 - 5t^2$

4. $ax + bx + 5a + 5b$

5. $2t + 2u + st + su$

6. $5y^2 + 13y - 6$

7. $3v^2 + 5v - 12$

8. $40a^3b^3 - 16a^2b^2 + 24a^3b$

9. $-c^2 - 10c - 25$

10. $-z^2 + 6z - 9$

Objective 1: Factoring a Sum or Difference of Cubes

11. Identify the expressions that are perfect cubes:
$$\{x^3, 8, 9, y^6, a^4, b^2, 3p^3, 27q^3, w^{12}, r^3s^6\}$$

12. Identify the expressions that are perfect cubes:
$$\{z^9, -81, 30, 8, 6x^3, y^{15}, 27a^3, b^2, p^3q^2, -1\}$$

13. From memory, write the formula to factor a sum of cubes:
$$a^3 + b^3 = \underline{\hspace{2cm}}$$

14. From memory, write the formula to factor a difference of cubes:
$$a^3 - b^3 = \underline{\hspace{2cm}}$$

For Exercises 15–30, factor the sums or differences of cubes. **(See Examples 1–2.)**

15. $y^3 - 8$

16. $x^3 + 27$

17. $1 - p^3$

18. $q^3 + 1$

19. $w^3 + 64$

20. $8 - t^3$

21. $x^3 - 1000$

22. $8y^3 - 27$

23. $64t^3 + 1$

24. $125r^3 + 1$

25. $1000a^3 + 27$

26. $216b^3 - 125$

27. $n^3 - \dfrac{1}{8}$

28. $\dfrac{8}{27} + m^3$

29. $125m^3 + 8$

30. $27p^3 - 64$

Objective 2: Factoring Binomials: A Summary

For Exercises 31–66, factor completely. **(See Examples 3–6.)**

31. $x^4 - 4$

32. $b^4 - 25$

33. $a^2 + 9$

34. $w^2 + 36$

35. $t^3 + 64$

36. $u^3 + 27$

37. $g^3 - 4$

38. $h^3 - 25$

39. $4b^3 + 108$

40. $3c^3 - 24$

41. $5p^2 - 125$

42. $2q^4 - 8$

43. $\dfrac{1}{64} - 8h^3$

44. $\dfrac{1}{125} + k^6$

45. $x^4 - 16$

46. $p^4 - 81$

47. $q^6 - 64$ **48.** $a^6 - 1$ **49.** $\dfrac{4x^2}{9} - w^2$ **50.** $\dfrac{16y^2}{25} - x^2$

51. $x^9 + 64y^3$ **52.** $125w^3 - z^9$ 💿 **53.** $2x^3 + 3x^2 - 2x - 3$ **54.** $3x^3 + x^2 - 12x - 4$

55. $16x^4 - y^4$ **56.** $1 - t^4$ 💿 **57.** $81y^4 - 16$ **58.** $u^5 - 256u$

59. $a^3 + b^6$ **60.** $u^6 - v^3$ **61.** $x^4 - y^4$ **62.** $a^4 - b^4$

63. $k^3 + 4k^2 - 9k - 36$ **64.** $w^3 - 2w^2 - 4w + 8$ **65.** $2t^3 - 10t^2 - 2t + 10$ **66.** $9a^3 + 27a^2 - 4a - 12$

Expanding Your Skills

For Exercises 67–70, factor completely.

67. $\dfrac{64}{125}p^3 - \dfrac{1}{8}q^3$ **68.** $\dfrac{1}{1000}r^3 + \dfrac{8}{27}s^3$ **69.** $a^{12} + b^{12}$ **70.** $a^9 - b^9$

Use Exercises 71–72 to investigate the relationship between division and factoring.

71. a. Use long division to divide $x^3 - 8$ by $(x - 2)$.

    b. Factor $x^3 - 8$.

72. a. Use long division to divide $y^3 + 27$ by $(y + 3)$.

    b. Factor $y^3 + 27$.

73. What trinomial multiplied by $(x - 4)$ gives a difference of cubes?

74. What trinomial multiplied by $(p + 5)$ gives a sum of cubes?

75. Write a binomial that when multiplied by $(4x^2 - 2x + 1)$ produces a sum of cubes.

76. Write a binomial that when multiplied by $(9y^2 + 15y + 25)$ produces a difference of cubes.

Problem Recognition Exercises

Factoring Strategy

PROCEDURE **Factoring Strategy**

Step 1 Factor out the GCF (Section 6.1).

Step 2 Identify whether the polynomial has two terms, three terms, or more than three terms.

Step 3 If the polynomial has more than three terms, try factoring by grouping (Section 6.1).

Step 4 If the polynomial has three terms, check first for a perfect square trinomial. Otherwise, factor the trinomial with the trial-and-error method or the ac-method (Sections 6.3 or 6.4).

Step 5 If the polynomial has two terms, determine if it fits the pattern for
- A difference of squares: $a^2 - b^2 = (a - b)(a + b)$ (Section 6.5)
- A sum of squares: $a^2 + b^2$ prime
- A difference of cubes: $a^3 - b^3 = (a - b)(a^2 + ab + b^2)$ (Section 6.6)
- A sum of cubes: $a^3 + b^3 = (a + b)(a^2 - ab + b^2)$ (Section 6.6)

Step 6 Be sure to factor the polynomial completely.

Step 7 Check by multiplying.

1. What is meant by a prime factor?

2. What is the first step in factoring any polynomial?

3. When factoring a binomial, what patterns can you look for?

4. What technique should be considered when factoring a four-term polynomial?

For Exercises 5–73,

a. Factor out the GCF from each polynomial. Then identify the category in which the polynomial best fits. Choose from
  - difference of squares
  - sum of squares
  - difference of cubes
  - sum of cubes
  - trinomial (perfect square trinomial)
  - trinomial (nonperfect square trinomial)
  - four terms-grouping
  - none of these

b. Factor the polynomial completely.

5. $2a^2 - 162$

6. $y^2 + 4y + 3$

7. $6w^2 - 6w$

8. $16z^4 - 81$

9. $3t^2 + 13t + 4$

10. $5r^3 + 5$

11. $3ac + ad - 3bc - bd$

12. $x^3 - 125$

13. $y^3 + 8$

14. $7p^2 - 29p + 4$

15. $3q^2 - 9q - 12$

16. $-2x^2 + 8x - 8$

17. $18a^2 + 12a$

18. $54 - 2y^3$

19. $4t^2 - 100$

20. $4t^2 - 31t - 8$

21. $10c^2 + 10c + 10$

22. $2xw - 10x + 3yw - 15y$

23. $x^3 + 0.001$

24. $4q^2 - 9$

25. $64 + 16k + k^2$

26. $s^2t + 5t + 6s^2 + 30$

27. $2x^2 + 2x - xy - y$

28. $w^3 + y^3$

29. $a^3 - c^3$

30. $3y^2 + y + 1$

31. $c^2 + 8c + 9$

32. $a^2 + 2a + 1$

33. $b^2 + 10b + 25$

34. $-t^2 - 4t + 32$

35. $-p^3 - 5p^2 - 4p$

36. $x^2y^2 - 49$

37. $6x^2 - 21x - 45$

38. $20y^2 - 14y + 2$

39. $5a^2bc^3 - 7abc^2$

40. $8a^2 - 50$

41. $t^2 + 2t - 63$

42. $b^2 + 2b - 80$

43. $ab + ay - b^2 - by$

44. $6x^3y^4 + 3x^2y^5$

45. $14u^2 - 11uv + 2v^2$

46. $9p^2 - 36pq + 4q^2$

47. $4q^2 - 8q - 6$

48. $9w^2 + 3w - 15$

49. $9m^2 + 16n^2$

50. $5b^2 - 30b + 45$

51. $6r^2 + 11r + 3$

52. $4s^2 + 4s - 15$

53. $16a^4 - 1$

54. $p^3 + p^2c - 9p - 9c$

55. $81u^2 - 90uv + 25v^2$

56. $4x^2 + 16$

57. $x^2 - 5x - 6$

58. $q^2 + q - 7$

59. $2ax - 6ay + 4bx - 12by$

60. $8m^3 - 10m^2 - 3m$

61. $21x^4y + 41x^3y + 10x^2y$

62. $2m^4 - 128$

63. $8uv - 6u + 12v - 9$

64. $4t^2 - 20t + st - 5s$

65. $12x^2 - 12x + 3$

66. $p^2 + 2pq + q^2$

67. $6n^3 + 5n^2 - 4n$

68. $4k^3 + 4k^2 - 3k$

69. $64 - y^2$

70. $36b - b^3$

71. $b^2 - 4b + 10$

72. $y^2 + 6y + 8$

73. $c^4 - 12c^2 + 20$

Section 6.7   Solving Equations Using the Zero Product Rule

Objectives

1. Definition of a Quadratic Equation
2. Zero Product Rule
3. Solving Equations by Factoring

1. Definition of a Quadratic Equation

In Section 2.1, we solved linear equations in one variable. These are equations of the form $ax + b = 0\,(a \neq 0)$. A linear equation in one variable is sometimes called a first-degree polynomial equation because the highest degree of all its terms is 1. A second-degree polynomial equation in one variable is called a quadratic equation.

> **DEFINITION**   **A Quadratic Equation in One Variable**
>
> If a, b, and c are real numbers such that $a \neq 0$, then a **quadratic equation** is an equation that can be written in the form
>
> $$ax^2 + bx + c = 0$$

The following equations are quadratic because they can each be written in the form $ax^2 + bx + c = 0, \quad (a \neq 0)$.

$$-4x^2 + 4x = 1 \qquad x(x - 2) = 3 \qquad (x - 4)(x + 4) = 9$$
$$-4x^2 + 4x - 1 = 0 \qquad x^2 - 2x = 3 \qquad x^2 - 16 = 9$$
$$x^2 - 2x - 3 = 0 \qquad x^2 - 25 = 0$$
$$x^2 + 0x - 25 = 0$$

2. Zero Product Rule

One method for solving a quadratic equation is to factor and apply the zero product rule. The **zero product rule** states that if the product of two factors is zero, then one or both of its factors is zero.

> **PROPERTY**   **Zero Product Rule**
>
> $$\text{If } ab = 0, \text{ then } a = 0 \text{ or } b = 0.$$

Example 1 **Using the Zero Product Rule**

Solve the equation by using the zero product rule. $(x - 4)(x + 3) = 0$

Solution:

$(x - 4)(x + 3) = 0$	Apply the zero product rule.
$x - 4 = 0$ or $x + 3 = 0$	Set each factor equal to zero.
$x = 4$ or $x = -3$	Solve each equation for x.

Check: $x = 4$ Check: $x = -3$

$(4 - 4)(4 + 3) \stackrel{?}{=} 0$ $(-3 - 4)(-3 + 3) \stackrel{?}{=} 0$

$\qquad (0)(7) \stackrel{?}{=} 0 ✔$ $\qquad (-7)(0) \stackrel{?}{=} 0 ✔$

The solutions are 4 and -3.

Skill Practice

1. Solve.
   $(x + 1)(x - 8) = 0$

Example 2 **Using the Zero Product Rule**

Solve the equation by using the zero product rule. $(2x + 1)(x - 5) = 0$

Solution:

$(2x + 1)(x - 5) = 0$	Apply the zero product rule.
$2x + 1 = 0$ or $x - 5 = 0$	Set each factor equal to zero.
$2x = -1$ or $x = 5$	Solve each equation for x.
$x = -\dfrac{1}{2}$ or $x = 5$	

The solutions $-\dfrac{1}{2}$ and 5 check in the original equation.

Skill Practice

2. Solve.
   $(4x - 5)(x + 6) = 0$

Example 3 **Using the Zero Product Rule**

Solve the equation using the zero product rule. $x(3x - 7) = 0$

Solution:

$x(3x - 7) = 0$	Apply the zero product rule.
$x = 0$ or $3x - 7 = 0$	Set each factor equal to zero.
$x = 0$ or $3x = 7$	Solve each equation for x.
$x = 0$ or $x = \dfrac{7}{3}$	

The solutions 0 and $\dfrac{7}{3}$ check in the original equation.

Skill Practice

3. Solve.
   $x(4x + 9) = 0$

Answers

1. $-1, 8$ 2. $\dfrac{5}{4}, -6$ 3. $0, -\dfrac{9}{4}$

3. Solving Equations by Factoring

Quadratic equations, like linear equations, arise in many applications in mathematics, science, and business. The following steps summarize the factoring method for solving a quadratic equation.

PROCEDURE Solving a Quadratic Equation by Factoring

Step 1 Write the equation in the form: $ax^2 + bx + c = 0$.

Step 2 Factor the quadratic expression completely.

Step 3 Apply the zero product rule. That is, set each factor equal to zero, and solve the resulting equations.

Note: The solution(s) found in step 3 may be checked by substitution in the original equation.

Skill Practice

Solve the equation.

4. $2y^2 + 19y = -24$

Example 4   Solving a Quadratic Equation

Solve the quadratic equation. $2x^2 - 9x = 5$

Solution:

$$2x^2 - 9x = 5$$

$$2x^2 - 9x - 5 = 0 \qquad \text{Write the equation in the form } ax^2 + bx + c = 0.$$

$$(2x + 1)(x - 5) = 0 \qquad \text{Factor the polynomial completely.}$$

$$2x + 1 = 0 \quad \text{or} \quad x - 5 = 0 \qquad \text{Set each factor equal to zero.}$$

$$2x = -1 \quad \text{or} \quad x = 5 \qquad \text{Solve each equation.}$$

$$x = -\frac{1}{2} \quad \text{or} \quad x = 5$$

$$\underline{\text{Check:}}\ x = -\frac{1}{2} \qquad\qquad \underline{\text{Check:}}\ x = 5$$

$$2x^2 - 9x = 5 \qquad\qquad 2x^2 - 9x = 5$$

$$2\left(-\frac{1}{2}\right)^2 - 9\left(-\frac{1}{2}\right) \overset{?}{=} 5 \qquad 2(5)^2 - 9(5) \overset{?}{=} 5$$

$$2\left(\frac{1}{4}\right) + \frac{9}{2} \overset{?}{=} 5 \qquad 2(25) - 45 \overset{?}{=} 5$$

$$\frac{1}{2} + \frac{9}{2} \overset{?}{=} 5 \qquad\qquad 50 - 45 \overset{?}{=} 5 ✔$$

$$\frac{10}{2} \overset{?}{=} 5 ✔$$

The solutions are $-\frac{1}{2}$ and 5.

Answer

4. $-8, -\dfrac{3}{2}$

Example 5 Solving a Quadratic Equation

Solve the quadratic equation. $4x^2 + 24x = 0$

Solution:

$4x^2 + 24x = 0$ The equation is already in the form $ax^2 + bx + c = 0$. (Note that $c = 0$.)

$4x(x + 6) = 0$ Factor completely.

$4x = 0$ or $x + 6 = 0$ Set each factor equal to zero.

$x = 0$ or $x = -6$

The solutions are 0 and -6. They both check in the original equation.

Skill Practice

Solve the equation.

5. $5s^2 = 45$

Example 6 Solving a Quadratic Equation

Solve the quadratic equation. $5x(5x + 2) = 10x + 9$

Solution:

$5x(5x + 2) = 10x + 9$

$25x^2 + 10x = 10x + 9$ Clear parentheses.

$25x^2 + 10x - 10x - 9 = 0$ Set the equation equal to zero.

$25x^2 - 9 = 0$ The equation is in the form $ax^2 + bx + c = 0$. (Note that $b = 0$.)

$(5x - 3)(5x + 3) = 0$ Factor completely.

$5x - 3 = 0$ or $5x + 3 = 0$ Set each factor equal to zero.

$5x = 3$ or $5x = -3$ Solve each equation.

$\dfrac{5x}{5} = \dfrac{3}{5}$ or $\dfrac{5x}{5} = \dfrac{-3}{5}$

$x = \dfrac{3}{5}$ or $x = -\dfrac{3}{5}$

The solutions are $\dfrac{3}{5}$ and $-\dfrac{3}{5}$. They both check in the original equation.

Skill Practice

Solve the equation.

6. $4z(z + 3) = 4z + 5$

The zero product rule can be used to solve higher degree polynomial equations provided the equations can be set to zero and written in factored form.

Answers

5. 3, -3 **6.** $-\dfrac{5}{2}, \dfrac{1}{2}$

Example 7   Solving Higher Degree Polynomial Equations

Solve the equation. $-6(y + 3)(y - 5)(2y + 7) = 0$

Solution:

$-6(y + 3)(y - 5)(2y + 7) = 0$ The equation is already in factored form and equal to zero.

Set each factor equal to zero.

Solve each equation for y.

$-6 \neq 0$	or	$y + 3 = 0$	or	$y - 5 = 0$	or	$2y + 7 = 0$
No solution,		$y = -3$	or	$y = 5$	or	$y = -\dfrac{7}{2}$

Notice that when the constant factor is set equal to zero, the result is a contradiction, $-6 = 0$. The constant factor does not produce a solution to the equation. Therefore, the only solutions are $-3, 5$, and $-\frac{7}{2}$. Each solution can be checked in the original equation.

Example 8   Solving a Higher Degree Polynomial Equation

Solve the equation. $w^3 + 5w^2 - 9w - 45 = 0$

Solution:

$w^3 + 5w^2 - 9w - 45 = 0$ This is a higher degree polynomial equation.

$w^3 + 5w^2 \mid - 9w - 45 = 0$ The equation is already set equal to zero. Now factor.

$w^2(w + 5) - 9(w + 5) = 0$ Because there are four terms, try factoring
$(w + 5)(w^2 - 9) = 0$ by grouping.

$(w + 5)(w - 3)(w + 3) = 0$ $w^2 - 9$ is a difference of squares and can be factored further.

$w + 5 = 0$	or	$w - 3 = 0$	or	$w + 3 = 0$	Set each factor equal to zero.
$w = -5$	or	$w = 3$	or	$w = -3$	Solve each equation.

The solutions are $-5, -3$ and 3. Each solution checks in the original equation.

Section 6.7 Practice Exercises

Study Skills Exercise

1. Define the key terms:

a. quadratic equation   **b. zero product rule**

Review Exercises

For Exercises 2–7, factor completely.

2. $6a - 8 - 3ab + 4b$

3. $4b^2 - 44b + 120$

4. $8u^2v^2 - 4uv$

5. $3x^2 + 10x - 8$

6. $3h^2 - 75$

7. $4x^2 + 16y^2$

Objective 1: Definition of a Quadratic Equation

For Exercises 8–13, identify the equations as linear, quadratic, or neither.

8. $4 - 5x = 0$

9. $5x^3 + 2 = 0$

10. $3x - 6x^2 = 0$

11. $1 - x + 2x^2 = 0$

12. $7x^4 + 8 = 0$

13. $3x + 2 = 0$

Objective 2: Zero Product Rule

For Exercises 14–22, solve the equations using the zero product rule. **(See Examples 1–2.)**

14. $(x - 5)(x + 1) = 0$

15. $(x + 3)(x - 1) = 0$

16. $(3x - 2)(3x + 2) = 0$

17. $(2x - 7)(2x + 7) = 0$

18. $2(x - 7)(x - 7) = 0$

19. $3(x + 5)(x + 5) = 0$

20. $(3x - 2)(2x - 3) = 0$

21. $x(5x - 1) = 0$

22. $x(3x + 8) = 0$

23. For a quadratic equation of the form $ax^2 + bx + c = 0$, what must be done before applying the zero product rule?

24. What are the requirements needed to use the zero product rule to solve a quadratic equation or higher degree polynomial equation?

Objective 3: Solving Equations by Factoring

For Exercises 25–72, solve the equations. **(See Examples 4–8.)**

25. $p^2 - 2p - 15 = 0$

26. $y^2 - 7y - 8 = 0$

27. $z^2 + 10z - 24 = 0$

28. $w^2 - 10w + 16 = 0$

29. $2q^2 - 7q = 4$

30. $4x^2 - 11x = 3$

31. $0 = 9x^2 - 4$

32. $4a^2 - 49 = 0$

33. $2k^2 - 28k + 96 = 0$

34. $0 = 2t^2 + 20t + 50$

35. $0 = 2m^3 - 5m^2 - 12m$

36. $3n^3 + 4n^2 + n = 0$

37. $5(3p + 1)(p - 3)(p + 6) = 0$

38. $4(2x - 1)(x - 10)(x + 7) = 0$

39. $x(x - 4)(2x + 3) = 0$

40. $x(3x + 1)(x + 1) = 0$

41. $-5x(2x + 9)(x - 11) = 0$

42. $-3x(x + 7)(3x - 5) = 0$

43. $x^3 - 16x = 0$

44. $t^3 - 36t = 0$

45. $3x^2 + 18x = 0$

46. $2y^2 - 20y = 0$

47. $16m^2 = 9$

48. $9n^2 = 1$

49. $2y^3 + 14y^2 = -20y$

50. $3d^3 - 6d^2 = 24d$

51. $5t - 2(t - 7) = 0$

52. $8h = 5(h - 9) + 6$

53. $2c(c - 8) = -30$

54. $3q(q - 3) = 12$

55. $b^3 = -4b^2 - 4b$

56. $x^3 + 36x = 12x^2$

57. $3(a^2 + 2a) = 2a^2 - 9$

58. $9(k - 1) = -4k^2$

59. $2n(n + 2) = 6$

60. $3p(p - 1) = 18$

61. $x(2x + 5) - 1 = 2x^2 + 3x + 2$

62. $3z(z - 2) - z = 3z^2 + 4$

63. $27q^2 = 9q$

64. $21w^2 = 14w$

65. $3(c^2 - 2c) = 0$

66. $2(4d^2 + d) = 0$

67. $y^3 - 3y^2 - 4y + 12 = 0$

68. $t^3 + 2t^2 - 16t - 32 = 0$

69. $(x - 1)(x + 2) = 18$

70. $(w + 5)(w - 3) = 20$

71. $(p + 2)(p + 3) = 1 - p$

72. $(k - 6)(k - 1) = -k - 2$

Problem Recognition Exercises

Polynomial Expressions and Polynomial Equations

For Exercises 1–36, factor each expression and solve each equation.

1. a. $x^2 + 6x - 7$
 b. $x^2 + 6x - 7 = 0$

2. a. $c^2 + 8c + 12$
 b. $c^2 + 8c + 12 = 0$

3. a. $2y^2 + 7y + 3$
 b. $2y^2 + 7y + 3 = 0$

4. a. $3x^2 - 8x + 5$
 b. $3x^2 - 8x + 5 = 0$

5. a. $5q^2 + q - 4 = 0$
 b. $5q^2 + q - 4$

6. a. $6a^2 - 7a - 3 = 0$
 b. $6a^2 - 7a - 3$

7. a. $a^2 - 64 = 0$
 b. $a^2 - 64$

8. a. $v^2 - 100 = 0$
 b. $v^2 - 100$

9. a. $4b^2 - 81$
 b. $4b^2 - 81 = 0$

10. a. $36t^2 - 49$
 b. $36t^2 - 49 = 0$

11. a. $8x^2 + 16x + 6 = 0$
 b. $8x^2 + 16x + 6$

12. a. $12y^2 + 40y + 32 = 0$
 b. $12y^2 + 40y + 32$

13. a. $x^3 - 8x^2 - 20x$
 b. $x^3 - 8x^2 - 20x = 0$

14. a. $k^3 + 5k^2 - 14k$
 b. $k^3 + 5k^2 - 14k = 0$

15. a. $b^3 + b^2 - 9b - 9 = 0$
 b. $b^3 + b^2 - 9b - 9$

16. a. $x^3 - 8x^2 - 4x + 32 = 0$
 b. $x^3 - 8x^2 - 4x + 32$

17. $2s^2 - 6s + rs - 3r$

18. $6t^2 + 3t + 10tu + 5u$

19. $8x^3 - 2x = 0$

20. $2b^3 - 50b = 0$

21. $2x^3 - 4x^2 + 2x = 0$

22. $3t^3 + 18t^2 + 27t = 0$

23. $7c^2 - 2c + 3 = 7(c^2 + c)$

24. $3z(2z + 4) = -7 + 6z^2$

25. $8w^3 + 27$

26. $1000q^3 - 1$

27. $5z^2 + 2z = 7$

28. $4h^2 + 25h = -6$

29. $3b(b + 6) = b - 10$

30. $3y^2 + 1 = y(y - 3)$

31. $5(2x - 3) - 2(3x + 1) = 4 - 3x$

32. $11 - 6a = -4(2a - 3) - 1$

33. $4s^2 = 64$

34. $81v^2 = 36$

35. $(x - 3)(x - 4) = 6$

36. $(x + 5)(x + 9) = 21$

Applications of Quadratic Equations

1. Applications of Quadratic Equations

Objectives

1. Applications of Quadratic Equations
2. Pythagorean Theorem

Example 1 Translating to a Quadratic Equation

The product of two consecutive integers is 48 more than the larger integer. Find the integers.

Solution:

Let x represent the first (smaller) integer.

Then $x + 1$ represents the second (larger) integer. — Label the variables.

(First integer)(second integer) = (second integer) + 48 — Verbal model

$$x(x + 1) = (x + 1) + 48 \qquad \text{Algebraic equation}$$

$$x^2 + x = x + 49 \qquad \text{Simplify.}$$

$$x^2 + x - x - 49 = 0 \qquad \text{Set one side of the equation equal to zero.}$$

$$x^2 - 49 = 0$$

$$(x - 7)(x + 7) = 0 \qquad \text{Factor.}$$

$$x - 7 = 0 \quad \text{or} \quad x + 7 = 0 \qquad \text{Set each factor equal to zero.}$$

$$x = 7 \quad \text{or} \quad x = -7 \qquad \text{Solve for } x.$$

Recall that x represents the smaller integer. Therefore, there are two possibilities for the pairs of consecutive integers.

If $x = 7$, then the larger integer is $x + 1$ or $7 + 1 = 8$.

If $x = -7$, then the larger integer is $x + 1$ or $-7 + 1 = -6$.

The integers are 7 and 8, or -7 and -6.

Skill Practice

1. The product of two consecutive odd integers is 9 more than 10 times the smaller integer. Find the pair of integers.

Example 2 Using a Quadratic Equation in a Geometry Application

A rectangular sign has an area of 40 ft². If the width is 3 feet shorter than the length, what are the dimensions of the sign?

Solution:

Let x represent the length of the sign. Then $x - 3$ represents the width (Figure 6-1).

Label the variables.

The problem gives information about the length of the sides and about the area. Therefore, we can form a relationship by using the formula for the area of a rectangle.

$x - 3$

x

Figure 6-1

Skill Practice

2. The length of a rectangle is 5 ft more than the width. The area is 36 ft². Find the length and width.

$A = l \cdot w$	Area equals length times width.
$40 = x(x - 3)$	Set up an algebraic equation.
$40 = x^2 - 3x$	Clear parentheses.
$0 = x^2 - 3x - 40$	Write the equation in the form, $ax^2 + bx + c = 0$.
$0 = (x - 8)(x + 5)$	Factor.
$0 = x - 8$ or $0 = x + 5$	Set each factor equal to zero.
$8 = x$ or $-5 \cancel{=} x$	Because x represents the length of a rectangle, reject the negative solution.

The variable x represents the length of the sign. The length is 8 ft.

The expression $x - 3$ represents the width. The width is 8 ft − 3 ft, or 5 ft.

Example 3   Using a Quadratic Equation in an Application

A stone is dropped off a 64-ft cliff and falls into the ocean below. The height of the stone above sea level is given by the equation

$$h = -16t^2 + 64 \qquad \text{where } h \text{ is the stone's height in feet, and } t \text{ is the time in seconds.}$$

Find the time required for the stone to hit the water.

64 ft

Solution:

When the stone hits the water, its height is zero. Therefore, substitute $h = 0$ into the equation.

$h = -16t^2 + 64$	The equation is quadratic.
$0 = -16t^2 + 64$	Substitute $h = 0$.
$0 = -16(t^2 - 4)$	Factor out the GCF.
$0 = -16(t - 2)(t + 2)$	Factor as a difference of squares.
$-16 \cancel{=} 0$ or $t - 2 = 0$ or $t + 2 = 0$	Set each factor to zero.
No solution, $t = 2$ or $t \cancel{=} -2$	Solve for t.

The negative value of t is rejected because the stone cannot fall for a negative time. Therefore, the stone hits the water after 2 seconds.

Answer

3. 9 seconds

In Example 3, we can analyze the path of the stone as it falls from the cliff. Use the equation $h = -16t^2 + 64$ to compute the height values at various times between 0 and 2 seconds (Table 6-1). The ordered pairs can be graphed where t is used in place of x and h is used in place of y. The graph of the height of the stone versus time is shown in Figure 6-2.

Table 6-1

Time, t (sec)	Height, h (ft)
0.0	64
0.5	60
1.0	48
1.5	28
2.0	0

Figure 6-2

2. Pythagorean Theorem

Recall that a right triangle is a triangle that contains a 90° angle. Furthermore, the sum of the squares of the two legs (the shorter sides) of a right triangle equals the square of the hypotenuse (the longest side). This important fact is known as the Pythagorean theorem. The Pythagorean theorem is an enduring landmark of mathematical history from which many mathematical ideas have been built. Although the theorem is named after Pythagoras (sixth century B.C.E.), a Greek mathematician and philosopher, it is thought that the ancient Babylonians were familiar with the principle more than a thousand years earlier.

For the right triangle shown in Figure 6-3, the **Pythagorean theorem** is stated as:

$$a^2 + b^2 = c^2$$

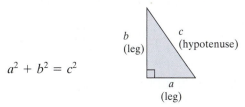

Figure 6-3

In this formula, a and b are the legs of the right triangle and c is the hypotenuse. Notice that the hypotenuse is the longest side of the right triangle and is opposite the 90° angle.

The triangle shown below is a right triangle. Notice that the lengths of the sides satisfy the Pythagorean theorem.

The shorter sides are labeled as a and b.

4 yd

5 yd   3 yd

Label the sides.

$a = 4$

$c = 5$ $b = 3$

The longest side is the hypotenuse. It is always labeled as c.

$a^2 + b^2 = c^2$ Apply the Pythagorean theorem.

$(4)^2 + (3)^2 = (5)^2$ Substitute $a = 4$, $b = 3$, and $c = 5$.

$16 + 9 = 25$

$25 = 25$ ✔

Skill Practice

4. Find the length of the missing side.

Example 4 Applying the Pythagorean Theorem

Find the length of the missing side of the right triangle.

Solution:

Label the triangle.

$$a^2 + b^2 = c^2$$ Apply the Pythagorean theorem.

$$a^2 + 6^2 = 10^2$$ Substitute $b = 6$ and $c = 10$.

$$a^2 + 36 = 100$$ Simplify.

The equation is quadratic. Set one side of the equation equal to zero.

$$a^2 + 36 - 100 = 100 - 100$$ Subtract 100 from both sides.

$$a^2 - 64 = 0$$

$$(a + 8)(a - 8) = 0$$ Factor.

$$a + 8 = 0 \quad \text{or} \quad a - 8 = 0$$ Set each factor equal to zero.

$$a \neq -8 \quad \text{or} \quad a = 8$$ Because x represents the length of a side of a triangle, reject the negative solution.

The third side is 8 ft.

Skill Practice

5. A 5-yd ladder leans against a wall. The distance from the bottom of the wall to the top of the ladder is 1 yd more than the distance from the bottom of the wall to the bottom of the ladder. Find both distances.

Example 5 Using a Quadratic Equation in an Application

A 13-ft board is used as a ramp to unload furniture off a loading platform. If the distance between the top of the board and the ground is 7 ft less than the distance between the bottom of the board and the base of the platform, find both distances.

Solution:

Let x represent the distance between the bottom of the board and the base of the platform. Then $x - 7$ represents the distance between the top of the board and the ground (Figure 6-4).

Figure 6-4

Answers

4. The length of the third side is 12 m.

5. The distance along the wall to the top of the ladder is 4 yd. The distance on the ground from the ladder to the wall is 3 yd.

$$a^2 + b^2 = c^2$$ Pythagorean theorem

$$x^2 + (x - 7)^2 = (13)^2$$

$$x^2 + (x)^2 - 2(x)(7) + (7)^2 = 169$$

$$x^2 + x^2 - 14x + 49 = 169$$

$$2x^2 - 14x + 49 = 169$$ Combine *like* terms.

$$2x^2 - 14x + 49 - 169 = 169 - 169$$ Set the equation equal to zero.

$$2x^2 - 14x - 120 = 0$$ Write the equation in the form $ax^2 + bx + c = 0$.

$$2(x^2 - 7x - 60) = 0$$ Factor.

$$2(x - 12)(x + 5) = 0$$

$2 \neq 0$ or $x - 12 = 0$ or $x + 5 = 0$ Set each factor equal to zero.

$x = 12$ or $x \neq -5$ Solve both equations for x.

> **Avoiding Mistakes**
>
> Recall that the square of a binomial results in a perfect square trinomial.
>
> $(a - b)^2 = a^2 - 2ab + b^2$
>
> Don't forget the middle term.

Recall that x represents the distance between the bottom of the board and the base of the platform. We reject the negative value of x because a distance cannot be negative. Therefore, the distance between the bottom of the board and the base of the platform is 12 ft. The distance between the top of the board and the ground is $x - 7 = 5$ ft.

Section 6.8 Practice Exercises

Boost *your* GRADE at ALEKS.com!   • Practice Problems   • e-Professors
• Self-Tests   • Videos
• NetTutor

Study Skills

1. Define the key term **Pythagorean theorem**.

Review Exercises

For Exercises 2–7, solve the quadratic equations.

2. $(6x + 1)(x + 4) = 0$

3. $9x(3x + 2) = 0$

4. $4x^2 - 1 = 0$

5. $x^2 - 5x = 6$

6. $x(x - 20) = -100$

7. $6x^2 - 7x - 10 = 0$

8. Explain what is wrong with the following problem: $(x - 3)(x + 2) = 5$
$$x - 3 = 5 \quad \text{or} \quad x + 2 = 5.$$

Objective 1: Applications of Quadratic Equations

9. If eleven is added to the square of a number, the result is sixty. Find all such numbers.

10. If a number is added to two times its square, the result is thirty-six. Find all such numbers.

11. If twelve is added to six times a number, the result is twenty-eight less than the square of the number. Find all such numbers.

12. The square of a number is equal to twenty more than the number. Find all such numbers.

13. The product of two consecutive odd integers is sixty-three. Find all such integers. **(See Example 1.)**

14. The product of two consecutive even integers is forty-eight. Find all such integers.

15. The sum of the squares of two consecutive integers is one more than ten times the larger number. Find all such integers.

16. The sum of the squares of two consecutive integers is nine less than ten times the sum of the integers. Find all such integers.

17. The length of a rectangular room is 5 yd more than the width. If 300 yd^2 of carpeting cover the room, what are the dimensions of the room? **(See Example 2.)**

x 300 yd^2

$x + 5$

18. The width of a rectangular painting is 2 in. less than the length. The area is 120 $in.^2$ Find the length and width.

x

$x - 2$

19. The width of a rectangular slab of concrete is 3 m less than the length. The area is 28 m^2.

  a. What are the dimensions of the rectangle?

  b. What is the perimeter of the rectangle?

20. The width of a rectangular picture is 7 in. less than the length. The area of the picture is 78 $in.^2$

  a. What are the dimensions of the picture?

  b. What is the perimeter of the picture?

21. The base of a triangle is 3 ft more than the height.

22. The height of a triangle is 15 cm more than the base.

23. In a physics experiment, a ball is dropped off a 144-ft platform. The height of the ball above the ground is given by the equation

$h = -16t^2 + 144$ where h is the ball's height in feet, and t is the time in seconds after the ball is dropped ($t \geq 0$).

Find the time required for the ball to hit the ground. (*Hint:* Let $h = 0$.) **(See Example 3.)**

24. A stone is dropped off a 256-ft cliff. The height of the stone above the ground is given by the equation

$h = -16t^2 + 256$ where h is the stone's height in feet, and t is the time in seconds after the stone is dropped ($t \geq 0$).

Find the time required for the stone to hit the ground.

25. An object is shot straight up into the air from ground level with an initial speed of 24 ft/sec. The height of the object (in feet) is given by the equation

$$h = -16t^2 + 24t \qquad \text{where } t \text{ is the time in seconds after launch } (t \geq 0).$$

Find the time(s) when the object is at ground level.

26. A rocket is launched straight up into the air from the ground with initial speed of 64 ft/sec. The height of the rocket (in feet) is given by the equation

$$h = -16t^2 + 64t \qquad \text{where } t \text{ is the time in seconds after launch } (t \geq 0).$$

Find the time(s) when the rocket is at ground level.

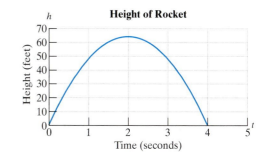

Objective 2: Pythagorean Theorem

27. Sketch a right triangle and label the sides with the words *leg* and *hypotenuse*.

28. State the Pythagorean theorem.

For Exercises 29–32, find the length of the missing side of the right triangle. **(See Example 4.)**

29.

30.

31.

32.

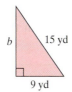

33. Find the length of the supporting brace.

34. Find the height of the airplane above the ground.

Section 6.8   Applications of Quadratic Equations

Key Concepts

Use the zero product rule to solve applications.

Some applications involve the Pythagorean theorem.

$$a^2 + b^2 = c^2$$

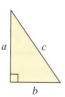

Examples

Example 1

Find two consecutive integers such that the sum of their squares is 61.

Let x represent one integer.
Let $x + 1$ represent the next consecutive integer.

$$x^2 + (x + 1)^2 = 61$$
$$x^2 + x^2 + 2x + 1 = 61$$
$$2x^2 + 2x - 60 = 0$$
$$2(x^2 + x - 30) = 0$$
$$2(x - 5)(x + 6) = 0$$
$$x = 5 \quad \text{or} \quad x = -6$$

If $x = 5$, then the next consecutive integer is 6.
If $x = -6$, then the next consecutive integer is -5.
The integers are 5 and 6, or -6 and -5.

Example 2

Find the length of the missing side.

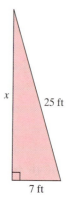

$$x^2 + (7)^2 = (25)^2$$
$$x^2 + 49 = 625$$
$$x^2 - 576 = 0$$
$$(x - 24)(x + 24) = 0$$
$$x = 24 \quad \text{or} \quad x = -24$$

The length of the side is 24 ft.

Chapter 6   Review Exercises

Section 6.1

For Exercises 1–4, identify the greatest common factor for each group of terms.

1. $15a^2b^4, 30a^3b, 9a^5b^3$ **2.** $3(x + 5), x(x + 5)$

3. $2c^3(3c - 5), 4c(3c - 5)$ **4.** $-2wyz, -4xyz$

For Exercises 5–10, factor out the greatest common factor.

5. $6x^2 + 2x^4 - 8x$ **6.** $11w^3y^3 - 44w^2y^5$

7. $-t^2 + 5t$ **8.** $-6u^2 - u$

9. $3b(b + 2) - 7(b + 2)$

10. $2(5x + 9) + 8x(5x + 9)$

For Exercises 11–14, factor by grouping.

11. $7w^2 + 14w + wb + 2b$

12. $b^2 - 2b + yb - 2y$

13. $60y^2 - 45y - 12y + 9$

14. $6a - 3a^2 - 2ab + a^2b$

Section 6.2

For Exercises 15–24, factor completely.

15. $x^2 - 10x + 21$ **16.** $y^2 - 19y + 88$

17. $-6z + z^2 - 72$ **18.** $-39 + q^2 - 10q$

19. $3p^2w + 36pw + 60w$ **20.** $2m^4 + 26m^3 + 80m^2$

21. $-t^2 + 10t - 16$ **22.** $-w^2 - w + 20$

23. $a^2 + 12ab + 11b^2$ **24.** $c^2 - 3cd - 18d^2$

Section 6.3

For Exercises 25–28, let a, b, and c represent positive integers.

25. When factoring a polynomial of the form $ax^2 - bx - c$, should the signs of the binomials be both positive, both negative, or different?

26. When factoring a polynomial of the form should the signs of the binomials be both positive, both negative, or different?

27. When factoring a polynomial of the form $ax^2 + bx + c$, should the signs of the binomials be both positive, both negative, or different?

28. When factoring a polynomial of the form $ax^2 + bx - c$, should the signs of the binomials be both positive, both negative, or different?

For Exercises 29–40, factor the trinomial using the trial-and-error method.

29. $2y^2 - 5y - 12$ **30.** $4w^2 - 5w - 6$

31. $10z^2 + 29z + 10$ **32.** $8z^2 + 6z - 9$

33. $2p^2 - 5p + 1$ **34.** $5r^2 - 3r + 7$

35. $10w^2 - 60w - 270$ **36.** $3y^2 - 18y - 48$

37. $9c^2 - 30cd + 25d^2$ **38.** $x^2 + 12x + 36$

39. $v^4 - 2v^2 - 3$ **40.** $x^4 + 7x^2 + 10$

41. In Exercises 29–40, which trinomials are perfect square trinomials?

Section 6.4

For Exercises 42–43, find a pair of integers whose product and sum are given.

42. Product: -5 sum: 4

43. Product: 15 sum: -8

For Exercises 44–57, factor the trinomial using the ac-method.

44. $3c^2 - 5c - 2$ **45.** $4y^2 + 13y + 3$

46. $t^2 + 13t + 12$ **47.** $4x^3 + 17x^2 - 15x$

48. $w^3 + 4w^2 - 5w$ **49.** $p^2 - 8pq + 15q^2$

50. $40v^2 + 22v - 6$ **51.** $40s^2 + 30s - 100$

Chapters 1–6 Cumulative Review Exercises

1. Simplify. $\dfrac{|4 - 25 \div (-5) \cdot 2|}{\sqrt{8^2 + 6^2}}$

2. Solve. $5 - 2(t + 4) = 3t + 12$

3. Solve for y. $3x - 2y = 8$

4. A child's piggy bank has $3.80 in quarters, dimes, and nickels. The number of nickels is two more than the number of quarters. The number of dimes is three less than the number of quarters. Find the number of each type of coin in the bank.

5. Solve the inequality. Graph the solution on a number line and write the solution set in interval notation.

$$-\frac{5}{12}x \le \frac{5}{3} \qquad \underrightarrow{\hspace{3cm}}$$

6. Given the equation $y = x + 4$

 a. Is the equation linear?

 b. Identify the slope.

 c. Identify the y-intercept.

 d. Identify the x-intercept.

 e. Graph the line.

7. Consider the equation, $x = 5$.

 a. Does the equation present a horizontal or vertical line?

 b. Determine the slope of the line, if it exists.

 c. Identify the x-intercept, if it exists.

 d. Identify the y-intercept, if it exists.

8. Find an equation of the line passing through the point $(-3, 5)$ and having a slope of 3. Write the final answer in slope-intercept form.

9. Solve the system. $\begin{aligned} 2x - 3y &= 4 \\ 5x - 6y &= 13 \end{aligned}$

For Exercises 10–12, perform the indicated operations.

10. $2\left(\dfrac{1}{3}y^3 - \dfrac{3}{2}y^2 - 7\right) - \left(\dfrac{2}{3}y^3 + \dfrac{1}{2}y^2 + 5y\right)$

11. $(4p^2 - 5p - 1)(2p - 3)$

12. $(2w - 7)^2$

13. Divide using long division:
$(r^4 + 2r^3 - 5r + 1) \div (r - 3)$

14. Simplify. $\dfrac{c^{12}c^{-5}}{c^3}$

15. Divide. Write the final answer in scientific notation: $\dfrac{8.0 \times 10^{-3}}{5.0 \times 10^{-6}}$

For Exercises 16–19, factor completely.

16. $w^4 - 16$

17. $2ax + 10bx - 3ya - 15yb$

18. $4x^2 - 8x - 5$

19. $y^3 - 27$

20. Solve. $4x(2x - 1)(x + 5) = 0$

Rational Expressions

CHAPTER OUTLINE

Chapter 7

In this chapter, we define a rational expression as a ratio of two polynomials. Then we perform operations on rational expressions and solve rational equations.

You may find it helpful to review operations with fractions before you begin this chapter. Try the matching puzzle, then record the letter in the spaces below to complete the sentence

1. Add. $\dfrac{7}{4} + \dfrac{5}{3} + \dfrac{1}{2}$ 2. Multiply. $\dfrac{5}{3} \cdot \dfrac{11}{4}$

3. Fraction that is not in lowest terms.

4. In this problem, a common denominator is not required.

5. Fractions can be eliminated from this problem by multiplying by the LCD.

a. $\dfrac{7}{8}x + \dfrac{2}{3} = \dfrac{1}{2}$ **t.** $\dfrac{45}{12}$

r. $\dfrac{47}{12}$ **p.** $\dfrac{34}{3} \cdot \dfrac{1}{4}$

n. $\dfrac{7}{8} + \dfrac{1}{3} - \dfrac{1}{2}$ **o.** $\dfrac{55}{12}$

The enemy of a good student is $\underline{}\ \underline{}\ \underline{}\ \underline{c}\ \underline{}\ \underline{}\ \underline{s}\ \underline{}\ \underline{i}\ \underline{n}\ \underline{}\ \underline{}\ \underline{i}\ \underline{}\ \underline{n}$.
$\ \ \ 4\ \ 1\ \ 2\ \ \ \ \ 1\ \ 5\ \ \ 3\ \ \ \ \ \ \ 5\ \ 3\ \ \ \ 2$

| Section 7.1 | **Introduction to Rational Expressions** |

Objectives

1. Definition of a Rational Expression
2. Evaluating Rational Expressions
3. Domain of a Rational Expression
4. Simplifying Rational Expressions to Lowest Terms
5. Simplifying a Ratio of −1

1. Definition of a Rational Expression

In Section 1.1, we defined a rational number as the ratio of two integers, $\frac{p}{q}$, where $q \neq 0$.

$$\text{Examples of rational numbers:}\quad \frac{2}{3}, -\frac{1}{5}, 9$$

In a similar way, we define a **rational expression** as the ratio of two polynomials, $\frac{p}{q}$, where $q \neq 0$.

$$\text{Examples of rational expressions:}\quad \frac{3x-6}{x^2-4},\quad \frac{3}{4},\quad \frac{6r^5+2r}{7}$$

2. Evaluating Rational Expressions

Skill Practice

Evaluate the expression for the given values of x.

$$\frac{x-3}{x+5}$$

1. $x = 2$
2. $x = 0$
3. $x = 3$
4. $x = -5$

| Example 1 | **Evaluating Rational Expressions** |

Evaluate the rational expression (if possible) for the given values of x: $\dfrac{12}{x-3}$

a. $x = 0$ **b.** $x = 1$ **c.** $x = -3$ **d.** $x = 3$

Solution:

Substitute the given value for the variable. Then use the order of operations to simplify.

a. $\dfrac{12}{x-3}$

$\dfrac{12}{(0)-3}$ Substitute $x = 0$.

$= \dfrac{12}{-3}$

$= -4$

b. $\dfrac{12}{x-3}$

$\dfrac{12}{(1)-3}$ Substitute $x = 1$.

$= \dfrac{12}{-2}$

$= -6$

c. $\dfrac{12}{x-3}$

$\dfrac{12}{(-3)-3}$ Substitute $x = -3$.

$= \dfrac{12}{-6}$

$= -2$

d. $\dfrac{12}{x-3}$

$\dfrac{12}{(3)-3}$ Substitute $x = 3$.

$= \dfrac{12}{0}$ Undefined.

Recall that division by zero is undefined.

Answers

1. $-\dfrac{1}{7}$ 2. $-\dfrac{3}{5}$
3. 0   4. Undefined

3. Domain of a Rational Expression

In Section 3.7, we presented the concept of the domain of a function.

> **DEFINITION Domain of a Function**
>
> The **domain** of a function is the set of all real numbers that when substituted into the function produces a real number.

For a rational expression, the domain will consist of the real numbers that when substituted into the expression do not make the denominator equal to zero. Therefore, in Example 1(d), because $\dfrac{12}{x-3}$ is undefined for $x = 3$, the domain is all real numbers except 3. We write this in set-builder notation as

$$\{x \mid x \text{ is a real number and } x \neq 3\}$$

Concept Connections

5. Fill in the blank.
The domain of a rational expression is all real numbers except those that make the _____ zero.

> **PROCEDURE Finding the Domain of a Rational Expression**
>
> **Step 1** Set the denominator equal to zero and solve the resulting equation.
> **Step 2** The domain is the set of real numbers *excluding* the values found in step 1.

Example 2 Finding the Domain of Rational Expressions

Find the domain of the expressions.

a. $\dfrac{y-3}{2y+7}$　　**b.** $\dfrac{-5}{x}$

Skill Practice

Find the domain.

6. $\dfrac{a+2}{2a-8}$

7. $\dfrac{2}{t}$

Solution:

a. $\dfrac{y-3}{2y+7}$

$2y + 7 = 0$　　Set the denominator equal to zero.

$2y = -7$　　Solve the equation.

$\dfrac{2y}{2} = \dfrac{-7}{2}$

$y = -\dfrac{7}{2}$　　The domain is the set of real numbers except $-\frac{7}{2}$.

Domain: $\{y \mid y \text{ is a real number and } y \neq -\frac{7}{2}\}$

b. $\dfrac{-5}{x}$

$x = 0$　　Set the denominator equal to zero.

Domain: $\{x \mid x \text{ is a real number and } x \neq 0\}$　　The domain is the set of real numbers except 0.

Answers

5. Denominator
6. $\{a \mid a \text{ is a real number and } a \neq 4\}$
7. $\{t \mid t \text{ is a real number and } t \neq 0\}$

Answers
8. $\{w \mid w$ is a real number and $w \neq 3,\ w \neq -3\}$
9. The set of all real numbers

Example 3 **Finding the Domain of Rational Expressions**

Find the domain of the expressions.

a. $\dfrac{a+10}{a^2-25}$ **b.** $\dfrac{2x^3+5}{x^2+9}$

Solution:

a. $\dfrac{a+10}{a^2-25}$

$a^2 - 25 = 0$	Set the denominator equal to zero. The equation is quadratic.
$(a-5)(a+5) = 0$	Factor.
$a-5=0$ or $a+5=0$	Set each factor equal to zero.
$a=5$ or $a=-5$	The domain is the set of real numbers except 5 and −5.

Domain: $\{a \mid a$ is a real number and $a \neq 5,\ a \neq -5\}$

b. $\dfrac{2x^3+5}{x^2+9}$

The quantity x^2 cannot be negative for any real number, x, so the denominator x^2+9 cannot equal zero. Therefore, no numbers are excluded from the domain.

The domain is the set of all real numbers.

4. Simplifying Rational Expressions to Lowest Terms

In many cases, it is advantageous to simplify or reduce a fraction to lowest terms. The same is true for rational expressions.

The method for simplifying rational expressions mirrors the process for simplifying fractions. In each case, factor the numerator and denominator. Common factors in the numerator and denominator form a ratio of 1 and can be reduced.

Simplifying a fraction: $\dfrac{21}{35} \xrightarrow{\text{Factor}} \dfrac{3 \cdot \overset{1}{\cancel{7}}}{5 \cdot \cancel{7}} = \dfrac{3}{5} \cdot (1) = \dfrac{3}{5}$

Simplifying a rational expression: $\dfrac{2x-6}{x^2-9} \xrightarrow{\text{Factor}} \dfrac{2\overset{1}{\cancel{(x-3)}}}{(x+3)\cancel{(x-3)}} = \dfrac{2}{(x+3)}(1) = \dfrac{2}{x+3}$

Informally, to simplify a rational expression to lowest terms, we simplify the ratio of common factors to 1. Formally, this is accomplished by applying the fundamental principle of rational expressions.

PROPERTY **Fundamental Principle of Rational Expressions**

Let p, q, and r represent polynomials where $q \neq 0$ and $r \neq 0$. Then

$$\frac{pr}{qr} = \frac{p}{q} \cdot \frac{r}{r} = \frac{p}{q} \cdot 1 = \frac{p}{q}$$

| Example 4 | **Simplifying a Rational Expression to Lowest Terms** |

Given the expression $\dfrac{2p - 14}{p^2 - 49}$

a. Factor the numerator and denominator.

b. Determine the domain of the expression.

c. Simplify the expression to lowest terms.

Solution:

a. $\dfrac{2p - 14}{p^2 - 49}$ Factor out the GCF in the numerator.

$= \dfrac{2(p - 7)}{(p + 7)(p - 7)}$ Factor the denominator as a difference of squares.

b. $(p + 7)(p - 7) = 0$ To find the domain restrictions, set the denominator equal to zero. The equation is quadratic.

$p + 7 = 0$ or $p - 7 = 0$ Set each factor equal to 0.

$p = -7$ or $p = 7$ The domain is all real numbers except -7 and 7.

Domain: $\{p \mid p \text{ is a real number and } p \neq -7, p \neq 7\}$

c. $\dfrac{2(\overset{1}{\cancel{p - 7}})}{(p + 7)(\cancel{p - 7})}$ Simplify the ratio of common factors to 1.

$= \dfrac{2}{p + 7}$ (provided $p \neq 7$ and $p \neq -7$)

In Example 4, it is important to note that the expressions

$$\frac{2p - 14}{p^2 - 49} \quad \text{and} \quad \frac{2}{p + 7}$$

are equal for all values of p that make each expression a real number. Therefore,

$$\frac{2p - 14}{p^2 - 49} = \frac{2}{p + 7}$$

for all values of p except $p = 7$ and $p = -7$. (At $p = 7$ and $p = -7$, the original expression is undefined.) This is why the domain of an expression is always determined before the expression is simplified.

From this point forward, we will write statements of equality between two rational expressions with the assumption that they are equal for all values of the variable for which each expression is defined.

Skill Practice

Simplify to lowest terms.

13. $\dfrac{15q^3}{9q^2}$

Example 5　Simplifying a Rational Expression to Lowest Terms

Simplify to lowest terms.　$\dfrac{18a^4}{9a^5}$

Solution:

$\dfrac{18a^4}{9a^5}$

$= \dfrac{2 \cdot 3 \cdot 3 \cdot a \cdot a \cdot a \cdot a}{3 \cdot 3 \cdot a \cdot a \cdot a \cdot a \cdot a}$　Factor the numerator and denominator.

$= \dfrac{2 \cdot (3 \cdot 3 \cdot a \cdot a \cdot a \cdot a)}{(3 \cdot 3 \cdot a \cdot a \cdot a \cdot a) \cdot a}$　Simplify common factors to lowest terms.

$= \dfrac{2}{a}$

> **TIP:** The expression $\dfrac{18a^4}{9a^5}$ can also be simplified using the properties of exponents.
>
> $$\dfrac{18a^4}{9a^5} = 2a^{4-5} = 2a^{-1} = \dfrac{2}{a}$$

Skill Practice

Simplify to lowest terms.

14. $\dfrac{x^2 - 1}{2x^2 - x - 3}$

Avoiding Mistakes

Given the expression

$$\dfrac{2c - 8}{10c^2 - 80c + 160}$$

do not be tempted to reduce before factoring. The terms $2c$ and $10c^2$ cannot be "canceled" because they are *terms* not factors.

The numerator and denominator must be in factored form before simplifying.

Example 6　Simplifying a Rational Expression to Lowest Terms

Simplify to lowest terms.　$\dfrac{2c - 8}{10c^2 - 80c + 160}$

Solution:

$\dfrac{2c - 8}{10c^2 - 80c + 160}$

$= \dfrac{2(c - 4)}{10(c^2 - 8c + 16)}$　Factor out the GCF.

$= \dfrac{2(c - 4)}{10(c - 4)^2}$　Factor the denominator.

$= \dfrac{2(c - 4)}{2 \cdot 5(c - 4)(c - 4)}$　Simplify the ratio of common factors to 1.

$= \dfrac{1}{5(c - 4)}$

The process to simplify a rational expression to lowest terms is based on the identity property of multiplication. Therefore, this process applies only to factors (remember that factors are multiplied). For example,

Answers

13. $\dfrac{5q}{3}$　**14.** $\dfrac{x - 1}{2x - 3}$

$$\frac{3x}{3y} = \frac{\overset{1}{\cancel{3}} \cdot x}{\underset{1}{\cancel{3}} \cdot y} = 1 \cdot \frac{x}{y} = \frac{x}{y}$$

Simplify

Terms that are added or subtracted cannot be reduced to lowest terms. For example,

$$\frac{x + 3}{y + 3}$$

Cannot be simplified

The objective of simplifying a rational expression to lowest terms is to create an equivalent expression that is simpler to use. Consider the rational expression from Example 6 in its original form and in its reduced form. If we choose an arbitrary value of c from the domain and substitute that value into each expression, we see that the reduced form is easier to evaluate. For example, substitute $c = 3$:

Original Expression   **Simplified Expression**

$$\frac{2c - 8}{10c^2 - 80c + 160} \qquad \frac{1}{5(c - 4)}$$

Substitute $c = 3$

$$= \frac{2(3) - 8}{10(3)^2 - 80(3) + 160} \qquad = \frac{1}{5(3 - 4)}$$

$$= \frac{6 - 8}{10(9) - 240 + 160} \qquad = \frac{1}{5(-1)}$$

$$= \frac{-2}{90 - 240 + 160} \qquad = -\frac{1}{5}$$

$$= \frac{-2}{10} \quad \text{or} \quad -\frac{1}{5}$$

5. Simplifying a Ratio of −1

When two factors are identical in the numerator and denominator, they form a ratio of 1 and can be reduced. Sometimes we encounter two factors that are opposites and form a ratio of −1. For example,

Simplified Form   **Details/Notes**

$$\frac{-5}{5} = -1 \qquad \text{The ratio of a number and its opposite is } -1.$$

$$\frac{100}{-100} = -1 \qquad \text{The ratio of a number and its opposite is } -1.$$

$$\frac{x + 7}{-x - 7} = -1 \qquad \frac{x + 7}{-x - 7} = \frac{x + 7}{-1(x + 7)} = \frac{\overset{1}{\cancel{x + 7}}}{-1(\cancel{x + 7})} = \frac{1}{-1} = -1$$

factor out −1

$$\frac{2 - x}{x - 2} = -1 \qquad \frac{2 - x}{x - 2} = \frac{-1(-2 + x)}{x - 2} = \frac{-1(\overset{1}{\cancel{x - 2}})}{\cancel{x - 2}} = \frac{-1}{1} = -1$$

> **Avoiding Mistakes**
>
> While the expression $2 - x$ and $x - 2$ are opposites, the expressions $2 - x$ and $2 + x$ are *not*.
>
> Therefore $\dfrac{2 - x}{2 + x}$ does not simplify to −1.

Recognizing factors that are opposites is useful when simplifying rational expressions.

Skill Practice

Simplify to lowest terms.

15. $\dfrac{2t - 12}{6 - t}$

Example 7 Simplifying a Rational Expression to Lowest Terms

Simplify to lowest terms. $\dfrac{3c - 3d}{d - c}$

Solution:

$\dfrac{3c - 3d}{d - c}$

$= \dfrac{3(c - d)}{d - c}$ Factor the numerator and denominator.

Notice that $(c - d)$ and $(d - c)$ are opposites and form a ratio of -1.

$= \dfrac{3(c \overset{-1}{\cancel{- d}})}{\cancel{d - c}}$ Details: $\dfrac{3(c - d)}{d - c} = \dfrac{3(c - d)}{-1(-d + c)} = \dfrac{3(c - d)}{-1(c - d)}$

$= 3(-1)$ $= \dfrac{3}{-1} = -3$

$= -3$

TIP: It is important to recognize that a rational expression can be written in several equivalent forms. In particular, two numbers with opposite signs form a negative quotient. Therefore, a number such as $-\frac{3}{4}$ can be written as:

$$-\dfrac{3}{4} \quad \text{or} \quad \dfrac{-3}{4} \quad \text{or} \quad \dfrac{3}{-4}$$

The negative sign can be written in the numerator, in the denominator, or out in front of the fraction. We demonstrate this concept in Example 8.

Skill Practice

Simplify to lowest terms.

16. $\dfrac{b - a}{a^2 - b^2}$

Example 8 Simplifying a Rational Expression to Lowest Terms

Simplify to lowest terms. $\dfrac{5 - y}{y^2 - 25}$

Solution:

$\dfrac{5 - y}{y^2 - 25}$

$= \dfrac{5 - y}{(y - 5)(y + 5)}$ Factor the numerator and denominator.

Notice that $5 - y$ and $y - 5$ are opposites and form a ratio of -1.

$= \dfrac{\overset{-1}{5 \cancel{- y}}}{\cancel{(y - 5)}(y + 5)}$ Details: $\dfrac{5 - y}{(y - 5)(y + 5)} = \dfrac{-1(-5 + y)}{(y - 5)(y + 5)}$

$\qquad\qquad\qquad\qquad\qquad\qquad = \dfrac{-1(y - 5)}{(y - 5)(y + 5)} = \dfrac{-1}{y + 5}$

$= \dfrac{-1}{y + 5} \quad \text{or} \quad \dfrac{1}{-(y + 5)} \quad \text{or} \quad -\dfrac{1}{y + 5}$

Answers

15. -2 **16.** $\dfrac{-1}{a + b}$

Section 7.1   Practice Exercises

Study Skills Exercises

1. Review Section R.2 in this text. Write an example of how to simplify (reduce) a fraction, multiply two fractions, divide two fractions, add two fractions, and subtract two fractions. Then as you learn about rational expressions, compare the operations on rational expressions with those on fractions. This is a great place to use 3×5 cards again. Write an example of an operation with fractions on one side and the same operation with rational expressions on the other side.

2. Define the key terms:

a. rational expression   **b. domain**

Objective 1: Definition of a Rational Expression

3. a. What is a rational number?

b. What is a rational expression?

4. a. Write an example of a rational number. (Answers will vary.)

b. Write an example of a rational expression. (Answers will vary.)

Objective 2: Evaluating Rational Expressions

For Exercises 5–10, substitute the given number into the expression and simplify (if possible). **(See Example 1.)**

5. $\dfrac{1}{x - 6}$; $x = -2$

6. $\dfrac{w - 10}{w + 6}$; $w = 0$

7. $\dfrac{w - 4}{2w + 8}$; $w = 0$

8. $\dfrac{y - 8}{2y^2 + y - 1}$; $y = 8$

9. $\dfrac{(a - 7)(a + 1)}{(a - 2)(a + 5)}$; $a = 2$

10. $\dfrac{(a + 4)(a + 1)}{(a - 4)(a - 1)}$; $a = 1$

11. A bicyclist rides 24 mi against a wind and returns 24 mi with the same wind. His average speed for the return trip traveling with the wind is 8 mph faster than his speed going out against the wind. If x represents the bicyclist's speed going out against the wind, then the total time, t, required for the round trip is given by

$$t = \frac{24}{x} + \frac{24}{x + 8}$$ where t is measured in hours.

a. Find the time required for the round trip if the cyclist rides 12 mph against the wind.

b. Find the time required for the round trip if the cyclist rides 24 mph against the wind.

12. The manufacturer of mountain bikes has a fixed cost of $56,000, plus a variable cost of $140 per bike. The average cost per bike, y (in dollars), is given by the equation:

$$y = \frac{56,000 + 140x}{x}$$ where x represents the number of bikes produced.

a. Find the average cost per bike if the manufacturer produces 1000 bikes.

b. Find the average cost per bike if the manufacturer produces 2000 bikes.

c. Find the average cost per bike if the manufacturer produces 10,000 bikes.

Objective 3: Domain of a Rational Expression

For Exercises 13–24, write the domain. **(See Examples 2–3.)**

13. $\dfrac{5}{k+2}$

14. $\dfrac{-3}{h-4}$

15. $\dfrac{x+5}{(2x-5)(x+8)}$

16. $\dfrac{4y+1}{(3y+7)(y+3)}$

17. $\dfrac{b+12}{b^2+5b+6}$

18. $\dfrac{c-11}{c^2-5c-6}$

19. $\dfrac{x-4}{x^2+9}$

20. $\dfrac{x+1}{x^2+4}$

21. $\dfrac{y^2-y-12}{12}$

22. $\dfrac{z^2+10z+9}{9}$

23. $\dfrac{t-5}{t}$

24. $\dfrac{2w+7}{w}$

25. Construct a rational expression that is undefined for $x=2$. (Answers will vary.)

26. Construct a rational expression that is undefined for $x=5$. (Answers will vary.)

27. Construct a rational expression that is undefined for $x=-3$ and $x=7$. (Answers will vary.)

28. Construct a rational expression that is undefined for $x=-1$ and $x=4$. (Answers will vary.)

29. Evaluate the expressions for $x=-1$.

   a. $\dfrac{3x^2-2x-1}{6x^2-7x-3}$ **b.** $\dfrac{x-1}{2x-3}$

30. Evaluate the expressions for $x=4$.

   a. $\dfrac{(x+5)^2}{x^2+6x+5}$ **b.** $\dfrac{x+5}{x+1}$

31. Evaluate the expressions for $x=1$.

   a. $\dfrac{5x+5}{x^2-1}$ **b.** $\dfrac{5}{x-1}$

32. Evaluate the expressions for $x=3$.

   a. $\dfrac{2x^2-4x-6}{2x^2-18}$ **b.** $\dfrac{x+1}{x+3}$

Objective 4: Simplifying Rational Expressions to Lowest Terms

For Exercises 33–42,

   a. Write the domain in set-builder notation.

   b. Simplify the expression to lowest terms. **(See Example 4.)**

33. $\dfrac{3y+6}{6y+12}$

34. $\dfrac{8x-8}{4x-4}$

35. $\dfrac{t^2-1}{t+1}$

36. $\dfrac{r^2-4}{r-2}$

37. $\dfrac{7w}{21w^2-35w}$

38. $\dfrac{12a^2}{24a^2-18a}$

39. $\dfrac{9x^2-4}{6x+4}$

40. $\dfrac{8b-20}{4b^2-25}$

41. $\dfrac{a^2+3a-10}{a^2+a-6}$

42. $\dfrac{t^2+3t-10}{t^2+t-20}$

For Exercises 43–84, simplify the expression to lowest terms. **(See Examples 5–6.)**

43. $\dfrac{7b^2}{21b}$

44. $\dfrac{15c^3}{3c^5}$

45. $\dfrac{18st^5}{12st^3}$

46. $\dfrac{20a^4b^2}{25ab^2}$

47. $\dfrac{-24x^2y^5z}{8xy^4z^3}$

48. $\dfrac{60rs^4t^2}{-12r^4s^2t^3}$

49. $\dfrac{3(y+2)}{6(y+2)}$

50. $\dfrac{8(x-1)}{4(x-1)}$

51. $\dfrac{(p-3)(p+5)}{(p+5)(p+4)}$

52. $\dfrac{(c+4)(c-1)}{(c+4)(c+2)}$

53. $\dfrac{(m+11)}{4(m+11)(m-11)}$

54. $\dfrac{(n-7)}{9(n+2)(n-7)}$

55. $\dfrac{x(2x+1)^2}{4x^3(2x+1)}$

56. $\dfrac{(p+2)(p-3)^4}{(p+2)^2(p-3)^2}$

57. $\dfrac{5}{20a-25}$

58. $\dfrac{7}{14c-21}$

59. $\dfrac{4w-8}{w^2-4}$

60. $\dfrac{3x+15}{x^2-25}$

61. $\dfrac{3x^2-6x}{9xy+18x}$

62. $\dfrac{6p^2+12p}{2pq-4p}$

63. $\dfrac{2x+4}{x^2-3x-10}$

64. $\dfrac{5z+15}{z^2-4z-21}$

65. $\dfrac{a^2-49}{a-7}$

66. $\dfrac{b^2-64}{b-8}$

67. $\dfrac{q^2+25}{q+5}$

68. $\dfrac{r^2+36}{r+6}$

69. $\dfrac{y^2+6y+9}{2y^2+y-15}$

70. $\dfrac{h^2+h-6}{h^2+2h-8}$

71. $\dfrac{3x^2+7x-6}{x^2+7x+12}$

72. $\dfrac{x^2-5x-14}{2x^2-x-10}$

73. $\dfrac{5q^2+5}{q^4-1}$

74. $\dfrac{4t^2+16}{t^4-16}$

75. $\dfrac{ac-ad+2bc-2bd}{2ac+ad+4bc+2bd}$ (*Hint:* Factor by grouping.)

76. $\dfrac{3pr-ps-3qr+qs}{3pr-ps+3qr-qs}$ (*Hint:* Factor by grouping.)

77. $\dfrac{2t^2-3t}{2t^4-13t^3+15t^2}$

78. $\dfrac{4m^3+3m^2}{4m^3+7m^2+3m}$

79. $\dfrac{49p^2-28pq+4q^2}{14p-4q}$

80. $\dfrac{3x-3y}{2x^2-4xy+2y^2}$

81. $\dfrac{5x^3+4x^2-45x-36}{x^2-9}$

82. $\dfrac{x^2-1}{ax^3-bx^2-ax+b}$

83. $\dfrac{2x^2-xy-3y^2}{2x^2-11xy+12y^2}$

84. $\dfrac{2c^2+cd-d^2}{5c^2+3cd-2d^2}$

Objective 5: Simplifying a Ratio of −1

85. What is the relationship between $x-2$ and $2-x$?

86. What is the relationship between $w+p$ and $-w-p$?

For Exercises 87–98, simplify to lowest terms. (See Examples 7–8.)

87. $\dfrac{x-5}{5-x}$

88. $\dfrac{8-p}{p-8}$

89. $\dfrac{-4-y}{4+y}$

90. $\dfrac{z+10}{-z-10}$

91. $\dfrac{3y-6}{12-6y}$

92. $\dfrac{4q-4}{12-12q}$

93. $\dfrac{k+5}{5-k}$

94. $\dfrac{2+n}{2-n}$

95. $\dfrac{10x-12}{10x+12}$

96. $\dfrac{4t-16}{16+4t}$

97. $\dfrac{x^2-x-12}{16-x^2}$

98. $\dfrac{49-b^2}{b^2-10b+21}$

Expanding Your Skills

For Exercises 99–102, factor and simplify to lowest terms.

99. $\dfrac{w^3-8}{w^2+2w+4}$

100. $\dfrac{y^3+27}{y^2-3y+9}$

101. $\dfrac{z^2-16}{z^3-64}$

102. $\dfrac{x^2-25}{x^3+125}$

Section 7.2 **Multiplication and Division of Rational Expressions**

Objectives

1. **Multiplication of Rational Expressions**
2. **Division of Rational Expressions**

1. Multiplication of Rational Expressions

Recall from Section R.2 that to multiply fractions, we multiply the numerators and multiply the denominators. The same is true for multiplying rational expressions.

> **PROPERTY Multiplication of Rational Expressions**
>
> Let p, q, r, and s represent polynomials, such that $q \neq 0$, $s \neq 0$. Then,
>
> $$\frac{p}{q} \cdot \frac{r}{s} = \frac{pr}{qs}$$

For example:

Multiply the Fractions

$$\frac{2}{3} \cdot \frac{5}{7} = \frac{10}{21}$$

Multiply the Rational Expressions

$$\frac{2x}{3y} \cdot \frac{5z}{7} = \frac{10xz}{21y}$$

Sometimes it is possible to simplify a ratio of common factors to 1 *before* multiplying. To do so, we must first factor the numerators and denominators of each fraction.

$$\frac{15}{14} \cdot \frac{21}{10} = \frac{3 \cdot \overset{1}{\cancel{5}}}{2 \cdot \cancel{7}} \cdot \frac{3 \cdot \overset{1}{\cancel{7}}}{2 \cdot \cancel{5}} = \frac{9}{4}$$

The same process is also used to multiply rational expressions.

> **PROCEDURE Multiplying Rational Expressions**
>
> **Step 1** Factor the numerators and denominators of all rational expressions.
> **Step 2** Simplify the ratios of common factors to 1.
> **Step 3** Multiply the remaining factors in the numerator, and multiply the remaining factors in the denominator.

Skill Practice

Multiply.

1. $\dfrac{7a}{3b} \cdot \dfrac{15b}{14a^2}$

Answer

1. $\dfrac{5}{2a}$

Example 1 **Multiplying Rational Expressions**

Multiply. $\dfrac{5a^2b}{2} \cdot \dfrac{6a}{10b}$

Solution:

$$\frac{5a^2b}{2} \cdot \frac{6a}{10b}$$

$$= \frac{5 \cdot a \cdot a \cdot b}{2} \cdot \frac{2 \cdot 3 \cdot a}{2 \cdot 5 \cdot b} \qquad \text{Factor into prime factors.}$$

$$= \frac{\overset{1}{\cancel{5}} \cdot a \cdot a \cdot \overset{1}{\cancel{b}}}{2} \cdot \frac{\overset{1}{\cancel{2}} \cdot 3 \cdot a}{\cancel{2} \cdot \cancel{5} \cdot \cancel{b}} \qquad \text{Simplify.}$$

$$= \frac{3a^3}{2} \qquad \text{Multiply remaining factors.}$$

Example 2 **Multiplying Rational Expressions**

Multiply. $\dfrac{3c - 3d}{6c} \cdot \dfrac{2}{c^2 - d^2}$

Solution:

$\dfrac{3c - 3d}{6c} \cdot \dfrac{2}{c^2 - d^2}$

$= \dfrac{3(c - d)}{2 \cdot 3 \cdot c} \cdot \dfrac{2}{(c - d)(c + d)}$ Factor.

$= \dfrac{\overset{1}{\cancel{3}}(\overset{1}{\cancel{c - d}})}{\underset{1}{\cancel{2}} \cdot \underset{1}{\cancel{3}} \cdot c} \cdot \dfrac{\overset{1}{\cancel{2}}}{(\cancel{c - d})(c + d)}$ Simplify.

$= \dfrac{1}{c(c + d)}$ Multiply remaining factors.

Skill Practice

Multiply.

2. $\dfrac{4x - 8}{x + 6} \cdot \dfrac{x^2 + 6x}{2x}$

Avoiding Mistakes

If all the factors in the numerator reduce to a ratio of 1, a factor of 1 is left in the numerator.

Example 3 **Multiplying Rational Expressions**

Multiply. $\dfrac{35 - 5x}{5x + 5} \cdot \dfrac{x^2 + 5x + 4}{x^2 - 49}$

Solution:

$\dfrac{35 - 5x}{5x + 5} \cdot \dfrac{x^2 + 5x + 4}{x^2 - 49}$

$= \dfrac{5(7 - x)}{5(x + 1)} \cdot \dfrac{(x + 4)(x + 1)}{(x - 7)(x + 7)}$ Factor the numerators and denominators completely.

$= \dfrac{\overset{1}{\cancel{5}}(\overset{-1}{\cancel{7 - x}})}{\cancel{5}(\cancel{x + 1})} \cdot \dfrac{(x + 4)(\overset{1}{\cancel{x + 1}})}{(\cancel{x - 7})(x + 7)}$ Simplify the ratios of common factors to 1 or −1.

$= \dfrac{-1(x + 4)}{x + 7}$ Multiply remaining factors.

$= \dfrac{-(x + 4)}{x + 7}$ or $\dfrac{x + 4}{-(x + 7)}$ or $-\dfrac{x + 4}{x + 7}$

Skill Practice

Multiply.

3. $\dfrac{p^2 + 4p + 3}{5p + 10} \cdot \dfrac{p^2 - p - 6}{9 - p^2}$

TIP: The ratio $\frac{7 - x}{x - 7} = -1$ because $7 - x$ and $x - 7$ are opposites.

---

2. Division of Rational Expressions

Recall that to divide fractions, multiply the first fraction by the reciprocal of the second.

$\dfrac{21}{10} \div \dfrac{49}{15}$ $\xrightarrow[\text{of the second fraction}]{\text{multiply by the reciprocal}}$ $\dfrac{21}{10} \cdot \dfrac{15}{49}$ $\xrightarrow{\text{factor}}$ $\dfrac{3 \cdot \overset{1}{\cancel{7}}}{2 \cdot \cancel{5}} \cdot \dfrac{3 \cdot \overset{1}{\cancel{5}}}{\cancel{7} \cdot 7} = \dfrac{9}{14}$

The same process is used to divide rational expressions.

PROPERTY **Division of Rational Expressions**

Let p, q, r, and s represent polynomials, such that $q \neq 0, r \neq 0, s \neq 0$. Then,

$$\dfrac{p}{q} \div \dfrac{r}{s} = \dfrac{p}{q} \cdot \dfrac{s}{r} = \dfrac{ps}{qr}$$

Answers

2. $2(x - 2)$

3. $\dfrac{-(p + 1)}{5}$ or $\dfrac{p + 1}{-5}$ or $-\dfrac{p + 1}{5}$

Skill Practice

Divide.

4. $\dfrac{7y - 14}{y + 1} \div \dfrac{y^2 + 2y - 8}{2y + 2}$

Example 4 Dividing Rational Expressions

Divide. $\dfrac{5t - 15}{2} \div \dfrac{t^2 - 9}{10}$

Solution:

$\dfrac{5t - 15}{2} \div \dfrac{t^2 - 9}{10}$

$= \dfrac{5t - 15}{2} \cdot \dfrac{10}{t^2 - 9}$ Multiply the first fraction by the reciprocal of the second.

$= \dfrac{5(t - 3)}{2} \cdot \dfrac{2 \cdot 5}{(t - 3)(t + 3)}$ Factor each polynomial.

$= \dfrac{5(t - 3)}{2} \cdot \dfrac{2 \cdot 5}{(t - 3)(t + 3)}$ Simplify the ratio of common factors to 1.

$= \dfrac{25}{t + 3}$

Avoiding Mistakes

When dividing rational expressions, take the reciprocal of the second fraction and change to multiplication *before* reducing like factors.

Skill Practice

Divide.

5. $\dfrac{4x^2 - 9}{2x^2 - x - 3} \div \dfrac{20x + 30}{x^2 + 7x + 6}$

Example 5 Dividing Rational Expressions

Divide. $\dfrac{p^2 - 11p + 30}{10p^2 - 250} \div \dfrac{30p - 5p^2}{2p + 4}$

Solution:

$\dfrac{p^2 - 11p + 30}{10p^2 - 250} \div \dfrac{30p - 5p^2}{2p + 4}$

$= \dfrac{p^2 - 11p + 30}{10p^2 - 250} \cdot \dfrac{2p + 4}{30p - 5p^2}$ Multiply the first fraction by the reciprocal of the second.

Factor the trinomial.
$p^2 - 11p + 30 = (p - 5)(p - 6)$

$= \dfrac{(p - 5)(p - 6)}{2 \cdot 5(p - 5)(p + 5)} \cdot \dfrac{2(p + 2)}{5p(6 - p)}$ Factor out the GCF.
$2p + 4 = 2(p + 2)$

Factor out the GCF. Then factor the difference of squares.
$10p^2 - 250 = 10(p^2 - 25)$
$\qquad\qquad = 2 \cdot 5(p - 5)(p + 5)$

Factor out the GCF.
$30p - 5p^2 = 5p(6 - p)$

$= \dfrac{(p - 5)(p - 6)}{2 \cdot 5(p - 5)(p + 5)} \cdot \dfrac{2(p + 2)}{5p(6 - p)}$ Simplify the ratio of common factors to 1 or −1.

$= -\dfrac{(p + 2)}{25p(p + 5)}$

Answers

4. $\dfrac{14}{y + 4}$ 5. $\dfrac{x + 6}{10}$

Example 6 Dividing Rational Expressions

Divide. $\dfrac{\dfrac{3x}{4y}}{\dfrac{5x}{6y}}$

Skill Practice

Divide.

6. $\dfrac{\dfrac{a^3b}{9c}}{\dfrac{4ab}{3c^3}}$

Solution:

$\dfrac{\dfrac{3x}{4y}}{\dfrac{5x}{6y}}$ ⟵ This fraction bar denotes division (÷). This expression is called a complex fraction because it has one or more rational expressions in its numerator or denominator.

$= \dfrac{3x}{4y} \div \dfrac{5x}{6y}$

$= \dfrac{3x}{4y} \cdot \dfrac{6y}{5x}$ — Multiply by the reciprocal of the second fraction.

$= \dfrac{3 \cdot \overset{1}{\cancel{x}}}{\cancel{2} \cdot 2 \cdot y} \cdot \dfrac{\overset{1}{\cancel{2}} \cdot 3 \cdot \overset{1}{\cancel{y}}}{5 \cdot \cancel{x}}$ — Simplify the ratio of common factors to 1.

$= \dfrac{9}{10}$

Sometimes multiplication and division of rational expressions appear in the same problem. In such a case, apply the order of operations by multiplying or dividing in order from left to right.

Example 7 Multiplying and Dividing Rational Expressions

Perform the indicated operations. $\dfrac{4}{c^2 - 9} \div \dfrac{6}{c - 3} \cdot \dfrac{3c}{8}$

Skill Practice

Perform the indicated operations.

7. $\dfrac{v}{v + 2} \div \dfrac{5v^2}{v^2 - 4} \cdot \dfrac{v}{10}$

Solution:

In this example, division occurs first, before multiplication. Parentheses may be inserted to reinforce the proper order.

$\left(\dfrac{4}{c^2 - 9} \div \dfrac{6}{c - 3} \right) \cdot \dfrac{3c}{8}$

$= \left(\dfrac{4}{c^2 - 9} \cdot \dfrac{c - 3}{6} \right) \cdot \dfrac{3c}{8}$ — Multiply the first fraction by the reciprocal of the second.

$= \left(\dfrac{2 \cdot 2}{(c - 3)(c + 3)} \cdot \dfrac{c - 3}{2 \cdot 3} \right) \cdot \dfrac{3 \cdot c}{2 \cdot 2 \cdot 2}$ — Now that each operation is written as multiplication, factor the polynomials and reduce the common factors.

$= \dfrac{\overset{1}{\cancel{2}} \cdot \overset{1}{\cancel{2}}}{\cancel{(c - 3)}(c + 3)} \cdot \dfrac{\overset{1}{(\cancel{c - 3})}}{2 \cdot \cancel{3}} \cdot \dfrac{\overset{1}{\cancel{3}} \cdot c}{2 \cdot 2 \cdot 2}$

$= \dfrac{c}{4(c + 3)}$ — Simplify.

Answers

6. $\dfrac{a^2c^2}{12}$ 7. $\dfrac{v - 2}{50}$

Section 7.2 Practice Exercises

Review Exercises

For Exercises 1–8, multiply or divide the fractions.

1. $\dfrac{3}{5} \cdot \dfrac{1}{2}$

2. $\dfrac{6}{7} \cdot \dfrac{5}{12}$

3. $\dfrac{3}{4} \div \dfrac{3}{8}$

4. $\dfrac{18}{5} \div \dfrac{2}{5}$

5. $6 \cdot \dfrac{5}{12}$

6. $\dfrac{7}{25} \cdot 5$

7. $\dfrac{\frac{21}{4}}{\frac{7}{5}}$

8. $\dfrac{\frac{9}{2}}{\frac{3}{4}}$

Objective 1: Multiplication of Rational Expressions

For Exercises 9–24, multiply. **(See Examples 1–3.)**

9. $\dfrac{2xy}{5x^2} \cdot \dfrac{15}{4y}$

10. $\dfrac{7s}{t^2} \cdot \dfrac{t^2}{14s^2}$

11. $\dfrac{6x^3}{9x^6y^2} \cdot \dfrac{18x^4y^7}{4y}$

12. $\dfrac{10a^2b}{15b^2} \cdot \dfrac{30b}{2a^3}$

13. $\dfrac{4x - 24}{20x} \cdot \dfrac{5x}{8}$

14. $\dfrac{5a + 20}{a} \cdot \dfrac{3a}{10}$

15. $\dfrac{3y + 18}{y^2} \cdot \dfrac{4y}{6y + 36}$

16. $\dfrac{2p - 4}{6p} \cdot \dfrac{4p^2}{8p - 16}$

17. $\dfrac{10}{2 - a} \cdot \dfrac{a - 2}{16}$

18. $\dfrac{b - 3}{6} \cdot \dfrac{20}{3 - b}$

19. $\dfrac{b^2 - a^2}{a - b} \cdot \dfrac{a}{a^2 - ab}$

20. $\dfrac{(x - y)^2}{x^2 + xy} \cdot \dfrac{x}{y - x}$

21. $\dfrac{y^2 + 2y + 1}{5y - 10} \cdot \dfrac{y^2 - 3y + 2}{y^2 - 1}$

22. $\dfrac{6a^2 - 6}{a^2 + 6a + 5} \cdot \dfrac{a^2 + 5a}{12a}$

23. $\dfrac{10x}{2x^2 + 3x + 1} \cdot \dfrac{x^2 + 7x + 6}{5x}$

24. $\dfrac{b - 3}{b^2 + b - 12} \cdot \dfrac{4b + 16}{b + 1}$

Objective 2: Division of Rational Expressions

For Exercises 25–38, divide. **(See Examples 4–6.)**

25. $\dfrac{4x}{7y} \div \dfrac{2x^2}{21xy}$

26. $\dfrac{6cd}{5d^2} \div \dfrac{8c^3}{10d}$

27. $\dfrac{\frac{8m^4n^5}{5n^6}}{\frac{24mn}{15m^3}}$

28. $\dfrac{\frac{10a^3b}{3a}}{\frac{5b}{9ab}}$

29. $\dfrac{4a + 12}{6a - 18} \div \dfrac{3a + 9}{5a - 15}$

30. $\dfrac{8b - 16}{3b + 3} \div \dfrac{5b - 10}{2b + 2}$

31. $\dfrac{3x - 21}{6x^2 - 42x} \div \dfrac{7}{12x}$

32. $\dfrac{4a^2 - 4a}{9a - 9} \div \dfrac{5}{12a}$

33. $\dfrac{m^2 - n^2}{9} \div \dfrac{3n - 3m}{27m}$

34. $\dfrac{9 - b^2}{15b + 15} \div \dfrac{b - 3}{5b}$

35. $\dfrac{3p + 4q}{p^2 + 4pq + 4q^2} \div \dfrac{4}{p + 2q}$

36. $\dfrac{x^2 + 2xy + y^2}{2x - y} \div \dfrac{x + y}{5}$

37. $\dfrac{p^2 - 2p - 3}{p^2 - p - 6} \div \dfrac{p^2 - 1}{p^2 + 2p}$

38. $\dfrac{4t^2 - 1}{t^2 - 5t} \div \dfrac{2t^2 + 5t + 2}{t^2 - 3t - 10}$

Mixed Exercises

For Exercises 39–64, multiply or divide as indicated.

39. $(w + 3) \cdot \dfrac{w}{2w^2 + 5w - 3}$

40. $\dfrac{5t + 1}{5t^2 - 31t + 6} \cdot (t - 6)$

41. $\dfrac{\frac{5t - 10}{12}}{\frac{4t - 8}{8}}$

42. $\dfrac{\frac{6m + 6}{5}}{\frac{3m + 3}{10}}$

43. $\dfrac{q + 1}{5q^2 - 28q - 12} \cdot (5q + 2)$

44. $(r - 5) \cdot \dfrac{4r}{2r^2 - 7r - 15}$

45. $\dfrac{2a^2 + 13a - 24}{8a - 12} \div (a + 8)$

46. $\dfrac{3y^2 + 20y - 7}{5y + 35} \div (3y - 1)$

47. $\dfrac{y^2 + 5y - 36}{y^2 - 2y - 8} \cdot \dfrac{y + 2}{y - 6}$

48. $\dfrac{z^2 - 11z + 28}{z - 1} \cdot \dfrac{z + 1}{z^2 - 6z - 7}$

49. $\dfrac{2t^2 + t - 1}{t^2 + 3t + 2} \cdot \dfrac{t + 4}{2t - 1}$

50. $\dfrac{3p^2 - 2p - 8}{3p^2 - 5p - 12} \cdot \dfrac{p + 1}{p - 2}$

51. $(5t - 1) \div \dfrac{5t^2 + 9t - 2}{3t + 8}$

52. $(2q - 3) \div \dfrac{2q^2 + 5q - 12}{q - 7}$

53. $\dfrac{x^2 + 2x - 3}{x^2 - 3x + 2} \cdot \dfrac{x^2 + 2x - 8}{x^2 + 4x + 3}$

54. $\dfrac{y^2 + y - 12}{y^2 - y - 20} \cdot \dfrac{y^2 + y - 30}{y^2 - 2y - 3}$

55. $\dfrac{\frac{w^2 - 6w + 9}{8}}{\frac{9 - w^2}{4w + 12}}$

56. $\dfrac{\frac{p^2 - 6p + 8}{24}}{\frac{16 - p^2}{6p + 6}}$

57. $\dfrac{5k^2 + 7k + 2}{k^2 + 5k + 4} \div \dfrac{5k^2 + 17k + 6}{k^2 + 10k + 24}$

58. $\dfrac{4h^2 - 5h + 1}{h^2 + h - 2} \div \dfrac{6h^2 - 7h + 2}{2h^2 + 3h - 2}$

59. $\dfrac{ax + a + bx + b}{2x^2 + 4x + 2} \cdot \dfrac{4x + 4}{a^2 + ab}$

60. $\dfrac{3my + 9m + ny + 3n}{9m^2 + 6mn + n^2} \cdot \dfrac{30m + 10n}{5y^2 + 15y}$

61. $\dfrac{y^4 - 1}{2y^2 - 3y + 1} \div \dfrac{2y^2 + 2}{8y^2 - 4y}$

62. $\dfrac{x^4 - 16}{6x^2 + 24} \div \dfrac{x^2 - 2x}{3x}$

63. $\dfrac{x^2 - xy - 2y^2}{x + 2y} \div \dfrac{x^2 - 4xy + 4y^2}{x^2 - 4y^2}$

64. $\dfrac{4m^2 - 4mn - 3n^2}{8m^2 - 18n^2} \div \dfrac{3m + 3n}{6m^2 + 15mn + 9n^2}$

For Exercises 65–70, multiply or divide as indicated. **(See Example 7.)**

65. $\dfrac{b^3 - 3b^2 + 4b - 12}{b^4 - 16} \cdot \dfrac{3b^2 + 5b - 2}{3b^2 - 10b + 3} \div \dfrac{3}{6b - 12}$

66. $\dfrac{x^2 - 25}{3x^2 + 3xy} \cdot \dfrac{x^2 + 4x + xy + 4y}{x^2 + 9x + 20} \div \dfrac{x - 5}{x}$

67. $\dfrac{a^2 - 5a}{a^2 + 7a + 12} \div \dfrac{a^3 - 7a^2 + 10a}{a^2 + 9a + 18} \div \dfrac{a + 6}{a + 4}$

68. $\dfrac{t^2 + t - 2}{t^2 + 5t + 6} \div \dfrac{t - 1}{t} \div \dfrac{5t - 5}{t + 3}$

69. $\dfrac{p^3 - q^3}{p - q} \cdot \dfrac{p + q}{2p^2 + 2pq + 2q^2}$

70. $\dfrac{r^3 + s^3}{r - s} \div \dfrac{r^2 + 2rs + s^2}{r^2 - s^2}$

For Exercises 3–4, write the domain in set-builder notation. Then reduce the expression to lowest terms.

3. $\dfrac{3x + 3}{5x^2 - 5}$

4. $\dfrac{x + 2}{x^2 - 3x - 10}$

For Exercises 5–8, multiply or divide as indicated.

5. $\dfrac{a + 3}{a + 7} \cdot \dfrac{a^2 + 3a - 10}{a^2 + a - 6}$

6. $\dfrac{6(a + 2b)}{2(a - 3b)} \cdot \dfrac{4(a + 3b)(a - 3b)}{9(a + 2b)(a - 2b)}$

7. $\dfrac{16y^2}{9y + 36} \div \dfrac{8y^3}{3y + 12}$

8. $\dfrac{5b^2 + 6b + 1}{b^2 + 5b + 6} \div (5b + 1)$

9. Which of the expressions are equivalent to $-\dfrac{5}{x - 3}$? Circle all that apply.

   a. $\dfrac{-5}{x - 3}$ **b.** $\dfrac{5}{-x + 3}$ **c.** $\dfrac{5}{3 - x}$ **d.** $\dfrac{5}{-(x - 3)}$

10. Which of the expressions are equivalent to $\dfrac{4 - a}{6}$? Circle all that apply.

   a. $\dfrac{a - 4}{-6}$ **b.** $\dfrac{a - 4}{6}$ **c.** $\dfrac{-(4 - a)}{-6}$ **d.** $-\dfrac{a - 4}{6}$

Objective 1: Least Common Denominator

11. Explain why the least common denominator of $\frac{1}{x^3}, \frac{1}{x^5}$, and $\frac{1}{x^4}$ is x^5.

12. Explain why the least common denominator of $\frac{2}{y^3}, \frac{9}{y^6}$, and $\frac{4}{y^5}$ is y^6.

For Exercises 13–30, identify the LCD. **(See Examples 1–2.)**

13. $\dfrac{4}{15}; \dfrac{5}{9}$

14. $\dfrac{7}{12}; \dfrac{1}{18}$

15. $\dfrac{1}{16}; \dfrac{1}{4}; \dfrac{1}{6}$

16. $\dfrac{1}{2}; \dfrac{11}{12}; \dfrac{3}{8}$

17. $\dfrac{1}{7}; \dfrac{2}{9}$

18. $\dfrac{2}{3}; \dfrac{5}{8}$

19. $\dfrac{1}{3x^2y}; \dfrac{8}{9xy^3}$

20. $\dfrac{5}{2a^4b^2}; \dfrac{1}{8ab^3}$

21. $\dfrac{6}{w^2}; \dfrac{7}{y}$

22. $\dfrac{2}{r}; \dfrac{3}{s^2}$

23. $\dfrac{p}{(p + 3)(p - 1)}; \dfrac{2}{(p + 3)(p + 2)}$

24. $\dfrac{6}{(q + 4)(q - 4)}; \dfrac{q^2}{(q + 1)(q + 4)}$

25. $\dfrac{7}{3t(t + 1)}; \dfrac{10t}{9(t + 1)^2}$

26. $\dfrac{13x}{15(x - 1)^2}; \dfrac{5}{3x(x - 1)}$

27. $\dfrac{y}{y^2 - 4}; \dfrac{3y}{y^2 + 5y + 6}$

28. $\dfrac{4}{w^2 - 3w + 2}; \dfrac{w}{w^2 - 4}$

29. $\dfrac{5}{3 - x}; \dfrac{7}{x - 3}$

30. $\dfrac{4}{x - 6}; \dfrac{9}{6 - x}$

31. Explain why a common denominator of

$$\frac{b+1}{b-1} \quad \text{and} \quad \frac{b}{1-b}$$

could be either $(b-1)$ or $(1-b)$.

32. Explain why a common denominator of

$$\frac{1}{6-t} \quad \text{and} \quad \frac{t}{t-6}$$

could be either $(6-t)$ or $(t-6)$.

Objective 2: Writing Rational Expressions with the Least Common Denominator

For Exercises 33–56, find the LCD. Then convert each expression to an equivalent expression with the denominator equal to the LCD. **(See Examples 3–5.)**

33. $\dfrac{6}{5x^2}; \dfrac{1}{x}$

34. $\dfrac{3}{y}; \dfrac{7}{9y^2}$

35. $\dfrac{4}{5x^2}; \dfrac{y}{6x^3}$

36. $\dfrac{3}{15b^2}; \dfrac{c}{3b^2}$

37. $\dfrac{5}{6a^2b}; \dfrac{a}{12b}$

38. $\dfrac{x}{15y^2}; \dfrac{y}{5xy}$

39. $\dfrac{6}{m+4}; \dfrac{3}{m-1}$

40. $\dfrac{3}{n-5}; \dfrac{7}{n+2}$

41. $\dfrac{6}{2x-5}; \dfrac{1}{x+3}$

42. $\dfrac{4}{m+3}; \dfrac{-3}{5m+1}$

43. $\dfrac{6}{(w+3)(w-8)}; \dfrac{w}{(w-8)(w+1)}$

44. $\dfrac{t}{(t+2)(t+12)}; \dfrac{18}{(t-2)(t+2)}$

45. $\dfrac{6p}{p^2-4}; \dfrac{3}{p^2+4p+4}$

46. $\dfrac{5}{q^2-6q+9}; \dfrac{q}{q^2-9}$

47. $\dfrac{1}{a-4}; \dfrac{a}{4-a}$

48. $\dfrac{3b}{2b-5}; \dfrac{2b}{5-2b}$

49. $\dfrac{4}{x-7}; \dfrac{y}{14-2x}$

50. $\dfrac{4}{3x-15}; \dfrac{z}{5-x}$

51. $\dfrac{1}{a+b}; \dfrac{6}{-a-b}$

52. $\dfrac{p}{-q-8}; \dfrac{1}{q+8}$

53. $\dfrac{-3}{24y+8}; \dfrac{5}{18y+6}$

54. $\dfrac{r}{10r+5}; \dfrac{2}{16r+8}$

55. $\dfrac{3}{5z}; \dfrac{1}{z+4}$

56. $\dfrac{-1}{4a-8}; \dfrac{5}{4a}$

Expanding Your Skills

For Exercises 57–60, find the LCD. Then convert each expression to an equivalent expression with the denominator equal to the LCD.

57. $\dfrac{z}{z^2+9z+14}; \dfrac{-3z}{z^2+10z+21}; \dfrac{5}{z^2+5z+6}$

58. $\dfrac{6}{w^2-3w-4}; \dfrac{1}{w^2+6w+5}; \dfrac{-9w}{w^2+w-20}$

59. $\dfrac{3}{p^3-8}; \dfrac{p}{p^2-4}; \dfrac{5p}{p^2+2p+4}$

60. $\dfrac{7}{q^3+125}; \dfrac{q}{q^2-25}; \dfrac{12}{q^2-5q+25}$

Section 7.4 Addition and Subtraction of Rational Expressions

1. Addition and Subtraction of Rational Expressions with the Same Denominator

To add or subtract rational expressions, the expressions must have the same denominator. As with fractions, we add or subtract rational expressions with the same denominator by combining the terms in the numerator and then writing the result over the common denominator. Then, if possible, we simplify the expression to lowest terms.

> **PROPERTY** Addition and Subtraction of Rational Expressions
>
> Let p, q, and r represent polynomials where $q \neq 0$. Then,
>
> **1.** $\dfrac{p}{q} + \dfrac{r}{q} = \dfrac{p + r}{q}$ **2.** $\dfrac{p}{q} - \dfrac{r}{q} = \dfrac{p - r}{q}$

Skill Practice

Add or subtract as indicated.

1. $\dfrac{3}{14} + \dfrac{4}{14}$

2. $\dfrac{2}{7d} - \dfrac{9}{7d}$

Example 1 Adding and Subtracting Rational Expressions with a Common Denominator

Add or subtract as indicated.

a. $\dfrac{1}{12} + \dfrac{7}{12}$ **b.** $\dfrac{2}{5p} - \dfrac{7}{5p}$

Solution:

a. $\dfrac{1}{12} + \dfrac{7}{12}$ The fractions have the same denominator.

$= \dfrac{1 + 7}{12}$ Add the terms in the numerators, and write the result over the common denominator.

$= \dfrac{8}{12}$

$= \dfrac{2}{3}$ Simplify to lowest terms.

b. $\dfrac{2}{5p} - \dfrac{7}{5p}$ The rational expressions have the same denominator.

$= \dfrac{2 - 7}{5p}$ Subtract the terms in the numerators, and write the result over the common denominator.

$= \dfrac{-5}{5p}$

$= \dfrac{(-\overset{-1}{\cancel{5}})}{\cancel{5}p}$ Simplify to lowest terms.

$= -\dfrac{1}{p}$

Answers

1. $\dfrac{1}{2}$ 2. $-\dfrac{1}{d}$

Example 2 **Adding and Subtracting Rational Expressions with a Common Denominator**

Add or subtract as indicated.

a. $\dfrac{2}{3d + 5} + \dfrac{7d}{3d + 5}$

b. $\dfrac{x^2}{x - 3} - \dfrac{-5x + 24}{x - 3}$

Solution:

a. $\dfrac{2}{3d + 5} + \dfrac{7d}{3d + 5}$

The rational expressions have the same denominator.

$= \dfrac{2 + 7d}{3d + 5}$

Add the terms in the numerators, and write the result over the common denominator.

$= \dfrac{7d + 2}{3d + 5}$

Because the numerator and denominator share no common factors, the expression is in lowest terms.

b. $\dfrac{x^2}{x - 3} - \dfrac{-5x + 24}{x - 3}$

The rational expressions have the same denominator.

$= \dfrac{x^2 - (-5x + 24)}{x - 3}$

Subtract the terms in the numerators, and write the result over the common denominator.

$= \dfrac{x^2 + 5x - 24}{x - 3}$

Simplify the numerator.

$= \dfrac{(x + 8)(x - 3)}{(x - 3)}$

Factor the numerator and denominator to determine if the rational expression can be simplified.

$= \dfrac{(x + 8)\cancel{(x - 3)}^{\;1}}{\cancel{(x - 3)}}$

Simplify to lowest terms.

$= x + 8$

Avoiding Mistakes

When subtracting rational expressions, use parentheses to group the terms in the numerator that follow the subtraction sign. This will help you remember to apply the distributive property.

2. Addition and Subtraction of Rational Expressions with Different Denominators

To add or subtract two rational expressions with unlike denominators, we must convert the expressions to equivalent expressions with the same denominator. For example, consider adding

$$\frac{1}{10} + \frac{12}{5y}$$

The LCD is $10y$. For each expression, identify the factors from the LCD that are missing from the denominator. Then multiply the numerator and denominator of the expression by the missing factor(s).

$$\underset{\substack{\text{Missing} \\ y}}{\frac{1}{10}} + \underset{\substack{\text{Missing} \\ 2}}{\frac{12}{5y}}$$

Answers

3. $x + 1$ **4.** $\dfrac{3t - 4}{2t + 1}$

$$= \frac{1 \cdot y}{10 \cdot y} + \frac{12 \cdot 2}{5y \cdot 2}$$

$$= \frac{y}{10y} + \frac{24}{10y} \qquad \text{The rational expressions now have the same denominators.}$$

$$= \frac{y + 24}{10y} \qquad \text{Add the numerators.}$$

Avoiding Mistakes

In the expression $\frac{y+24}{10y}$, notice that you cannot reduce the 24 and 10 because 24 is not a factor in the numerator, it is a term. Only factors can be reduced—not terms.

After successfully adding or subtracting two rational expressions, always check to see if the final answer is simplified. If necessary, factor the numerator and denominator, and reduce common factors. The expression

$$\frac{y + 24}{10y}$$

is in lowest terms because the numerator and denominator do not share any common factors.

PROCEDURE **Adding or Subtracting Rational Expressions**

Step 1 Factor the denominators of each rational expression.

Step 2 Identify the LCD.

Step 3 Rewrite each rational expression as an equivalent expression with the LCD as its denominator.

Step 4 Add or subtract the numerators, and write the result over the common denominator.

Step 5 Simplify to lowest terms.

Skill Practice

Add.

5. $\dfrac{4}{3x} + \dfrac{1}{2x^2}$

Avoiding Mistakes

Do not reduce after rewriting the fractions with the LCD. You will revert back to the original expression.

Answer

5. $\dfrac{8x + 3}{6x^2}$

Example 3 **Subtracting Rational Expressions with Different Denominators**

Subtract. $\quad \dfrac{4}{7k} - \dfrac{3}{k^2}$

Solution:

$$\frac{4}{7k} - \frac{3}{k^2}$$

Step 1: The denominators are already factored.

Step 2: The LCD is $7k^2$.

$$= \frac{4 \cdot k}{7k \cdot k} - \frac{3 \cdot 7}{k^2 \cdot 7} \qquad \textbf{Step 3:} \quad \text{Write each expression with the LCD.}$$

$$= \frac{4k}{7k^2} - \frac{21}{7k^2}$$

$$= \frac{4k - 21}{7k^2} \qquad \textbf{Step 4:} \quad \text{Subtract the numerators, and write the result over the LCD.}$$

Step 5: The expression is in lowest terms because the numerator and denominator share no common factors.

Example 4 Subtracting Rational Expressions with Different Denominators

Subtract. $\dfrac{2q - 4}{3} - \dfrac{q + 1}{2}$

Solution:

$\dfrac{2q - 4}{3} - \dfrac{q + 1}{2}$

Step 1: The denominators are already factored.

Step 2: The LCD is 6.

$= \dfrac{2(2q - 4)}{2 \cdot 3} - \dfrac{3(q + 1)}{3 \cdot 2}$

Step 3: Write each expression with the LCD.

$= \dfrac{2(2q - 4) - 3(q + 1)}{6}$

Step 4: Subtract the numerators, and write the result over the LCD.

$= \dfrac{4q - 8 - 3q - 3}{6}$

$= \dfrac{q - 11}{6}$

Step 5: The expression is in lowest terms because the numerator and denominator share no common factors.

Skill Practice

Subtract.

6. $\dfrac{q}{12} - \dfrac{q - 2}{4}$

Example 5 Adding Rational Expressions with Different Denominators

Add. $\dfrac{1}{x - 5} + \dfrac{-10}{x^2 - 25}$

Solution:

$\dfrac{1}{x - 5} + \dfrac{-10}{x^2 - 25}$

$= \dfrac{1}{x - 5} + \dfrac{-10}{(x - 5)(x + 5)}$

Step 1: Factor the denominators.

Step 2: The LCD is $(x - 5)(x + 5)$.

$= \dfrac{1(x + 5)}{(x - 5)(x + 5)} + \dfrac{-10}{(x - 5)(x + 5)}$

Step 3: Write each expression with the LCD.

$= \dfrac{1(x + 5) + (-10)}{(x - 5)(x + 5)}$

Step 4: Add the numerators, and write the result over the LCD.

$= \dfrac{x + 5 - 10}{(x - 5)(x + 5)}$

$= \dfrac{\overset{1}{\cancel{x - 5}}}{\cancel{(x - 5)}(x + 5)}$

Step 5: Simplify.

$= \dfrac{1}{x + 5}$

Skill Practice

Add.

7. $\dfrac{1}{x - 4} + \dfrac{-8}{x^2 - 16}$

Answers

6. $\dfrac{-q + 3}{6}$ **7.** $\dfrac{1}{x + 4}$

Skill Practice

Subtract.

8. $\dfrac{2y}{y-1} - \dfrac{1}{y} - \dfrac{2y+1}{y^2-y}$

Example 6   **Adding and Subtracting Rational Expressions with Different Denominators**

Add or subtract as indicated. $\dfrac{p+2}{p-1} - \dfrac{2}{p+6} - \dfrac{14}{p^2+5p-6}$

Solution:

$$\dfrac{p+2}{p-1} - \dfrac{2}{p+6} - \dfrac{14}{p^2+5p-6}$$

$$= \dfrac{p+2}{p-1} - \dfrac{2}{p+6} - \dfrac{14}{(p-1)(p+6)}$$ **Step 1:** Factor the denominators.

Step 2: The LCD is $(p-1)(p+6)$.

Step 3: Write each expression with the LCD.

$$= \dfrac{(p+2)(p+6)}{(p-1)(p+6)} - \dfrac{2(p-1)}{(p+6)(p-1)} - \dfrac{14}{(p-1)(p+6)}$$

$$= \dfrac{(p+2)(p+6) - 2(p-1) - 14}{(p-1)(p+6)}$$ **Step 4:** Combine the numerators, and write the result over the LCD.

$$= \dfrac{p^2+6p+2p+12-2p+2-14}{(p-1)(p+6)}$$ **Step 5:** Clear parentheses in the numerator.

$$= \dfrac{p^2+6p}{(p-1)(p+6)}$$ Combine *like* terms.

$$= \dfrac{p(p+6)}{(p-1)(p+6)}$$ Factor the numerator to determine if the expression is in lowest terms.

$$= \dfrac{p(\overset{1}{\cancel{p+6}})}{(p-1)(\cancel{p+6})}$$ Simplify to lowest terms.

$$= \dfrac{p}{p-1}$$

When the denominators of two rational expressions are opposites, we can produce identical denominators by multiplying one of the expressions by the ratio $\frac{-1}{-1}$. This is demonstrated in the next example.

Answer

8. $\dfrac{2y-3}{y-1}$

Example 7 Adding Rational Expressions with Different Denominators

Add the rational expressions. $\dfrac{1}{d-7} + \dfrac{5}{7-d}$

Skill Practice

Add or subtract as indicated.

9. $\dfrac{3}{p-8} + \dfrac{1}{8-p}$

Solution:

$\dfrac{1}{d-7} + \dfrac{5}{7-d}$

The expressions $d-7$ and $7-d$ are opposites and differ by a factor of -1. Therefore, multiply the numerator and denominator of *either* expression by -1 to obtain a common denominator.

$= \dfrac{1}{d-7} + \dfrac{(-1)5}{(-1)(7-d)}$ Note that $-1(7-d) = -7 + d$ or $d-7$.

$= \dfrac{1}{d-7} + \dfrac{-5}{d-7}$ Simplify.

$= \dfrac{1+(-5)}{d-7}$ Add the terms in the numerators, and write the result over the common denominator.

$= \dfrac{-4}{d-7}$

3. Using Rational Expressions in Translations

Example 8 Using Rational Expressions in Translations

Translate the English phrase into a mathematical expression. Then simplify by combining the rational expressions.

The difference of the reciprocal of n and the quotient of n and 3

Skill Practice

Translate the English phrase into a mathematical expression. Then simplify by combining the rational expressions.

10. The sum of 1 and the quotient of 2 and a

Solution:

The difference of the reciprocal of n and the quotient of n and 3

$$\underbrace{\left(\dfrac{1}{n}\right)}_{\text{The reciprocal of } n} \overset{\text{The difference of}}{-} \underbrace{\left(\dfrac{n}{3}\right)}_{\text{The quotient of } n \text{ and 3}}$$

$\dfrac{1}{n} - \dfrac{n}{3}$ The LCD is $3n$.

$= \dfrac{3 \cdot 1}{3 \cdot n} - \dfrac{n \cdot n}{3 \cdot n}$ Write each expression over the LCD.

$= \dfrac{3 - n^2}{3n}$ Subtract the numerators.

Answers

9. $\dfrac{2}{p-8}$ or $\dfrac{-2}{8-p}$

10. $1 + \dfrac{2}{a}$; $\dfrac{a+2}{a}$

Section 7.4 Practice Exercises

Review Exercises

1. For the rational expression $\dfrac{x^2 - 4x - 5}{x^2 - 7x + 10}$

   a. Find the value of the expression (if possible) when $x = 0, 1, -1, 2,$ and 5.

   b. Factor the denominator and identify the domain. Write the domain in set-builder notation.

   c. Reduce the expression to lowest terms.

2. For the rational expression $\dfrac{a^2 + a - 2}{a^2 - 4a - 12}$

   a. Find the value of the expression (if possible) when $a = 0, 1, -2, 2,$ and 6.

   b. Factor the denominator, and identify the domain. Write the domain in set-builder notation.

   c. Reduce the expression to lowest terms.

For Exercises 3–4, multiply or divide as indicated.

3. $\dfrac{2b^2 - b - 3}{2b^2 - 3b - 9} \div \dfrac{b^2 - 1}{4b + 6}$

4. $\dfrac{6t - 1}{5t - 30} \cdot \dfrac{10t - 25}{2t^2 - 3t - 5}$

Objective 1: Addition and Subtraction of Rational Expressions with the Same Denominator

For Exercises 5–26, add or subtract the expressions with like denominators as indicated. **(See Examples 1–2.)**

5. $\dfrac{7}{8} + \dfrac{3}{8}$

6. $\dfrac{1}{3} + \dfrac{7}{3}$

7. $\dfrac{9}{16} - \dfrac{3}{16}$

8. $\dfrac{14}{15} - \dfrac{4}{15}$

9. $\dfrac{5a}{a + 2} - \dfrac{3a - 4}{a + 2}$

10. $\dfrac{2b}{b - 3} - \dfrac{b - 9}{b - 3}$

11. $\dfrac{5c}{c + 6} + \dfrac{30}{c + 6}$

12. $\dfrac{12}{2 + d} + \dfrac{6d}{2 + d}$

13. $\dfrac{5}{t - 8} - \dfrac{2t + 1}{t - 8}$

14. $\dfrac{7p + 1}{2p + 1} - \dfrac{p - 4}{2p + 1}$

15. $\dfrac{9x^2}{3x - 7} - \dfrac{49}{3x - 7}$

16. $\dfrac{4w^2}{2w - 1} - \dfrac{1}{2w - 1}$

17. $\dfrac{m^2}{m + 5} + \dfrac{10m + 25}{m + 5}$

18. $\dfrac{k^2}{k - 3} - \dfrac{6k - 9}{k - 3}$

19. $\dfrac{2a}{a + 2} + \dfrac{4}{a + 2}$

20. $\dfrac{5b}{b + 4} + \dfrac{20}{b + 4}$

21. $\dfrac{x^2}{x + 5} - \dfrac{25}{x + 5}$

22. $\dfrac{y^2}{y - 7} - \dfrac{49}{y - 7}$

23. $\dfrac{r}{r^2 + 3r + 2} + \dfrac{2}{r^2 + 3r + 2}$

24. $\dfrac{x}{x^2 - x - 12} - \dfrac{4}{x^2 - x - 12}$

25. $\dfrac{1}{3y^2 + 22y + 7} - \dfrac{-3y}{3y^2 + 22y + 7}$

26. $\dfrac{5}{2x^2 + 13x + 20} + \dfrac{2x}{2x^2 + 13x + 20}$

For Exercises 27–28, find an expression that represents the perimeter of the figure (assume that $x > 0$, $y > 0$, and $t > 0$).

27.

28.

Objective 2: Addition and Subtraction of Rational Expressions with Different Denominators

For Exercises 29–70, add or subtract the expressions with unlike denominators as indicated. **(See Examples 3–7.)**

29. $\dfrac{5}{4} + \dfrac{3}{2a}$

30. $\dfrac{11}{6p} + \dfrac{-7}{4p}$

31. $\dfrac{4}{5xy^3} + \dfrac{2x}{15y^2}$

32. $\dfrac{5}{3a^2b} + \dfrac{-7}{6b^2}$

33. $\dfrac{2}{s^3t^3} - \dfrac{3}{s^4t}$

34. $\dfrac{1}{p^2q} - \dfrac{2}{pq^3}$

35. $\dfrac{z}{3z - 9} - \dfrac{z - 2}{z - 3}$

36. $\dfrac{3w - 8}{2w - 4} - \dfrac{w - 3}{w - 2}$

37. $\dfrac{5}{a + 1} + \dfrac{4}{3a + 3}$

38. $\dfrac{2}{c - 4} + \dfrac{1}{5c - 20}$

39. $\dfrac{k}{k^2 - 9} - \dfrac{4}{k - 3}$

40. $\dfrac{7}{h + 2} - \dfrac{2h - 3}{h^2 - 4}$

41. $\dfrac{3a - 7}{6a + 10} - \dfrac{10}{3a^2 + 5a}$

42. $\dfrac{k + 2}{8k} - \dfrac{3 - k}{12k}$

43. $\dfrac{x}{x - 4} + \dfrac{3}{x + 1}$

44. $\dfrac{4}{y - 3} + \dfrac{y}{y - 5}$

45. $\dfrac{6a}{a^2 - b^2} + \dfrac{2a}{a^2 + ab}$

46. $\dfrac{7x}{x^2 + 2xy + y^2} + \dfrac{3x}{x^2 + xy}$

47. $\dfrac{p}{3} - \dfrac{4p - 1}{-3}$

48. $\dfrac{r}{7} - \dfrac{r - 5}{-7}$

49. $\dfrac{4n}{n - 8} - \dfrac{2n - 1}{8 - n}$

50. $\dfrac{m}{m - 2} - \dfrac{3m + 1}{2 - m}$

51. $\dfrac{5}{x} + \dfrac{3}{x + 2}$

52. $\dfrac{6}{y - 1} + \dfrac{9}{y}$

53. $\dfrac{5}{p - 3} - \dfrac{2}{p - 1}$

54. $\dfrac{1}{7x} + \dfrac{5}{2y^2}$

55. $\dfrac{y}{4y + 2} + \dfrac{3y}{6y + 3}$

56. $\dfrac{4}{q^2 - 2q} - \dfrac{5}{3q - 6}$

57. $\dfrac{4w}{w^2 + 2w - 3} + \dfrac{2}{1 - w}$

58. $\dfrac{z - 23}{z^2 - z - 20} - \dfrac{2}{5 - z}$

59. $\dfrac{3a - 8}{a^2 - 5a + 6} + \dfrac{a + 2}{a^2 - 6a + 8}$

60. $\dfrac{3b + 5}{b^2 + 4b + 3} + \dfrac{-b + 5}{b^2 + 2b - 3}$

61. $\dfrac{3x}{x^2 + x - 6} + \dfrac{x}{x^2 + 5x + 6}$

62. $\dfrac{x}{x^2 + 5x + 4} - \dfrac{2x}{x^2 - 2x - 3}$

63. $\dfrac{3y}{2y^2 - y - 1} - \dfrac{4y}{2y^2 - 7y - 4}$

64. $\dfrac{5}{6y^2 - 7y - 3} + \dfrac{4y}{3y^2 + 4y + 1}$

65. $\dfrac{3}{2p - 1} - \dfrac{4p + 4}{4p^2 - 1}$

66. $\dfrac{1}{3q - 2} - \dfrac{6q + 4}{9q^2 - 4}$

67. $\dfrac{m}{m + n} - \dfrac{m}{m - n} + \dfrac{1}{m^2 - n^2}$

68. $\dfrac{x}{x + y} - \dfrac{2xy}{x^2 - y^2} + \dfrac{y}{x - y}$

69. $\dfrac{2}{a + b} + \dfrac{2}{a - b} - \dfrac{4a}{a^2 - b^2}$

70. $\dfrac{-2x}{x^2 - y^2} + \dfrac{1}{x + y} - \dfrac{1}{x - y}$

For Exercises 71–72, find an expression that represents the perimeter of the figure (assume that $x > 0$ and $t > 0$).

71.

$\dfrac{2}{x+3}$

$\dfrac{1}{x+2}$

72.

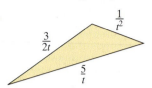

$\dfrac{3}{2t}$

$\dfrac{1}{t^2}$

$\dfrac{5}{t}$

Objective 3: Using Rational Expressions in Translations

73. Let a number be represented by n. Write the reciprocal of n.

74. Write the reciprocal of the sum of a number and 6.

75. Write the quotient of 5 and the sum of a number and 2.

76. Let a number be represented by p. Write the quotient of 12 and p.

For Exercises 77–80, translate the English phrases into algebraic expressions. Then simplify by combining the rational expressions. **(See Example 8.)**

77. The sum of a number and the quantity seven times the reciprocal of the number.

78. The sum of a number and the quantity five times the reciprocal of the number.

79. The difference of the reciprocal of n and the quotient of 2 and n.

80. The difference of the reciprocal of m and the quotient of $3m$ and 7.

Expanding Your Skills

For Exercises 81–86, perform the indicated operations.

81. $\dfrac{-3}{w^3 + 27} - \dfrac{1}{w^2 - 9}$

82. $\dfrac{m}{m^3 - 1} + \dfrac{1}{(m-1)^2}$

83. $\dfrac{2p}{p^2 + 5p + 6} - \dfrac{p+1}{p^2 + 2p - 3} + \dfrac{3}{p^2 + p - 2}$

84. $\dfrac{3t}{8t^2 + 2t - 1} - \dfrac{5t}{2t^2 - 9t - 5} + \dfrac{2}{4t^2 - 21t + 5}$

85. $\dfrac{3m}{m^2 + 3m - 10} + \dfrac{5}{4 - 2m} - \dfrac{1}{m + 5}$

86. $\dfrac{2n}{3n^2 - 8n - 3} + \dfrac{1}{6 - 2n} - \dfrac{3}{3n + 1}$

For Exercises 87–90, simplify by applying the order of operations.

87. $\left(\dfrac{2}{k+1} + 3\right)\left(\dfrac{k+1}{4k+7}\right)$

88. $\left(\dfrac{p+1}{3p+4}\right)\left(\dfrac{1}{p+1} + 2\right)$

89. $\left(\dfrac{1}{10a} - \dfrac{b}{10a^2}\right) \div \left(\dfrac{1}{10} - \dfrac{b}{10a}\right)$

90. $\left(\dfrac{1}{2m} + \dfrac{n}{2m^2}\right) \div \left(\dfrac{1}{4} + \dfrac{n}{4m}\right)$

Problem Recognition Exercises

Operations on Rational Expressions

In Sections 7.1–7.4, we learned how to simplify, add, subtract, multiply, and divide rational expressions. The procedure for each operation is different, and it takes considerable practice to determine the correct method to apply for a given problem. The following review exercises give you the opportunity to practice the specific techniques for simplifying rational expressions.

For Exercises 1–20, perform any indicated operations and simplify the expression.

1. $\dfrac{5}{3x + 1} - \dfrac{2x - 4}{3x + 1}$

2. $\dfrac{\dfrac{w + 1}{w^2 - 16}}{\dfrac{w + 1}{w + 4}}$

3. $\dfrac{3}{y} \cdot \dfrac{y^2 - 5y}{6y - 9}$

4. $\dfrac{-1}{x + 3} + \dfrac{2}{2x - 1}$

5. $\dfrac{x - 9}{9x - x^2}$

6. $\dfrac{1}{p} - \dfrac{3}{p^2 + 3p} + \dfrac{p}{3p + 9}$

7. $\dfrac{c^2 + 5c + 6}{c^2 + c - 2} \div \dfrac{c}{c - 1}$

8. $\dfrac{2x^2 - 5x - 3}{x^2 - 9} \cdot \dfrac{x^2 + 6x + 9}{10x + 5}$

9. $\dfrac{6a^2b^3}{72ab^7c}$

10. $\dfrac{2a}{a + b} - \dfrac{b}{a - b} - \dfrac{-4ab}{a^2 - b^2}$

11. $\dfrac{p^2 + 10pq + 25q^2}{p^2 + 6pq + 5q^2} \div \dfrac{10p + 50q}{2p^2 - 2q^2}$

12. $\dfrac{3k - 8}{k - 5} + \dfrac{k - 12}{k - 5}$

13. $\dfrac{20x^2 + 10x}{4x^3 + 4x^2 + x}$

14. $\dfrac{w^2 - 81}{w^2 + 10w + 9} \cdot \dfrac{w^2 + w + 2zw + 2z}{w^2 - 9w + zw - 9z}$

15. $\dfrac{8x^2 - 18x - 5}{4x^2 - 25} \div \dfrac{4x^2 - 11x - 3}{3x - 9}$

16. $\dfrac{xy + 7x + 5y + 35}{x^2 + ax + 5x + 5a}$

17. $\dfrac{a}{a^2 - 9} - \dfrac{3}{6a - 18}$

18. $\dfrac{4}{y^2 - 36} + \dfrac{2}{y^2 - 4y - 12}$

19. $(t^2 + 5t - 24)\left(\dfrac{t + 8}{t - 3}\right)$

20. $\dfrac{6b^2 - 7b - 10}{b - 2}$

Section 7.5 Complex Fractions

Objectives

1. Simplifying Complex Fractions (Method I)
2. Simplifying Complex Fractions (Method II)

1. Simplifying Complex Fractions (Method I)

A **complex fraction** is a fraction whose numerator or denominator contains one or more rational expressions. For example,

$$\frac{\dfrac{1}{ab}}{\dfrac{2}{b}} \qquad \text{and} \qquad \frac{1 + \dfrac{3}{4} - \dfrac{1}{6}}{\dfrac{1}{2} + \dfrac{1}{3}}$$

are complex fractions.

Two methods will be presented to simplify complex fractions. The first method (Method I) follows the order of operations to simplify the numerator and denominator separately before dividing. The process is summarized as follows.

> **PROCEDURE Simplifying a Complex Fraction (Method I)**
>
> **Step 1** Add or subtract expressions in the numerator to form a single fraction. Add or subtract expressions in the denominator to form a single fraction.
>
> **Step 2** Divide the rational expressions from Step 1 by multiplying the numerator of the complex fraction by the reciprocal of the denominator of the complex fraction.
>
> **Step 3** Simplify to lowest terms if possible.

Skill Practice

Simplify the expression.

1. $\dfrac{\dfrac{6x}{y}}{\dfrac{9}{2y}}$

Example 1 Simplifying Complex Fractions (Method I)

Simplify the expression. $\dfrac{\dfrac{1}{ab}}{\dfrac{2}{b}}$

Solution:

Step 1: The numerator and denominator of the complex fraction are already single fractions.

$$\dfrac{\dfrac{1}{ab}}{\dfrac{2}{b}} \longleftarrow \qquad \text{This fraction bar denotes division } (\div).$$

$$= \frac{1}{ab} \div \frac{2}{b}$$

$$= \frac{1}{ab} \cdot \frac{b}{2} \qquad \textbf{Step 2:} \quad \text{Multiply the numerator of the complex fraction by the reciprocal of } \tfrac{2}{b}, \text{ which is } \tfrac{b}{2}.$$

$$= \frac{1}{ab} \cdot \frac{\overset{1}{\cancel{b}}}{2} \qquad \textbf{Step 3:} \quad \text{Reduce common factors and simplify.}$$

$$= \frac{1}{2a}$$

Answer

1. $\dfrac{4x}{3}$

Sometimes it is necessary to simplify the numerator and denominator of a complex fraction before the division can be performed. This is illustrated in Example 2.

| **Example 2** | **Simplifying Complex Fractions (Method I)** |

Simplify the expression.
$$\frac{1 + \dfrac{3}{4} - \dfrac{1}{6}}{\dfrac{1}{2} + \dfrac{1}{3}}$$

Skill Practice

Simplify the expression.

2. $\dfrac{\dfrac{3}{4} - \dfrac{1}{6} + 2}{\dfrac{1}{3} + \dfrac{1}{2}}$

Solution:

$$\frac{1 + \dfrac{3}{4} - \dfrac{1}{6}}{\dfrac{1}{2} + \dfrac{1}{3}}$$

Step 1: Combine fractions in the numerator and denominator separately.

$$= \frac{1 \cdot \dfrac{12}{12} + \dfrac{3}{4} \cdot \dfrac{3}{3} - \dfrac{1}{6} \cdot \dfrac{2}{2}}{\dfrac{1}{2} \cdot \dfrac{3}{3} + \dfrac{1}{3} \cdot \dfrac{2}{2}}$$

The LCD in the numerator is 12.
The LCD in the denominator is 6.

$$= \frac{\dfrac{12}{12} + \dfrac{9}{12} - \dfrac{2}{12}}{\dfrac{3}{6} + \dfrac{2}{6}}$$

$$= \frac{\dfrac{19}{12}}{\dfrac{5}{6}}$$

Form a single fraction in the numerator and in the denominator.

$$= \frac{19}{\overset{}{\underset{2}{12}}} \cdot \frac{\overset{1}{6}}{5}$$

Step 2: Multiply by the reciprocal of $\frac{5}{6}$, which is $\frac{6}{5}$.

$$= \frac{19}{10}$$

Step 3: Simplify.

Answer

2. $\dfrac{31}{10}$

Example 3 Simplifying Complex Fractions (Method I)

Simplify the expression. $\dfrac{\dfrac{1}{x} + \dfrac{1}{y}}{x - \dfrac{y^2}{x}}$

Solution:

$\dfrac{\dfrac{1}{x} + \dfrac{1}{y}}{x - \dfrac{y^2}{x}}$

The LCD in the numerator is xy. The LCD in the denominator is x.

$= \dfrac{\dfrac{1 \cdot y}{x \cdot y} + \dfrac{1 \cdot x}{y \cdot x}}{\dfrac{x \cdot x}{1 \cdot x} - \dfrac{y^2}{x}}$

Rewrite the expressions using common denominators.

$= \dfrac{\dfrac{y}{xy} + \dfrac{x}{xy}}{\dfrac{x^2}{x} - \dfrac{y^2}{x}}$

$= \dfrac{\dfrac{y + x}{xy}}{\dfrac{x^2 - y^2}{x}}$

Form single fractions in the numerator and denominator.

$= \dfrac{y + x}{xy} \cdot \dfrac{x}{x^2 - y^2}$

Multiply by the reciprocal of the denominator.

$= \dfrac{\overset{1}{\cancel{y + x}}}{xy} \cdot \dfrac{\overset{1}{\cancel{x}}}{\cancel{(x + y)}(x - y)}$

Factor and reduce. Note that $(y + x) = (x + y)$.

$= \dfrac{1}{y(x - y)}$

Simplify.

2. Simplifying Complex Fractions (Method II)

We will now simplify the expressions from Examples 2 and 3 again using a second method to simplify complex fractions (Method II). Recall that multiplying the numerator and denominator of a rational expression by the same quantity does not change the value of the expression because we are multiplying by a number equivalent to 1. This is the basis for Method II.

> **PROCEDURE Simplifying a Complex Fraction (Method II)**
>
> **Step 1** Multiply the numerator and denominator of the complex fraction by the LCD of *all* individual fractions within the expression.
>
> **Step 2** Apply the distributive property, and simplify the numerator and denominator.
>
> **Step 3** Simplify to lowest terms if possible.

Example 4   Simplifying Complex Fractions (Method II)

Simplify the expression.
$$\frac{1 + \dfrac{3}{4} - \dfrac{1}{6}}{\dfrac{1}{2} + \dfrac{1}{3}}$$

Skill Practice

Simplify the expression.

4. $\dfrac{1 - \dfrac{3}{5}}{\dfrac{1}{4} - \dfrac{7}{10} + 1}$

Solution:

$$\frac{1 + \dfrac{3}{4} - \dfrac{1}{6}}{\dfrac{1}{2} + \dfrac{1}{3}}$$

The LCD of the expressions $1, \frac{3}{4}, \frac{1}{6}, \frac{1}{2}$, and $\frac{1}{3}$ is 12.

TIP: In step 1, we are multiplying the original expression by $\frac{12}{12}$, which equals 1.

$$= \frac{12\left(1 + \dfrac{3}{4} - \dfrac{1}{6}\right)}{12\left(\dfrac{1}{2} + \dfrac{1}{3}\right)}$$

Step 1: Multiply the numerator and denominator of the complex fraction by 12.

$$= \frac{12 \cdot 1 + 12 \cdot \dfrac{3}{4} - 12 \cdot \dfrac{1}{6}}{12 \cdot \dfrac{1}{2} + 12 \cdot \dfrac{1}{3}}$$

Step 2: Apply the distributive property.

$$= \frac{12 \cdot 1 + \overset{3}{\cancel{12}} \cdot \dfrac{3}{\cancel{4}} - \overset{2}{\cancel{12}} \cdot \dfrac{1}{\cancel{6}}}{\overset{6}{\cancel{12}} \cdot \dfrac{1}{\cancel{2}} + \overset{4}{\cancel{12}} \cdot \dfrac{1}{\cancel{3}}}$$

Simplify each term.

$$= \frac{12 + 9 - 2}{6 + 4}$$

$$= \frac{19}{10}$$

Step 3: Simplify.

Example 5   Simplifying a Complex Fraction (Method II)

Simplify the expression.
$$\frac{\dfrac{1}{x} + \dfrac{1}{y}}{x - \dfrac{y^2}{x}}$$

Skill Practice

Simplify the expressions.

5. $\dfrac{\dfrac{z}{3} - \dfrac{3}{z}}{1 + \dfrac{3}{z}}$

Solution:

$$\frac{\dfrac{1}{x} + \dfrac{1}{y}}{x - \dfrac{y^2}{x}}$$

The LCD of the expressions $\frac{1}{x}, \frac{1}{y}, x$, and $\frac{y^2}{x}$ is xy.

$$= \frac{xy\left(\dfrac{1}{x} + \dfrac{1}{y}\right)}{xy\left(x - \dfrac{y^2}{x}\right)}$$

Step 1: Multiply numerator and denominator of the complex fraction by xy.

Answers

4. $\dfrac{8}{11}$ **5.** $\dfrac{z-3}{3}$

$$= \frac{xy \cdot \dfrac{1}{x} + xy \cdot \dfrac{1}{y}}{xy \cdot x - xy \cdot \dfrac{y^2}{x}}$$

Step 2: Apply the distributive property, and simplify each term.

$$= \frac{y + x}{x^2 y - y^3}$$

$$= \frac{y + x}{y(x^2 - y^2)}$$

Step 3: Factor completely, and reduce common factors.

$$= \frac{\overset{1}{\cancel{y + x}}}{y\cancel{(x + y)}(x - y)}$$

Note that $(y + x) = (x + y)$.

$$= \frac{1}{y(x - y)}$$

Skill Practice

Simplify the expression.

6. $\dfrac{\dfrac{4}{p - 3} + 1}{1 + \dfrac{2}{p - 3}}$

Example 6 Simplifying a Complex Fraction (Method II)

Simplify the expression.

$$\frac{\dfrac{1}{k + 1} - 1}{\dfrac{1}{k + 1} + 1}$$

Solution:

$$\frac{\dfrac{1}{k + 1} - 1}{\dfrac{1}{k + 1} + 1}$$

The LCD of $\dfrac{1}{k + 1}$ and 1 is $(k + 1)$.

$$= \frac{(k + 1)\left(\dfrac{1}{k + 1} - 1\right)}{(k + 1)\left(\dfrac{1}{k + 1} + 1\right)}$$

Step 1: Multiply numerator and denominator of the complex fraction by $(k + 1)$.

$$= \frac{\overset{1}{\cancel{(k + 1)}} \cdot \dfrac{1}{\cancel{(k + 1)}} - (k + 1) \cdot 1}{\overset{1}{\cancel{(k + 1)}} \cdot \dfrac{1}{\cancel{(k + 1)}} + (k + 1) \cdot 1}$$

Step 2: Apply the distributive property.

$$= \frac{1 - (k + 1)}{1 + (k + 1)}$$

Simplify.

$$= \frac{1 - k - 1}{1 + k + 1}$$

$$= \frac{-k}{k + 2}$$

Step 3: The expression is already in lowest terms.

Answer

6. $\dfrac{p + 1}{p - 1}$

Section 7.5 Practice Exercises

Study Skills Exercise

1. Define the key term **complex fraction**.

Review Exercises

For Exercises 2–3, write the domain in set-builder notation, and simplify the expression.

2. $\dfrac{y(2y + 9)}{y^2(2y + 9)}$

3. $\dfrac{a + 5}{2a^2 + 7a - 15}$

For Exercises 4–6, perform the indicated operations.

4. $\dfrac{2}{w - 2} + \dfrac{3}{w}$

5. $\dfrac{6}{5} - \dfrac{3}{5k - 10}$

6. $\dfrac{x^2 - 2xy + y^2}{x^4 - y^4} \div \dfrac{3x^2y - 3xy^2}{x^2 + y^2}$

Objectives 1–2: Simplifying Complex Fractions (Methods I and II)

For Exercises 7–34, simplify the complex fractions. **(See Examples 1–6.)**

7. $\dfrac{\dfrac{7}{18y}}{\dfrac{2}{9}}$

8. $\dfrac{\dfrac{a^2}{2a - 3}}{\dfrac{5a}{8a - 12}}$

9. $\dfrac{\dfrac{3x + 2y}{2y}}{\dfrac{6x + 4y}{2}}$

10. $\dfrac{\dfrac{2x - 10}{4}}{\dfrac{x^2 - 5x}{3x}}$

11. $\dfrac{\dfrac{8a^4b^3}{3c}}{\dfrac{a^7b^2}{9c}}$

12. $\dfrac{\dfrac{12x^2}{5y}}{\dfrac{8x^6}{9y^2}}$

13. $\dfrac{\dfrac{4r^3s}{t^5}}{\dfrac{2s^7}{r^2t^9}}$

14. $\dfrac{\dfrac{5p^4q}{w^4}}{\dfrac{10p^2}{qw^2}}$

15. $\dfrac{\dfrac{1}{8} + \dfrac{4}{3}}{\dfrac{1}{2} - \dfrac{5}{12}}$

16. $\dfrac{\dfrac{8}{9} - \dfrac{1}{3}}{\dfrac{7}{6} + \dfrac{1}{9}}$

17. $\dfrac{\dfrac{1}{h} + \dfrac{1}{k}}{\dfrac{1}{hk}}$

18. $\dfrac{\dfrac{1}{b} + 1}{\dfrac{1}{b}}$

19. $\dfrac{\dfrac{n + 1}{n^2 - 9}}{\dfrac{2}{n + 3}}$

20. $\dfrac{\dfrac{5}{k - 5}}{\dfrac{k + 1}{k^2 - 25}}$

21. $\dfrac{2 + \dfrac{1}{x}}{4 + \dfrac{1}{x}}$

22. $\dfrac{6 + \dfrac{6}{k}}{1 + \dfrac{1}{k}}$

23. $\dfrac{\dfrac{m}{7} - \dfrac{7}{m}}{\dfrac{1}{7} + \dfrac{1}{m}}$

24. $\dfrac{\dfrac{2}{p} + \dfrac{p}{2}}{\dfrac{p}{3} - \dfrac{3}{p}}$

25. $\dfrac{\dfrac{1}{5} - \dfrac{1}{y}}{\dfrac{7}{10} + \dfrac{1}{y^2}}$

26. $\dfrac{\dfrac{1}{m^2} + \dfrac{2}{3}}{\dfrac{1}{m} - \dfrac{5}{6}}$

27. $\dfrac{\dfrac{8}{a + 4} + 2}{\dfrac{12}{a + 4} - 2}$

28. $\dfrac{\dfrac{2}{w + 1} + 3}{\dfrac{3}{w + 1} + 4}$

29. $\dfrac{1 - \dfrac{4}{t^2}}{1 - \dfrac{2}{t} - \dfrac{8}{t^2}}$

30. $\dfrac{1 - \dfrac{9}{p^2}}{1 - \dfrac{1}{p} - \dfrac{6}{p^2}}$

31. $\dfrac{t + 4 + \dfrac{3}{t}}{t - 4 - \dfrac{5}{t}}$

32. $\dfrac{\dfrac{9}{4m} + \dfrac{9}{2m^2}}{\dfrac{3}{2} + \dfrac{3}{m}}$

33. $\dfrac{\dfrac{1}{k - 6} - 1}{\dfrac{2}{k - 6} - 2}$

34. $\dfrac{\dfrac{3}{y - 3} + 4}{8 + \dfrac{6}{y - 3}}$

For Exercises 35–38, translate the English phrases into algebraic expressions. Then simplify the expressions.

35. The sum of one-half and two-thirds, divided by five.

36. The quotient of ten and the difference of two-fifths and one-fourth.

37. The quotient of three and the sum of two-thirds and three-fourths.

38. The difference of three-fifths and one-half, divided by four.

39. In electronics, resistors oppose the flow of current. For two resistors in parallel, the total resistance is given by

$$R = \dfrac{1}{\dfrac{1}{R_1} + \dfrac{1}{R_2}}$$

a. Find the total resistance if $R_1 = 2\,\Omega$ (ohms) and $R_2 = 3\,\Omega$.

b. Find the total resistance if $R_1 = 10\,\Omega$ and $R_2 = 15\,\Omega$.

40. Suppose that Joëlle makes a round trip to a location that is d miles away. If the average rate going to the location is r_1 and the average rate on the return trip is given by r_2, the average rate of the entire trip, R, is given by

$$R = \dfrac{2d}{\dfrac{d}{r_1} + \dfrac{d}{r_2}}$$

a. Find the average rate of a trip to a destination 30 mi away when the average rate going there was 60 mph and the average rate returning home was 45 mph. (Round to the nearest tenth of a mile per hour.)

b. Find the average rate of a trip to a destination that is 50 mi away if the driver travels at the same rates as in part (a). (Round to the nearest tenth of a mile per hour.)

c. Compare your answers from parts (a) and (b) and explain the results in the context of the problem.

Expanding Your Skills

For Exercises 41–48, simplify the complex fractions using either method.

41. $\dfrac{2x^{-1} + 8y^{-1}}{4x^{-1}}$

$\left(Hint: 2x^{-1} = \dfrac{2}{x}\right)$

42. $\dfrac{6a^{-1} + 4b^{-1}}{8b^{-1}}$

43. $\dfrac{(mn)^{-2}}{m^{-2} + n^{-2}}$

44. $\dfrac{(xy)^{-1}}{2x^{-1} + 3y^{-1}}$

45. $\dfrac{\dfrac{1}{z^2 - 9} + \dfrac{2}{z + 3}}{\dfrac{3}{z - 3}}$

46. $\dfrac{\dfrac{5}{w^2 - 25} - \dfrac{3}{w + 5}}{\dfrac{4}{w - 5}}$

47. $\dfrac{\dfrac{2}{x - 1} + 2}{\dfrac{2}{x + 1} - 2}$

48. $\dfrac{\dfrac{1}{y - 3} + 1}{\dfrac{2}{y + 3} - 1}$

For Exercises 49–50, simplify the complex fractions (*Hint:* Use the order of operations and begin with the fraction on the lower right.)

49. $1 + \dfrac{1}{1 + 1}$

50. $1 + \dfrac{1}{1 + \dfrac{1}{1 + 1}}$

Rational Equations

1. Introduction to Rational Equations

Thus far we have studied two specific types of equations in one variable: linear equations and quadratic equations. Recall,

$$ax + b = 0, \text{ where } a \neq 0, \text{ is a } \textbf{linear equation}$$

$$ax^2 + bx + c = 0, \text{ where } a \neq 0, \text{ is a } \textbf{quadratic equation}.$$

We will now study another type of equation called a rational equation.

Objectives

1. **Introduction to Rational Equations**
2. **Solving Rational Equations**
3. **Solving Formulas Involving Rational Expressions**

DEFINITION Rational Equation

An equation with one or more rational expressions is called a **rational equation**.

The following equations are rational equations:

$$\frac{y}{2} + \frac{y}{4} = 6 \qquad \frac{1}{x} + \frac{1}{3} = \frac{5}{6} \qquad \frac{6}{t^2 - 7t + 12} + \frac{2t}{t - 3} = \frac{3t}{t - 4}$$

To understand the process of solving a rational equation, first review the process of clearing fractions from Section 2.3. We can clear the fractions in an equation by multiplying both sides of the equation by the LCD of all terms.

Example 1 Solving a Rational Equation

Solve. $\quad \dfrac{y}{2} + \dfrac{y}{4} = 6$

Solution:

$$\frac{y}{2} + \frac{y}{4} = 6 \qquad \text{The LCD of all terms in the equation is 4.}$$

$$4\left(\frac{y}{2} + \frac{y}{4}\right) = 4(6) \qquad \text{Multiply both sides of the equation by 4 to clear fractions.}$$

$$\overset{2}{\cancel{4}} \cdot \frac{y}{2} + \overset{1}{\cancel{4}} \cdot \frac{y}{4} = 4(6) \qquad \text{Apply the distributive property.}$$

$$2y + y = 24 \qquad \text{Clear fractions.}$$

$$3y = 24 \qquad \text{Solve the resulting equation (linear).}$$

$$y = 8$$

Check: $\quad \dfrac{y}{2} + \dfrac{y}{4} = 6$

$$\frac{(8)}{2} + \frac{(8)}{4} \overset{?}{=} 6$$

$$4 + 2 \overset{?}{=} 6$$

The solution is 8. $\qquad\qquad 6 \overset{?}{=} 6 ✔ \text{ (True)}$

Skill Practice

Solve the equations.

1. $\dfrac{t}{5} - \dfrac{t}{4} = 2$

Answer

1. -40

2. Solving Rational Equations

The same process of clearing fractions is used to solve rational equations when variables are present in the denominator. Variables in the denominator make it necessary to take note of **domain restrictions**. Values for the variable that make any denominator zero are restricted from the domain.

Example 2 **Solving a Rational Equation**

Solve the equation. $\dfrac{x+1}{x} + \dfrac{1}{3} = \dfrac{5}{6}$

Solution:

$$\dfrac{x+1}{x} + \dfrac{1}{3} = \dfrac{5}{6}$$

The LCD of all the expressions is $6x$. The domain restriction is $x \neq 0$.

$$6x \cdot \left(\dfrac{x+1}{x} + \dfrac{1}{3} \right) = 6x \cdot \left(\dfrac{5}{6} \right)$$

Multiply by the LCD.

$$\overset{1}{6x} \cdot \left(\dfrac{x+1}{x} \right) + \overset{2}{6x} \cdot \left(\dfrac{1}{3} \right) = \overset{1}{6x} \cdot \left(\dfrac{5}{6} \right)$$

Apply the distributive property.

$$6(x+1) + 2x = 5x$$

Clear fractions.

$$6x + 6 + 2x = 5x$$

Solve the resulting equation.

$$8x + 6 = 5x$$

$$3x = -6$$

$$x = -2$$

-2 is not a restricted value.

Check: $\dfrac{x+1}{x} + \dfrac{1}{3} = \dfrac{5}{6}$

$$\dfrac{(-2)+1}{(-2)} + \dfrac{1}{3} \overset{?}{=} \dfrac{5}{6}$$

$$\dfrac{-1}{-2} + \dfrac{1}{3} \overset{?}{=} \dfrac{5}{6}$$

$$\dfrac{1}{2} + \dfrac{1}{3} \overset{?}{=} \dfrac{5}{6}$$

$$\dfrac{3}{6} + \dfrac{2}{6} \overset{?}{=} \dfrac{5}{6}$$

The solution is -2.

$$\dfrac{5}{6} \overset{?}{=} \dfrac{5}{6} \quad \checkmark \text{ (True)}$$

Example 3 **Solving a Rational Equation**

Skill Practice

Solve the equations.

3. $\dfrac{x}{x+1} - 2 = \dfrac{-1}{x+1}$

Solve the equation. $1 + \dfrac{3a}{a-2} = \dfrac{6}{a-2}$

Solution:

$$1 + \frac{3a}{a-2} = \frac{6}{a-2}$$

The LCD of all the expressions is $a - 2$. The domain restriction is $a \neq 2$.

$$(a-2)\left(1 + \frac{3a}{a-2}\right) = (a-2)\left(\frac{6}{a-2}\right)$$

Multiply by the LCD.

$$(a-2)1 + (a-2)^1\left(\frac{3a}{a-2}\right) = (a-2)^1\left(\frac{6}{a-2}\right)$$

Apply the distributive property.

$$a - 2 + 3a = 6$$

Solve the resulting equation (linear).

$$4a - 2 = 6$$

$$4a = 8$$

$$a = 2$$

2 is a restricted value.

Check: $1 + \dfrac{3a}{a-2} = \dfrac{6}{a-2}$

$$1 + \frac{3(2)}{(2)-2} \stackrel{?}{=} \frac{6}{(2)-2}$$

$$1 + \frac{6}{0} \stackrel{?}{=} \frac{6}{0}$$

The denominator is 0 when $a = 2$.

Because the value $a = 2$ makes the denominator zero in one (or more) of the rational expressions within the equation, the equation is undefined for $a = 2$. That is, 2 is not in the domain. No other potential solutions exist for the equation.

The equation $1 + \dfrac{3a}{a-2} = \dfrac{6}{a-2}$ has no solution.

Examples 1–3 show that the steps to solve a rational equation mirror the process of clearing fractions from Section 2.3. However, there is one significant difference. The solutions of a rational equation must not make the denominator equal to zero for any expression within the equation. When $a = 2$ is substituted into the expression

$$\frac{3a}{a-2} \quad \text{or} \quad \frac{6}{a-2}$$

the denominator is zero and the expression is undefined. Therefore, 2 cannot be a solution to the equation

$$1 + \frac{3a}{a-2} = \frac{6}{a-2}$$

Answer

3. No solution (the value -1 does not check.)

The steps to solve a rational equation are summarized as follows.

> **PROCEDURE** Solving a Rational Equation
> **Step 1** Factor the denominators of all rational expressions. Determine the domain restrictions.
> **Step 2** Identify the LCD of all expressions in the equation.
> **Step 3** Multiply both sides of the equation by the LCD.
> **Step 4** Solve the resulting equation.
> **Step 5** Check potential solutions in the original equation.

After multiplying by the LCD and then simplifying, the rational equation will be either a linear equation or higher degree equation.

Skill Practice

Solve the equation.

4. $\dfrac{z}{2} - \dfrac{1}{2z} = \dfrac{12}{z}$

Example 4 Solving a Rational Equation

Solve the equation. $\quad 1 - \dfrac{4}{p} = -\dfrac{3}{p^2}$

Solution:

$1 - \dfrac{4}{p} = -\dfrac{3}{p^2}$ — **Step 1:** The denominators are already factored. The domain restriction is $p \neq 0$.

Step 2: The LCD of all expressions is p^2.

$p^2\left(1 - \dfrac{4}{p}\right) = p^2\left(-\dfrac{3}{p^2}\right)$ — **Step 3:** Multiply by the LCD.

$p^2(1) - \overset{p}{\cancel{p^2}}\left(\dfrac{4}{\cancel{p}}\right) = \overset{1}{\cancel{p^2}}\left(-\dfrac{3}{\cancel{p^2}}\right)$ — Apply the distributive property.

$p^2 - 4p = -3$ — **Step 4:** Solve the resulting quadratic equation.

$p^2 - 4p + 3 = 0$ — Set the equation equal to zero and factor.

$(p - 3)(p - 1) = 0$

$p - 3 = 0 \quad \text{or} \quad p - 1 = 0$ — Set each factor equal to zero.

$p = 3 \quad \text{or} \quad p = 1$ — **Step 5:** Check: $p = 3$ $\qquad$ Check: $p = 1$

3 and 1 are not restricted values.

$1 - \dfrac{4}{p} = -\dfrac{3}{p^2} \qquad 1 - \dfrac{4}{p} = -\dfrac{3}{p^2}$

$1 - \dfrac{4}{(3)} \overset{?}{=} -\dfrac{3}{(3)^2} \qquad 1 - \dfrac{4}{(1)} \overset{?}{=} -\dfrac{3}{(1)^2}$

$\dfrac{3}{3} - \dfrac{4}{3} \overset{?}{=} -\dfrac{3}{9} \qquad 1 - 4 \overset{?}{=} -3$

Both solutions 3 and 1 check.

$-\dfrac{1}{3} \overset{?}{=} -\dfrac{1}{3} ✔ \qquad -3 \overset{?}{=} -3 ✔$

Answer

4. $5, -5$

Example 5 Solving a Rational Equation

Solve the equation. $\quad \dfrac{6}{t^2 - 7t + 12} + \dfrac{2t}{t - 3} = \dfrac{3t}{t - 4}$

Solution:

$$\dfrac{6}{t^2 - 7t + 12} + \dfrac{2t}{t - 3} = \dfrac{3t}{t - 4}$$

$$\dfrac{6}{(t - 3)(t - 4)} + \dfrac{2t}{t - 3} = \dfrac{3t}{t - 4}$$

Step 1: Factor the denominators. The domain restrictions are $t \neq 3$ and $t \neq 4$.

Step 2: The LCD is $(t - 3)(t - 4)$.

Step 3: Multiply by the LCD on both sides.

$$(t - 3)(t - 4)\left(\dfrac{6}{(t - 3)(t - 4)} + \dfrac{2t}{t - 3}\right) = (t - 3)(t - 4)\left(\dfrac{3t}{t - 4}\right)$$

$$(t - 3)(t - 4)\left(\dfrac{6}{(t - 3)(t - 4)}\right) + (t - 3)(t - 4)\left(\dfrac{2t}{t - 3}\right) = (t - 3)(t - 4)\left(\dfrac{3t}{t - 4}\right)$$

$$6 + 2t(t - 4) = 3t(t - 3)$$

$6 + 2t^2 - 8t = 3t^2 - 9t$ **Step 4:** Solve the resulting equation.

$0 = 3t^2 - 2t^2 - 9t + 8t - 6$

$0 = t^2 - t - 6$ Because the resulting equation is quadratic, set the equation equal to zero and factor.

$0 = (t - 3)(t + 2)$

$t - 3 = 0 \quad \text{or} \quad t + 2 = 0$ Set each factor equal to zero.

$\quad\quad t = 3 \quad \text{or} \quad t = -2$

3 is a restricted value, but -2 is not restricted.   **Step 5:** Check the potential solutions in the original equation.

<u>Check:</u> $t = 3$

3 cannot be a solution to the equation because it will make the denominator zero in the original equation.

$$\dfrac{6}{t^2 - 7t + 12} + \dfrac{2t}{t - 3} = \dfrac{3t}{t - 4}$$

$$\dfrac{6}{(3)^2 - 7(3) + 12} + \dfrac{2(3)}{(3) - 3} \overset{?}{=} \dfrac{3(3)}{(3) - 4}$$

$$\dfrac{6}{0} + \dfrac{6}{0} \overset{?}{=} \dfrac{9}{-1}$$

Zero in the denominator

The only solution is -2.

<u>Check:</u> $t = -2$

$$\dfrac{6}{t^2 - 7t + 12} + \dfrac{2t}{t - 3} = \dfrac{3t}{t - 4}$$

$$\dfrac{6}{(-2)^2 - 7(-2) + 12} + \dfrac{2(-2)}{(-2) - 3} \overset{?}{=} \dfrac{3(-2)}{(-2) - 4}$$

$$\dfrac{6}{4 + 14 + 12} + \dfrac{-4}{-5} \overset{?}{=} \dfrac{-6}{-6}$$

$$\dfrac{6}{30} + \dfrac{4}{5} \overset{?}{=} 1$$

$$\dfrac{1}{5} + \dfrac{4}{5} = 1 \; ✔ \; \text{(True)}$$

$t = -2$ is a solution.

Concept Connections

6. Explain why it is important to check the solution(s) to a rational equation.

Answers

5. 4; (the value -4 does not check.)
6. It is important to check that the solutions are in the domain of each expression in the equation.

TIP: The solution to Example 8 can be written in several forms. The quantity

$$\frac{2A - hB}{h}$$

can be left as a single rational expression or can be split into two fractions and simplified.

$$b = \frac{2A - hB}{h} = \frac{2A}{h} - \frac{hB}{h} = \frac{2A}{h} - B$$

Skill Practice

10. Solve for h.

$$\frac{b}{x} = \frac{a}{h} + 1$$

Example 9 Solving Formulas Involving Rational Expressions

Solve for z. $\quad y = \dfrac{x - z}{x + z}$

Solution:

To solve for z, we must clear fractions so that z appears in the numerator.

$$y = \frac{x - z}{x + z} \qquad \text{LCD is } (x + z).$$

$$y(x + z) = \left(\frac{x - z}{x + z}\right)(x + z) \qquad \text{Multiply both sides of the equation by the LCD.}$$

$$yx + yz = x - z \qquad \text{Apply the distributive property.}$$

$$yz + z = x - yx \qquad \text{Collect } z \text{ terms on one side of the equation and collect terms not containing } z \text{ on the other side.}$$

$$z(y + 1) = x - yx \qquad \text{Factor out } z.$$

$$z = \frac{x - yx}{y + 1} \qquad \text{Divide by } y + 1 \text{ to solve for } z.$$

Answer

10. $h = \dfrac{ax}{b - x}$ or $\dfrac{-ax}{x - b}$

Section 7.6 Practice Exercises

Boost your GRADE at ALEKS.com!

ALEKS®
version 3.0

- Practice Problems
- Self-Tests
- NetTutor
- e-Professors
- Videos

Study Skills Exercise

1. Define the key terms:

   a. linear equation   **b.** quadratic equation   **c.** rational equation   **d.** domain restrictions

Review Exercises

For Exercises 2–7, perform the indicated operations.

2. $\dfrac{2}{x - 3} - \dfrac{3}{x^2 - x - 6}$

3. $\dfrac{2x - 6}{4x^2 + 7x - 2} \div \dfrac{x^2 - 5x + 6}{x^2 - 4}$

4. $\dfrac{2y}{y - 3} + \dfrac{4}{y^2 - 9}$

5. $\dfrac{h - \dfrac{1}{h}}{\dfrac{1}{5} - \dfrac{1}{5h}}$

6. $\dfrac{w - 4}{w^2 - 9} \cdot \dfrac{w - 3}{w^2 - 8w + 16}$

7. $1 + \dfrac{1}{x} - \dfrac{12}{x^2}$

Objective 1: Introduction to Rational Equations

For Exercises 8–13, solve the equations by first clearing the fractions. **(See Example 1.)**

8. $\dfrac{1}{3}z + \dfrac{2}{3} = -2z + 10$

9. $\dfrac{5}{2} + \dfrac{1}{2}b = 5 - \dfrac{1}{3}b$

10. $\dfrac{3}{2}p + \dfrac{1}{3} = \dfrac{2p - 3}{4}$

11. $\dfrac{5}{3} - \dfrac{1}{6}k = \dfrac{3k + 5}{4}$

12. $\dfrac{2x - 3}{4} + \dfrac{9}{10} = \dfrac{x}{5}$

13. $\dfrac{4y + 2}{3} - \dfrac{7}{6} = -\dfrac{y}{6}$

Objective 2: Solving Rational Equations

14. For the equation

$$\dfrac{1}{w} - \dfrac{1}{2} = -\dfrac{1}{4}$$

    a. Identify the domain.

    b. Identify the LCD of the fractions in the equation.

    c. Solve the equation.

15. For the equation

$$\dfrac{3}{z} - \dfrac{4}{5} = -\dfrac{1}{5}$$

    a. Identify the domain.

    b. Identify the LCD of the fractions in the equation.

    c. Solve the equation.

16. Identify the LCD of all the denominators in the equation.

$$\dfrac{x + 1}{x^2 + 2x - 3} = \dfrac{1}{x + 3} - \dfrac{1}{x - 1}$$

For Exercises 17–46, solve the equations. **(See Examples 2–5.)**

17. $\dfrac{1}{8} = \dfrac{3}{5} + \dfrac{5}{y}$

18. $\dfrac{2}{7} - \dfrac{1}{x} = \dfrac{2}{3}$

19. $\dfrac{4}{t} = \dfrac{3}{t} + \dfrac{1}{8}$

20. $\dfrac{9}{b} - \dfrac{8}{b} = \dfrac{1}{4}$

21. $\dfrac{5}{6x} + \dfrac{7}{x} = 1$

22. $\dfrac{14}{3x} - \dfrac{5}{x} = 2$

23. $1 - \dfrac{2}{y} = \dfrac{3}{y^2}$

24. $1 - \dfrac{2}{m} = \dfrac{8}{m^2}$

25. $\dfrac{a + 1}{a} = 1 + \dfrac{a - 2}{2a}$

26. $\dfrac{7b - 4}{5b} = \dfrac{9}{5} - \dfrac{4}{b}$

27. $\dfrac{w}{5} - \dfrac{w + 3}{w} = -\dfrac{3}{w}$

28. $\dfrac{t}{12} + \dfrac{t + 3}{3t} = \dfrac{1}{t}$

29. $\dfrac{2}{m + 3} = \dfrac{5}{4m + 12} - \dfrac{3}{8}$

30. $\dfrac{2}{4n - 4} - \dfrac{7}{4} = \dfrac{-3}{n - 1}$

31. $\dfrac{p}{p - 4} - 5 = \dfrac{4}{p - 4}$

32. $\dfrac{-5}{q + 5} = \dfrac{q}{q + 5} + 2$

33. $\dfrac{2t}{t + 2} - 2 = \dfrac{t - 8}{t + 2}$

34. $\dfrac{4w}{w - 3} - 3 = \dfrac{3w - 1}{w - 3}$

35. $\dfrac{x^2 - x}{x - 2} = \dfrac{12}{x - 2}$

36. $\dfrac{x^2 + 9}{x + 4} = \dfrac{-10x}{x + 4}$

37. $\dfrac{x^2 + 3x}{x - 1} = \dfrac{4}{x - 1}$

38. $\dfrac{2x^2 - 21}{2x - 3} = \dfrac{-11x}{2x - 3}$

39. $\dfrac{2x}{x + 4} - \dfrac{8}{x - 4} = \dfrac{2x^2 + 32}{x^2 - 16}$

40. $\dfrac{4x}{x + 3} - \dfrac{12}{x - 3} = \dfrac{4x^2 + 36}{x^2 - 9}$

41. $\dfrac{x}{x+6} = \dfrac{72}{x^2-36} + 4$

42. $\dfrac{y}{y+4} = \dfrac{32}{y^2-16} + 3$

43. $\dfrac{5}{3x-3} - \dfrac{2}{x-2} = \dfrac{7}{x^2-3x+2}$

44. $\dfrac{6}{5a+10} - \dfrac{1}{a-5} = \dfrac{4}{a^2-3a-10}$

45. $\dfrac{y-2}{y-3} = \dfrac{11}{y^2-7y+12} + \dfrac{y}{y-4}$

46. $\dfrac{6}{w+1} - \dfrac{3}{w+5} = \dfrac{18}{w^2+6w+5}$

For Exercises 47–50, translate to a rational equation and solve. **(See Example 6.)**

47. The reciprocal of a number is added to three. The result is the quotient of 25 and the number. Find the number.

48. The difference of three and the reciprocal of a number is equal to the quotient of 20 and the number. Find the number.

49. If a number added to five is divided by the difference of the number and two, the result is three-fourths. Find the number.

50. If twice a number added to three is divided by the number plus one, the result is three-halves. Find the number.

Objective 3: Solving Formulas Involving Rational Expressions

For Exercises 51–68, solve for the indicated variable. **(See Examples 7–9.)**

51. $K = \dfrac{ma}{F}$ for m

52. $K = \dfrac{ma}{F}$ for a

53. $K = \dfrac{IR}{E}$ for E

54. $K = \dfrac{IR}{E}$ for R

55. $I = \dfrac{E}{R+r}$ for R

56. $I = \dfrac{E}{R+r}$ for r

57. $h = \dfrac{2A}{B+b}$ for B

58. $\dfrac{C}{\pi r} = 2$ for r

59. $\dfrac{V}{\pi h} = r^2$ for h

60. $\dfrac{V}{lw} = h$ for w

61. $x = \dfrac{at+b}{t}$ for t

62. $\dfrac{T+mf}{m} = g$ for m

63. $\dfrac{x-y}{xy} = z$ for x

64. $\dfrac{w-n}{wn} = P$ for w

65. $a+b = \dfrac{2A}{h}$ for h

66. $1+rt = \dfrac{A}{P}$ for P

67. $\dfrac{1}{R} = \dfrac{1}{R_1} + \dfrac{1}{R_2}$ for R

68. $\dfrac{b+a}{ab} = \dfrac{1}{f}$ for b

Problem Recognition Exercises

Comparing Rational Equations and Rational Expressions

Often adding or subtracting rational expressions is confused with solving rational equations. When adding rational expressions, we combine the terms to simplify the expression. When solving an equation, we clear the fractions and find numerical solutions, if possible. Both processes begin with finding the LCD, but the LCD is used differently in each process. Compare these two examples.

Example 1:

Add. $\dfrac{4}{x} + \dfrac{x}{3}$ (The LCD is $3x$.)

$$= \frac{3}{3} \cdot \left(\frac{4}{x}\right) + \left(\frac{x}{3}\right) \cdot \frac{x}{x}$$

$$= \frac{12}{3x} + \frac{x^2}{3x}$$

$$= \frac{12 + x^2}{3x} \quad \text{The final answer is a}$$
rational expression.

Example 2:

Solve. $\dfrac{4}{x} + \dfrac{x}{3} = -\dfrac{8}{3}$ (The LCD is $3x$.)

$$\frac{3x}{1}\left(\frac{4}{x} + \frac{x}{3}\right) = \frac{3x}{1}\left(-\frac{8}{3}\right)$$

$$12 + x^2 = -8x$$

$$x^2 + 8x + 12 = 0$$

$$(x + 2)(x + 6) = 0$$

$$x + 2 = 0 \text{ or } x + 6 = 0$$

$x = -2$ or $x = -6$ The final answers are numbers. The solutions -2 and -6 both check in the original equation.

For Exercises 1–12, solve the equation or simplify the expression by combining the terms.

1. $\dfrac{y}{2y + 4} - \dfrac{2}{y^2 + 2y}$

2. $\dfrac{1}{x + 2} + 2 = \dfrac{x + 11}{x + 2}$

3. $\dfrac{5t}{2} - \dfrac{t - 2}{3} = 5$

4. $3 - \dfrac{2}{a - 5}$

5. $\dfrac{7}{6p^2} + \dfrac{2}{9p} + \dfrac{1}{3p^2}$

6. $\dfrac{3b}{b + 1} - \dfrac{2b}{b - 1}$

7. $4 + \dfrac{2}{h - 3} = 5$

8. $\dfrac{2}{w + 1} + \dfrac{3}{(w + 1)^2}$

9. $\dfrac{1}{x - 6} - \dfrac{3}{x^2 - 6x} = \dfrac{4}{x}$

10. $\dfrac{3}{m} - \dfrac{6}{5} = -\dfrac{3}{m}$

11. $\dfrac{7}{2x + 2} + \dfrac{3x}{4x + 4}$

12. $\dfrac{10}{2t - 1} - 1 = \dfrac{t}{2t - 1}$

Applications of Rational Equations and Proportions

Objectives

1. **Solving Proportions**
2. **Applications of Proportions and Similar Triangles**
3. **Distance, Rate, and Time Applications**
4. **Work Applications**

1. Solving Proportions

In this section, we look at how rational equations can be used to solve a variety of applications. The first type of rational equation that will be applied is called a proportion.

> **DEFINITION** Proportion
>
> An equation that equates two ratios or rates is called a **proportion**. Thus, for $b \neq 0$ and $d \neq 0$, $\frac{a}{b} = \frac{c}{d}$ is a proportion.

A proportion can be solved by multiplying both sides of the equation by the LCD and clearing fractions.

Skill Practice

Solve the proportion.

1. $\dfrac{10}{b} = \dfrac{2}{33}$

Example 1 Solving a Proportion

Solve the proportion. $\dfrac{3}{11} = \dfrac{123}{w}$

Solution:

$$\frac{3}{11} = \frac{123}{w} \qquad \text{The LCD is } 11w.$$

$$11w\left(\frac{3}{11}\right) = 11w\left(\frac{123}{w}\right) \qquad \text{Multiply by the LCD and clear fractions.}$$

$$3w = 11 \cdot 123 \qquad \text{Solve the resulting equation (linear).}$$

$$3w = 1353$$

$$\frac{3w}{3} = \frac{1353}{3}$$

$$w = 451$$

Check: $w = 451$

$$\frac{3}{11} = \frac{123}{w}$$

$$\frac{3}{11} \overset{?}{=} \frac{123}{(451)}$$

The solution is 451.

$$\frac{3}{11} \overset{?}{=} \frac{3}{11} \checkmark \text{ (True)} \qquad \text{Simplify to lowest terms.}$$

Answer

1. 165

TIP: The cross products of any proportion are equal. That is, for $b \neq 0$ and $d \neq 0$, the proportion $\frac{a}{b} = \frac{c}{d}$ is equivalent to $ad = bc$. Some rational equations are proportions and can be solved by equating the cross products. Consider the proportion from Example 1:

$$\frac{3}{11} \diagdown \!\!\!\!\! \diagup \frac{123}{w}$$

$3 \cdot w = 11 \cdot 123$	Equate the cross products.
$3w = 1353$	Solve the resulting equation.
$\dfrac{3w}{3} = \dfrac{1353}{3}$	
$w = 451$	The solution is 451.

2. Applications of Proportions and Similar Triangles

Example 2 Using a Proportion in an Application

For a recent year, the population of Alabama was approximately 4.2 million. At that time, Alabama had seven representatives in the U.S. House of Representatives. In the same year, North Carolina had a population of approximately 7.2 million. If representation in the House is based on population in equal proportions for each state, how many representatives did North Carolina have?

Solution:

Let x represent the number of representatives for North Carolina.

Set up a proportion by writing two equivalent ratios.

$$\boxed{\frac{\text{Population of Alabama}}{\text{number of representatives}}} \rightarrow \frac{4.2}{7} = \frac{7.2}{x} \leftarrow \boxed{\frac{\text{Population of North Carolina}}{\text{number of representatives}}}$$

$$\frac{4.2}{7} = \frac{7.2}{x}$$

$7x \cdot \dfrac{4.2}{7} = 7x \cdot \dfrac{7.2}{x}$	Multiply by the LCD, $7x$.
$4.2x = (7.2)(7)$	Solve the resulting linear equation.
$4.2x = 50.4$	
$\dfrac{4.2x}{4.2} = \dfrac{50.4}{4.2}$	
$x = 12$	North Carolina had 12 representatives.

TIP: The equation from Example 2 could have been solved by first equating the cross products:

$$\frac{4.2}{7} \diagdown \!\!\!\!\! \diagup \frac{7.2}{x}$$

$$4.2x = (7.2)(7)$$
$$4.2x = 50.4$$
$$x = 12$$

Proportions are used in geometry with **similar triangles**. Two triangles are similar if their angles have equal measure. In such a case, the lengths of the corresponding sides are proportional. The triangles in Figure 7-1 are similar. Therefore, the following ratios are equivalent.

$$\frac{a}{x} = \frac{b}{y} = \frac{c}{z}$$

Answer

2. 300

Figure 7-1

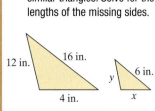
Example 3 **Using Similar Triangles to Find an Unknown Side in a Triangle**

The triangles in Figure 7-2 are similar.

Figure 7-2

a. Solve for x.   **b.** Solve for y.

Solution:

a. The lengths of the upper right sides of the triangles are given. These form a known ratio of $\frac{10}{6}$. Because the triangles are similar, the ratio of the other corresponding sides must be equal to $\frac{10}{6}$. To solve for x, we have:

Bottom side from large triangle	➝	x	$=$	10 in.	←	Right side from large triangle
bottom side from small triangle	➝	9 in.		6 in.	←	right side from small triangle

$$\frac{x}{9} = \frac{10}{6} \qquad \text{The LCD is 18.}$$

$$\overset{2}{\cancel{18}} \cdot \left(\frac{x}{\cancel{9}}\right) = \overset{3}{\cancel{18}} \cdot \left(\frac{10}{\cancel{6}}\right) \qquad \text{Multiply by the LCD.}$$

$$2x = 30 \qquad \text{Clear fractions.}$$

$$x = 15 \qquad \text{Divide by 2.}$$

The length of side x is 15 in.

b. To solve for y, the ratio of the upper left sides of the triangles must equal $\frac{10}{6}$.

Left side from large triangle	➝	8 in.	$=$	10 in.	←	Right side from large triangle
left side from small triangle	➝	y		6 in.	←	right side from small triangle

$$\frac{8}{y} = \frac{10}{6} \qquad \text{The LCD is } 6y.$$

$$6\cancel{y} \cdot \left(\frac{8}{\cancel{y}}\right) = \cancel{6}y \cdot \left(\frac{10}{\cancel{6}}\right) \qquad \text{Multiply by the LCD.}$$

$$48 = 10y \qquad \text{Clear fractions.}$$

$$\frac{48}{10} = \frac{10y}{10}$$

$$4.8 = y$$

The length of side y is 4.8 in.

| Example 4 | **Using Similar Triangles in an Application** |

The shadow cast by a yardstick is 2 ft long. The shadow cast by a tree is 11 ft long. Find the height of the tree.

Solution:

Step 1: Read the problem.

Let x represent the height of the tree.   **Step 2:** Label the variables.

We will assume that the measurements were taken at the same time of day. Therefore, the angle of the Sun is the same on both objects, and we can set up similar triangles (Figure 7-3).

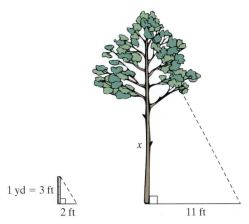

1 yd = 3 ft

2 ft

x

11 ft

Figure 7-3

Step 3: Create a verbal model.

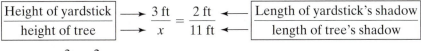

| Height of yardstick | $\longrightarrow$ | $\dfrac{3\ \text{ft}}{x}$ | $=$ | $\dfrac{2\ \text{ft}}{11\ \text{ft}}$ | $\longleftarrow$ | Length of yardstick's shadow |
| height of tree | $\longrightarrow$ | | | | $\longleftarrow$ | length of tree's shadow |

$$\frac{3}{x} = \frac{2}{11}$$

Step 4: Write a mathematical equation.

$$11x\left(\frac{3}{x}\right) = \left(\frac{2}{11}\right)11x$$

Step 5: Multiply by the LCD.

$$33 = 2x$$

Solve the equation.

$$\frac{33}{2} = \frac{2x}{2}$$

$$16.5 = x$$

Step 6: Interpret the results and write the answer in words.

The tree is 16.5 ft high.

Skill Practice

4. The Sun casts a 3.2-ft shadow of a 6 ft man. At the same time, the Sun casts a 80-ft shadow of a building. How tall is the building?

3. Distance, Rate, and Time Applications

In Sections 2.7 and 4.4 we presented applications involving the relationship among the variables distance, rate, and time. Recall that $d = rt$.

Answer

4. The building is 150 ft tall.

Skill Practice

5. Alison paddles her kayak in a river where the current of the water is 2 mph. She can paddle 20 mi with the current in the same time that she can paddle 10 mi against the current. Find the speed of the kayak in still water.

Example 5 Using a Rational Equation in a Distance, Rate, and Time Application

A small plane flies 440 mi with the wind from Memphis, TN, to Oklahoma City, OK. In the same amount of time, the plane flies 340 miles against the wind from Oklahoma City to Little Rock, AR (see Figure 7-4). If the wind speed is 30 mph, find the speed of the plane in still air.

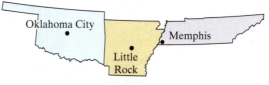

Figure 7-4

Solution:

Let x represent the speed of the plane in still air.

Then $x + 30$ is the speed of the plane with the wind.

$x - 30$ is the speed of the plane against the wind.

Organize the given information in a chart.

	Distance	Rate	Time
With the wind	440	$x + 30$	$\dfrac{440}{x + 30}$
Against the wind	340	$x - 30$	$\dfrac{340}{x - 30}$

Because $d = rt$, then $t = \dfrac{d}{r}$

TIP: The equation

$$\frac{440}{x + 30} = \frac{340}{x - 30}$$

is a proportion. The fractions can also be cleared by equating the cross products.

$$\frac{440}{x + 30} \diagup\kern-1em\diagdown \frac{340}{x - 30}$$

$$440(x - 30) = 340(x + 30)$$

The plane travels with the wind for the same amount of time as it travels against the wind, so we can equate the two expressions for time.

$$\left(\begin{array}{c} \text{Time with} \\ \text{the wind} \end{array} \right) = \left(\begin{array}{c} \text{time against} \\ \text{the wind} \end{array} \right)$$

$$\frac{440}{x + 30} = \frac{340}{x - 30}$$

The LCD is $(x + 30)(x - 30)$.

$$(x + 30)(x - 30) \cdot \frac{440}{x + 30} = (x + 30)(x - 30) \cdot \frac{340}{x - 30}$$

$$440(x - 30) = 340(x + 30)$$

$$440x - 13{,}200 = 340x + 10{,}200$$

Solve the resulting linear equation.

$$100x = 23{,}400$$

$$x = 234$$

The plane's speed in still air is 234 mph.

Answer

5. 6 mph

Example 6 Using a Rational Equation in a Distance, Rate, and Time Application

A motorist drives 100 mi between two cities in a bad rainstorm. For the return trip in sunny weather, she averages 10 mph faster and takes $\frac{1}{2}$ hr less time. Find the average speed of the motorist in the rainstorm and in sunny weather.

Solution:

Let x represent the motorist's speed during the rain.

Then $x + 10$ represents the speed in sunny weather.

	Distance	Rate	Time
Trip during rainstorm	100	x	$\dfrac{100}{x}$
Trip during sunny weather	100	$x + 10$	$\dfrac{100}{x + 10}$

Because $d = rt$, then $t = \dfrac{d}{r}$

Because the same distance is traveled in $\frac{1}{2}$ hr less time, the difference between the time of the trip during the rainstorm and the time during sunny weather is $\frac{1}{2}$ hr.

$$\left(\begin{array}{c}\text{Time during}\\\text{the rainstorm}\end{array}\right) - \left(\begin{array}{c}\text{time during}\\\text{sunny weather}\end{array}\right) = \left(\frac{1}{2}\text{ hr}\right) \qquad \text{Verbal model}$$

$$\frac{100}{x} - \frac{100}{x + 10} = \frac{1}{2} \qquad \begin{array}{l}\text{Mathematical}\\\text{equation}\end{array}$$

$$2x(x + 10)\left(\frac{100}{x} - \frac{100}{x + 10}\right) = 2x(x + 10)\left(\frac{1}{2}\right) \qquad \begin{array}{l}\text{Multiply by}\\\text{the LCD.}\end{array}$$

$$2x(x + 10)\left(\frac{100}{x}\right) - 2x(x + 10)\left(\frac{100}{x + 10}\right) = 2x(x + 10)\left(\frac{1}{2}\right) \qquad \begin{array}{l}\text{Apply the}\\\text{distributive}\\\text{property.}\end{array}$$

$$200(x + 10) - 200x = x(x + 10) \qquad \text{Clear fractions.}$$

$$200x + 2000 - 200x = x^2 + 10x \qquad \begin{array}{l}\text{Solve the}\\\text{resulting}\\\text{equation}\\\text{(quadratic).}\end{array}$$

$$2000 = x^2 + 10x$$

$$0 = x^2 + 10x - 2000 \qquad \begin{array}{l}\text{Set the}\\\text{equation equal}\\\text{to zero.}\end{array}$$

$$0 = (x - 40)(x + 50) \qquad \text{Factor.}$$

$$x = 40 \qquad \text{or} \qquad x = -50$$

Because a rate of speed cannot be negative, reject $x = -50$. Therefore, the speed of the motorist in the rainstorm is 40 mph. Because $x + 10 = 40 + 10 = 50$, the average speed for the return trip in sunny weather is 50 mph.

Skill Practice

6. Harley rode his mountain bike 12 mi to the top of the mountain and the same distance back down. His speed going up was 8 mph slower than coming down. The ride up took 2 hr longer than the ride coming down. Find his speeds.

Avoiding Mistakes

The equation

$$\frac{100}{x} - \frac{100}{x + 10} = \frac{1}{2}$$

is not a proportion because the left-hand side has more than one fraction. Do not try to multiply the cross products. Instead, multiply by the LCD to clear fractions.

Answer

6. Uphill speed was 4 mph; downhill speed was 12 mph.

4. Work Applications

Example 7 demonstrates how work rates are related to a portion of a job that can be completed in one unit of time.

Example 7 Using a Rational Equation in a Work Problem

A new printing press can print the morning edition in 2 hr, whereas the old printer required 4 hr. How long would it take to print the morning edition if both printers were working together?

Solution:

Let x represent the time required for both printers working together to complete the job.

One method to approach this problem is to determine the portion of the job that each printer can complete in 1 hr and extend that rate to the portion of the job completed in x hr.

- The old printer can perform the job in 4 hr. Therefore, it completes $\frac{1}{4}$ of the job in 1 hr and $\frac{1}{4}x$ jobs in x hr.
- The new printer can perform the job in 2 hr. Therefore, it completes $\frac{1}{2}$ of the job in 1 hr and $\frac{1}{2}x$ jobs in x hr.

	Work Rate	Time	Portion of Job Completed
Old printer	$\dfrac{1 \text{ job}}{4 \text{ hr}}$	x hours	$\dfrac{1}{4}x$
New printer	$\dfrac{1 \text{ job}}{2 \text{ hr}}$	x hours	$\dfrac{1}{2}x$

The sum of the portions of the job completed by each printer must equal one whole job.

$$\begin{pmatrix} \text{Portion of job} \\ \text{completed by} \\ \text{old printer} \end{pmatrix} + \begin{pmatrix} \text{portion of job} \\ \text{completed by} \\ \text{new printer} \end{pmatrix} = \begin{pmatrix} 1 \\ \text{whole} \\ \text{job} \end{pmatrix}$$

$\dfrac{1}{4}x + \dfrac{1}{2}x = 1$ The LCD is 4.

$4\left(\dfrac{1}{4}x + \dfrac{1}{2}x\right) = 4(1)$ Multiply by the LCD.

$\overset{1}{4} \cdot \dfrac{1}{\overset{}{4}}x + \overset{2}{4} \cdot \dfrac{1}{2}x = 4 \cdot 1$ Apply the distributive property.

$x + 2x = 4$ Solve the resulting linear equation.

$3x = 4$

$x = \dfrac{4}{3}$ or $x = 1\dfrac{1}{3}$ The time required to print the morning edition using both printers is $1\frac{1}{3}$ hr.

Answer

7. 12 hr

TIP: An alternative approach to solving a "work" problem is to add rates of speed. In Example 7, we could have set up an equation as follows.

$$\left(\begin{matrix}\text{Rate of speed} \\ \text{of old printer}\end{matrix}\right) + \left(\begin{matrix}\text{rate of speed} \\ \text{of new printer}\end{matrix}\right) = \left(\begin{matrix}\text{rate of speed of} \\ \text{both working together}\end{matrix}\right)$$

$$\frac{1 \text{ job}}{4 \text{ hr}} + \frac{1 \text{ job}}{2 \text{ hr}} = \frac{1 \text{ job}}{x \text{ hr}}$$

$$\frac{1}{4} + \frac{1}{2} = \frac{1}{x}$$

$$4x\left(\frac{1}{4} + \frac{1}{2}\right) = 4x\left(\frac{1}{x}\right) \qquad \text{Multiply by the LCD, } 4x.$$

$$x + 2x = 4$$

$$3x = 4$$

$$x = \frac{4}{3} \qquad \text{The time required for both printers working together is } 1\frac{1}{3} \text{ hr.}$$

Section 7.7   Practice Exercises

Boost your GRADE at ALEKS.com!

- Practice Problems
- Self-Tests
- NetTutor
- e-Professors
- Videos

Study Skills Exercise

1. Define the key terms:

    a. proportion   **b. similar triangles**

Review Exercises

For Exercises 2–7, determine whether each of the following is an equation or an expression. If it is an equation, solve it. If it is an expression, perform the indicated operation.

2. $\dfrac{b}{5} + 3 = 9$

3. $\dfrac{m}{m-1} - \dfrac{2}{m+3}$

4. $\dfrac{2}{a+5} + \dfrac{5}{a^2-25}$

5. $\dfrac{3y+6}{20} \div \dfrac{4y+8}{8}$

6. $\dfrac{z^2+z}{24} \cdot \dfrac{8}{z+1}$

7. $\dfrac{3}{p+3} = \dfrac{12p+19}{p^2+7p+12} - \dfrac{5}{p+4}$

8. Determine whether 1 is a solution to the equation. $\dfrac{1}{x-1} + \dfrac{1}{2} = \dfrac{2}{x^2-1}$

Objective 1: Solving Proportions

For Exercises 9–22, solve the proportions. **(See Example 1.)**

9. $\dfrac{8}{5} = \dfrac{152}{p}$

10. $\dfrac{6}{7} = \dfrac{96}{y}$

11. $\dfrac{19}{76} = \dfrac{z}{4}$

12. $\dfrac{15}{135} = \dfrac{w}{9}$

13. $\dfrac{5}{3} = \dfrac{a}{8}$

14. $\dfrac{b}{14} = \dfrac{3}{8}$

15. $\dfrac{2}{1.9} = \dfrac{x}{38}$

16. $\dfrac{16}{1.3} = \dfrac{30}{p}$

17. $\dfrac{y + 1}{2y} = \dfrac{2}{3}$

18. $\dfrac{w - 2}{4w} = \dfrac{1}{6}$

19. $\dfrac{9}{2z - 1} = \dfrac{3}{z}$

20. $\dfrac{1}{t} = \dfrac{1}{4 - t}$

21. $\dfrac{8}{9a - 1} = \dfrac{5}{3a + 2}$

22. $\dfrac{4p + 1}{3} = \dfrac{2p - 5}{6}$

23. Charles' law describes the relationship between the initial and final temperature and volume of a gas held at a constant pressure.

$$\frac{V_i}{V_f} = \frac{T_i}{T_f}$$

  a. Solve the equation for V_f.

  b. Solve the equation for T_f.

24. The relationship between the area, height, and base of a triangle is given by the proportion

$$\frac{A}{b} = \frac{h}{2}$$

  a. Solve the equation for A.

  b. Solve the equation for b.

Objective 2: Applications of Proportions and Similar Triangles

For Exercises 25–32, solve using proportions.

 25. Toni drives her Honda Civic 132 mi on the highway on 4 gal of gas. At this rate how many miles can she drive on 9 gal of gas? **(See Example 2.)**

26. Tim takes his pulse for 10 sec and counts 12 beats. How many beats per minute is this?

27. Suppose a household of 4 people produces 128 lb of garbage in one week. At this rate, how many pounds will 48 people produce in 1 week?

28. Property tax on a $180,000 house is $4000. At this rate, how much tax would be paid on a $216,000 home?

29. Andrew is on a low-carbohydrate diet. If his diet book tells him that an 8-oz serving of pineapple contains 19.2 g of carbohydrate, how many grams of carbohydrate does a 5-oz serving contain?

30. Cooking oatmeal requires 1 cup of water for every $\frac{1}{2}$ cup of oats. How many cups of water will be required for $\frac{3}{4}$ cup of oats?

31. A map has a scale of 75 mi/in. If two cities measure 3.5 in. apart, how many miles does this represent?

32. A map has a scale of 50 mi/in. If two cities measure 6.5 in. apart, how many miles does this represent?

For Exercises 33–36, the figures are similar. Solve for x and y. **(See Example 3.)**

33.

34.

35.

36.

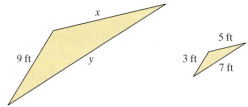

37. To estimate the height of a light pole, a mathematics student measures the length of a shadow cast by a meterstick and the length of the shadow cast by the light pole. Find the height of the light pole. **(See Example 4.)**

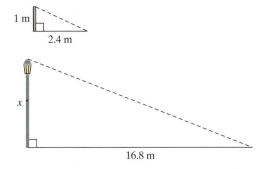

38. To estimate the height of a building, a student measures the length of a shadow cast by a yardstick and the length of the shadow cast by the building. Find the height of the building.

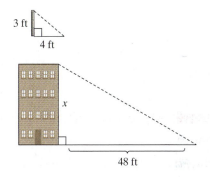

39. A 6-ft-tall man standing 54 ft from a light post casts an 18-ft shadow. What is the height of the light post?

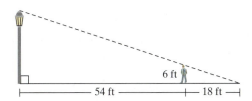

40. For a science project at school, a student must measure the height of a tree. The student measures the length of the shadow of the tree and then measures the length of the shadow cast by a yardstick. Use similar triangles to find the height of the tree.

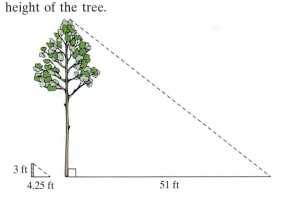

Objective 3: Distance, Rate, and Time Applications

41. A boat travels 54 mi upstream against the current in the same amount of time it takes to travel 66 mi downstream with the current. If the current is 2 mph, what is the speed of the boat in still water? (Use $t = \frac{d}{r}$ to complete the table.) **(See Example 5.)**

	Distance	Rate	Time
With the current (downstream)			
Against the current (upstream)			

42. A plane flies 630 mi with the wind in the same time that it takes to fly 455 mi against the wind. If this plane flies at the rate of 217 mph in still air, what is the speed of the wind? (Use $t = \frac{d}{r}$ to complete the table.)

	Distance	Rate	Time
With the wind			
Against the wind			

43. The jet stream is a fast flowing air current found in the atmosphere at around 36,000 ft above the surface of the Earth. During one summer day, the speed of the jet stream is 35 mph. A plane flying with the jet stream can fly 700 mi in the same amount of time that it would take to fly 500 mi against the jet stream. What is the speed of the plane in still air?

44. A fisherman travels 9 mi downstream with the current in the same time that he travels 3 mi upstream against the current. If the speed of the current is 6 mph, what is the speed at which the fisherman travels in still water?

45. An athlete in training rides his bike 20 mi and then immediately follows with a 10-mi run. The total workout takes him 2.5 hr. He also knows that he bikes about twice as fast as he runs. Determine his biking speed and his running speed.

46. Devon can cross-country ski 5 km/hr faster than his sister Shanelle. Devon skis 45 km in the same time Shanelle skis 30 km. Find their speeds.

47. Floyd can walk 2 mph faster than his wife, Rachel. It takes Rachel 3 hr longer than Floyd to hike a 12-mi trail through the park. Find their speeds. **(See Example 6.)**

48. Janine bikes 3 mph faster than her sister, Jessica. Janine can ride 36 mi in 1 hr less time than Jessica can ride the same distance. Find each of their speeds.

49. Sergio rode his bike 4 mi. Then he got a flat tire and had to walk back 4 mi. It took him 1 hr longer to walk than it did to ride. If his rate walking was 9 mph less than his rate riding, find the two rates.

50. Amber jogs 10 km in $\frac{3}{4}$ hr less than she can walk the same distance. If her walking rate is 3 km/hr less than her jogging rate, find her rates jogging and walking (in km/hr).

Objective 4: Work Applications

51. If it takes a person 2 hr to paint a room, what fraction of the room would be painted in 1 hr?

52. If it takes a copier 3 hr to complete a job, what fraction of the job would be completed in 1 hr?

53. If the cold-water faucet is left on, the sink will fill in 10 min. If the hot-water faucet is left on, the sink will fill in 12 min. How long would it take the sink to fill if both faucets are left on? **(See Example 7.)**

54. The CUT-IT-OUT lawn mowing company consists of two people: Tina and Bill. If Tina cuts a lawn by herself, she can do it in 4 hr. If Bill cuts the same lawn himself, it takes him an hour longer than Tina. How long would it take them if they worked together?

55. A manuscript needs to be printed. One printer can do the job in 50 min, and another printer can do the job in 40 min. How long would it take if both printers were used?

56. A pump can empty a small pond in 4 hr. Another more efficient pump can do the job in 3 hr. How long would it take to empty the pond if both pumps were used?

57. A pipe can fill a reservoir in 16 hr. A drainage pipe can drain the reservoir in 24 hr. How long would it take to fill the reservoir if the drainage pipe were left open by mistake? (*Hint:* The rate at which water drains should be negative.)

58. A hole in the bottom of a child's plastic swimming pool can drain the pool in 60 min. If the pool had no hole, a hose could fill the pool in 40 min. How long would it take the hose to fill the pool with the hole?

59. Tim and Al are bricklayers. Tim can construct an outdoor grill in 5 days. If Al helps Tim, they can build it in only 2 days. How long would it take Al to build the grill alone?

60. Norma is a new and inexperienced secretary. It takes her 3 hr to prepare a mailing. If her boss helps her, the mailing can be completed in 1 hr. How long would it take the boss to do the job by herself?

Expanding Your Skills

For Exercises 61–64, solve using proportions.

61. The ratio of smokers to nonsmokers in a restaurant is 2 to 7. There are 100 more nonsmokers than smokers. How many smokers and nonsmokers are in the restaurant?

62. The ratio of fiction to nonfiction books sold in a bookstore is 5 to 3. One week there were 180 more fiction books sold than nonfiction. Find the number of fiction and nonfiction books sold during that week.

63. There are 440 students attending a biology lecture. The ratio of male to female students at the lecture is 6 to 5. How many men and women are attending the lecture?

64. The ratio of dogs to cats at the humane society is 5 to 8. The total number of dogs and cats is 650. How many dogs and how many cats are at the humane society?

Variation

Section 7.8

1. Definition of Direct and Inverse Variation

In this section, we introduce the concept of variation. Direct and inverse variation models can show how one quantity varies in proportion to another.

Objectives

1. **Definition of Direct and Inverse Variation**
2. **Translations Involving Variation**
3. **Applications of Variation**

DEFINITION   Direct and Inverse Variation

Let k be a nonzero constant real number. Then the following statements are equivalent:

1. y varies **directly** as x.
   y is directly proportional to x. $\Big\}$ $y = kx$

2. y varies **inversely** as x.
   y is inversely proportional to x. $\Big\}$ $y = \dfrac{k}{x}$

Note: The value of k is called the constant of variation.

For a car traveling 30 mph, the equation $d = 30t$ indicates that the distance traveled is *directly proportional* to the time of travel. For positive values of k, when two variables are directly related, as one variable increases, the other variable will also increase. Likewise, if one variable decreases, the other will decrease. In the equation $d = 30t$, the longer the time of the trip, the greater the distance traveled. The shorter the time of the trip, the shorter the distance traveled.

For positive values of k, when two variables are *inversely related*, as one variable increases, the other will decrease, and vice versa. Consider a car traveling between Toronto and Montreal, a distance of 500 km. The time required to make the trip is inversely proportional to the speed of travel: $t = 500/r$. As the rate of speed, r, increases, the quotient $500/r$ will decrease. Thus the time will decrease. Similarly, as the rate of speed decreases, the trip will take longer.

2. Translations Involving Variation

The first step in using a variation model is to translate an English phrase into an equivalent mathematical equation.

Example 1 **Translating to a Variation Model**

Translate each expression into an equivalent mathematical model.

a. The circumference of a circle varies directly as the radius.

b. At a constant temperature, the volume of a gas varies inversely as the pressure.

c. The length of time of a meeting is directly proportional to the *square* of the number of people present.

Solution:

a. Let C represent circumference and r represent radius. The variables are directly related, so use the model $C = kr$.

b. Let V represent volume and P represent pressure. Because the variables are inversely related, use the model $V = k/P$.

c. Let t represent time, and let N be the number of people present at a meeting. Because t is directly related to N^2, use the model $t = kN^2$.

Sometimes a variable varies directly as the product of two or more other variables. In this case, we have joint variation.

DEFINITION **Joint Variation**

Let k be a nonzero constant real number. Then the following statements are equivalent:

y varies **jointly** as w and z.

y is jointly proportional to w and z. $\left.\right\}$ $y = kwz$

Example 2 **Translating to a Variation Model**

Translate each expression into an equivalent mathematical model.

a. y varies jointly as u and the square root of v.

b. The gravitational force of attraction between two planets varies jointly as the product of their masses and inversely as the square of the distance between them.

Solution:

a. $y = ku\sqrt{v}$

b. Let m_1 and m_2 represent the masses of the two planets. Let F represent the gravitational force of attraction and d represent the distance between the planets.

The variation model is $F = \dfrac{km_1m_2}{d^2}$

3. Applications of Variation

Consider the variation models $y = kx$ and $y = k/x$. In either case, if values for x and y are known, we can solve for k. Once k is known, we can use the variation equation to find y if x is known, or to find x if y is known. This concept is the basis for solving many problems involving variation.

PROCEDURE Finding a Variation Model
Step 1 Write a general variation model that relates the variables given in the problem. Let k represent the constant of variation.
Step 2 Solve for k by substituting known values of the variables into the model from step 1.
Step 3 Substitute the value of k into the original variation model from step 1.

Example 3 — Solving an Application Involving Direct Variation

The variable z varies directly as w. When w is 16, z is 56.

a. Write a variation model for this situation. Use k as the constant of variation.

b. Solve for the constant of variation.

c. Find the value of z when w is 84.

Solution:

a. $z = kw$

b. $z = kw$

$56 = k(16)$ Substitute known values for z and w. Then solve for the unknown value of k.

$\dfrac{56}{16} = \dfrac{k(16)}{16}$ To isolate k, divide both sides by 16.

$\dfrac{7}{2} = k$ Simplify $\dfrac{56}{16}$ to $\dfrac{7}{2}$.

c. With the value of k known, the variation model can now be written as

$z = \dfrac{7}{2}w$.

$z = \dfrac{7}{2}(84)$ To find z when $w = 84$, substitute $w = 84$ into the equation.

$z = 294$

Example 4 — Solving an Application Involving Direct Variation

The speed of a racing canoe in still water varies directly as the square root of the length of the canoe.

a. If a 16-ft canoe can travel 6.2 mph in still water, find a variation model that relates the speed of a canoe to its length.

b. Find the speed of a 25-ft canoe.

Solution:

a. Let s represent the speed of the canoe and L represent the length. The general variation model is $s = k\sqrt{L}$. To solve for k, substitute the known values for s and L.

$$s = k\sqrt{L}$$

$$6.2 = k\sqrt{16} \qquad \text{Substitute } s = 6.2 \text{ mph and } L = 16 \text{ ft.}$$

$$6.2 = k \cdot 4$$

$$\frac{6.2}{4} = \frac{\cancel{4}k}{\cancel{4}} \qquad \text{Solve for } k.$$

$$k = 1.55$$

$$s = 1.55\sqrt{L} \qquad \text{Substitute } k = 1.55 \text{ into the model } s = k\sqrt{L}.$$

b. $s = 1.55\sqrt{L}$

$$= 1.55\sqrt{25} \qquad \text{Find the speed when } L = 25 \text{ ft.}$$

$$= 7.75 \text{ mph} \qquad \text{The speed is 7.75 mph.}$$

Skill Practice

10. The yield on a bond varies inversely as the price. The yield on a particular bond is 5% when the price is $100. Find the yield when the price is $125.

Example 5 Solving an Application Involving Inverse Variation

The loudness of sound measured in decibels varies inversely as the square of the distance between the listener and the source of the sound. If the loudness of sound is 17.92 decibels at a distance of 10 ft from a stereo speaker, what is the decibel level 20 ft from the speaker?

Solution:

Let L represent the loudness of sound in decibels and d represent the distance in feet. The inverse relationship between decibel level and the square of the distance is modeled by

$$L = \frac{k}{d^2}$$

$$17.92 = \frac{k}{(10)^2} \qquad \text{Substitute } L = 17.92 \text{ decibels and } d = 10 \text{ ft.}$$

$$17.92 = \frac{k}{100}$$

$$(17.92)100 = \frac{k}{100} \cdot 100 \qquad \text{Solve for } k \text{ (clear fractions).}$$

$$k = 1792$$

$$L = \frac{1792}{d^2} \qquad \text{Substitute } k = 1792 \text{ into the original model } L = \frac{k}{d^2}.$$

With the value of k known, we can find L for any value of d.

$$L = \frac{1792}{(20)^2} \qquad \text{Find the loudness when } d = 20 \text{ ft.}$$

$$= 4.48 \text{ decibels} \qquad \text{The loudness is 4.48 decibels.}$$

Notice that the loudness of sound is 17.92 decibels at a distance 10 ft from the speaker. When the distance from the speaker is increased to 20 ft, the decibel level decreases to 4.48 decibels. This is consistent with an inverse relationship. For $k > 0$, as one variable is increased, the other is decreased. It also seems reasonable that the further one moves away from the source of a sound, the softer the sound becomes.

Answer

10. 4%

Example 6 Solving an Application Involving Joint Variation

The kinetic energy of an object varies jointly as the weight of the object at sea level and as the square of its velocity. During a hurricane, a 0.5-lb stone traveling at 60 mph has 81 J (joules) of kinetic energy. Suppose the wind speed doubles to 120 mph. Find the kinetic energy.

Solution:

Let E represent the kinetic energy, let w represent the weight, and let v represent the velocity of the stone. The variation model is

$E = kwv^2$

$81 = k(0.5)(60)^2$ Substitute $E = 81$ J, $w = 0.5$ lb, and $v = 60$ mph.

$81 = k(0.5)(3600)$ Simplify exponents.

$81 = k(1800)$

$\dfrac{81}{1800} = \dfrac{k(1800)}{1800}$ Divide by 1800.

$0.045 = k$ Solve for k.

With the value of k known, the model $E = kwv^2$ can now be written as $E = 0.045wv^2$. We now find the kinetic energy of a 0.5-lb stone traveling at 120 mph.

$E = 0.045(0.5)(120)^2$

$\quad = 324$

The kinetic energy of a 0.5-lb stone traveling at 120 mph is 324 J.

In Example 6, when the velocity increased by 2 times, the kinetic energy increased by 4 times (note that 324 J = 4 · 81 J). This factor of 4 occurs because the kinetic energy is proportional to the *square* of the velocity. When the velocity increased by 2 times, the kinetic energy increased by 2^2 times.

Skill Practice

11. The amount of simple interest earned in an account varies jointly as the interest rate and time of the investment. An account earns \$72 in 4 yr at 2% interest. How much interest would be earned in 3 yr at a rate of 5%?

Answer

11. \$135

Section 7.8 Practice Exercises

Boost your GRADE at ALEKS.com!

ALEKS®
version 3.0

• Practice Problems
• Self-Tests
• NetTutor

• e-Professors
• Videos

Study Skills Exercise

1. Define the key terms:

   a. **direct variation**   b. **inverse variation**   c. **joint variation**

Review Exercises

For Exercises 2–7, perform the indicated operation, or solve the equation.

2. $\dfrac{5p}{p+2} + \dfrac{10}{p+2}$

3. $\dfrac{2y}{3} - \dfrac{3y-1}{5} = 1$

4. $\dfrac{3}{q-1} \cdot \dfrac{2q^2+3q-5}{6q+24}$

5. $\dfrac{a}{4} + \dfrac{3}{a} = 2$

6. $\dfrac{3}{b^2 + 5b - 14} - \dfrac{2}{b^2 - 49}$

7. $\dfrac{a + \dfrac{a}{b}}{\dfrac{a}{b} - a}$

Objective 1: Definition of Direct and Inverse Variation

8. In the equation $r = kt$, does r vary directly or inversely with t?

9. In the equation $w = \dfrac{k}{v}$, does w vary directly or inversely with v?

10. In the equation $P = \dfrac{k \cdot c}{v}$, does P vary directly or inversely as v?

Objective 2: Translations Involving Variation

For Exercises 11–22, write a variation model. Use k as the constant of variation. **(See Examples 1–2.)**

11. T varies directly as q.

12. W varies directly as z.

13. b varies inversely as c.

14. m varies inversely as t.

15. Q is directly proportional to x and inversely proportional to y.

16. d is directly proportional to p and inversely proportional to n.

17. c varies jointly as s and t.

18. w varies jointly as p and f.

19. L varies jointly as w and the square root of v.

20. q varies jointly as v and the square root of w.

21. x varies directly as the square of y and inversely as z.

22. a varies directly as n and inversely as the square of d.

For Exercises 23–28, find the constant of variation, k. **(See Example 3.)**

23. y varies directly as x and when x is 4, y is 18.

24. m varies directly as x and when x is 8, m is 22.

25. p varies inversely as q and when q is 16, p is 32.

26. T varies inversely as x and when x is 40, T is 200.

27. y varies jointly as w and v. When w is 50 and v is 0.1, y is 8.75.

28. N varies jointly as t and p. When t is 1 and p is 7.5, N is 330.

Objective 3: Applications of Variation

Solve Exercises 29–40 using the steps found on page 561. **(See Example 3.)**

29. x varies directly as p. If $x = 50$ when $p = 10$, find x when p is 14.

30. y is directly proportional to z. If $y = 12$ when $z = 36$, find y when z is 21.

31. b is inversely proportional to c. If b is 4 when c is 3, find b when $c = 2$.

32. q varies inversely as w. If q is 8 when w is 50, find q when w is 125.

33. Z varies directly as the square of w. If $Z = 14$ when $w = 4$, find Z when $w = 8$.

34. m varies directly as the square of x. If $m = 200$ when $x = 20$, find m when x is 32.

35. Q varies inversely as the square of p. If $Q = 4$ when $p = 3$, find Q when $p = 2$.

36. z is inversely proportional to the square of t. If $z = 15$ when $t = 4$, find z when $t = 10$.

37. L varies jointly as a and the square root of b. If $L = 72$ when $a = 8$ and $b = 9$, find L when $a = \frac{1}{2}$ and $b = 36$.

38. Y varies jointly as the cube of x and the square root of w. $Y = 128$ when $x = 2$ and $w = 16$. Find Y when $x = \frac{1}{2}$ and $w = 64$.

39. B varies directly as m and inversely as n. $B = 20$ when $m = 10$ and $n = 3$. Find B when $m = 15$ and $n = 12$.

40. R varies directly as s and inversely as t. $R = 14$ when $s = 2$ and $t = 9$. Find R when $s = 4$ and $t = 3$.

For Exercises 41–56, use a variation model to solve for the unknown value. (See Examples 4–6.)

41. The amount of medicine that a physician prescribes for a patient varies directly as the weight of the patient. A physician prescribes 3 g of a medicine for a 150-lb person.

   a. How many grams should be prescribed for a 180-lb person?

   b. How many grams should be prescribed for a 225-lb person?

   c. How many grams should be prescribed for a 120-lb person?

42. The number of turkeys needed for a banquet is directly proportional to the number of guests that must be fed. Master Chef Rico knows that he needs to cook 3 turkeys to feed 42 guests.

   a. How many turkeys should he cook to feed 70 guests?

   b. How many turkeys should he cook to feed 140 guests?

   c. How many turkeys should be cooked to feed 700 guests at an inaugural ball?

   d. How many turkeys should be cooked for a wedding with 100 guests?

43. The unit cost of producing CDs is inversely proportional to the number of CDs produced. If 5000 CDs are produced, the cost per CD is $0.48.

   a. What would be the unit cost if 6000 CDs were produced?

   b. What would be the unit cost if 8000 CDs were produced?

   c. What would be the unit cost if 2400 CDs were produced?

44. An author self-publishes a book and finds that the number of books she can sell per month varies inversely as the price of the book. The author can sell 1500 books per month when the price is set at $8 per book.

   a. How many books would she expect to sell if the price were $12?

   b. How many books would she expect to sell if the price were $15?

   c. How many books would she expect to sell if the price were $6?

45. The amount of pollution entering the atmosphere over a given time varies directly as the number of people living in an area. If 80,000 people cause 56,800 tons of pollutants, how many tons enter the atmosphere in a city with a population of 500,000?

46. The area of a picture projected on a wall varies directly as the square of the distance from the projector to the wall. If a 10-ft distance produces a 16-ft² picture, what is the area of a picture produced when the projection unit is moved to a distance 20 ft from the wall?

47. The stopping distance of a car varies directly as the square of the speed of the car. If a car traveling at 40 mph has a stopping distance of 109 ft, find the stopping distance of a car that is traveling at 25 mph. (Round your answer to one decimal place.)

48. The intensity of a light source varies inversely as the square of the distance from the source. If the intensity is 48 lumens at a distance of 5 ft, what is the intensity when the distance is 8 ft?

49. The power in an electric circuit varies jointly as the current and the square of the resistance. If the power is 144 W (watts) when the current is 4 A and the resistance is 6 Ω, find the power when the current is 3 A and the resistance is 10 Ω.

50. The current in a wire varies directly as the voltage and inversely as the resistance. If the current is 9 A (amperes) when the voltage is 90 V (volts) and the resistance is 10 Ω (ohms), find the current when the voltage is 185 V and the resistance is 10 Ω.

51. The resistance of a wire varies directly as its length and inversely as the square of its diameter. A 40-ft wire 0.1 in. in diameter has a resistance of 4 Ω. What is the resistance of a 50-ft wire with a diameter of 0.2 in.?

52. Some body-builders claim that, within safe limits, the number of repetitions that a person can complete on a given weight-lifting exercise is inversely proportional to the amount of weight lifted. Roxanne can bench press 45 lb for 15 repetitions.

  a. How many repetitions can Roxanne bench with 60 lb of weight?

  b. How many repetitions can Roxanne bench with 75 lb of weight?

  c. How many repetitions can Roxanne bench with 100 lb of weight?

53. The weight of a medicine ball varies directly as the cube of its radius. A ball with a radius of 3 in. weighs 4.32 lb. How much would a medicine ball weigh if its radius is 5 in.?

54. The surface area of a cube varies directly as the square of the length of an edge. The surface area is 24 ft² when the length of an edge is 2 ft. Find the surface area of a cube with an edge that is 5 ft.

55. The amount of simple interest earned in an account varies jointly as the amount of principal invested and the amount of time the money is invested. If $2500 in principal earns $500 in interest after 4 yr, then how much interest will be earned on $7000 invested for 10 yr?

56. The amount of simple interest earned in an account varies jointly as the amount of principal invested and the amount of time the money is invested. If $6000 in principal earns $840 in interest after 2 yr, then how much interest will be earned on $4500 invested for 8 yr?

Group Activity

Computing Monthly Mortgage Payments

Materials: A calculator

Estimated Time: 15–20 minutes

Group Size: 3

When a person borrows money to buy a house, the bank usually requires a down payment of between 0% and 20% of the cost of the house. The bank then issues a loan for the remaining balance on the house. The loan to buy a house is called a *mortgage*. Monthly payments are made to pay off the mortgage over a period of years.

A formula to calculate the monthly payment, P, for a loan is given by the complex fraction:

$$P = \dfrac{\dfrac{Ar}{12}}{1 - \dfrac{1}{\left(1 + \dfrac{r}{12}\right)^{12t}}} \qquad \text{where} \quad \begin{array}{l} P \text{ is the monthly payment} \\ A \text{ is the original amount of the mortgage} \\ r \text{ is the annual interest rate} \\ t \text{ is the term of the loan in years} \end{array}$$

Suppose a person wants to buy a $200,000 house. The bank requires a down payment of 20%, and the loan is issued for 30 yr at 7.5% interest for 30 yr.

1. Find the amount of the down payment. _____

2. Find the amount of the mortgage. _____

3. Find the monthly payment (to the nearest cent). _____

4. Multiply the monthly payment found in question 3 by the total number of months in a 30-yr period. Interpret what this value means in the context of the problem.

5. How much total interest was paid on the loan for the house? _____

6. What was the total amount paid to the bank (include the down payment). _____

Chapter 7 Summary

Section 7.1 Introduction to Rational Expressions

Key Concepts

A **rational expression** is a ratio of the form $\frac{p}{q}$ where p and q are polynomials and $q \neq 0$.

The **domain** of an algebraic expression is the set of real numbers that when substituted for the variable makes the expression equal a real number. For a rational expression, the domain is all real numbers except those that make the denominator zero.

Simplifying a Rational Expression to Lowest Terms

Factor the numerator and denominator completely, and reduce factors whose ratio is equal to 1 or to −1. A rational expression written in lowest terms will still have the same restrictions on the domain as the original expression.

Examples

Example 1

$$\frac{x + 2}{x^2 - 5x - 14} \quad \text{is a rational expression.}$$

Example 2

To find the domain of $\dfrac{x + 2}{x^2 - 5x - 14}$ factor the denominator: $\dfrac{x + 2}{(x + 2)(x - 7)}$

The domain is $\{x \mid x$ is a real number and $x \neq -2, x \neq 7\}$.

Example 3

Simplify to lowest terms. $\dfrac{x + 2}{x^2 - 5x - 14}$

$$\frac{\overset{1}{\cancel{x + 2}}}{\cancel{(x + 2)}(x - 7)} \quad \text{Simplify.}$$

$$= \frac{1}{x - 7} \quad \text{(provided } x \neq 7, x \neq -2\text{)}.$$

Section 7.2 — Multiplication and Division of Rational Expressions

Key Concepts

Multiplying Rational Expressions

Multiply the numerators and multiply the denominators. That is, if $q \neq 0$ and $s \neq 0$, then

$$\frac{p}{q} \cdot \frac{r}{s} = \frac{pr}{qs}$$

Factor the numerator and denominator completely. Then reduce factors whose ratio is 1 or -1.

Dividing Rational Expressions

Multiply the first expression by the reciprocal of the second expression. That is, for $q \neq 0$, $r \neq 0$, and $s \neq 0$,

$$\frac{p}{q} \div \frac{r}{s} = \frac{p}{q} \cdot \frac{s}{r} = \frac{ps}{qr}$$

Examples

Example 1

Multiply. $\dfrac{b^2 - a^2}{a^2 - 2ab + b^2} \cdot \dfrac{a^2 - 3ab + 2b^2}{2a + 2b}$

$$= \frac{\overset{-1}{(b - a)}\overset{1}{(b + a)}}{(a - b)(a - b)} \cdot \frac{(a - 2b)\overset{1}{(a - b)}}{2(a + b)}$$

$$= -\frac{a - 2b}{2} \quad \text{or} \quad \frac{2b - a}{2}$$

Example 2

Divide. $\dfrac{x - 2}{15} \div \dfrac{x^2 + 2x - 8}{20x}$

$$= \frac{x - 2}{15} \cdot \frac{20x}{x^2 + 2x - 8}$$

$$= \frac{\overset{1}{(x - 2)}}{\underset{3}{15}} \cdot \frac{\overset{4}{20x}}{(x - 2)(x + 4)\underset{1}{}}$$

$$= \frac{4}{3(x + 4)}$$

Section 7.3   Least Common Denominator

Key Concepts

Converting a Rational Expression to an Equivalent Expression with a Different Denominator

Multiply numerator and denominator of the rational expression by the missing factors necessary to create the desired denominator.

Examples

Example 1

Convert $\dfrac{-3}{x-2}$ to an equivalent expression with the indicated denominator:

$$\frac{-3}{x-2} = \frac{}{5(x-2)(x+2)}$$

Multiply numerator and denominator by the missing factors from the denominator.

$$\frac{-3 \cdot 5(x+2)}{(x-2) \cdot 5(x+2)} = \frac{-15x - 30}{5(x-2)(x+2)}$$

Finding the Least Common Denominator (LCD) of Two or More Rational Expressions

1. Factor all denominators completely.
2. The LCD is the product of unique factors from the denominators, where each factor is raised to its highest power.

Example 2

Identify the LCD. $\dfrac{1}{8x^3y^2z}; \dfrac{5}{6xy^4}$

1. Write the denominators as a product of prime factors:

$$\frac{1}{2^3x^3y^2z}; \frac{5}{2 \cdot 3xy^4}$$

2. The LCD is $2^3 3 x^3 y^4 z$ or $24x^3y^4z$

Section 7.4   Addition and Subtraction of Rational Expressions

Key Concepts

To add or subtract rational expressions, the expressions must have the same denominator.

Steps to Add or Subtract Rational Expressions

1. Factor the denominators of each rational expression.
2. Identify the LCD.
3. Rewrite each rational expression as an equivalent expression with the LCD as its denominator.
4. Add or subtract the numerators, and write the result over the common denominator.
5. Simplify.

Examples

Example 1

Subtract. $\dfrac{c}{c^2 - c - 12} - \dfrac{1}{2c - 8}$

$$= \frac{c}{(c-4)(c+3)} - \frac{1}{2(c-4)}$$

The LCD is $2(c-4)(c+3)$.

$$= \frac{2c}{2(c-4)(c+3)} - \frac{1(c+3)}{2(c-4)(c+3)}$$

$$= \frac{2c - (c+3)}{2(c-4)(c+3)}$$

$$= \frac{2c - c - 3}{2(c-4)(c+3)} = \frac{c-3}{2(c-4)(c+3)}$$

| Section 7.5 | Complex Fractions |

Key Concepts

Complex fractions can be simplified by using Method I or Method II.

Method I

1. Add or subtract expressions in the numerator to form a single fraction. Add or subtract expressions in the denominator to form a single fraction.
2. Divide the rational expressions from step 1 by multiplying the numerator of the complex fraction by the reciprocal of the denominator of the complex fraction.
3. Simplify to lowest terms, if possible.

Method II

1. Multiply the numerator and denominator of the complex fraction by the LCD of all individual fractions within the expression.
2. Apply the distributive property, and simplify the result.
3. Simplify to lowest terms, if possible.

Examples

Example 1

Simplify.

$$\frac{1 - \dfrac{4}{w^2}}{1 - \dfrac{1}{w} - \dfrac{6}{w^2}} = \frac{\dfrac{w^2}{w^2} - \dfrac{4}{w^2}}{\dfrac{w^2}{w^2} - \dfrac{w}{w^2} - \dfrac{6}{w^2}}$$

$$= \frac{\dfrac{w^2 - 4}{w^2}}{\dfrac{w^2 - w - 6}{w^2}} = \frac{w^2 - 4}{w^2} \cdot \frac{w^2}{w^2 - w - 6}$$

$$= \frac{(w - 2)\overset{1}{\cancel{(w + 2)}}}{w^2} \cdot \frac{\overset{1}{w^2}}{(w - 3)\cancel{(w + 2)}}$$

$$= \frac{w - 2}{w - 3}$$

Example 2

Simplify.

$$\frac{1 - \dfrac{4}{w^2}}{1 - \dfrac{1}{w} - \dfrac{6}{w^2}} = \frac{w^2\left(1 - \dfrac{4}{w^2}\right)}{w^2\left(1 - \dfrac{1}{w} - \dfrac{6}{w^2}\right)}$$

$$= \frac{w^2 - 4}{w^2 - w - 6} = \frac{(w - 2)\overset{1}{\cancel{(w + 2)}}}{(w - 3)\cancel{(w + 2)}}$$

$$= \frac{w - 2}{w - 3}$$

Section 7.6 Rational Equations

Key Concepts

An equation with one or more rational expressions is called a **rational equation**.

Steps to Solve a Rational Equation

1. Factor the denominators of all rational expressions.
2. Identify the LCD of all expressions in the equation.
3. Multiply both sides of the equation by the LCD.
4. Solve the resulting equation.
5. Check each potential solution in the original equation.

Examples

Example 1

Solve. $\dfrac{1}{w} - \dfrac{1}{2w-1} = \dfrac{-2w}{2w-1}$

The LCD is $w(2w-1)$.

$$w(2w-1)\frac{1}{w} - w(2w-1)\frac{1}{2w-1}$$
$$= w(2w-1)\frac{-2w}{2w-1}$$

$$(2w-1)(1) - w(1) = w(-2w)$$

$$2w - 1 - w = -2w^2 \qquad \text{Quadratic equation}$$

$$2w^2 + w - 1 = 0$$

$$(2w-1)(w+1) = 0$$

$$w \ne \tfrac{1}{2} \qquad \text{or} \qquad w = -1$$
$$\text{Does not check.} \qquad\qquad \text{Checks.}$$

Example 2

Solve for I. $q = \dfrac{VQ}{I}$

$$I \cdot q = \frac{VQ}{I} \cdot I$$

$$Iq = VQ$$

$$I = \frac{VQ}{q}$$

Section 7.7	**Applications of Rational Equations and Proportions**

Key Concepts

Solving Proportions

An equation that equates two rates or ratios is called a **proportion**:

$$\frac{a}{b} = \frac{c}{d} \quad (b \neq 0, d \neq 0)$$

To solve a proportion, multiply both sides of the equation by the LCD.

Examples 2 and 3 give applications of rational equations.

Example 2

Two cars travel from Los Angeles to Las Vegas. One car travels an average of 8 mph faster than the other car. If the faster car travels 189 mi in the same time as the slower car travels 165 mi, what is the average speed of each car?

Let r represent the speed of the slower car.
Let $r + 8$ represent the speed of the faster car.

	Distance	**Rate**	**Time**
Slower car	165	r	$\dfrac{165}{r}$
Faster car	189	$r + 8$	$\dfrac{189}{r + 8}$

$$\frac{165}{r} = \frac{189}{r + 8}$$

$$165(r + 8) = 189r$$

$$165r + 1320 = 189r$$

$$1320 = 24r$$

$$55 = r$$

The slower car travels 55 mph, and the faster car travels $55 + 8 = 63$ mph.

Examples

Example 1

A 90-g serving of a particular ice cream contains 10 g of fat. How much fat does 400 g of the same ice cream contain?

$$\frac{10 \text{ g fat}}{90 \text{ g ice cream}} = \frac{x \text{ grams fat}}{400 \text{ g ice cream}}$$

$$\frac{10}{90} = \frac{x}{400}$$

$$\overset{40}{\cancel{3600}} \cdot \left(\frac{10}{\cancel{90}}\right) = \left(\frac{x}{\cancel{400}}\right) \cdot \overset{9}{\cancel{3600}}$$

$$400 = 9x$$

$$x = \frac{400}{9} \approx 44.4 \text{ g}$$

Example 3

Beth and Cecelia have a house cleaning business. Beth can clean a particular house in 5 hr by herself. Cecelia can clean the same house in 4 hr. How long would it take if they cleaned the house together?

Let x be the number of hours it takes for both Beth and Cecelia to clean the house.

Beth can clean $\frac{1}{5}$ of the house in an hour and $\frac{1}{5}x$ of the house in x hr.
Cecelia can clean $\frac{1}{4}$ of the house in an hour and $\frac{1}{4}x$ of the house in x hr.

$$\frac{1}{5}x + \frac{1}{4}x = 1 \quad \text{Together they clean one whole house.}$$

$$20\left(\frac{1}{5}x + \frac{1}{4}x\right) = (1)20$$

$$4x + 5x = 20$$

$$9x = 20$$

$$x = \frac{20}{9}, \text{ or } 2\frac{2}{9} \text{ hr working together.}$$

Section 7.8 Variation

Key Concepts

Direct Variation

y varies directly as x.
y is directly proportional to x. $\left.\right\}$ $y = kx$

Inverse Variation

y varies inversely as x.
y is inversely proportional to x. $\left.\right\}$ $y = \dfrac{k}{x}$

Joint Variation

y varies jointly as w and z.
y is jointly proportional to w and z. $\left.\right\}$ $y = kwz$

Steps to Find a Variation Model

1. Write a general variation model that relates the variables given in the problem. Let k represent the constant of variation.
2. Solve for k by substituting known values of the variables into the model from step 1.
3. Substitute the value of k into the original variation model from step 1.

Examples

Example 1

t varies directly as the square root of x.

$t = k\sqrt{x}$

Example 2

W is inversely proportional to the cube of x.

$W = \dfrac{k}{x^3}$

Example 3

y is jointly proportional to x and the square of z.

$y = kxz^2$

Example 4

C varies directly as the square root of d and inversely as t. If $C = 12$ when d is 9 and t is 6, find C if d is 16 and t is 12.

Step 1. $C = \dfrac{k\sqrt{d}}{t}$

Step 2. $12 = \dfrac{k\sqrt{9}}{6} \Rightarrow 12 = \dfrac{k \cdot 3}{6} \Rightarrow k = 24$

Step 3. $C = \dfrac{24\sqrt{d}}{t} \Rightarrow C = \dfrac{24\sqrt{16}}{12} \Rightarrow C = 8$

Chapter 7 Review Exercises

Section 7.1

1. For the rational expression $\dfrac{t - 2}{t + 9}$

   a. Evaluate the expression (if possible) for $t = 0, 1, 2, -3, -9$

   b. Write the domain of the expression in set-builder notation.

2. For the rational expression $\dfrac{k + 1}{k - 5}$

   a. Evaluate the expression for $k = 0, 1, 5, -1, -2$

   b. Write the domain of the expression in set-builder notation.

3. Which of the rational expressions are equal to -1?

a. $\dfrac{2 - x}{x - 2}$

b. $\dfrac{x - 5}{x + 5}$

c. $\dfrac{-x - 7}{x + 7}$

d. $\dfrac{x^2 - 4}{4 - x^2}$

For Exercises 4–13, write the domain in set-builder notation. Then simplify the expressions to lowest terms.

4. $\dfrac{x - 3}{(2x - 5)(x - 3)}$

5. $\dfrac{h + 7}{(3h + 1)(h + 7)}$

6. $\dfrac{4a^2 + 7a - 2}{a^2 - 4}$

7. $\dfrac{2w^2 + 11w + 12}{w^2 - 16}$

8. $\dfrac{z^2 - 4z}{8 - 2z}$

9. $\dfrac{15 - 3k}{2k^2 - 10k}$

10. $\dfrac{2b^2 + 4b - 6}{4b + 12}$

11. $\dfrac{3m^2 - 12m - 15}{9m + 9}$

12. $\dfrac{n + 3}{n^2 + 6n + 9}$

13. $\dfrac{p + 7}{p^2 + 14p + 49}$

Section 7.2

For Exercises 14–27, multiply or divide as indicated.

14. $\dfrac{3y^3}{3y - 6} \cdot \dfrac{y - 2}{y}$

15. $\dfrac{2u + 10}{u} \cdot \dfrac{u^3}{4u + 20}$

16. $\dfrac{11}{v - 2} \cdot \dfrac{2v^2 - 8}{22}$

17. $\dfrac{8}{x^2 - 25} \cdot \dfrac{3x + 15}{16}$

18. $\dfrac{4c^2 + 4c}{c^2 - 25} \div \dfrac{8c}{c^2 - 5c}$

19. $\dfrac{q^2 - 5q + 6}{2q + 4} \div \dfrac{2q - 6}{q + 2}$

20. $\left(\dfrac{-2t}{t + 1}\right)(t^2 - 4t - 5)$

21. $(s^2 - 6s + 8)\left(\dfrac{4s}{s - 2}\right)$

22. $\dfrac{\dfrac{a^2 + 5a + 1}{7a - 7}}{\dfrac{a^2 + 5a + 1}{a - 1}}$

23. $\dfrac{\dfrac{n^2 + n + 1}{n^2 - 4}}{\dfrac{n^2 + n + 1}{n + 2}}$

24. $\dfrac{5h^2 - 6h + 1}{h^2 - 1} \div \dfrac{16h^2 - 9}{4h^2 + 7h + 3} \cdot \dfrac{3 - 4h}{30h - 6}$

25. $\dfrac{3m - 3}{6m^2 + 18m + 12} \cdot \dfrac{2m^2 - 8}{m^2 - 3m + 2} \div \dfrac{m + 3}{m + 1}$

26. $\dfrac{x - 2}{x^2 - 3x - 18} \cdot \dfrac{6 - x}{x^2 - 4}$

27. $\dfrac{4y^2 - 1}{1 + 2y} \div \dfrac{y^2 - 4y - 5}{5 - y}$

Section 7.3

For Exercises 28–33, identify the LCD. Then write each fraction as an equivalent fraction with the LCD as its denominator.

28. $\dfrac{2}{5a}; \dfrac{3}{10b}$

29. $\dfrac{7}{4x}; \dfrac{11}{6y}$

30. $\dfrac{1}{x^2y^4}; \dfrac{3}{xy^5}$

31. $\dfrac{5}{ab^3}; \dfrac{3}{ac^2}$

32. $\dfrac{5}{p + 2}; \dfrac{p}{p - 4}$

33. $\dfrac{6}{q}; \dfrac{1}{q + 8}$

34. Determine the LCD.

$$\dfrac{6}{n^2 - 9}; \dfrac{5}{n^2 - n - 6}$$

35. Determine the LCD.

$$\dfrac{8}{m^2 - 16}; \dfrac{7}{m^2 - m - 12}$$

36. State two possible LCDs that could be used to add the fractions.

$$\dfrac{7}{c - 2} + \dfrac{4}{2 - c}$$

37. State two possible LCDs that could be used to subtract the fractions.

$$\dfrac{10}{3 - x} - \dfrac{5}{x - 3}$$

Section 7.4

For Exercises 38–49, add or subtract as indicated.

38. $\dfrac{h + 3}{h + 1} + \dfrac{h - 1}{h + 1}$

39. $\dfrac{b - 6}{b - 2} + \dfrac{b + 2}{b - 2}$

40. $\dfrac{a^2}{a - 5} - \dfrac{25}{a - 5}$

41. $\dfrac{x^2}{x + 7} - \dfrac{49}{x + 7}$

42. $\dfrac{y}{y^2 - 81} + \dfrac{2}{9 - y}$

43. $\dfrac{3}{4 - t^2} + \dfrac{t}{2 - t}$

44. $\dfrac{4}{3m} - \dfrac{1}{m + 2}$

45. $\dfrac{5}{2r + 12} - \dfrac{1}{r}$

46. $\dfrac{4p}{p^2 + 6p + 5} - \dfrac{3p}{p^2 + 5p + 4}$

47. $\dfrac{3q}{q^2 + 7q + 10} - \dfrac{2q}{q^2 + 6q + 8}$

48. $\dfrac{1}{h} + \dfrac{h}{2h + 4} - \dfrac{2}{h^2 + 2h}$

49. $\dfrac{x}{3x + 9} - \dfrac{3}{x^2 + 3x} + \dfrac{1}{x}$

Section 7.5

For Exercises 50–57, simplify the complex fractions.

50. $\dfrac{\frac{a - 4}{3}}{\frac{a - 2}{3}}$

51. $\dfrac{\frac{z + 5}{z}}{\frac{z - 5}{3}}$

52. $\dfrac{\frac{2 - 3w}{2}}{\frac{2}{w} - 3}$

53. $\dfrac{\frac{2}{y} + 6}{\frac{3y + 1}{4}}$

54. $\dfrac{\frac{y}{x} - \frac{x}{y}}{\frac{1}{x} + \frac{1}{y}}$

55. $\dfrac{\frac{b}{a} - \frac{a}{b}}{\frac{1}{b} - \frac{1}{a}}$

56. $\dfrac{\frac{6}{p + 2} + 4}{\frac{8}{p + 2} - 4}$

57. $\dfrac{\frac{25}{k + 5} + 5}{\frac{5}{k + 5} - 5}$

Section 7.6

For Exercises 58–65, solve the equations.

58. $\dfrac{2}{x} + \dfrac{1}{2} = \dfrac{1}{4}$

59. $\dfrac{1}{y} + \dfrac{3}{4} = \dfrac{1}{4}$

60. $\dfrac{2}{h - 2} + 1 = \dfrac{h}{h + 2}$

61. $\dfrac{w}{w - 1} = \dfrac{3}{w + 1} + 1$

62. $\dfrac{t + 1}{3} - \dfrac{t - 1}{6} = \dfrac{1}{6}$

63. $\dfrac{4p - 4}{p^2 + 5p - 14} + \dfrac{2}{p + 7} = \dfrac{1}{p - 2}$

64. $\dfrac{1}{z + 2} = \dfrac{4}{z^2 - 4} - \dfrac{1}{z - 2}$

65. $\dfrac{y + 1}{y + 3} = \dfrac{y^2 - 11y}{y^2 + y - 6} - \dfrac{y - 3}{y - 2}$

66. Four times a number is added to 5. The sum is then divided by 6. The result is $\frac{7}{2}$. Find the number.

67. Solve the formula $\dfrac{V}{h} = \dfrac{\pi r^2}{3}$ for h.

68. Solve the formula $\dfrac{A}{b} = \dfrac{h}{2}$ for b.

Section 7.7

For Exercises 69–70, solve the proportions.

69. $\dfrac{m + 2}{8} = \dfrac{m}{3}$

70. $\dfrac{12}{a} = \dfrac{5}{8}$

71. A bag of popcorn states that it contains 4 g of fat per serving. If a serving is 2 oz, how many grams of fat are in a 5-oz bag?

72. Bud goes 10 mph faster on his Harley Davidson motorcycle than Ed goes on his Honda motorcycle. If Bud travels 105 mi in the same time that Ed travels 90 mi, what are the rates of the two bikers?

73. There are two pumps set up to fill a small swimming pool. One pump takes 24 min by itself to fill the pool, but the other takes 56 min by itself. How long would it take if both pumps work together?

74 Consider the similar triangles shown here. Find the values of *x* and *y*.

76. Suppose *y* varies inversely with the cube of *x*, and $y = 9$ when $x = 2$. Find *y* when $x = 3$.

77. Suppose *y* varies jointly with *x* and the square root of *z*, and $y = 3$ when $x = 3$ and $z = 4$. Find *y* when $x = 8$ and $z = 9$.

 78. The distance, *d*, that one can see to the horizon varies directly as the square root of the height above sea level. If a person 25 m above sea level can see 30 km, how far can a person see if she is 64 m above sea level?

Section 7.8

75. The force applied to a spring varies directly with the distance that the spring is stretched.

 a. Write a variation model using *k* as the constant of variation.

 b. When 6 lb of force is applied, the spring stretches 2 ft. Find *k*.

 c. How much force is required to stretch the spring 4.2 ft?

Chapter 7   Test

For Exercises 1–2,

 a. Write the domain of the rational expressions in set-builder notation.

 b. Reduce the rational expression to lowest terms.

1. $\dfrac{5(x - 2)(x + 1)}{30(2 - x)}$ **2.** $\dfrac{7a^2 - 42a}{a^3 - 4a^2 - 12a}$

3. Identify the rational expressions that are equal to −1.

 a. $\dfrac{x + 4}{x - 4}$ **b.** $\dfrac{7 - 2x}{2x - 7}$

 c. $\dfrac{9x^2 + 16}{-9x^2 - 16}$ **d.** $-\dfrac{x + 5}{x + 5}$

For Exercises 4–9, perform the indicated operation.

4. $\dfrac{2}{y^2 + 4y + 3} + \dfrac{1}{3y + 9}$

5. $\dfrac{9 - b^2}{5b + 15} \div \dfrac{b - 3}{b + 3}$

6. $\dfrac{w^2 - 4w}{w^2 - 8w + 16} \cdot \dfrac{w - 4}{w^2 + w}$

7. $\dfrac{t}{t - 2} - \dfrac{8}{t^2 - 4}$

8. $\dfrac{1}{x + 4} + \dfrac{2}{x^2 + 2x - 8} + \dfrac{x}{x - 2}$

9. $\dfrac{1 - \dfrac{4}{m}}{m - \dfrac{16}{m}}$

For Exercises 10–13, solve the equation.

10. $\dfrac{3}{a} + \dfrac{5}{2} = \dfrac{7}{a}$

11. $\dfrac{p}{p - 1} + \dfrac{1}{p} = \dfrac{p^2 + 1}{p^2 - p}$

12. $\dfrac{3}{c - 2} - \dfrac{1}{c + 1} = \dfrac{7}{c^2 - c - 2}$

13. $\dfrac{4x}{x - 4} = 3 + \dfrac{16}{x - 4}$

14. Solve the formula $\dfrac{C}{2} = \dfrac{A}{r}$ for r.

15. If $\frac{3}{2}$ is added to the reciprocal of a number the result is $\frac{2}{5}$ times the reciprocal of that number. Find the number.

16. Solve the proportion.

$$\frac{y+7}{-4} = \frac{1}{4}$$

17. A recipe for vegetable soup calls for $\frac{1}{2}$ cup of carrots for six servings. How many cups of carrots are needed to prepare 15 servings?

18. A motorboat can travel 28 mi downstream in the same amount of time as it can travel 18 mi upstream. Find the speed of the current if the boat can travel 23 mph in still water.

19. Two printers working together can complete a job in 2 hr. If one printer requires 6 hr to do the job alone, how many hours would the second printer need to complete the job alone?

20. Consider the similar triangles shown here. Find the values of a and b.

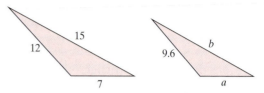

21. Find the LCD of the following pairs of rational expressions.

a. $\dfrac{x}{3(x+3)}; \dfrac{7}{5(x+3)}$ **b.** $\dfrac{-2}{3x^2y}; \dfrac{4}{xy^2}$

22. The amount of medication prescribed for a patient varies directly as the patient's weight. If a 160-lb person is prescribed 6 mL of a medicine, then how much medicine would be prescribed to a 220-lb person?

23. The number of drinks sold at a concession stand varies inversely as price. If the price is set at \$1.25 per drink, then 400 drinks are sold. If the price is set at \$2.50 per drink, then how many drinks are sold?

Chapters 1–7   Cumulative Review Exercises

For Exercises 1–2, simplify completely.

1. $\left(\dfrac{1}{2}\right)^{-4} + 2^4$ **2.** $|3-5| + |-2+7|$

3. Solve. $\dfrac{1}{2} - \dfrac{3}{4}(y-1) = \dfrac{5}{12}$

4. Complete the table.

Set-Builder Notation	Graph	Interval Notation
$\{x \mid x \geq -1\}$	→	
	→	$(-\infty, 5)$

5. The perimeter of a rectangular swimming pool is 104 m. The length is 1 m more than twice the width. Find the length and width.

6. The height of a triangle is 2 in. less than the base. The area is 40 in.^{2} Find the base and height of the triangle.

7. Simplify. $\left(\dfrac{4x^{-1}y^{-2}}{z^4}\right)^{-2}(2y^{-1}z^3)^3$

8. The length and width of a rectangle are given in terms of x.

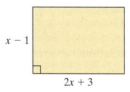

$x - 1$

$2x + 3$

a. Write a polynomial that represents the perimeter of the rectangle.

b. Write a polynomial that represents the area of the rectangle.

9. Factor completely: $25x^2 - 30x + 9$

10. Factor. $10cd + 5d - 6c - 3$

11. Determine the domain of the expression.

$$\frac{x+3}{(x-5)(2x+1)}$$

12. Simplify to lowest terms.

$$\frac{x^2 - 9}{x^2 + 8x + 15}$$

13. Divide. $\dfrac{2x - 6}{x^2 - 16} \div \dfrac{10x^2 - 90}{x^2 - x - 12}$

14. Simplify.

$$\frac{\dfrac{3}{4} - \dfrac{1}{x}}{\dfrac{1}{3x} - \dfrac{1}{4}}$$

15. Solve.

$$\frac{7}{y^2 - 4} = \frac{3}{y - 2} + \frac{2}{y + 2}$$

16. Solve the proportion.

$$\frac{2b - 5}{6} = \frac{4b}{7}$$

17. Determine the x- and y-intercepts.

    a. $-2x + 4y = 8$ **b.** $y = 5x$

18. Determine the slope

    a. of the line containing the points $(0, -6)$ and $(-5, 1)$

    b. of the line $y = -\dfrac{2}{3}x - 6$

    c. of a line parallel to a line having a slope of 4.

    d. of a line perpendicular to a line having a slope of 4.

19. Find an equation of a line passing through the point $(1, 2)$ and having a slope of 5. Write the answer in slope-intercept form.

20. A group of teenagers buys 2 large popcorns and 6 drinks at the movie theater for $16. A couple buys 1 large popcorn and 2 drinks for $6.50. Find the price for 1 large popcorn and the price for 1 drink.

Radicals

8

Chapter 8

Chapter 8 is devoted to the study of radicals and their applications. We first present the techniques to add, subtract, multiply, and divide radical expressions. This is followed by radical equations and their applications.

Several of the key terms from the chapter can be found in the word search puzzle. Try to locate the words from the list in the puzzle.

square root   cube   radicand

Pythagorean   index   radical

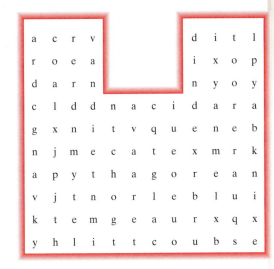

Section 8.1 Introduction to Roots and Radicals

Objectives

1. Definition of a Square Root
2. Definition of an *n*th-Root
3. Translations Involving *n*th-Roots
4. Pythagorean Theorem

1. Definition of a Square Root

Recall that to square a number means to multiply the number by itself: $b^2 = b \cdot b$. The reverse operation to squaring a number is to find its square roots. For example, finding a square root of 49 is equivalent to asking: "What number when squared equals 49?"

One obvious answer to this question is 7 because $(7)^2 = 49$. But -7 will also work because $(-7)^2 = 49$.

DEFINITION Square Root

b is a **square root** of a if $b^2 = a$.

Skill Practice

Identify the square roots of the numbers.

1. 64
2. -36
3. 36
4. $\dfrac{25}{16}$

TIP: All positive real numbers have two real-valued square roots: one positive and one negative. Zero has only one square root, which is 0 itself. Finally, for any negative real number, there are no real-valued square roots.

Example 1 Identifying the Square Roots of a Number

Identify the square roots of

a. 9　　　b. 121　　　c. 0　　　d. -4

Solution:

a. 3 is a square root of 9 because $(3)^2 = 9$.
   -3 is a square root of 9 because $(-3)^2 = 9$.

b. 11 is a square root of 121 because $(11)^2 = 121$.
   -11 is a square root of 121 because $(-11)^2 = 121$.

c. 0 is a square root of 0 because $(0)^2 = 0$.

d. There are no real numbers that when squared will equal a negative number. Therefore, there are no real-valued square roots of -4.

Recall from Section 1.2 that the positive square root of a real number can be denoted with a radical sign, $\sqrt{\ }$.

DEFINITION Notation for Positive and Negative Square Roots

Let a represent a positive real number. Then,

1. $\sqrt{a}$ is the **positive square root** of a. The positive square root is also called the **principal square root**.
2. $-\sqrt{a}$ is the **negative square root** of a.
3. $\sqrt{0} = 0$

Answers

1. $8; -8$
2. There are no real-valued square roots.
3. $6; -6$　　4. $\dfrac{5}{4}; -\dfrac{5}{4}$

Example 2 Simplifying Square Roots

Simplify the square roots.

a. $\sqrt{36}$ b. $\sqrt{225}$ c. $\sqrt{1}$ d. $\sqrt{\dfrac{9}{4}}$ e. $\sqrt{0.49}$

Solution:

a. $\sqrt{36}$ denotes the positive square root of 36. $\sqrt{36} = 6$

b. $\sqrt{225}$ denotes the positive square root of 225. $\sqrt{225} = 15$

c. $\sqrt{1}$ denotes the positive square root of 1. $\sqrt{1} = 1$

d. $\sqrt{\dfrac{9}{4}}$ denotes the positive square root of $\dfrac{9}{4}$. $\sqrt{\dfrac{9}{4}} = \dfrac{3}{2}$

e. $\sqrt{0.49}$ denotes the positive square root. $\sqrt{0.49} = 0.7$

The numbers 36, 225, 1, $\frac{9}{4}$, and 0.49 are **perfect squares** because their square roots are rational numbers. Radicals that cannot be simplified to rational numbers are irrational numbers. Recall that an irrational number cannot be written as a terminating or repeating decimal. For example, the symbol $\sqrt{13}$ is used to represent the exact value of the square root of 13. The symbol $\sqrt{42}$ is used to represent the exact value of the square root of 42. These values are irrational numbers but can be approximated by rational numbers by using a calculator.

$$\sqrt{13} \approx 3.605551275 \qquad \sqrt{42} \approx 6.480740698$$

Note: The only way to denote the *exact* values of the square root of 13 and the square root of 42 is $\sqrt{13}$ and $\sqrt{42}$.

A negative number cannot have a real number as a square root because no real number when squared is negative. For example, $\sqrt{-25}$ is *not a real number* because there is no real number, b, for which $(b)^2 = -25$.

> **TIP:** Before using a calculator to evaluate a square root, try estimating the value first.
>
> $\sqrt{13}$ must be a number between 3 and 4 because $\sqrt{9} < \sqrt{13} < \sqrt{16}$.
>
> $\sqrt{42}$ must be a number between 6 and 7 because $\sqrt{36} < \sqrt{42} < \sqrt{49}$.

Example 3 Evaluating Square Roots if Possible

Simplify the square roots if possible.

a. $\sqrt{-100}$ b. $-\sqrt{100}$ c. $\sqrt{-64}$

Solution:

a. $\sqrt{-100}$ Not a real number

b. $-\sqrt{100}$

   $-1 \cdot \sqrt{100}$ The expression $-\sqrt{100}$ is equivalent to $-1 \cdot \sqrt{100}$.

   $-1 \cdot 10 = -10$

c. $\sqrt{-64}$ Not a real number

2. Definition of an *n*th-Root

Finding a square root of a number is the reverse process of squaring a number. This concept can be extended to finding a third root (called a cube root), a fourth root, and in general, an *n*th-root.

> **DEFINITION**   *n*th-Root
>
> b is an **nth-root** of a if $b^n = a$.

Concept Connections

13. Identify the index and radicand.
$$\sqrt[3]{2xy}$$

The radical sign, $\sqrt{}$, is used to denote the principal square root of a number. The symbol, $\sqrt[n]{}$, is used to denote the principal *n*th-root of a number.

In the expression $\sqrt[n]{a}$, n is called the **index** of the radical, and a is called the **radicand**. For a square root, the index is 2, but it is usually not written ($\sqrt[2]{a}$ is denoted simply as $\sqrt{a}$). A radical with an index of 3 is called a **cube root**, $\sqrt[3]{a}$.

> **DEFINITION** $\sqrt[n]{a}$
>
> 1. If n is a positive *even* integer and $a > 0$, then $\sqrt[n]{a}$ is the principal (positive) *n*th-root of a.
> 2. If $n > 1$ is a positive *odd* integer, then $\sqrt[n]{a}$ is the *n*th-root of a.
> 3. If $n > 1$ is a positive integer, then $\sqrt[n]{0} = 0$.

For the purpose of simplifying radicals, it is helpful to know the following patterns:

Perfect cubes	Perfect fourth powers	Perfect fifth powers
$1^3 = 1$	$1^4 = 1$	$1^5 = 1$
$2^3 = 8$	$2^4 = 16$	$2^5 = 32$
$3^3 = 27$	$3^4 = 81$	$3^5 = 243$
$4^3 = 64$	$4^4 = 256$	$4^5 = 1024$
$5^3 = 125$	$5^4 = 625$	$5^5 = 3125$

Skill Practice

Simplify the expressions if possible.

14. $\sqrt[3]{27}$ 15. $\sqrt[4]{1}$
16. $\sqrt[3]{216}$ 17. $\sqrt[5]{-32}$
18. $\sqrt[4]{\dfrac{16}{625}}$ 19. $\sqrt{0.25}$
20. $\sqrt[4]{-1}$

> **Example 4**   **Simplifying *n*th-Roots**
>
> Simplify the expressions, if possible.
>
> **a.** $\sqrt[3]{8}$ **b.** $\sqrt[4]{16}$ **c.** $\sqrt[5]{32}$ **d.** $\sqrt[3]{-64}$
>
> **e.** $\sqrt[3]{\dfrac{125}{27}}$ **f.** $\sqrt{0.01}$ **g.** $\sqrt[4]{-81}$
>
> **Solution:**
>
> **a.** $\sqrt[3]{8} = 2$ Because $(2)^3 = 8$
>
> **b.** $\sqrt[4]{16} = 2$ Because $(2)^4 = 16$
>
> **c.** $\sqrt[5]{32} = 2$ Because $(2)^5 = 32$
>
> **d.** $\sqrt[3]{-64} = -4$ Because $(-4)^3 = -64$
>
> **e.** $\sqrt[3]{\dfrac{125}{27}} = \dfrac{5}{3}$ Because $\left(\dfrac{5}{3}\right)^3 = \dfrac{125}{27}$

Answers

13. Index: 3; radicand: $2xy$
14. 3   15. 1   16. 6
17. -2 18. $\dfrac{2}{5}$ 19. 0.5
20. Not a real number.

f. $\sqrt{0.01} = 0.1$ Because $(0.1)^2 = 0.01$

Note: $\sqrt{0.01}$ is equivalent to $\sqrt{\dfrac{1}{100}} = \dfrac{1}{10}$, or 0.1.

g. $\sqrt[4]{-81}$ is not a real number because no real number raised to the fourth power equals -81.

Example 4(g) illustrates that an *n*th-root of a negative number is not a real number if the index is even because no real number raised to an even power is negative.

Finding an *n*th-root of a variable expression is similar to finding an *n*th-root of a numerical expression. However, for roots with an even index, particular care must be taken to obtain a nonnegative solution.

DEFINITION $\sqrt[n]{a^n}$

1. If n is a positive odd integer, then $\sqrt[n]{a^n} = a$
2. If n is a positive even integer, then $\sqrt[n]{a^n} = |a|$

The absolute value bars are necessary for roots with an even index because the variable, a, may represent a positive quantity or a negative quantity. By using absolute value bars, we ensure that $\sqrt[n]{a^n} = |a|$ is nonnegative and represents the principal *n*th-root of a.

Example 5   Simplifying Expressions of the Form $\sqrt[n]{a^n}$

Simplify the expressions.

a. $\sqrt{(-3)^2}$ **b.** $\sqrt{x^2}$ **c.** $\sqrt[3]{x^3}$ **d.** $\sqrt[4]{x^4}$ **e.** $\sqrt[5]{x^5}$

Solution:

a. $\sqrt{(-3)^2} = |-3| = 3$ Because the index is *even*, absolute value bars are necessary to make the answer nonnegative.

b. $\sqrt{x^2} = |x|$ Because the index is *even*, absolute value bars are necessary to make the answer nonnegative.

c. $\sqrt[3]{x^3} = x$ Because the index is *odd*, no absolute value bars are necessary.

d. $\sqrt[4]{x^4} = |x|$ Because the index is *even*, absolute value bars are necessary to make the answer nonnegative.

e. $\sqrt[5]{x^5} = x$ Because the index is *odd*, no absolute value bars are necessary.

If n is an even integer, then $\sqrt[n]{a^n} = |a|$. However, if the variable a is assumed to be nonnegative, then the absolute value bars may be omitted, that is, $\sqrt[n]{a^n} = a$ provided $a \geq 0$. In many examples and exercises, we will make the assumption that the variables within a radical expression are positive real numbers. In such a case, the absolute value bars are not needed to evaluate $\sqrt[n]{a^n}$.

It is helpful to become familiar with the patterns associated with perfect squares and perfect cubes involving variable expressions.

The following powers of x are perfect squares:

Perfect squares

$(x^1)^2 = x^2$
$(x^2)^2 = x^4$
$(x^3)^2 = x^6$
$(x^4)^2 = x^8$
$\ldots$

TIP: Any expression raised to an even power (multiple of 2) is a perfect square.

The following powers of x are perfect cubes:

Perfect cubes

$(x^1)^3 = x^3$
$(x^2)^3 = x^6$
$(x^3)^3 = x^9$
$(x^4)^3 = x^{12}$
$\ldots$

TIP: Any expression raised to a power that is a multiple of 3 is a perfect cube.

Skill Practice

Simplify the expressions. Assume the variables represent positive real numbers.

26. $\sqrt{y^{10}}$ **27.** $\sqrt[3]{x^{12}}$

28. $\sqrt{x^4 y^2}$ **29.** $\sqrt{25c^4}$

Example 6   Simplifying nth-Roots

Simplify the expressions. Assume that the variables are positive real numbers.

a. $\sqrt{c^6}$ **b.** $\sqrt[3]{d^{15}}$ **c.** $\sqrt{a^2 b^2}$ **d.** $\sqrt[3]{64 z^6}$

Solution:

a. $\sqrt{c^6}$ The expression c^6 is a perfect square.

     $\sqrt{c^6} = c^3$ This is because $\sqrt{(c^3)^2} = c^3$.

b. $\sqrt[3]{d^{15}}$ The expression d^{15} is a perfect cube.

     $\sqrt[3]{d^{15}} = d^5$ This is because $\sqrt[3]{(d^5)^3} = d^5$.

c. $\sqrt{a^2 b^2} = ab$ This is because $\sqrt{a^2 b^2} = \sqrt{(ab)^2} = ab$.

d. $\sqrt[3]{64 z^6} = 4z^2$ This is because $\sqrt[3]{(4z^2)^3} = 4z^2$.

TIP: A perfect nth-root can be simplified by dividing the power of the radicand by the index.

$$\sqrt[3]{d^{15}} = d^{15/3} = d^5$$

---

3. Translations Involving nth-Roots

It is important to understand the vocabulary and language associated with nth-roots. For instance, you must be able to distinguish between the square of a number and the square *root* of a number. The following example offers practice translating between English form and algebraic form.

Answers

26. y^5 **27.** x^4

28. $x^2 y$ **29.** $5c^2$

Example 7 **Translating from English Form to Algebraic Form**

Translate each English phrase into an algebraic expression.

a. The difference of the square of x and the principal square root of 7

b. The quotient of 1 and the cube root of z

Solution:

a. $x^2 - \sqrt{7}$

b. The quotient of $\rightarrow \dfrac{1}{\sqrt[3]{z}}$

4. Pythagorean Theorem

Recall that the **Pythagorean theorem** relates the lengths of the three sides of a right triangle (Figure 8-1).

$$a^2 + b^2 = c^2$$

The principal square root can be used to solve for an unknown side of a right triangle if the lengths of the other two sides are known.

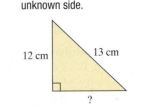

Figure 8-1

Example 8 **Applying the Pythagorean Theorem**

Use the Pythagorean theorem and the definition of the principal square root of a number to find the length of the unknown side.

Solution:

Label the sides of the triangle.

$a^2 + b^2 = c^2$

$a^2 + (8)^2 = (10)^2$ Apply the Pythagorean theorem.

$a^2 + 64 = 100$ Simplify.

$a^2 = 100 - 64$

$a^2 = 36$ This equation is quadratic. One method for solving the equation is to set the equation equal to zero, factor, and apply the zero product rule. However, we can also use the definition of a square root to solve for a.

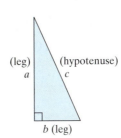

$a = \sqrt{36}$ or $a = -\sqrt{36}$ By definition, a must be one of the square roots of 36 (either 6 or -6). However, because a represents a distance, choose the *positive* (principal) square root of 36.

$a = 6$

The third side is 6 in. long.

Skill Practice

33. A wire is attached to the top of a 20-ft pole. How long is the wire if it reaches a point on the ground 15 ft from the base of the pole?

?   20 ft

15 ft

Example 9   Applying the Pythagorean Theorem

A bridge across a river is 600 yd long. A boat ramp at point R is 200 yd due north of point P on the bridge, such that the line segments $\overline{PQ}$ and $\overline{PR}$ form a right angle (Figure 8-2). How far does a kayak travel if it leaves from the boat ramp and paddles to point Q? Use a calculator and round the answer to the nearest yard.

Figure 8-2

Solution:

Label the triangle:

$$a^2 + b^2 = c^2$$

$$(200)^2 + (600)^2 = c^2 \qquad \text{Apply the Pythagorean theorem.}$$

$$40{,}000 + 360{,}000 = c^2 \qquad \text{Simplify.}$$

$$400{,}000 = c^2$$

By definition, c must be one of the square roots of 400,000. Because the value of c is a distance, choose the positive square root of 400,000.

$$c = \sqrt{400{,}000}$$

$$c \approx 632$$

Use a calculator to approximate the positive square root of 400,000.

The kayak must travel approximately 632 yd.

c $a = 200$ yd

$b = 600$ yd

Answer

33. The wire is 25 ft long.

Calculator Connections

Topic: Evaluating Square Roots and Higher Order Roots on a Calculator

A calculator can be used to approximate the value of a radical expression. To evaluate a square root, use the $\sqrt{}$ key. For example, evaluate: $\sqrt{25}$, $\sqrt{60}$, $\sqrt{\frac{13}{3}}$

Scientific Calculator

Enter: 25 $\boxed{\sqrt{x}}$ **Result:** $\boxed{ 5}$

Enter: 60 $\boxed{\sqrt{x}}$ **Result:** $\boxed{7.745966692}$

Enter: 13 $\boxed{\div}$ 3 $\boxed{=}$ $\boxed{\sqrt{x}}$ **Result:** $\boxed{2.081665999}$

Graphing Calculator

On the graphing calculator, the radicand is enclosed in parentheses.

```
√(25)
              5
√(60)
      7.745966692
√(13/3)
      2.081665999
```

TIP: The values $\sqrt{60}$ and $\sqrt{\frac{13}{3}}$ are approximated on the calculator to 10 digits. However, $\sqrt{60}$ and $\sqrt{\frac{13}{3}}$ are actually irrational numbers. Their decimal forms are nonterminating and nonrepeating. The only way to represent the exact answers is by writing the radical forms, $\sqrt{60}$ and $\sqrt{\frac{13}{2}}$.

To evaluate cube roots, your calculator may have a $\sqrt[3]{}$ key. Otherwise, for cube roots and roots of higher index (fourth roots, fifth roots, and so on), try using the $\sqrt[x]{y}$ key or $\sqrt[x]{}$ key. For example, evaluate $\sqrt[3]{64}$, $\sqrt[4]{81}$, and $\sqrt[3]{162}$:

Scientific Calculator

Enter: 64 `2nd` `√ʸ` 3 `=`	**Result:**		4
Enter: 81 `2nd` `√ʸ` 4 `=`	**Result:**		3
Enter: 162 `2nd` `√ʸ` 3 `=`	**Result:**	5.451361778	

Graphing Calculator

On a graphing calculator, the index is usually entered first.

```
3*√(64)
              4
4*√(81)
              3
3*√(162)
      5.451361778
```

Calculator Exercises

Estimate the value of each radical. Then use a calculator to approximate the radical to three decimal places. **(See the Tip on page 583.)**

1. $\sqrt{5}$ **2.** $\sqrt{17}$ **3.** $\sqrt{50}$ **4.** $\sqrt{96}$

5. $\sqrt{33}$ **6.** $\sqrt{145}$ **7.** $\sqrt{80}$ **8.** $\sqrt{170}$

9. $\sqrt[3]{7}$ **10.** $\sqrt[3]{28}$ **11.** $\sqrt[3]{65}$ **12.** $\sqrt[3]{124}$

Section 8.1   Practice Exercises

Boost *your* GRADE at ALEKS.com!

- Practice Problems
- Self-Tests
- NetTutor
- e-Professors
- Videos

Study Skill Exercise

1. Define the key terms:

     a. square root   **b. positive square root**   **c. principal square root**   **d. negative square root**

     e. perfect square   **f. *n*th-root**   **g. index**   **h. radicand**

     i. cube root   **j. Pythagorean theorem**

Objective 1: Definition of a Square Root

For Exercises 2–9, determine the square roots. **(See Example 1.)**

2. 4   **3.** 144   **4.** -64 **5.** -49

6. 81   **7.** 0   **8.** $\dfrac{16}{9}$ **9.** $\dfrac{1}{25}$

10. a. What is the principal square root of 64?

    b. What is the negative square root of 64?

11. a. What is the principal square root of 169?

    b. What is the negative square root of 169?

12. Does every number have two square roots? Explain.

13. Which number has only one square root?

14. Which of the following are perfect squares?

    $\{0, 1, 4, 15, 30, 49, 72, 81, 144, 300, 625, 900\}$

15. Which of the following are perfect squares?

    $\{8, 9, 12, 16, 25, 36, 42, 64, 95, 121, 140, 169\}$

For Exercises 16–31, evaluate the square roots. **(See Example 2.)**

16. $\sqrt{16}$

17. $\sqrt{4}$

18. $\sqrt{81}$

19. $\sqrt{49}$

20. $\sqrt{0.25}$

21. $\sqrt{0.16}$

22. $\sqrt{0.64}$

23. $\sqrt{0.09}$

24. $\sqrt{\dfrac{1}{9}}$

25. $\sqrt{\dfrac{25}{16}}$

26. $\sqrt{\dfrac{49}{121}}$

27. $\sqrt{\dfrac{1}{144}}$

28. $\sqrt{64 + 36}$

29. $\sqrt{16 + 9}$

30. $\sqrt{169 - 144}$

31. $\sqrt{225 - 144}$

32. Explain the difference between $\sqrt{-16}$ and $-\sqrt{16}$.

33. Using the definition of a square root, explain why $\sqrt{-16}$ does not have a real-valued square root.

34. Evaluate. $-\sqrt{|-25|}$

For Exercises 35–46, evaluate the square roots, if possible. **(See Example 3.)**

35. $-\sqrt{4}$

36. $-\sqrt{1}$

37. $\sqrt{-4}$

38. $\sqrt{-1}$

39. $\sqrt{-\dfrac{4}{49}}$

40. $-\sqrt{-\dfrac{9}{25}}$

41. $-\sqrt{-\dfrac{1}{36}}$

42. $-\sqrt{\dfrac{1}{36}}$

43. $-\sqrt{400}$

44. $-\sqrt{121}$

45. $\sqrt{-900}$

46. $\sqrt{-169}$

Objective 2: Definition of an *n*th-Root

47. Which of the following are perfect cubes?

    $\{0, 1, 3, 9, 27, 36, 42, 90, 125\}$

48. Which of the following are perfect cubes?

    $\{6, 8, 16, 20, 30, 64, 111, 150, 216\}$

49. Does $\sqrt[3]{-27}$ have a real-valued cube root?

50. Does $\sqrt[3]{-8}$ have a real-valued cube root?

For Exercises 51–66, evaluate the *n*th roots, if possible. **(See Example 4.)**

51. $\sqrt[3]{27}$

52. $\sqrt[3]{-27}$

53. $\sqrt[3]{64}$

54. $\sqrt[3]{-64}$

55. $-\sqrt[4]{16}$

56. $-\sqrt[4]{81}$

57. $\sqrt[4]{-1}$

58. $\sqrt[4]{0}$

59. $\sqrt[4]{-256}$

60. $\sqrt[4]{-625}$

61. $\sqrt[5]{-32}$

62. $-\sqrt[5]{32}$

63. $-\sqrt[6]{1}$

64. $\sqrt[6]{64}$

65. $\sqrt[6]{0}$

66. $\sqrt[6]{-1}$

For Exercises 67–86, simplify the expressions. **(See Example 5.)**

67. $\sqrt{(4)^2}$

68. $\sqrt{(8)^2}$

69. $\sqrt{(-4)^2}$

70. $\sqrt{(-8)^2}$

71. $\sqrt[3]{(5)^3}$

72. $\sqrt[3]{(7)^3}$

73. $\sqrt[3]{(-5)^3}$

74. $\sqrt[3]{(-7)^3}$

75. $\sqrt[4]{(2)^4}$

76. $\sqrt[4]{(10)^4}$

77. $\sqrt[4]{(-2)^4}$

78. $\sqrt[4]{(-10)^4}$

79. $\sqrt{a^2}$

80. $\sqrt{b^2}$

81. $\sqrt[3]{y^3}$

82. $\sqrt[3]{z^3}$

83. $\sqrt[4]{w^4}$

84. $\sqrt[4]{p^4}$

85. $\sqrt[5]{x^5}$

86. $\sqrt[5]{y^5}$

87. Determine which of the expressions are perfect squares. Then state a rule for determining perfect squares based on the exponent of the expression.

$\{x^2, a^3, y^4, z^5, (ab)^6, (pq)^7, w^8 x^8, c^9 d^9, m^{10}, n^{11}\}$

88. Determine which of the expressions are perfect cubes. Then state a rule for determining perfect cubes based on the exponent of the expression.

$\{a^2, b^3, c^4, d^5, e^6, (xy)^7, (wz)^8, (pq)^9, t^{10} s^{10}, m^{11} n^{11}, u^{12} v^{12}\}$

89. Determine which of the expressions are perfect fourth powers. Then state a rule for determining perfect fourth powers based on the exponent of the expression.

$\{m^2, n^3, p^4, q^5, r^6, s^7, t^8, u^9, v^{10}, (ab)^{11}, (cd)^{12}\}$

90. Determine which of the expressions are perfect fifth powers. Then state a rule for determining perfect fifth powers based on the exponent of the expression.

$\{a^2, b^3, c^4, d^5, e^6, k^7, w^8, x^9, y^{10}, z^{11}\}$

For Exercises 91–106, simplify the expressions. Assume the variables represent positive real numbers. **(See Example 6.)**

91. $\sqrt{y^{12}}$

92. $\sqrt{z^{20}}$

93. $\sqrt{a^8 b^{30}}$

94. $\sqrt{t^{50} s^{60}}$

95. $\sqrt[3]{q^{24}}$

96. $\sqrt[3]{x^{33}}$

97. $\sqrt[3]{8w^6}$

98. $\sqrt[3]{-27x^{27}}$

99. $\sqrt{(5x)^2}$

100. $\sqrt{(6w)^2}$

101. $-\sqrt{25x^2}$

102. $-\sqrt{36w^2}$

103. $\sqrt[3]{(5p^2)^3}$

104. $\sqrt[3]{(2k^4)^3}$

105. $\sqrt[3]{125p^6}$

106. $\sqrt[3]{8k^{12}}$

Objective 3: Translations Involving *n*th-Roots

For Exercises 107–110, translate the English phrase to an algebraic expression. **(See Example 7.)**

107. The sum of the principal square root of q and the square of p

108. The product of the principal square root of 11 and the cube of x

109. The quotient of 6 and the principal fourth root of x

110. The difference of the square of y and 1

Objective 4: Pythagorean Theorem

For Exercises 111–116, find the length of the third side of each triangle using the Pythagorean theorem. Round the answer to the nearest tenth if necessary. **(See Example 8.)**

111.

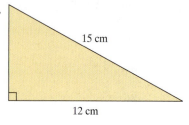

15 cm

12 cm

112.

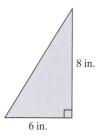

8 in.

6 in.

113.

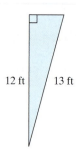

12 ft   13 ft

114.

5 m

4 m

115.

6.5 cm

2.4 cm

116.

14.8 ft

9.2 ft

117. Find the length of the diagonal of the square tile shown in the figure. Round the answer to the nearest tenth of an inch.

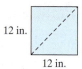

12 in.

12 in.

118. A baseball diamond is 90 ft on a side. Find the distance between home plate and second base. Round the answer to the nearest tenth of a foot.

Second base

90 ft

?

90 ft

Home plate

119. A new plasma television is listed as being 42 in. This distance is the diagonal distance across the screen. If the screen measures 28 inches in height, what is the actual width of the screen? Round to the nearest tenth of an inch.
(See Example 9.)

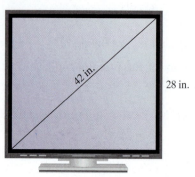

42 in.

28 in.

120. A marine biologist wants to track the migration of a pod of whales. He receives a radio signal from a tagged humpback whale and determines that the whale is 21 mi east and 37 mi north of his laboratory. Find the direct distance between the whale and the laboratory. Round to the nearest tenth of a mile.

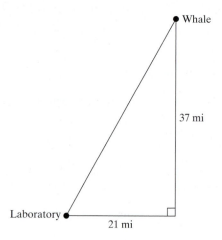

Whale

37 mi

Laboratory

21 mi

121. On a map, the cities Asheville, North Carolina, Roanoke, Virginia, and Greensboro, North Carolina, form a right triangle (see the figure). The distance between Asheville and Roanoke is 300 km. The distance between Roanoke and Greensboro is 134 km. How far is it from Greensboro to Asheville? Round the answer to the nearest kilometer.

122. Jackson, Mississippi, is west of Meridian, Mississippi, a distance of 141 km. Tupelo, Mississippi, is north of Meridian, a distance of 209 km. How far is it from Jackson to Tupelo? Round the answer to the nearest kilometer.

Expanding Your Skills

123. For what values of x will $\sqrt{x}$ be a real number?

124. For what values of x will $\sqrt{-x}$ be a real number?

Simplifying Radicals

Section 8.2

1. Multiplication Property of Radicals

You may have already recognized certain properties of radicals involving a product.

Objectives

1. Multiplication Property of Radicals
2. Simplifying Radicals Using the Multiplication Property of Radicals
3. Simplifying Radicals Using the Order of Operations
4. Simplifying Cube Roots

> **PROPERTY Multiplication Property of Radicals**
>
> Let a and b represent real numbers such that $\sqrt[n]{a}$ and $\sqrt[n]{b}$ are both real. Then,
>
> $$\sqrt[n]{ab} = \sqrt[n]{a} \cdot \sqrt[n]{b} \quad \textbf{Multiplication property of radicals}$$

The multiplication property of radicals indicates that a product within a radicand can be written as a product of radicals provided the roots are real numbers.

$$\sqrt{100} = \sqrt{25} \cdot \sqrt{4}$$

The reverse process is also true. A product of radicals can be written as a single radical provided the roots are real numbers and they have the same indices.

$$\overset{\text{Same index}}{\sqrt[n]{2} \cdot \sqrt[n]{18}} = \sqrt{36}$$

2. Simplifying Radicals Using the Multiplication Property of Radicals

In algebra, it is customary to simplify radical expressions as much as possible.

DEFINITION   Simplified Form of a Radical

Consider any radical expression where the radicand is written as a product of prime factors. The expression is in **simplified form** if all of the following conditions are met:

1. The radicand has no factor raised to a power greater than or equal to the index.
2. There are no radicals in the denominator of a fraction.
3. The radicand does not contain a fraction.

The expression $\sqrt{x^2}$ is not simplified because it fails condition 1. Because x^2 is a perfect square, $\sqrt{x^2}$ is easily simplified.

$$\sqrt{x^2} = x \quad \text{(for } x \geq 0\text{)}$$

However, how is an expression such as $\sqrt{x^7}$ simplified? This and many other radical expressions are simplified using the multiplication property of radicals. Examples 1–3 illustrate how nth powers can be removed from the radicands of nth-roots.

Example 1   Using the Multiplication Property to Simplify a Radical Expression

Use the multiplication property of radicals to simplify the expression $\sqrt{x^7}$. Assume $x \geq 0$.

Solution:

The expression $\sqrt{x^7}$ is equivalent to $\sqrt{x^6 \cdot x}$. By applying the multiplication property of radicals, we have

$$\sqrt{x^6 \cdot x} = \sqrt{x^6} \cdot \sqrt{x} \qquad x^6 \text{ is a perfect square because } (x^3)^2 = x^6$$
$$= x^3 \cdot \sqrt{x} \qquad \text{Simplify.}$$
$$= x^3\sqrt{x}$$

In Example 1, the expression x^7 is not a perfect square. Therefore, to simplify $\sqrt{x^7}$, it was necessary to write the expression as the product of the largest perfect square and a remaining, or "leftover," factor: $\sqrt{x^7} = \sqrt{x^6 \cdot x}$.

Example 2   Using the Multiplication Property to Simplify Radicals

Use the multiplication property of radicals to simplify the expressions. Assume the variables represent positive real numbers.

a. $\sqrt{a^{15}}$ **b.** $\sqrt{x^2 y^5}$ **c.** $\sqrt{s^9 t^{11}}$

Solution:

The goal is to rewrite each radicand as the product of the largest perfect square and a leftover factor.

a. $\sqrt{a^{15}}$

$$= \sqrt{a^{14} \cdot a} \qquad a^{14} \text{ is the largest perfect square in the radicand.}$$
$$= \sqrt{a^{14}} \cdot \sqrt{a} \qquad \text{Apply the multiplication property of radicals.}$$
$$= a^7\sqrt{a} \qquad \text{Simplify.}$$

b. $\sqrt{x^2 y^5}$

$\quad = \sqrt{x^2 y^4 \cdot y}$ $x^2 y^4$ is the largest perfect square in the radicand.

$\quad = \sqrt{x^2 y^4} \cdot \sqrt{y}$ Apply the multiplication property of radicals.

$\quad = xy^2 \sqrt{y}$ Simplify.

c. $\sqrt{s^9 t^{11}}$

$\quad = \sqrt{s^8 t^{10} \cdot st}$ $s^8 t^{10}$ is the largest perfect square in the radical.

$\quad = \sqrt{s^8 t^{10}} \cdot \sqrt{st}$ Apply the multiplication property of radicals.

$\quad = s^4 t^5 \sqrt{st}$ Simplify.

Each expression in Example 2 involves a radicand that is a product of variable factors. If a numerical factor is present, sometimes it is necessary to factor the coefficient before simplifying the radical.

Example 3 **Using the Multiplication Property to Simplify Radicals**

Use the multiplication property of radicals to simplify the expressions. Assume the variables represent positive real numbers.

a. $\sqrt{50}$ **b.** $5\sqrt{24a^6}$ **c.** $\sqrt{81x^4 y^3}$

Solution:

The goal is to rewrite each radicand as the product of the largest perfect square and a leftover factor.

a. Write the radicand as a product of prime factors. From the prime factorization, the largest perfect square is easily identified.

$\sqrt{50} = \sqrt{5^2 \cdot 2}$ Factor the radicand. $2\overline{)50}$
                            5^2 is the largest perfect square. $5\overline{)25}$
                                                                              5

$\quad = \sqrt{5^2} \cdot \sqrt{2}$ Apply the multiplication property of radicals.

$\quad = 5\sqrt{2}$ Simplify.

b. $5\sqrt{24a^6} = 5\sqrt{2^3 \cdot 3 \cdot a^6}$ Write the radicand as a product of prime factors: $24 = 2^3 \cdot 3$.

$\quad = 5\sqrt{2^2 a^6 \cdot 2 \cdot 3}$ $2^2 a^6$ is the largest perfect square in the radicand.

$\quad = 5\sqrt{2^2 a^6} \cdot \sqrt{2 \cdot 3}$ Apply the multiplication property of radicals.

$\quad = 5 \cdot 2a^3 \sqrt{6}$ Simplify the radical.

$\quad = 10a^3 \sqrt{6}$ Simplify the coefficient of the radical.

c. $\sqrt{81x^4y^3} = \sqrt{3^4x^4y^3}$ Write the radical as a product of prime factors. *Note:* $81 = 3^4$.

$= \sqrt{3^4x^4y^2 \cdot y}$ $3^4x^4y^2$ is the largest square in the radicand.

$= \sqrt{3^4x^4y^2} \cdot \sqrt{y}$ Apply the multiplication property of radicals.

$= 3^2x^2y \cdot \sqrt{y}$ Simplify the radical.

$= 9x^2y\sqrt{y}$ Simplify the coefficient of the radical.

Avoiding Mistakes

The multiplication property of radicals allows us to simplify a product of factors within a radical. For example,

$$\sqrt{x^2y^2} = \sqrt{x^2} \cdot \sqrt{y^2} = xy \quad \text{(for } x \geq 0 \text{ and } y \geq 0)$$

However, this rule does not apply to *terms* that are added or subtracted *within* the radical. For example,

$$\sqrt{x^2 + y^2} \quad \text{and} \quad \sqrt{x^2 - y^2}$$

cannot be simplified.

---

3. Simplifying Radicals Using the Order of Operations

Often a radical can be simplified by applying the order of operations. In Example 4, the first step will be to simplify the expression within the radicand.

Example 4 **Simplifying Radicals Using the Order of Operations**

Simplify the expressions. Assume the variables represent positive real numbers.

a. $\sqrt{\dfrac{a^5}{a^3}}$ **b.** $\sqrt{\dfrac{6}{96}}$ **c.** $\sqrt{\dfrac{27x^5}{3x}}$

Solution:

a. $\sqrt{\dfrac{a^5}{a^3}}$ The radical contains a fraction. However, the fraction can be simplified.

$= \sqrt{a^2}$ Reduce the fraction to lowest terms.

$= a$ Simplify the radical.

b. $\sqrt{\dfrac{6}{96}}$ The radical contains a fraction that can be simplified.

$= \sqrt{\dfrac{1}{16}}$ Reduce the fraction to lowest terms.

$= \dfrac{1}{4}$ Simplify.

c. $\sqrt{\dfrac{27x^5}{3x}}$ The fraction within the radicand can be simplified.

$= \sqrt{9x^4}$ Reduce to lowest terms.

$= 3x^2$ Simplify.

Example 5 Simplifying Radical Expressions

Simplify the expressions.

a. $\dfrac{5\sqrt{20}}{2}$ **b.** $\dfrac{2 - \sqrt{36}}{12}$

Solution:

a. $\dfrac{5\sqrt{20}}{2} = \dfrac{5\sqrt{2^2 \cdot 5}}{2}$ Following the order of operations, first simplify the radical. 2^2 is the largest perfect square in the radicand.

$= \dfrac{5\sqrt{2^2} \cdot \sqrt{5}}{2}$ Apply the multiplication property of radicals.

$= \dfrac{5 \cdot 2\sqrt{5}}{2}$ Simplify the radical.

$= \dfrac{5 \cdot 2\sqrt{5}}{2}$ Simplify to lowest terms.

$= 5\sqrt{5}$

b. $\dfrac{2 - \sqrt{36}}{12}$

$= \dfrac{2 - 6}{12}$ Following the order of operations, first simplify the radical.

$= \dfrac{-4}{12}$ Next, simplify the numerator.

$= -\dfrac{1}{3}$ Reduce the fraction to lowest terms.

Avoiding Mistakes

$\dfrac{5\sqrt{20}}{2}$ cannot be simplified as written because 20 is under the radical and 2 is not. To reduce to lowest terms, the radical must be simplified first, $\dfrac{5 \cdot 2\sqrt{5}}{2}$. Then factors outside the radical can be simplified.

4. Simplifying Cube Roots

Example 6 Simplifying Cube Roots

Use the multiplication property of radicals to simplify the expressions.

a. $\sqrt[3]{z^5}$ **b.** $\sqrt[3]{-80}$

Solution:

a. $\sqrt[3]{z^5}$

$= \sqrt[3]{z^3 \cdot z^2}$ z^3 is the largest perfect cube in the radicand.

$= \sqrt[3]{z^3} \cdot \sqrt[3]{z^2}$ Apply the multiplication property of radicals.

$= z\sqrt[3]{z^2}$ Simplify.

b. $\sqrt[3]{-80}$

$= \sqrt[3]{-1 \cdot 2^4 \cdot 5}$ Factor the radicand.

$= \sqrt[3]{-1 \cdot 2^3 \cdot 2 \cdot 5}$ -1 and 2^3 are perfect cubes.

$= \sqrt[3]{-1} \cdot \sqrt[3]{2^3} \cdot \sqrt[3]{2 \cdot 5}$ Apply the multiplication property of radicals.

$= -1 \cdot 2 \cdot \sqrt[3]{10}$ Simplify.

$= -2\sqrt[3]{10}$

$2)\overline{80}$
$2)\overline{40}$
$2)\overline{20}$
$2)\overline{10}$
5

Skill Practice

Simplify.

18. $\sqrt[3]{\dfrac{x^{12}}{y^{6}}}$ **19.** $\sqrt[3]{\dfrac{81}{3}}$

Example 7 Simplifying Cube Roots

Simplify the expressions.

a. $\sqrt[3]{\dfrac{a^{16}}{a}}$ **b.** $\sqrt[3]{\dfrac{2}{16}}$

Solution:

a. $\sqrt[3]{\dfrac{a^{16}}{a}}$ The radical contains a fraction that can be simplified.

$= \sqrt[3]{a^{15}}$ Reduce to lowest terms.

$= a^{5}$ Simplify.

b. $\sqrt[3]{\dfrac{2}{16}}$ The radical contains a fraction that can be simplified.

$= \sqrt[3]{\dfrac{1}{8}}$ Reduce to lowest terms.

$= \dfrac{1}{2}$ Simplify.

Answers

18. $\dfrac{x^{4}}{y^{2}}$ **19.** 3

Calculator Connections

Topic: Verifying Simplified Radicals

A calculator can support the multiplication property of radicals. For example, use a calculator to evaluate $\sqrt{50}$ and its simplified form $5\sqrt{2}$.

Scientific Calculator

Enter: 50 $\boxed{\sqrt{x}}$ **Result:** $\boxed{7.071067812}$

Enter: 2 $\boxed{\sqrt{x}}$ $\boxed{\times}$ 5 $\boxed{=}$ **Result:** $\boxed{7.071067812}$

Graphing Calculator

```
√(50)
          7.071067812
5*√(2)
          7.071067812
```

TIP: The decimal approximation for $\sqrt{50}$ and $5\sqrt{2}$ agree for the first 10 digits. This in itself does not make $\sqrt{50} = 5\sqrt{2}$. It is the multiplication property of radicals that guarantees that the expressions are equal.

Calculator Exercises

Simplify the radical expressions algebraically. Then use a calculator to approximate the original expression and its simplified form.

1. $\sqrt{125}$ **2.** $\sqrt{18}$ **3.** $\sqrt[3]{54}$ **4.** $\sqrt[3]{108}$

Section 8.2   Practice Exercises

Study Skills Exercise

1. Define the key terms:

a. simplified form of a radical   **b. multiplication property of radicals**

Review Exercises

2. Which of the following are perfect squares? $\{2, 4, 6, 16, 20, 25, x^2, x^3, x^{15}, x^{20}, x^{25}\}$

3. Which of the following are perfect cubes? $\{3, 6, 8, 9, 12, 27, y^3, y^8, y^9, y^{12}, y^{27}\}$

4. Which of the following are perfect fourth powers? $\{4, 16, 20, 25, 81, w^4, w^{16}, w^{20}, w^{25}, w^{81}\}$

For Exercises 5–12, simplify the expressions, if possible. Assume the variables represent positive real numbers.

5. $-\sqrt{25}$ **6.** $\sqrt{-25}$ **7.** $-\sqrt[3]{27}$ **8.** $\sqrt[3]{-27}$

9. $\sqrt[4]{a^8}$ **10.** $\sqrt[5]{b^{15}}$ **11.** $\sqrt{4x^2y^4}$ **12.** $\sqrt{9p^{10}}$

 13. On a map, Seattle, Washington, is 378 km west of Spokane, Washington. Portland, Oregon, is 236 km south of Seattle. Approximate the distance between Portland and Spokane to the nearest kilometer.

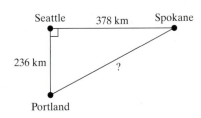

14. A new roof is needed on a shed. How many square feet of tar paper would be needed to cover the top of the roof?

Objective 2: Simplifying Radicals Using the Multiplication Property of Radicals

For Exercises 15–50, use the multiplication property of radicals to simplify the expressions. Assume the variables represent positive real numbers. **(See Examples 1–3.)**

15. $\sqrt{18}$ **16.** $\sqrt{75}$ **17.** $\sqrt{28}$ **18.** $\sqrt{40}$

19. $6\sqrt{20}$ **20.** $10\sqrt{27}$ **21.** $-2\sqrt{50}$ **22.** $-11\sqrt{8}$

23. $\sqrt{a^5}$ **24.** $\sqrt{b^9}$ **25.** $\sqrt{w^{22}}$ **26.** $\sqrt{p^{18}}$

27. $\sqrt{m^4n^5}$ **28.** $\sqrt{c^2d^9}$ **29.** $x\sqrt{x^{13}y^{10}}$ **30.** $v\sqrt{u^{10}v^7}$

31. $3\sqrt{t^{10}}$ **32.** $-4\sqrt{m^8n^4}$ **33.** $\sqrt{8x^3}$ **34.** $\sqrt{27y^5}$

35. $\sqrt{16z^3}$ **36.** $\sqrt{9y^5}$ **37.** $-\sqrt{45w^6}$ **38.** $-\sqrt{56v^8}$

39. $\sqrt{z^{25}}$ **40.** $\sqrt{25p^{49}}$ **41.** $-\sqrt{15z^{11}}$ **42.** $-\sqrt{6k^{15}}$

43. $5\sqrt{104a^2b^7}$ **44.** $3\sqrt{88m^4n^{11}}$ **45.** $\sqrt{26pq}$ **46.** $\sqrt{15a}$

47. $m\sqrt{m^{10}n^{16}}$ **48.** $c^2\sqrt{c^4d^{12}}$ **49.** $\sqrt{48a^3b^5c^4}$ **50.** $\sqrt{18xy^4z^3}$

Objective 3: Simplifying Radicals Using the Order of Operations

For Exercises 51–70, use the order of operations, if necessary, to simplify the expressions. Assume the variables represent positive real numbers. **(See Examples 4–5.)**

51. $\sqrt{\dfrac{a^9}{a}}$ **52.** $\sqrt{\dfrac{x^5}{x}}$ **53.** $\sqrt{\dfrac{y^{15}}{y^5}}$ **54.** $\sqrt{\dfrac{c^{31}}{c^{11}}}$

55. $\sqrt{\dfrac{5}{20}}$ **56.** $\sqrt{\dfrac{3}{75}}$ **57.** $\sqrt{\dfrac{40}{10}}$ **58.** $\sqrt{\dfrac{80}{5}}$

59. $\sqrt{\dfrac{32x^3}{8x}}$ **60.** $\sqrt{\dfrac{200b^{11}}{2b^5}}$ **61.** $\sqrt{\dfrac{50p^7}{2p}}$ **62.** $\sqrt{\dfrac{45t^9}{5t^5}}$

63. $\dfrac{3\sqrt{20}}{2}$ **64.** $\dfrac{5\sqrt{18}}{3}$ **65.** $\dfrac{5\sqrt{24}}{10}$ **66.** $\dfrac{2\sqrt{27}}{6}$

67. $\dfrac{10+\sqrt{4}}{3}$ **68.** $\dfrac{-1+\sqrt{25}}{4}$ **69.** $\dfrac{20-\sqrt{36}}{2}$ **70.** $\dfrac{3-\sqrt{81}}{3}$

For Exercises 71–74, find the exact length of the third side of each triangle using the Pythagorean theorem. Write the answer as a simplified radical.

71.
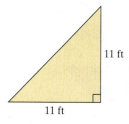
11 ft

11 ft

72.

21 m

20 m

73.
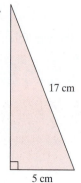
17 cm

5 cm

74.
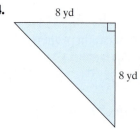
8 yd

8 yd

Objective 4: Simplifying Cube Roots

For Exercises 75–86, simplify the cube roots. **(See Examples 6–7.)**

75. $\sqrt[3]{a^8}$

76. $\sqrt[3]{8v^3}$

77. $7\sqrt[3]{16z^3}$

78. $5\sqrt[3]{54t^6}$

79. $\sqrt[3]{16a^5b^6}$

80. $\sqrt[3]{81p^9q^{11}}$

81. $\sqrt[3]{\dfrac{z^4}{z}}$

82. $\sqrt[3]{\dfrac{w^8}{w^2}}$

83. $\sqrt[3]{-\dfrac{32}{4}}$

84. $\sqrt[3]{-\dfrac{128}{2}}$

85. $\sqrt[3]{40}$

86. $\sqrt[3]{54}$

Mixed Exercises

For Exercises 87–110, simplify the expressions. Assume the variables represent positive real numbers.

87. $\sqrt{\dfrac{3}{27}}$

88. $\sqrt{\dfrac{5}{125}}$

89. $\sqrt{16a^3}$

90. $\sqrt{125x^6}$

91. $\sqrt{\dfrac{4x^3}{x}}$

92. $\sqrt{\dfrac{9z^5}{z}}$

93. $\sqrt{8p^2q}$

94. $\sqrt{6cd^3}$

95. $\sqrt{32}$

96. $\sqrt{64}$

97. $\sqrt{52u^4v^7}$

98. $\sqrt{44p^8q^{10}}$

99. $\sqrt{216}$

100. $\sqrt{250}$

101. $\sqrt[3]{216}$

102. $\sqrt[3]{250}$

103. $\sqrt[3]{16a^3}$

104. $\sqrt[3]{125x^6}$

105. $\sqrt[3]{\dfrac{x^5}{x^2}}$

106. $\sqrt[3]{\dfrac{y^{11}}{y^2}}$

107. $\dfrac{-6\sqrt{20}}{12}$

108. $\dfrac{-5\sqrt{32}}{10}$

109. $\dfrac{-4-\sqrt{25}}{18}$

110. $\dfrac{8-\sqrt{100}}{2}$

Expanding Your Skills

For Exercises 111–114, simplify the expressions. Assume the variables represent positive real numbers.

111. $\sqrt{(-2-5)^2+(-4+3)^2}$

112. $\sqrt{(-1-7)^2+[1-(-1)]^2}$

113. $\sqrt{x^2+10x+25}$

114. $\sqrt{x^2+6x+9}$

Addition and Subtraction of Radicals

Section 8.3

1. Definition of *Like* Radicals

Objectives

1. Definition of *Like* Radicals
2. Addition and Subtraction of Radicals

DEFINITION *Like* Radicals

Two radical terms are said to be *like* radicals if they have the same index and the same radicand.

Section 8.4 **Multiplication of Radicals**

Objectives

1. **Multiplication Property of Radicals**
2. **Expressions of the Form ($\sqrt[n]{a}$)^n**
3. **Special Case Products**
4. **Multiplying Conjugate Radical Expressions**

1. Multiplication Property of Radicals

In this section, we will learn how to multiply radicals that have the same index. Recall from Section 8.2 the multiplication property of radicals.

> **PROPERTY** **Multiplication Property of Radicals**
>
> Let a and b represent real numbers such that $\sqrt[n]{a}$ and $\sqrt[n]{b}$ are both real. Then,
> $$\sqrt[n]{a} \cdot \sqrt[n]{b} = \sqrt[n]{ab}$$

To multiply two radical expressions, use the multiplication property of radicals along with the commutative and associative properties of multiplication.

Skill Practice

Multiply the expressions and simplify the result. Assume the variables represent positive real numbers.

1. $\sqrt{2} \cdot \sqrt{5}$
2. $(-5z\sqrt{6})(4z\sqrt{2})$
3. $(9y\sqrt{x})\left(\dfrac{1}{3}y\sqrt{xy}\right)$

Example 1 **Multiplying Radical Expressions**

Multiply the expressions and simplify the result. Assume the variables represent positive real numbers.

a. $\sqrt{3} \cdot \sqrt{2}$ **b.** $(5\sqrt{3})(2\sqrt{15})$ **c.** $(6a\sqrt{ab})\left(\dfrac{1}{3}a\sqrt{a}\right)$

Solution:

a. $\sqrt{3} \cdot \sqrt{2} = \sqrt{6}$ Multiplication property of radicals

b. $(5\sqrt{3})(2\sqrt{15}) = (5 \cdot 2)(\sqrt{3} \cdot \sqrt{15})$ Commutative and associative properties of multiplication

$= 10\sqrt{45}$ Multiplication property of radicals

$= 10\sqrt{3^2 \cdot 5}$ Simplify the radical.

$= 10 \cdot 3\sqrt{5}$

$= 30\sqrt{5}$

c. $(6a\sqrt{ab})\left(\dfrac{1}{3}a\sqrt{a}\right) = \left(6a \cdot \dfrac{1}{3}a\right)(\sqrt{ab} \cdot \sqrt{a})$ Commutative and associative properties of multiplication

$= 2a^2\sqrt{a^2b}$ Multiplication property of radicals

$= 2a^2 \cdot a\sqrt{b}$ Simplify the radical.

$= 2a^3\sqrt{b}$

When multiplying radical expressions with more than one term, we use the distributive property.

Answers

1. $\sqrt{10}$ 2. $-40z^2\sqrt{3}$
3. $3xy^2\sqrt{y}$

Example 2 **Multiplying Radical Expressions with Multiple Terms**

Multiply the expressions. Assume the variables represent positive real numbers.

a. $\sqrt{5}(4 + 3\sqrt{5})$ **b.** $(\sqrt{x} - 10)(\sqrt{y} + 4)$ **c.** $(2\sqrt{3} - \sqrt{5})(\sqrt{3} + 6\sqrt{5})$

Solution:

a. $\sqrt{5}(4 + 3\sqrt{5})$

$= \sqrt{5}(4) + \sqrt{5}(3\sqrt{5})$ Apply the distributive property.

$= 4\sqrt{5} + 3\sqrt{5^2}$ Multiplication property of radicals

$= 4\sqrt{5} + 3 \cdot 5$ Simplify the radical.

$= 4\sqrt{5} + 15$

b. $(\sqrt{x} - 10)(\sqrt{y} + 4)$

$= \sqrt{x}(\sqrt{y}) + \sqrt{x}(4) - 10(\sqrt{y}) - 10(4)$ Apply the distributive property.

$= \sqrt{xy} + 4\sqrt{x} - 10\sqrt{y} - 40$ Simplify.

c. $(2\sqrt{3} - \sqrt{5})(\sqrt{3} + 6\sqrt{5})$

$= 2\sqrt{3}(\sqrt{3}) + 2\sqrt{3}(6\sqrt{5}) - \sqrt{5}(\sqrt{3}) - \sqrt{5}(6\sqrt{5})$ Apply the distributive property.

$= 2\sqrt{3^2} + 12\sqrt{15} - \sqrt{15} - 6\sqrt{5^2}$ Multiplication property of radicals

$= 2 \cdot 3 + 11\sqrt{15} - 6 \cdot 5$ Simplify radicals. Combine *like* radicals.

$= 6 + 11\sqrt{15} - 30$

$= -24 + 11\sqrt{15}$ Combine *like* terms.

Skill Practice

Multiply the expressions and simplify the result. Assume the variables represent positive real numbers.

4. $\sqrt{7}(2\sqrt{7} - 4)$
5. $(\sqrt{x} + 2)(\sqrt{x} - 3)$
6. $(2\sqrt{a} + 4\sqrt{6}) \cdot (\sqrt{a} - 3\sqrt{6})$

2. Expressions of the Form $(\sqrt[n]{a})^n$

The multiplication property of radicals can be used to simplify an expression of the form $(\sqrt{a})^2$, where $a \geq 0$.

$$(\sqrt{a})^2 = \sqrt{a} \cdot \sqrt{a} = \sqrt{a^2} = a$$

This logic can be applied to nth-roots. If $\sqrt[n]{a}$ is a real number, then $(\sqrt[n]{a})^n = a$.

Skill Practice

Simplify the expressions.

7. $(\sqrt{13})^2$
8. $(\sqrt[3]{x})^3$
9. $(2\sqrt{11})^2$

Example 3 **Simplifying Radical Expressions**

Simplify the expressions. Assume $x \geq 0$.

a. $(\sqrt{7})^2$ **b.** $(\sqrt[4]{x})^4$ **c.** $(3\sqrt{2})^2$

Solution:

a. $(\sqrt{7})^2 = 7$ **b.** $(\sqrt[4]{x})^4 = x$ **c.** $(3\sqrt{2})^2 = 3^2 \cdot (\sqrt{2})^2 = 9 \cdot 2 = 18$

Answers

4. $14 - 4\sqrt{7}$
5. $x - \sqrt{x} - 6$
6. $2a - 2\sqrt{6a} - 72$
7. 13   **8.** x **9.** 44

Section 8.4 Practice Exercises

Study Skills Exercise

1. When writing a radical expression, be sure to note the difference between an exponent on a coefficient and an index to a radical. Write an algebraic expression for each of the following:

   x cubed times the square root of y

   x times the cube root of y

Review Exercises

For Exercises 2–5, perform the indicated operations and simplify. Assume the variables represent positive real numbers.

2. $\sqrt{25} + \sqrt{16} - \sqrt{36}$ **3.** $\sqrt{100} - \sqrt{4} + \sqrt{9}$ **4.** $6x\sqrt{18} + 2\sqrt{2x^2}$ **5.** $10\sqrt{zw^4} - w^2\sqrt{49z}$

6. What three conditions are needed for a radical expression to be in simplified form?

Objective 1: Multiplication Property of Radicals

For Exercises 7–26, multiply the expressions. **(See Example 1.)**

7. $\sqrt{5} \cdot \sqrt{3}$ **8.** $\sqrt{7} \cdot \sqrt{6}$ **9.** $\sqrt{47} \cdot \sqrt{47}$ **10.** $\sqrt{59} \cdot \sqrt{59}$

11. $\sqrt{b} \cdot \sqrt{b}$ **12.** $\sqrt{t} \cdot \sqrt{t}$ **13.** $(2\sqrt{15})(3\sqrt{p})$ **14.** $(4\sqrt{2})(5\sqrt{q})$

15. $\sqrt{10} \cdot \sqrt{5}$ **16.** $\sqrt{2} \cdot \sqrt{10}$ **17.** $(-\sqrt{7})(-2\sqrt{14})$ **18.** $(-6\sqrt{2})(-\sqrt{22})$

19. $(3x\sqrt{2})(\sqrt{14})$ **20.** $(4y\sqrt{3})(\sqrt{6})$ **21.** $\left(\frac{1}{6}x\sqrt{xy}\right)(24x\sqrt{x})$ **22.** $\left(\frac{1}{4}u\sqrt{uv}\right)(8u\sqrt{v})$

23. $(6w\sqrt{5})(w\sqrt{8})$ **24.** $(t\sqrt{2})(5\sqrt{6t})$ **25.** $(-2\sqrt{3})(4\sqrt{5})$ **26.** $(-\sqrt{7})(2\sqrt{3})$

For Exercises 27–28, find the exact perimeter and exact area of the rectangles.

27.

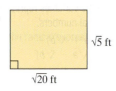

$\sqrt{5}$ ft

$\sqrt{20}$ ft

28.

$\sqrt{8}$ in.

$\sqrt{2}$ in.

For Exercises 29–30, find the exact area of the triangles.

29.

$\sqrt{12}$ cm

$\sqrt{3}$ cm

30.

$\sqrt{28}$ m

$\sqrt{7}$ m

For Exercises 31–44, multiply the expressions. Assume the variables represent positive real numbers. **(See Example 2.)**

31. $\sqrt{3w} \cdot \sqrt{3w}$

32. $\sqrt{6p} \cdot \sqrt{6p}$

33. $(8\sqrt{5y})(-2\sqrt{2})$

34. $(4\sqrt{5x})(7\sqrt{3})$

35. $\sqrt{2}(\sqrt{6} - \sqrt{3})$

36. $\sqrt{5}(\sqrt{10} + \sqrt{7})$

37. $4\sqrt{x}(\sqrt{x} + 5)$

38. $2\sqrt{y}(3 - \sqrt{y})$

39. $(\sqrt{3} + 2\sqrt{10})(4\sqrt{3} - \sqrt{10})$

40. $(8\sqrt{7} - \sqrt{5})(\sqrt{7} + 3\sqrt{5})$

41. $(\sqrt{a} - 3b)(9\sqrt{a} - b)$

42. $(11\sqrt{m} + 4n)(\sqrt{m} + n)$

43. $(p + 2\sqrt{p})(8p + 3\sqrt{p} - 4)$

44. $(5s - \sqrt{s})(s + 5\sqrt{s} + 6)$

Objective 2: Expressions of the Form $(\sqrt[n]{a})^n$

For Exercises 45–52, simplify the expressions. Assume the variables represent positive real numbers. **(See Example 3.)**

45. $(\sqrt{10})^2$

46. $(\sqrt{23})^2$

47. $(\sqrt[3]{4})^3$

48. $(\sqrt[3]{29})^3$

49. $(\sqrt[4]{t})^4$

50. $(\sqrt[4]{xy})^4$

51. $(4\sqrt{c})^2$

52. $(10\sqrt{2pq})^2$

Objective 3: Special Case Products

For Exercises 53–60, multiply the radical expressions. Assume the variables represent positive real numbers. **(See Examples 4–5.)**

53. $(\sqrt{13} + 4)^2$

54. $(6 - \sqrt{11})^2$

55. $(\sqrt{a} - 2)^2$

56. $(\sqrt{p} + 3)^2$

57. $(2\sqrt{a} - 3)^2$

58. $(3\sqrt{w} + 4)^2$

59. $(\sqrt{10} - \sqrt{11})^2$

60. $(\sqrt{3} - \sqrt{2})^2$

Objective 4: Multiplying Conjugate Radical Expressions

For Exercises 61–72, multiply the radical expressions. Assume the variables represent positive real numbers. **(See Examples 6–7.)**

61. $(\sqrt{5} + 2)(\sqrt{5} - 2)$

62. $(\sqrt{3} - 4)(\sqrt{3} + 4)$

63. $(\sqrt{x} + \sqrt{y})(\sqrt{x} - \sqrt{y})$

64. $(\sqrt{a} + \sqrt{b})(\sqrt{a} - \sqrt{b})$

65. $(\sqrt{10} - \sqrt{11})(\sqrt{10} + \sqrt{11})$

66. $(\sqrt{3} - \sqrt{2})(\sqrt{3} + \sqrt{2})$

67. $(\sqrt{6} + \sqrt{2})(\sqrt{6} - \sqrt{2})$

68. $(\sqrt{15} - \sqrt{5})(\sqrt{15} + \sqrt{5})$

69. $(8\sqrt{x} - 2\sqrt{y})(8\sqrt{x} + 2\sqrt{y})$

70. $(4\sqrt{s} + 11\sqrt{t})(4\sqrt{s} - 11\sqrt{t})$

71. $(5\sqrt{3} - \sqrt{2})(5\sqrt{3} + \sqrt{2})$

72. $(2\sqrt{7} - 4\sqrt{3})(2\sqrt{7} + 4\sqrt{3})$

Mixed Exercises

For Exercises 73–84, multiply the expressions in parts (a) and (b) and compare the process used. Assume the variables represent positive real numbers.

73. a. $3(x + 2)$
   b. $\sqrt{3}(\sqrt{x} + \sqrt{2})$

74. a. $-5(6 + y)$
   b. $-\sqrt{5}(\sqrt{6} + \sqrt{y})$

75. a. $(2a + 3)^2$
   b. $(2\sqrt{a} + 3)^2$

76. a. $(6 - z)^2$
   b. $(\sqrt{6} - z)^2$

77. a. $(b - 5)(b + 5)$
   b. $(\sqrt{b} - 5)(\sqrt{b} + 5)$

78. a. $(3w - 1)(3w + 1)$
   b. $(3\sqrt{w} - 1)(3\sqrt{w} + 1)$

79. a. $(x - 2y)^2$
   b. $(\sqrt{x} - 2\sqrt{y})^2$

80. a. $(5c + 2d)^2$
   b. $(5\sqrt{c} + 2\sqrt{d})^2$

81. a. $(p - q)(p + q)$
   b. $(\sqrt{p} - \sqrt{q})(\sqrt{p} + \sqrt{q})$

82. a. $(t - 3)(t + 3)$
   b. $(\sqrt{t} - \sqrt{3})(\sqrt{t} + \sqrt{3})$

83. a. $(y - 3)^2$
   b. $(\sqrt{x - 2} - 3)^2$

84. a. $(p + 4)^2$
   b. $(\sqrt{x + 1} + 4)^2$

Section 8.5 | Division of Radicals and Rationalization

1. Simplified Form of a Radical

Recall the conditions for a radical to be simplified.

> **DEFINITION** Simplified Form of a Radical
>
> Consider any radical expression where the radicand is written as a product of prime factors. The expression is in simplified form if all of the following conditions are met:
>
> 1. The radicand has no factor raised to a power greater than or equal to the index.
> 2. There are no radicals in the denominator of a fraction.
> 3. The radicand does not contain a fraction.

The basis of the second and third conditions, which restrict radicals from the denominator of an expression, are largely historical. In some cases, removing a radical from the denominator of a fraction will create an expression that is computationally simpler.

The process to remove a radical from the denominator is called **rationalizing the denominator**. In this section, we will show three approaches that can be used to achieve the second and third conditions of a simplified radical.

1. Rationalizing by applying the division property of radicals.
2. Rationalizing when the denominator contains a singe radical term.
3. Rationalizing when the denominator contains two terms involving square roots.

2. Division Property of Radicals

The multiplication property of radicals enables a product within a radical to be separated and written as a product of radicals. We now state a similar property for radicals involving quotients.

> **PROPERTY** Division Property of Radicals
>
> Let a and b represent real numbers such that $\sqrt[n]{a}$ and $\sqrt[n]{b}$ are both real. Then,
>
> $$\sqrt[n]{\frac{a}{b}} = \frac{\sqrt[n]{a}}{\sqrt[n]{b}} \qquad b \neq 0$$

The division property of radicals indicates that a quotient within a radicand can be written as a quotient of radicals provided the roots are real numbers. For example:

$$\sqrt{\frac{4}{9}} = \frac{\sqrt{4}}{\sqrt{9}}$$

The reverse process is also true. A quotient of radicals can be written as a single radical provided that the roots are real numbers and they have the same indices.

$$\text{same index} \left[\frac{\sqrt[3]{125}}{\sqrt[3]{8}} = \sqrt[3]{\frac{125}{8}} \right.$$

In Examples 1 and 2, we will apply the division property of radicals to simplify radical expressions.

Example 1   **Using the Division Property to Simplify Radicals**

Use the division property of radicals to simplify the expressions. Assume the variables represent positive real numbers.

a. $\sqrt{\dfrac{a^{10}}{b^4}}$ **b.** $\sqrt{\dfrac{20x^3}{9}}$

Skill Practice

Simplify the expressions.

1. $\sqrt{\dfrac{c^4}{49}}$

2. $\sqrt{\dfrac{12b^5}{25}}$

Solution:

a. $\sqrt{\dfrac{a^{10}}{b^4}}$ The radicand contains an irreducible fraction.

$= \dfrac{\sqrt{a^{10}}}{\sqrt{b^4}}$ Apply the division property to rewrite as a quotient of radicals.

$= \dfrac{a^5}{b^2}$ Simplify the radicals.

b. $\sqrt{\dfrac{20x^3}{9}}$ The radicand contains an irreducible fraction.

$= \dfrac{\sqrt{20x^3}}{\sqrt{9}}$ Apply the division property to rewrite as a quotient of radicals.

$= \dfrac{\sqrt{2^2 \cdot 5 \cdot x^2 \cdot x}}{\sqrt{9}}$ Factor the radicand in the numerator to simplify the radical.

$= \dfrac{2x\sqrt{5x}}{3}$ Simplify the radicals in the numerator and denominator. The expression is simplified since it now satisfies all conditions.

Answers

1. $\dfrac{c^2}{7}$ **2.** $\dfrac{2b^2\sqrt{3b}}{5}$

Skill Practice

Simplify the expressions.

3. $\dfrac{\sqrt[3]{250}}{\sqrt[3]{2}}$

4. $\dfrac{\sqrt{10z^9}}{\sqrt{z}}$

Example 2 **Using the Division Property to Simplify Radicals**

Use the division property of radicals to simplify the expressions. Assume the variables represent positive real numbers.

a. $\dfrac{\sqrt[3]{9}}{\sqrt[3]{72}}$ **b.** $\dfrac{\sqrt{7y^3}}{\sqrt{y}}$

Solution:

a. $\dfrac{\sqrt[3]{9}}{\sqrt[3]{72}}$ There is a radical in the denominator of the fraction.

$= \sqrt[3]{\dfrac{9}{72}}$ Apply the division property to write the quotient under a single radical.

$= \sqrt[3]{\dfrac{1}{8}}$ Reduce to lowest terms.

$= \dfrac{1}{2}$ Simplify the radical.

b. $\dfrac{\sqrt{7y^3}}{\sqrt{y}}$ There is a radical in the denominator of the fraction.

$= \sqrt{\dfrac{7y^3}{y}}$ Apply the division property to write the quotient under a single radical.

$= \sqrt{7y^2}$ Simplify the fraction.

$= y\sqrt{7}$ Simplify the radical.

Examples 1 and 2 show that radical expressions can sometimes be simplified by using the division property of radicals. However, there are cases where other methods are needed. For example:

$\dfrac{2}{\sqrt{2}}$ and $\dfrac{2}{\sqrt{5}+\sqrt{3}}$ are two such cases.

3. Rationalizing the Denominator: One Term

To begin the first case, recall that the nth-root of a perfect nth power is easily simplified. For example:

$$\sqrt{x^2} = x \quad x \geq 0$$

Skill Practice

Simplify the expression by rationalizing the denominator.

5. $\dfrac{3}{\sqrt{5}}$

Example 3 **Rationalizing the Denominator: One Term**

Simplify the expression. $\dfrac{2}{\sqrt{2}}$

Solution:

A square root of a perfect square is needed in the denominator to remove the radical.

$\dfrac{2}{\sqrt{2}} = \dfrac{2}{\sqrt{2}} \cdot \dfrac{\sqrt{2}}{\sqrt{2}}$ Multiply the numerator and denominator by $\sqrt{2}$ because $\sqrt{2} \cdot \sqrt{2} = \sqrt{2^2}$.

$= \dfrac{2\sqrt{2}}{\sqrt{2^2}}$ Multiply the radicals.

Answers

3. 5 **4.** $z^4\sqrt{10}$ **5.** $\dfrac{3\sqrt{5}}{5}$

$$= \frac{2\sqrt{2}}{2} \qquad \text{Simplify.}$$

$$= \frac{2\sqrt{2}}{2} \qquad \text{Reduce the fraction to lowest terms.}$$

$$= \sqrt{2}$$

Example 4 **Rationalizing the Denominator: One Term**

Simplify the expression. Assume x represents a positive real number. $\sqrt{\dfrac{x}{5}}$

Solution:

$$\sqrt{\frac{x}{5}} \qquad \text{The radicand contains an irreducible fraction.}$$

$$= \frac{\sqrt{x}}{\sqrt{5}} \qquad \text{Apply the division property of radicals.}$$

$$= \frac{\sqrt{x}}{\sqrt{5}} \cdot \frac{\sqrt{5}}{\sqrt{5}} \qquad \begin{array}{l}\text{Multiply the numerator and denominator by } \sqrt{5} \\ \text{because } \sqrt{5} \cdot \sqrt{5} = \sqrt{5^2}.\end{array}$$

$$= \frac{\sqrt{5x}}{\sqrt{5^2}} \qquad \text{Multiply the radicals.}$$

$$= \frac{\sqrt{5x}}{5} \qquad \text{Simplify the radicals.}$$

Avoiding Mistakes

In the expression $\dfrac{\sqrt{5x}}{5}$, do not try to "cancel" the factor of $\sqrt{5}$ from the numerator with the factor of 5 in the denominator. $\sqrt{5}$ and 5 are not equal.

Example 5 **Rationalizing the Denominator: One Term**

Simplify the expression. Assume w represents a positive real number.

$$\frac{14\sqrt{w}}{\sqrt{7}}$$

Solution:

$$\frac{14\sqrt{w}}{\sqrt{7}} \qquad \text{Fraction contains a radical in the denominator.}$$

$$= \frac{14\sqrt{w}}{\sqrt{7}} \cdot \frac{\sqrt{7}}{\sqrt{7}} \qquad \begin{array}{l}\text{Multiply the numerator and denominator by } \sqrt{7} \\ \text{because } \sqrt{7} \cdot \sqrt{7} = \sqrt{7^2}.\end{array}$$

$$= \frac{14\sqrt{7w}}{\sqrt{7^2}} \qquad \text{Multiply the radicals.}$$

$$= \frac{14\sqrt{7w}}{7} \qquad \text{Simplify.}$$

$$= \frac{\overset{2}{14}\sqrt{7w}}{\underset{1}{7}} \qquad \text{Reduce to lowest terms.}$$

$$= 2\sqrt{7w}$$

TIP: In the expression

$$\frac{14\sqrt{7w}}{7}$$

the factor of 14 and the factor of 7 may be reduced because both factors are outside the radical.

$$\frac{14\sqrt{7w}}{7} = \frac{14}{7} \cdot \sqrt{7w}$$

$$= 2\sqrt{7w}$$

Skill Practice

Simplify the expression by rationalizing the denominator.

8. $\sqrt{\dfrac{z}{18}}$

Example 6 **Rationalizing the Denominator: One Term**

Simplify the expression. Assume w represents a positive real number.

$$\sqrt{\frac{w}{12}}$$

Solution:

$\sqrt{\dfrac{w}{12}}$ The radical contains an irreducible fraction.

$= \dfrac{\sqrt{w}}{\sqrt{12}}$ Apply the division property of radicals.

$= \dfrac{\sqrt{w}}{\sqrt{2^2 \cdot 3}}$ Factor 12 to simplify the radical.

$= \dfrac{\sqrt{w}}{2\sqrt{3}}$ The $\sqrt{3}$ in the denominator needs to be rationalized.

$= \dfrac{\sqrt{w}}{2\sqrt{3}} \cdot \dfrac{\sqrt{3}}{\sqrt{3}}$ Multiply the numerator and denominator by $\sqrt{3}$ because $\sqrt{3} \cdot \sqrt{3} = \sqrt{3^2}$.

$= \dfrac{\sqrt{3w}}{2\sqrt{3^2}}$ Multiply the radicals.

$= \dfrac{\sqrt{3w}}{2 \cdot 3}$ Simplify.

$= \dfrac{\sqrt{3w}}{6}$ This cannot be simplified further because 3 is inside the radical and 6 is not.

4. Rationalizing the Denominator: Two Terms

Recall from the multiplication of polynomials that the product of two conjugates results in a difference of squares.

$$(a + b)(a - b) = a^2 - b^2$$

If either a or b has a square root factor, the expression will simplify without a radical; that is, the expression is rationalized. For example,

$$(\sqrt{5} - \sqrt{3})(\sqrt{5} + \sqrt{3}) = (\sqrt{5})^2 - (\sqrt{3})^2$$
$$= 5 - 3$$
$$= 2$$

Multiplying a binomial by its conjugate is the basis for rationalizing a denominator with two terms involving square roots.

Answer

8. $\dfrac{\sqrt{2z}}{6}$

Example 7 **Rationalizing the Denominator: Two Terms**

Simplify the expression by rationalizing the denominator.

$$\frac{2}{\sqrt{5} + \sqrt{3}}$$

Solution:

$\dfrac{2}{\sqrt{5} + \sqrt{3}}$ To rationalize a denominator with two terms, multiply the numerator and denominator by the conjugate of the denominator.

$= \dfrac{2}{(\sqrt{5} + \sqrt{3})} \cdot \dfrac{(\sqrt{5} - \sqrt{3})}{(\sqrt{5} - \sqrt{3})}$

Conjugates

The denominator is in the form $(a + b)(a - b)$, where $a = \sqrt{5}$ and $b = \sqrt{3}$.

$= \dfrac{2(\sqrt{5} - \sqrt{3})}{(\sqrt{5})^2 - (\sqrt{3})^2}$ In the denominator, apply the formula $(a + b)(a - b) = a^2 - b^2$.

$= \dfrac{2(\sqrt{5} - \sqrt{3})}{5 - 3}$ Simplify.

$= \dfrac{2(\sqrt{5} - \sqrt{3})}{2}$

$= \dfrac{2(\sqrt{5} - \sqrt{3})}{2}$ Reduce to lowest terms.

$= \sqrt{5} - \sqrt{3}$

Example 8 **Rationalizing the Denominator: Two Terms**

Simplify the expression by rationalizing the denominator.

$$\frac{\sqrt{x} + \sqrt{2}}{\sqrt{x} - \sqrt{2}}$$

Solution:

$\dfrac{\sqrt{x} + \sqrt{2}}{\sqrt{x} - \sqrt{2}} = \dfrac{(\sqrt{x} + \sqrt{2})}{(\sqrt{x} - \sqrt{2})} \cdot \dfrac{(\sqrt{x} + \sqrt{2})}{(\sqrt{x} + \sqrt{2})}$ Multiply the numerator and denominator by the conjugate of the denominator.

Conjugates

$= \dfrac{(\sqrt{x} + \sqrt{2})^2}{(\sqrt{x} - \sqrt{2})(\sqrt{x} + \sqrt{2})}$

$= \dfrac{(\sqrt{x})^2 + 2(\sqrt{x})(\sqrt{2}) + (\sqrt{2})^2}{(\sqrt{x})^2 - (\sqrt{2})^2}$ Simplify using special case products.

$= \dfrac{x + 2\sqrt{2x} + 2}{x - 2}$ Simplify the radicals.

5. Simplifying Quotients That Contain Radicals

Sometimes a radical expression within a quotient must be reduced to lowest terms. This is demonstrated in the next example.

Example 9   Simplifying a Radical Quotient to Lowest Terms

Simplify the expression $\dfrac{4 - \sqrt{20}}{10}$.

Solution:

$$\frac{4 - \sqrt{20}}{10}$$ First simplify $\sqrt{20}$ by writing the radicand as a product of prime factors.

$$= \frac{4 - \sqrt{2^2 \cdot 5}}{10}$$

$$= \frac{4 - 2\sqrt{5}}{10}$$ Simplify the radical.

$$= \frac{2(2 - \sqrt{5})}{2 \cdot 5}$$ Factor out the GCF.

$$= \frac{2(2 - \sqrt{5})}{2 \cdot 5}$$ Reduce to lowest terms.

$$= \frac{2 - \sqrt{5}}{5}$$

Avoiding Mistakes

Remember that it is not correct to reduce *terms* within a rational expression. In the expression

$$\frac{4 - 2\sqrt{5}}{10}$$

do not try to reduce the 4 and the 10. Only common *factors* can be canceled.

Answer

11. $\dfrac{3 - \sqrt{6}}{6}$

Section 8.5   Practice Exercises

Study Skills Exercise

1. Define the key term **rationalizing the denominator**.

Review Exercises

For Exercises 2–10, perform the indicated operations. Assume the variables represent positive real numbers.

2. $x\sqrt{45} + 4\sqrt{20x^2}$ **3.** $(2\sqrt{y} + 3)(3\sqrt{y} + 7)$ **4.** $(4\sqrt{w} - 2)(2\sqrt{w} - 4)$

5. $4\sqrt{3} + \sqrt{5} \cdot \sqrt{15}$ **6.** $\sqrt{7} \cdot \sqrt{21} + 2\sqrt{27}$ **7.** $(5 - \sqrt{a})^2$

8. $(\sqrt{z} + 3)^2$ **9.** $(\sqrt{2} + \sqrt{7})(\sqrt{2} - \sqrt{7})$ **10.** $(\sqrt{3} + 5)(\sqrt{3} - 5)$

Objective 2: Division Property of Radicals

For Exercises 11–30, use the division property of radicals, if necessary, to simplify the expressions. Assume the variables represent positive real numbers. **(See Examples 1–2.)**

11. $\sqrt{\dfrac{3}{16}}$

12. $\sqrt{\dfrac{7}{25}}$

13. $\sqrt{\dfrac{a^4}{b^4}}$

14. $\sqrt{\dfrac{y^6}{z^2}}$

15. $\sqrt{\dfrac{c^3}{4}}$

16. $\sqrt{\dfrac{d^5}{9}}$

17. $\sqrt[3]{\dfrac{x^2}{27}}$

18. $\sqrt[3]{\dfrac{c^2}{8}}$

19. $\sqrt[3]{\dfrac{y^5}{27y^3}}$

20. $\sqrt[3]{\dfrac{7ac}{64c^4}}$

21. $\sqrt{\dfrac{200}{81}}$

22. $\sqrt{\dfrac{80}{49}}$

23. $\dfrac{\sqrt{8}}{\sqrt{50}}$

24. $\dfrac{\sqrt{21}}{\sqrt{12}}$

25. $\dfrac{\sqrt{p}}{\sqrt{4p^3}}$

26. $\dfrac{\sqrt{9t}}{\sqrt{t^5}}$

27. $\dfrac{\sqrt[3]{z^5}}{\sqrt[3]{z^2}}$

28. $\dfrac{\sqrt[3]{a^7}}{\sqrt[3]{a}}$

29. $\dfrac{\sqrt[3]{24x^5}}{\sqrt[3]{3x^4}}$

30. $\dfrac{\sqrt[3]{2y^8}}{\sqrt[3]{54y^7}}$

Objective 3: Rationalizing the Denominator: One Term

For Exercises 31–50, rationalize the denominators. Assume the variable expressions represent positive real numbers. **(See Examples 3–6.)**

31. $\dfrac{1}{\sqrt{6}}$

32. $\dfrac{5}{\sqrt{2}}$

33. $\dfrac{15}{\sqrt{5}}$

34. $\dfrac{14}{\sqrt{7}}$

35. $\dfrac{6}{\sqrt{x+1}}$

36. $\dfrac{8}{\sqrt{y-3}}$

37. $\sqrt{\dfrac{6}{x}}$

38. $\sqrt{\dfrac{8}{y}}$

39. $\sqrt{\dfrac{3}{7}}$

40. $\sqrt{\dfrac{5}{11}}$

41. $\dfrac{10}{\sqrt{6y}}$

42. $\dfrac{15}{\sqrt{3w}}$

43. $\dfrac{9}{2\sqrt{6}}$

44. $\dfrac{15}{4\sqrt{10}}$

45. $\sqrt{\dfrac{p}{27}}$

46. $\sqrt{\dfrac{x}{32}}$

47. $\dfrac{5}{\sqrt{20}}$

48. $\dfrac{8}{\sqrt{24}}$

49. $\sqrt{\dfrac{x^2}{y^3}}$

50. $\sqrt{\dfrac{a}{b^5}}$

Objective 4: Rationalizing the Denominator: Two Terms

For Exercises 51–52, multiply the conjugates.

51. $(\sqrt{2}+3)(\sqrt{2}-3)$

52. $(\sqrt{3}+\sqrt{7})(\sqrt{3}-\sqrt{7})$

53. What is the conjugate of $\sqrt{5}-\sqrt{3}$? Multiply $\sqrt{5}-\sqrt{3}$ by its conjugate.

54. What is the conjugate of $\sqrt{7}+\sqrt{2}$? Multiply $\sqrt{7}+\sqrt{2}$ by its conjugate.

55. What is the conjugate of $\sqrt{x}+10$? Multiply $\sqrt{x}+10$ by its conjugate.

56. What is the conjugate of $12-\sqrt{y}$? Multiply $12-\sqrt{y}$ by its conjugate.

For Exercises 57–68, rationalize the denominators. Assume the variable expressions represent positive real numbers. **(See Examples 7–8.)**

57. $\dfrac{4}{\sqrt{2}+3}$

58. $\dfrac{6}{4-\sqrt{3}}$

59. $\dfrac{1}{\sqrt{5}-\sqrt{2}}$

60. $\dfrac{2}{\sqrt{3}+\sqrt{7}}$

61. $\dfrac{\sqrt{8}}{\sqrt{3}+1}$

62. $\dfrac{\sqrt{18}}{1-\sqrt{2}}$

63. $\dfrac{1}{\sqrt{x}-\sqrt{3}}$

64. $\dfrac{1}{\sqrt{y}+\sqrt{5}}$

65. $\dfrac{2-\sqrt{3}}{2+\sqrt{3}}$

66. $\dfrac{\sqrt{3}-\sqrt{2}}{\sqrt{3}+\sqrt{2}}$

67. $\dfrac{\sqrt{5}+4}{2-\sqrt{5}}$

68. $\dfrac{3+\sqrt{2}}{\sqrt{2}-5}$

Objective 5: Simplifying Quotients That Contain Radicals

For Exercises 69–76, simplify the expression. **(See Example 9.)**

69. $\dfrac{10-\sqrt{50}}{5}$

70. $\dfrac{4+\sqrt{12}}{2}$

71. $\dfrac{21+\sqrt{98}}{14}$

72. $\dfrac{3-\sqrt{18}}{6}$

73. $\dfrac{2-\sqrt{28}}{2}$

74. $\dfrac{5+\sqrt{75}}{5}$

75. $\dfrac{14+\sqrt{72}}{6}$

76. $\dfrac{15-\sqrt{125}}{10}$

Recall that a radical is simplified if

1. The radicand has no factor raised to a power greater than or equal to the index.

2. There are no radicals in the denominator of a fraction.

3. The radicand does not contain a fraction.

For Exercises 77–80, state which condition(s) fails. Then simplify the radical.

77. a. $\sqrt{8x^9}$

b. $\dfrac{5}{\sqrt{5x}}$

c. $\sqrt{\dfrac{1}{3}}$

78. a. $\sqrt{\dfrac{7}{2}}$

b. $\sqrt{18y^6}$

c. $\dfrac{2}{\sqrt{4x}}$

79. a. $\dfrac{3}{\sqrt{x+1}}$

b. $\sqrt{\dfrac{9w^2}{t}}$

c. $\sqrt{24a^5b^9}$

80. a. $\sqrt{\dfrac{12}{z^3}}$

b. $\dfrac{4}{\sqrt{a}-\sqrt{b}}$

c. $\sqrt[3]{27m^3n^7}$

Mixed Exercises

For Exercises 81–96, simplify the radical expressions, if possible. Assume the variables represent positive real numbers.

81. $\sqrt{45}$

82. $-\sqrt{108y^4}$

83. $-\sqrt{\dfrac{18w^2}{25}}$

84. $\sqrt{\dfrac{8a^2}{7}}$

85. $\sqrt{-36}$

86. $\sqrt{54b^5}$

87. $\sqrt{\dfrac{s^2}{t}}$

88. $\dfrac{x+\sqrt{y}}{x-\sqrt{y}}$

89. $\dfrac{\sqrt{2m^5}}{\sqrt{8m}}$

90. $\dfrac{\sqrt{10w}}{\sqrt{5w^3}}$

91. $\sqrt{\dfrac{81}{t^3}}$

92. $-\sqrt{a^3bc^6}$

93. $\dfrac{3}{\sqrt{11}+\sqrt{5}}$

94. $\dfrac{4}{\sqrt{10}+\sqrt{2}}$

95. $\dfrac{\sqrt{a}+\sqrt{b}}{\sqrt{a}-\sqrt{b}}$

96. $\dfrac{\sqrt{x}+1}{\sqrt{x}-1}$

Expanding Your Skills

97. Find the slope of the line through the points: $(5\sqrt{2}, 3)$ and $(\sqrt{2}, 6)$.

98. Find the slope of the line through the points: $(4\sqrt{5}, -1)$ and $(6\sqrt{5}, -5)$.

99. Find the slope of the line through the points: $(\sqrt{3}, -1)$ and $(4\sqrt{3}, 0)$.

100. Find the slope of the line through the points: $(-2\sqrt{7}, -5)$ and $(\sqrt{7}, 2)$.

Problem Recognition Exercises

Operations on Radicals

Perform the indicated operations and simplify if possible. Assume that all variables represent positive real numbers.

1. $(\sqrt{3})(\sqrt{6})$

2. $(\sqrt{2})(\sqrt{14})$

3. $\sqrt{3} + \sqrt{6}$

4. $\sqrt{2} + \sqrt{14}$

5. $\dfrac{\sqrt{6}}{\sqrt{3}}$

6. $\dfrac{\sqrt{14}}{\sqrt{2}}$

7. $(3 + \sqrt{z})(3 - \sqrt{z})$

8. $(4 - \sqrt{y})(4 + \sqrt{y})$

9. $(2\sqrt{5} + 1)(\sqrt{5} - 2)$

10. $(4\sqrt{3} - 5)(\sqrt{3} + 4)$

11. $2\sqrt{x^2 y} - 3x\sqrt{y}$

12. $8\sqrt{a^3 b^2} + 3a\sqrt{ab^2}$

13. $-3\sqrt{2}\,(4\sqrt{2} + 2\sqrt{3} + 1)$

14. $-8\sqrt{5}\,(2\sqrt{5} - \sqrt{3} - 2)$

15. $\dfrac{2}{\sqrt{x} - 7}$

16. $\dfrac{5}{\sqrt{y} + 4}$

17. $\dfrac{9}{\sqrt{3}}$

18. $\dfrac{15}{\sqrt{5}}$

19. $\sqrt{\dfrac{7}{x}}$

20. $\sqrt{\dfrac{11}{y}}$

21. $\sqrt{y^4 z^{11}}$

22. $\sqrt{8q^6}$

23. $\sqrt[3]{27p^8}$

24. $\sqrt[3]{125u^{11}v^{12}}$

25. $\dfrac{\sqrt{10x^3}}{\sqrt{x}}$

26. $\dfrac{\sqrt{15y^3}}{\sqrt{5y}}$

27. $6\sqrt{75} - 5\sqrt{12}$

28. $\sqrt{90} - \sqrt{40}$

29. $(\sqrt{2} + 7)^2$

30. $(\sqrt{3} + \sqrt{5})^2$

31. $\dfrac{x - 5}{\sqrt{x} + \sqrt{5}}$

32. $\dfrac{y - 7}{\sqrt{y} + \sqrt{7}}$

33. $(4\sqrt{x} + \sqrt{y})(\sqrt{x} - 3\sqrt{y})$

34. $\sqrt[4]{\dfrac{1}{81}}$

35. $\sqrt[3]{\dfrac{125}{27}}$

36. $(\sqrt{5} - \sqrt{11})^2$

37. $(\sqrt{x} - 6)^2$

38. $2\sqrt{6} - 5\sqrt{6}$

39. $5\sqrt{a} + 7\sqrt{a} - \sqrt{a}$

40. $(2\sqrt{3} - 10)(2\sqrt{3} + 10)$

41. $(\sqrt{u} - 3\sqrt{v})(\sqrt{u} + 3\sqrt{v})$

42. $x\sqrt{18} + \sqrt{2x^2}$

43. $4\sqrt{75} - 20\sqrt{3}$

44. $\sqrt{5}(\sqrt{5} + \sqrt{7})$

45. $\sqrt{a}(\sqrt{a} + 2)$

46. $(3\sqrt{2} - 4)(5\sqrt{2} + 1)$

Section 8.6 | Radical Equations

1. Solving Radical Equations

> **DEFINITION** Radical Equation
>
> An equation with one or more radicals containing a variable is called a **radical equation**.

For example, $\sqrt{x} = 5$ is a radical equation. Recall that $(\sqrt[n]{a})^n = a$ provided $\sqrt[n]{a}$ is a real number. The basis to solve a radical equation is to eliminate the radical by raising both sides of the equation to a power equal to the index of the radical.

To solve the equation $\sqrt{x} = 5$, square both sides of the equation.

$$\sqrt{x} = 5$$
$$(\sqrt{x})^2 = (5)^2$$
$$x = 25$$

By raising each side of a radical equation to a power equal to the index of the radical, a new equation is produced. However, it is important to note that the new equation may have **extraneous solutions**; that is, some or all of the solutions to the new equation may *not* be solutions to the original radical equation. For this reason, it is necessary to check *all* potential solutions in the original equation. For example, consider the equation $x = 4$. By squaring both sides we produce a quadratic equation.

Square both sides.
$$x = 4$$
$$(x)^2 = (4)^2 \qquad \text{Squaring both sides produces a quadratic equation.}$$
$$x^2 = 16$$
$$x^2 - 16 = 0$$
$$(x - 4)(x + 4) = 0 \qquad \text{Solving this equation, we find two solutions.}$$
$$\qquad\qquad\qquad\qquad\qquad \text{However, } x = -4 \text{ does not check. The value}$$
$$x = 4 \quad \text{or} \quad x \ne -4 \qquad -4 \text{ is an extraneous solution because it is not a}$$
$$\qquad\qquad\qquad\qquad\qquad \text{solution to the original equation, } x = 4.$$

> **PROCEDURE** Solving a Radical Equation
>
> **Step 1** Isolate the radical. If an equation has more than one radical, choose one of the radicals to isolate.
>
> **Step 2** Raise each side of the equation to a power equal to the index of the radical.
>
> **Step 3** Solve the resulting equation.
>
> **Step 4** Check the potential solutions in the original equation.*
>
> *Extraneous solutions can only arise when both sides of the equation are raised to an *even power*. Therefore, an equation with odd-index roots will not have an extraneous solution. However, it is still recommended that you check *all* potential solutions regardless of the type of root.

Example 1 Solving a Radical Equation

Solve the equation. $\sqrt{2x + 1} + 5 = 8$

Solution:

$$\sqrt{2x + 1} + 5 = 8$$

$$\sqrt{2x + 1} = 8 - 5 \qquad \text{Isolate the radical.}$$

$$\sqrt{2x + 1} = 3$$

$$(\sqrt{2x + 1})^2 = (3)^2 \qquad \text{Raise both sides to a power equal to the index of the radical.}$$

$$2x + 1 = 9 \qquad \text{Simplify both sides.}$$

$$2x = 8 \qquad \text{Solve the resulting equation (the equation is linear).}$$

$$x = 4$$

Check: Check 4 as a potential solution.

$$\sqrt{2x + 1} + 5 = 8$$

$$\sqrt{2(4) + 1} + 5 \overset{?}{=} 8$$

$$\sqrt{8 + 1} + 5 \overset{?}{=} 8$$

$$\sqrt{9} + 5 \overset{?}{=} 8$$

$$3 + 5 \overset{?}{=} 8 \ ✔ \qquad \text{The answer checks.}$$

The solution is 4.

Skill Practice

Solve the equation.

1. $\sqrt{p - 4} - 2 = 4$

Example 2 Solving a Radical Equation

Solve the equation. $8 + \sqrt{x + 2} = 7$

Solution:

$$8 + \sqrt{x + 2} = 7$$

$$\sqrt{x + 2} = 7 - 8 \qquad \text{Isolate the radical.}$$

$$\sqrt{x + 2} = -1$$

$$(\sqrt{x + 2})^2 = (-1)^2 \qquad \text{Raise both sides to a power equal to the index of the radical.}$$

$$x + 2 = 1 \qquad \text{Simplify.}$$

$$x = -1 \qquad \text{Solve the resulting equation.}$$

Check: Check -1 as a potential solution.

$$8 + \sqrt{x + 2} = 7$$

$$8 + \sqrt{(-1) + 2} \overset{?}{=} 7$$

$$8 + \sqrt{1} \overset{?}{=} 7$$

$$8 + 1 \neq 7 \qquad \text{The value } -1 \text{ does not check. It is an extraneous solution.}$$

There is no solution to the equation.

Skill Practice

Solve the equation.

2. $\sqrt{2y + 5} + 7 = 4$

TIP: After isolating the radical in Example 2, the equation shows a square root equated to a negative number.

$$\sqrt{x + 2} = -1$$

By definition, a principal square root of any real number must be nonnegative. Therefore, there can be no solution to this equation.

Answers

1. 40

2. No solution (the value 2 does not check.)

73. $(5a^2c^{-1/2}d^{1/2})^2$ **74.** $(2x^{-1/3}y^2z^{5/3})^3$ **75.** $\left(\dfrac{x^{-2/3}}{y^{-3/4}}\right)^{12}$ **76.** $\left(\dfrac{m^{-1/4}}{n^{-1/2}}\right)^{-4}$

77. $\left(\dfrac{16w^{-2}z}{2wz^{-8}}\right)^{1/3}$ **78.** $\left(\dfrac{50p^{-1}q}{2pq^{-3}}\right)^{1/2}$ **79.** $(25x^2y^4z^3)^{1/2}$ **80.** $(8a^6b^3c^2)^{2/3}$

Objective 5: Applications of Rational Exponents

 81. a. If the area, A, of a square is known, then the length of its sides, s, can be computed by the formula: $s = A^{1/2}$. Compute the length of the sides of a square having an area of 100 in.^{2}

  b. Compute the length of the sides of a square having an area of 72 in.^{2} Round your answer to the nearest 0.01 in.

82. The radius, r, of a sphere of volume, V, is given by
$$r = \left(\dfrac{3V}{4\pi}\right)^{1/3}$$

Find the radius of a spherical ball having a volume of 55 in.^{3} Round your answer to the nearest 0.01 in.

For Exercises 83–84, use the following information.

If P dollars in principal grows to A dollars after t years with annual interest, then the rate of return is given by

$$r = \left(\dfrac{A}{P}\right)^{1/t} - 1$$

83. a. In one account, $10,000 grows to $16,802 after 5 years. Compute the interest rate to the nearest tenth of a percent. **(See Example 6.)**

  b. In another account $10,000 grows to $18,000 after 7 years. Compute the interest rate to the nearest tenth of a percent.

  c. Which account produced a higher average yearly return?

84. a. In one account, $5000 grows to $23,304.79 in 20 years. Compute the interest rate to the nearest whole percent.

  b. In another account, $6000 grows to $34,460.95 in 30 years. Compute the interest rate to the nearest whole percent.

  c. Which account produced a higher average yearly return?

Expanding Your Skills

85. Is $(a + b)^{1/2}$ the same as $a^{1/2} + b^{1/2}$? Explain why or why not by giving an example.

For Exercises 86–91, simplify the expressions. Write the final answer with positive exponents only.

86. $\left(\dfrac{1}{8}\right)^{2/3} + \left(\dfrac{1}{4}\right)^{1/2}$ **87.** $\left(\dfrac{1}{8}\right)^{-2/3} + \left(\dfrac{1}{4}\right)^{-1/2}$ **88.** $\left(\dfrac{1}{16}\right)^{-1/4} - \left(\dfrac{1}{49}\right)^{-1/2}$

89. $\left(\dfrac{1}{16}\right)^{1/4} - \left(\dfrac{1}{49}\right)^{1/2}$ **90.** $\left(\dfrac{x^2y^{-1/3}z^{2/3}}{x^{2/3}y^{1/4}z}\right)^{12}$ **91.** $\left(\dfrac{a^2b^{1/2}c^{-2}}{a^{-3/4}b^0c^{1/8}}\right)^8$

Group Activity

Approximating Square Roots

Materials: A calculator

Estimated Time: 15 minutes

Group Size: 2

Suppose that n represents a positive real number that is *not* a perfect square. To approximate $\sqrt{n}$, we will make repeated use of the formula:

$$\sqrt{n} \approx \frac{1}{2}\left(x + \frac{n}{x}\right) \quad \text{where } x \text{ is an approximation of the square root of } n.$$

We will outline the steps to use this formula and demonstrate by approximating $\sqrt{8}$.

Step 1: Begin by letting x be the nonzero whole number that is closest to the square root of n.

Step 1: To approximate $\sqrt{8}$ we begin with $x = 3$, because 3 is equal to $\sqrt{9}$ which is close to $\sqrt{8}$.

Step 2: Substitute the starting value of x and the value of n into the formula.

Step 2: $\sqrt{n} \approx \dfrac{1}{2}\left(x + \dfrac{n}{x}\right)$

$$\sqrt{8} \approx \frac{1}{2}\left(3 + \frac{8}{3}\right)$$

$$\approx 2.83333333333$$

Step 3: Replace x by the answer obtained in step 2. Then apply the formula again, using the new value of x.

Step 3: New value of $x = 2.8\overline{3}$

$$\sqrt{8} \approx \frac{1}{2}\left(2.8\overline{3} + \frac{8}{2.8\overline{3}}\right)$$

$$\approx 2.828431373$$

Step 4: Repeatedly apply step 3, each time using the new value of x from the previous step. You can repeat this process until two consecutive answers differ by less than the desired level of accuracy you want.

TIP: You can check to determine if your answer is reasonable by squaring the result.

$$(2.828431373)^2 \approx 8.000024029$$

1. Use the process outlined to approximate $\sqrt{28}$ accurate to 0.0001.

2. Use the process outlined to approximate $\sqrt{104}$ accurate to 0.00001.

Chapter 8   Summary

Section 8.1   Introduction to Roots and Radicals

Key Concepts

b is a **square root** of a if $b^2 = a$.

The expression $\sqrt{a}$ represents the **principal square root** of a.

b is an nth-root of a if $b^n = a$.

1. If n is a positive *even* integer and $a > 0$, then $\sqrt[n]{a}$ is the principal (positive) nth-root of a.
2. If $n > 1$ is a positive *odd* integer, then $\sqrt[n]{a}$ is the nth-root of a.
3. If $n > 1$ is any positive integer, then $\sqrt[n]{0} = 0$.

$\sqrt[n]{a^n} = |a|$ if n is even.

$\sqrt[n]{a^n} = a$ if n is odd.

$\sqrt[n]{a}$ is not a real number if a is *negative* and n is even.

Pythagorean Theorem

$a^2 + b^2 = c^2$

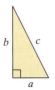

Examples

Example 1

The square roots of 16 are 4 and -4 because $(4)^2 = 16$ and $(-4)^2 = 16$.

$\sqrt{16} = 4$ Because $4^2 = 16$

$\sqrt[4]{16} = 2$ Because $2^4 = 16$

$\sqrt[3]{125} = 5$ Because $5^3 = 125$

Example 2

$\sqrt{y^2} = |y|$ $\sqrt[3]{y^3} = y$ $\sqrt[4]{y^4} = |y|$

Example 3

$\sqrt[4]{-16}$ is not a real number.

Example 4

Find the length of the unknown side.

$a^2 + b^2 = c^2$

$(8)^2 + b^2 = (17)^2$

$64 + b^2 = 289$

$b^2 = 289 - 64$

$b^2 = 225$

$b = \sqrt{225}$ Because b denotes a length,
b must be the positive square
$b = 15$ root of 225.

The third side is 15 cm.

b $c = 17$ cm

$a = 8$ cm

Section 8.2   Simplifying Radicals

Key Concepts

Multiplication Property of Radicals

If $\sqrt[n]{a}$ and $\sqrt[n]{b}$ are both real, then

$$\sqrt[n]{ab} = \sqrt[n]{a} \cdot \sqrt[n]{b}$$

Simplifying Radicals

Consider a radical expression whose radicand is written as a product of prime factors. Then the radical is in simplified form if each of the following criteria are met:

1. The radicand has no factor raised to a power greater than or equal to the index.
2. There are no radicals in the denominator of a fraction.
3. The radicand does not contain a fraction.

Examples

Example 1

$$\sqrt{3} \cdot \sqrt{5} = \sqrt{3 \cdot 5} = \sqrt{15}$$

Example 2

$$\sqrt{\frac{b^7}{b^3}} = \sqrt{b^4} = b^2$$

Example 3

$$\sqrt[3]{16x^5y^7} = \sqrt[3]{2^4x^5y^7}$$
$$= \sqrt[3]{2^3x^3y^6 \cdot 2x^2y}$$
$$= \sqrt[3]{2^3x^3y^6} \cdot \sqrt[3]{2x^2y}$$
$$= 2xy^2\sqrt[3]{2x^2y}$$

Section 8.3   Addition and Subtraction of Radicals

Key Concepts

Two radical terms are *like* radicals if they have the same index and the same radicand.

Use the distributive property to add or subtract *like* radicals.

Examples

Example 1

Like radicals. $\sqrt[3]{5z}, \quad 6\sqrt[3]{5z}$

Example 2

$$3\sqrt{7} - 10\sqrt{7} + \sqrt{7}$$
$$= (3 - 10 + 1)\sqrt{7}$$
$$= -6\sqrt{7}$$

Section 8.4 Multiplication of Radicals

Key Concepts

Multiplication Property of Radicals

$\sqrt[n]{a} \cdot \sqrt[n]{b} = \sqrt[n]{ab}$ provided $\sqrt[n]{a}$ and $\sqrt[n]{b}$ are both real.

Examples

Example 1

$$(6\sqrt{5})(4\sqrt{3}) = 6 \cdot 4\sqrt{5 \cdot 3}$$
$$= 24\sqrt{15}$$

Example 2

$$3\sqrt{2}(\sqrt{2} + 5\sqrt{7} - \sqrt{6}) = 3\sqrt{4} + 15\sqrt{14} - 3\sqrt{12}$$
$$= 3\sqrt{2^2} + 15\sqrt{14} - 3\sqrt{2^2 \cdot 3}$$
$$= 3 \cdot 2 + 15\sqrt{14} - 3 \cdot 2\sqrt{3}$$
$$= 6 + 15\sqrt{14} - 6\sqrt{3}$$

Special Case Products

$(a + b)(a - b) = a^2 - b^2$

$(a + b)^2 = a^2 + 2ab + b^2$

$(a - b)^2 = a^2 - 2ab + b^2$

Example 3

$$(4\sqrt{x} + \sqrt{2})(4\sqrt{x} - \sqrt{2}) = (4\sqrt{x})^2 - (\sqrt{2})^2$$
$$= 16x - 2$$

Example 4

$$(\sqrt{x} - \sqrt{5y})^2 = (\sqrt{x})^2 - 2(\sqrt{x})(\sqrt{5y}) + (\sqrt{5y})^2$$
$$= x - 2\sqrt{5xy} + 5y$$

Section 8.5 Division of Radicals and Rationalization

Key Concepts

Division Property of Radicals

If $\sqrt[n]{a}$ and $\sqrt[n]{b}$ are both real, then

$$\sqrt[n]{\frac{a}{b}} = \frac{\sqrt[n]{a}}{\sqrt[n]{b}} \quad b \neq 0$$

Rationalizing the Denominator with One Term

Multiply the numerator and denominator by an appropriate expression to create an nth-root of an nth-power in the denominator.

Rationalizing a Two-Term Denominator Involving Square Roots

Multiply the numerator and denominator by the conjugate of the denominator.

Examples

Example 1

$$\frac{\sqrt{x^{11}}}{\sqrt{x^3}} = \sqrt{\frac{x^{11}}{x^3}} = \sqrt{x^8} = x^4$$

Example 2

$$\frac{10}{\sqrt{5}} = \frac{10}{\sqrt{5}} \cdot \frac{\sqrt{5}}{\sqrt{5}} = \frac{10\sqrt{5}}{\sqrt{5^2}} = \frac{10\sqrt{5}}{5} = 2\sqrt{5}$$

Example 3

$$\frac{\sqrt{2}}{\sqrt{x} - \sqrt{3}} = \frac{\sqrt{2}}{(\sqrt{x} - \sqrt{3})} \cdot \frac{(\sqrt{x} + \sqrt{3})}{(\sqrt{x} + \sqrt{3})}$$
$$= \frac{\sqrt{2x} + \sqrt{6}}{x - 3}$$

Section 8.6   Radical Equations

Key Concepts

An equation with one or more radicals containing a variable is a **radical equation**.

Steps for Solving a Radical Equation

1. Isolate the radical. If an equation has more than one radical, choose one of the radicals to isolate.
2. Raise each side of the equation to a power equal to the index of the radical.
3. Solve the resulting equation.
4. Check the potential solutions in the original equation.

Note: Raising both sides of an equation to an even power may result in extraneous solutions.

Examples

Example 1

Solve. $\sqrt{2x-4} + 3 = 7$

Step 1: $\qquad \sqrt{2x-4} = 4$

Step 2: $\quad (\sqrt{2x-4})^2 = (4)^2$

Step 3: $\qquad\qquad 2x - 4 = 16$

$\qquad\qquad\qquad 2x = 20$

$\qquad\qquad\qquad\quad x = 10$

Step 4:

Check:

$$\sqrt{2x-4} + 3 = 7$$
$$\sqrt{2(10)-4} + 3 \overset{?}{=} 7$$
$$\sqrt{20-4} + 3 \overset{?}{=} 7$$
$$\sqrt{16} + 3 \overset{?}{=} 7$$
$$4 + 3 \overset{?}{=} 7 \checkmark \qquad \text{The solution checks.}$$

The solution is 10.

Section 8.7   Rational Exponents

Key Concepts

If $\sqrt[n]{a}$ is a real number, then

$$a^{1/n} = \sqrt[n]{a}$$
$$a^{m/n} = (\sqrt[n]{a})^m = \sqrt[n]{a^m}$$

Examples

Example 1

$121^{1/2} = \sqrt{121} = 11$

Example 2

$27^{2/3} = (\sqrt[3]{27})^2 = (3)^2 = 9$

Example 3

$8^{-1/3} = \dfrac{1}{8^{1/3}} = \dfrac{1}{\sqrt[3]{8}} = \dfrac{1}{2}$

Chapter 8   Review Exercises

Section 8.1

For Exercises 1–4, state the principal square root and the negative square root.

1. 196

2. 1.44

3. 0.64

4. 225

5. Explain why $\sqrt{-64}$ is *not* a real number.

6. Explain why $\sqrt[3]{-64}$ *is* a real number.

For Exercises 7–18, simplify the expressions, if possible.

7. $-\sqrt{144}$

8. $-\sqrt{25}$

9. $\sqrt{-144}$

10. $\sqrt{-25}$

11. $\sqrt{y^2}$

12. $\sqrt[3]{y^3}$

13. $\sqrt[4]{y^4}$

14. $-\sqrt[3]{125}$

15. $-\sqrt[4]{625}$

16. $\sqrt[3]{p^{12}}$

17. $\sqrt[4]{\dfrac{81}{t^8}}$

18. $\sqrt[3]{\dfrac{-27}{w^3}}$

 19. The radius, r, of a circle can be found from the area of the circle according to the formula:

$$r = \sqrt{\dfrac{A}{\pi}}$$

a. What is the radius of a circular garden whose area is 160 m^2? Round to the nearest tenth of a meter.

b. What is the radius of a circular fountain whose area is 1600 ft^2? Round to the nearest tenth of a foot.

20. Suppose a ball is thrown with an initial velocity of 76 ft/sec at an angle of 30° (see figure). Then the horizontal position of the ball, x (measured in feet), depends on the number of seconds, t, after the ball is thrown according to the equation:

$$x = 38\sqrt{3}t$$

a. What is the horizontal position of the ball after 1 sec? Round your answer to the nearest tenth of a foot.

b. What is the horizontal position of the ball after 2 sec? Round your answer to the nearest tenth of a foot.

For Exercises 21–22, translate the English phrases into algebraic expressions.

21. The square of b plus the principal square root of 5.

22. The difference of the cube root of y and the fourth root of x.

For Exercises 23–24, translate the algebraic expressions into English phrases. (Answers may vary.)

23. $\dfrac{2}{\sqrt{p}}$

24. $8\sqrt{q}$

25. A hedge extends 5 ft from the wall of a house. A 13-ft ladder is placed at the edge of the hedge. How far up the house is the tip of the ladder?

13 ft   ?

5 ft

26. Nashville, TN, is north of Birmingham, Alabama, a distance of 182 miles. Augusta, Georgia, is east of Birmingham, a distance of 277 miles. How far is it from Augusta to Nashville? Round the answer to the nearest mile.

Nashville

182 miles

Birmingham   277 miles   Augusta

Section 8.2

For Exercises 27–32, use the multiplication property of radicals to simplify. Assume the variables represent positive real numbers.

27. $\sqrt{x^{17}}$ **28.** $\sqrt[3]{40}$ **29.** $\sqrt{28}$

30. $5\sqrt{18x^3}$ **31.** $\sqrt[3]{27y^{10}}$ **32.** $2\sqrt{27y^{10}}$

For Exercises 33–42, use order of operations to simplify. Assume the variables represent positive real numbers.

33. $\sqrt{\dfrac{c^5}{c^3}}$ **34.** $\sqrt{\dfrac{t^9}{t^3}}$

35. $\sqrt{\dfrac{200y^5}{2y}}$ **36.** $\sqrt{\dfrac{18x^3}{2x}}$

37. $\sqrt[3]{\dfrac{48x^4}{6x}}$ **38.** $\sqrt[3]{\dfrac{128a^{17}}{2a^2}}$

39. $\dfrac{5\sqrt{12}}{2}$ **40.** $\dfrac{2\sqrt{45}}{6}$

41. $\dfrac{12 - \sqrt{49}}{5}$ **42.** $\dfrac{20 + \sqrt{100}}{5}$

Section 8.3

For Exercises 43–48, add or subtract as indicated. Assume the variables represent positive real numbers.

43. $8\sqrt{6} - \sqrt{6}$

44. $1.6\sqrt{y} - 1.4\sqrt{y} + 0.6\sqrt{y}$

45. $x\sqrt{20} - 2\sqrt{45x^2}$

46. $y\sqrt{64y} + 3\sqrt{y^3}$

47. $3\sqrt{75} - 4\sqrt{28} + \sqrt{7}$

48. $2\sqrt{50} - 4\sqrt{20} - 6\sqrt{2}$

For Exercises 49–50, translate the English phrases into algebraic expressions and simplify.

49. The sum of the principal square root of the fourth power of x and the square of $5x$.

50. The difference of the principal square root of 128 and the square root of 2.

51. Find the exact perimeter of the triangle.

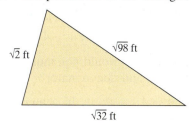

52. Find the exact perimeter of a square whose sides are $3\sqrt{48}$ m.

Section 8.4

For Exercises 53–62, multiply the expressions. Assume the variables represent positive real numbers.

53. $\sqrt{5} \cdot \sqrt{125}$ **54.** $\sqrt{10p} \cdot \sqrt{6}$

55. $(5\sqrt{6})(7\sqrt{2x})$ **56.** $(3\sqrt{y})(-2z\sqrt{11y})$

57. $8\sqrt{m}(\sqrt{m} + 3)$ **58.** $\sqrt{2}(\sqrt{7} + 8)$

59. $(5\sqrt{2} + \sqrt{13})(-\sqrt{2} - 3\sqrt{13})$

60. $(\sqrt{p} + 2\sqrt{q})(4\sqrt{p} - \sqrt{q})$

61. $(8\sqrt{w} - \sqrt{z})(8\sqrt{w} + \sqrt{z})$

62. $(2x - \sqrt{y})^2$

63. Find the exact volume of the box.

Section 8.5

For Exercises 64–67, use the division property of radicals to write the radicals in simplified form. Assume all variables are positive real numbers.

64. $\dfrac{\sqrt[3]{x^7}}{\sqrt[3]{x^4}}$ **65.** $\dfrac{\sqrt{a^{11}}}{\sqrt{a}}$ **66.** $\dfrac{\sqrt{250c}}{\sqrt{10}}$ **67.** $\dfrac{\sqrt{96y^3}}{\sqrt{6y^2}}$

68. To rationalize the denominator in the expression

$$\frac{6}{\sqrt{a} + 5}$$

which quantity would you multiply by in the numerator and denominator?

  a. $\sqrt{a} + 5$ **b.** $\sqrt{a} - 5$ **c.** $\sqrt{a}$ **d.** -5

69. To rationalize the denominator in the expression

$$\frac{w}{\sqrt{w} - 4}$$

which quantity would you multiply by in the numerator and denominator?

  a. $\sqrt{w} - 4$ **b.** $\sqrt{w} + 4$

  c. $\sqrt{w}$ **d.** 4

For Exercises 70–75, rationalize the denominators. Assume the variables represent positive real numbers.

70. $\dfrac{11}{\sqrt{7}}$ **71.** $\sqrt{\dfrac{18}{y}}$ **72.** $\dfrac{\sqrt{24}}{\sqrt{6x^7}}$

73. $\dfrac{10}{\sqrt{7} - \sqrt{2}}$ **74.** $\dfrac{6}{\sqrt{w} + 2}$ **75.** $\dfrac{\sqrt{7} + 3}{\sqrt{7} - 3}$

76. The velocity of an object, v (in meters per second: m/sec) depends on the kinetic energy, E (in joules: J), and mass, m (in kilograms: kg), of the object according to the formula:

$$v = \sqrt{\frac{2E}{m}}$$

  a. What is the exact velocity of a 3-kg object whose kinetic energy is 100 J?

  b. What is the exact velocity of a 5-kg object whose kinetic energy is 162 J?

Section 8.6

For Exercises 77–85, solve the equations. Be sure to check the potential solutions.

77. $\sqrt{p + 6} = 12$ **78.** $\sqrt{k + 1} = -7$

79. $\sqrt{3x - 17} - 10 = 0$

80. $\sqrt{14n + 10} = 4\sqrt{n}$

81. $\sqrt{2z + 2} = \sqrt{3z - 5}$

82. $\sqrt{5y - 5} - \sqrt{4y + 1} = 0$

83. $\sqrt{2m + 5} = m + 1$

84. $\sqrt{3n - 8} - n + 2 = 0$

85. $\sqrt[3]{2y + 13} = -5$

 86. The length of the sides of a cube is related to the volume of the cube according to the formula: $x = \sqrt[3]{V}$.

  a. What is the volume of the cube if the side length is 21 in.?

  b. What is the volume of the cube if the side length is 15 cm?

Section 8.7

For Exercises 87–92, simplify the expressions.

87. $(-27)^{1/3}$ **88.** $121^{1/2}$ **89.** $-16^{1/4}$

90. $(-16)^{1/4}$ **91.** $4^{-3/2}$ **92.** $\left(\dfrac{1}{9}\right)^{-3/2}$

For Exercises 93–96, write the expression in radical notation. Assume the variables represent positive real numbers.

93. $z^{1/5}$ **94.** $q^{2/3}$

95. $(w^3)^{1/4}$ **96.** $\left(\dfrac{b}{121}\right)^{1/2}$

For Exercises 97–100, write the expression using rational exponents rather than radical notation. Assume the variables represent positive real numbers.

97. $\sqrt[5]{a^2}$ **98.** $5\sqrt[3]{m^2}$

99. $\sqrt[5]{a^2b^4}$ **100.** $\sqrt{6}$

For Exercises 101–106, simplify using the properties of rational exponents. Write the answer with positive exponents only. Assume the variables represent positive real numbers.

101. $y^{2/3}y^{4/3}$ **102.** $a^{1/3}a^{1/2}$

103. $\dfrac{6^{4/5}}{6^{1/5}}$ **104.** $\left(\dfrac{b^4b^0}{b^{1/4}}\right)^4$

105. $(64a^3b^6)^{1/3}$ **106.** $(5^{1/2})^{3/2}$

107. The radius, r, of a right circular cylinder can be found if the volume, V, and height, h, are known. The radius is given by

$$r = \left(\frac{V}{\pi h}\right)^{1/2}$$

Find the radius of a right circular cylinder whose volume is 150.8 cm³ and whose height is 12 cm. Round the answer to the nearest tenth of a centimeter.

Chapter 8   Test

1. State the conditions for a radical expression to be in simplified form.

For Exercises 2–7, simplify the radicals, if possible. Assume the variables represent positive real numbers.

2. $\sqrt{242x^2}$ 3. $\sqrt[3]{48y^4}$ 4. $\sqrt{-64}$

5. $\sqrt{\dfrac{5a^6}{81}}$ 6. $\dfrac{9}{\sqrt{6}}$ 7. $\dfrac{2}{\sqrt{5}+6}$

8. Translate the English phrases into algebraic expressions and simplify.

   a. The sum of the principal square root of twenty-five and the cube of five.

   b. The difference of the square of four and the principal square root of 16.

9. A baseball player hits the ball at an angle of 30° with an initial velocity of 112 ft/sec. The horizontal position of the ball, x (measured in feet), depends on the number of seconds, t, after the ball is struck according to the equation

$$x = 56\sqrt{3}t$$

   a. What is the horizontal position of the ball after 1 sec? Round the answer to the nearest foot.

   b. What is the horizontal position of the ball after 3.5 sec? Round the answer to the nearest foot.

For Exercises 10–19, perform the indicated operations. Assume the variables represent positive real numbers.

10. $6\sqrt{z} - 3\sqrt{z} + 5\sqrt{z}$

11. $\sqrt{3}(4\sqrt{2} - 5\sqrt{3})$

12. $\sqrt{50t^2} - t\sqrt{288}$

13. $\sqrt{360} + \sqrt{250} - \sqrt{40}$

14. $(3\sqrt{5} - 1)^2$

15. $(6\sqrt{2} - \sqrt{5})(\sqrt{2} + 4\sqrt{5})$ 16. $\dfrac{\sqrt{2m^3n}}{\sqrt{72m^5}}$

17. $(4 - 3\sqrt{x})(4 + 3\sqrt{x})$ 18. $\sqrt{\dfrac{2}{11}}$

19. $\dfrac{6}{\sqrt{7} - \sqrt{3}}$

20. A triathlon consists of a swim, followed by a bike ride, followed by a run. The swim begins on a beach at point A. The swimmers must swim 50 yd to a buoy at point B, then 200 yd to a buoy at point C, and then return to point A on the beach. How far is the distance from point C to point A? (Round to the nearest yard.)

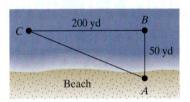

For Exercises 21–23, solve the equations.

21. $\sqrt{2x + 7} + 6 = 2$

22. $\sqrt{1 - 7x} = 1 - x$

23. $\sqrt[3]{x + 6} = \sqrt[3]{2x - 8}$

24. The height, y (in inches), of a tomato plant can be approximated by the time, t (in weeks), after the seed has germinated according to the equation

$$y = 6\sqrt{t}$$

   a. Use the equation to find the height of the plant after 4 weeks.

   b. Use the equation to find the height of the plant after 9 weeks.

Example 3 Solving a Quadratic Equation Using the Quadratic Formula

Solve the quadratic equation using the quadratic formula.

$$\frac{1}{4}w^2 - \frac{1}{2}w - \frac{5}{4} = 0$$

Solution:

$$\frac{1}{4}w^2 - \frac{1}{2}w - \frac{5}{4} = 0$$ To simplify the equation, multiply both sides by 4.

$$4\left(\frac{1}{4}w^2 - \frac{1}{2}w - \frac{5}{4}\right) = 4(0)$$ Clear fractions.

$$w^2 - 2w - 5 = 0$$ The equation is in the form $ax^2 + bx + c = 0$.

$$a = 1, b = -2, c = -5$$ Identify a, b, and c.

$$w = \frac{-b \pm \sqrt{b^2 - 4ac}}{2a}$$

$$w = \frac{-(-2) \pm \sqrt{(-2)^2 - 4(1)(-5)}}{2(1)}$$ Apply the quadratic formula.

$$= \frac{2 \pm \sqrt{4 + 20}}{2}$$ Simplify.

$$= \frac{2 \pm \sqrt{24}}{2}$$

$$= \frac{2 \pm 2\sqrt{6}}{2}$$ The solutions are irrational numbers.

$$= \frac{2(1 \pm \sqrt{6})}{2}$$ Factor and simplify.

$$= 1 \pm \sqrt{6}$$

The solutions are $1 + \sqrt{6}$ and $1 - \sqrt{6}$.

3. Review of the Methods for Solving a Quadratic Equation

Three methods have been presented for solving quadratic equations.

> **PROCEDURE** Methods for Solving a Quadratic Equation
> - Factor and use the zero product rule (Section 6.7).
> - Use the square root property. Complete the square if necessary (Sections 9.1 and 9.2).
> - Use the quadratic formula (Section 9.3).

Using the zero product rule only works if you can factor the equation. The square root property and the quadratic formula can be used to solve any quadratic equation. Before solving a quadratic equation, take a minute to analyze it. Each problem must be evaluated individually before choosing the most efficient method to find its solutions.

Example 4 **Solving Quadratic Equations Using Any Method**

Solve the quadratic equations using any method.

a. $(x + 1)^2 = 5$ **b.** $t^2 - t - 30 = 0$ **c.** $2x^2 + 5x + 1 = 0$

Solution:

a. $(x + 1)^2 = 5$ Because the equation is the square of a binomial equal to a constant, the square root property can be applied easily.

$x + 1 = \pm\sqrt{5}$ Apply the square root property.

$x = -1 \pm \sqrt{5}$ Isolate x.

The solutions are $-1 + \sqrt{5}$ and $-1 - \sqrt{5}$.

b. $t^2 - t - 30 = 0$ This equation factors.

$(t - 6)(t + 5) = 0$ Factor and apply the zero product rule.

$t = 6$ or $t = -5$

The solutions are 6 and -5.

c. $2x^2 + 5x + 1 = 0$ The equation does not factor. Because it is already in the form $ax^2 + bx + c = 0$, use the quadratic formula.

$a = 2, b = 5, c = 1$ Identify a, b, and c.

$x = \dfrac{-(5) \pm \sqrt{(5)^2 - 4(2)(1)}}{2(2)}$ Apply the quadratic formula.

$x = \dfrac{-5 \pm \sqrt{25 - 8}}{4}$ Simplify.

$x = \dfrac{-5 \pm \sqrt{17}}{4}$

The solutions are $\dfrac{-5 + \sqrt{17}}{4}$ and $\dfrac{-5 - \sqrt{17}}{4}$.

Skill Practice

Solve the equations using any method.

7. $p^2 + 7p + 12 = 0$

8. $5y^2 + 7y + 1 = 0$

9. $(w - 8)^2 = 3$

Answers

7. $-3, -4$

8. $\dfrac{-7 \pm \sqrt{29}}{10}$

9. $8 \pm \sqrt{3}$

4. Applications of Quadratic Equations

Skill Practice

10. The length of a rectangle is 2 in. longer than the width. The area is 10 in.² Find the exact values of the length and width. Then use a calculator to approximate the dimensions to the nearest tenth of an inch.

Example 5 Solving a Quadratic Equation in an Application

The length of a box is 2 in. longer than the width. The height of the box is 4 in. and the volume of the box is 200 in.³ Find the exact dimensions of the box. Then use a calculator to approximate the dimensions to the nearest tenth of an inch.

Solution:

Label the box as follows (Figure 9-1):

Width $= x$

Length $= x + 2$

Height $= 4$

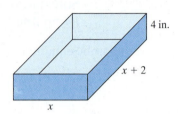

4 in.

$x + 2$

x

Figure 9-1

The volume of a box is given by the formula: $V = lwh$

$$V = l \cdot w \cdot h$$

$$200 = (x + 2)(x)(4)$$

Substitute $V = 200$, $l = x + 2$, $w = x$, and $h = 4$.

$$200 = (x + 2)4x$$

$$200 = 4x^2 + 8x$$

$$0 = 4x^2 + 8x - 200$$

$$4x^2 + 8x - 200 = 0$$

The equation is in the form $ax^2 + bx + c = 0$.

$$\frac{4x^2}{4} + \frac{8x}{4} - \frac{200}{4} = \frac{0}{4}$$

The coefficients are all divisible by 4. Dividing by 4 will create smaller values of a, b, and c to be used in the quadratic formula.

$$x^2 + 2x - 50 = 0$$

$a = 1$, $b = 2$, $c = -50$

$$x = \frac{-2 \pm \sqrt{(2)^2 - 4(1)(-50)}}{2(1)}$$

Apply the quadratic formula.

$$= \frac{-2 \pm \sqrt{4 + 200}}{2}$$

Simplify.

$$= \frac{-2 \pm \sqrt{204}}{2}$$

$$= \frac{-2 \pm 2\sqrt{51}}{2}$$

Simplify the radical. $\sqrt{204} = \sqrt{2^2 \cdot 51} = 2\sqrt{51}$

$$= \frac{\overset{1}{2}(-1 \pm \sqrt{51})}{\underset{1}{2}}$$

Factor and simplify.

$$= -1 \pm \sqrt{51}$$

Avoiding Mistakes

We do not use the solution $x = -1 - \sqrt{51}$ because it is a negative number, that is,

$-1 - \sqrt{51} \approx -8.1$

The width of an object cannot be negative.

Because the width of the box must be positive, use $x = -1 + \sqrt{51}$.

The width is $(-1 + \sqrt{51})$ in. ≈ 6.1 in.

The length is $x + 2$: $(-1 + \sqrt{51} + 2)$ in. or $(1 + \sqrt{51})$ in. ≈ 8.1 in.

The height is 4 in.

Answer

10. The width is $(-1 + \sqrt{11})$ in. or approximately 2.3 in. The length is $(1 + \sqrt{11})$ in. or approximately 4.3 in.

Calculator Connections

Use the quadratic formula to verify that the solutions to the equation $x(x + 7) + 4 = 0$ are

$$x = \frac{-7 + \sqrt{33}}{2} \quad \text{and} \quad x = \frac{-7 - \sqrt{33}}{2}$$

A calculator can be used to obtain decimal approximations for the irrational solutions of a quadratic equation.

Scientific Calculator

Enter: 7 +/- + 33 √ = ÷ 2 =   **Result:** -0.6277186767

Enter: 7 +/- − 33 √ = ÷ 2 =   **Result:** -6.372281323

Graphing Calculator

```
(-7+√(33))/2
          -.6277186767
(-7-√(33))/2
          -6.372281323
```

Calculator Exercises

Use a calculator to obtain a decimal approximation of each expression.

1. $\dfrac{-5 + \sqrt{17}}{4}$ and $\dfrac{-5 - \sqrt{17}}{4}$

2. $\dfrac{-40 + \sqrt{1920}}{-32}$ and $\dfrac{-40 - \sqrt{1920}}{-32}$

Section 9.3 Practice Exercises

Review Exercises

For Exercises 1–4, apply the square root property to solve the equation.

1. $z^2 = 169$ **2.** $p^2 = 1$ **3.** $(x - 4)^2 = 28$ **4.** $(y + 3)^2 = 7$

For Exercises 5–6, solve the equations by completing the square.

5. $3a^2 - 12a - 12 = 0$ **6.** $x^2 - 5x + 1 = 0$

Objective 1: Derivation of the Quadratic Formula

7. State the quadratic formula from memory.

8. Can all quadratic equations be solved by using the quadratic formula?

For Exercises 9–14, write the equations in the form $ax^2 + bx + c = 0$. Then identify the values of a, b, and c.

9. $2x^2 - x = 5$ **10.** $5(x^2 + 2) = -3x$ **11.** $-3x(x - 4) = -2x$

12. $x(x - 2) = 3(x + 1)$ **13.** $x^2 - 9 = 0$ **14.** $x^2 + 25 = 0$

Objective 2: Solving Quadratic Equations Using the Quadratic Formula

For Exercises 15–32, solve the equations using the quadratic formula. **(See Examples 1–3.)**

15. $t^2 + 16t + 64 = 0$

16. $y^2 - 10y + 25 = 0$

17. $6k^2 - k - 2 = 0$

18. $3n^2 + 5n - 2 = 0$

19. $5t^2 - t = 3$

20. $2a^2 + 5a = 1$

21. $x(x - 2) = 1$

22. $2y(y - 3) = -1$

23. $2p^2 = -10p - 11$

24. $z^2 = 4z + 1$

25. $-4y^2 - y + 1 = 0$

26. $-5z^2 - 3z + 4 = 0$

27. $2x(x + 1) = 3 - x$

28. $3m(m - 2) = -m + 1$

29. $0.2y^2 = -1.5y - 1$

30. $0.2t^2 = t + 0.5$

31. $\dfrac{2}{3}x^2 + \dfrac{4}{9}x = \dfrac{1}{3}$

32. $\dfrac{1}{2}x^2 + \dfrac{1}{6}x = 1$

Objective 3: Review of the Methods for Solving a Quadratic Equation

For Exercises 33–56, choose any method to solve the quadratic equations. **(See Example 4.)**

33. $16x^2 - 9 = 0$

34. $\dfrac{1}{4}x^2 + 5x + 13 = 0$

35. $(x - 5)^2 = -21$

36. $2x^2 + x + 5 = 0$

37. $\dfrac{1}{9}x^2 + \dfrac{8}{3}x + 11 = 0$

38. $7x^2 = 12x$

39. $2x^2 - 6x - 3 = 0$

40. $4(x + 1)^2 = -15$

41. $9x^2 = 11x$

42. $25x^2 - 4 = 0$

43. $(2y - 3)^2 = 5$

44. $(6z + 1)^2 = 7$

45. $0.4x^2 = 0.2x + 1$

46. $0.6x^2 = 0.1x + 0.8$

47. $9z^2 - z = 0$

48. $16p^2 - p = 0$

49. $r^2 - 52 = 0$

50. $y^2 - 32 = 0$

51. $-2.5t(t - 4) = 1.5$

52. $1.6p(p - 2) = 0.8$

53. $(m - 3)(m + 2) = 9$

54. $(h - 6)(h - 1) = 12$

55. $x^2 + x + 3 = 0$

56. $3x^2 - 20x + 12 = 0$

Objective 4: Applications of Quadratic Equations

57. In a rectangle, the length is 1 m less than twice the width and the area is 100 m². Approximate the dimensions to the nearest tenth of a meter. **(See Example 5.)**

58. In a triangle, the height is 2 cm more than the base. The area is 72 cm². Approximate the base and height to the nearest tenth of a centimeter.

59. The volume of a rectangular storage area is 240 ft³. The length is 2 ft more than the width. The height is 6 ft. Approximate the dimensions to the nearest tenth of a foot.

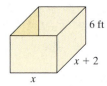

60. In a right triangle, one leg is 2 ft shorter than the other leg. The hypotenuse is 12 ft. Approximate the lengths of the legs to the nearest tenth of a foot.

61. In a rectangle, the length is 4 ft longer than the width. The area is 72 ft². Approximate the dimensions to the nearest tenth of a foot.

62. In a triangle, the base is 4 cm less than twice the height. The area is 60 cm². Approximate the base and height to the nearest tenth of a centimeter.

63. In a right triangle, one leg is 3 m longer than the other leg. The hypotenuse is 13 m. Approximate the lengths of the legs to the nearest tenth of a meter.

Problem Recognition Exercises

Solving Quadratic Equations

For Exercises 1–2, solve the equations using each of the three methods.

   a. Factoring and applying the zero product rule

   b. Completing the square and applying the square root property

   c. Applying the quadratic formula

1. $6x^2 + 7x - 3 = 0$

2. $y^2 + 14y + 49 = 0$

For Exercises 3–22, solve using any method.

3. $x(x - 8) = 6$

4. $2 - 6y = -y^2$

5. $(t - 3)^2 - 8 = 0$

6. $(p + 11)^2 - 12 = 0$

7. $6m^2 = 5m$

8. $12n^2 = 6n$

9. $\dfrac{1}{36}w^2 - 1 = 0$

10. $\dfrac{25}{49}y^2 - 1 = 0$

11. $8x^2 - 22x + 5 = 0$

12. $9w^2 - 15w + 4 = 0$

13. $p^2 + 20 = 0$

14. $m^2 + 15 = 0$

15. $y(y + 8) = -1$

16. $t(t + 6) = -3$

17. $\dfrac{1}{2}a^2 - 4a + 8 = 0$

18. $\dfrac{1}{3}b^2 + 2b + 3 = 0$

19. $2x^2 - 14x = -20$

20. $3n^2 - 6n = 9$

21. $(w + 1)^2 = 100$

22. $(u - 5)^2 = 64$

Skill Practice

3. Given $y = -x^2 - 4x$, perform parts (a)–(g), as in Example 1.

Example 1 **Analyzing a Quadratic Function**

Given the function defined by $y = -x^2 + 4x - 3$,

   a. Determine whether the parabola opens upward or downward.

   b. Find the vertex of the parabola.

   c. Find the x-intercept(s).

   d. Find the y-intercept.

   e. Sketch the parabola.

   f. Write the domain of the function in interval notation.

   g. Write the range of the function in interval notation.

Solution:

   a. The function $y = -x^2 + 4x - 3$ is written in the form $y = ax^2 + bx + c$, where $a = -1, b = 4$, and $c = -3$. Because the value of a is negative, the parabola opens *downward*.

   b. The x-coordinate of the vertex is given by $x = \dfrac{-b}{2a}$.

$$x = \frac{-b}{2a} = \frac{-(4)}{2(-1)}$$ Substitute $b = 4$ and $a = -1$.

$$= \frac{-4}{-2}$$ Simplify.

$$= 2$$

The y-coordinate of the vertex is found by substituting $x = 2$ into the equation and solving for y.

$$y = -x^2 + 4x - 3$$

$$= -(2)^2 + 4(2) - 3$$ Substitute $x = 2$.

$$= -4 + 8 - 3$$

$$= 1$$

The vertex is $(2, 1)$.   Because the parabola opens downward, the vertex is the maximum point on the graph of the parabola.

   c. To find the x-intercept(s), substitute $y = 0$ and solve for x.

$$y = -x^2 + 4x - 3$$

$$0 = -x^2 + 4x - 3$$ Substitute $y = 0$. The resulting equation is quadratic.

$$0 = -1(x^2 - 4x + 3)$$ Factor out -1.

$$0 = -1(x - 3)(x - 1)$$ Factor the trinomial.

Answers

3. **a.** downward
  b. $(-2, 4)$
  c. $(0, 0)$ and $(-4, 0)$
  d. $(0, 0)$
  e.

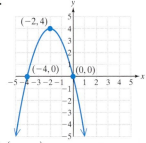

  f. $(-\infty, \infty)$
  g. $(-\infty, 4]$

$x - 3 = 0$ or $x - 1 = 0$ Apply the zero product rule.

$x = 3$ or $x = 1$

The x-intercepts are $(3, 0)$ and $(1, 0)$.

d. To find the y-intercept, substitute $x = 0$ and solve for y.

$$y = -x^2 + 4x - 3$$
$$= -(0)^2 + 4(0) - 3 \qquad \text{Substitute } x = 0.$$
$$= -3$$

The y-intercept is $(0, -3)$.

e. Using the results of parts (a)–(d), we have a parabola that opens downward with vertex $(2, 1)$, x-intercepts at $(3, 0)$ and $(1, 0)$, and y-intercept at $(0, -3)$ (Figure 9-3).

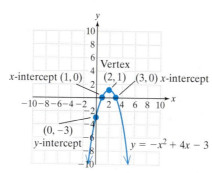

Figure 9-3

f. Because any real number, x, when substituted into the equation $y = -x^2 + 4x - 3$ produces a real number, the domain is all real numbers.

Domain: $(-\infty, \infty)$

g. From the graph, we see that the vertex is the maximum point on the graph. Therefore, the maximum y-value is $y = 1$. The range is restricted to $y \leq 1$.

Range: $(-\infty, 1]$

3. Graphing a Parabola

Example 1 illustrates a process to sketch a quadratic function by finding the location of the defining characteristics of the function. These include the vertex and the x- and y-intercepts. Furthermore, notice that the parabola defining the graph of a quadratic function is symmetric with respect to the axis of symmetry.

To analyze the graph of a parabola, we recommend the following guidelines.

4. The vertex is $(0, -4)$. This is also the y-intercept.

5. Sketch the parabola through the y-intercept and vertex (Figure 9-5).

To verify the proper shape of the graph, find additional points to the right or left of the vertex and use the symmetry of the parabola to sketch the curve.

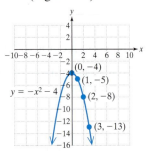

x	y
1	-5
2	-8
3	-13

Figure 9-5

4. Applications of Quadratic Functions

Skill Practice

6. A basketball player shoots a basketball at an angle of 45°. The height of the ball y (in feet) is given by $y = -16t^2 + 40t + 6$ where t is time in seconds. Find the maximum height of the ball and the time required to reach that height.

Example 4 Using a Quadratic Function in an Application

A golfer hits a ball at an angle of 30°. The height of the ball y (in feet) can be represented by

$$y = -16x^2 + 60x$$ where x is the time in seconds after the ball was hit (Figure 9-6).

Find the maximum height of the ball. In how many seconds will the ball reach its maximum height?

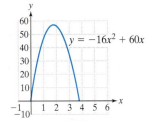

Figure 9-6

Solution:

The function is written in the form $y = ax^2 + bx + c$, where $a = -16$, $b = 60$, and $c = 0$. Because a is negative, the parabola opens downward. Therefore, the maximum height of the ball occurs at the vertex of the parabola.

The x-coordinate of the vertex is given by

$$x = \frac{-b}{2a} = \frac{-(60)}{2(-16)} = \frac{-60}{-32} = \frac{15}{8}$$

Substituting $x = \frac{15}{8}$ into the equation, we have

$$y = -16\left(\frac{15}{8}\right)^2 + 60\left(\frac{15}{8}\right)$$

$$= -16\left(\frac{225}{64}\right) + \frac{900}{8}$$

$$= -\frac{225}{4} + \frac{450}{4}$$

$$= \frac{225}{4}$$

The vertex is $\left(\frac{15}{8}, \frac{225}{4}\right)$ or equivalently $(1.875, 56.25)$.

The ball reaches its maximum height of 56.25 ft after 1.875 sec.

Answer

6. The ball reaches a maximum height of 31 ft in 1.25 sec.

Calculator Connections

Some graphing calculators have *Minimum* and *Maximum* features that enable the user to approximate the minimum and maximum values of a function. Otherwise, *Zoom* and *Trace* can be used.

For example, the maximum value of the function from Example 4, $y = -16x^2 + 60x$, can be found using the *Maximum* feature.

The minimum value of the function from Example 2, $y = x^2 - 6x + 9$, can be found using the *Minimum* feature.

Calculator Exercises

Without using a calculator, find the vertex of each parabola. Then verify the answer using a graphing calculator.

1. $y = x^2 + 4x + 7$

2. $y = x^2 - 20x + 105$

3. $y = -x^2 - 3x - 4.85$

4. $y = -x^2 + 3.5x - 0.5625$

5. $y = 2x^2 - 10x + \dfrac{25}{2}$

6. $y = 3x^2 + 16x + \dfrac{64}{3}$

7. Use a graphing calculator to graph the functions on the same screen. Describe the relationship between the graph of $y = x^2$ and the graphs in parts (b) and (c).

a. $y = x^2$ **b.** $y = x^2 + 4$ **c.** $y = x^2 - 3$

8. Use a graphing calculator to graph the functions on the same screen. Describe the relationship between the graph of $y = x^2$ and the graphs in parts (b) and (c).

a. $y = x^2$ **b.** $y = (x - 3)^2$ **c.** $y = (x + 2)^2$

9. Use a graphing calculator to graph the functions on the same screen. Describe the relationship between the graph of $y = x^2$ and the graphs in parts (b) and (c).

a. $y = x^2$ **b.** $y = 2x^2$ **c.** $y = \frac{1}{2}x^2$

10. Use a graphing calculator to graph the functions on the same screen. Describe the relationship between the graph of $y = x^2$ and the graphs in parts (b) and (c).

a. $y = x^2$ **b.** $y = -2x^2$ **c.** $y = -\frac{1}{2}x^2$

Section 9.4 Practice Exercises

Study Skills Exercise

1. Define the key terms.

    a. quadratic function   **b. parabola**   **c. vertex of a parabola**   **d. axis of symmetry**

Review Exercises

For Exercises 2–8, solve the quadratic equations using any one of the following methods: factoring, the square root property, or the quadratic formula.

2. $3(y^2 + 1) = 10y$ **3.** $3 + a(a + 2) = 18$ **4.** $4t^2 - 7 = 0$ **5.** $2z^2 + 4z - 10 = 0$

6. $(b + 1)^2 = 6$ **7.** $(x - 2)^2 = 8$ **8.** $3p^2 - 12p - 12 = 0$

Objective 1: Definition of a Quadratic Function

For Exercises 9–20, identify the equations as linear, quadratic, or neither.

9. $y = -8x + 3$ **10.** $y = 5x - 12$ **11.** $y = 4x^2 - 8x + 22$ **12.** $y = x^2 + 10x - 3$

13. $y = -5x^3 - 8x + 14$ **14.** $y = -3x^4 + 7x - 11$ **15.** $y = 15x$ **16.** $y = -9x$

17. $y = -21x^2$ **18.** $y = 3x^2$ **19.** $y = -x^3 + 1$ **20.** $y = 7x^4 - 4$

Objective 2: Vertex of a Parabola

21. How do you determine whether the graph of a function $y = ax^2 + bx + c\ (a \neq 0)$ opens upward or downward?

For Exercises 22–25, identify a and determine if the parabola opens upward or downward. **(See Example 1.)**

22. $y = x^2 - 15$ **23.** $y = 2x^2 + 23$ **24.** $y = -3x^2 + x - 18$ ⊙**25.** $y = -10x^2 - 6x - 20$

26. How do you find the vertex of a parabola?

For Exercises 27–34, find the vertex of the parabola. **(See Example 1.)**

⊙**27.** $y = 2x^2 + 4x - 6$ **28.** $y = x^2 - 4x - 4$ **29.** $y = -x^2 + 2x - 5$ **30.** $y = 2x^2 - 4x - 6$

31. $y = x^2 - 2x + 3$ **32.** $y = -x^2 + 4x - 2$ **33.** $y = 3x^2 - 4$ **34.** $y = 4x^2 - 1$

Objective 3: Graphing a Parabola

For Exercises 35–38, find the *x*- and *y*-intercepts of the function. Then match each function with a graph. **(See Example 1.)**

35. $y = x^2 - 7$ **36.** $y = x^2 - 9$ **37.** $y = (x + 3)^2 - 4$ **38.** $y = (x - 2)^2 - 1$

a.

b.

c.

d.
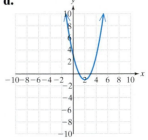

For Exercises 39–50, **(See Examples 1–3.)**

    a. Determine whether the graph of the parabola opens upward or downward.

    b. Find the vertex.

    c. Find the *x*-intercept(s), if possible.

    d. Find the *y*-intercept.

    e. Sketch the function.

    f. Identify the domain of the function.

    g. Identify the range of the function.

39. $y = x^2 - 9$

40. $y = x^2 - 4$

41. $y = x^2 - 2x - 8$

42. $y = x^2 + 2x - 24$

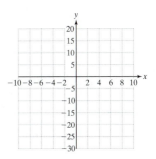

43. $y = -x^2 + 6x - 9$

44. $y = -x^2 + 10x - 25$

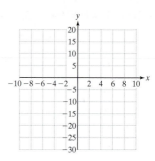

45. $y = -x^2 + 8x - 15$

46. $y = -x^2 - 4x + 5$

47. $y = x^2 + 6x + 10$

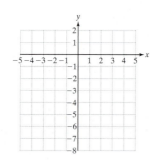

48. $y = x^2 + 4x + 5$

49. $y = -2x^2 - 2$

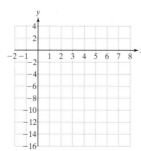

50. $y = -x^2 - 5$

51. True or False: The function $y = -5x^2$ has a maximum value but no minimum value.

52. True or False: The graph of $y = -4x^2 + 9x - 6$ opens upward.

53. True or False: The graph of $y = 1.5x^2 - 6x - 3$ opens downward.

54. True or False: The function $y = 2x^2 - 5x + 4$ has a maximum value but no minimum value.

Objective 4: Applications of Quadratic Functions

55. A child kicks a ball into the air, and the height of the ball, y (in feet), can be approximated by

$$y = -16t^2 + 40t + 3 \qquad \text{where } t \text{ is the number of seconds after the ball was kicked.}$$

a. Find the maximum height of the ball. **(See Example 4.)**

b. How long will it take the ball to reach its maximum height?

56. A concession stand at the Arthur Ashe Tennis Center sells a hamburger/drink combination dinner for $5. The profit, y (in dollars), can be approximated by

$$y = -0.001x^2 + 3.6x - 400 \qquad \text{where } x \text{ is the number of dinners prepared.}$$

a. Find the number of dinners that should be prepared to maximize profit.

b. What is the maximum profit?

57. For a fund raising activity, a charitable organization produces calendars to sell in the community. The profit, y (in dollars), can be approximated by

$$y = -\frac{1}{40}x^2 + 10x - 500 \qquad \text{where } x \text{ is the number of calendars produced.}$$

a. Find the number of calendars that should be produced to maximize the profit.

b. What is the maximum profit?

58. The pressure, x, in an automobile tire can affect its wear. Both over-inflated and under-inflated tires can lead to poor performance and poor mileage. For one particular tire, the number of miles that a tire lasts, y (in thousands), is given by

$$y = -0.875x^2 + 57.25x - 900 \qquad \text{where } x \text{ is the tire pressure in pounds per square inch (psi).}$$

a. Find the tire pressure that will yield the maximum number of miles that a tire will last. Round to the nearest whole unit.

b. Find the maximum number of miles that a tire will last if the proper tire pressure is maintained. Round to the nearest thousand miles.

Group Activity

Maximizing Volume

Materials: A calculator and a sheet of $8\frac{1}{2}$ by 11 in. paper.

Estimated Time: 25–30 minutes

Group Size: 3

Antonio is going to build a custom gutter system for his house. He plans to use rectangular strips of aluminum that are $8\frac{1}{2}$ in. wide and 72 in. long. Each piece of aluminum will be turned up at a distance of x in. from the sides to form a gutter. See Figure 9-7.

Figure 9-7

Antonio wants to maximize the volume of water that the gutters can hold. To do this, he must determine the distance, x, that should be turned up to form the height of the gutter.

1. To familiarize yourself with this problem, we will simulate the gutter problem using an $8\frac{1}{2}$ by 11 in. piece of paper. For each value of x in the table, turn up the sides of the paper x in. from the edge. Then measure the base, the height, and the length of the paper "gutter," and calculate the volume.

Height, x	Base	Length	Volume
0.5 in.		11 in.	
1.0 in.		11 in.	
1.5 in.		11 in.	
2.0 in.		11 in.	
2.5 in.		11 in.	
3.0 in.		11 in.	
3.5 in.		11 in.	

2. Now follow these steps to find the optimal distance, x, that you should fold the paper to make the greatest volume within the paper gutter.

   a. If the height of the paper gutter is x in., write an expression for the base of the paper gutter.

   b. Write a function for the volume of the paper gutter.

   c. Find the vertex of the parabola defined by the function in part (b).

   d. Interpret the meaning of the vertex from part (c).

3. Use the concept from the paper gutter to write a function for the volume of Antonio's aluminum gutter that is 72 in. long.

4. Now find the optimal distance, x, that he should fold the aluminum sheet to make the greatest volume within the 72-in.-long aluminum gutter. What is the maximum volume?

Chapter 9   Summary

Section 9.1   The Square Root Property

Key Concepts

Square Root Property

If $x^2 = k$, then $x = \pm\sqrt{k}$.

The square root property can be used to solve a quadratic equation written as a square of a binomial equal to a constant.

Example

Example 1

$(x - 5)^2 = 13$

$x - 5 = \pm\sqrt{13}$ Square root property

$x = 5 \pm \sqrt{13}$ Solve for x.

Section 9.2   Completing the Square

Key Concepts

Solving a Quadratic Equation of the Form
$ax^2 + bx + c = 0\ (a \neq 0)$ **by Completing the Square and Applying the Square Root Property**

1. Divide both sides by a to make the leading coefficient 1.
2. Isolate the variable terms on one side of the equation.
3. Complete the square by adding the square of $\frac{1}{2}$ the linear term coefficient to both sides of the equation. Then factor the resulting perfect square trinomial.
4. Apply the square root property and solve for x.

Example

Example 1

$2x^2 - 8x - 6 = 0$

Step 1: $\dfrac{2x^2}{2} - \dfrac{8x}{2} - \dfrac{6}{2} = \dfrac{0}{2}$

$x^2 - 4x - 3 = 0$

Step 2: $x^2 - 4x = 3$

Step 3: $x^2 - 4x + 4 = 3 + 4$ Note that

$(x - 2)^2 = 7$ $[\frac{1}{2}(-4)]^2 = (-2)^2 = 4$

Step 4: $x - 2 = \pm\sqrt{7}$

$x = 2 \pm \sqrt{7}$

Section 9.3   Quadratic Formula

Key Concepts

The solutions to a quadratic equation of the form $ax^2 + bx + c = 0$ $(a \neq 0)$ are given by the **quadratic formula**:

$$x = \frac{-b \pm \sqrt{b^2 - 4ac}}{2a}$$

Three Methods for Solving a Quadratic Equation

1. Factoring
2. Completing the square and applying the square root property
3. Using the quadratic formula

Example

Example 1

$$3x^2 = 2x + 4$$

$$3x^2 - 2x - 4 = 0 \qquad a = 3, b = -2, c = -4$$

$$x = \frac{-(-2) \pm \sqrt{(-2)^2 - 4(3)(-4)}}{2(3)}$$

$$= \frac{2 \pm \sqrt{4 + 48}}{6}$$

$$= \frac{2 \pm \sqrt{52}}{6}$$

$$= \frac{2 \pm 2\sqrt{13}}{6} \qquad \text{Simplify the radical.}$$

$$= \frac{2(1 \pm \sqrt{13})}{6} \qquad \text{Factor.}$$

$$= \frac{1 \pm \sqrt{13}}{3} \qquad \text{Simplify.}$$

Section 9.4 Graphing Quadratic Functions

Key Concepts

Let a, b, and c represent real numbers such that $a \neq 0$. Then a function defined by $y = ax^2 + bx + c$ is called a **quadratic function**.

The graph of a quadratic function is called a **parabola**.

The leading coefficient, a, of a quadratic function, $y = ax^2 + bx + c$, determines if the parabola will open upward or downward. If $a > 0$, then the parabola opens upward. If $a < 0$, then the parabola opens downward.

Finding the Vertex of a Parabola

1. The x-coordinate of the vertex of the parabola with equation $y = ax^2 + bx + c$ $(a \neq 0)$ is given by

$$x = \frac{-b}{2a}$$

2. To find the corresponding y-coordinate of the vertex, substitute the value of the x-coordinate found in step 1 and solve for y.

If a parabola opens upward, the vertex is the lowest point on the graph. If a parabola opens downward, the vertex is the highest point on the graph.

Examples

Example 1

$y = x^2 - 4x - 3$ is a quadratic function. Its graph is in the shape of a parabola.

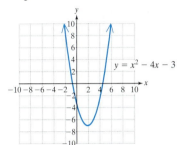

Example 2

Find the vertex of the parabola defined by $y = 3x^2 + 6x - 1$.

$$x = \frac{-b}{2a} = \frac{-6}{2 \cdot 3} = -1$$

$$y = 3(-1)^2 + 6(-1) - 1 = -4$$

The vertex is $(-1, -4)$.

For the equation $y = 3x^2 + 6x - 1$, $a = 3 > 0$. Therefore, the parabola opens upward. The vertex $(-1, -4)$ represents the lowest point of the graph.

Chapter 9   Review Exercises

Section 9.1

For Exercises 1–4, identify the equations as linear or quadratic.

1. $5x - 10 = 3x - 6$ **2.** $(x + 6)^2 = 6$

3. $x(x - 4) = 5x - 2$ **4.** $3(x + 6) = 18(x - 1)$

For Exercises 5–12, solve the equations using the square root property.

5. $x^2 = 25$ **6.** $x^2 - 19 = 0$

7. $x^2 + 49 = 0$ **8.** $x^2 = -48$

9. $(x + 1)^2 = 14$ **10.** $(x - 2)^2 = 60$

11. $\left(x - \frac{1}{8}\right)^2 = \frac{3}{64}$ **12.** $(2x - 3)^2 = 20$

Section 9.2

For Exercises 13–16, find the constant that should be added to each expression to make it a perfect square trinomial.

13. $x^2 + 12x$ **14.** $x^2 - 18x$

15. $x^2 - 5x$ **16.** $x^2 + 7x$

For Exercises 17–20, solve the quadratic equations by completing the square and applying the square root property.

17. $x^2 + 8x + 3 = 0$ **18.** $x^2 - 2x - 4 = 0$

19. $2x^2 - 6x - 6 = 0$ **20.** $3x^2 - 7x - 3 = 0$

21. A right triangle has legs of equal length. If the hypotenuse is 15 ft long, find the length of each leg. Round the answer to the nearest tenth of a foot.

22. A can in the shape of a right circular cylinder holds approximately 362 cm³ of liquid. If the height of the can is 12.1 cm, find the radius of the can. Round to the nearest tenth of a centimeter. (*Hint:* The volume of a right circular cylinder is given by: $V = \pi r^2 h$)

$r = ?$

12.1 cm

Section 9.3

23. Write the quadratic formula from memory.

For Exercises 24–33, solve the quadratic equations using the quadratic formula.

24. $5x^2 + x - 7 = 0$ **25.** $x^2 + 4x + 4 = 0$

26. $3x^2 - 2x + 2 = 0$ **27.** $2x^2 - x - 3 = 0$

28. $\frac{1}{8}x^2 + x = \frac{5}{2}$ **29.** $\frac{1}{6}x^2 + x + \frac{1}{3} = 0$

30. $1.2x^2 + 6x = 7.2$

31. $0.01x^2 - 0.02x - 0.04 = 0$

32. $(x + 6)(x + 2) = 10$

33. $(x - 1)(x - 7) = -18$

34. One number is two more than another number. Their product is 11.25. Find the numbers.

35. The base of a parallelogram is 1 cm longer than the height, and the area is 24 cm². Find the exact values of the base and height of the parallelogram. Then use a calculator to approximate the values to the nearest tenth of a centimeter.

36. An astronaut on the moon tosses a rock upward with an initial velocity of 25 ft/sec. The height of the rock, $h(t)$ (in feet), is determined by the number of seconds, t, after the rock is released according to the equation:

$$h(t) = -2.7t^2 + 25t + 5$$

Find the time required for the rock to hit the ground. [*Hint:* At ground level, $h(t) = 0$.] Round to the nearest tenth of a second.

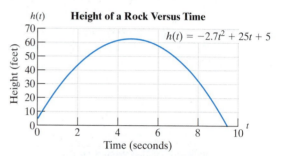

Height of a Rock Versus Time

$h(t) = -2.7t^2 + 25t + 5$

Height (feet) vs Time (seconds)

Section 9.4

For Exercises 37–40, identify a and determine if the parabola opens upward or downward.

37. $y = x^2 - 3x + 1$ **38.** $y = -x^2 + 8x + 2$

39. $y = -2x^2 + x - 12$ **40.** $y = 5x^2 - 2x - 6$

For Exercises 41–44, find the vertex for each parabola.

41. $y = 3x^2 + 6x + 4$ **42.** $y = -x^2 + 8x + 3$

43. $y = -2x^2 + 12x - 5$ **44.** $y = 2x^2 + 2x - 1$

For Exercises 45–48,

   a. Determine whether the graph of the parabola opens upward or downward.

   b. Find the vertex.

   c. Find the x-intercept(s) if possible.

   d. Find the y-intercept.

   e. Sketch the function.

   f. Write the domain in interval notation.

   g. Write the range in interval notation.

45. $y = 3x^2 + 12x + 9$

46. $y = -3x^2 + 12x - 10$

47. $y = -8x^2 - 16x - 12$

48. $y = 2x^2 + 4x - 1$

49. An object is launched into the air from ground level with an initial velocity of 256 ft/sec. The height of the object, y (in feet), can be approximated by the function

$$y = -16t^2 + 256t$$ where t is the number of seconds after launch.

   a. Find the maximum height of the object.

   b. Find the time required for the object to reach its maximum height.

Chapter 9   Test

1. Solve the equation by applying the square root property.

$$(3x + 1)^2 = -14$$

2. Solve the equation by completing the square and applying the square root property.

$$x^2 - 8x - 5 = 0$$

3. Solve the equation by using the quadratic formula.

$$3x^2 - 5x = -1$$

For Exercises 4–10, solve the equations using any method.

4. $5x^2 + x - 2 = 0$

5. $(c - 12)^2 = 12$

6. $y^2 + 14y - 1 = 0$

7. $3t^2 = 30$

8. $4x(3x + 2) = 15$

9. $6p^2 - 11p = 0$

10. $\dfrac{1}{4}x^2 - \dfrac{3}{2}x = \dfrac{11}{4}$

11. The surface area, S, of a sphere is given by the formula $S = 4\pi r^2$, where r is the radius of the sphere. Find the radius of a sphere whose surface area is 201 in.^{2} Round to the nearest tenth of an inch.

$S = 201$ in.^{2}

12. The height of a triangle is 2 m longer than twice the base, and the area is 24 m^2. Find the exact values of the base and height. Then use a calculator to approximate the base and height to the nearest tenth of a meter.

13. Explain how to determine if a parabola opens upward or downward.

For Exercises 14–16, find the vertex of the parabola.

14. $y = x^2 - 10x + 25$ **15.** $y = 3x^2 - 6x + 8$

16. $y = -x^2 - 16$

17. Suppose a parabola opens upward and the vertex is located at $(-4, 3)$. How many x-intercepts does the parabola have?

18. Given the parabola, $y = x^2 + 6x + 8$.

   a. Determine whether the parabola opens upward or downward.

   b. Find the vertex of the parabola.

   c. Find the x-intercepts.

   d. Find the y-intercept.

   e. Graph the parabola.

19. Graph the parabola and label the vertex, x-intercepts, and y-intercept.

$$y = -x^2 + 25$$

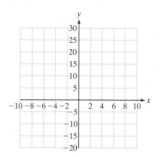

20. The Phelps Arena in Atlanta holds 20,000 seats. If the Atlanta Hawks charge x dollars per ticket for a game, then the total revenue, y (in dollars), can be approximated by

$$y = -400x^2 + 20,000x$$

   where x is the price per ticket.

   a. Find the ticket price that will produce the maximum revenue.

   b. What is the maximum revenue?

Chapters 1–9   Cumulative Review Exercises

1. Solve. $3x - 5 = 2(x - 2)$

2. Solve for h. $A = \frac{1}{2}bh$

3. Solve. $\frac{1}{2}y - \frac{5}{6} = \frac{1}{4}y + 2$

4. Determine whether 2 is a solution to the inequality. $-3x + 4 < x + 8$

5. Graph the solution to the inequality: $-3x + 4 < x + 8$. Then write the solution in set-builder notation and in interval notation.

_____

6. The graph depicts the death rate from 60 U.S. cities versus the median education level of the people living in that city. The death rate, y, is measured in number of deaths per 100,000 people. The median education level, x, is a type of "average" and is measured by grade level. (*Source*: U.S. Bureau of the Census)

The death rate can be predicted from the median education level according to the equation:

$$y = -37.6x + 1353 \qquad \text{where } 9 \le x \le 13.$$

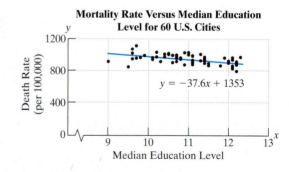

   a. From the graph, does it appear that the death rate increases or decreases as the median education level increases?

   b. What is the slope of the line? Interpret the slope in the context of the death rate and education level.

c. For a city in the United States with a median education level of 12, what would be the expected death rate?

d. If the death rate of a certain city is 977 per 100,000 people, what would be the approximate median education level?

7. Simplify completely. Write the final answer with positive exponents only.

$$\left(\frac{2a^2b^{-3}}{c}\right)^{-1} \cdot \left(\frac{4a^{-1}}{b^2}\right)^2$$

8. Approximately 5.2×10^7 disposable diapers are thrown into the trash each day in the United States and Canada. How many diapers are thrown away each year?

9. In 1989, the Hipparcos satellite found the distance between Earth and the star, Polaris, to be approximately 2.53×10^{15} mi. If 1 light-year is approximately 5.88×10^{12} miles, how many light-years is Polaris from Earth?

10. Perform the indicated operation.
$(2x - 3)^2 - 4(x - 1)$

11. Divide using long division.
$(2y^4 - 4y^3 + y - 5) \div (y - 2)$

12. Factor. $2x^2 - 9x - 35$

13. Factor completely. $2xy + 8xa - 3by - 12ab$

14. The base of a triangle is 1 m more than the height. If the area is 36 m², find the base and height.

15. Multiply. $\dfrac{x^2 + 10x + 9}{x^2 - 81} \cdot \dfrac{18 - 2x}{x^2 + 2x + 1}$

16. Reduce to lowest terms.

$$\frac{5x + 10}{x^2 - 4}$$

17. Perform the indicated operations.

$$\frac{x^2}{x - 5} - \frac{10x - 25}{x - 5}$$

18. Simplify completely.

$$\frac{\dfrac{1}{x + 1} - \dfrac{1}{x - 1}}{\dfrac{x}{x^2 - 1}}$$

19. Solve.

$$1 - \frac{1}{y} = \frac{12}{y^2}$$

20. At Lake George, a fishing enthusiast must release any fish that is less than 8 inches in length. If a fish is caught and measured at 20 cm, must the fish be released back to the lake? (*Hint:* 1 in. = 2.54 cm)

21. Write an equation of the line passing through the point $(-2, 3)$ and having a slope of $\frac{1}{2}$. Write the final answer in slope-intercept form.

For Exercises 22–23,

a. Find the x-intercept (if it exists).

b. Find the y-intercept (if it exists).

c. Find the slope (if it exists).

d. Graph the line.

22. $2x - 4y = 12$

23. $4x + 12 = 0$

24. Solve the system by using the addition method. If the system has no solution or infinitely many solutions, so state.

$$\frac{1}{2}x - \frac{1}{4}y = \frac{1}{6}$$

$$12x - 3y = 8$$

25. Solve the system by using the substitution method. If the system has no solution or infinitely many solutions, so state.

$$2x - y = 8$$

$$4x - 4y = 3x - 3$$

26. In a right triangle, one acute angle is 2° more than three times the other acute angle. Find the measure of each angle.

27. A bank of 27 coins contains only dimes and quarters. The total value of the coins is $4.80. Find the number of dimes and the number of quarters.

28. Sketch the inequality, $x - y \leq 4$.

29. Which of the following are irrational numbers?
$\{0, -\frac{2}{3}, \pi, \sqrt{7}, 1.2, \sqrt{25}\}$

For Exercises 30–31, simplify the radicals.

30. $\sqrt{\frac{1}{7}}$

31. $\dfrac{\sqrt{16x^4}}{\sqrt{2x}}$

32. Perform the indicated operation. $(4\sqrt{3} + \sqrt{x})^2$

33. Add the radicals. $-3\sqrt{2x} + \sqrt{50x}$

34. Rationalize the denominator.

$$\frac{4}{2 - \sqrt{a}}$$

35. Solve. $\sqrt{x + 11} = x + 5$

36. Factor completely. $8c^3 - y^3$

37. Which graph defines y as a function of x?

a. **b.**

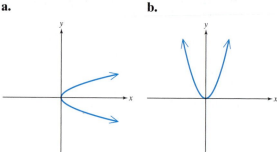

38. Given the functions defined by $f(x) = -\frac{1}{2}x + 4$ and $g(x) = x^2$, find

a. $f(6)$ **b.** $g(-2)$ **c.** $f(0) + g(3)$

39. Find the domain and range of the function.
$\{(2, 4), (-1, 3), (9, 2), (-6, 8)\}$

40. Find the slope of the line passing through the points $(3, -1)$ and $(-4, -6)$.

41. Find the slope of the line defined by $-4x - 5y = 10$.

42. What value of k would make the expression a perfect square trinomial?

$$x^2 + 10x + k$$

43. Solve the quadratic equation by completing the square and applying the square root property. $2x^2 + 12x + 6 = 0$.

44. Solve the quadratic equation by using the quadratic formula. $2x^2 + 12x + 6 = 0$.

45. Graph the parabola defined by the equation. Label the vertex, x-intercepts, and y-intercept.

$$y = x^2 + 4x + 4$$

Student Answer Appendix

Chapter R

Chapter Opener Puzzles

$\dfrac{d}{1}\ \dfrac{e}{2}\ \dfrac{c}{3}\ \dfrac{i}{4}\ \dfrac{m}{5}\ \dfrac{a}{6}\ \dfrac{l}{7}\ \dfrac{l}{7}$.

Section R.2 Practice Exercises, pp. 17–20

1. Numerator: 7; denominator: 8; proper
3. Numerator: 9; denominator: 5; improper
5. Numerator: 6; denominator: 6; improper
7. Numerator: 12; denominator: 1; improper
9. $\dfrac{3}{4}$ 11. $\dfrac{4}{3}$ 13. $\dfrac{1}{6}$ 15. $\dfrac{2}{2}$
17. $\dfrac{5}{2}$ or $2\tfrac{1}{2}$ 19. $\dfrac{6}{2}$ or 3
21. The set of whole numbers includes the number 0 and the set of natural numbers does not.
23. For example: $\dfrac{2}{4}$ 25. Prime   27. Composite
29. Composite   31. Prime   33. $2 \times 2 \times 3 \times 3$
35. $2 \times 3 \times 7$ 37. $2 \times 5 \times 11$ 39. $3 \times 3 \times 3 \times 5$
41. $\dfrac{1}{5}$ 43. $\dfrac{3}{8}$ 45. $\dfrac{7}{8}$ 47. $\dfrac{3}{4}$ 49. $\dfrac{5}{8}$ 51. $\dfrac{3}{4}$
53. False: When adding or subtracting fractions, it is necessary to have a common denominator.
55. $\dfrac{4}{3}$ 57. $\dfrac{2}{3}$ 59. $\dfrac{9}{2}$ 61. $\dfrac{3}{5}$ 63. $\dfrac{5}{3}$
65. $\dfrac{90}{13}$ 67. $1050   69. 4 graduated with honors.
71. 8 aprons   73. 8 jars   75. $\dfrac{3}{7}$ 77. $\dfrac{1}{2}$ 79. 30
81. 40   83. $\dfrac{7}{8}$ 85. $\dfrac{3}{40}$ 87. $\dfrac{3}{26}$ 89. $\dfrac{29}{36}$
91. $\dfrac{7}{10}$ 93. $\dfrac{35}{48}$ 95. $\dfrac{7}{24}$ 97. $\dfrac{51}{28}$ or $1\tfrac{23}{28}$ 99. 46
101. $\dfrac{14}{5}$ or $2\tfrac{4}{5}$ 103. $\dfrac{46}{5}$ or $9\tfrac{1}{5}$ 105. $\dfrac{1}{6}$ 107. $\dfrac{11}{54}$
109. $\dfrac{7}{2}$ or $3\tfrac{1}{2}$ 111. $\dfrac{13}{8}$ or $1\tfrac{5}{8}$ 113. $\dfrac{59}{12}$ or $4\tfrac{11}{12}$ 115. $\dfrac{1}{8}$
117. $8\tfrac{19}{24}$ in.   119. $1\tfrac{7}{12}$ hr   121. $2\tfrac{1}{4}$ lb   123. 25 in.

Section R.3 Calculator Connections, p. 26

1. $0.\overline{4}$ 2. $0.\overline{63}$ 3. $0.1\overline{36}$ 4. $0.\overline{384615}$

1.–2.
```
4/9
          .4444444444
7/11
          .6363636364
```

3.–4.
```
3/22
          .1363636364
5/13
          .3846153846
```

Section R.3 Practice Exercises, pp. 27–29

1. Tens   3. Hundreds   5. Tenths   7. Hundredths
9. No, the symbols I, V, X, and so on each represent certain values but the values are not dependent on the position of the symbol within the number.   11. 0.7   13. 0.36
15. $1.\overline{2}$ 17. $0.\overline{21}$ 19. 214.1   21. 39.268
23. 40,000   25. 0.73   27. $\dfrac{9}{20}$ 29. $\dfrac{181}{1000}$
31. $\dfrac{51}{25}$ or $2\tfrac{1}{25}$ 33. $\dfrac{13,007}{1000}$ or $13\tfrac{7}{1000}$ 35. $\dfrac{5}{9}$
37. $\dfrac{10}{9}$ or $1\tfrac{1}{9}$ 39. $0.3, \dfrac{3}{10}$ 41. $0.75, \dfrac{3}{4}$ 43. $0.0375, \dfrac{3}{80}$
45. $0.157, \dfrac{157}{1000}$ 47. $2.7, \dfrac{27}{10}$ 49. Multiply by 100%.
51. 5%   53. 90%   55. 120%   57. 750%
59. 13.5%   61. 0.3%   63. 6%   65. 450%
67. 62.5%   69. 31.25%   71. $83.\overline{3}$%   73. $93.\overline{3}$%
75. $42   77. $3375   79. 7%   81. $792
83. $192   85. $67,500

Section R.4 Practice Exercises, pp. 37–43

1. b, e, i   3. 32 m   5. 17.2 mi   7. $11\tfrac{1}{2}$ in.
9. 31.4 ft   11. a, f, g   13. 33 cm^2 15. 16.81 m^2
17. 84 in.^{2} 19. 10.12 km^2 21. 13.8474 ft^2
23. 66 in.^{2} 25. 31.5 ft^2 27. c, d, h   29. 307.72 cm^3
31. 39 in.^{3} 33. 113.04 cm^3 35. 1695.6 cm^3
37. 3052.08 in.^{3} 39. 113.04 cm^3
41. a. $0.25/ft^2 b. $104   43. Perimeter   45. 54 ft
47. a. 57,600 ft^2 b. 19,200 pieces
49. a. 50.24 in.^{2} b. 113.04 in.^{2} c. One 12-in. pizza
51. 289.3824 cm^3 53. True   55. True   57. True
59. Not possible   61. For example: 100°, 80°
63. a. $\angle 1$ and $\angle 3$, $\angle 2$ and $\angle 4$
b. $\angle 1$ and $\angle 2$, $\angle 2$ and $\angle 3$, $\angle 3$ and $\angle 4$, $\angle 1$ and $\angle 4$
c. $m(\angle 1) = 100°, m(\angle 2) = 80°, m(\angle 3) = 100°$
65. 57°   67. 78°   69. 147°   71. 58°
73. 7   75. 1   77. 1   79. 5
81. $m(\angle a) = 45°, m(\angle b) = 135°, m(\angle c) = 45°$, $m(\angle d) = 135°, m(\angle e) = 45°, m(\angle f) = 135°, m(\angle g) = 45°$
83. Scalene   85. Isosceles   87. True   89. 45°
91. No, a 90° angle plus an angle greater than 90° would make the sum of the angles greater than 180°.   93. 40°
95. 37°   97. $m(\angle a) = 80°, m(\angle b) = 80°, m(\angle c) = 100°$, $m(\angle d) = 100°, m(\angle e) = 65°, m(\angle f) = 115°, m(\angle g) = 115°$, $m(\angle h) = 35°, m(\angle i) = 145°, m(\angle j) = 145°$
99. $m(\angle a) = 70°, m(\angle b) = 65°, m(\angle c) = 65°$, $m(\angle d) = 110°, m(\angle e) = 70°, m(\angle f) = 110°, m(\angle g) = 115°$, $m(\angle h) = 115°, m(\angle i) = 65°, m(\angle j) = 70°, m(\angle k) = 65°$
101. 82 ft   103. 36 in.^{2} 105. 15.2464 cm^2

Chapter 1
Chapter Opener Puzzle

-10	$+$	2	$+$	-3	$=$	-11
$+$		$\cdot$		$\cdot$		$-$
12	$\div$	-2	$\cdot$	3	$=$	-18
$+$		$+$		$+$		$+$
6	$-$	4	$-$	10	$=$	-8
$=$		$=$		$=$		$=$
8	$\cdot$	0	$-$	1	$=$	-1

Section 1.1 Calculator Connections, p. 52

1. ≈ 3.464101615 **2.** ≈ 9.949874371
3. ≈ 12.56637061 **4.** ≈ 1.772453851

Section 1.1 Practice Exercises, pp. 53–55

3.

5. a; rational   **7.** b; rational   **9.** a; rational
11. c; irrational   **13.** a; rational   **15.** a; rational
17. b; rational   **19.** c; irrational
21. For example: $\pi, -\sqrt{2}, \sqrt{3}$
23. For example: $-4, -1, 0$
25. For example: $-\dfrac{3}{4}, \dfrac{1}{2}, 0.206$ **27.** $-\dfrac{3}{2}, -4, 0.\overline{6}, 0, 1$
29. 1   **31.** $-4, 0, 1$ **33. a.** $>$ **b.** $>$ **c.** $<$ **d.** $>$
35. -18 **37.** 6.1   **39.** $\dfrac{5}{8}$ **41.** $-\dfrac{7}{3}$ **43.** 3
45. $-\dfrac{7}{3}$ **47.** 8   **49.** -72.1 **51.** 2   **53.** 1.5
55. -1.5 **57.** $\dfrac{3}{2}$ **59.** -10 **61.** $-\dfrac{1}{2}$
63. False, $|n|$ is never negative.   **65.** True   **67.** False
69. True   **71.** False   **73.** False   **75.** False
77. True   **79.** True   **81.** False   **83.** True
85. True   **87.** True   **89.** For all $a < 0$

Section 1.2 Calculator Connections, p. 62

1. 2   **2.** 91   **3.** 84   **4.** 12   **5.** 49   **6.** 18
1.–3.   **4.–6.**

```
(4+6)/(8-3)
                    2
110-5*(2+1)-4
                   91
100-2*(5-3)^3
                   84
```

```
3+(4-1)²
                   12
(12-6+1)²
                   49
3*8-√(32+2²)
                   18
```

7. 4   **8.** 27   **9.** 0.5   **7.–9.**

```
√(18-2)
                    4
(4*3-3*3)^3
                   27
(20-3²)/(26-2²)
                  .5
```

Section 1.2 Practice Exercises, pp. 63–65

3. $-4, 5.\overline{6}, 0, 4.02, \dfrac{7}{9}$ **5.** 9.2   **7.** -19 **9.** 15
11. 3   **13.** 9   **15.** $\left(\dfrac{1}{6}\right)^4$ **17.** a^3b^2 **19.** $(5c)^5$
21. a. x **b.** Yes, 1   **23.** $x \cdot x \cdot x$ **25.** $2b \cdot 2b \cdot 2b$
27. $10 \cdot y \cdot y \cdot y \cdot y \cdot y$ **29.** $2 \cdot w \cdot z \cdot z$ **31.** 36
33. $\dfrac{1}{49}$ **35.** 0.008   **37.** 64   **39.** 9   **41.** 2
43. 12   **45.** 4   **47.** $\dfrac{1}{3}$ **49.** $\dfrac{5}{9}$ **51.** 20   **53.** 60
55. 8   **57.** 78   **59.** 0   **61.** $\dfrac{7}{6}$ **63.** 45   **65.** 16
67. 15   **69.** 19   **71.** 3   **73.** 39   **75.** 26   **77.** $\dfrac{5}{12}$
79. $\dfrac{5}{2}$ **81.** 57,600 ft^2 **83.** 21 ft^2 **85.** $3x$
87. $\dfrac{x}{7}$ or $x \div 7$ **89.** $2 - a$ **91.** $2y + x$
93. $4(x + 12)$ **95.** $3 - Q$ **97.** $2y^3$; 16
99. $|z - 8|$; 2   **101.** $5\sqrt{x}$; 10   **103.** $yz - x$; 16
105. 1   **107.** $\dfrac{1}{4}$
109. a. $36 \div 4 \cdot 3$ Division must be performed before
      $= 9 \cdot 3$ multiplication.
      $= 27$
b. $36 - 4 + 3$ Subtraction must be performed
      $= 32 + 3$ before addition.
      $= 35$
111. This is acceptable, provided division and multiplication are performed in order from left to right, and subtraction and addition are performed in order from left to right.

Section 1.3 Practice Exercises, pp. 71–73

3. $>$ **5.** $>$ **7.** $>$
9. -6 **11.** 3   **13.** 3   **15.** -3 **17.** -17
19. 7   **21.** -19 **23.** -23 **25.** -5 **27.** -3
29. 0   **31.** 0   **33.** -5 **35.** -3 **37.** 0
39. -23 **41.** -6 **43.** -3 **45.** 21.3
47. $-\dfrac{3}{14}$ **49.** $-\dfrac{1}{6}$ **51.** $-\dfrac{15}{16}$ **53.** $\dfrac{1}{20}$
55. -2.4 or $-\dfrac{12}{5}$ **57.** $\dfrac{1}{4}$ or 0.25   **59.** 0   **61.** $-\dfrac{7}{8}$
63. -1 **65.** $\dfrac{11}{9}$ **67.** -23.08 **69.** -0.002117
71. To add two numbers with different signs, subtract the smaller absolute value from the larger absolute value and apply the sign of the number with the larger absolute value.
73. -1 **75.** 10   **77.** 5   **79.** 1

81. $-6 + (-10); -16$ **83.** $-3 + 8; 5$
85. $-21 + 17; -4$ **87.** $3(-14 + 20); 18$
89. $(-7 + (-2)) + 5; -4$ **91.** $-5 + 13 + (-11); -3°F$
93. $-2 + 6 + (-5); -1$ yd or 1-yd loss
95. a. $52.23 + (-52.95)$ **b.** Yes
97. a. $100 + 200 + (-500) + 300 + 100 + (-200)$ **b.** $0

Section 1.4 Calculator Connections, pp. 78–79

1. -13 **2.** -2 **3.** 711
1.–3.

```
-8+(-5)
              -13
4+(-5)+(-1)
              -2
627-(-84)
              711
```

4. -0.18 **5.** 11.3 **6.** -990 **7.** -17 **8.** 38
4.–6.   7.–8.

```
-0.06-0.12
            -.18
-3.2+(-14.5)
            -17.7
-472+(-518)
            -990
```

```
-12-9+4
            -17
209-108+(-63)
            38
```

Section 1.4 Practice Exercises, pp. 79–81

3. x^2 **5.** $-b + 2$ **7.** 9 **9.** -3 **11.** -12
13. 4 **15.** -2 **17.** 8 **19.** -8 **21.** 2 **23.** 6
25. 40 **27.** -40 **29.** 0 **31.** -20 **33.** -24
35. 25 **37.** -5 **39.** $-\frac{3}{2}$ **41.** $\frac{41}{24}$ **43.** $\frac{2}{5}$ **45.** $-\frac{2}{3}$
47. 9.2 **49.** -5.72 **51.** -10 **53.** -14 **55.** -51
57. -173.188 **59.** 3.243 **61.** $6 - (-7); 13$
63. $3 - 18; -15$ **65.** $-5 - (-11); 6$
67. $-1 - (-13); 12$ **69.** $-32 - 20; -52$
71. $200 + 400 + 600 + 800 - 1000; 1000 **73.** $152°F$
75. $19,881$ m   **77.** 13 **79.** -9 **81.** 5 **83.** -25
85. -2 **87.** $-\frac{11}{30}$ **89.** $-\frac{29}{9}$ **91.** -2 **93.** -11
95. 2 **97.** -7 **99.** 5 **101.** 5 **103.** 3

Chapter 1 Problem Recognition Exercises, p. 82

1. Add their absolute values and apply a negative sign.
2. Subtract the smaller absolute value from the larger absolute value. Apply the sign of the number with the larger absolute value.
3. 41 **4.** 13 **5.** 31 **6.** 46 **7.** -1.3 **8.** -3.6
9. -16 **10.** -7 **11.** $-\frac{1}{12}$ **12.** $\frac{7}{24}$ **13.** -36
14. -59 **15.** -12 **16.** -50 **17.** $-\frac{19}{6}$ **18.** $-\frac{8}{5}$
19. -5 **20.** -32 **21.** 0 **22.** 0 **23.** -7.7
24. -10.5 **25.** -114 **26.** -56 **27.** -32
28. -46 **29.** 0 **30.** 0 **31.** -30 **32.** -400
33. -57 **34.** -26 **35.** -1 **36.** -6 **37.** 4
38. -1 **39.** $\frac{1}{36}$ **40.** $\frac{1}{64}$

Section 1.5 Calculator Connections, p. 89

1. -30 **2.** -2 **3.** 625 **4.** 625 **5.** -625 **6.** -5.76
1.–3.   4.–6.

```
-6(5)
            -30
-5.2/2.6
            -2
(-5)(-5)(-5)(-5)
            625
```

```
(-5)^4
            625
-5^4
            -625
-2.4²
            -5.76
```

7. 5.76 **8.** -1 **9.** 4 **10.** -36
7.–8.   9.–10.

```
(-2.4)²
            5.76
(-1)(-1)(-1)
            -1
```

```
-8.4/-2.1
            4
90/(-5)(2)
            -36
```

Section 1.5 Practice Exercises, pp. 89–92

3. True   **5.** False   **7.** -56 **9.** -42 **11.** 143
13. 240 **15.** -12.76 **17.** $\frac{3}{4}$ **19.** 36 **21.** -36
23. $-\frac{27}{125}$ **25.** 0.0016 **27.** -6 **29.** 10 **31.** 2
33. $-\frac{1}{5}$ **35.** $(-2)(-7) = 14$ **37.** $-5 \cdot 0 = 0$
39. No number multiplied by zero equals 6.
41. $(-6)(4) = -24$ **43.** 6 **45.** -6 **47.** -6
49. 6 **51.** 8 **53.** -8 **55.** -8 **57.** 8 **59.** 0
61. Undefined   **63.** 0 **65.** 0 **67.** $-\frac{3}{2}$ **69.** $\frac{3}{10}$
71. -2 **73.** -7.912 **75.** 0.092 **77.** -6 **79.** 2.1
81. 9 **83.** -9 **85.** $-\frac{64}{27}$ **87.** -0.008
89. -0.0016 **91.** -29 **93.** 48 **95.** -14.28
97. 340 **99.** $-\frac{10}{9}$ **101.** $\frac{14}{9}$ **103.** -30 **105.** 96
107. 2 **109.** -1 **111.** $-\frac{4}{33}$ **113.** $-\frac{4}{7}$ **115.** -24
117. $-\frac{1}{20}$ **119.** -23 **121.** 12 **123.** $\frac{9}{7}$
125. Undefined   **127.** -48 **129.** -6 **131.** -1
133. 7 **135.** 12 **137.** -40 **139.** $\frac{7}{2}$
141. No. The first expression is equivalent to $10 \div (5x)$. The second is $10 \div 5 \cdot x$.
143. $-3.75(0.3); -1.125$ **145.** $\frac{16}{5} \div \left(-\frac{8}{9}\right); -\frac{18}{5}$
147. $-0.4 + 6(-0.42); -2.92$ **149.** $-\frac{1}{4} - 6\left(-\frac{1}{3}\right); \frac{7}{4}$
151. $3(-2) + 3 = -3;$ loss of $3
153. a. -10 **b.** 24 **c.** In part (a), we subtract; in part (b), we multiply.

Section 1.6 Practice Exercises, pp. 102–105

3. 8 **5.** -8 **7.** $-\dfrac{9}{2}$ or -4.5 **9.** 0 **11.** $\dfrac{7}{8}$

13. $-\dfrac{4}{45}$ **15.** $-8 + 5$ **17.** $x + 8$ **19.** $4(5)$

21. $-12x$ **23.** $x + (-3); -3 + x$

25. $4p + (-9); -9 + 4p$ **27.** $x + (4 + 9); x + 13$

29. $(-5 \cdot 3)x; -15x$ **31.** $\left(\dfrac{6}{11} \cdot \dfrac{11}{6}\right)x; x$

33. $\left(-4 \cdot -\dfrac{1}{4}\right)t; t$ **35.** Reciprocal   **37.** 0

39. $30x + 6$ **41.** $-2a - 16$ **43.** $15c - 3d$

45. $-7y + 14$ **47.** $-\dfrac{2}{3}x + 4$ **49.** $\dfrac{1}{3}m - 1$

51. $-2p - 10$ **53.** $6w + 10z - 16$ **55.** $4x + 8y - 4z$

57. $6w - x + 3y$ **59.** $6 + 2x$ **61.** $24z$ **63.** b

65. i   **67.** g   **69.** d   **71.** h

73.

Term	Coefficient
$2x$	2
$-y$	-1
$18xy$	18
5	5

75.

Term	Coefficient
$-x$	-1
$8y$	8
$-9x^2y$	-9
-3	-3

77. The variable factors are different.

79. The variables are the same and raised to the same power.   **81.** For example: $5y, -2x, 6$ **83.** $2p - 12$

85. $-6y^2 - 7y$ **87.** $7x^3y + xy - 4$ **89.** $3t - \dfrac{7}{5}$

91. $-6x + 22$ **93.** $4w$ **95.** $-3x + 17$

97. $10t - 44$ **99.** -18 **101.** $-2t + 7$

103. $51a - 27$ **105.** -6 **107.** $4q - \dfrac{1}{3}$

109. $6n$ **111.** $2x + 18$ **113.** $32.33z - 30.81$

115. $-2x - 34$ **117.** $9z - 35$ **119.** Equivalent

121. Not equivalent. The terms are not *like* terms and cannot be combined.   **123.** Not equivalent; subtraction is not commutative.   **125.** Equivalent   **127.** $14\frac{2}{7} + (2\frac{1}{3} + \frac{2}{3})$ is easier.   **129. a.** 55   **b.** 210

Chapter 1 Review Exercises, pp. 111–113

1. a. $7, 1$ **b.** $7, -4, 0, 1$ **c.** $7, 0, 1$ **d.** $7, \frac{1}{3}, -4, 0, -0.\overline{2}, 1$ **e.** $-\sqrt{3}, \pi$ **f.** $7, \frac{1}{3}, -4, 0, -\sqrt{3}, -0.\overline{2}, \pi, 1$

3. 6   **5.** 0   **7.** False   **9.** True   **11.** True

13. True   **15.** $\dfrac{7}{y}$ or $7 \div y$ **17.** $a - 5$ **19.** $13z - 7$

21. 60   **23.** 4   **25.** 225   **27.** $\dfrac{1}{10}$ **29.** $\dfrac{27}{8}$

31. 11   **33.** 10   **35.** 4   **37.** -17 **39.** $-\dfrac{5}{22}$

41. $-\dfrac{27}{10}$ **43.** -4.28 **45.** 8

47. When a and b are both negative or when a and b have different signs and the number with the larger absolute value is negative.

49. -12 **51.** -1 **53.** $-\dfrac{29}{18}$ **55.** -1.2

57. -10.2 **59.** $\dfrac{10}{3}$ **61.** -1 **63.** $-7 - (-18); 11$

65. $7 - 13; -6$ **67.** $(6 + (-12)) - 21; -27$

69. -170 **71.** -2 **73.** $-\dfrac{1}{6}$ **75.** 0   **77.** 0

79. $-\dfrac{3}{2}$ **81.** -30 **83.** $\dfrac{1}{4}$ **85.** -2 **87.** 17

89. $-\dfrac{7}{120}$ **91.** $-\dfrac{1}{3}$ **93.** -2 **95.** -23 **97.** 70.6

99. False, any nonzero real number raised to an even power is positive.   **101.** True   **103.** True   **105.** For example: $2 + 3 = 3 + 2$ **107.** For example: $5 + (-5) = 0$ **109.** For example: $5 \cdot 2 = 2 \cdot 5$

111. For example: $3 \cdot \dfrac{1}{3} = 1$

113. $5x - 2y = 5x + (-2y)$, then use the commutative property of addition.

115. $3y, 10x, -12, xy$ **117.** $8a - b - 10$

119. $-8z - 18$ **121.** $p - 2$ **123.** $-14q - 1$

125. $4x + 24$

Chapter 1 Test, pp. 113–114

1. Rational, all repeating decimals are rational numbers.

2.

3. a. False   **b.** True   **c.** True   **d.** True

4. a. $(4x)(4x)(4x)$ **b.** $4 \cdot x \cdot x \cdot x$

5. a. Twice the difference of a and b

b. b subtracted from twice a

6. $\dfrac{\sqrt{c}}{d^2}$ or $\sqrt{c} \div d^2$ **7.** 6   **8.** -19 **9.** -12

10. 28   **11.** $-\dfrac{7}{8}$ **12.** 4.66   **13.** -32 **14.** -12

15. Undefined   **16.** -28 **17.** 0   **18.** 96   **19.** $\dfrac{2}{3}$

20. -8 **21.** 9   **22.** $\dfrac{1}{3}$ **23.** $-\dfrac{3}{5}$

24. a. Commutative property of multiplication   **b.** Identity property of addition   **c.** Associative property of addition   **d.** Inverse property of multiplication   **e.** Associative property of multiplication

25. $-12x + 2y - 4$ **26.** $-12m - 24p + 21$

27. $-6k - 8$ **28.** $-4p - 23$ **29.** $4p - \dfrac{4}{3}$

30. 5   **31.** 18   **32.** -6 **33.** -32

34. $12 - (-4); 16$ **35.** $6 - 8; -2$

Chapter 2
Chapter Opener Puzzle

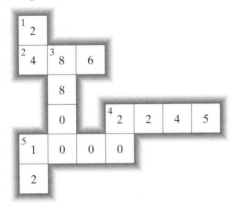

Section 2.1 Practice Exercises, pp. 124–126

3. Expression   **5.** Equation   **7.** Substitute the value into the equation and determine if the right-hand side is equal to the left-hand side.   **9.** No   **11.** Yes
13. Yes   **15.** -1 **17.** 20   **19.** -17 **21.** -16
23. 0   **25.** 1.3   **27.** $\frac{11}{2}$ or $5\frac{1}{2}$ **29.** -2 **31.** -2.13
33. -3.2675 **35.** 9   **37.** -4 **39.** 0   **41.** -15
43. $-\frac{4}{5}$ **45.** -10 **47.** 4   **49.** 41   **51.** -127
53. -2.6 **55.** $-8 + x = 42$; The number is 50.
57. $x - (-6) = 18$; The number is 12.
59. $x \cdot 7 = -63$ or $7x = -63$; The number is -9.
61. $\frac{x}{12} = \frac{1}{3}$; The number is 4.
63. $x + \frac{5}{8} = \frac{13}{8}$; The number is 1.   **65.** 10   **67.** $-\frac{1}{9}$
69. -12 **71.** $\frac{22}{3}$ **73.** -36 **75.** 16   **77.** 2
79. $-\frac{7}{4}$ **81.** 11   **83.** -36 **85.** $\frac{7}{2}$ **87.** 4
89. 3.6   **91.** 0.4084   **93.** Yes   **95.** No   **97.** Yes
99. Yes   **101.** For example: $y + 9 = 15$ **103.** For example: $2p = -8$ **105.** For example: $5a + 5 = 5$
107. -1 **109.** 7

Section 2.2 Practice Exercises, pp. 133–135

3. $-5z + 2$ **5.** $10p - 10$ **7.** To simplify an expression, clear parentheses and combine *like* terms. To solve an equation, use the addition, subtraction, multiplication, and division properties of equality to isolate the variable.   **9.** -3
11. -5 **13.** 2   **15.** 6   **17.** $\frac{5}{2}$ **19.** -42
21. $-\frac{3}{4}$ **23.** 5   **25.** -4 **27.** -26 **29.** 10
31. -8 **33.** $-\frac{7}{3}$ **35.** 0   **37.** -3 **39.** -2
41. $\frac{9}{2}$ **43.** $-\frac{1}{3}$ **45.** 10   **47.** -6 **49.** 0
51. -2 **53.** $-\frac{25}{4}$ **55.** $\frac{10}{3}$ **57.** -0.25

59. Contradiction; no solution   **61.** Conditional equation; -15 **63.** Identity; all real numbers
65. One solution   **67.** Infinitely many solutions
69. 7   **71.** $\frac{1}{2}$ **73.** 0   **75.** All real numbers
77. -46 **79.** 2   **81.** $\frac{13}{2}$ **83.** -5
85. No solution   **87.** 2.205   **89.** 10   **91.** -1
93. $a = 15$ **95.** $a = 4$
97. For example: $5x + 2 = 2 + 5x$

Section 2.3 Practice Exercises, pp. 140–142

3. -2 **5.** -5 **7.** No solution   **9.** 18, 36
11. 100; 1000; 10,000   **13.** 30, 60   **15.** 4   **17.** $-\frac{19}{2}$
19. $-\frac{15}{4}$ **21.** 8   **23.** 3   **25.** 15
27. No solution   **29.** All real numbers   **31.** 5
33. 2   **35.** -15 **37.** 6   **39.** 3
41. All real numbers   **43.** 67   **45.** 90   **47.** 4
49. -3.8 **51.** No solution   **53.** -0.25 **55.** -6
57. $\frac{8}{3}$ or $2\frac{2}{3}$ **59.** -9 **61.** $\frac{1}{10}$ **63.** -2
65. -1 **67.** 2

Chapter 2 Problem Recognition Exercises, p. 142

1. Expression; $-4b + 18$ **2.** Expression; $20p - 30$
3. Equation; -8 **4.** Equation; -14 **5.** Equation; $\frac{1}{3}$
6. Equation; $-\frac{4}{3}$ **7.** Expression; $6z - 23$
8. Expression; $-x - 9$ **9.** Equation; $\frac{7}{9}$
10. Equation; $-\frac{13}{10}$ **11.** Equation; 20
12. Equation; -3 **13.** Equation; $\frac{1}{2}$ **14.** Equation; -6
15. Expression; $\frac{5}{8}x + \frac{7}{4}$ **16.** Expression; $-26t + 18$
17. Equation; no solution   **18.** Equation; no solution
19. Equation; $\frac{23}{12}$ **20.** Equation; $\frac{5}{8}$
21. Equation; all real numbers
22. Equation; all real numbers
23. Equation; $\frac{1}{2}$ **24.** Equation; 0   **25.** Expression; 0
26. Expression; -1 **27.** Expression; $2a + 13$
28. Expression; $8q + 3$ **29.** Equation; 10
30. Equation; $-\frac{1}{20}$

Section 2.4 Practice Exercises, pp. 149–152

3. $x + 5$ **5.** $3x$ **7.** $3x + 20$ **9.** The number is -4.
11. The number is -3.   **13.** The number is 5.
15. The number is -5.   **17.** The number is 9.
19. a. $x + 1, x + 2$ **b.** $x - 1, x - 2$ **21.** The integers are -34 and -33.   **23.** The integers are 13 and 15.

25. The sides are 14 in., 15 in., 16 in., 17 in., and 18 in.
27. The integers are 42, 44, and 46.   **29.** The integers are 13, 15, and 17.   **31.** The lengths of the pieces are 33 cm and 53 cm.   **33.** Karen's age is 35, and Clarann's age is 23.
35. There were 201 Republicans and 232 Democrats.
37. 4.698 million watch *The Dr. Phil Show.*   **39.** The Congo River is 4370 km long, and the Nile River is 6825 km.
41. The area of Africa is 30,065,000 km^2. The area of Asia is 44,579,000 km^2.   **43.** They walked 12.3 mi on the first day and 8.2 mi on the second.   **45.** The pieces are 6 in., 18 in., and 24 in.   **47.** 42, 43, and 44   **49.** Jennifer Lopez made $37 million, and U2 made $69 million.   **51.** The number is 11.
53. The page numbers are 470 and 471.   **55.** The number is 10.   **57.** The deepest point in the Arctic Ocean is 5122 m.
59. The number is $\frac{7}{16}$.   **61.** The number is 2.5.

Section 2.5 Practice Exercises, pp. 157–159

3. The numbers are 21 and 22.   **5.** 12.5%   **7.** 85%
9. 0.75   **11.** 1050.8   **13.** 885   **15.** 2200
17. Molly will have to pay $106.99.   **19.** Approximately 231,000 cases   **21.** 2%   **23.** Javon's taxable income was $84,000.   **25.** Pam will earn $420.
27. Bob borrowed $1200.   **29.** The rate is 6%.
31. Penny needs to invest $3302.   **33. a.** $20.40
b. $149.60   **35.** The original price was $470.59.
37. The discount rate is 12%.   **39.** The original cost was $60.   **41.** The tax rate is 5%.   **43.** The ticket price was $67.   **45.** The original price was $210,000.
47. Alina made $4600 that month.   **49.** Diane sold $645 over $200 worth of merchandise.

Section 2.6 Calculator Connections, p. 164

1. 140.056   **2.** 31.831   **3.** 1.273   **4.** 0.455

1–2.
```
880/(2π)
            140.0563499
1600/(π*(4)²)
            31.83098862
```
3–4.
```
20/(5π)
            1.273239545
10/(7π)
            .4547284088
```

Section 2.6 Practice Exercises, pp. 165–168

3. −5   **5.** 0   **7.** −2   **9.** $a = P - b - c$
11. $y = x + z$ **13.** $q = p - 250$ **15.** $b = \dfrac{A}{h}$
17. $t = \dfrac{PV}{nr}$ **19.** $x = 5 + y$ **21.** $y = -3x - 19$
23. $y = \dfrac{-2x + 6}{3}$ or $y = -\dfrac{2}{3}x + 2$
25. $x = \dfrac{y + 9}{-2}$ or $x = -\dfrac{1}{2}y - \dfrac{9}{2}$
27. $y = \dfrac{-4x + 12}{-3}$ or $y = \dfrac{4}{3}x - 4$
29. $y = \dfrac{-ax + c}{b}$ or $y = -\dfrac{a}{b}x + \dfrac{c}{b}$
31. $t = \dfrac{A - P}{Pr}$ or $t = \dfrac{A}{Pr} - \dfrac{1}{r}$

33. $c = \dfrac{a - 2b}{2}$ or $c = \dfrac{a}{2} - b$ **35.** $y = 2Q - x$
37. $a = MS$ **39.** $R = \dfrac{P}{I^2}$
41. The length is 7 ft, and the width is 5 ft.
43. The length is 120 yd and the width is 30 yd.
45. The length is 195 m, and the width is 100 m.
47. The sides are 22 m, 22 m, and 27 m.
49. "Adjacent supplementary angles form a straight angle." The words *Supplementary* and *Straight* both begin with the same letter.   **51.** The angles are 55° and 35°.
53. The angles are 45° and 135°.
55. $x = 20$; the vertical angles measure 37°.
57. The measures of the angles are 30°, 60°, and 90°.
59. The measures of the angles are 42°, 54°, and 84°.
61. $x = 17$; the measures of the angles are 34° and 56°.
63. a. $A = lw$ **b.** $w = \dfrac{A}{l}$ **c.** The width is 29.5 ft.
65. a. $P = 2l + 2w$ **b.** $l = \dfrac{P - 2w}{2}$ **c.** The length is 103 m.
67. a. $C = 2\pi r$ **b.** $r = \dfrac{C}{2\pi}$ **c.** The radius is approximately 140 ft.   **69. a.** 415.48 m^2 **b.** 10,386.89 m^3

Section 2.7 Practice Exercises, pp. 173–177

3. $c = \dfrac{r}{d}$ **5.** 4   **7.** $200 - t$ **9.** $100 - x$
11. $3000 - y$ **13.** 53 tickets were sold at $3 and 28 tickets were sold at $2.   **15.** Josh downloaded 17 songs for $0.90 and 8 songs for $1.50   **17.** Angelina purchased 25 bottles of soda and 20 bottles of flavored water.
19. $x + 7$ **21.** $d + 2000$
23. 20 oz of 50% antifreeze solution
25. The pharmacist needs to use 21 mL of the 1% saline solution.   **27.** The contractor needs to mix 6.75 oz of 50% acid solution.   **29. a.** 300 mi   **b.** $5x$ **c.** $5(x + 2)$ or $5x + 60$
31. She walks 4 mph to the lake.   **33.** Bryan hiked 6 mi up the canyon.   **35.** The plane travels 600 mph in still air.
37. The slower car travels 48 mph and the faster car travels 52 mph.   **39.** The speeds of the vehicles are 40 mph and 50 mph.   **41.** The rates of the boats are 20 mph and 40 mph.
43. a. 2 lb   **b.** $0.10x$ **c.** $0.10(x + 3) = 0.10x + 0.30$
45. 10 lb of coffee sold at $12 per pound and 40 lb of coffee sold at $8 per pound.   **47.** The boats will meet in $\frac{3}{4}$ hr (45 min).
49. Sam purchased 16 packages of wax and 5 bottles of sunscreen.   **51.** 2.5 quarts of 85% chlorine solution
53. 20 L of water   **55.** The Japanese bullet train travels 300 km/hr and the Acela Express travels 240 km/hr.

Section 2.8 Practice Exercises, pp. 187–192

3. −3

Set-Builder Notation	Graph	Interval Notation
5. $\{x \mid x \geq 6\}$	(graph at 6)	$[6, \infty)$
7. $\{x \mid x \leq 2.1\}$	(graph at 2.1)	$(-\infty, 2.1]$
9. $\{x \mid -2 < x \leq 7\}$	(graph from −2 to 7)	$(-2, 7]$

Set-Builder Notation	Graph	Interval Notation
11. $\left\{x \mid x > \dfrac{3}{4}\right\}$		$\left(\dfrac{3}{4}, \infty\right)$
13. $\{x \mid -1 < x < 8\}$		$(-1, 8)$
15. $\{x \mid x \leq -14\}$		$(-\infty, -14]$

Set-Builder Notation	Graph	Interval Notation
17. $\{x \mid x \geq 18\}$		$[18, \infty)$
19. $\{x \mid x < -0.6\}$		$(-\infty, -0.6)$
21. $\{x \mid -3.5 \leq x < 7.1\}$		$[-3.5, 7.1)$

23. a. $x = 3$ **b.** $x > 3$

25. a. $p = 13$ **b.** $p \leq 13$

27. a. $c = -3$ **b.** $c < -3$

29. a. $z = -\dfrac{3}{2}$ **b.** $z \geq -\dfrac{3}{2}$

31.

33. $-3 < x < 5$

35. $2 \leq x \leq 6$

37. $(-\infty, 1]$

39. $(10, \infty)$

41. $(3, \infty)$

43. $(-\infty, 8]$

45. $(2, \infty)$

47. $(-\infty, -2)$

49. $[14, \infty)$

51. $[-24, \infty)$

53. $[-3, 3)$

55. $\left(0, \dfrac{5}{2}\right)$

57. $(10, 12)$

59. $[-1, 4)$

61. $[90, \infty)$

63. $(-9, \infty)$

65. $\left[-\dfrac{15}{2}, \infty\right)$

67. $(-\infty, -3)$

69. $\left[-\dfrac{1}{3}, \infty\right)$

71. $(-3, \infty)$

73. $(-\infty, 7)$

75. $(-\infty, -5]$

77. $\left(-\infty, \dfrac{15}{4}\right]$

79. $(-\infty, -9)$

81. $(-3, \infty)$

83. $(-\infty, 0]$

85. No   **87.** Yes
89. a. A $[93, 100]$; B+ $[89, 93)$; B $[84, 89)$; C+ $[80, 84)$; C $[75, 80)$; F $[0, 75)$ **b.** B **c.** C   **91.** $L \geq 10$
93. $w > 75$ **95.** $t \leq 72$ **97.** $L \geq 8$ **99.** $2 < h < 5$
101. More than 10.2 in. of rain is needed.
103. a. \$1539   **b.** 200 birdhouses cost \$1440. It is cheaper to purchase 200 birdhouses because the discount is greater.
105. a. $2.00x > 75 + 0.17x$ **b.** $x \geq 41$; profit occurs when 41 or more lemonades are sold.
107. $[13, \infty)$

109. $[-4, \infty)$

111. $(14.5, \infty)$

Chapter 2 Review Exercises, pp. 199–202

1. a. Equation **b.** Expression **c.** Equation
d. Equation   **3. a.** Nonlinear **b.** Linear
c. Nonlinear **d.** Linear   **5.** -8

7. $\dfrac{21}{4}$ **9.** $-\dfrac{21}{5}$ **11.** $-\dfrac{10}{7}$ **13.** The number is 60.

15. The number is -8.   **17.** 1   **19.** $\dfrac{9}{4}$

21. -3 **23.** $\dfrac{3}{4}$ **25.** 0   **27.** $\dfrac{3}{8}$

29. A contradiction has no solution and an identity is true for all real numbers.   **31.** 6   **33.** 13   **35.** -10

37. $\dfrac{5}{3}$ **39.** 2.5   **41.** -4.2 **43.** -312

45. No solution   **47.** All real numbers
49. The number is 30.   **51.** The number is -7.
53. The integers are 66, 68, and 70.
55. The sides are 25 in., 26 in., and 27 in.
57. The minimum salary was \$30,000 in 1980.
59. 23.8   **61.** 12.5%   **63.** 160   **65.** The novel originally cost \$29.50.   **67. a.** \$840 **b.** \$3840

69. $K = C + 273$ **71.** $s = \dfrac{P}{4}$ **73.** $x = \dfrac{y - b}{m}$

75. $y = \dfrac{-2x - 2}{5}$ **77. a.** $h = \dfrac{3V}{\pi r^2}$ **b.** The height is 5.1 in.

79. The angles are 22°, 78°, and 80°.   **81.** The length is 5 ft, and the width is 4 ft.   **83.** The measure of angle y is 53°.   **85.** Gus rides 15 mph and Winston rides 18 mph.
87. They meet in 2.25 hr (2 hr and 15 min).
89. 20 lb of the 40% solder should be used.

91. $\left(-\infty, \dfrac{1}{2}\right]$

93. a. $637　**b.** 300 plants cost $1410, and 295 plants cost $1416. 300 plants cost less.

95. $\left(-\dfrac{1}{3}, \infty\right)$

97. $\left(-\infty, -\dfrac{14}{5}\right]$

99. $(-\infty, 34.5)$

101. $\left(-\infty, \dfrac{5}{2}\right]$

103. $[-6, 5]$

105. a. $1.50x > 33 + 0.4x$　**b.** $x > 30$; a profit is realized if more than 30 hot dogs are sold.

Chapter 2 Test, pp. 202–203

1. a. b, d　**2. a.** $5x + 7$　**b.** 9　**3.** -16　**4.** 12

5. $-\dfrac{16}{9}$　**6.** $\dfrac{7}{3}$　**7.** 15　**8.** $\dfrac{13}{4}$　**9.** $\dfrac{20}{21}$

10. No solution　**11.** -3　**12.** -47

13. All real numbers　**14.** $y = -3x - 4$　**15.** $r = \dfrac{C}{2\pi}$

16. 90　**17.** The numbers are 18 and 13.
18. The sides are 61 in., 62 in., 63 in., 64 in., and 65 in.
19. The cost was $82.00.　**20.** Each basketball ticket was $36.32, and each hockey ticket was $40.64.
21. Clarita originally borrowed $5000.　**22.** The field is 110 m long and 75 m wide.　**23.** $y = 30$; The measures of the angles are $30°$, $39°$, and $111°$.
24. Paula needs 30 lb of macadamia nuts.
25. One family travels 55 mph and the other travels 50 mph.
26. The measures of the angles are $32°$ and $58°$.
27. a. $(-\infty, 0)$　　　　**b.** $[-2, 5)$

28. $(-2, \infty)$　　　　**29.** $(-\infty, -4]$

30. $\left(-\dfrac{3}{2}, \infty\right)$　　　**31.** $[-5, 1]$

32. More than 26.5 in. is required.

Chapters 1–2 Cumulative Review Exercises, pp. 203–204

1. $\dfrac{1}{2}$　**2.** -7　**3.** $-\dfrac{5}{12}$　**4.** 16　**5.** 4

6. $\sqrt{5^2 - 9}$; 4　**7.** $-14 + 12$; -2　**8.** $-7x^2y$, $4xy$, -6
9. $9x + 13$　**10.** 4　**11.** -7.2

12. All real numbers　**13.** -8　**14.** $-\dfrac{4}{7}$　**15.** -80

16. The numbers are 77 and 79.　**17.** The cost before tax was $350.00.　**18.** The height is $\frac{41}{6}$ cm or $6\frac{5}{6}$ cm.
19. $(-2, \infty)$　　　　**20.** $[-1, 9]$

Chapter 3

Chapter Opener Puzzle

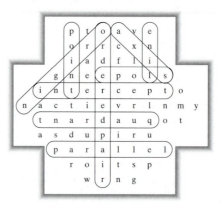

Section 3.1 Practice Exercises, pp. 210–214

3. a. Month 10　**b.** 30　**c.** Between months 3 and 5 and between months 10 and 12　**d.** Months 8 and 9
e. Month 3　**f.** 80
5. a. On day 1 the price per share was $89.25.
b. $1.75　**c.** $-$2.75
7.

9.

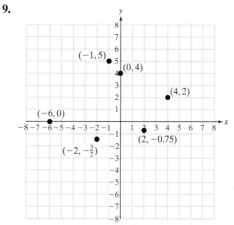

11. IV　**13.** II　**15.** III　**17.** I
19. $(0, -5)$ lies on the y-axis.
21. $\left(\frac{7}{8}, 0\right)$ is located on the x-axis.
23. $A(-4, 2)$, $B(\frac{1}{2}, 4)$, $C(3, -4)$, $D(-3, -4)$, $E(0, -3)$, $F(5, 0)$
25. a. $A(400, 200)$, $B(200, -150)$, $C(-300, -200)$, $D(-300, 250)$, $E(0, 450)$　**b** 450 m

27. a. $(250, 225), (175, 193), (315, 330), (220, 209), (450,$
$570), (400, 480), (190, 185);$ the ordered pair $(250, 225)$ means
that 250 people produce $225 in popcorn sales.

b.

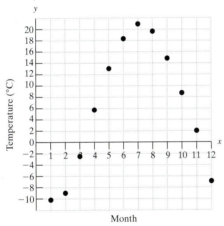

29. a. $(10, 6800)$ means that in 1990, the poverty level
was $6800.

b.

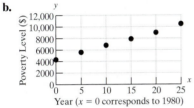

31. a. $(1, -10.2), (2, -9.0), (3, -2.5), (4, 5.7), (5, 13.0),$
$(6, 18.3), (7, 20.9), (8, 19.6), (9, 14.8), (10, 8.7), (11, 2.0),$
$(12, -6.9).$

b.

33. a.

b. Increasing   **c.** Increasing

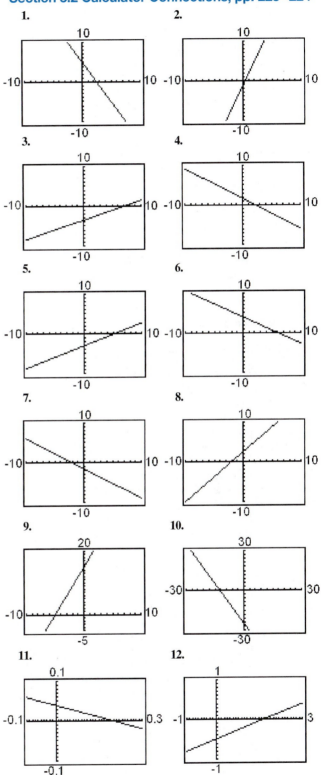

Section 3.2 Practice Exercises, pp. 224–230

3. $(2, 4);$ quadrant I   **5.** $(0, -1);$ y-axis
7. $(3, -4);$ quadrant IV   **9.** Yes   **11.** Yes   **13.** No
15. No   **17.** Yes

19.

x	y
1	-3
-2	0
-3	1
-4	2

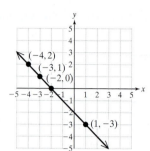

21.

x	y
-2	3
-1	0
-4	9

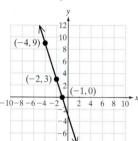

23.

x	y
0	4
2	0
3	-2

25.

x	y
0	-2
5	-5
10	-8

27.

x	y
0	-2
-3	-2
5	-2

29.

x	y
3/2	-1
3/2	2
3/2	-3

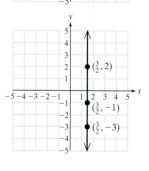

31.

x	y
0	4.6
1	3.4
2	2.2

33.

35.

37.

39.

41.

43.

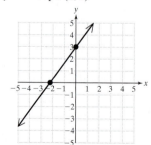

45. y-axis   **47.** x-intercept: $(-1, 0)$; y-intercept: $(0, -3)$
49. x-intercept: $(-4, 0)$; y-intercept: $(0, 1)$

51. x-intercept: $\left(-\dfrac{9}{4}, 0\right)$;   **53.** x-intercept: $\left(\dfrac{8}{3}, 0\right)$;

y-intercept: $(0, 3)$ y-intercept: $(0, 2)$

55. x-intercept: $(-4, 0)$;
y-intercept: $(0, 8)$

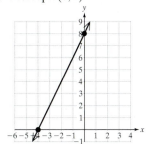

57. x-intercept: $(0, 0)$;
y-intercept: $(0, 0)$

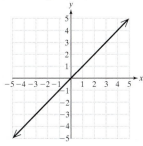

Section 3.3 Practice Exercises, pp. 237–243

3. x-intercept: $(6, 0)$;
y-intercept: $(0, -2)$

5. x-intercept: $(0, 0)$;
y-intercept: $(0, 0)$

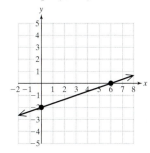

59. x-intercept: $(10, 0)$;
y-intercept: $(0, 5)$

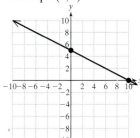

61. x-intercept: $(0, 0)$;
y-intercept: $(0, 0)$

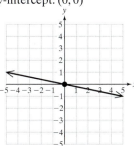

7. x-intercept: $(2, 0)$;
y-intercept: $(0, 8)$

63. True
67. a. Horizontal
c. no x-intercept;
y-intercept: $(0, -1)$

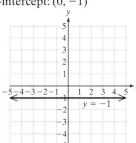

65. True
69. a. Vertical
c. x-intercept: $(4, 0)$;
no y-intercept

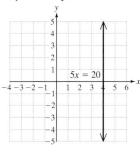

9. $m = \dfrac{1}{3}$ **11.** $m = \dfrac{6}{11}$ **13.** Undefined

15. Positive   **17.** Negative   **19.** Zero
21. Undefined   **23.** Positive   **25.** Negative

27. $m = \dfrac{1}{2}$ **29.** $m = -3$ **31.** $m = 0$

33. The slope is undefined.   **35.** $\dfrac{1}{3}$ **37.** -3

39. $\dfrac{3}{5}$ **41.** Zero   **43.** Undefined   **45.** $\dfrac{28}{5}$ **47.** $-\dfrac{7}{8}$

49. -0.45 or $-\dfrac{9}{20}$ **51.** -0.15 or $-\dfrac{3}{20}$

53. a. -2 **b.** $\dfrac{1}{2}$ **55. a.** 0 **b.** undefined

57. a. $\dfrac{4}{5}$ **b.** $-\dfrac{5}{4}$ **59. a.** undefined   **b.** 0

61. Perpendicular   **63.** Parallel   **65.** Neither
67. l_1: $m = 2$, l_2: $m = 2$; parallel
69. l_1: $m = 5$, l_2: $m = -\dfrac{1}{5}$; perpendicular

71. a. Horizontal
c. no x-intercept;
y-intercept $(0, -5)$

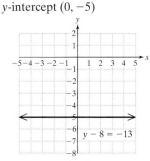

73. a. Vertical
c. All points on the y-axis are
y-intercepts; x-intercept: $(0, 0)$

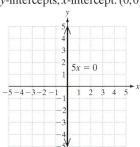

71. l_1: $m = \dfrac{1}{4}$, l_2: $m = 4$; neither

73. a. $m = 47$ **b.** The number of male prisoners increased at a rate of 47 thousand per year during this time period.
75. a. $m \approx 0.009$ **b.** The slope $m = 0.009$ means that postage increased by about $0.009 per year (or equivalently 0.9¢ per year) during this time.

77. $m = \dfrac{3}{4}$ **79.** $m = 0$

75. A horizontal line may not have an x-intercept. A vertical line may not have a y-intercept.   **77.** a b, d
79. a. $y = 10{,}068$ **b.** $x = 3$ **c.** $(1, 10068)$ One year after purchase the value of the car is $10,068. $(3, 7006)$ Three years after purchase the value of the car is $7006.

81. For example:
$(4, 0)$ and $(-2, -4)$

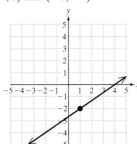

83. For example:
$(3, -1)$ and $(1, 5)$

85.

87.

89.

91. $\dfrac{3m - 3n}{2b}$ or $\dfrac{-3m + 3n}{-2b}$ **93.** $\left(\dfrac{c}{a}, 0\right)$

95. For example: $(7, 1)$

Section 3.4 Calculator Connections, p. 247

1. Perpendicular

2. Parallel

3. Neither

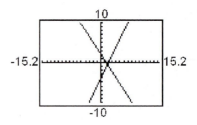

4. The lines may appear parallel; however, they are not parallel because the slopes are different.

5. The lines may appear to coincide on a graph; however, they are not the same line because the y-intercepts are different.

6. The line may appear to be horizontal, but it is not. The slope is 0.001 rather than 0.

Section 3.4 Practice Exercises, pp. 248–251

3. x-intercept: $(10, 0)$; y-intercept: $(0, -2)$

5. x-intercept: none; y-intercept: $(0, -3)$

7. x-intercept: $(0, 0)$; y-intercept: $(0, 0)$

9. x-intercept: $(4, 0)$; ; y-intercept: none

11. $m = -2$; y-intercept: $(0, 3)$ **13.** $m = 1$; y-intercept: $(0, -2)$ **15.** $m = -1$; y-intercept: $(0, 0)$

17. $m = \dfrac{3}{4}$; y-intercept: $(0, -1)$ **19.** $m = \dfrac{2}{5}$; y-intercept: $\left(0, -\dfrac{4}{5}\right)$ **21.** $m = 3$; y-intercept: $(0, -5)$

23. $m = -1$; y-intercept: $(0, 6)$ **25.** Undefined slope; no y-intercept   **27.** $m = 0$; y-intercept: $\left(0, -\dfrac{1}{4}\right)$

29. $m = \dfrac{2}{3}$; y-intercept: $(0, 0)$

31.

33.

35. b   **37.** e   **39.** c

41. $y = \dfrac{1}{2}x - 3$ **43.** $y = -2x + 9$

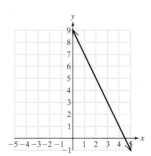

45. $y = -\frac{1}{2}x + \frac{3}{2}$

47. $y = -x$

49. $y = \frac{4}{5}x$

51. $y = -\frac{2}{3}$

53. Perpendicular   **55.** Neither   **57.** Parallel
59. Perpendicular   **61.** Parallel   **63.** Perpendicular

65. Neither   **67.** Parallel   **69.** $y = -\frac{1}{3}x + 2$

71. $y = 10x - 19$ **73.** $y = -11$ **75.** $y = 5x$
77. $y = 6x - 2$ **79. a.** $m = 49.95$. The cost increases
$49.95 per day.   **b.** $(0, 31.95)$. The base fee for renting a car
is $31.95.   **c.** $381.60   **81.** $y = -\frac{A}{B}x + \frac{C}{B}$; the slope is $-\frac{A}{B}$.

83. $m = -\frac{6}{7}$ **85.** $m = \frac{11}{8}$

Chapter 3 Problem Recognition Exercises, p. 252

1. a, c, d   **2.** b, f, h   **3.** a   **4.** f   **5.** b, f   **6.** c
7. c, d   **8.** f   **9.** e   **10.** g   **11.** b   **12.** h
13. g   **14.** e   **15.** c   **16.** b, h   **17.** e   **18.** e
19. b, f, h   **20.** c, d

Section 3.5 Practice Exercises, pp. 257–260

3.

5.

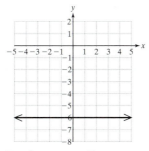

7. 9   **9.** 0   **11.** $y = 3x + 7$ or $3x - y = -7$

13. $y = -4x - 14$ or $4x + y = -14$ **15.** $y = -\frac{1}{2}x - \frac{1}{2}$

or $x + 2y = -1$ **17.** $y = \frac{1}{4}x + 8$ or $x - 4y = -32$

19. $y = 4.5x - 25.6$ or $45x - 10y = 256$ **21.** $y = -2$
23. $y = 2x - 2$ or $2x - y = 2$ **25.** $y = -x - 4$ or
$x + y = -4$ **27.** $y = -0.2x - 2.86$ or
$20x + 100y = -286$ **29.** $y = -2x + 1$

31. $y = 2x + 4$ **33.** $y = \frac{1}{2}x - 1$ **35.** $y = 4x + 13$ or

$4x - y = -13$ **37.** $y = -\frac{3}{2}x + 6$ or $3x + 2y = 12$

39. $y = -2x - 8$ or $2x + y = -8$ **41.** $y = -\frac{1}{5}x - 6$

or $x + 5y = -30$ **43.** $y = 3x - 8$ or $3x - y = 8$
45. iv   **47.** vi   **49.** iii   **51.** $y = 1$ **53.** $x = 2$
55. $y = 2$ **57.** $x = -6$ **59.** $x = -4$

Section 3.6 Calculator Connections, p. 263

1. 13.3

2. −42.3

3. 345

4. 95

Section 3.6 Practice Exercises, pp. 264–267

3. x-intercept: $(6, 0)$; y-intercept: $(0, 5)$ **5.** x-intercept:
$(-2, 0)$; y-intercept: $(0, -4)$ **7.** x-intercept: none;
y-intercept: $(0, -9)$ **9. a.** $3.00   **b.** $7.20
c. The y-intercept is $(0, 1.6)$. This indicates that the minimum
wage was $1.60 per hour in the year 1970.   **d.** The slope is
0.14. This indicates that the minimum wage has risen
approximately $0.14 per year during this period.
11. a. $30.7°$ **b.** $13.4°$ **c.** $m = -2.333$. The average
temperature in January decreases at a rate of 2.333° per 1°
of latitude.   **d.** $(53.2, 0)$. At 53.2° latitude, the average
temperature in January is 0°.
13. a. $95   **b.** $190   **c.** $(0, 0)$. For 0 kilowatt-hours used,
the cost is $0.   **d.** $m = 0.095$. The cost increases by $0.095
for each kilowatt-hour used.   **15. a.** $y = -0.08x + 7.6$
b. 5 days   **17. a.** $y = 2.5x + 31$ **b.** 43.5 in.
19. a. $y = 0.25x + 20$ **b.** $84.50   **21. a.** $y = 90x + 105$
b. $1185.00   **23. a.** $y = 0.8x + 100$ **b.** $260.00

Section 3.7 Calculator Connections, p. 275

1.

2.

3.   **4.**

Section 3.7 Practice Exercises, pp. 275–279

3. a. x-intercept: $\left(\dfrac{5}{3}, 0\right)$; y-intercept: $(0, -5)$ **b.** 3

c.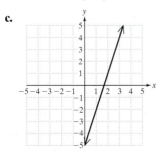

5. Domain: $\{4, 3, 0\}$; range: $\{2, 7, 1, 6\}$
7. Domain: $\{\frac{1}{2}, 0, 1\}$; range: $\{3\}$ **9.** Domain: $\{0, 5, -8, 8\}$;
range: $\{0, 2, 5\}$ **11.** Domain: $\{\text{Atlanta, Macon, Pittsburgh}\}$;
range: $\{\text{GA, PA}\}$ **13.** Domain: $\{\text{New York, California}\}$;
range: $\{\text{Albany, Los Angeles, Buffalo}\}$ **15.** The relation is
a function if each element in the domain has exactly one
corresponding element in the range.
17. The relations in Exercises 7, 9, and 11 are functions.
19. Yes   **21.** No   **23.** No   **25.** Yes   **27.** Yes
29. $-5, -1, -11$ **31.** $\frac{1}{5}, \frac{1}{4}, \frac{1}{2}$ **33.** $7, 2, 3$ **35.** $0, 1, 2$
37. b   **39.** c   **41.** The function value at $x = 6$ is 2.
43. The function value at $x = \frac{1}{2}$ is $\frac{1}{4}$.   **45.** $(2, 7)$
47. $(0, 8)$ **49. a.** $s(1) = 32$. The speed of an object
1 sec after being dropped is 32 ft/sec.   **b.** $s(2) = 64$. The
speed of an object 2 sec after being dropped is 64 ft/sec.
c. $s(10) = 320$. The speed of an object 10 sec after being
dropped is 320 ft/sec.   **d.** 294.4 ft/sec
51. a. $h(0) = 3$. The initial height of the ball is 3 ft.
b. $h(1) = 51$. The height of the ball 1 sec after being kicked
is 51 ft.   **c.** $h(2) = 67$. The height of the ball 2 sec after being
kicked is 67 ft.   **d.** $h(4) = 3$. The height of the ball 4 sec after
being kicked is 3 ft.

Chapter 3 Review Exercises, pp. 286–290

1.

3. III   **5.** IV   **7.** IV   **9.** x-axis
11. a. On day 1, the price was $26.25.   **b.** Day 2   **c.** $2.25
13. No   **15.** Yes

17.

x	y
2	1
3	4
1	-2

19.

x	y
0	-1
3	1
-6	-5

21.   **23.**

25. Vertical   **27.** Horizontal

    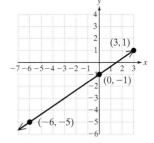

29. x-intercept: $(-3, 0)$; y-intercept: $\left(0, \dfrac{3}{2}\right)$
31. x-intercept: $(0, 0)$; y-intercept: $(0, 0)$
33. x-intercept: none; y-intercept: $(0, -4)$
35. x-intercept: $\left(-\dfrac{5}{2}, 0\right)$; y-intercept: none
37. $m = \dfrac{12}{5}$ **39.** $-\dfrac{2}{3}$ **41.** Undefined
43. a. -5 **b.** $\dfrac{1}{5}$ **45.** $m_1 = \dfrac{2}{3}; m_2 = \dfrac{2}{3}$; parallel
47. $m_1 = -\dfrac{5}{12}; m_2 = \dfrac{12}{5}$; perpendicular   **49.** $y = \dfrac{5}{2}x - 5$;
$m = \dfrac{5}{2}$; y-intercept: $(0, -5)$ **51.** $y = \dfrac{1}{3}x$;
$m = \dfrac{1}{3}$; y-intercept: $(0, 0)$ **53.** $y = -\dfrac{5}{2}$; $m = 0$;
y-intercept: $\left(0, -\dfrac{5}{2}\right)$ **55.** Neither   **57.** Parallel
59. Perpendicular   **61.** $y = 5x$ or $5x - y = 0$
63. For example: $5x + 2y = -4$ **65.** $y - y_1 = m(x - x_1)$

67. For example: $y = -5$ **69.** $y = \frac{2}{3}x + \frac{5}{3}$ or

$2x - 3y = -5$ **71.** $y = -5$

73. $y = 4x + 31$ or $4x - y = -31$

75. a. $m = 137$ **b.** The number of prescriptions increased by 137 million per year during this time period.
c. $y = 137x + 2140$ **d.** 4195 million

77. a. $y = 8x + 700$ **b.** $1340   **79.** Domain: {2}; range: {0, 1, −5, 2}; not a function

81. Domain: $(-\infty, \infty)$; range: $[-2, \infty)$; function

83. Domain: {3, −4, 0, 2}; range: $\{0, \frac{1}{2}, 3, -12\}$; function

85. a. 0   **b.** 4   **c.** $-\frac{1}{6}$ **d.** $\frac{3}{2}$ **e.** $-\frac{1}{2}$

Chapter 3 Test, pp. 290–292

1. a. II   **b.** IV   **c.** III   **2.** 0   **3.** 0

4. a. (5, 46) At age 5 the boy's height was 46 in. (7, 50) (9, 55) (11, 60)

b.

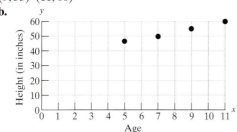

c. 57.5 in.   **d.** No, his height will maximize in his teen years.

5. a. No   **b.** Yes   **c.** Yes   **d.** Yes

6.

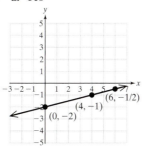

x	y
0	−2
4	−1
6	$-\frac{1}{2}$

7. a. 202 beats per minute   **b.** (20, 200), (30, 190), (40, 180) (50, 170), (60, 160)

8. Horizontal   **9.** Vertical

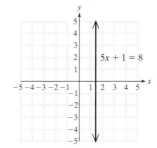

10. x-intercept: $\left(-\frac{3}{2}, 0\right)$; y-intercept: (0, 2)

11. x-intercept: (0, 0); y-intercept: (0, 0)

12. x-intercept: (4, 0); y-intercept: none

13. x-intercept: none; y-intercept: (0, 3)   **14.** $\frac{2}{5}$

15. a. $\frac{1}{3}$ **b.** $\frac{4}{3}$ **16. a.** $-\frac{1}{4}$ **b.** 4

17. a. Undefined   **b.** 0

18.

19.

20. Perpendicular   **21.** $y = \frac{1}{4}x + \frac{1}{2}$ or $x - 4y = -2$

22. $y = -\frac{7}{2}x + 15$ or $7x + 2y = 30$ **23.** $y = -6$

24. $y = -\frac{1}{3}x + 1$ or $x + 3y = 3$ **25.** $y = 3x + 8$ or

$3x - y = -8$ **26. a.** $\frac{3}{4}$ **b.** $\frac{1}{2}$ **27. a.** $y = 1.5x + 10$

b. $25   **28. a.** $m = 20$; The slope indicates that there is an increase of 20 thousand medical doctors per year.
b. $y = 20x + 414$ **c.** 1014 thousand or, equivalently, 1,014,000   **29. a.** {0, 2, −15, 4, 9}   **b.** {−1, 3, −8, 4}
c. Function   **30. a.** 6   **b.** 12

Chapters 1–3 Cumulative Review Exercises, p. 293

1. a. Rational   **b.** Rational   **c.** Irrational   **d.** Rational

2. a. $\frac{2}{3}, \frac{2}{3}$ **b.** −5.3; 5.3   **3.** 69   **4.** −13   **5.** 18

6. $\frac{3}{4} \div -\frac{7}{8}; -\frac{6}{7}$ **7.** (−2.1)(−6); 12.6

8. The associative property of addition   **9.** 4   **10.** 5

11. $-\frac{9}{2}$ **12.** −2   **13.** 9241 mi^2 **14.** $a = \frac{c - b}{3}$

15.

16. x-intercept: (−2, 0); y-intercept: (0, 1)

17. $y = -\frac{3}{2}x - 6$;   slope: $-\frac{3}{2}$; y-intercept: (0, −6)

18. $2x + 3 = 5$ can be written as $x = 1$, which represents a vertical line. A vertical line of the form $x = k$ ($k \neq 0$) has an x-intercept of $(k, 0)$ and no y-intercept.

19. $y = -3x + 1$ or $3x + y = 1$ **20.** $y = \frac{2}{3}x + 6$ or $2x - 3y = -18$

represent intersecting lines.   **5.** No solution. The equations represent parallel lines.   **6.** Infinitely many solutions. The equations represent the same line.   **7.** $(5, 0)$
8. $(1, -7)$ **9.** $(4, -5)$ **10.** $(2, 3)$ **11.** $(2, 0)$
12. $(8, 10)$ **13.** $\left(2, -\dfrac{5}{7}\right)$ **14.** $\left(-\dfrac{14}{3}, -4\right)$
15. No solution; inconsistent system   **16.** No solution; inconsistent system   **17.** $(-1, 0)$ **18.** $(5, 0)$
19. Infinitely many solutions; $\{(x, y) \mid y = 2x - 14\}$; dependent system   **20.** Infinitely many solutions; $\{(x, y) \mid x = 5y - 9\}$; dependent system
21. $(2200, 1000)$ **22.** $(3300, 1200)$ **23.** $(5, -7)$
24. $(2, -1)$ **25.** $\left(\dfrac{2}{3}, \dfrac{1}{2}\right)$ **26.** $\left(\dfrac{1}{4}, -\dfrac{3}{2}\right)$

Section 4.4 Practice Exercises, pp. 331–335

1. $(-1, 4)$ **3.** $\left(\dfrac{5}{2}, 1\right)$ **5.** The numbers are 4 and 16.
7. The angles are 80° and 10°.   **9.** DVDs are $10.50 each, and CDs are $15.50 each.   **11.** Technology stock costs $16 per share, and the mutual fund costs $11 per share.
13. Patricia bought forty 42¢ stamps and ten 59¢ stamps.
15. Shanelle invested $3500 in the 10% account and $6500 in the 7% account.   **17.** $9000 was borrowed at 6%, and $3000 was borrowed at 9%.   **19.** Invest $12,000 in the bond fund and $18,000 in the stock fund.   **21.** 15 gal of the 50% mixture should be mixed with 10 gal of the 40% mixture.   **23.** 12 gal of the 45% disinfectant solution should be mixed with 8 gal of the 30% disinfectant solution.
25. She should mix 20 mL of the 13% solution with 30 mL of the 18% solution.   **27.** The speed of the boat in still water is 6 mph, and the speed of the current is 2 mph.   **29.** The speed of the plane in still air is 300 mph, and the wind is 20 mph.   **31.** The speed of the plane in still air is 525 mph and the speed of the wind is 75 mph.   **33.** There are 17 dimes and 22 nickels.   **35. a.** 835 free throws and 1597 field goals   **b.** 4029 points   **c.** Approximately 50 points per game   **37.** The speed of the plane in still air is 160 mph, and the wind is 40 mph.   **39.** 12 lb of candy should be mixed with 8 lb of nuts.   **41.** $15,000 was invested in the 5.5% account, and $45,000 was invested in the 6.5% account.
43. Dallas scored 30 points, and Buffalo scored 13 points.
45. There were 300 women and 200 men in the survey.

Section 4.5 Practice Exercises, pp. 341–345

3.

5. When the inequality symbol is $\leq$ or $\geq$
7. All of the points in the shaded region are solutions to the inequality.   **9.** a

11. For example:
$(0, 5)\ (2, 7)\ (-1, 8)$

13. For example:
$(1, -1)\ (3, 0)\ (-2, -9)$

15. For example:
$(0, 0)\ (0, 2)\ (-1, -3)$

17.

19.

21.

23.

25.

27.

29.

31.

33.

53.

35. a. The set of ordered pairs above the line $x + y = 4$, for example, $(6, 3)(-2, 8)(0, 5)$ **b.** The set of ordered pairs on the line $x + y = 4$, for example, $(0, 4)(4, 0)(2, 2)$ **c.** The set of ordered pairs below the line $x + y = 4$, for example, $(0, 0)(-2, 1)(3, 0)$

Chapter 4 Review Exercises, pp. 352–355

1. Yes   **3.** No   **5.** Intersecting lines (the lines have different slopes)   **7.** Parallel lines (the lines have the same slope but different y-intercepts)   **9.** Coinciding lines (the lines have the same slope and same y-intercept)
11. $(0, -2)$

37.

39.

13. Infinitely many solutions; $\{(x, y)\,|\,2x + y = 5\}$; dependent system

41.

43.

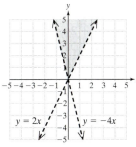

15. $(1, -1)$ **17.** No solution; inconsistent system

45.

47.

49.

51.

19. 200 mi   **21.** $(-4, 1)$ **23.** Infinitely many solutions; $\{(x, y)\,|\,y = -2x + 2\}$; dependent system   **25. a.** y in the second equation is easiest to isolate because its coefficient is 1. **b.** $\left(\dfrac{9}{2}, 3\right)$ **27.** $(0, 4)$ **29.** No solution; inconsistent system   **31.** The angles are $42°$ and $48°$.
33. See page 317.   **35. b.** $(2, 2)$ **37.** $(-6, 2)$

69. $\dfrac{c^{27}d^{31}}{2}$ **71.** $-\dfrac{27a^9b^3}{c^6}$ **73.** $16b^{26}$ **75.** x^{2m}

77. $125a^{6n}$ **79.** $\dfrac{m^{2b}}{n^{3b}}$ **81.** $\dfrac{3^n a^{3n}}{5^n b^{4n}}$

Section 5.3 Practice Exercises, pp. 381–383

3. c^9 **5.** y **7.** 3^6 or 729 **9.** $7^4 w^{28}z^8$ or $2401w^{28}z^8$
11. a. 1 **b.** 1 **13.** 1 **15.** 1 **17.** -1 **19.** 1
21. 1 **23.** -7 **25. a.** $\dfrac{1}{t^5}$ **b.** $\dfrac{1}{t^5}$ **27.** $\dfrac{343}{8}$ **29.** 25
31. $\dfrac{1}{a^3}$ **33.** $\dfrac{1}{12}$ **35.** $\dfrac{1}{16b^2}$ **37.** $\dfrac{6}{x^2}$ **39.** $\dfrac{1}{64}$
41. $-\dfrac{3}{y^4}$ **43.** $-\dfrac{1}{t^3}$ **45.** a^5 **47.** $\dfrac{x^4}{x^{-6}} = x^{4-(-6)} = x^{10}$
49. $2a^{-3} = 2 \cdot \dfrac{1}{a^3} = \dfrac{2}{a^3}$ **51.** $\dfrac{1}{x^4}$ **53.** 1 **55.** y^4
57. $\dfrac{n^{27}}{m^{18}}$ **59.** $\dfrac{81k^{24}}{j^{20}}$ **61.** $\dfrac{1}{p^6}$ **63.** $\dfrac{1}{r^3}$ **65.** a^8
67. $\dfrac{1}{y^8}$ **69.** $\dfrac{1}{7^7}$ **71.** 1 **73.** $\dfrac{1}{a^4 b^6}$ **75.** $\dfrac{1}{w^{21}}$
77. $\dfrac{1}{27}$ **79.** 1 **81.** $\dfrac{1}{64x^6}$ **83.** $-\dfrac{16y^4}{z^2}$ **85.** $-\dfrac{a^{12}}{6}$
87. $80c^{21}d^{24}$ **89.** $\dfrac{p^{27}}{8}$ **91.** $\dfrac{4d^{16}}{c^2}$ **93.** $\dfrac{9y^2}{2x}$
95. $\dfrac{9}{20}$ **97.** $\dfrac{9}{10}$ **99.** $\dfrac{5}{4}$ **101.** $\dfrac{10}{3}$

Section 5.4 Calculator Connections, p. 386

1.–2.

```
(5.2E6)*(4.6E-3)
                23920
(2.19E-8)*(7.84E
-4)
           1.71696E-11
```

3.–4.

```
(4.76E-5)/(2.38E
9)
                2E-14
(8.5E4)/(4.0E-1)
               212500
```

5.

```
((9.6E7)*(4.0E-3
))/(2.0E-2)
            19200000
```

6.

```
((5.0E-12)*(6.4E
-5))/((1.6E-8)*(
4.0E2))
                5E-11
```

Section 5.4 Practice Exercises, pp. 387–389

3. b^{13} **5.** 10^{13} **7.** $\dfrac{1}{y^5}$ **9.** $\dfrac{x^{20}}{y^{12}}$ **11.** w^4 **13.** 10^4

15. Move the decimal point between the 2 and 3 and
multiply by 10^{-10}; 2.3×10^{-10}.   **17.** 5×10^4
19. 2.08×10^5 **21.** 6.01×10^6 **23.** 8×10^{-6}
25. 1.25×10^{-4} **27.** 6.708×10^{-3} **29.** 1.7×10^{-24} g
31. $\$2.7 \times 10^{10}$ **33.** 6.8×10^7 gal; 1.0×10^2 miles
35. Move the decimal point nine places to the left;
0.000 000 0031.   **37.** 0.00005   **39.** 2800   **41.** 0.000603
43. 2,400,000   **45.** 0.019   **47.** 7032
49. 0.000 000 000 001 g   **51.** 1600 calories and 2800 calories

53. 5.0×10^4 **55.** 3.6×10^{11} **57.** 2.2×10^4
59. 2.25×10^{-13} **61.** 3.2×10^{14} **63.** 2.432×10^{-10}
65. 3.0×10^{13} **67.** 6.0×10^5 **69.** 1.38×10^1
71. 5.0×10^{-14} **73.** 3.75 in.   **75.** $\$2.97 \times 10^{10}$
77. a. 6.5×10^7 **b.** 2.3725×10^{10} days
c. 5.694×10^{11} hr   **d.** 2.04984×10^{15} sec

Chapter 5 Problem Recognition Exercises, p. 389

1. t^8 **2.** 2^8 or 256 **3.** y^5 **4.** p^6 **5.** $r^4 s^8$
6. $a^3 b^9 c^6$ **7.** w^6 **8.** $\dfrac{1}{m^{16}}$ **9.** $\dfrac{x^4 z^3}{y^7}$ **10.** $\dfrac{a^3 c^8}{b^6}$
11. 1.25×10^3 **12.** 1.24×10^5 **13.** 8.0×10^8
14. 6.0×10^{-9} **15.** p^{15} **16.** p^{15} **17.** $\dfrac{1}{v^2}$
18. $c^{50}d^{40}$ **19.** 3 **20.** -4 **21.** $\dfrac{b^9}{2^{15}}$ **22.** $\dfrac{81}{y^6}$
23. $\dfrac{16y^4}{81x^4}$ **24.** $\dfrac{25d^6}{36c^2}$ **25.** $3a^7 b^5$ **26.** $64x^7 y^{11}$
27. $\dfrac{y^4}{x^8}$ **28.** $\dfrac{1}{a^{10}b^{10}}$ **29.** $\dfrac{1}{t^2}$ **30.** $\dfrac{1}{p^7}$ **31.** $\dfrac{8w^6 x^9}{27}$
32. $\dfrac{25b^8}{16c^6}$ **33.** $\dfrac{q^3 s}{r^2 t^5}$ **34.** $\dfrac{m^2 p^3 q}{n^3}$ **35.** $\dfrac{1}{y^{13}}$ **36.** w^{10}
37. $-\dfrac{1}{8a^{18}b^6}$ **38.** $\dfrac{4x^{18}}{9y^{10}}$ **39.** $\dfrac{k^8}{5h^6}$ **40.** $\dfrac{6n^{10}}{m^{12}}$

Section 5.5 Practice Exercises, pp. 395–398

3. $\dfrac{45}{x^2}$ **5.** $\dfrac{2}{t^4}$ **7.** $\dfrac{1}{3^{12}}$
9. 4.0×10^{-2} is in scientific notation in which 10 is raised
to the -2 power. 4^{-2} is not in scientific notation and 4 is being
raised to the -2 power.   **11.** $-7x^4 + 7x^2 + 9x + 6$
13. Binomial; $10; 2$ **15.** Monomial; $6; 2$
17. Binomial; $-1; 4$ **19.** Trinomial; $12; 4$
21. Monomial; $23; 0$ **23.** Monomial; $-32; 3$
25. The exponents on the x-factors are different.
27. $35x^2 y$ **29.** $10y$ **31.** $8b^5 d^2 - 9d$
33. $4y^2 + y - 9$ **35.** $4a - 8c$ **37.** $a - \dfrac{1}{2}b - 2$
39. $\dfrac{4}{3}z^2 - \dfrac{5}{3}$ **41.** $7.9t^3 - 3.4t^2 + 6t - 4.2$ **43.** $-4h + 5$
45. $2m^2 - 3m + 15$ **47.** $-3v^3 - 5v^2 - 10v - 22$
49. $-8a^3 b^2$ **51.** $-53x^3$ **53.** $-5a - 3$ **55.** $16k + 9$
57. $2s - 14$ **59.** $3t^2 - 4t - 3$ **61.** $-2r - 3s + 3t$
63. $\dfrac{3}{4}x + \dfrac{1}{3}y - \dfrac{3}{10}$ **65.** $-\dfrac{2}{3}h^2 + \dfrac{3}{5}h - \dfrac{5}{2}$
67. $2.4x^4 - 3.1x^2 - 4.4x - 6.7$
69. $4b^3 + 12b^2 - 5b - 12$ **71.** $-3x^3 - 2x^2 + 11x - 31$
73. $4y^3 + 2y^2 + 2$ **75.** $3a^2 - 3a + 5$
77. $9ab^2 - 3ab + 16a^2 b$ **79.** $4z^5 + z^4 + 9z^3 - 3z - 2$
81. $2x^4 + 11x^3 - 3x^2 + 8x - 4$ **83.** $-2w^2 - 7w + 18$
85. $-p^2 q - 4pq^2 + 3pq$ **87.** 0 **89.** $-5ab + 6ab^2$
91. $11y^2 - 10y - 4$ **93.** For example, $x^3 + 6$
95. For example, $8x^5$ **97.** For example, $-6x^2 + 2x + 5$

Section 5.6 Practice Exercises, pp. 404–407

3. $-2y^2$ **5.** $-8y^4$ **7.** $8uvw^2$ **9.** $7u^2v^2w^4$
11. $-12y$ **13.** $21p$ **15.** $12a^{14}b^8$ **17.** $-2c^{10}d^{12}$
19. $16p^2q^2 - 24p^2q + 40pq^2$ **21.** $-4k^3 + 52k^2 + 24k$
23. $-45p^3q - 15p^4q^3 + 30pq^2$ **25.** $y^2 - y - 90$
27. $m^2 - 14m + 24$ **29.** $12p^2 - 5p - 2$
31. $12w^2 - 32w + 16$ **33.** $p^2 - 14pw + 33w^2$
35. $12x^2 + 28x - 5$ **37.** $8a^2 - 22a + 9$
39. $9t^2 - 18t - 7$ **41.** $3m^2 + 28mn + 32n^2$
43. $5s^3 + 8s^2 - 7s - 6$ **45.** $27w^3 - 8$
47. $p^4 + 5p^3 - 2p^2 - 21p + 5$
49. $6a^3 - 23a^2 + 38a - 45$
51. $8x^3 - 36x^2y + 54xy^2 - 27y^3$ **53.** $9a^2 - 16b^2$
55. $81k^2 - 36$ **57.** $\frac{1}{4} - t^2$ **59.** $u^6 - 25v^2$
61. $4 - 9a^2$ **63.** $\frac{4}{9} - p^2$ **65.** $a^2 + 10a + 25$
67. $x^2 - 2xy + y^2$ **69.** $4c^2 + 20c + 25$
71. $9t^4 - 24st^2 + 16s^2$ **73.** $t^2 - 14t + 49$
75. $16q^2 + 24q + 9$ **77. a.** 36 **b.** 20
c. $(a + b)^2 \neq a^2 + b^2$ in general. **79. a.** $9x^2 + 6xy + y^2$
b. $9x^2y^2$ **81.** $36 - y^2$ **83.** $49q^2 - 42q + 9$
85. $r^3 - 15r^2 + 63r - 49$ **87.** $3t^3 - 12t$
89. $81w^2 - 16z^2$ **91.** $25s^2 - 30st + 9t^2$
93. $10a^2 - 13ab + 4b^2$ **95.** $s^2 + \frac{147}{5}s - 18$
97. $4k^3 - 4k^2 - 5k - 25$ **99.** $u^3 + u^2 - 10u + 8$
101. $w^3 + w^2 + 4w + 12$ **103.** $\frac{4}{25}p^2 - \frac{4}{5}pq + q^2$
105. $4v^4 + 48v^3$ **107.** $4h^2 - 7.29$ **109.** $5k^6 - 19k^3 + 18$
111. $6.25y^2 + 5.5y + 1.21$ **113.** $h^3 + 9h^2 + 27h + 27$
115. $8a^3 - 48a^2 + 96a - 64$ **117.** $6w^4 + w^3 - 15w^2 - 11w - 5$
119. $30x^3 + 55x^2 - 10x$ **121.** $2y^3 - y^2 - 15y + 18$
123. $x + 6$ **125.** $k = 6$ or -6

Section 5.7 Practice Exercises, pp. 412–414

1. $6z^5 - 10z^4 - 4z^3 - z^2 - 6$ **3.** $10x^2 - 29xy - 3y^2$
5. $11x - 2y$ **7.** $y^2 - \frac{3}{4}y + \frac{1}{2}$ **9.** $a^3 + 27$
11. Use long division when the divisor is a polynomial with two or more terms. **13.** $5t^2 + 6t$ **15.** $3a^2 + 2a - 7$
17. $x^2 + 4x - 1$ **19.** $3p^2 - p$ **21.** $1 + \frac{2}{m}$
23. $-2y^2 + y - 3$ **25.** $x^2 - 6x - \frac{1}{4} + \frac{2}{x}$
27. $a - 1 + \frac{b}{a}$ **29.** $3t - 1 + \frac{3}{2t} - \frac{1}{2t^2} + \frac{2}{t^3}$
31. a. $z + 2 + \frac{1}{z + 5}$ **33.** $t + 3 + \frac{2}{t + 1}$
35. $7b + 4$ **37.** $k - 6$ **39.** $2p^2 + 3p - 4$
41. $k - 2 + \frac{-4}{k + 1}$ **43.** $2x^2 - x + 6 + \frac{2}{2x - 3}$
45. $y^2 + 2y + 1 + \frac{2}{3y - 1}$ **47.** $a - 3 + \frac{18}{a + 3}$
49. $4x^2 + 8x + 13$ **51.** $w^2 + 5w - 2 + \frac{1}{w^2 - 3}$

53. $n^2 + n - 6$ **55.** $x - 1 + \frac{-8}{5x^2 + 5x + 1}$
57. Multiply $(x - 2)(x^2 + 4) = x^3 - 2x^2 + 4x - 8$, which does not equal $x^3 - 8$. **59.** Monomial division; $3a^2 + 4a$
61. Long division; $p + 2$ **63.** Long division;
$t^3 - 2t^2 + 5t - 10 + \frac{4}{t + 2}$ **65.** Long division;
$w^2 + 3 + \frac{1}{w^2 - 2}$ **67.** Long division; $n^2 + 4n + 16$
69. Monomial division; $-3r + 4 - \frac{3}{r^2}$ **71.** $x + 1$
73. $x^3 + x^2 + x + 1$ **75.** $x + 1 + \frac{1}{x - 1}$
77. $x^3 + x^2 + x + 1 + \frac{1}{x - 1}$

Chapter 5 Problem Recognition Exercises, p. 415

1. $2x^3 - 8x^2 + 14x - 12$ **2.** $-3y^4 - 20y^2 - 32$
3. $x^2 - 1$ **4.** $4y^2 + 12$ **5.** $36y^2 - 84y + 49$
6. $9z^2 + 12z + 4$ **7.** $36y^2 - 49$ **8.** $9z^2 - 4$
9. $16x^2 + 8xy + y^2$ **10.** $4a^2 + 4ab + b^2$ **11.** $16x^2y^2$
12. $4a^2b^2$ **13.** $-x^2 - 3x + 4$ **14.** $5m^2 - 4m + 1$
15. $-7m^2 - 16m$ **16.** $-4n^5 + n^4 + 6n^2 - 7n + 2$
17. $8x^2 + 16x + 34 + \frac{74}{x - 2}$
18. $-4x^2 - 10x - 30 + \frac{-95}{x - 3}$
19. $6x^3 + 5x^2y - 6xy^2 + y^3$ **20.** $6a^3 - a^2b + 5ab^2 + 2b^3$
21. $x^3 + y^6$ **22.** $m^6 + 1$ **23.** $4b$ **24.** $-12z$
25. $a^4 - 4b^2$ **26.** $y^6 - 36z^2$ **27.** $64u^2 + 48uv + 9v^2$
28. $4p^2 - 4pt + t^2$ **29.** $4p + 4 + \frac{-2}{2p - 1}$
30. $2v - 7 + \frac{29}{2v + 3}$ **31.** $4x^2y^2$ **32.** $-9pq$
33. $10a^2 - 57a + 54$ **34.** $28a^2 - 17a - 3$
35. $\frac{9}{49}x^2 - \frac{1}{4}$ **36.** $\frac{4}{25}y^2 - \frac{16}{9}$
37. $-\frac{11}{9}x^3 + \frac{5}{9}x^2 - \frac{1}{2}x - 4$ **38.** $-\frac{13}{10}y^2 - \frac{9}{10}y + \frac{4}{15}$
39. $1.3x^2 - 0.3x - 0.5$ **40.** $5w^3 - 4.1w^2 + 2.8w - 1.2$

Chapter 5 Review Exercises, pp. 420–423

1. Base: 5; exponent: 3 **3.** Base: -2; exponent: 0
5. a. 36 **b.** 36 **c.** -36 **7.** 5^{13} **9.** x^9 **11.** 10^3
13. b^8 **15.** k **17.** 2^8 **19.** Exponents are added only when multiplying factors with the same base. In such a case, the base does not change. **21.** \$7146.10
23. 7^{12} **25.** p^{18} **27.** $\frac{a^2}{b^2}$ **29.** $\frac{5^2}{c^4d^{10}}$ **31.** $2^4a^4b^8$
33. $-\frac{3^3x^9}{5^3y^6z^3}$ **35.** a^{11} **37.** $4h^{14}$ **39.** $\frac{x^6y^2}{4}$ **41.** 1
43. -1 **45.** 2 **47.** $\frac{1}{z^5}$ **49.** $\frac{1}{36a^2}$ **51.** $\frac{17}{16}$

14. a. Nonperfect square trinomial   **b.** $(7p - 1)(p - 4)$
15. a. Nonperfect square trinomial   **b.** $3(q - 4)(q + 1)$
16. a. Perfect square trinomial   **b.** $-2(x - 2)^2$
17. a. None of these   **b.** $6a(3a + 2)$
18. a. Difference of cubes   **b.** $2(3 - y)(9 + 3y + y^2)$
19. a. Difference of squares   **b.** $4(t - 5)(t + 5)$
20. a. Nonperfect square trinomial   **b.** $(4t + 1)(t - 8)$
21. a. Nonperfect square trinomial   **b.** $10(c^2 + c + 1)$
22. a. Four terms-grouping   **b.** $(w - 5)(2x + 3y)$
23. a. Sum of cubes   **b.** $(x + 0.1)(x^2 - 0.1x + 0.01)$
24. a. Difference of squares   **b.** $(2q - 3)(2q + 3)$
25. a. Perfect square trinomial   **b.** $(8 + k)^2$
26. a. Four terms-grouping   **b.** $(t + 6)(s^2 + 5)$
27. a. Four terms-grouping   **b.** $(x + 1)(2x - y)$
28. a. Sum of cubes   **b.** $(w + y)(w^2 - wy + y^2)$
29. a. Difference of cubes   **b.** $(a - c)(a^2 + ac + c^2)$
30. a. Nonperfect square trinomial   **b.** Prime
31. a. Nonperfect square trinomial   **b.** Prime
32. a. Perfect square trinomial   **b.** $(a + 1)^2$
33. a. Perfect square trinomial   **b.** $(b + 5)^2$
34. a. Nonperfect square trinomial   **b.** $-1(t + 8)(t - 4)$
35. a. Nonperfect square trinomial   **b.** $-p(p + 4)(p + 1)$
36. a. Difference of squares   **b.** $(xy - 7)(xy + 7)$
37. a. Nonperfect square trinomial   **b.** $3(2x + 3)(x - 5)$
38. a. Nonperfect square trinomial   **b.** $2(5y - 1)(2y - 1)$
39. a. None of these   **b.** $abc^2(5ac - 7)$
40. a. Difference of squares   **b.** $2(2a - 5)(2a + 5)$
41. a. Nonperfect square trinomial   **b.** $(t + 9)(t - 7)$
42. a. Nonperfect square trinomial   **b.** $(b + 10)(b - 8)$
43. a. Four terms-grouping   **b.** $(b + y)(a - b)$
44. a. None of these   **b.** $3x^2y^4(2x + y)$
45. a. Nonperfect square trinomial   **b.** $(7u - 2v)(2u - v)$
46. a. Nonperfect square trinomial   **b.** Prime
47. a. Nonperfect square trinomial   **b.** $2(2q^2 - 4q - 3)$
48. a. Nonperfect square trinomial   **b.** $3(3w^2 + w - 5)$
49. a. Sum of squares   **b.** Prime
50. a. Perfect square trinomial   **b.** $5(b - 3)^2$
51. a. Nonperfect square trinomial   **b.** $(3r + 1)(2r + 3)$
52. a. Nonperfect square trinomial   **b.** $(2s - 3)(2s + 5)$
53. a. Difference of squares   **b.** $(2a - 1)(2a + 1)(4a^2 + 1)$
54. a. Four terms-grouping   **b.** $(p + c)(p - 3)(p + 3)$
55. a. Perfect square trinomial   **b.** $(9u - 5v)^2$
56. a. Sum of squares   **b.** $4(x^2 + 4)$
57. a. Nonperfect square trinomial   **b.** $(x - 6)(x + 1)$
58. a. Nonperfect square trinomial   **b.** Prime
59. a. a. Four terms-grouping   **b.** $2(x - 3y)(a + 2b)$
60. a. Nonperfect square trinomial
b. $m(4m + 1)(2m - 3)$ **61. a.** Nonperfect square
trinomial   **b.** $x^2y(3x + 5)(7x + 2)$
62. a. Difference of squares   **b.** $2(m^2 - 8)(m^2 + 8)$
63. a. Four terms-grouping   **b.** $(4v - 3)(2u + 3)$
64. a. Four terms-grouping   **b.** $(t - 5)(4t + s)$
65. a. Perfect square trinomial   **b.** $3(2x - 1)^2$
66. a. Perfect square trinomial   **b.** $(p + q)^2$
67. a. Nonperfect square trinomial   **b.** $n(2n - 1)(3n + 4)$
68. a. Nonperfect square trinomial   **b.** $k(2k - 1)(2k + 3)$
69. a. Difference of squares   **b.** $(8 - y)(8 + y)$
70. a. Difference of squares   **b.** $b(6 - b)(6 + b)$
71. a. Nonperfect square trinomial   **b.** Prime
72. a. Nonperfect square trinomial   **b.** $(y + 4)(y + 2)$
73. a. Nonperfect square trinomial   **b.** $(c^2 - 10)(c^2 - 2)$

Section 6.7 Practice Exercises, pp. 476–478

3. $4(b - 5)(b - 6)$ **5.** $(3x - 2)(x + 4)$ **7.** $4(x^2 + 4y^2)$
9. Neither   **11.** Quadratic   **13.** Linear   **15.** $-3, 1$
17. $\dfrac{7}{2}, -\dfrac{7}{2}$ **19.** -5 **21.** $0, \dfrac{1}{5}$ **23.** The polynomial
must be factored completely.   **25.** $5, -3$ **27.** $-12, 2$
29. $4, -\dfrac{1}{2}$ **31.** $\dfrac{2}{3}, -\dfrac{2}{3}$ **33.** $6, 8$ **35.** $0, -\dfrac{3}{2}, 4$
37. $-\dfrac{1}{3}, 3, -6$ **39.** $0, 4, -\dfrac{3}{2}$ **41.** $0, -\dfrac{9}{2}, 11$
43. $0, 4, -4$ **45.** $-6, 0$ **47.** $\dfrac{3}{4}, -\dfrac{3}{4}$ **49.** $0, -5, -2$
51. $-\dfrac{14}{3}$ **53.** $5, 3$ **55.** $0, -2$ **57.** -3 **59.** $-3, 1$
61. $\dfrac{3}{2}$ **63.** $0, \dfrac{1}{3}$ **65.** $0, 2$ **67.** $3, -2, 2$ **69.** $-5, 4$
71. $-5, -1$

Chapter 6 Problem Recognition Exercises, p. 478

1. a. $(x + 7)(x - 1)$ **b.** $-7, 1$ **2. a.** $(c + 6)(c + 2)$
b. $-6, -2$ **3. a.** $(2y + 1)(y + 3)$ **b.** $-\dfrac{1}{2}, -3$
4. a. $(3x - 5)(x - 1)$ **b.** $\dfrac{5}{3}, 1$ **5. a.** $\dfrac{4}{5}, -1$
b. $(5q - 4)(q + 1)$ **6. a.** $-\dfrac{1}{3}, \dfrac{3}{2}$ **b.** $(3a + 1)(2a - 3)$
7. a. $-8, 8$ **b.** $(a + 8)(a - 8)$ **8. a.** $-10, 10$
b. $(v + 10)(v - 10)$ **9. a.** $(2b + 9)(2b - 9)$ **b.** $-\dfrac{9}{2}, \dfrac{9}{2}$
10. a. $(6t + 7)(6t - 7)$ **b.** $-\dfrac{7}{6}, \dfrac{7}{6}$
11. a. $-\dfrac{3}{2}, -\dfrac{1}{2}$ **b.** $2(2x + 3)(2x + 1)$
12. a. $-\dfrac{4}{3}, -2$ **b.** $4(3y + 4)(y + 2)$
13. a. $x(x - 10)(x + 2)$ **b.** $0, 10, -2$
14. a. $k(k + 7)(k - 2)$ **b.** $0, -7, 2$
15. a. $-1, 3, -3$ **b.** $(b + 1)(b - 3)(b + 3)$
16. a. $8, -2, 2$ **b.** $(x - 8)(x + 2)(x - 2)$
17. $(s - 3)(2s + r)$ **18.** $(2t + 1)(3t + 5u)$
19. $-\dfrac{1}{2}, 0, \dfrac{1}{2}$ **20.** $-5, 0, 5$ **21.** $0, 1$ **22.** $-3, 0$
23. $\dfrac{1}{3}$ **24.** $-\dfrac{7}{12}$ **25.** $(2w + 3)(4w^2 - 6w + 9)$
26. $(10q - 1)(100q^2 + 10q + 1)$ **27.** $-\dfrac{7}{5}, 1$
28. $-6, -\dfrac{1}{4}$ **29.** $-\dfrac{2}{3}, -5$ **30.** $-1, -\dfrac{1}{2}$ **31.** 3
32. 0 **33.** $-4, 4$ **34.** $-\dfrac{2}{3}, \dfrac{2}{3}$ **35.** $1, 6$ **36.** $-2, -12$

Section 6.8 Practice Exercises, pp. 483–486

3. $0, -\dfrac{2}{3}$ **5.** $6, -1$ **7.** $-\dfrac{5}{6}, 2$
9. The numbers are 7 and -7.   **11.** The numbers are 10
and -4.   **13.** The numbers are -9 and -7, or 7 and 9.

15. The numbers are 5 and 6, or -1 and 0.   **17.** The room is 15 yd by 20 yd.   **19. a.** The slab is 7 m by 4 m. **b.** 22 m   **21.** The base is 7 ft and the height is 4 ft. **23.** 3 sec   **25.** 0 sec and 1.5 sec

27.

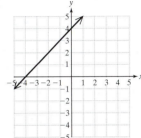

29. $c = 25$ cm   **31.** $a = 15$ in.

33. The brace is 20 in. long.   **35.** The kite is 19 yd high. **37.** The bottom of the ladder is 8 ft from the house. The distance from the top of the ladder to the ground is 15 ft. **39.** The hypotenuse is 10 m.

Chapter 6 Review Exercises, pp. 493–495

1. $3a^2b$ **3.** $2c(3c - 5)$ **5.** $2x(3x + x^3 - 4)$ **7.** $t(-t + 5)$ or $-t(t - 5)$ **9.** $(b + 2)(3b - 7)$ **11.** $(w + 2)(7w + b)$ **13.** $3(4y - 3)(5y - 1)$ **15.** $(x - 3)(x - 7)$ **17.** $(z - 12)(z + 6)$ **19.** $3w(p + 10)(p + 2)$ **21.** $-(t - 8)(t - 2)$ **23.** $(a + b)(a + 11b)$ **25.** Different   **27.** Both positive   **29.** $(2y + 3)(y - 4)$ **31.** $(2z + 5)(5z + 2)$ **33.** Prime   **35.** $10(w - 9)(w + 3)$ **37.** $(3c - 5d)^2$ **39.** $(v^2 + 1)(v^2 - 3)$ **41.** The trinomials in Exercises 37 and 38.   **43.** $-3, -5$ **45.** $(y + 3)(4y + 1)$ **47.** $x(x + 5)(4x - 3)$ **49.** $(p - 3q)(p - 5q)$ **51.** $10(4s - 5)(s + 2)$ **53.** $2z^4(z + 7)(z - 3)$ **55.** $(3p + 2)(2p + 5)$ **57.** $(3w - z)^2$ **59.** $(a - b)(a + b)$ **61.** $(a - 7)(a + 7)$ **63.** $(10 - 9t)(10 + 9t)$ **65.** Prime   **67.** $(y + 6)^2$ **69.** $(3a - 2)^2$ **71.** $-3(v + 2)^2$ **73.** $2(c^2 - 3)(c^2 + 3)$ **75.** $(p + 3)(p - 4)(p + 4)$ **77.** $(a + b)(a^2 - ab + b^2)$ **79.** $(4 + a)(16 - 4a + a^2)$ **81.** $(p^2 + 2)(p^4 - 2p^2 + 4)$ **83.** $6(x - 2)(x^2 + 2x + 4)$ **85.** $x(x - 6)(x + 6)$ **87.** $4(2h^2 + 5)$ **89.** $(x + 4)(x + 1)(x - 1)$ **91.** $n(2 + n)(4 - 2n + n^2)$ **93.** $(x - 3)(2x + 1) = 0$ can be solved directly by the zero product rule because it is a product of factors set equal to zero.

95. $9, \dfrac{1}{2}$ **97.** $0, 7, \dfrac{9}{4}$ **99.** $-\dfrac{1}{4}, 6$ **101.** $5, -5$

103. $4, 2$ **105.** $-\dfrac{2}{3}$ **107.** $\dfrac{11}{2}, -12$ **109.** $0, 2, -2$

111. The ball is at ground level at 0 and 1 sec. **113.** The legs are 6 ft and 8 ft; the hypotenuse is 10 ft. **115.** The numbers are 29 and 30, or -2 and -1.

Chapter 6 Test, p. 495

1. $3x(5x^3 - 1 + 2x^2)$ **2.** $(a - 5)(7 - a)$ **3.** $(6w - 1)(w - 7)$ **4.** $(13 - p)(13 + p)$ **5.** $(q - 8)^2$ **6.** $(2 + t)(4 - 2t + t^2)$ **7.** $(a + 4)(a + 8)$ **8.** $(x + 7)(x - 6)$ **9.** $(2y - 1)(y - 8)$ **10.** $(2z + 1)(3z + 8)$ **11.** $(3t - 10)(3t + 10)$ **12.** $(v + 9)(v - 9)$ **13.** $3(a + 6b)(a + 3b)$ **14.** $(c - 1)(c + 1)(c^2 + 1)$ **15.** $(y - 7)(x + 3)$ **16.** Prime   **17.** $-10(u - 2)(u - 1)$ **18.** $3(2t - 5)(2t + 5)$ **19.** $5(y - 5)^2$ **20.** $7q(3q + 2)$ **21.** $(2x + 1)(x - 2)(x + 2)$ **22.** $(y - 5)(y^2 + 5y + 25)$ **23.** $(mn - 9)(mn + 9)$ **24.** $16(a - 2b)(a + 2b)$ **25.** $(4x - 3y^2)(16x^2 + 12xy^2 + 9y^4)$ **26.** $3y(x - 4)(x + 2)$

27. $\dfrac{3}{2}, -5$ **28.** $0, 7$ **29.** $8, -2$ **30.** $\dfrac{1}{5}, -1$

31. $3, -3, -10$ **32.** The tennis court is 12 yd by 26 yd. **33.** The two integers are 5 and 7, or -5 and -7. **34.** The base is 12 in., and the height is 7 in. **35.** The shorter leg is 5 ft.

Chapters 1–6 Cumulative Review Exercises, p. 496

1. $\dfrac{7}{5}$ **2.** -3 **3.** $y = \dfrac{3}{2}x - 4$

4. There are 10 quarters, 12 nickels, and 7 dimes. **5.** $[-4, \infty)$ **6. a.** Yes   **b.** 1
c. $(0, 4)$ **d.** $(-4, 0)$ **e.**

7. a. Vertical line   **b.** Undefined   **c.** $(5, 0)$ **d.** Does not exist   **8.** $y = 3x + 14$ **9.** $(5, 2)$

10. $-\dfrac{7}{2}y^2 - 5y - 14$ **11.** $8p^3 - 22p^2 + 13p + 3$

12. $4w^2 - 28w + 49$ **13.** $r^3 + 5r^2 + 15r + 40 + \dfrac{121}{r - 3}$

14. c^4 **15.** 1.6×10^3 **16.** $(w - 2)(w + 2)(w^2 + 4)$ **17.** $(a + 5b)(2x - 3y)$ **18.** $(2x - 5)(2x + 1)$

19. $(y - 3)(y^2 + 3y + 9)$ **20.** $0, \dfrac{1}{2}, -5$

Chapter 7

Chapter Opener Puzzle

$\underset{4}{\text{p}}\ \underset{1}{\text{r}}\ \underset{2}{\text{o}}\ \underset{}{\text{c}}\ \underset{1}{\text{r}}\ \underset{5}{\text{a}}\ \underset{3}{\text{s}}\ \underset{}{\text{t}}\ \underset{}{\text{i}}\ \underset{5}{\text{n}}\ \underset{3}{\text{a}}\ \underset{2}{\text{t}}\ \underset{}{\text{i}}\ \underset{}{\text{o}}\ \underset{}{\text{n}}$

Section 7.1 Practice Exercises, pp. 505–507

3. a. A number $\dfrac{p}{q}$, where p and q are integers and $q \neq 0$

b. An expression $\dfrac{p}{q}$, where p and q are polynomials and

$q \neq 0$ **5.** $-\dfrac{1}{8}$ **7.** $-\dfrac{1}{2}$ **9.** Undefined

11. a. $3\frac{1}{5}$ hr or 3.2 hr   **b.** $1\frac{3}{4}$ hr or 1.75 hr **13.** $\{k \mid k$ is a real number and $k \neq -2\}$

15. $\left\{ x \mid x \text{ is a real number and } x \neq \dfrac{5}{2}, x \neq -8 \right\}$

17. $\{b \mid b$ is a real number and $b \neq -2, b \neq -3\}$ **19.** The set of all real numbers   **21.** The set of all real numbers   **23.** $\{t \mid t$ is a real number and $t \neq 0\}$

25. For example: $\dfrac{1}{x - 2}$ **27.** For example: $\dfrac{1}{(x + 3)(x - 7)}$

29. a. $\dfrac{2}{5}$ **b.** $\dfrac{2}{5}$ **31. a.** Undefined   **b.** Undefined

33. a. $\{y \mid y$ is a real number and $y \neq -2\}$ **b.** $\dfrac{1}{2}$

35. a. $\{t \mid t \text{ is a real number and } t \neq -1\}$ **b.** $t - 1$

37. a. $\{w \mid w \text{ is a real number and } w \neq 0, \ w \neq \frac{5}{3}\}$

b. $\dfrac{1}{3w - 5}$ **39. a.** $\{x \mid x \text{ is a real number and } x \neq -\frac{2}{3}\}$

b. $\dfrac{3x - 2}{2}$ **41. a.** $\{a \mid a \text{ is a real number and } a \neq -3, \ a \neq 2\}$

b. $\dfrac{a + 5}{a + 3}$ **43.** $\dfrac{b}{3}$ **45.** $\dfrac{3t^2}{2}$ **47.** $-\dfrac{3xy}{z^2}$

49. $\dfrac{1}{2}$ **51.** $\dfrac{p - 3}{p + 4}$ **53.** $\dfrac{1}{4(m - 11)}$ **55.** $\dfrac{2x + 1}{4x^2}$

57. $\dfrac{1}{4a - 5}$ **59.** $\dfrac{4}{w + 2}$ **61.** $\dfrac{x - 2}{3(y + 2)}$ **63.** $\dfrac{2}{x - 5}$

65. $a + 7$ **67.** Cannot simplify   **69.** $\dfrac{y + 3}{2y - 5}$

71. $\dfrac{3x - 2}{x + 4}$ **73.** $\dfrac{5}{(q + 1)(q - 1)}$ **75.** $\dfrac{c - d}{2c + d}$

77. $\dfrac{1}{t(t - 5)}$ **79.** $\dfrac{7p - 2q}{2}$ **81.** $5x + 4$ **83.** $\dfrac{x + y}{x - 4y}$

85. They are opposites.   **87.** -1 **89.** -1 **91.** $-\dfrac{1}{2}$

93. Cannot simplify   **95.** $\dfrac{5x - 6}{5x + 6}$ **97.** $-\dfrac{x + 3}{4 + x}$

99. $w - 2$ **101.** $\dfrac{z + 4}{z^2 + 4z + 16}$

Section 7.2 Practice Exercises, pp. 512–513

1. $\dfrac{3}{10}$ **3.** 2 **5.** $\dfrac{5}{2}$ **7.** $\dfrac{15}{4}$ **9.** $\dfrac{3}{2x}$ **11.** $3xy^4$

13. $\dfrac{x - 6}{8}$ **15.** $\dfrac{2}{y}$ **17.** $-\dfrac{5}{8}$ **19.** $-\dfrac{b + a}{a - b}$

21. $\dfrac{y + 1}{5}$ **23.** $\dfrac{2(x + 6)}{2x + 1}$ **25.** 6 **27.** $\dfrac{m^6}{n^2}$ **29.** $\dfrac{10}{9}$

31. $\dfrac{6}{7}$ **33.** $-m(m + n)$ **35.** $\dfrac{3p + 4q}{4(p + 2q)}$ **37.** $\dfrac{p}{p - 1}$

39. $\dfrac{w}{2w - 1}$ **41.** $\dfrac{5}{6}$ **43.** $\dfrac{q + 1}{q - 6}$ **45.** $\dfrac{1}{4}$

47. $\dfrac{y + 9}{y - 6}$ **49.** $\dfrac{t + 4}{t + 2}$ **51.** $\dfrac{3t + 8}{t + 2}$ **53.** $\dfrac{x + 4}{x + 1}$

55. $-\dfrac{w - 3}{2}$ **57.** $\dfrac{k + 6}{k + 3}$ **59.** $\dfrac{2}{a}$ **61.** $2y(y + 1)$

63. $x + y$ **65.** 2 **67.** $\dfrac{1}{a - 2}$ **69.** $\dfrac{p + q}{2}$

Section 7.3 Practice Exercises, pp. 517–519

3. $\{x \mid x \text{ is a real number and } x \neq 1, x \neq -1\}$; $\dfrac{3}{5(x - 1)}$

5. $\dfrac{a + 5}{a + 7}$ **7.** $\dfrac{2}{3y}$ **9.** a, b, c, d

11. x^5 is the greatest power of x that appears in any denominator.   **13.** 45 **15.** 48 **17.** 63 **19.** $9x^2y^3$
21. w^2y **23.** $(p + 3)(p - 1)(p + 2)$ **25.** $9t(t + 1)^2$
27. $(y - 2)(y + 2)(y + 3)$ **29.** $3 - x \text{ or } x - 3$
31. Because $(b - 1)$ and $(1 - b)$ are opposites; they differ by a factor of -1.

33. $\dfrac{6}{5x^2}$; $\dfrac{5x}{5x^2}$ **35.** $\dfrac{24x}{30x^3}$; $\dfrac{5y}{30x^3}$ **37.** $\dfrac{10}{12a^2b}$; $\dfrac{a^3}{12a^2b}$

39. $\dfrac{6m - 6}{(m + 4)(m - 1)}$; $\dfrac{3m + 12}{(m + 4)(m - 1)}$

41. $\dfrac{6x + 18}{(2x - 5)(x + 3)}$; $\dfrac{2x - 5}{(2x - 5)(x + 3)}$

43. $\dfrac{6w + 6}{(w + 3)(w - 8)(w + 1)}$; $\dfrac{w^2 + 3w}{(w + 3)(w - 8)(w + 1)}$

45. $\dfrac{6p^2 + 12p}{(p - 2)(p + 2)^2}$; $\dfrac{3p - 6}{(p - 2)(p + 2)^2}$

47. $\dfrac{1}{a - 4}$; $\dfrac{-a}{a - 4}$ or $\dfrac{-1}{4 - a}$; $\dfrac{a}{4 - a}$

49. $\dfrac{8}{2(x - 7)}$; $\dfrac{-y}{2(x - 7)}$ or $\dfrac{-8}{2(7 - x)}$; $\dfrac{y}{2(7 - x)}$

51. $\dfrac{1}{a + b}$; $\dfrac{-6}{a + b}$ or $\dfrac{-1}{-a - b}$; $\dfrac{6}{-a - b}$

53. $\dfrac{-9}{24(3y + 1)}$; $\dfrac{20}{24(3y + 1)}$ **55.** $\dfrac{3z + 12}{5z(z + 4)}$; $\dfrac{5z}{5z(z + 4)}$

57. $\dfrac{z^2 + 3z}{(z + 2)(z + 7)(z + 3)}$; $\dfrac{-3z^2 - 6z}{(z + 2)(z + 7)(z + 3)}$; $\dfrac{5z + 35}{(z + 2)(z + 7)(z + 3)}$

59. $\dfrac{3p + 6}{(p - 2)(p^2 + 2p + 4)(p + 2)}$; $\dfrac{p^3 + 2p^2 + 4p}{(p - 2)(p^2 + 2p + 4)(p + 2)}$; $\dfrac{5p^3 - 20p}{(p - 2)(p^2 + 2p + 4)(p + 2)}$

Section 7.4 Practice Exercises, pp. 526–528

1. a. $-\dfrac{1}{2}$, -2, 0, undefined, undefined

b. $(x - 5)(x - 2)$; $\{x \mid x \text{ is a real number and } x \neq 5, x \neq 2\}$

c. $\dfrac{x + 1}{x - 2}$ **3.** $\dfrac{2(2b - 3)}{(b - 3)(b - 1)}$ **5.** $\dfrac{5}{4}$ **7.** $\dfrac{3}{8}$

9. 2 **11.** 5 **13.** $\dfrac{-2(t - 2)}{t - 8}$ **15.** $3x + 7$

17. $m + 5$ **19.** 2 **21.** $x - 5$ **23.** $\dfrac{1}{r + 1}$

25. $\dfrac{1}{y + 7}$ **27.** $\dfrac{15x}{y}$ **29.** $\dfrac{5a + 6}{4a}$ **31.** $\dfrac{2(6 + x^2y)}{15xy^3}$

33. $\dfrac{2s - 3t^2}{s^4t^3}$ **35.** $-\dfrac{2}{3}$ **37.** $\dfrac{19}{3(a + 1)}$

39. $\dfrac{-3(k + 4)}{(k - 3)(k + 3)}$ **41.** $\dfrac{a - 4}{2a}$ **43.** $\dfrac{(x + 6)(x - 2)}{(x - 4)(x + 1)}$

45. $\dfrac{2(4a - b)}{(a + b)(a - b)}$ **47.** $\dfrac{5p - 1}{3}$ or $\dfrac{-5p + 1}{-3}$

49. $\dfrac{6n - 1}{n - 8}$ or $\dfrac{-6n + 1}{8 - n}$ **51.** $\dfrac{2(4x + 5)}{x(x + 2)}$

53. $\dfrac{3p + 1}{(p - 3)(p - 1)}$ **55.** $\dfrac{3y}{2(2y + 1)}$ **57.** $\dfrac{2(w - 3)}{(w + 3)(w - 1)}$

59. $\dfrac{4a - 13}{(a - 3)(a - 4)}$ **61.** $\dfrac{4x(x + 1)}{(x + 3)(x - 2)(x + 2)}$

63. $\dfrac{-y(y + 8)}{(2y + 1)(y - 1)(y - 4)}$ **65.** $\dfrac{1}{2p + 1}$

67. $\dfrac{-2mn + 1}{(m + n)(m - n)}$ **69.** 0 **71.** $\dfrac{2(3x + 7)}{(x + 3)(x + 2)}$

73. $\dfrac{1}{n}$ **75.** $\dfrac{5}{n + 2}$ **77.** $n + \left(7 \cdot \dfrac{1}{n}\right); \dfrac{n^2 + 7}{n}$

79. $\dfrac{1}{n} - \dfrac{2}{n}; -\dfrac{1}{n}$ **81.** $\dfrac{-w^2}{(w + 3)(w - 3)(w^2 - 3w + 9)}$

83. $\dfrac{p^2 - 2p + 7}{(p + 2)(p + 3)(p - 1)}$

85. $\dfrac{-m - 21}{2(m + 5)(m - 2)}$ or $\dfrac{m + 21}{2(m + 5)(2 - m)}$ **87.** $\dfrac{3k + 5}{4k + 7}$

89. $\dfrac{1}{a}$

Chapter 7 Problem Recognition Exercises, p. 529

1. $\dfrac{-2x + 9}{3x + 1}$ **2.** $\dfrac{1}{w - 4}$ **3.** $\dfrac{y - 5}{2y - 3}$

4. $\dfrac{7}{(x + 3)(2x - 1)}$ **5.** $-\dfrac{1}{x}$ **6.** $\dfrac{1}{3}$ **7.** $\dfrac{c + 3}{c}$

8. $\dfrac{x + 3}{5}$ **9.** $\dfrac{a}{12b^4 c}$ **10.** $\dfrac{2a - b}{a - b}$ **11.** $\dfrac{p - q}{5}$

12. 4 **13.** $\dfrac{10}{2x + 1}$ **14.** $\dfrac{w + 2z}{w + z}$ **15.** $\dfrac{3}{2x + 5}$

16. $\dfrac{y + 7}{x + a}$ **17.** $\dfrac{1}{2(a + 3)}$ **18.** $\dfrac{2(3y + 10)}{(y - 6)(y + 6)(y + 2)}$

19. $(t + 8)^2$ **20.** $6b + 5$

Section 7.5 Practice Exercises, pp. 535–536

3. $\left\{a \mid a \text{ is a real number and } a \neq \dfrac{3}{2}, a \neq -5\right\}; \dfrac{1}{2a - 3}$

5. $\dfrac{3(2k - 5)}{5(k - 2)}$ **7.** $\dfrac{7}{4y}$ **9.** $\dfrac{1}{2y}$ **11.** $\dfrac{24b}{a^3}$ **13.** $\dfrac{2r^5 t^4}{s^6}$

15. $\dfrac{35}{2}$ **17.** $k + h$ **19.** $\dfrac{n + 1}{2(n - 3)}$ **21.** $\dfrac{2x + 1}{4x + 1}$

23. $m - 7$ **25.** $\dfrac{2y(y - 5)}{7y^2 + 10}$ **27.** $-\dfrac{a + 8}{a - 2}$ or $\dfrac{a + 8}{2 - a}$

29. $\dfrac{t - 2}{t - 4}$ **31.** $\dfrac{t + 3}{t - 5}$ **33.** $\dfrac{1}{2}$ **35.** $\dfrac{\frac{1}{2} + \frac{2}{3}}{5}; \dfrac{7}{30}$

37. $\dfrac{3}{\frac{2}{3} + \frac{3}{4}}; \dfrac{36}{17}$ **39. a.** $\dfrac{6}{5}\,\Omega$ **b.** $6\,\Omega$ **41.** $\dfrac{y + 4x}{2y}$

43. $\dfrac{1}{n^2 + m^2}$ **45.** $\dfrac{2z - 5}{3(z + 3)}$ **47.** $-\dfrac{x + 1}{x - 1}$ or $\dfrac{x + 1}{1 - x}$

49. $\dfrac{3}{2}$

Section 7.6 Practice Exercises, pp. 544–546

3. $\dfrac{2}{4x - 1}$ **5.** $5(h + 1)$ **7.** $\dfrac{(x + 4)(x - 3)}{x^2}$

9. 3 **11.** $\dfrac{5}{11}$ **13.** $\dfrac{1}{3}$

15. a. $\{z \mid z \text{ is a real number and } z \neq 0\}$ **b.** $5z$ **c.** 5

17. $-\dfrac{200}{19}$ **19.** 8 **21.** $\dfrac{47}{6}$ **23.** 3, -1 **25.** 4

27. 5; (the value 0 does not check.) **29.** -5
31. No solution; (the value 4 does not check.) **33.** 4
35. 4, -3 **37.** -4; (the value 1 does not check.)
39. No solution; (the value -4 does not check.)
41. 4; (the value -6 does not check.) **43.** -25
45. -1 **47.** The number is 8. **49.** The number is -26.

51. $m = \dfrac{FK}{a}$ **53.** $E = \dfrac{IR}{K}$

55. $R = \dfrac{E - Ir}{I}$ or $R = \dfrac{E}{I} - r$

57. $B = \dfrac{2A - hb}{h}$ or $B = \dfrac{2A}{h} - b$ **59.** $h = \dfrac{V}{r^2 \pi}$

61. $t = \dfrac{b}{x - a}$ or $t = \dfrac{-b}{a - x}$ **63.** $x = \dfrac{y}{1 - yz}$ or $x = \dfrac{-y}{yz - 1}$ **65.** $h = \dfrac{2A}{a + b}$ **67.** $R = \dfrac{R_1 R_2}{R_2 + R_1}$

Chapter 7 Problem Recognition Exercises, p. 547

1. $\dfrac{y - 2}{2y}$ **2.** 6 **3.** 2 **4.** $\dfrac{3a - 17}{a - 5}$ **5.** $\dfrac{4p + 27}{18p^2}$

6. $\dfrac{b(b - 5)}{(b - 1)(b + 1)}$ **7.** 5 **8.** $\dfrac{2w + 5}{(w + 1)^2}$ **9.** 7

10. 5 **11.** $\dfrac{3x + 14}{4(x + 1)}$ **12.** $\dfrac{11}{3}$

Section 7.7 Practice Exercises, pp. 555–559

3. Expression; $\dfrac{m^2 + m + 2}{(m - 1)(m + 3)}$ **5.** Expression; $\dfrac{3}{10}$

7. Equation; 2 **9.** 95 **11.** 1 **13.** $\dfrac{40}{3}$ **15.** 40

17. 3 **19.** -1 **21.** 1 **23. a.** $V_f = \dfrac{V_i T_f}{T_i}$

b. $T_f = \dfrac{T_i V_f}{V_i}$ **25.** Toni can drive 297 mi on 9 gal of gas.
27. They would produce 1536 lb. **29.** 5 oz contains 12 g of carbohydrate. **31.** This represents 262.5 mi.
33. $x = 4$ cm; $y = 5$ cm **35.** $x = 3.75$ cm; $y = 4.5$ cm
37. The height of the pole is 7 m. **39.** The light post is 24 ft high. **41.** The speed of the boat is 20 mph.
43. The plane flies 210 mph in still air. **45.** He runs 8 mph and bikes 16 mph. **47.** Floyd walks 4 mph and Rachel walks 2 mph. **49.** Sergio rode 12 mph and walked 3 mph. **51.** $\dfrac{1}{2}$ of the room **53.** $5\frac{5}{11}$ $(5.\overline{45})$ min
55. $22\frac{2}{9}$ $(22.\overline{2})$ min **57.** 48 hr **59.** $3\frac{1}{3}$ $(3.\overline{3})$ days
61. There are 40 smokers and 140 nonsmokers.
63. There are 240 men and 200 women.

Section 7.8 Practice Exercises, pp. 563–566

3. $y = 12$ **5.** $a = 6; a = 2$ **7.** $\dfrac{b + 1}{1 - b}$ **9.** Inversely

11. $T = kq$ **13.** $b = \dfrac{k}{c}$ **15.** $Q = \dfrac{kx}{y}$ **17.** $c = kst$

19. $L = kw\sqrt{v}$ **21.** $x = \dfrac{ky^2}{z}$ **23.** $k = \dfrac{9}{2}$

25. $k = 512$ **27.** $k = 1.75$ **29.** $x = 70$ **31.** $b = 6$

33. $Z = 56$ **35.** $Q = 9$ **37.** $L = 9$ **39.** $B = \dfrac{15}{2}$

41. a. 3.6 g **b.** 4.5 g **c.** 2.4 g   **43. a.** \$0.40 **b.** \$0.30 **c.** \$1.00   **45.** 355,000 tons   **47.** 42.6 ft **49.** 300 W   **51.** 1.25 Ω   **53.** 20 lb   **55.** \$3500

Chapter 7 Review Exercises pp. 574–577

1. a. $-\dfrac{2}{9}, -\dfrac{1}{10}, 0, -\dfrac{5}{6}$, undefined

b. $\{t \mid t$ is a real number and $t \neq -9\}$ **3.** a, c, d
5. $\{h \mid h$ is a real number and $h \neq -\frac{1}{3}, h \neq -7\}$; $\frac{1}{3h + 1}$
7. $\{w \mid w$ is a real number and $w \neq 4, w \neq -4\}$; $\frac{2w + 3}{w - 4}$
9. $\{k \mid k$ is a real number and $k \neq 0, k \neq 5\}$; $-\frac{3}{2k}$
11. $\{m \mid m$ is a real number and $m \neq -1\}$; $\frac{m - 5}{3}$

13. $\{p \mid p$ is a real number and $p \neq -7\}$; $\frac{1}{p + 7}$ **15.** $\dfrac{u^2}{2}$

17. $\dfrac{3}{2(x - 5)}$ **19.** $\dfrac{q - 2}{4}$ **21.** $4s(s - 4)$

23. $\dfrac{1}{n - 2}$ **25.** $\dfrac{1}{m + 3}$ **27.** $-\dfrac{2y - 1}{y + 1}$

29. LCD $= 12xy$; $\dfrac{21y}{12xy}, \dfrac{22x}{12xy}$

31. LCD $= ab^3c^2$; $\dfrac{5c^2}{ab^3c^2}, \dfrac{3b^3}{ab^3c^2}$

33. LCD $= q(q + 8)$; $\dfrac{6q + 48}{q(q + 8)}, \dfrac{q}{q(q + 8)}$

35. LCD $= (m + 4)(m - 4)(m + 3)$
37. $3 - x$ or $x - 3$ **39.** 2   **41.** $x - 7$

43. $\dfrac{t^2 + 2t + 3}{(2 - t)(2 + t)}$ **45.** $\dfrac{3(r - 4)}{2r(r + 6)}$ **47.** $\dfrac{q}{(q + 5)(q + 4)}$

49. $\dfrac{1}{3}$ **51.** $\dfrac{3(z + 5)}{z(z - 5)}$ **53.** $\dfrac{8}{y}$ **55.** $-(b + a)$

57. $-\dfrac{k + 10}{k + 4}$ **59.** -2 **61.** 2   **63.** 3   **65.** $-11, 1$

67. $h = \dfrac{3V}{\pi r^2}$ **69.** $\dfrac{6}{5}$ **71.** 10 g   **73.** Together the pumps would fill the pool in 16.8 min.   **75. a.** $F = kd$ **b.** $k = 3$ **c.** 4.5 lb   **77.** $y = 12$

Chapter 7 Test pp. 577–578

1. a. $\{x \mid x$ is a real number and $x \neq 2\}$ **b.** $-\dfrac{x + 1}{6}$

2. a. $\{a \mid a$ is a real number and $a \neq 0, a \neq 6, a \neq -2\}$
b. $\dfrac{7}{a + 2}$ **3.** b, c, d   **4.** $\dfrac{y + 7}{3(y + 3)(y + 1)}$

5. $-\dfrac{b + 3}{5}$ **6.** $\dfrac{1}{w + 1}$ **7.** $\dfrac{t + 4}{t + 2}$

8. $\dfrac{x(x + 5)}{(x + 4)(x - 2)}$ **9.** $\dfrac{1}{m + 4}$ **10.** $\dfrac{8}{5}$ **11.** 2
12. 1   **13.** No solution. (the value 4 does not check.)

14. $r = \dfrac{2A}{C}$ **15.** The number is $-\dfrac{2}{5}$.   **16.** -8

17. $1\frac{1}{4}$ (1.25) cups of carrots   **18.** The speed of the current is 5 mph.   **19.** It would take the second printer 3 hr to do the job working alone.   **20.** $a = 5.6; b = 12$
21. a. $15(x + 3)$ **b.** $3x^2y^2$ **22.** 8.25 mL
23. 200 drinks are sold.

Chapters 1–7 Cumulative Review Exercises pp. 578–579

1. 32   **2.** 7   **3.** $\dfrac{10}{9}$

4.

Set-Builder Notation	Graph	Interval Notation
$\{x \mid x \geq -1\}$	→ -1	$[-1, \infty)$
$\{x \mid x < 5\}$	← 5	$(-\infty, 5)$

5. The width is 17 m and the length is 35 m.
6. The base is 10 in. and the height is 8 in.   **7.** $\dfrac{x^2yz^{17}}{2}$

8. a. $6x + 4$ **b.** $2x^2 + x - 3$ **9.** $(5x - 3)^2$
10. $(2c + 1)(5d - 3)$

11. $\left\{x \mid x$ is a real number and $x \neq 5, x \neq -\dfrac{1}{2}\right\}$

12. $\dfrac{x - 3}{x + 5}$ **13.** $\dfrac{1}{5(x + 4)}$ **14.** -3 **15.** 1   **16.** $-\dfrac{7}{2}$
17. a. x-intercept: $(-4, 0)$; y-intercept: $(0, 2)$
b. x-intercept: $(0, 0)$; y-intercept: $(0, 0)$

18. a. $m = -\dfrac{7}{5}$ **b.** $m = -\dfrac{2}{3}$ **c.** $m = 4$ **d.** $m = -\dfrac{1}{4}$
19. $y = 5x - 3$ **20.** One large popcorn costs \$3.50, and one drink costs \$1.50.

Chapter 8
Chapter Opener Puzzle

Section 8.1 Calculator Connections, p. 589

1. 2.236   **3.** 7.071   **5.** 5.745   **7.** 8.944
9. 1.913   **11.** 4.021

Section 8.1 Practice Exercises, pp. 589–593

3. 12, -12 **5.** There are no real-valued square roots of -49.   **7.** 0   **9.** $\dfrac{1}{5}, -\dfrac{1}{5}$ **11. a.** 13 **b.** -13 **13.** 0
15. 9, 16, 25, 36, 64, 121, 169   **17.** 2   **19.** 7   **21.** 0.4

23. 0.3   **25.** $\dfrac{5}{4}$ **27.** $\dfrac{1}{12}$ **29.** 5   **31.** 9

33. There is no real value of b for which $b^2 = -16$.
35. -2 **37.** Not a real number   **39.** Not a real number
41. Not a real number   **43.** -20 **45.** Not a real number
47. $0, 1, 27, 125$ **49.** Yes, -3 **51.** 3   **53.** 4
55. -2 **57.** Not a real number   **59.** Not a real number
61. -2 **63.** -1 **65.** 0   **67.** 4   **69.** 4   **71.** 5
73. -5 **75.** 2   **77.** 2   **79.** $|a|$ **81.** y **83.** $|w|$
85. x **87.** $x^2, y^4, (ab)^6, w^8x^8, m^{10}$. The expression is a
perfect square if the exponent is even.   **89.** $p^4, t^8, (cd)^{12}$.
The expression is a perfect fourth power if the exponent is a
multiple of 4.   **91.** y^6 **93.** a^4b^{15} **95.** q^8 **97.** $2w^2$
99. $5x$ **101.** $-5x$ **103.** $5p^2$ **105.** $5p^2$

107. $\sqrt{q} + p^2$ **109.** $\dfrac{6}{\sqrt[4]{x}}$ **111.** 9 cm   **113.** 5 ft

115. 6.9 cm   **117.** 17.0 in.   **119.** 31.3 in.
121. 268 km   **123.** $x \geq 0$

Section 8.2 Calculator Connections, p. 598

1.
```
√(125)
       11.18033989
5*√(5)
       11.18033989
```
2.
```
√(18)
       4.242640687
3*√(2)
       4.242640687
```
3.
```
3√(54)
       3.77976315
3*3√(2)
       3.77976315
```
4.
```
3√(108)
       4.762203156
3*3√(4)
       4.762203156
```

Section 8.2 Practice Exercises, pp. 599–601

3. $8, 27, y^3, y^9, y^{12}, y^{27}$ **5.** -5 **7.** -3 **9.** a^2
11. $2xy^2$ **13.** 446 km   **15.** $3\sqrt{2}$ **17.** $2\sqrt{7}$
19. $12\sqrt{5}$ **21.** $-10\sqrt{2}$ **23.** $a^2\sqrt{a}$ **25.** w^{11}
27. $m^2n^2\sqrt{n}$ **29.** $x^7y^5\sqrt{x}$ **31.** $3t^5$ **33.** $2x\sqrt{2x}$
35. $4z\sqrt{z}$ **37.** $-3w^3\sqrt{5}$ **39.** $z^{12}\sqrt{z}$ **41.** $-z^5\sqrt{15z}$
43. $10ab^3\sqrt{26b}$ **45.** $\sqrt{26pq}$ **47.** m^6n^8

49. $4ab^2c^2\sqrt{3ab}$ **51.** a^4 **53.** y^5 **55.** $\dfrac{1}{2}$

57. 2   **59.** $2x$ **61.** $5p^3$ **63.** $3\sqrt{5}$ **65.** $\sqrt{6}$
67. 4   **69.** 7   **71.** $11\sqrt{2}$ ft   **73.** $2\sqrt{66}$ cm
75. $a^2\sqrt[3]{a^2}$ **77.** $14z\sqrt[3]{2}$ **79.** $2ab^2\sqrt[3]{2a^2}$ **81.** z

83. -2 **85.** $2\sqrt[3]{5}$ **87.** $\dfrac{1}{3}$ **89.** $4a\sqrt{a}$

91. $2x$ **93.** $2p\sqrt{2q}$ **95.** $4\sqrt{2}$ **97.** $2u^2v^3\sqrt{13v}$
99. $6\sqrt{6}$ **101.** 6   **103.** $2a\sqrt[3]{2}$ **105.** x

107. $-\sqrt{5}$ **109.** $-\dfrac{1}{2}$ **111.** $5\sqrt{2}$ **113.** $x + 5$

Section 8.3 Practice Exercises, pp. 605–607

3. $2y$ **5.** $6x\sqrt{x}$ **7.** $\dfrac{5c^3}{4}$ **9.** Not a real number

11. For example, $2\sqrt{3}, 6\sqrt[3]{3}$ **13.** c **15.** $8\sqrt{2}$
17. $4\sqrt{7}$ **19.** $2\sqrt{10}$ **21.** $11\sqrt{y}$ **23.** 0
25. $5y\sqrt{15}$ **27.** $x\sqrt{y} - y\sqrt{x}$ **29.** $3\sqrt[3]{6} + 8\sqrt[3]{6}; 11\sqrt[3]{6}$

31. $4\sqrt{5} - 6\sqrt{5}; -2\sqrt{5}$ **33.** $8\sqrt{3}$ **35.** 0   **37.** $2\sqrt{2}$
39. $16p^2\sqrt{5}$ **41.** $10\sqrt{2k}$ **43.** $a^2\sqrt{b}$ **45.** $3\sqrt{5}$
47. $\dfrac{29}{18}z\sqrt{6}$ **49.** $-1.7\sqrt{10}$ **51.** $2x\sqrt{x}$ **53.** $3\sqrt{7}$
55. $4\sqrt{w} + 2\sqrt{6w} + 2\sqrt{10w}$ **57.** $6x^3\sqrt{y}$
59. $2\sqrt{3} - 4\sqrt{6}$ **61.** $-4x\sqrt{2} + \sqrt{2x}$ **63.** $9\sqrt{2}$ m
65. $16\sqrt{3}$ in.   **67.** Radicands are not the same.
69. One term has a radical. One does not.

71. The indices are different.   **73.** $\dfrac{\sqrt{3}}{3}$ **75. a.** 80 m
b. 159 m

Section 8.4 Practice Exercises, pp. 612–613

3. 11   **5.** $3w^2\sqrt{z}$ **7.** $\sqrt{15}$ **9.** 47   **11.** b
13. $6\sqrt{15p}$ **15.** $5\sqrt{2}$ **17.** $14\sqrt{2}$ **19.** $6x\sqrt{7}$
21. $4x^3\sqrt{y}$ **23.** $12w^2\sqrt{10}$ **25.** $-8\sqrt{15}$
27. Perimeter: $6\sqrt{5}$ ft; area: 10 ft^2 **29.** 3 cm^2 **31.** $3w$
33. $-16\sqrt{10y}$ **35.** $2\sqrt{3} - \sqrt{6}$ **37.** $4x + 20\sqrt{x}$
39. $-8 + 7\sqrt{30}$ **41.** $9a - 28b\sqrt{a} + 3b^2$
43. $8p^2 + 19p\sqrt{p} + 2p - 8\sqrt{p}$ **45.** 10   **47.** 4
49. t **51.** $16c$ **53.** $29 + 8\sqrt{13}$ **55.** $a - 4\sqrt{a} + 4$
57. $4a - 12\sqrt{a} + 9$ **59.** $21 - 2\sqrt{110}$ **61.** 1
63. $x - y$ **65.** -1 **67.** 4   **69.** $64x - 4y$ **71.** 73
73. a. $3x + 6$ **b.** $\sqrt{3x} + \sqrt{6}$ **75. a.** $4a^2 + 12a + 9$
b. $4a + 12\sqrt{a} + 9$ **77. a.** $b^2 - 25$ **b.** $b - 25$
79. a. $x^2 - 4xy + 4y^2$ **b.** $x - 4\sqrt{xy} + 4y$
81. a. $p^2 - q^2$ **b.** $p - q$ **83. a.** $y^2 - 6y + 9$
b. $x - 6\sqrt{x - 2} + 7$

Section 8.5 Practice Exercises, pp. 620–623

3. $6y + 23\sqrt{y} + 21$ **5.** $9\sqrt{3}$ **7.** $25 - 10\sqrt{a} + a$
9. -5 **11.** $\dfrac{\sqrt{3}}{4}$ **13.** $\dfrac{a^2}{b^2}$ **15.** $\dfrac{c\sqrt{c}}{2}$ **17.** $\dfrac{\sqrt[3]{x^2}}{3}$
19. $\dfrac{\sqrt[3]{y^2}}{3}$ **21.** $\dfrac{10\sqrt{2}}{9}$ **23.** $\dfrac{2}{5}$ **25.** $\dfrac{1}{2p}$ **27.** z
29. $2\sqrt[3]{x}$ **31.** $\dfrac{\sqrt{6}}{6}$ **33.** $3\sqrt{5}$ **35.** $\dfrac{6\sqrt{x + 1}}{x + 1}$
37. $\dfrac{\sqrt{6x}}{x}$ **39.** $\dfrac{\sqrt{21}}{7}$ **41.** $\dfrac{5\sqrt{6y}}{3y}$ **43.** $\dfrac{3\sqrt{6}}{4}$
45. $\dfrac{\sqrt{3p}}{9}$ **47.** $\dfrac{\sqrt{5}}{2}$ **49.** $\dfrac{x\sqrt{y}}{y^2}$ **51.** -7
53. $\sqrt{5} + \sqrt{3}; 2$ **55.** $\sqrt{x} - 10; x - 100$
57. $\dfrac{4\sqrt{2} - 12}{-7}$ or $\dfrac{12 - 4\sqrt{2}}{7}$ **59.** $\dfrac{\sqrt{5} + \sqrt{2}}{3}$
61. $\sqrt{6} - \sqrt{2}$ **63.** $\dfrac{\sqrt{x} + \sqrt{3}}{x - 3}$ **65.** $7 - 4\sqrt{3}$
67. $-13 - 6\sqrt{5}$ **69.** $2 - \sqrt{2}$ **71.** $\dfrac{3 + \sqrt{2}}{2}$
73. $1 - \sqrt{7}$ **75.** $\dfrac{7 + 3\sqrt{2}}{3}$ **77. a.** Condition 1 fails;
$2x^4\sqrt{2x}$ **b.** Condition 2 fails; $\dfrac{\sqrt{5x}}{x}$ **c.** Condition 3 fails;
$\dfrac{\sqrt{3}}{3}$ **79. a.** Condition 2 fails; $\dfrac{3\sqrt{x} - 3}{x - 1}$ **b.** Conditions
1 and 3 fail; $\dfrac{3w\sqrt{t}}{t}$ **c.** Condition 1 fails; $2a^2b^4\sqrt{6ab}$
81. $3\sqrt{5}$ **83.** $-\dfrac{3w\sqrt{2}}{5}$ **85.** Not a real number

87. $\dfrac{s\sqrt{t}}{t}$ **89.** $\dfrac{m^2}{2}$ **91.** $\dfrac{9\sqrt{t}}{t^2}$ **93.** $\dfrac{\sqrt{11}-\sqrt{5}}{2}$

95. $\dfrac{a+2\sqrt{ab}+b}{a-b}$ **97.** $-\dfrac{3\sqrt{2}}{8}$ **99.** $\dfrac{\sqrt{3}}{9}$

Chapter 8 Problem Recognition Exercises, p. 623

1. $3\sqrt{2}$ **2.** $2\sqrt{7}$ **3.** Cannot be simplified further
4. Cannot be simplified further   **5.** $\sqrt{2}$ **6.** $\sqrt{7}$
7. $9-z$ **8.** $16-y$ **9.** $8-3\sqrt{5}$
10. $-8+11\sqrt{3}$ **11.** $-x\sqrt{y}$ **12.** $11ab\sqrt{a}$
13. $-24-6\sqrt{6}-3\sqrt{2}$ **14.** $-80+8\sqrt{15}+16\sqrt{5}$
15. $\dfrac{2\sqrt{x}+14}{x-49}$ **16.** $\dfrac{5\sqrt{y}-20}{y-16}$ **17.** $3\sqrt{3}$
18. $3\sqrt{5}$ **19.** $\dfrac{\sqrt{7x}}{x}$ **20.** $\dfrac{\sqrt{11y}}{y}$ **21.** $y^2z^5\sqrt{z}$
22. $2q^3\sqrt{2}$ **23.** $3p^2\sqrt[3]{p^2}$ **24.** $5u^3v^4\sqrt[3]{u^2}$ **25.** $x\sqrt{10}$
26. $y\sqrt{3}$ **27.** $20\sqrt{3}$ **28.** $\sqrt{10}$ **29.** $51+14\sqrt{2}$
30. $8+2\sqrt{15}$ **31.** $\sqrt{x}-\sqrt{5}$ **32.** $\sqrt{y}-\sqrt{7}$
33. $4x-11\sqrt{xy}-3y$ **34.** $\dfrac{1}{3}$ **35.** $\dfrac{5}{3}$
36. $16-2\sqrt{55}$ **37.** $x-12\sqrt{x}+36$ **38.** $-3\sqrt{6}$
39. $11\sqrt{a}$ **40.** -88 **41.** $u-9v$ **42.** $4x\sqrt{2}$
43. 0 **44.** $5+\sqrt{35}$ **45.** $a+2\sqrt{a}$
46. $26-17\sqrt{2}$

Section 8.6 Practice Exercises, pp. 628–630

3. $\dfrac{\sqrt{2}-\sqrt{10}}{-8}$ or $\dfrac{\sqrt{10}-\sqrt{2}}{8}$ **5.** $\dfrac{2\sqrt{6}}{3}$

7. $x^2+8x+16$ **9.** $x+8\sqrt{x}+16$ **11.** $2x-3$
13. t^2+2t+1 **15.** 36 **17.** 15 **19.** No solution
(the value 29 does not check)   **21.** 5 **23.** $-\dfrac{1}{2}$
25. 6 **27.** 8 **29.** No solution (the value $\frac{19}{2}$ does
not check)   **31.** 1 **33.** $4, -3$ **35.** 0
37. No solution (the value -4 does not check)
39. 4, (the value -1 does not check)   **41.** $0, -1$
43. 12, (the value 4 does not check)   **45.** -6 **47.** -1
49. $\sqrt{x+10}=1$; -9 **51.** $\sqrt{2x}=x-4$; 8
53. $\sqrt[3]{x+1}=2$; 7 **55. a.** 80 ft/sec   **b.** 289 ft
57. a. 16 in.   **b.** 25 weeks   **59.** $\dfrac{9}{5}$

61. $\dfrac{3}{2}$ (the value -1 does not check)

Section 8.7 Practice Exercises, pp. 634–636

1. a. 3   **b.** 125   **3.** 27   **5.** $a+1$ **7.** 9   **9.** 5

11. 3   **13.** -2 **15.** -2 **17.** $\dfrac{1}{6}$ **19.** $\sqrt[3]{x}$

21. $\sqrt{4a}$ or $2\sqrt{a}$ **23.** $\sqrt[5]{yz}$ **25.** $\sqrt[3]{u^2}$ **27.** $5\sqrt{q}$

29. $\sqrt{\dfrac{x}{9}}$ or $\dfrac{\sqrt{x}}{3}$ **31.** $a^{m/n}=\sqrt[n]{a^m}$ or $(\sqrt[n]{a})^m$, provided the

roots exist.   **33.** 8   **35.** $\dfrac{1}{9}$ **37.** -32 **39.** 2

41. $(\sqrt{y})^9$ **43.** $\sqrt[3]{c^5d}$ **45.** $\dfrac{1}{\sqrt[5]{qr}}$

47. $6\sqrt[3]{y^2}$ **49.** $x^{1/3}$ **51.** $(xy)^{1/3}$ **53.** $5x^{1/2}$
55. $y^{2/3}$ **57.** $(m^3n)^{1/4}$ **59.** $4k^{3/3}$ or $4k$ **61.** x
63. y^2 **65.** 6   **67.** a^7 **69.** $y^{4/3}$ **71.** 2
73. $\dfrac{25a^4d}{c}$ **75.** $\dfrac{y^9}{x^8}$ **77.** $\dfrac{2z^3}{w}$ **79.** $5xy^2z^{3/2}$
81. a. 10 in.   **b.** 8.49 in.   **83. a.** 10.9%   **b.** 8.8%
c. The account in part (a)   **85.** No, for example,
$(36+64)^{1/2}\neq 36^{1/2}+64^{1/2}$ **87.** 6   **89.** $\dfrac{5}{14}$ **91.** $\dfrac{a^{22}b^4}{c^{17}}$

Chapter 8 Review Exercises, pp. 642–644

1. Principal square root: 14; negative square root: -14
3. Principal square root: 0.8; negative square root: -0.8
5. There is no real number b such that $b^2=-64$.   **7.** -12
9. Not a real number   **11.** $|y|$ **13.** $|y|$ **15.** -5
17. $\dfrac{3}{t^2}$ **19. a.** 7.1 m   **b.** 22.6 ft   **21.** $b^2+\sqrt{5}$
23. The quotient of 2 and the principal square root of p
25. 12 ft   **27.** $x^8\sqrt{x}$ **29.** $2\sqrt{7}$ **31.** $3y^3\sqrt[3]{y}$
33. c   **35.** $10y^2$ **37.** $2x$ **39.** $5\sqrt{3}$ **41.** 1
43. $7\sqrt{6}$ **45.** $-4x\sqrt{5}$ **47.** $15\sqrt{3}-7\sqrt{7}$
49. $\sqrt{x^4+(5x)^2}$; $26x^2$ **51.** $12\sqrt{2}$ ft   **53.** 25
55. $70\sqrt{3x}$ **57.** $8m+24\sqrt{m}$ **59.** $-49-16\sqrt{26}$
61. $64w-z$ **63.** $10\sqrt{3}\ m^3$ **65.** a^5 **67.** $4\sqrt{y}$
69. b   **71.** $\dfrac{3\sqrt{2y}}{y}$ **73.** $2\sqrt{7}+2\sqrt{2}$
75. $-8-3\sqrt{7}$ **77.** 138   **79.** 39   **81.** 7
83. 2, (the value -2 does not check)
85. -69 **87.** -3 **89.** -2 **91.** $\dfrac{1}{8}$
93. $\sqrt[5]{z}$ **95.** $\sqrt[4]{w^3}$ **97.** $a^{2/5}$ **99.** $(a^2b^4)^{1/5}$
101. y^2 **103.** $6^{3/5}$ **105.** $4ab^2$ **107.** 2.0 cm

Chapter 8 Test, pp. 645–646

1. 1. The radicand has no factor raised to a power greater
than or equal to the index. 2. There are no radicals in the
denominator of a fraction. 3. The radicand does not contain a
fraction.   **2.** $11x\sqrt{2}$ **3.** $2y\sqrt[3]{6y}$ **4.** Not a real number
5. $\dfrac{a^3\sqrt{5}}{9}$ **6.** $\dfrac{3\sqrt{6}}{2}$ **7.** $\dfrac{2\sqrt{5}-12}{-31}$ or $\dfrac{12-2\sqrt{5}}{31}$
8. a. $\sqrt{25}+5^3$; 130   **b.** $4^2-\sqrt{16}$; 12   **9. a.** 97 ft
b. 339 ft   **10.** $8\sqrt{z}$ **11.** $4\sqrt{6}-15$ **12.** $-7t\sqrt{2}$
13. $9\sqrt{10}$ **14.** $46-6\sqrt{5}$ **15.** $-8+23\sqrt{10}$
16. $\dfrac{\sqrt{n}}{6m}$ **17.** $16-9x$ **18.** $\dfrac{\sqrt{22}}{11}$ **19.** $\dfrac{3\sqrt{7}+3\sqrt{3}}{2}$
20. 206 yd   **21.** No solution (the value $\frac{9}{2}$ does not check)
22. $0, -5$ **23.** 14   **24. a.** 12 in.   **b.** 18 in.   **c.** 25 weeks
25. 1000   **26.** 2   **27.** $\sqrt[5]{x^3}$ or $(\sqrt[5]{x})^3$ **28.** $5\sqrt{y}$
29. $(ab^3)^{1/4}$ **30.** $p^{11/12}$ **31.** $5^{3/5}$ **32.** $3mn^2$

Chapters 1–8 Cumulative Review Exercises, pp. 646–647

1. 1   **2.** -2 **3.** -15 **4.** $-9x^2+2x+10$
5. $\dfrac{2x}{y}-1+\dfrac{4}{x}$ **6.** $2(5c+2)^2$ **7.** $-\dfrac{2}{5},\dfrac{1}{2}$ **8.** 1
9. No solution (the value 5 does not check)

10. $\dfrac{x(5x+8)}{16(x-1)}$ **11.**

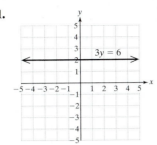

12. a. $y = 880$; the cost of renting the office space for 3 months is \$880.   **b.** $x = 12$; the cost of renting office space for 12 months is \$2770.   **c.** $m = 210$; the cost increases at a rate of \$210 per month.   **d.** $(0, 250)$; the down payment of renting the office space is \$250.   **13.** $y = -x + 1$

14. $(1, -4)$ **15.**

16. 8 L of 20% solution should be mixed with 4 L of 50% solution.   **17.** $3\sqrt{11}$ **18.** $7x\sqrt{3}$

19. $\dfrac{x + \sqrt{xy}}{x - y}$ **20.** 5

Chapter 9

Chapter Opener Puzzle

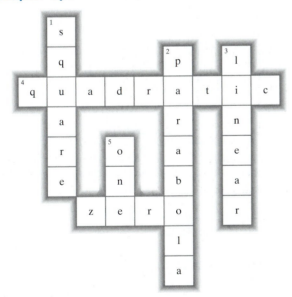

Section 9.1 Practice Exercises, pp. 653–655

3. a. Linear   **b.** Quadratic   **c.** Linear   **5.** $-5, \dfrac{1}{2}$

7. $7, -5$ **9.** $-2, -\dfrac{1}{6}$ **11.** $-7, -\dfrac{3}{2}$ **13.** $12, -12$

15. $8, -2$ **17.** $\dfrac{1}{4}, -2$ **19.** $-1, -7$ **21.** ± 7

23. ± 10 **25.** There are no real-valued solutions.

27. $\pm\sqrt{3}$ **29.** $9, 1$ **31.** $11, -1$ **33.** $11 \pm \sqrt{5}$

35. $-1 \pm 3\sqrt{2}$ **37.** $\dfrac{1}{4} \pm \dfrac{\sqrt{7}}{4}$ **39.** $\dfrac{1}{2} \pm \sqrt{15}$

41. There are no real-valued solutions.   **43.** $\pm\dfrac{5}{2}$

45. The solution checks.   **47.** False. -8 is also a solution.

49. a. 64 ft   **b.** 3.5 sec   **c.** 8.8 sec   **51.** 7.1 m

53. 8.0 ft

Section 9.2 Practice Exercises, pp. 659–661

3. $5 \pm \sqrt{21}$ **5.** $4; y^2 + 4y + 4 = (y + 2)^2$

7. $36; p^2 - 12p + 36 = (p - 6)^2$

9. $\dfrac{81}{4}; x^2 - 9x + \dfrac{81}{4} = \left(x - \dfrac{9}{2}\right)^2$

11. $\dfrac{25}{36}; d^2 + \dfrac{5}{3}d + \dfrac{25}{36} = \left(d + \dfrac{5}{6}\right)^2$

13. $\dfrac{1}{100}; m^2 - \dfrac{1}{5}m + \dfrac{1}{100} = \left(m - \dfrac{1}{10}\right)^2$

15. $\dfrac{1}{4}; u^2 + u + \dfrac{1}{4} = \left(u + \dfrac{1}{2}\right)^2$ **17.** $x = 2, x = -6$

19. $y = -1, y = -5$ **21.** $x = 1 \pm \sqrt{2}$

23. $x = 1 \pm \sqrt{6}$ **25.** $-2 \pm \sqrt{3}$ **27.** $-\dfrac{1}{2} \pm \dfrac{\sqrt{13}}{2}$

29. $-1 \pm \sqrt{41}$ **31.** $2 \pm \sqrt{5}$ **33.** $-2, -4$

35. $3, 8$ **37.** ± 11 **39.** $-2 \pm \sqrt{2}$ **41.** $-13, 5$

43. 13 **45.** $10, -2$ **47.** $7, -1$ **49.** $4 \pm \sqrt{15}$

51. $-1 \pm \sqrt{6}$ **53.** $11, -2$ **55.** $0, 7$ **57.** $\dfrac{1}{2}, -\dfrac{3}{4}$

59. $\pm\sqrt{14}$ **61.** There are no real-valued solutions.

63. 1 **65.** The suitcase is 10 in. by 14 in. by 30 in. The bag must be checked because 10 in. + 14 in. + 30 in. = 54 in., which is greater than 45 in.

Section 9.3 Calculator Connections, p. 667

1.
```
(-5+√(17))/4
     -.2192235936
(-5-√(17))/4
     -2.280776406
```

2.
```
(-40+√(1920))/-3
2
     -.1193063938
(-40-√(1920))/-3
2
     2.619306394
```

Section 9.3 Practice Exercises, pp. 667–669

1. ± 13 **3.** $4 \pm 2\sqrt{7}$ **5.** $2 \pm 2\sqrt{2}$

7. For $ax^2 + bx + c = 0, x = \dfrac{-b \pm \sqrt{b^2 - 4ac}}{2a}$

9. $2x^2 - x - 5 = 0; a = 2, b = -1, c = -5$

11. $-3x^2 + 14x + 0 = 0; a = -3, b = 14, c = 0$

13. $x^2 + 0x - 9 = 0; a = 1, b = 0, c = -9$

15. -8 **17.** $\dfrac{2}{3}, -\dfrac{1}{2}$ **19.** $\dfrac{1 \pm \sqrt{61}}{10}$ **21.** $1 \pm \sqrt{2}$

23. $\dfrac{-5 \pm \sqrt{3}}{2}$ **25.** $\dfrac{1 \pm \sqrt{17}}{-8}$ or $\dfrac{-1 \pm \sqrt{17}}{8}$

27. $\dfrac{-3 \pm \sqrt{33}}{4}$ **29.** $\dfrac{-15 \pm \sqrt{145}}{4}$

31. $\dfrac{-2 \pm \sqrt{22}}{6}$ **33.** $\dfrac{3}{4}, -\dfrac{3}{4}$

35. There are no real-valued solutions.   **37.** $-12 \pm 3\sqrt{5}$

39. $\dfrac{3 \pm \sqrt{15}}{2}$ **41.** $0, \dfrac{11}{9}$ **43.** $\dfrac{3 \pm \sqrt{5}}{2}$

45. $\dfrac{1 \pm \sqrt{41}}{4}$ **47.** $0, \dfrac{1}{9}$ **49.** $\pm 2\sqrt{13}$

51. $\dfrac{-10 \pm \sqrt{85}}{-5}$ or $\dfrac{10 \pm \sqrt{85}}{5}$ **53.** $\dfrac{1 \pm \sqrt{61}}{2}$

55. There are no real-valued solutions.   **57.** The width is 7.3 m. The length is 13.7 m   **59.** The length is 7.4 ft. The width is 5.4 ft. The height is 6 ft.   **61.** The width is 6.7 ft. The length is 10.7 ft.   **63.** The legs are 10.6 m and 7.6 m.

Chapter 9 Problem Recognition Exercises, p. 669

1. $\dfrac{1}{3}, -\dfrac{3}{2}$ **2.** -7 **3.** $4 \pm \sqrt{22}$ **4.** $3 \pm \sqrt{7}$

5. $3 \pm 2\sqrt{2}$ **6.** $-11 \pm 2\sqrt{3}$ **7.** $0, \dfrac{5}{6}$ **8.** $0, \dfrac{1}{2}$

9. ± 6 **10.** $\pm \dfrac{7}{5}$ **11.** $\dfrac{5}{2}, \dfrac{1}{4}$ **12.** $\dfrac{4}{3}, \dfrac{1}{3}$

13. There are no real-valued solutions.   **14.** There are no real-valued solutions.   **15.** $-4 \pm \sqrt{15}$ **16.** $-3 \pm \sqrt{6}$

17. 4 **18.** -3 **19.** $5, 2$ **20.** $3, -1$ **21.** $9, -11$

22. $13, -3$

Section 9.4 Calculator Connections, p. 677

1. $(-2, 3)$

2. $(10, 5)$

3. $(-1.5, -2.6)$

4. $(1.75, 2.5)$

5. $1 \pm \sqrt{6}$

6. $\left(-\dfrac{8}{3}, 0\right)$

7. The graph in part (b) is shifted up 4 units. The graph in part (c) is shifted down 3 units.

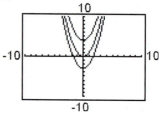

8. In part (b), the graph is shifted to the right 3 units. In part (c), the graph is shifted to the left 2 units.

9. In part (b), the graph is stretched vertically by a factor of 2. In part (c), the graph is shrunk vertically by a factor of $\frac{1}{2}$.

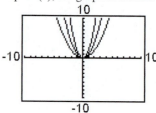

10. In part (b), the graph has been stretched vertically and reflected across the x-axis. In part (c), the graph has been shrunk vertically and reflected across the x-axis.

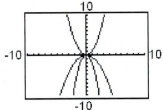

Section 9.4 Practice Exercises, pp. 678–681

3. $-5, 3$ **5.** $-1 \pm \sqrt{6}$ **7.** $2 \pm 2\sqrt{2}$
9. Linear   **11.** Quadratic   **13.** Neither
15. Linear   **17.** Quadratic   **19.** Neither
21. If $a > 0$ the graph opens upward; if $a < 0$ the graph opens downward.   **23.** $a = 2$; upward   **25.** $a = -10$; downward   **27.** $(-1, -8)$ **29.** $(1, -4)$ **31.** $(1, 2)$
33. $(0, -4)$ **35.** c; x-intercepts: $(\sqrt{7}, 0)(-\sqrt{7}, 0)$; y-intercept: $(0, -7)$ **37.** a; x-intercepts: $(-1, 0)(-5, 0)$; y-intercept: $(0, 5)$
39. a. Upward   **b.** $(0, -9)$ **c.** $(3, 0)(-3, 0)$ **d.** $(0, -9)$
e.   **f.** $(-\infty, \infty)$ **g.** $[-9, \infty)$

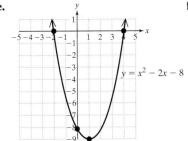

41. a. Upward   **b.** $(1, -9)$ **c.** $(4, 0)(-2, 0)$ **d.** $(0, -8)$
e.   **f.** $(-\infty, \infty)$ **g.** $[-9, \infty)$

43. a. Downward   **b.** $(3, 0)$ **c.** $(3, 0)$ **d.** $(0, -9)$
e.   **f.** $(-\infty, \infty)$ **g.** $(-\infty, 0]$

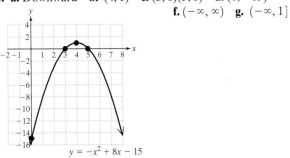

45. a. Downward   **b.** $(4, 1)$ **c.** $(3, 0)(5, 0)$ **d.** $(0, -15)$
e.   **f.** $(-\infty, \infty)$ **g.** $(-\infty, 1]$

$y = -x^2 + 8x - 15$

47. a. Upward   **b.** $(-3, 1)$ **c.** none   **d.** $(0, 10)$
e.   **f.** $(-\infty, \infty)$ **g.** $[1, -\infty)$

$y = x^2 + 6x + 10$

49. a. Downward   **b.** $(0, -2)$ **c.** none   **d.** $(0, -2)$
e.   **f.** $(-\infty, \infty)$ **g.** $(-\infty, -2]$

$y = -2x^2 - 2$

51. True   **53.** False   **55. a.** 28 ft   **b.** 1.25 sec
57. a. 200 calendars   **b.** $500

Chapter 9 Review Exercises, pp. 685–687

1. Linear   **3.** Quadratic   **5.** ± 5 **7.** The equation has no real-valued solutions.   **9.** $-1 \pm \sqrt{14}$
11. $\dfrac{1}{8} \pm \dfrac{\sqrt{3}}{8}$ **13.** 36   **15.** $\dfrac{25}{4}$ **17.** $-4 \pm \sqrt{13}$
19. $\dfrac{3}{2} \pm \dfrac{\sqrt{21}}{2}$ **21.** 10.6 ft   **23.** For $ax^2 + bx + c = 0$, $x = \dfrac{-b \pm \sqrt{b^2 - 4ac}}{2a}$ **25.** -2 **27.** $\dfrac{3}{2}, -1$
29. $-3 \pm \sqrt{7}$ **31.** $1 \pm \sqrt{5}$
33. The equation has no real-valued solutions.
35. The height is $\dfrac{-1 + \sqrt{97}}{2} \approx 4.4$ cm.   The base is $\dfrac{1 + \sqrt{97}}{2} \approx 5.4$ cm.   **37.** $a = 1$; upward   **39.** $a = -2$; downward   **41.** Vertex: $(-1, 1)$ **43.** Vertex: $(3, 13)$
45. a. Upward   **b.** $(-2, -3)$ **c.** $(-3, 0)(-1, 0)$ **d.** $(0, 9)$

e.

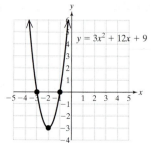

f. $(-\infty, \infty)$ **g.** $[-3, \infty)$

19. Vertex: $(0, 25)$; x-intercepts: $(-5, 0)(5, 0)$; y-intercept: $(0, 25)$

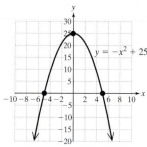

47. a. Downward   **b.** $(-1, -4)$ **c.** No x-intercepts
d. $(0, -12)$

e.

f. $(-\infty, \infty)$ **g.** $(-\infty, -4]$

20. a. $25 per ticket   **b.** $250,000

49. a. 1024 ft   **b.** 8 sec

Chapter 9 Test, pp. 687–688

1. The equation has no real-valued solutions.

2. $4 \pm \sqrt{21}$ **3.** $\dfrac{5 \pm \sqrt{13}}{6}$ **4.** $\dfrac{-1 \pm \sqrt{41}}{10}$

5. $12 \pm 2\sqrt{3}$ **6.** $-7 \pm 5\sqrt{2}$ **7.** $\pm\sqrt{10}$

8. $\dfrac{5}{6}, -\dfrac{3}{2}$ **9.** $0, \dfrac{11}{6}$ **10.** $3 \pm 2\sqrt{5}$ **11.** 4.0 in.

12. The base is $\dfrac{-1 + \sqrt{97}}{2} \approx 4.4$ m. The height is

$1 + \sqrt{97} \approx 10.8$ m.   **13.** For $y = ax^2 + bx + c$, if $a > 0$ the parabola opens upward, if $a < 0$ the parabola opens downward.   **14.** $(5, 0)$ **15.** $(1, 5)$ **16.** $(0, -16)$
17. The parabola has no x-intercepts.
18. a. Opens upward   **b.** Vertex: $(-3, -1)$
c. x-intercepts: $(-2, 0)$ and $(-4, 0)$ **d.** y-intercept: $(0, 8)$
e.

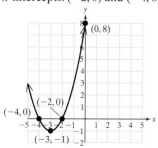

Chapters 1–9 Cumulative Review Exercises, pp. 688–689

1. 1   **2.** $h = \dfrac{2A}{b}$ **3.** $\dfrac{34}{3}$ **4.** Yes, 2 is a solution.

5. $\{x \mid x > -1\}$; $(-1, \infty)$ ⟵——————→
$\qquad\qquad\qquad\qquad\qquad -1$

6. a. Decreases   **b.** $m = -37.6$. For each additional increase in education level, the death rate decreases by approximately 38 deaths per 100,000 people.
c. 901.8 per 100,000   **d.** 10th grade

7. $\dfrac{8c}{a^4 b}$ **8.** 1.898×10^{10} diapers

9. Approximately 430 light-years   **10.** $4x^2 - 16x + 13$

11. $2y^3 + 1 - \dfrac{3}{y - 2}$ **12.** $(2x + 5)(x - 7)$

13. $(y + 4a)(2x - 3b)$ **14.** The base is 9 m, and the

height is 8 m.   **15.** $-\dfrac{2}{x + 1}$ **16.** $\dfrac{5}{x - 2}$ **17.** $x - 5$

18. $-\dfrac{2}{x}$ **19.** $4, -3$ **20.** 20 cm ≈ 7.9 in. The fish

must be released.   **21.** $y = \dfrac{1}{2}x + 4$

22. a. $(6, 0)$ **b.** $(0, -3)$ **c.** $\frac{1}{2}$
d.

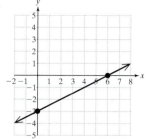

23. a. $(-3, 0)$ **b.** No y-intercept   **c.** Slope is undefined.
d.

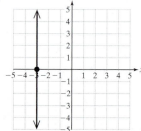

24. $\left(1, \dfrac{4}{3}\right)$ **25.** $(5, 2)$ **26.** The angles are $22°$ and $68°$.

27. There are 13 dimes and 14 quarters.

28.

29. $\pi, \sqrt{7}$ **30.** $\dfrac{\sqrt{7}}{7}$ **31.** $2x\sqrt{2x}$

32. $48 + 8\sqrt{3x} + x$ **33.** $2\sqrt{2x}$ **34.** $\dfrac{8 + 4\sqrt{a}}{4 - a}$

35. -2, (the value -7 does not check)

36. $(2c - y)(4c^2 + 2cy + y^2)$ **37.** b

38. a. 1 **b.** 4 **c.** 13

39. Domain: $\{2, -1, 9, -6\}$; range: $\{4, 3, 2, 8\}$

40. $m = \dfrac{5}{7}$ **41.** $m = -\dfrac{4}{5}$ **42.** 25

43. $-3 \pm \sqrt{6}$ **44.** $-3 \pm \sqrt{6}$

45.

Credits

Application Index

Subject Index

Perimeter and Circumference

Rectangle

$P = 2l + 2w$

Square

$P = 4s$

Triangle

$P = a + b + c$

Circle

Circumference: $C = 2\pi r$

Area

Rectangle

$A = lw$

Square

$A = s^2$

Parallelogram

$A = bh$

Triangle

$A = \frac{1}{2}bh$

Trapezoid

$A = \frac{1}{2}(b_1 + b_2)h$

Circle

$A = \pi r^2$

Volume

Rectangular Solid

$V = lwh$

Cube

$V = s^3$

Right Circular Cylinder

$V = \pi r^2 h$

Right Circular Cone

$V = \frac{1}{3}\pi r^2 h$

Sphere

$V = \frac{4}{3}\pi r^3$

Angles

- Two angles are **complementary** if the sum of their measures is 90°.

- Two angles are **supplementary** if the sum of their measures is 180°.

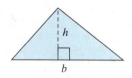

- $\angle a$ and $\angle c$ are vertical angles, and $\angle b$ and $\angle d$ are vertical angles. The measures of vertical angles are equal.

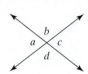

- The sum of the measures of the angles of a triangle is 180°.

$x° + y° + z° = 180°$